Lecture Notes in Artificial Intelligence 5736

Edited by R. Goebel, J. Siekmann, and W. Wahlster

Subseries of Lecture Notes in Computer Science

Miltiadis D. Lytras Ernesto Damiani
John M. Carroll Robert D. Tennyson
David Avison Ambjörn Naeve
Adrian Dale Paul Lefrere Felix Tan
Janice Sipior Gottfried Vossen (Eds.)

Visioning and Engineering the Knowledge Society

A Web Science Perspective

Second World Summit
on the Knowledge Society, WSKS 2009
Chania, Crete, Greece, September 16-18, 2009
Proceedings

Volume Editors

Miltiadis D. Lytras, mlytras@acgmail.gr

Ernesto Damiani, damiani@dti.unimi.it

John M. Carroll, jcarroll@ist.psu.edu

Robert D. Tennyson, rtenny@umn.edu

David Avison, avison@essec.fr

Ambjörn Naeve, amb@nada.kth.se

Adrian Dale, adrian.dale@creatifica.com

Paul Lefrere, plefrere@gltweb.com

Felix Tan, felix.tan@aut.ac.nz

Janice Sipior, janice.sipior@villanova.edu

Gottfried Vossen, vossen@uni-muenster.de

Library of Congress Control Number: 2009935332

CR Subject Classification (1998): I.2.6, I.2.4, K.4, K.3, J.1, H.2.8

LNCS Sublibrary: SL 7 – Artificial Intelligence

ISSN 0302-9743

ISBN 978-3-642-04753-4 Springer Berlin Heidelberg New York

springer.com

Typesetting: Camera-ready by author, data conversion by Scientific Publishing Services, Chennai, India
Printed on acid-free paper SPIN: 12767915 06/3180 5 4 3 2 1 0

Preface

It is a great pleasure to share with you the Springer LNCS proceedings of the Second World Summit on the Knowledge Society, WSKS 2009, organized by the Open Research Society, Ngo, http://www.open-knowledge-society.org, and held in Samaria Hotel, in the beautiful city of Chania in Crete, Greece, September 16–18, 2009.

The 2nd World Summit on the Knowledge Society (WSKS 2009) was an international scientific event devoted to promoting dialogue on the main aspects of the knowledge society towards a better world for all. The multidimensional economic and social crisis of the last couple of years has brought to the fore the need to discuss in depth new policies and strategies for a human centric developmental processes in the global context.

This annual summit brings together key stakeholders involved in the worldwide development of the knowledge society, from academia, industry, and government, including policy makers and active citizens, to look at the impact and prospects of information technology, and the knowledge-based era it is creating, on key facets of living, working, learning, innovating, and collaborating in today's hyper-complex world.

The summit provides a distinct, unique forum for cross-disciplinary fertilization of research, favoring the dissemination of research on new scientific ideas relevant to international research agendas such as the EU (FP7), OECD, or UNESCO. We focus on the key aspects of a new sustainable deal for a bold response to the multidimensional crisis of our times.

Eleven general pillars provide the constitutional elements of the summit:

- Pillar 1. Information Technologies – Knowledge Management Systems – E-business and Business, and Organizational and Inter-organizational Information Systems for the Knowledge Society
- Pillar 2. Knowledge, Learning, Education, Learning Technologies, and E-learning for the Knowledge Society
- Pillar 3. Social and Humanistic Computing for the Knowledge Society – Emerging Technologies for Society and Humanity
- Pillar 4. Culture and Cultural Heritage – Technology for Culture Management – Management of Tourism and Entertainment – Tourism Networks in the Knowledge Society
- Pillar 5. E-government and E-democracy in the Knowledge Society
- Pillar 6. Innovation, Sustainable Development, and Strategic Management for the Knowledge Society
- Pillar 7. Service Science, Management, Engineering, and Technology
- Pillar 8. Intellectual and Human Capital Development in the Knowledge Society
- Pillar 9. Advanced Applications for Environmental Protection and Green Economy Management
- Pillar 10. Future Prospects for the Knowledge Society: from Foresight Studies to Projects and Public Policies
- Pillar 11. Technologies and Business Models for the Creative Industries

In the 2nd World Summit on the Knowledge Society, six main tracks and three workshops were organized. This volume includes 60 full research articles, selected after a double blind review process from 256 submissions, contributed by 480 co-authors.

We are very happy, because in this volume of LNAI you will find excellent quality research giving sound propositions for advanced systems towards the knowledge society.

In the next figure we summarize the context of the research contributions presented at WSKS 2009.

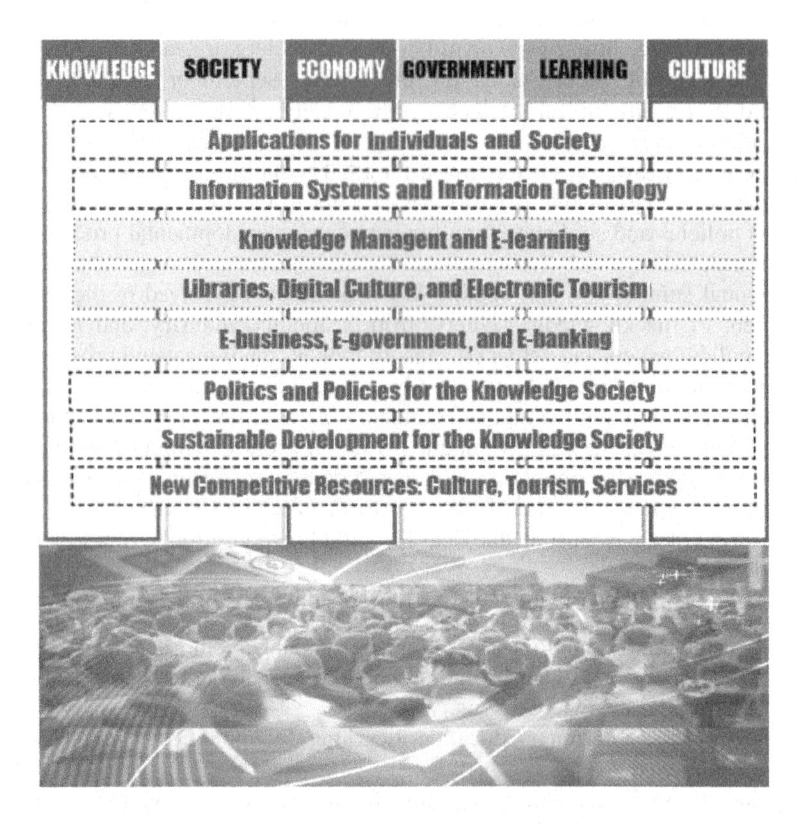

I would like to thank the more than 480 co-authors from 59 countries for their submissions, the Program Committee members and their subreviewers for the thoroughness of their reviews and the colleagues in the Open Research Society for the great support they offered during the organization of the event in Chania.

We are honored by the support and encouragement of the Editors-in-Chief of the five ISI SCI/SSCI listed journals that agreed to publish special issues from extended versions of papers presented at the summit:

- Robert Tennyson, Editor-in-Chief of Computers in Human Behaviour
- Amit Sheth, Editor-in-Chief of the International Journal of Semantic Web and Information Systems
- Adrian Dale, Editor-in-Chief of the Journal of Information Science
- Felix Tan, Editor-in-Chief of the Journal of Global Information Management
- Janice Sipior, Editor-in-Chief of Information Systems Management

A great thank you also to Alfred Hofmann from Springer and his staff for the excellent support during the publication of LNCS/LNAI 5736 and CCIS 49.

Last but not least, I would like to express my gratitude to the staff and members of the Open Research Society for their hard work throughout the organization of the summit and their efforts to promote a better world for all based on knowledge and learning.

We need a better world. We contribute with our sound voices to the agenda, policies, and actions. We invite you to join your voice with ours and all together to shape a new deal for our world: education, sustainable development, health, opportunities for well being, culture, collaboration, peace, democracy, and technology for all.

Looking forward to seeing you at the third event in the series, about which you can find more information at: http://www.open-knowledge-society.org/summit.htm.

With 30 special issues already agreed for WSKS 2010, and 6 main tracks planned, we would like to ask for your involvement, and we would be happy to see you joining us.

THANK YOU – Efharisto Poli!!

July 2009 Miltiadis D. Lytras

Organization

WSKS 2009 was organized by the International Scientific Council for the Knowledge Society and supported by the Open Research Society, Ngo, http://www.open-knowledge-society.org and the International Journal of the Knowledge Society Research, http://www.igi-global.com/ijksr

Executive Committee

General Chair of WSKS 2009

Professor Miltiadis D. Lytras

President, Open Research Society, Ngo

Miltiadis D. Lytras is the President and Founder of the Open Research Society, NGO. His research focuses on semantic web, knowledge management, and e-learning, with more than 100 publications in these areas. He has co-edited / co-edits 25 special issues in International Journals (e.g., IEEE Transaction on Knowledge and Data Engineering, IEEE Internet Computing, IEEE Transactions on Education, Computers in Human Behavior, Interactive Learning Environments, the Journal of Knowledge Management, the Journal of Computer Assisted Learning, etc.) and has authored/ (co-)edited 25 books (e.g., Open Source for Knowledge and Learning Management, Ubiquitous and Pervasive Knowledge Management, Intelligent Learning Infrastructures for Knowledge Intensive Organizations, Semantic Web Based Information Systems, China Information Technology Hanbook, Real World Applications of Semantic Web and Ontologies, Web 2.0: The Business Model, etc.). He is the founder and officer of the Semantic Web and Information Systems Special Interest Group of the Association for Information Systems (http://www.sigsemis.org). He serves as the (Co-) Editor-in-Chief of 12 international journals (e.g., the International Journal of Knowledge and Learning, the International Journal of Technology Enhanced Learning, the International Journal on Social and Humanistic Computing, the International Journal on Semantic Web and Information Systems, the International Journal on Digital Culture and Electronic Tourism, the International Journal of Electronic Democracy, the International Journal of Electronic Banking, and the International Journal of Electronic Trade, etc.) while he is associate editor or editorial board member in seven more.

WSKS 2009 Co-chairs

Professor Ernesto Damiani

University of Milan, Italy

Ernesto Damiani is a professor at the Dept. of Information Technology, University of Milan, where he leads the Software Architectures Lab. Prof. Damiani holds/has held visiting positions at several international institutions, including George Mason University (Fairfax, VA, USA) and LaTrobe University (Melbourne, Australia). Prof. Damiani is an Adjunct Professor at the Sydney University of Technology (Australia). He has written several books and filed international patents; also, he has co-authored more than two hundred research papers on advanced secure service-oriented architectures, open source software and business process design, software reuse, and Web data semantics. Prof. Damiani is the Vice Chair of IFIP WG 2.12 on Web Data Semantics and the secretary of IFIP WG 2.13 on Open Source Software Development. He coordinates several research projects funded by the Italian Ministry of Research and by private companies, including Siemens Mobile, Cisco Systems, ST Microelectronics, BT Exact, Engineering, Telecom Italy, and others.

Professor John M. Carroll

The Pennsylvania State University, USA

John M. Carroll was a founder of human-computer interaction, the youngest of the 9 core areas of computer science identified by the Association for Computing Machinery (ACM). He served on the program committee of the 1982 Bureau of Standards Conference on the Human Factors of Computing Systems that, in effect, inaugurated the field, and was the direct predecessor of the field's flagship conference series, the ACM CHI conferences. Through the past two decades, Carroll has been a leader in the development of the field of Human-Computer Interaction. In 1984 he founded the User Interface Institute at the IBM Thomas J. Watson Research Center, the most influential corporate research laboratory during the latter 1980s. In 1994, he joined Virginia Tech as Department Head of Computer Science where he established an internationally renowned HCI focus in research and teaching. Carroll has served on the editorial boards of every major HCI journal – the International Journal of Human-Computer Interaction, the International Journal of Human-Computer Studies, Human-Computer Interaction, Transactions on Computer-Human Interaction, Transactions on Information Systems, Interacting with Computers, and Behavior and Information Technology. He was a founding associate editor of the field's premier journal, ACM Transactions on Computer-Human Interaction, and a founding member of the editorial boards of Transactions on Information Systems, Behavior and Information Technology, and the International Journal of Human-Computer Interaction. He is currently on the Senior Editorial Advisory Board for the field's oldest journal, the International Journal of Human-Computer Systems. He served on the editorial board of all three editions of the Handbook of Human-Computer Interaction, and was associate editor for the section on Human-Computer Interaction in the Handbook of Computer Science and Engineering. He has served on more than 50 program committees for international HCI conferences, serving as chair or associate chair 12 times. He has been nominated to become the next Editor-in-Chief of the ACM Transactions on Computer-Human Interaction. He is currently serving his second term on the National Research Council's Committee on Human Factors. Carroll has published 13 books and more than 250 technical papers, and has produced more than 70 miscellaneous reports (videotapes, workshops, tutorials, conference demonstrations and discussant talks). He has presented more than 30 plenary or distinguished lectures.

Professor Robert Tennyson

University of Minnesota, USA

Robert Tennyson is currently a professor of educational psychology and technology in learning and cognition. In addition to his faculty position, he is a program coordinator for the psychological foundations of education. His published works range from basic theoretical articles on human learning to applied books on instructional design and technology. He serves as the Editor-in-Chief of the scientific journal, Computers in Human Behavior, published by Elsevier Science and now in its 17th year, as well as serving on several editorial boards for professional journals. His research and publications include topics such as cognitive learning and complex cognitive processes, intelligent systems, complex-dynamic simulations, testing and measurement, instructional design, and advanced learning technologies. His international activities include directing a NATO-sponsored advanced research workshop in Barcelona and a NATO advanced study institute in Grimstad, Norway – both on the topic of automated instructional design and delivery. He has recently directed an institute on technology in Athens and Kuala Lumpur. His other international activities include twice receiving a Fulbright Research Award in Germany and once in Russia. His teaching interests include psychology of learning, technology-based systems design, evaluation, and management systems.

Professor David Avison

ESSEC Business School, France

David Avison is Distinguished Professor of Information Systems at ESSEC Business School, near Paris, France, after being Professor at the School of Management at Southampton University for nine years. He has also held posts at Brunel and Aston Universities in England, and the University of Technology Sydney and University of New South Wales in Australia, and elsewhere. He is President-elect of the Association of Information Systems (AIS). He is joint editor of Blackwell Science's Information Systems Journal now in its eighteenth volume, rated as a 'core' international journal. So far, 25 books are to his credit including the fourth edition of the well-used text Information Systems Development: Methodologies, Techniques and Tools (jointly authored with Guy Fitzgerald). He has published a large number of research papers in learned journals, and has edited texts and conference papers. He was Chair of the International Federation of Information Processing (IFIP) 8.2 group on the impact of IS/IT on organizations and society and is now vice chair of IFIP technical committee 8. He was past President of the UK Academy for Information Systems and also chair of the UK Heads and Professors of IS and is presently member of the IS

Senior Scholars Forum. He was joint program chair of the International Conference in Information Systems (ICIS) in Las Vegas (previously also research program stream chair at ICIS Atlanta), joint program chair of IFIP TC8 conference at Santiago Chile, program chair of the IFIP WG8.2 conference in Amsterdam, panels chair for the European Conference in Information Systems at Copenhagen and publicity chair for the entity-relationship conference in Paris, and chair of several other UK and European conferences. He will be joint program chair of the IFIP TC8 conference in Milan, Italy in 2008. He also acts as consultant and has most recently worked with a leading manufacturer developing their IT/IS strategy. He researches in the area of information systems development and more generally on information systems in their natural organizational setting, in particular using action research, though he has also used a number of other qualitative research approaches.

Dr. Ambjorn Naeve

KTH-Royal Institute of Technology, Sweden

Ambjörn Naeve (www.nada.kth.se/~amb) has a background in mathematics and computer science and received his Ph.D. in computer science from KTH in 1993. He is presently coordinator of research on Interactive Learning Environments and the Semantic Web at the Centre for user-oriented Information technology Design (CID: http://cid.nada.kth.se) at the Royal Institute of Technology (KTH: www.kth.se) in Stockholm, where he heads the Knowledge Management Research group (KMR: http://kmr.nada.kth.se).

Professor Adrian Dale

Creatifica Associates, UK

Adrian Dale is renowned as a radical thinker in the information and knowledge management fields. He has managed the information, records, and knowledge functions of several UK organizations in the public and private sector, driving the shift from the paper to electronic worlds. He coaches the knowledge management teams of a number of public and private sector organizations, helping them to create the radical change necessary today. Current clients include the Cabinet Office; Shell Exploration and Production; CBI; the Department for Children, Schools and Families; Scottish Enterprise; Health Scotland; NICE; and the National Library for Health. Past clients have included Health London, HM Treasury, the Ministry of Defence, the House of Commons, DeFRA, Environment Agency, the Competition Commission, KPMG, Christian Aid, Motor Neurone Disease Associate, the Learning and Skills Council, and Bedford Hospital NHS Trust. He has 21 years of experience in the fields of IT, Knowledge, and Information Management. Before becoming independent, he was Chief Knowledge Officer for Unilever Research with responsibility for IT & Knowledge Management in their Research and Innovation Programs. Adrian is Chairman of the International Online Information Conference, Editor of the Journal of Information Science, and a fellow of the Chartered Institute of Library and Information Professionals.

Professor Paul Lefrere

University of Tampere, Finland

Paul Lefrere is a professor of eLearning at the University of Tampere, Finland (Hypermedia Lab; Vocational Learning and e-skills Centre). Before that, he was Senior Lecturer at the UK Open University's Institute of Educational Technology. From 2003-2005, he was on leave of absence as Microsoft's Executive Director for eLearning, in which role he served on various European and national advisory groups concerned with professional learning and related topics, and also on Europe's e-learning Industry Group, eLIG. Until 2003 he was Policy Adviser at the Open University, where he was also Academic Director of a number of international multi-university projects concerned with e-skills and knowledge management. Since he returned to the OU he has been engaged in a number of development and consultancy activities in many countries of the world including the Middle East and Pakistan.

Professor Felix Tan

Auckland University of Technology, New Zealand

Dr. Felix B. Tan is Professor of Information Systems and Chair of the Business Information Systems discipline. He is also Director of the Centre for Research on Information Systems Management (CRISM). He serves as the Editor-in-Chief of the Journal of Global Information Management. He is a Fellow of the Information Resources Management Association as well as the New Zealand Computer Society. He also served on the Council of the Association for Information Systems from 2003-2005. He has held visiting positions with the National University of Singapore, The University of Western Ontario, Canada, and was a visiting professor at Georgia State University, USA, in May/June 2005 and the University of Hawaii at Manoa in January 2006. Dr. Tan is internationally known for his work in the global IT field. Dr. Tan's current research interests are in electronic commerce, global information management, business-IT alignment, and the management of IT. He actively uses the repertory grid and narrative inquiry methods in his research. Dr. Tan has published in MIS Quarterly, Information Management, the Journal of Information Technology, the Information Systems Journal, IEEE Transactions on Engineering Management, IEEE Transactions on Personal Communications, the International Journal of HCI, and the International Journal of Electronic Commerce, as well as other journals, and has refereed a number of conference proceedings.

Professor Janice Sipior

School of Business, Villanova University, USA

Janice C. Sipior is Associate Professor of Management Information Systems at Villanova University, an Augustinian university located in Pennsylvania, USA. Her academic experience also includes faculty positions at Canisius College, USA; the University of North Carolina, USA; Moscow State Linguistic University, Russia; and the University of Warsaw, Poland. She was previously employed in computer planning at HSBC (Hong Kong-Shanghai Bank Corporation). Her research interests include ethical and legal aspects of information technology, system development strategies, and knowledge management. Her research has been published in over 65 refereed journals, international conference proceedings, and books. She is Chair of the Association for Computing Machinery, Special Interest Group on Management Information Systems (ACM-SIGMIS), and serves as a Senior Editor of Data Base, an Associate Editor of the Information Resources Management Journal, and Editorial Board Member of Information Systems Management.

Professor Gottfried Vossen

University of Muenster, Germany

Gottfried Vossen is Professor of Computer Science in the Department of Information Systems at the University of Muenster in Germany. He is the European Editor-in-Chief of Elsevier's Information Systems and Director of the European Research Center for Information Systems (ERCIS) in Muenster. His research interests include conceptual as well as application-oriented problems concerning databases, information systems, electronic learning, and the Web.

Program Chairs

Miltiadis D. Lytras	American College of Greece, Greece
Patricia Ordonez De Pablos	University of Oviedo, Spain
Miguel Angel Sicilia	University of Alcala, Spain

Knowledge Management and E-Learning Symposium Chairs

Ambjorn Naeve	Royal Institute of Technology, Sweden
Miguel Angel Sicilia	University of Alcala, Spain

Publicity Chair

Ekaterini Pitsa	Open Research Society, Greece

Exhibition Chair

Efstathia Pitsa	University of Cambridge, UK

Sponsoring Organizations

Gold

Inderscience Publishers, http://www.inderscience.com

Program and Scientific Committee Members (Serving also as Reviewers)

Adrian Paschke	Technical University Dresden, Germany
Adriana Schiopoiu Burlea	University of Craiova, Romania
Agnes Kukulska-Hulme	The Open University, UK
Ahmad Syamil	Arkansas State University, USA
Aimé Lay-Ekuakille	University of Salento, Italy
Alan Chamberlain	University of Nottingham, UK
Alejandro Diaz-Morcillo	University of Cartagena, Spain
Alok Mishra	Atilim University, Turkey
Alyson Gill	Arkansas State University, USA
Ambjörn Naeve	Royal Institute of Technology, Sweden
Ana Isabel Jiménez-Zarco	Open University of Catalonia, Spain
Anas Tawileh	Cardiff University, UK
Anastasia Petrou	University of Peloponnese, Greece
Anastasios A. Economides	University of Macedonia, Greece
Andreas Holzinger	Medical University Graz, Austria
Andy Dearden	Sheffield Hallam University, UK
Ane Troger	Aston Business School, UK
Angela J. Daniels	Arkansas State University, USA
Anna Lisa Guido	University of Salento, Italy
Anna Maddalena	DISI, University of Genoa, Italy
Anna Maria Tammaro	University of Parma, Italy
Ansgar Scherp	OFFIS - Multimedia and Internet Information Services, Germany
Antonio Cartelli	University of Cassino, Italy
Antonio Tomeu	University of Cadiz, Spain
Riccardo Lancellotti	University of Modena and Reggio Emilia, Italy
Apostolos Gkamas	University of Peloponnese, Greece
Arianna D'Ulizia	National Research Council, Italy
Aristomenis Macris	University of Piraeus, Greece
Badr Al-Daihani	Cardiff University, UK
Beatriz Fariña	University of Valladolid, Spain
Berardina Nadja De Carolis	University of Bari, Italy
Bob Folden	Texas A&M University-Commerce, USA
Bodil Nistrup Madsen	Copenhagen Business School, Denmark
Bradley Moore	University of West Alabama, USA
Campbell R. Harvey	Duke University, USA
Carla Limongelli	Università "Roma Tre", Italy
Carlos Bobed	University of Zaragoza, Spain
Carlos Ferran	Penn State Great Valley University, USA
Carmen Costilla	Technical University of Madrid, Spain
Carolina Lopez Nicolas	University of Murcia, Spain
Charles A. Morrissey	Pepperdine University, USA
Chengbo Wang	Glasgow Caledonian University, UK
Chengcui Zhang	University of Alabama at Birmingham, USA

Christian Wagner	City University of Hong Kong, Hong Kong (China)
Christos Bouras	University of Patras, Greece
Chunzhao Liu	Chinese Academy of Sciences, China
Claire Dormann	Carleton University, Canada
Claus Pahl	Dublin City University, Ireland
Cui Tao	Brigham Young University, USA
Damaris Fuentes-Lorenzo	IMDEA Networks, Spain
Daniel R. Fesenmaier	Temple University, USA
Daniela Leal Musa	Federal University of Sao Paulo, Brazil
Daniela Tsaneva	Cardiff School of Computer Science,UK
Darijus Strasunskas	Norwegian University of Science and Technology (NTNU), Norway
David O'Donnell	Intellectual Capital Research Institute of Ireland, Ireland
David R. Harding	Jr. Arkansas State University, USA
Dawn Jutla	Saint Mary's University, Canada
Denis Gillet	Swiss Federal Institute of Technology in Lausanne (EPFL), Switzerland
Diane H. Sonnenwald	Göteborg University and University College of Borås, Sweden
Dimitri Konstantas	University of Geneva, Switzerland
Dimitris N. Chryssochoou	University of Crete, Greece
Douglas L. Micklich	Illinois State University, USA
Dusica Novakovic	London Metropolitan University, USA
Edward Dieterle	Harvard Graduate School of Education, USA
Ejub Kajan	High School of Applied Studies, Serbia
Elena García-Barriocanal	University of Alcalá, Spain
Emma O'Brien	University of Limerick, Ireland
Eric Tsui	The Hong Kong Polytechnic University, Hong Kong (China)
Eva Rimbau-Gilabert	Open University of Catalonia, Spain
Evanegelia Karagiannopoulou	University of Ioannina, Greece
Evangelos Sakkopoulos	University of Patras, Greece
Fernanda Lima	Universidade Catolica de Brasilia, Brazil
Filippo Sciarrone	Università "Roma Tre", Italy
Francesc Burrull	Polytechnic University of Cartagena, Spain
Francesca Lonetti	ISTI - Area della Ricerca CNR, Italy
Francisco Palomo Lozano	University of Cádiz, Spain
Gang Wu	Tsinghua University, China
Gavin McArdle	University College Dublin, Ireland
George A. Jacinto	Arkansas State University, USA
Georgios Skoulas	University of Macedonia, Greece
Gianluca Elia	University of Salento, Italy
Gianluigi Viscusi	University of Milano-Bicocca, Italy
Giovanni Vincenti	S.r.l. Rome, Italy
Giuseppe Pirrò	University of Calabria, Italy
Giuseppe Vendramin	University of Salento, Italy

Vincenza Pelillo	University of Salento, Italy
Gregg Janie	University of West Alabama, USA
Guillermo Ibañez	Universidad de Alcalá, Spain
Hai Jiang	Arkansas State University, USA
Haim Kilov	Stevens Institute of Technology, USA
Hanh H. Hoang	Hue University, Vietnam
Hanne Erdman Thomsen	Copenhagen Business School, Denmark
Hanno Schauer	Universität Duisburg-Essen, Germany
Heinz V. Dreher	Curtin University of Technology, Australia
Helena Corrales Herrero	University of Valladolid, Spain
Helena Villarejo	University of Valladolid. Spain
Hyggo Almeida	University of Campina Grande, Brazil
Inma Rodríguez-Ardura	Open University of Catalonia, Spain
Ino Martínez León	Universidad Politécnica de Cartagena, Spain
Ioan Marius Bilasco	Laboratoire Informatique de Grenoble (LIG), France
Ioanna Constantiou	Copenhagen Business School, Denmark
Ioannis Papadakis	Ionian University, Greece
Ioannis Stamelos	AUTH, Greece
Irene Daskalopoulou	University of Peloponnese, Greece
Isabel Ramos	University of Minho, Portugal
James Braman	Towson University, USA
Jan-Willem Strijbos	Leiden University, The Netherlands
Javier De Andrés	University of Oviedo, Spain
Javier Fabra	University of Zaragoza, Spain
Jeanne D. Maes	University of South Alabama, USA
Jens O. Meissner	Lucerne School of Business, Switzerland
Jerome Darmont	University of Lyon (ERIC Lyon 2), France
Jesus Contreras	ISOCO, Spain
Jesús Ibáñez	University Pompeu Fabra, Spain
Jianhan Zhu	The Open University, UK
Johann Gamper	Free University of Bozen-Bolzano, Italy
Jon A. Preston	Clayton State University, USA
Jorge Gracia	University of Zaragoza, Spain
Jose Jesus García Rueda	Carlos III University of Madrid, Spain
José Luis García-Lapresta	Universidad de Valladolid, Spain
Jose Luis Isla Montes	University of Cadiz, Spain
Josemaria Maalgosa Sanahuja	Polytechnic University of Cartagena, Spain
Joseph C. Paradi	University of Toronto, Canada
Joseph Feller	University College Cork, Ireland
Joseph Hardin	University of Michigan, USA
Joze Gricar	University of Maribor, Slovenia
Juan Gabriel Cegarra Navarro	Universidad Politécnica de Cartagena, Spain
Juan Manuel Dodero	University of Cádiz, Spain
Juan Miguel Gómez Berbís	Univesidad Carlos III de Madrid, Spain
Juan Pablo de Castro Fernández	University of Valladolid, Spain
Juan Vicente Perdiz	University of Valladolid, Spain

Juan Ye	University College Dublin, Ireland
Julià Minguillón	Universitat Oberta de Catalunya (UOC), Spain
Jyotishman Pathak	Mayo Clinic College of Medicine, USA
Karim Mohammed Rezaul	University of Wales, UK
Karl-Heinz Pognmer	Copenhagen Business School, Denmark
Katerina Pastra	Institute for Language and Speech Processing, Greece
Ken Fisher	London Metropolitan University, UK
Kleanthis Thramboulidis	University of Patras, Greece
Konstantinos Tarabanis	University of Macedonia, Greece
Kylie Hsu	California State University at Los Angeles, USA
Laura Papaleo	DISI, Italy
Laura Sanchez Garcia	Universidade Federal do Parana, Brazil
Laurel D. Riek	Cambridge University, UK
Lazar Rusu	Royal Institute of Technology (KTH), Sweden
Leonel Morgado	University of Trás-os-Montes e Alto Douro, Portugal
Leyla Zhuhadar	Western Kentucky University, USA
Lily Diaz-Kommonen	University of Art and Design Helsinki, Finland
Linda A. Jackson	Michigan State University, USA
Liqiong Deng	University of West Georgia, USA
Lori L. Scarlatos	Stony Brook University, USA
Lubomir Stanchev	Indiana University - Purdue University Fort Wayne, USA
Luis Angel Galindo Sanchez	Universidad Carlos III de Madrid, Spain
Luis Iribarne	Departamento de Lenguajes y Computacion, Universidad de Almeria, Spain
Luke Tredinnick	London Metropolitan University, UK
Lynne Nikolychuk	King's College London, UK
M. Carmen de Castro	Universidad de Cádiz, Spain
M. Latif	Manchester Metropolitan University, UK
Mahmoud Youssef	Arab Academy for Science and Technology, Egypt
Maiga Chang	Athabasca University, Canada
Manolis Vavalis	University of Thessaly, Greece
Manuel Rubio-Sanchez	Rey Juan Carlos University, Spain
Marco Temperini	Università La Sapienza, Italy
Marcos Castilho	Departamento de Informática da Universidade Federal Paraná, Brazil
Marcos Martin-Fernandez	Associate Valladolid University, Spain
Maria Chiara Caschera	IRPPS-CNR, Rome, Italy
Maria Grazia Gnoni	University of Salento, Lecce, Italy
Maria Helena Braz	Technical University of Lisbon, Portugal
Maria Jesús Martinez-Argüelles	Open University of Catalonia, Spain
María Jesús Segovia Vargas	Universidad Complutense de Madrid, Spain
María Jesús Verdú Pérez	University of Valladolid, Spain
Maria Joao Ferreira	Universidade Portucalense, Portugal
Maria Papadaki	University of Ioannina, Greece

Paola Mello	University of Bologna, Italy
Paolo Toth	University of Bologna, Italy
Patricia A. Walls	Arkansas State University, USA
Paul A. Kirschner	Open University of the Netherlands, The Netherlands
Paul G. Mezey	Memorial University, Canada
Pedro J. Muñoz Merino	Universidad Carlos III de Madrid, Spain
Pedro Soto-Acosta	University of Murcia, Spain
Peisheng Zhao	George Mason University, USA
Pekka Muukkonen	University of Turku, Finland
Per Anker Jensen	Copenhagen Business School, Denmark
Peter Gomber	Johann Wolfgang Goethe-Universität Frankfurt, Germany
Phil Fitzsimmons	University of Wollongong, Australia
Pierre Deransart	INRIA-Rocquencourt, France
Pilar Manzanares-Lopez	Technical University of Cartagena, Spain
Pirkko Walden	Abo Akademi University, Finland
Ralf Klamma	RWTH Aachen University, Germany
Raquel Hijón Neira	Universidad Rey Juan Carlos, Spain
Raymond Y.K. Lau	City University of Hong Kong, Hong Kong SAR
Razaq Raj	Leeds Metropolitan University, UK
Razvan Daniel Zota	Academy of Economic Studies Bucharest, Romania
Ricardo Colomo Palacios	Universidad Carlos III de Madrid, Spain
Ricardo Lopez-Ruiz	University of Zaragoza, Spain.
Rob Potharst	Erasmus University, The Netherlands
Robert Fullér	Åbo Akademi University, Finland
Roberto García	Universitat de Lleida, Spain
Roberto Paiano	University of Salento, Italy
Roman Povalej	University of Karlsruhe, Germany
Rushed Kanawati	LIPN – CNRS, France
Russel Pears	Auckland University of Technology, New Zealand
Ryan Robeson	Arkansas State University, USA
Sabine H. Hoffmann	Macquarie University, Australia
Sadat Shami	Cornell University, USA
Salam Abdallah	Abu Dhabi University, United Arab Emirates
Samiaji Sarosa	Atma Jaya Yogyakarta University, Indonesia
Sean Mehan	University of the Highlands and Islands, Scotland, UK
Sean Wolfgand M. Siqueira	Federal University of the State of Rio de Janeiro (UNIRIO), Brazil
Sebastian Matyas	Otto-Friedrich-Universität Bamberg, Germany
Sergio Ilarri	University of Zaragoza, Spain
Shantha Liyanage	Macquarie University, Australia
Shaoyi He	California State University, USA
She-I Chang	National Chung Cheng University, Taiwan (China)
Sherif Sakr	University of New South Wales, Australia
Sijung Hu	Loughborough University, UK
Silvia Rita Viola	Università Politecnica delle Marche - Ancona, Italy

Silvia Schiaffino	Univ. Nac. del Centro de la Provincia de Buenos Aives, Argentina
Sima Yazdani	Cisco Systems, USA
Sinuhe Arroyo	University of Alcalá de Henares, Spain
Sonia Berman	University of Cape Town, South Africa
Soror Sahri	Université Paris-Dauphine, France
Spyros D. Arsenis	University of Thessaly, Greece
Staffan Elgelid	Arizona State University, USA
Stefan Hrastinski	Uppsala University, Sweden
Stephan Lukosch	FernUniversität in Hagen, Germany
Steve Barker	King's College London, UK
Steve Creason	Metropolitan State University, USA
Stone Schmidt	Arkansas State University, USA
Suneeti Rekhari	University of Wollongong, Australia
Taha Osman	Nottingham Trent University, UK
Takaharu Kameoka	Mie Univeristy, Japan
Teck Yong Eng	University of London, UK
Terrill L. Frantz	Carnegie Mellon University, USA
Thanos C. Papadopoulos	The University of Warwick, UK
Tobias Ley	Know-Center, Austria
Toyohide Watanabe	Nagoya University, Japan
	Université Libre de Bruxelles , Belgium
Upasana Singh	University of KwaZulu, South Africa
Vaclav Snasel	VSB-Technical University of Ostrava, Czech Republic
Vagan Terziyan	University of Jyvaskyla , Finland
Val Clulow	Monash University, Australia
Véronique Heiwy	IUT de Paris, France
Vincenzo Ciancia	University of Pisa, Italy
Violeta Damjanovic	Salzburg Research, Austria
Virginie Sans	Université de Cergy-Pontoise, France
Vlasios Voudouris	London Metropolitan University, UK
Walt Scacchi	University of California, USA
Weili Wu	University of Texas at Dallas, USA
Xavier Calbet	EUMETSAT, Germany
Xiaohua (Tony) Hu	Drexel University, USA
Xihui (Paul) Zhang	University of Memphis, USA
Yihong Ding	Brigham Young University, USA
Yolaine Bourda	SUPELEC, France
Yuan An	Drexel University, USA
Yun Shen	University of Bristol, UK
Yvon Kermarrec	Institut Télécom / Télécom Bretagne, France
Ziqi Liao	Hong Kong Baptist University, Hong Kong (China)

Table of Contents

Knowledge, Learning, Education, Learning Technologies and E-Learning for the Knowledge Society

Information Technologies - Knowledge Management Systems - E-Business and Business, Organizational and Inter-organizational Information Systems for the Knowledge Society

Engineering the Knowledge Society through Web Science: Advanced Systems, Semantics, and Social Networks

E-Government and E-Democracy for the Knowledge Society

Software Engineering for the Knowledge Society

Promoting a Humanistic Perspective of Creativity by Interpersonal Qualities and Web-Based Tools

Renate Motschnig[1,2] and Tomáš Pitner[1,2,*]

[1] University of Vienna, Faculty of Computer Science,
Rathausstrasse 19/9, 1010 Vienna, Austria
renate.motschnig@univie.ac.at
http://www.pri.univie.ac.at

[2] Masaryk University, Faculty of Informatics,
Botanická 68a, 60200 Brno, Czech Republic
tomp@fi.muni.cz
http://www.fi.muni.cz

Abstract. The rapid rise of web-based services and tools for communication, collaboration and learning confronts us with the question, whether and under what conditions these tools have the potential to promote human creativity. In this paper we approach this question by first elaborating a person-centered and relational notion of creativity. Based on this perspective on creativity, we investigate the potential of web-based tools and usage contexts in order to find out in which ways they are likely to promote creativity. As a result it will be argued that creativity — in a humanistic sense — can be fostered by web-based tools only if certain conditions are met. These include specific personal (and interpersonal) qualities of the participating persons, such as openness to experience and a non-judgmental attitude, as well as requirements on web-based tools. The paper is intended to expose the dimensions and conditions under which web-based tools can be used in ways most likely promote creativity.

Keywords: human creativity, Person-Centered Approach, web-based services.

1 Introduction

In a time with constant and rapid change we are often faced with new situations, i.e. situations that we encounter for the first time and for which previous learning is inadequate to provide a solution [3], pp. 211—215. We cannot believe that any portion of static knowledge, how big so ever, will suffice to deal with the novel and uncertain, dynamic conditions of the future [10], p. 120. Hence, dealing with new situations constructively, in other words most appropriately or "creatively",

* This research has been supported by the Faculty of Computer Science, University of Vienna and the Faculty of Informatics, Masaryk University.

M.D. Lytras et al. (Eds.): WSKS 2009, LNAI 5736, pp. 1–12, 2009.

certainly constitutes an essential goal for present and future education and life. Since the web and the tools operating on it have become our (most?) widely used companions it is more than justified to ask the question: *Under what conditions do web-based tools promote or even impede our creativity?*

We assume that many of the readers of this paper, like ourselves, have had experiences in which they felt that web-based activities have had most fruitful effects on creating outcomes and meaning. For example, mailing, chatting, collaborating on a text, or producing/sharing some artifact with others was perceived as successful. It was rewarded with a feeling of expansion, joy, pleasure, richness of meaning or any other positive sensation making us perceive that something constructive had been formed that did not exist before. We equally assume that any reader has had a distinctly negative experience with web-based tools, evoking responses such as: "what a waste of time", "how complicated is it to express my thought by typing text only", "what endless forms/steps do I need to follow in order to achieve the most simple transaction", etc.

The primary goal of this paper is to illuminate some conditions for designing and using web-based tools in ways that are most likely to promote creativity. For this purpose, chapter two first elaborates on a notion of creativity based on the person-centered approach [8,9] and proceeds by extending it by relational aspects [5,7]. Based on the individual facets of a humanistic perspective on creativity and the creative process, in chapter three the potential of web-based tools is investigated in order to find out in which ways and under what conditions they can support the individual facets of creativity. The general considerations are confirmed by examples of selected Web 2.0 services for supporting creativity. The final chapter summarizes and discusses the findings and points to further research.

Initial results indicate that, from a personal (and interpersonal) perspective, creativity will emerge if the participating persons are sufficiently free to choose their way of involvement, are not judged prematurely, feel safe to express themselves and are sufficiently open to a wide range of aspects of their experience. From the software perspective some preconditions are that the software must be easy and straightforward to use, must allow one to produce artifacts effectively, has to be appealing to the users, and must make it easy to establish and maintain relationships with persons as well as artifacts.

2 The Notion of Creativity in the Person-Centered Approach

The Person-Centered Approach (PCA) is a branch of humanistic psychology founded by Carl R. Rogers (1902—1987), one of the most renowned American psychologists of the 20th century. Originating in psychology, the PCA has spread to disciplines such as education, social science, international communication, management, conflict resolution, health care and others. In a nutshell, the basic assumption underlying the PCA is that human beings, like all living organisms, have the natural tendency to actualize, i.e. to maintain and to enhance their organisms. The tendency, furthermore, is directed, amongst others, towards differentiation of organs, the use of tools, and socialization. It can unfold best in a climate in which a person experiences, at least to some degree, the genuineness or congruence, unconditional positive regard, and empathic understanding of (at least one) other person or persons. According to Rogers [9], the actualizing tendency "is the primary motivator for creativity as the organism forms new relationships to the environment in its endeavor most fully to be itself".

Given, human beings have an innate tendency towards enhancement, what is it that fosters creativity and how can web-based tools influence that process? In order to respond to these questions let us first illuminate the notion of creativity from a PCA perspective[1] [8,9].

2.1 Aspects and Conditions for Creativity from a Person- and Relationship-Centered Perspective

The creative process and product. In a rapidly changing world there is a genuine need for creative adaptation to cope with the ever changing demands of our environment. Any creative product is a novel construction such that the novelty grows out of the unique qualities of *a person in his or her interaction* with some entity of the environment. "... the creative process is [...] the emergence in action of a novel relational product, growing out of the uniqueness of the individual on the one hand, and the materials, events, people, or circumstances of his life on the other" [9]. The product must be acceptable to some group at some point of time. However, this fact is not helpful to our definition because of fluctuating valuations and the fact that the individual creates because it is satisfying him or her. Still, it has been confirmed by research in the PCA that when a person is open to all of his or her experience, their behavior will be creative and their creativity may be trusted to be essentially constructive. This can be explained by assuming by positing that when a person is open to all aspects of his or her experience and has available to their awareness all the varied sensings and perceivings which are going on in their organism, then the novel products of their interaction with the environment will tend to be constructive for him-/herself and others and his/her behavior will tend in the direction of constructively social living. This appears to be consistent with what Senge [12] claims for a thorough "sensing" phase of the U-process that is designed to bring about organizational change.

Inner conditions for constructive creativity. Rogers identified three inner conditions for constructive creativity. The first one, as already been mentioned above, is openness to experience or extensionality. It means a lack of rigidity, and permeability of boundaries in concepts, beliefs, perceptions, and hypotheses. It means a tolerance for ambiguity, where ambiguity exists. It also means the ability to

[1] The rest of this paper draws heavily from [9], pp. 350—359 using several original wordings that have just slightly been adapted for brevity and gender-sensitive language.

receive much conflicting information without forcing closure upon the situation. The second condition is that the source or locus of evaluation is internal. For the creative person, the value of their creation is established not by praise or criticism of others but by satisfaction to himself or herself. This doesn't mean that the judgments of others are oblivious, it just says that the primary feeling comes from "me in action" with something emerging into existence. The third condition is the ability to toy with elements and concepts. It includes the ability to play spontaneously with ideas, colors, shapes, relationships — to jungle elements into impossible juxtapositions, to translate from one form to another, etc. This can lead to exploration and seeing from innumerable possibilities new options that lead to evolutionary forms with better meeting some inner need and/or more permanent value.

The creative act. The selection of a "product" which is more satisfying and/or forms a more effective relationship with its environment is referred to as the creative act. There is one quality of the creative act that can be described: its selectivity, or emphasis, or attempt to bring out the essence. I bring structure into my relationship to reality until it feels: "This is it!" For example, a writer selects those words which give unity to his expression. Typically, a concomitant to the creative act is anxiety of separateness on the one hand and the desire to communicate and share one's creation on the other hand. We wish to share this new aspect of "me-in-relationship-to-my-environment" with others.

Conditions fostering constructive creativity. From the nature of the inner conditions of creativity it is clear that they cannot be forced, but must be admitted to emerge. The likelihood of emergence is maximized by providing a climate of psychological safety and freedom. Safety is achieved if a person is accepted as someone of unconditional worth. In order for this attitude to be genuine, we need to have an unconditional faith in the other person. If he or she apprehends this attitude, he/she has less need of rigidity, senses safety, and can be more spontaneous, actualizing. To feel safe it also needs a climate in which external judgment is absent and there is no need for defensiveness. Only then can I recognize the locus of evaluation within myself.

The process which provides the ultimate in psychological safety is empathic understanding. If I accept you but know nothing about you, the acceptance is shallow and it might change if come to know you. But if I enter your inner world and see it from your perspective, and still accept what you are feeling and doing from your point of view, this will provide additional safety and will permit your real self to emerge and to express itself in known and novel formings.

Psychological freedom is present when a teacher or facilitative person allows the individual a complete freedom of *symbolic* expression. This permissiveness gives the individual complete freedom to think, to feel, to be, whatever is most inward within him-/herself. It fosters the openness and playful juggling of percepts, concepts, and meanings which is part of creativity.

Rogers' daughter, Natalie Rogers adds a third condition, namely the *offering of stimulating and challenging experiences* [11]. This criterion appears to be particularly relevant for web-based tools since it needs to be explicitly considered in the design of web technology.

Moving from the point of view of an individual to that of a relationship and recalling that creativity comes from *forming relationships with the environment*, the authors claim that creativity, in particular, springs from our striving for forming/cultivating *constructive relationships* with social others. Living in such relationships equally satisfies our desire to communicate that has been identified as a concomitant of the creative act. Motivation for the creative act often lies in an interpersonal relationship that is reciprocally enhancing and forming itself. Constructive creativity in this case emerges from dialogue and potential transcendence and may be "documented" or conserved in a creative product.

We summarize the inner and environmental conditions fostering creativity from a person and relationship centered perspective in Figure 1.

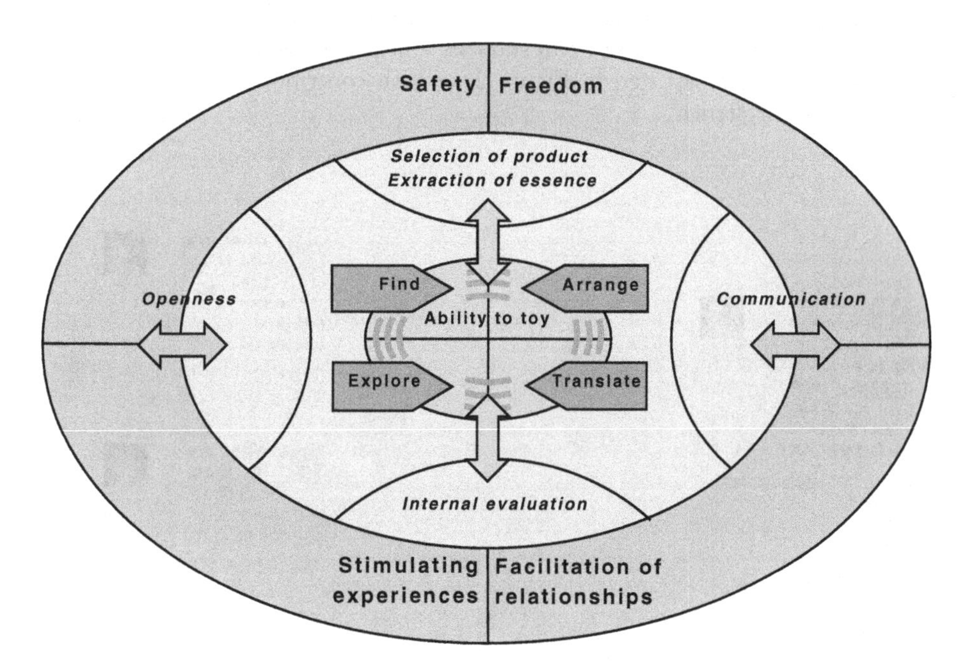

Fig. 1. Conditions fostering creativity from a person- and relationship-centered perspective

2.2 Web-Based Services and Scenarios in Light of Creativity

It appears that web-based tools cannot establish the initial repertoire of a person's inner conditions of constructive creativity, namely his or her openness to experience, inner locus of evaluation, and the ability to toy with elements and concepts. However, given the inner conditions are met to a sufficient degree, our hypothesis is that the use of web-based tools has — under specific circumstances to be elaborated later, the potential to influence the inner as well as environmental conditions enabling a person's or group's constructive creativity (the capacity to create) and their engagement in inventing creative products. This is because web-based tools, along with their usage contexts, both technological and social, influence a person's or group's interaction with their environment. Let us investigate this potential in a general way and illustrate our arguments by examples.

Web-based tools and inner conditions for constructive creativity. The first of the three inner conditions was identified as openness to experience, meaning the ability to receive much conflicting information without forcing closure upon the situation. For sure, web–based tools, with their enormous capacities for interconnection anytime and anywhere support the fact that we receive lots of conflicting information. For example, the various "voices" represented in a forum's entries represent a variety of opinions. If we are not driven by the need to select one of them for strategic reasons, they can show us a broad scope of meanings existing side-by-side — as illustrated in Figure 2 — each contributing its share to the whole picture or "truth".

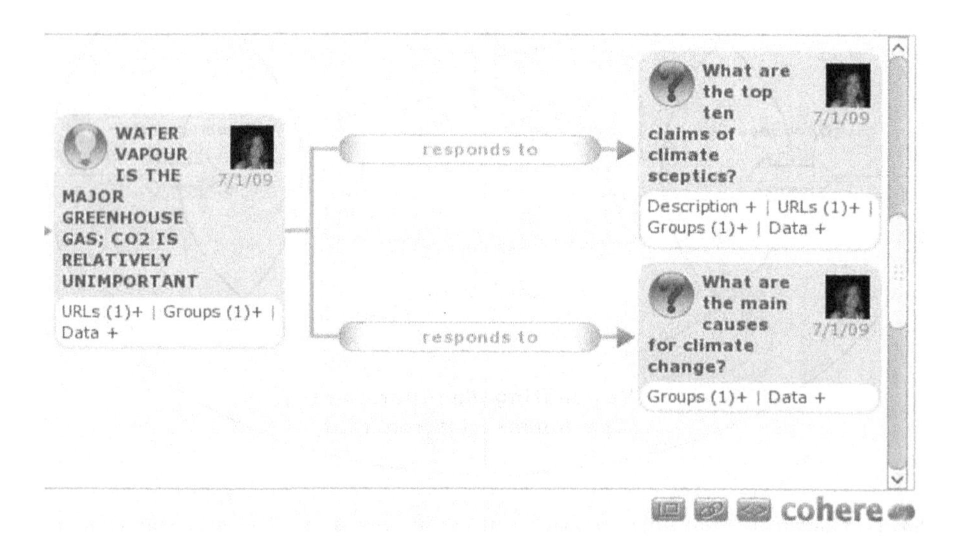

Fig. 2. Variety of opinions in a creative process boosted by a Web 2.0 service (Cohere)

The second inner condition requires that the source of evaluation be internal rather than established by others' judgment. In this respect, web-based tools often motivate one to share his or her ideas/texts/work with others, but only after one's conscious decision on what to share. This means that some form of inner judgment precedes the potential judgment by others. Also, we can choose with whom we share. Furthermore, delayed (as an opposite to instant) responses from others will "teach" us to stay with our own perception for some time prior to getting responses from others.

The third condition, the ability to toy with elements or concepts will be strengthened by web-based tools whenever "toying" is experienced as straightforward and rewarding in some sense. Let us recall that "toying" is aimed at exploration and seeing several possibilities from which new option arise that can lead to new forms, some of which may turn out to be superior in some respect than others. In this view, toying supported by web-based tools can be seen as vastly extending our possibilities of toying in the real world. The effectiveness of our toying will be increased if it is fast and easy to produce, search/find, revise, combine, see from different perspectives, visualize, as well as communicate and share the new elements and concepts. In this respect, web-based tools do have the potential to vastly extend human capacities by offering additional storage and processing power. This is particularly the case with the broadly acknowledged concept of *mashups* bringing added value by creative integration and reuse of existing services. However, there is also the risk of web-based tools being too attractive to humans in so far as they tie too much attention that may be missing form other real-world activities required for healthy, balanced human development.

Web-based tools and psychological safety. Acceptance of a person and empathic understanding, in our view, cannot be achieved by tools per se. However, the persons with whom we interconnect using the tools can provide these conditions. In this context, web-based tools, can be seen as extending the reach or pool of persons who might provide these conditions. We can interconnect with someone by whom we think we or my product will be received with acceptance and understanding. Still it has to be taken into account that the communication of acceptance and empathic understanding over the web will be only partial, cutting out several bodily and feeling-level messages. Often the message will arrive with a time lag making it harder to "tune in" as a full person with feelings and spontaneous meanings. Yet, another aspect that has been subsumed under psychological safety is — in our view — significantly influenced by web-based communication, namely the aspect of judgment. Whereas in face-to-face contact others often impose evaluation or judgment of ideas, thoughts, concepts prematurely, and we can hardly escape these judgments, the time-lag introduced by web-based tools may be used to escape premature evaluation. We can form our own perception of what we have created before turning to any responses we receive via the web. Another difference can be put as follows. Basically, the computer keyboard accepts everything and we can form our own relationship to our expressed ideas prior to sending them off. Hence, the condition of absence of external judgment can significantly and willingly be influenced by the use of computerized tools such as those providing support for professional social networks as shown in Figure 3. They appear to let us be ourselves for an initial period of time. This is not necessarily the case with close social others, who tend to judge and to compete rather than facilitate our becoming most fully ourselves.

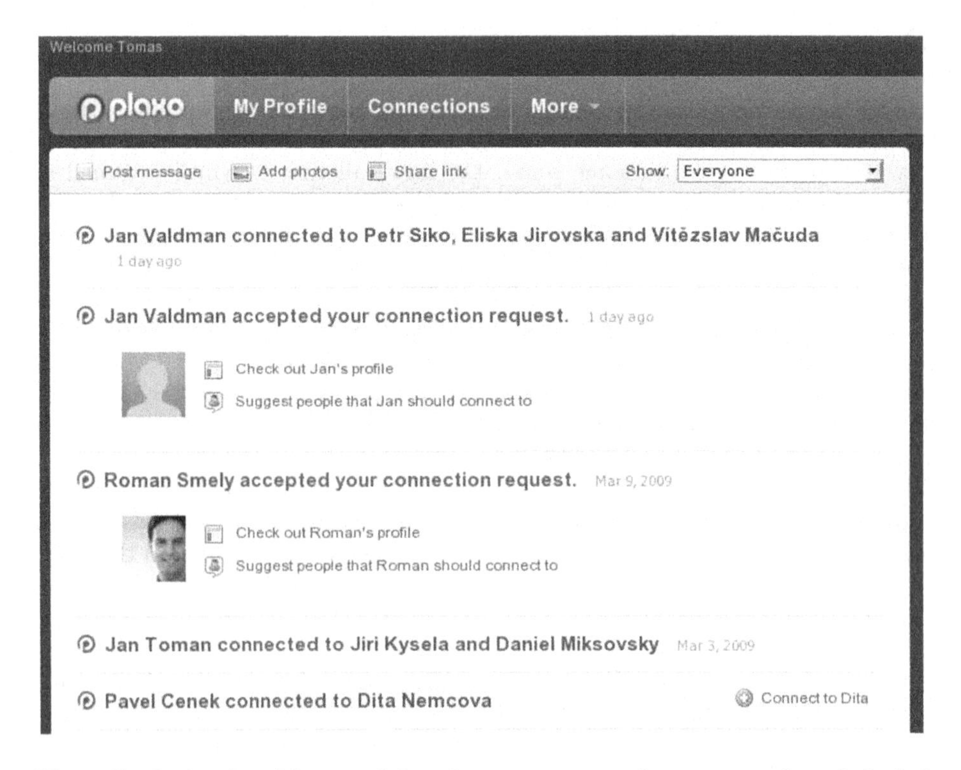

Fig. 3. Professional social network based on trust approaches aspects of psychological safety (Plaxo)

Psychological freedom in the context of web-based tools. This condition emphasizes the complete freedom of symbolic expression, meaning the full freedom to think, to feel, to be whatever one is. In this respect, web-based tools that are easy to use for one's desired expression, e.g. in words, in drawings, in tagging (as illustrated in Figure 4), in making a collage can be seen as promoting symbolic expression. In fact, they make some ways of symbolic expression easier and faster to create and adapt than physical tools. However, if the tools are complicated, counterintuitive, and hard to use, they may discourage symbolic expression. Therefore the usability of tools is a major concern if creativity is to be influenced positively. From a different perspective, web-based tools put a particular façade on some symbolic expression and in this sense are not the same as full symbolic expression such as in drama, dancing, handwriting on a paper or flipchart instead of typing, etc. Thus, web-based expression, in general, does not involve as many motor skills and physical properties and states as are involved in direct contact with other media. Web-based symbolic expression is often faster and can be communicated/spread more easily, but it is equally more abstract. Consequently, in each usage-scenario, it needs to be carefully weighed whether a more alienated but probably faster and farther reaching virtual medium shall be traded for a slower but more realistic production, involving different skills,

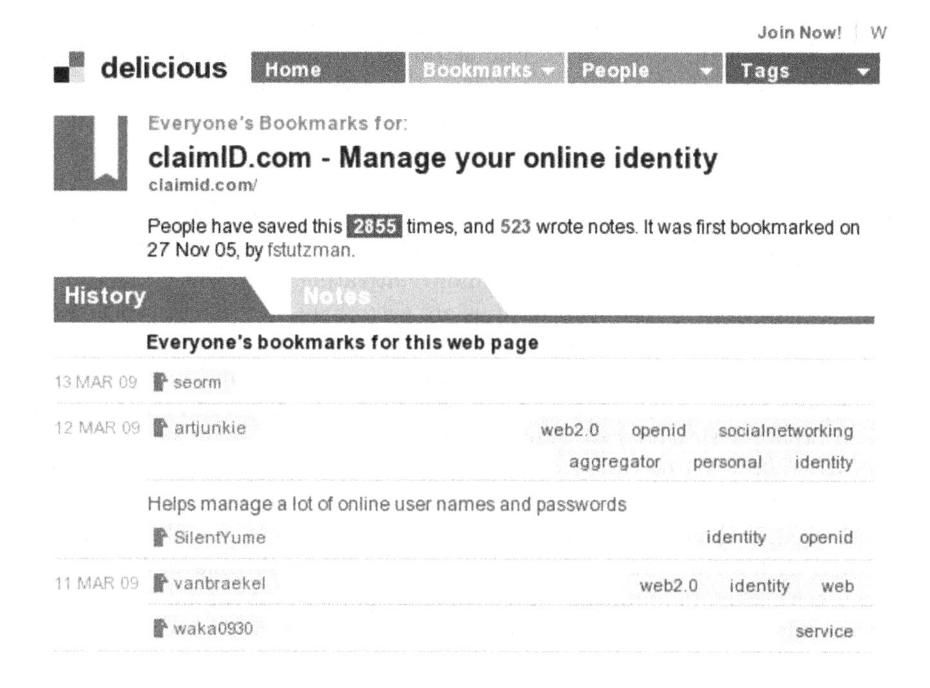

Fig. 4. Informal nature of tagging brings the freedom of expression (Delicious)

opportunities and challenges. Summarizing, web-based tools, at the same time, enhance and reduce our symbolic expression. Only when used in combination with immediate expression can the whole repertoire of symbolic expression be enhanced. Still, some web-based tools make playing and juggling of entities particularly straightforward and communicable to others. By this very feature they offer a degree of freedom unmet by other tools.

Web-based tools, in addition to being designed to offer a large amount of freedom of symbolic expression, offer a further dimension of psychological freedom. They appear to increase the number of possibilities and choices humans need to make for spending their time in general and learning in particular. More often than in face-to-face situations, we need to choose which medium to use for communication, when to communicate, with whom to connect, when/whether to respond. Watzlawick's axiom "We cannot not communicate" takes on an extended scope and meaning. With all these additional choices to make, we undoubtedly have extended potentialities. It must not be overlooked, however, that we have equally more occupation, effort and responsibility, resulting in the fact that our inner resources may be taken up for activities that we somehow need to accomplish but do not consider creative and extending ourselves at all.

Web-based tools in light of offering stimulation and challenging experiences. There is no doubt that web-based tools can offer stimulating as well as challenging experiences, e.g. by allowing for simulations, video-transmission, interaction, communication, etc. in ways hardly possible to be achieved in the physical world. However, the virtual world and way of interaction is still different to the real world such that the virtual challenges, while potentially highly instructive, should, in my view, stay subordinate to real challenges or help to resolve problems coming from the real world.

Web-based tools as channels for interpersonal contact. The notion of swarm creativity confirms that a major potential for web-based tools to foster creativity comes from their facilitation of forming/cultivating constructive relationships with social others. These relationships are ever changing, creative constructions par excellence that allow us for taking on perspectives that otherwise would stay concealed due to our innate constraints of our sensing organisms as rendering ourselves as the center of our experience. In this respect, the scope of social others with whom we can share ideas, thoughts, meanings, artifacts can be vastly extended by the use of simple, universally accessible web-based tools. A considerable constraint for interpersonal contact, however, that must not be overlooked, is the reduced transmission channel that acts as a filter with regard to the very human condition as a conscious and spiritual being with feelings and meanings arising in immediate interpersonal contact that cannot be substituted by virtual means. Consequently, any taking over of immediate contact by virtual tools needs to be considered with utmost thoughtfulness and care, in order not to cut off the very basic of the human species, our need for intimate contact and experience of relationships. In case that this experience insistently cuts off some of our senses, in other words, comes through reduced channels, it fails to provide stimulation to the cut-off features of our experience. As a consequence, the postulated condition of offering experiences is violated for particular channels. This again makes the inner condition of openness to experience obsolete, since, although openness may be given, the respective sensatory channel will be void not rendering any experience. According to a humanistic, person-centered theory of creativity, this would result in a decline, if not extinction (due to the violation of an inner condition) of creativity and consequently in adverse effects on the human condition. — Again, a balanced use of web-based tools and our immediate capabilities appears to have the potential of increasing creativity rather than extinguishing it by one-sided extremes.

3 Conclusion

In this paper a humanistic perspective of creativity has been described and investigated in light of web-based tools. It has been argued that web-based tools, per se, do not foster creativity: Their contribution to creativity primarily depends on the capacities of persons who use them. A major criterion is whether these persons provide each other with a constructive, non-judgmental atmosphere. In particular, it is important to provide the inner preconditions for constructive creativity, such as openness to experience, internal rather than external evaluation, and the joyful creation and juggling of "pieces" until something new emerges that takes on new qualities. When these inner conditions are present, usable web-based tools can be applied in a way most likely to contribute and strengthen outer conditions fostering creativity. These have been identified as a safe, resourceful, and understanding environment that not only provides freedom and variety of symbolic expression but also offers stimulating and challenging experiences and facilitates the forming and exploring of various relationships.

Other criteria fostering creativity are the particular usage scenarios of tools and the authentic purposes for employing tools. Importantly, added value can be achieved by a thoughtful blending of face-to-face and online activities that extend the potential for and the repertoire of environmental and social relationships. This is because, in sum, they address more channels of expression than any singe medium including immediacy and thus offer a richer basis for creating new concepts, forms and products.

Further research will substantiate the analytical investigation in this paper by deriving a simple framework with dimensions and associated aspects of web-based tools for supporting creativity. A complementary research thread will address the use of novel Web 2.0 services and their mashups in adapted learning scenarios. Also, interpersonal attitudes of facilitators will be assessed and checked for correlation with students' perceptions and/or their creative outcomes. Case studies, action research procedures and design-based research are intended to be employed to find out more about the conditions, scenarios, tool- and interpersonal properties that help humans to optimize their creative potentials.

References

1. Chatti, M.A., Jarke, M., Frosch-Wilke, D.: The future of e-learning: a shift to knowledge networking and social software. Int. J. Knowledge and Learning 3(4-5), 404–420 (2007)
2. Fischer, G.: Social creativity: turning barriers into opportunities for collaborative design. In: Proceedings of the Eighth Conference on Participatory Design: Artful integration: interweaving Media, Materials and Practices, Toronto, Ontario, Canada, July 27 - 31, vol. 1, pp. 152–161. ACM, New York (2004)
3. Holzkamp, K.L.: Eine subjektwissenschaftliche Grundlegung. Campus, New York (1995)
4. Jonassen, D.: Accommodating ways of human knowing in the design of information and instruction. Int. J. Knowledge and Learning 2(3-4), 181–190 (2006)
5. Motschnig-Pitrik, R.: Significant Learning Communities — A Humanistic Approach to Knowledge and Human Resource Development in the Age of the Internet. In: Proceedings of the WSKS (World Summit of the Knowledge Society), Athens, Greece, September 23-26. CCIS, pp. 1–10. Springer, Heidelberg (2008)
6. Motschnig-Pitrik, R.: Co-Actualization in Person Centered Relationships: The Construct, Initial Evidence in Education, Implications. In: Proceedings of the 8th World Conference for PCE 2008 (abstract)
7. Barrett-Lennard, G.T.: Relationship at the Centre: Healing in a Troubled World. Whurr Publishers, Philadelphia (2005)
8. Rogers, C.R.: Towards a theory of creativity. ETC: A Review of General Semantics 11, 249–260 (1954); In: [9]

9. Rogers, C.R.: On becoming a person: A therapist view of psychotherapy. Houghton Mifflin, Boston (1961)
10. Rogers, C.R.: Freedom To Learn for the 80s. C.E. Merrill Publ., Columbus (1983)
11. Rogers, N.: The Creative Connection: Expressive arts as healing. Science and Behavior Books, Palo Alto (1993)
12. Senge, P.M.: The Fifth Discipline, The Art and Practice of the Learning Organization. Currency Doubleday, USA (2006)
13. Sigala, M.: Integrating Web 2.0 in e-learning environments: a socio-technical approach. Int. J. Knowledge and Learning 3(6), 628–648 (2007)

TSW: A Web-Based Automatic Correction System for C Programming Exercises

Pietro Longo[1], Andrea Sterbini[1], and Marco Temperini[2]

[1] DI, University of Rome "La Sapienza", via Salaria 113, 00198 Rome, Italy
sterbini@di.uniroma1.it
[2] DIS, University of Rome "La Sapienza", Via Ariosto 25, 00185 Rome, Italy
marte@dis.uniroma1.it

Abstract. We present the TSW system (TestSystem Web), a web-based environment currently developed at the Rome 1 University, for the delivery of C programming exercises and their automatic correction.

The core of the correction system automatically tests the student's programs by applying unit-tests and/or by comparing the behaviour of the student's code to a reference implementation. Care is taken to avoid error propagation from a function to other functionally depending procedures by redirecting the failing calls to the corresponding reference implementation. The system "instruments" the student's code by using a code analyser and rewriter to handle instruction tracing and function calls redirection. The rewriter can be easily extended to develop other analysis instruments. As an example, we have developed: a **code coverage tool** that reports how much of the student's code has been visited during the test, a **cyclomatic complexity evaluator** to compare the number of different logic paths in the code, a **tracker for stack depth usage** to check for proper implementation of recursive functions, a **function/loop execution counter** for the evaluation of the execution complexity. Additional care is taken to capture disruptive errors that would abort the program: "segmentation faults" caused by wrong pointer dereferentiation, and time-out caused by run-away processes.

With these tools, the teacher can write rich unit tests that can either compare the behaviour of the function under analysis with a reference implementation (e.g. by generating random input and comparing the results), or by submitting well-crafted special inputs to elicit special cases or by comparing the complexity and/or stack depth counters. Each test applied will then explain to the student what was the problem found.

TSW is the core component of a future larger social knowledge project, in which students will cooperatively/competitively participate to the definition and test of each-other's programs, sharing ideas and learning from each other.

Keywords: automatic grading, automatic correction, programming exercises.

1 Motivation

In sciences and engineering faculties, such as the ones we belong, computer science foundations and programming are topics normally thought in one or more

M.D. Lytras et al. (Eds.): WSKS 2009, LNAI 5736, pp. 13–21, 2009.

courses, during the first year of every study programmes. Our courses are normally populated by 90 to 150 students, and managed by one teachers and a tutor. Often, in our experience, students that do not attend the labs and homework activities and then try the exam, fail and should repeat the exam at the next semester. The environment we are confronting at present is that of courses attended by 130 students (some of which are worker-students that never attend face-to-face classes). We are giving homeworks every two weeks, both to analyse their progress, and to give them useful feedback through the homework correction. While we see that such activity could strengthen the students' motivation and satisfaction in programming (not to mention increasing their performance as a final result), we see the effort needed on our side as well: an overall number of 650 student's homeworks to be corrected. This can be managed only by using a framework for automatic correction and feedback generation.

So far we have been experimenting, a bit roughly, with black-box-testing of the student's programs, by checking on several input sets the conformance of its output with the expected ones. This approach is helpful, though in some cases - let's say 15% of them - we had to go over a direct hand correction of the code to cope with very simple errors. Such an approach is too simple minded indeed and could lower the student's motivations (e.g. because of "insignificant" output differences) as normally the students are more focused on the logic of the exercise, than on its I/O functions. Providing the student with a whole library containing the proper I/O functions, would heal some of these errors, yet we think this would move the frontier of the testable problems just a little further. Is it possible to provide the teacher with a sufficiently simple framework to support the definition of exercises, the specification of the various detailed tests, and their final automated correction?

In the development of a framework to answer to the above question we have the following goals:

- make the correction completely automatic. This way the corrector can be used also as a self-evaluation tool in distance-learning settings, letting the students test their code and resubmit, in an iterative improvement process;
- give the student a detailed correction report that explains, where possible, what the error was and how to avoid it;
- avoid the propagation of errors from a function to others, to capture good work that normally fails in black-box tests;
- collect evidence of side-effect related errors (e.g. improper use of global variables or of references pointers)

2 Related Work

A similar problem has been solved by Morris [1] for programming courses in Java. In his work he uses Java reflection to analyse the student code and to replace failing functions with his reference implementation.

In C a similar technique is harder to apply because reflection is not available. Yet, a viable alternative to reflection can be found in Aspect Programming [2].

Aspect Programming allows the definition of "aspects", i.e. the specification of code that should be weaved through the program to add some functionality (e.g. counting the number of calls of a given function). With Aspect programming we can easily instrument the student code so to:

- test each function against a set of unit-tests
- replace failing functions with a reference implementation
- count function calls to check algorithm complexity

Tests are written by following the software engineering Unit-test methodology, where the smallest possible tests (unit tests) are defined and then collected in test suites.

In previous work [4] we have followed this approach, finding that the aspect-C weaver and compiler is rather hard to use on general C exercises. In particular the compiler was a little too picky respect to the normal coding errors common in student's code. Thus we have built a code transformation system that is sufficiently general and robust respect to the student's code.

In this paper we present the Test System (TS), that transforms the submitted code to be tested, the Test Builder (TB), a Java application used to prepare the C templates needed to write an exercise with its set of tests, and the Test System Web (TSW), the web-interface used by teachers and students.

The TWS system is thought as the initial core component of a larger (social) system where students will cooperate and participate to a social game, based on reputation and tournaments, by sharing their ideas and competing at a peer level. E.g., students will be able to play against each other, either by writing programs that win against many of the peers' tests, or by writing tests that catch many of the peers' bugs. This will be obtained (at the core level) by making easy to participate to the definition of tests and by introducing automatic comparison and classification of programs coming from different students. The teacher itself will be able to classify clusters of similar solutions and design better tests to catch common errors suggested by the students' programs. This will help her/him to produce more detailed explanation to be presented as feedback.

3 The Core Test System

The core TestSystem is a Java application that takes a C programming exercise (a single file containing a set of functions without the main program) and "instruments" the program to test it against a set of unit-tests.

The unit tests are written in C++, under the framework of the CppUnit library, with the help of a set of libraries and C macros that we have developed to:

- prepare the feedback messages for failed tests,
- capture fatal errors related to invalid pointers and wrong memory accesses,
- check for the correct usage of assertions,
- check for the maximum stack depth used by the functions,
- check for the iteration (loop, do, while) counters,

- check for the function call counters,
- compute the instruction coverage (i.e. count how many of the instructions have been executed),
- provide dummy versions of each function to avoid link errors from missing implementations,
- automatically switch to the reference implementation of a function whenever it fails its tests, to avoid error-propagation to other tests.

The TestSystem parses the student's code and applies a set of transformations that add the machinery needed to support the above mentioned macros. The code rewriter is very general, and can be easily extended to add other probes. E.g. we have added a **cyclomatic complexity analyzer** that shows how many different execution paths are present. The cyclomatic complexity is very useful to show which correct solutions are simpler (and thus to assign them higher grades).

The report produced by running the tests is made as a set of XML files, that can be further analyzed or transformed through XSL transformations. The default report shows a compact table collecting for each function the number of tests passed, its cyclomatic complexity and its statement coverage (see figure 1). Separate pages are generated with all the test details and the feedbacks.

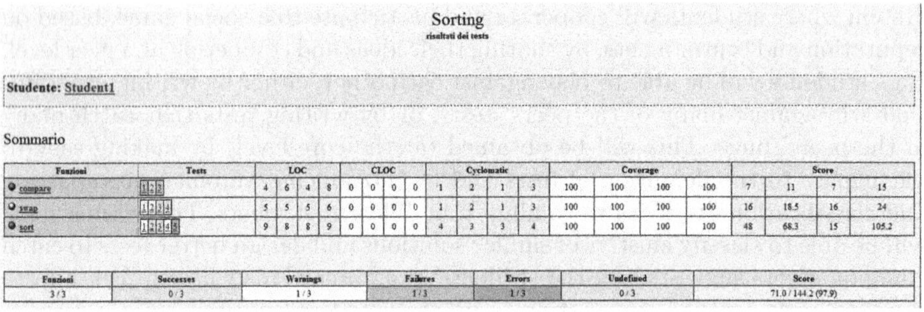

Sorting
risultati dei tests

Studente: Student1

Sommario

Funzioni	Tests	LOC				CLOC				Cyclomatic				Coverage				Score			
compare	1 2 3	4	6	4	8	0	0	0	0	1	2	1	3	100	100	100	100	7	11	7	15
swap	1 2 3 4	5	5	5	6	0	0	0	0	1	2	1	3	100	100	100	100	16	18.5	16	24
sort	1 2 3 4 5	9	8	8	9	0	0	0	0	4	3	3	4	100	100	100	100	48	68.3	15	105.2

Funzioni	Successes	Warnings	Failures	Errors	Undefined	Score
3 / 3	0 / 3	1 / 3	1 / 3	1 / 3	0 / 3	71.0 / 144.2 (97.9)

Fig. 1. The test report for the sorting exercise

3.1 Writing a Test

Writing a test, with the Test Builder that (as we will see) generates all the surrounding machinery, means normally just:

- defining the involved data-types
- writing the reference implementations
- filling the test function templates

A simple test example for a **sort** function is shown in listing 1.1. Here we can see that the messages to be shown if the test fails are prepared before the test to avoid loosing the output in case a segmentation fault appears.

```
void test_sort(){
    // define the input array
    int input[] = {2, 6, 3, 8, 9, 0, 9, 4, 5, 7};
    int len = 10;
    // Prepare the input/output description messages used in
       the test reports
    in << "The array input[] = {2, 6, 3, 8, 9, 0, 9, 4, 5, 7}
        with len = 10";
    out << "The sorted array: input[] = {0, 1, 2, 3, 4, 5 ,6
       , 7, 8, 9}";
    // invoke the function
    sort(input, len);
    // verify that the output is correct
    for(int i = 0; i < 10; i++)
        // throw an exception (with default message) if an
           element is out of place
        MY_ASSERT(input[i] == i);
}
```

Listing 1.1. A simple test just checking the output of the sort function

A more complex example that checks for the number of loops executed and compare it to the reference implementation is shown in listing 1.2. Here we can see how both the tested and the reference functions are called, and then the counters are compared.

```
void test_sort_10000(){
    int a[10000];
    int b[10000];
    // fill two arrays with the same 10000 random numbers
    srand(time(NULL));
    for(int i = 0; i < 10000; i++){
        a[i] = rand();
        b[i] = a[i];
    }
    in << "An array of 10000 random integers";
    out << "The array efficiently sorted";
    // we add a hint that will appear if the function goes
       timeout
    ADD_TIMEOUT_MESSAGE(
        HINT("To sort efficiently use the quicksort or
            mergesort algorithm"));
    PROTECT();
    // call the reference version to compute her complexity
    sort_Ok(a, 0, 9999);
    // call the tested version to compute her complexity
    sort(b, 0, 9999);
    // check the counters and fail if the sort is slower than
        10 times sort_OK
```

```
    COMPARE_COMPLEXITY_MESSAGE(sort, sort_Ok, 10 * COMP,
        HINT("To sort efficiently use the quicksort or
            mergesort algorithm"));
}
```

Listing 1.2. A more interesting example, checking the complexity of the implemented solution against the reference implementation

4 Building a New Exercise

Functions definition. The construction of the set of files, tests and directories that makes all the machinery to test an exercise is very complex (see the directory structure to the right).

To ease the task, a Java-based application (the TestBuilder) is used to prepare all the C++ templates needed to write the tests. The interface allows the definition of prototypes of the functions to be implemented in the exercise, their order of testing, and the possible existence of functional dependencies (see Fig.2).

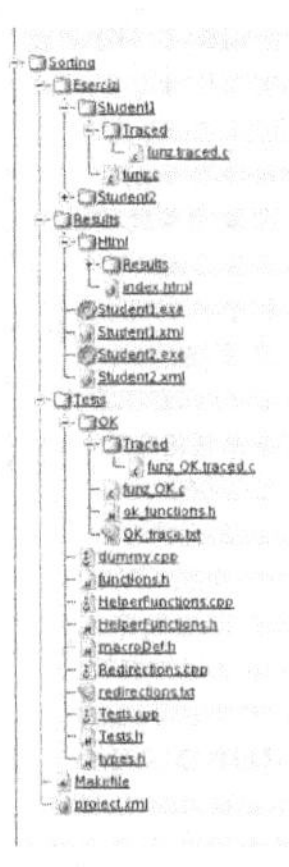

Directory structure of an exercise.

Fig. 2. Creation of an exercise: definition of function prototypes, test order and functional dependencies

Functional dependencies are used to enable the automatic replacement of a submitted function with its reference implementation when tests fail. This way the tests of the following (functionally-dependent) procedures can be done safely, without error propagation. In the figure we see that the three functions `compare`, `swap` and `sort` should be tested in this order, and that the first two should be replaced with the reference version if they fail a test.

Tests definition For each function a set of test names can be defined, to produce the corresponding code templates and test coordination machinery (see in figure 3 the interface to define the set of test names).

When the project is built all the needed files are generated.

Tests

add a test ok

Functions to test

name	score	warning	failure	warning	timeout	
test_sort_Null	5	1	1	☑	☐	
test_sort_5	20	10	5	☐	☐	
test_sort_10	20	10	5	☐	☐	
test_sort_1000	30	5	2	☐	☑	5
test_sort_10000	30	5	2	☐	☑	5

Fig. 3. Creation of the template of the unit tests

5 The Web-Based Interface

The system is made available through a web-based interface, the TSW (Test System Web), see figure 4. The simple interface allows three types of users: administrator, teacher, and a student.

The administrator can mainly define new teachers.

Fig. 4. TSW: the home page

A teacher can:

- add new exercises to a course
- browse all the student's tests
- handle students
- check the wall of fame to compare the student's solutions

A student can:

- submit new solutions to an exercise (even more than once)
- browse his personal tests
- check the wall of fame to compare his rank to the others

6 Conclusions

We have presented the TWS system for the automatic correction of C programming exercises.

The direct application of the system described is of course related to the support of distance, asyncronous learning activities aiming at the development of programming abilities. We see some relevance of our effort, though, in the area of development and management of socially based knowledge. In particular we see a natural evolution of the system towards a reputation based application, allowing to the system users to "add" knowledge in the system and to participate in the construction of the technical and methodological knowledge that is used to evaluate the exercises. Such social level, which is still to be developed for tsw, yet can enjoy authors' experiences in social knowledge systems, can give a social dimension to the presently narrow applicability of the system.

The most labour-intensive part of using the system is the construction of new unit-tests for the submitted solutions. To further lower the work needed to produce a new exercise we are planning the following enhancements:

leverage the reference solution: We already use a "reference" version of the solution to replace failing functions. The reference function can be leveraged to make easier writing black-box tests. The idea is that a test could just compare the output of the two solutions (the "reference" and the student one) to discover if there is any difference.

random data structures: Thus we are extending TWS to allow the specification of randomly generated input data for the functions to be tested. The output's correctness is checked by means of a set of classes that either test the isomorphism of the resulting data structures (the student's one and the teacher's one) or transform the data structures to a "canonical" form and then check for equality.

classification of common errors: As soon the random generation and comparison of data structures is in place, any pair of solutions can be compared. This allows us to classify the student's solutions by clustering together the solutions that behaves in similar ways. Thus the teacher can use the student's solutions as reference examples of common errors.

recycling the student's solutions: On a parallel track, we are designing a simple work flow and an easier test definition to enable recycling (reuse) of the student's submissions. In our future experiments we are going to ask the students to write both code and tests for their code (thus teaching the usefulness of the "write-test-then-the-code" style of programming). Then we will run tournaments where the exercises implemented by each student (and by the teacher) will be tested against tests implemented by the teacher and by the other students. The outcomes will elicit the best implementations of both the exercises and the tests. Our final goals are both: to raise the interest of the students, and to "recycle" the tests to incrementally enhance the automatic corrector.

social games: The TSW system will be used as part of a more general cooperative learning activities management system. In particular, we are planning its integration within our SocialX system for the management of general cooperative activities [5,6].

knowledge based metadata: Moreover, the TSW exercises' metadata should be enriched to allow their automatic retrieval, to become part of an automatically-generated course, as we do in our Lecomps system for the construction of adaptive courses [8,7].

References

1. Morris, D.S.: Automatically Grading Java Programming Assignments via Reflection, Inheritance, and Regular Expressions. In: Proc. Frontiers in Education 2002, Boston, USA (2002)
2. `http://www.aspectc.org`
3. `http://cppunit.sourceforge.net`
4. Sterbini, A., Temperini, M.: Automatic correction of C programming exercises through Unit-Testing and Aspect-Programming. EISTA (2004)
5. Sterbini, A., Temperini, M.: Good students help each other: improving knowledge sharing through reputation systems. In: ITHET 2007, Kumamoto, Japan, July 10-13 (2007)
6. Sterbini, A., Temperini, M.: Learning from peers: motivating students through reputation systems. In: SPeL 2008, Turku, Finland, July 28 (2008)
7. Fernandez, G., Sterbini, A., Temperini, M.: Learning Objects: a Metadata Approach. In: Eurocon 2007, Varsaw, September 9-12 (2007)
8. Limongelli, C., Sterbini, A., Temperini, M.: Automated course configuration based on automated planning: framework and first experiments. In: International Conference Methods and Technologies for Learning (ICMTL 2005), Palermo, Italy (2005)

Taking Up the Challenge of Evaluating Research Results in the Field of Technology Enhanced Learning

Rosa Maria Bottino[1], Michela Ott[1], and Francesca Pozzi[1]

[1] Istituto Tecnologie Didattiche – CNR
Via De Marini, 6, Genoa Italy
{bottino,ott,pozzi}@itd.cnr.it

Abstract. The present contribution tackles the issue of the evaluation of research results in the field of Technology Enhanced Learning (TEL). It presents an evaluation model designed to evaluate a web based Pedagogical Planner and the theoretical model underpinning it. The evaluation model, designed and tested within a European project, proved to be able to highlight the strengths and weaknesses of the TEL system at hand, by distinguishing between theory and practice, between the advantages and disadvantages derived from the theoretical model and/or those referable to the implemented tool.

Keywords: Research Evaluation, Technology Enhanced Learning, Educational Innovation, Pedagogical Planning.

1 Introduction

The evaluation of educational research results is now considered an important issue in order to ensure "quality and sustainability in the development and delivery of courses, programmes and services…" [1] in one word, in order to improve the quality of teaching and learning; it should also be regarded as a key aspect to guarantee a well grounded, unquestionable basis for the educational developments of today's Knowledge Society.

In the field of Educational Technology, a field where theory and practice are by definition very strictly intertwined, a wide number of studies have been published over the years, which focus on the evaluation of the usability and of the educational effectiveness of the tools and systems produced and adopted [2,3] including, more recently, (e-)learning environments [4]. Even in those contexts where Knowledge Management (KM) practices are increasingly being adopted, the evaluation of the KM systems is becoming an emergent issue [5]. As pointed out by Kim et al. [6], while the research community shows keen interest in proposing and testing evaluation theories and methods regarding the available ICT tools, very few studies focus on the evaluation of innovative educational systems seen in their double-sided aspect: theory and practice, educational ideas and their actualization, educational models and developed tools. To date, very few attempts have been made, in fact, to design and set up comprehensive evaluation systems taking into account both the actual educational ICT tools and the underpinning theoretical models. The recent valuable effort of Hlapanis

M.D. Lytras et al. (Eds.): WSKS 2009, LNAI 5736, pp. 22–30, 2009.

and Dimitrakopoulou [7], who addressed the issue of evaluating both a course model concerning teachers' education and the related web-based environment, confirmed the need to assume both perspectives, so as to better understand the role played by the conceptual design of an educational resource with respect to its concrete actualization (the implemented tool).

This paper reports on the efforts made by the authors in this direction in the framework of the European project ReMATH[1] – Representing Mathematics with Digital Media. ReMATH was a research and development project co-funded by the European Commission (IST-4-26751) which had two main goals: the development of ICT-based tools for mathematics education at secondary school level and the design and experimentation, in different school contexts, of learning activities involving the use of such tools. This latter goal implied the necessity of having a common resource to build and share learning activities among the research teams involved in the project (including the teachers who were in charge of carrying out the classroom experiments). The design of such a resource was based on the idea of "Pedagogical Plan" for representing teaching and learning activities. A conceptual model of "Pedagogical Plan" was, then, developed within ReMATH [8]. On the basis of such model, a web-based tool, called the Pedagogical Plan Manager (PPM), aimed at the production and sharing of pedagogical plans, was then worked out. In the framework of the ReMATH project, an evaluation was carried out, focusing on both the conceptual model of "Pedagogical Plans" and its implementation (PPM), namely the web-based environment instantiating the Pedagogical Plan concept.

The paper, then, explains the methodology adopted for the evaluation of the work done by the authors in the framework of the above mentioned project and it also summarizes the evaluation results shedding light on how the overall evaluation process proved to be highly effective in suggesting possible improvements to both the model that had been conceived and the ICT system that had been implemented.

2 The Context: The ReMATH Project and the "Pedagogical Plan" Model and Tool

The ReMATH project had the primary aim of building an integrated theoretical and operative framework for mathematics learning. The project gathered several research teams throughout Europe who, despite the fact that they all worked in the field of technology enhanced learning in mathematics, based their work on different theories and on different approaches [9]. As mentioned before, during the project a number of classroom experiments were carried out according to a cross – experimentation approach [10], that is, it was decided that each team participating in the ReMATH consortium, would carry out two different classroom experiments using respectively the ICT-based tool they themselves had implemented ("familiar"), and the ICT tool produced by another team ("alien"). Due to the substantial differences among the partners' theoretical perspectives and actual working methodologies, the need to establish a common language to describe pedagogical ideas and theories, as well as practicalities and specific activities to be carried out, emerged from the very beginning of the

[1] More info about the ReMATH project: http://remath.cti.gr/default_remath.asp

project. The key to the mutual comprehension of the research teams, to the feasibility of cross-experiments and, finally, to the definition of a shared framework, was, then, the building up of a common resource aimed at enabling the partners to share pedagogical knowledge, compare approaches and foster pedagogical design reuse.

In the light of the above mentioned objectives and needs, the project required the realization of a specific system to support the process of pedagogical design, namely the description of learning activities to be enacted during cross-experiments (thus also enabling and fostering their reusability). In response to such a specific need, a Pedagogical Plan model was then devised. A Pedagogical Plan was conceived as a description of pedagogical activities to be carried out in a real context (e.g. a classroom, a laboratory, etc.) where a number of different descriptors should be made explicit, at different level of details [11]: educational target (What learning outcomes? What learning contexts? Who are the target learners?); pedagogical rationale (Why those learning outcomes? Why applying a certain strategy? Why using a given tool?); Specifications (Which activities are to be carried out? Which roles are to be assumed by the different actors? Which resources and tools are to be used? etc.). The Pedagogical Plan model was conceived as a flexible, modular entity capable of embracing both simple plans for single activities and very complex structures with activities of different type arranged into different levels. Such flexibility was achieved by defining the plan as a tree-like hierarchy made up of simpler, elementary plans. Thus the PPM prototype allows the representation of Pedagogical Plans as hierarchical entities which can be built and read at different levels of detail. This structure supports both "authors" of Pedagogical Plans (providing them with the possibility of working with a top-down structure) and "readers", who in top-down organizations have a facilitating factor to navigate from the general to the particular and vice versa, and to explicitly select the descriptors they want to focus on, thus contributing to a better understanding of complex plans (grasping the general structure, and relating rationales with concrete details, etc.)

The PPM has been designed as a wholly web-based system, accessible via standard web browser, whose interface has been designed so as to allow both authors (PPM Editor) and readers (PPM Viewer) to deal easily and naturally with the hierarchical structure.

3 A Methodology to Evaluate the Pedagogical Plan Model and System

In the following the methodology adopted within ReMATH to evaluate the Pedagogical Plan (model and tool) is described.

3.1 General Approach to Evaluation

The need for "identifying and making explicit the evaluation criteria for Pedagogical Plans" was set as a priority from the very beginning in the ReMATH project proposal but it was also pointed out as a key point during the annual reviews of the project. Within the project, evaluation was regarded in a "formative perspective", so as to enable subsequent improvements of the system. In this view, some of the evaluation means produced (which are detailed in the following paragraphs) were used starting from the very beginning of the project, in order to allow an "in itinere" evaluation of the work in progress.

The evaluation techniques adopted basically followed the paradigms of "mixed research methodology" [12], they made, in fact, a combined use of quantitative and qualitative approaches: both quantitative data coming from questionnaires and qualitative data coming from direct observation, individual interviews and free-style opinions reported in questionnaires were, in fact, gathered and considered for the evaluation of the system, even though, due to the limited number of people involved in the process, the qualitative part assumed a more crucial role than the quantitative one.

A global evaluation model was, then, defined, which is explained hereunder, having the main aim of verifying on the one hand the soundness of the pedagogical plan model and, on the other hand, the usefulness of the actual tool developed in the project.

3.2 Evaluation Focus: Different "Elements" under the Lens

In the framework of the ReMATH project, the need for a separate evaluation of both the model and the implemented tool, emerged clearly from the substantial interdependence between the subsumed theoretical model and the PPM features; this approach was also meant to provide the PPM designers with sufficient feedback to grant the possibility of carrying out significant improvements on the work done. Furthermore, in addition to the two above mentioned research products (the model and the PPM), during the project lifespan a number of actual Pedagogical Plans were produced and subsequently experimented in real educational contexts. This implied that a thorough evaluation of the research results could not avoid consideration of these artifacts (actual Pedagogical Plans), that, in fact, provide an instantiation of the global working methodology adopted and the functionalities of the developed system.

3.3 Evaluation Indicators

Three main categories of indicators were recognized as crucial to shed light on the general suitability and effectiveness of the Pedagogical Plan: perceived usefulness, perceived ease of use [13, 14], and (re-)usability.

In particular, the "perceived usefulness" focuses on the possibility offered by the Pedagogical Plan model to serve the users' needs, in terms, for example, of suitability/comprehensiveness of the pedagogical descriptors, ability of the model to present the information effectively, etc. Indicators pertaining to this category mainly address the evaluation of the pedagogical plan model, while indicators of the other two categories mainly focus on more concrete aspects (namely PPM and actual Pedagogical Plans). The "perceived ease of use" deals with the effort required by the user to use the PPM, and should be regarded in terms of interface quality, adequacy of support provided, general understandability of the system, etc... Finally, indicators related to "(re-)usability" besides considering the clarity of the implemented Pedagogical Plans and both the understandability and pedagogical soundness of their contents, also refer to their ability to provide enough information to support classroom enactment of the envisaged activities.

3.4 Evaluation Perspectives

When dealing with the evaluation of the Pedagogical Plan in its complexity (namely: model + PPM + plans), three different users' perspectives should be assumed: the author's perspective, the reader's perspective and the experimenter's perspective (Fig. 1).

In particular, the *Author* uses the PPM Editor to elaborate a Pedagogical Plan expressing not only the activities to be carried out with students, but also the educational objectives, the theoretical assumptions that have inspired the design of the plan, the roles to be enacted by the different actors (students, teachers…) and a number of other significant aspects involving the educational activities. The *Reader* is the one who reads the plans through the PPM Viewer with the aim of better understanding how they can be used in school practice, and, eventually (not necessarily), of adopting them. The *Experimenter* uses the plans to practically enact them in a classroom context. Readers and Experimenters, despite significant commonalities, when approaching the reading of the plans have very different objectives and their viewpoints may, therefore, differ significantly: while the final goal of the experimenter is that of actually using the written plan as a support for the classroom enactment of the envisaged activities, the reader simply reads the plan without necessarily having in mind to use it, he has the broader aim of grasping ideas, assuming new or different points of view, approaching and understanding new working methodologies etc…

The three different actors (Authors, Readers and Experimenters) also access the system from different entry points (Fig. 1): while authors typically "enter" from the PPM Editor (which is still content – free) and need to have a clear understanding of the subsumed theoretical model, both readers and experimenters take the "way in", starting from the actual implemented plans and may have a different level of interest in fully understanding the theoretical model underpinning the plans.

The three users, in theory, can be both researchers and teachers and the Pedagogical Plan was conceived to hopefully serve the needs of both of them.

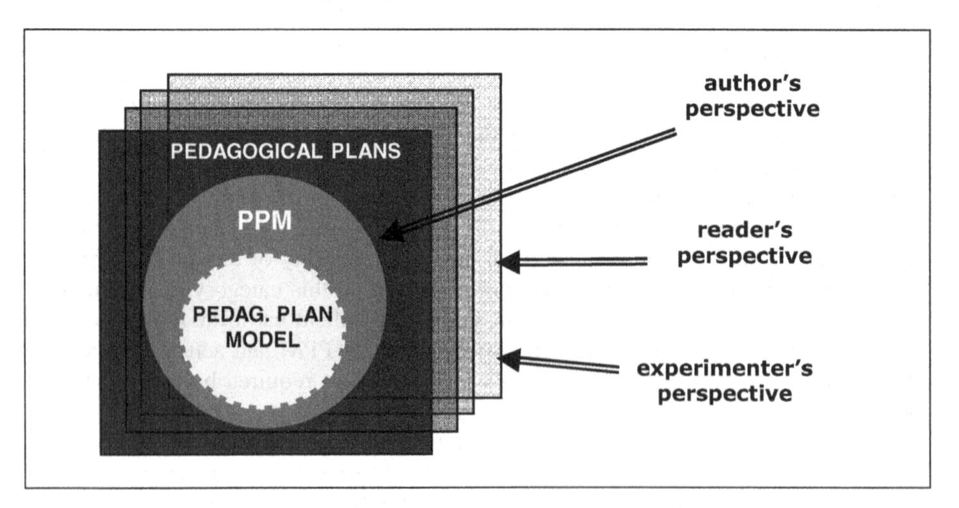

Fig. 1. Perspectives considered in the evaluation

In ReMATH, nevertheless, the authors were generally researchers while the readers and the experimenters were both teachers and researchers: the presence of different expertise in the same role provided the actual evaluation process with significant added value.

In order to finalize the evaluation model, the need also emerged to analyze the use made by the ReMATH partners of the Pedagogical Plan.

A fourth perspective was then considered, in order to highlight the strengths and weaknesses of the Pedagogical Plan; this was the role of its designers and developers (i.e. the authors of this paper) who, from their view point (actually the fourth perspective), carried out a thorough investigation of the actual use made of the implemented system by the ReMATH users.

Table 1. Synthesis of the evaluation focuses, objectives and means

	evaluation focus	**objectives of evaluation**	**evaluation means**
Author's perspective	PPM Editor	to assess the soundness of the subsumed theoretical model and its ability to allow the full expression of ideas, working methods, educational theories and principles	questionnaire informal interviews
Reader's perspective	PPM Viewer	to assess the perceived ease of use of the system	case-study approach (test group composed of two mathematics teachers and one mathematics researcher) questionnaires and individual interviews
Experimenter's perspective	PPM Viewer	to judge the pedagogical soundness of the plan and to verify whether experimenters were able to find out sufficient information to carry out the envisaged activities in the classroom. (plan re-usability and understandability, which implies completeness and clearness of plans)	two questionnaires (one devoted to the "experimenting researcher" and the other to the "experimenting teacher")
Designer's perspective	pedagogical plans	to evaluate how ReMATH partners had really used the most relevant PPM features	direct observation of the various versions of the implemented pedagogical plans

4 Results

As already mentioned, the evaluation model was applied and used within ReMATH and has provided a series of results, in terms of indications concerning the model, the PPM and the realized plans. In the following, a synthesis of these results is provided to give an idea of the kind of feedback that the evaluation model, as it is conceived, may provide.

The evaluation carried out in ReMATH from the authors' perspective, provided useful information mainly about the PPM Editor. The experience of use of such a system was judged by the ReMATH plan authors as largely positive; they found that the PPM Editor had provided a satisfactory level of support for the task of designing Pedagogical Plans and that the conceptual model underpinning the PPM had provided them with concrete help in expressing their ideas; most of them also noticed the use of the PPM Editor had allowed them to bring into sharper focus their initial ideas. While no major problems were reported, some authors did express the impression that constructing their plans required a considerable effort and some of them suggested diminishing the amount of descriptors, which is a useful indication that may be considered to modify the model.

The evaluation carried out by the readers shed light on the strong points and weaknesses of the PPM Viewer. The ReMATH readers/teachers appeared not to be fully satisfied with the PPM Viewer and mostly asked for a simplified interface of the system, with less, more practical information, so as to be able to get an immediate, complete picture of the plans. From the interviews with the readers it also emerged that, while the structural information of plans was adequately conveyed by the PPM Viewer, some of the contents of plans needed, instead, to be filtered, so as to allow readers to immediately find practical information.

Results of the evaluation from the experimenters' perspective brought to light considerable differences between teachers' and researchers' points of view. Most experimenting teachers declared that they never acted as the main users of the PPM, but had sporadic access to the system; they preferred to have their own work "mediated", suggested and guided by the researchers. On the contrary, it seems that the PPM constituted an important tool for the experimenting researchers, who used it extensively and appreciated it very much. Most experimenting researchers also found that the PPM was an appropriate tool to allow communication between them and with the experimenting teachers and judged plans to be clear and exhaustive enough to be used and experimented in real contexts.

The analysis of the process of construction of the different plans (fourth perspective) showed that the hierarchical organization and the descriptors of the Pedagogical Plan model had been used differently, in order to achieve different levels of detail/generality in the overall description of the plan. Common patterns of behavior among the PPM users did emerge, but the variety of approaches and modalities of use basically confirmed the flexibility of the model, which appeared to be suitable to meet different needs. This is particularly important, because in some situations it may be appropriate to convey the overall gist of a plan and thus focus at a general, abstract level while on other occasions, attention may need to be directed towards very concrete aspects mainly considering details.

All in all, researchers were, no doubt, the ones that appeared to benefit from the PPM system facilities at a higher level; they were satisfied with the plan model and appeared to be almost completely satisfied with the use of the PPM tool which, in their words, fully met their expectations and allowed them both to bring into sharper focus their ideas and to easily communicate with each other. On the contrary, the ReMATH teachers (both the readers and the experimenters) considered the effort required to access the pedagogical planning system a little too high and found considerable difficulties in approaching the PPM system (basically the Viewer); they also found some of the descriptors redundant.

These basic considerations emerging from the evaluation process consent, on the one hand, to appreciate the Pedagogical Plan in its complexity, and, on the other one, to separate the different elements (plan model, PPM Editor, PPM Viewer, actual plans) so as to identify the strong points and weaknesses of each single element. In particular, while the Pedagogical Plan model was largely appreciated, its main quality lying in its flexibility, it also appeared that the PPM Viewer presented many weak points, thus requiring substantial revisions.

5 Discussion and Conclusions

The evaluation of research results in the field of Educational Technology is *per se* a research issue, given that the rapidly growing use of technology is changing the way in which knowledge is produced, stored and distributed [15].

The experience reported in this paper is an example of a methodology developed to evaluate a system in the field of TEL. Even if the Pedagogical Planner developed in the framework of the ReMATH project had specificities and requirements of its own that make the methodology not completely exportable as it is, there are still a number of elements of such a methodology that we think are worth sharing and discussing, in the light of possible reuse for evaluating other TEL systems and – more in general - KM systems. In particular, the experience showed that there are at least four key points to be considered when taking up the challenge of evaluting TEL systems: the *evaluation process,* the *evaluation focus,* the *evaluation method and means* and the *evaluators.*

As to the *evaluation process,* in ReMATH this was carried out both "a posteriori" and "in itinere". In particular, the monitoring of the work in progress and the continuous interactions between the designers and the actors in ReMATH, helped to better tune the system at the various development steps, thus having the effect of better finalizing the research efforts. Such an approach allowed us to refine the conceptual model according to the emerging user's needs before building up the PPM system, so as to avoid significant lacks, inconsistencies and/or redundancies. As to the *evaluation focus* (namely the elements to be evaluated) the ReMATH experience showed that, in order to assess an innovative system, it is worth evaluating the conceptual model underpinning it, together with its main functionalities and the outputs it may provide. Such a complex evaluation should be done globally, also taking into account the possible different viewpoints. This allows one to "locate" exactly the strengths and weaknesses of each research output. In ReMATH this kind of approach helped, for instance, to shed light on the PPM Viewer weaknesses with respect to the conceptual model strengths, thus allowing for a well focused revision of the former. As to *the evaluation methods and means* the ReMATH experience underlined the importance of gathering both quantitive and qualitative data. As a matter of fact, due to the complexity of the aspects to be evaluated, a quantitative approach would miss some "hot" aspects, that on the contrary may be well caught thanks to a qualitative evaluation. Thus, a mixed – approach appears to be the most appropriate. Moreover, the ReMATH experience suggests that the wider the number of the evaluation means used is, the more we can have a thorough and deep insight into the key aspects of the research outputs and activities. As to the *evaluators* (namely, the type of persons/users to involve in the evaluation process), it should be noted that very often a system may have different kinds of users. The multi-evaluator approach may provide a significant added value to the evaluation process, because it allows one to take into account a more ample spectrum of viewpoints and focused opinions, thus helping in understanding whether the developed system really satisfies all the users' needs.

References

1. Latchem, C. (ed.): A content analysis of the British Journal of Educational Technology. British Journal of Educational Technology 37(4), 503–511 (2006)
2. Jacobs, G.: Evaluating courseware: Some critical questions. Innovations in Education and Training International 35, 3–8 (1998)
3. Crowther, M.S., Keller, C.C., Waddoups Gregory, L.: Improving the quality and effectiveness of computer-mediated instruction through usability evaluations. British Journal of Educational Technology 35(3), 289–303 (2004)
4. Bligh, B.: Towards a compendium of learning space evaluations, `http://www.lsri.nottingham.ac.uk/JELS/JELS_newsletter1_Dec08.pdf` (accessed December 08, 2008)
5. Alexandropoulou, D.A., Angelis, V.A., Mavri, M.: Knowledge management and higher education: present state and future trends. Int. Journal of Knowledge and Learning 5(1), 96–106 (2009)
6. Kim, S., Brock, D.M., Orkand, A., Astion, M.L.: Design implications from a usability study of GramStain-TutorTM. British Journal of Educational Technology 32(5), 595–605 (2001)
7. Hlapanis, G., Dimitrakopoulou, A.: A course model implemented in a teacher's learning community context: issues of course assessment. Behaviour & Information Technology 26(6), 561–578 (2007)
8. Bottino, R.M., Earp, J., Olimpo, G., Ott, M., Pozzi, F., Tavella, M.: Supporting the Design of Pilot Learning Activities with the Pedagogical Plan Manager. In: Learning to live in the knowledge society. IFIP – International Federation for Information Processing - Book Series, pp. 37–44. Springer, Boston (2008)
9. Artigue, M.: Digital technologies: A window on theoretical issues in Mathematics education. In: Pitta-Pantazi, D., Philippou, G. (eds.) Proceedings of the V Congress of the European Society for Research in Mathematics Education CERME, vol. 5, pp. 68–82 (2007)
10. Bottino, R.M., Artigue, M., Noss, R.: Building European Collaboration in Technology-Enhanced Learning in Mathematics. In: Balacheff, N., et al. (eds.) Technology Enhanced Learning, pp. 73–87. Springer, Heidelberg (2009)
11. Bottino, R.M., Earp, J., Olimpo, G., Ott, M., Pozzi, F., Tavella, M.: Pedagogical Plans as Communication Oriented objects. Poster presented at Cal 2009 Conference, Brighton, March 23-24 (2009)
12. Johnson, R.B., Onwuegbuzie, A.J., Turner, L.A.: Toward a Definition of Mixed Methods Research. Journal of Mixed Methods Research 1, 112–133 (2007)
13. Bagozzi, R.P., Davis, F.D., Warshaw, P.R.: Development and test of a theory of technology learning and usage. Human Relations 45(7), 660–686 (1992)
14. Davis, F.D.: Perceived usefulness, perceived ease of use and user acceptance of information technology. MIS Quarterly 13(3), 319 (1989)
15. Sammour, G., Schreurs, J., Al-Zoubi, A.Y., Vanhoof, K.: The role of knowledge management and e-learning in professional development. Int. Journal of Knowledge and Learning 4(5), 465–477 (2009)

Project Management Competences Development Using an Ontology-Based e-Learning Platform

Constanta-Nicoleta Bodea

The Academy of Economic Studies,
Calea Dorobanţi, no.15-17, Cladirea Centrului de Calcul,
Bucharest, Romania
bodea@ase.ro

Abstract. The paper presents a web based learning environment in project management, capable of building and conducting a complete and personalized training cycle from the definition of the learning objectives to the assessment of the learning results for each learner. This paper focuses on the organization of the learning content according to the course ontology developed by the author.

Keywords: project management, ontology, learning technologies, knowledge management.

1 Introduction

There is a unanimous view among politicians, economists and researchers that knowledge is an important driving force in our economy. Researches were done on the nature of the knowledge society and many models have been developed, as societal transitions models ([7], [10]). The Finnish model, for example, is presented in [7] in contrast to the Silicon Valley model. "The Finnish model is based on high-level basic education and strong commitment of all citizens to lifelong learning. The state has used incentives, strategic planning and participatory mechanisms. The state has acted as a promoter of technological and social innovations, as public venture capitalist and producer of knowledge labor, thus creating the conditions under which Finnish business could restructure itself and compete globally". This might be an example for the rest of Europe, especially when taking into account the difficulties in implementing the Lisbon strategy in different countries.

The key to success in implementing knowledge-based society lies in structural changes in education through effective lifelong learning and e-learning technologies. ([5]). To be a knowledge worker, "a person must be capable of exploiting the knowledge and expertise produced by other people, be able to develop his or her core competence on a continual basis, know how to operate in networks, master ICT, and be able to build such a space where he or she can co-operate with others" ([7]). Competence development has become an increasingly important issue today.

Competence development is achieved through formal/informal education and training, using different learning technologies. The e-learning systems are widely accepted and used today on the competences development processes. In the lifelong

M.D. Lytras et al. (Eds.): WSKS 2009, LNAI 5736, pp. 31–39, 2009.

learning, one of the main requirements for the e-learning systems is the content personalization, because each person/user has a specific target in the competence development process.

The e-learning system which we are proposing will solve some present limitations of the e-learning systems, especially those referring to the flexibility/adaptability of the learning process assisted by the computer and to the promotion of the traditional didactical method. The system development objective is to ensure the interoperability with the e-learning systems at national and international level, by the evaluation and adopting some standards dedicated to the e-learning domain (IMS, IEEE, SCORM etc.) referring to the learning process design, content structuring, learning resources descriptive metadata, learner and competences modeling [11].

These two main objectives are fulfilled bringing toghether new conceptual approaches and technical solutions for three basic elements: *the teaching-learning process* (e.g. learning and support activities flow, delivery conditions, triggers - notifications or timed events), *the learning content* (e.g. learning ontology, learning object and metadata) and *the actors-roles model* (e.g. learner model).

2 The System Architecture

The system architecture (see fig 1) was built on the base of the following elements: infrastructure, services and offered content ([9]). The e-learning *infrastructure*

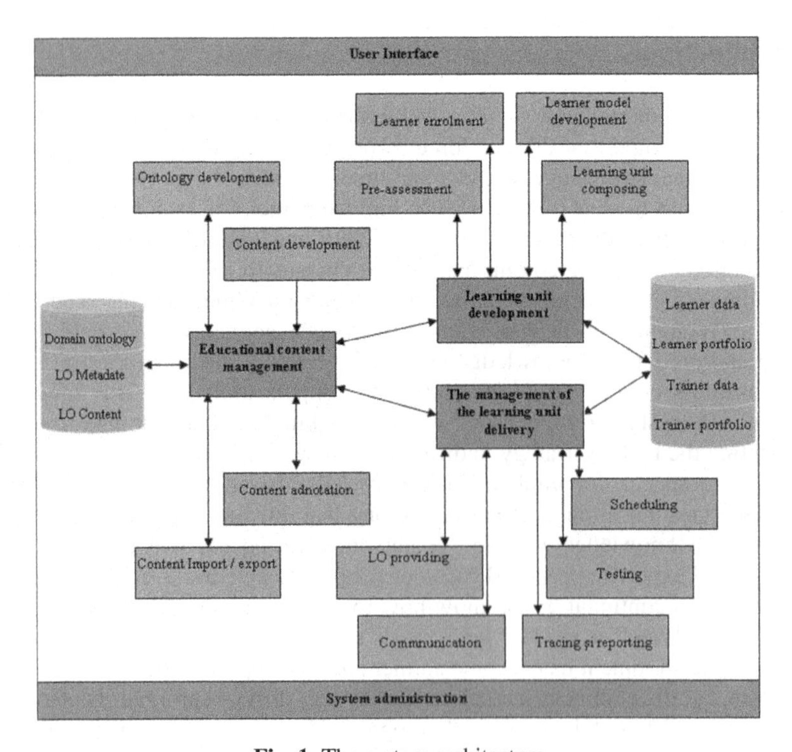

Fig. 1. The system architecture

includes two major software categories of applications, clearly distinguished and complementary:

1. LCMS – Learning Content Management System
2. LMS – Learning Management System

The *services* assure the successfully implementation of a e-learning technology, such as efficient resource planning, adapting to the organizational context and/or to the individual preferences of the student, application integration and management, expert consultancy.

Referring to the *content*, it has observed that the current organizations prefer more and more to purchase a preexisting content delivered by a third party or to reuse some parts by import and export. This is way the tendency of specialization of some content suppliers and the creation and maintenance of the big collections/warehouses of digital educational content. These warehouses interoperability need respecting some standards, from which the project team selected the recommendations of IMS Digital Repository.

3 The Learning Ontology Approach

A learning ontology is an explicit formal specification of how to represent the learning objects, learning concepts (classes) and other entities and the relationships among them ([4]). It describes the learning terms and the relationships between them and provides a clear definition of each term used. Ontologies are created using ontology editors, such as Protégé.

An interesting guide to develop a learning ontology is given in [4]. The proposed methodology for developing learning ontology include the following steps: identifying the purpose (why is the ontology being built), ontology capture mechanism (identifying all the key concepts and relationships), coding (representing the ontology in a formal language, using a suitable editor), refinement, testing and maintenance of the ontology.

A precise and formal description of the course content will be made by explicit references to the learning ontology, using semantic annotations. The modeling of an ontology-based course can be accomplished on two levels of knowledge organization (figure 2):

- the upper level: the concepts set of the course topic selected form the ontological domain concepts
- the lower level: learning resources (books, web presentations, movies) associated with the upper level concepts; the ontology may be used as a semantic index for accessing the resources.

In the course development phase, learning paths can be created at the conceptual level based on semantic relations between the concepts (see figure 3). In this phase, it is considered:

- a sequence of concepts obtained by browsing of the domain ontology, which give the access order to the learning objects
- the corresponding learning objects sequence, which is associated to the ontology concepts and which constitute the personalized course.

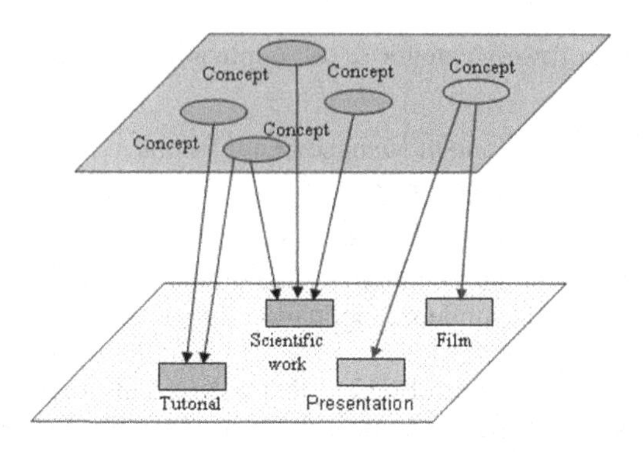

Fig. 2. The content semantic annotation

At the conceptual level, the learning paths can be developed based on semantic relations between the concepts, on two dimensions: the horizontal dimension and the vertical dimension (see figure 4).

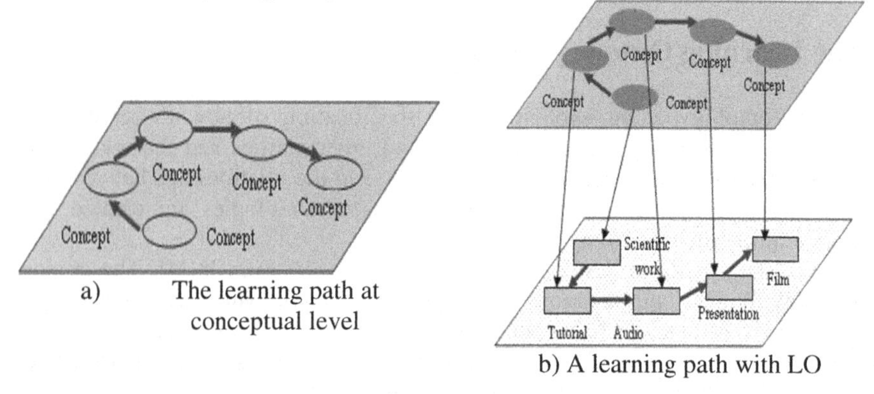

a) The learning path at conceptual level

b) A learning path with LO

Fig. 3. The learning path

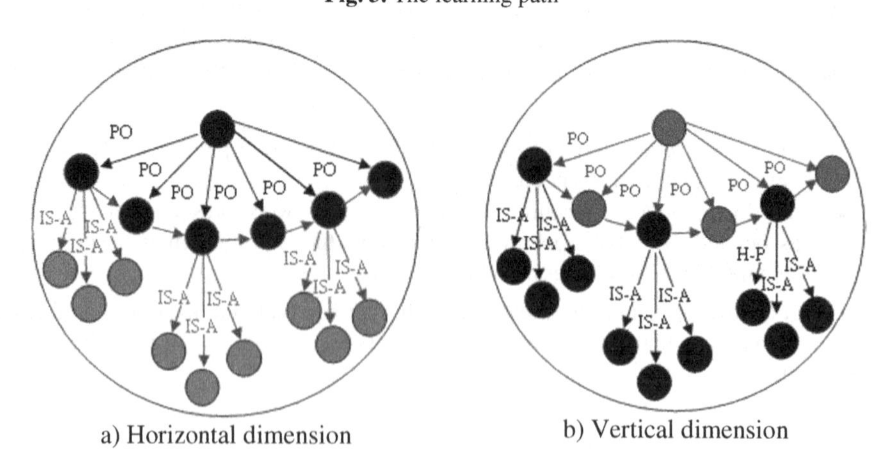

a) Horizontal dimension

b) Vertical dimension

Fig. 4. The learning path development

On the horizontal dimension, the learning sequence is established by moving from a given concept (the main subject), the ontology is browsed by following the decomposition relations (PO – Part Of relation). On the vertical dimension, the ontology is browsed on the specialized connections (the IS-A relationship) with different results base on the direction: from down to up (synthesis and topic completion) and from up to down (topics development).

4 The Examples of the Ontology-Based Learning Systems

The following systems are based on ontologies and standards that have an important role in the representation of learning objects and repositories ([4], [8]):

CIPHER (http://www.cipherweb.org): The system supports the exploration of national and regional heritage resources.

Connexions (http://cnx.rice.edu): It is an open source project that provides learning objects, a repository, a markup language and a set of tools for authoring, composing modules into courses and navigating through these courses.

Conzilla (http: www.conzilla.org/): Conzilla is being developed as part of the PADLR project as a means of accessing and annotating learning objects. It is a concept browser that allows the user to navigate through a space of context maps to access associated content. While the context maps are not reffered to as ontologies, they may be regarded as equivalent.

Edutella (http://edutella.jxta.org): This project provides an infrastructure for Peer-to-Peer systems for exchanging educational resources. Edutella uses metadata based on standards such as IEEE LOM to decsribe resources.

EML (Educational Modelling Language) (http://eml.ou.nl/introduction/explanation.htm): It is a notational system developed at the Open University of the Netherlands as a means of representing the content of a study unit and the students and teachers roles, relations, interactions and activities. It now forms the basis for the IMS Learning Design Specification. As with many XML based approaches ontologies are not mentioned. However, the study units, domain and learning theory models can be contructed as a set of ontologies.

5 The Educational Ontology

Applying the ontology learning approach for the project management domain requires adopting a standard for the domain concepts and project managers competencies. This standard is ICB – International Competence Baseline of the IPMA – International Project Management Association.

The training material is structured on indexed learning objects (LO). For the specification of the relations and interdependencies between the elements, the learning system uses ontology ([2], [4]); these allow the abstracting, definition and inter-correlation of the training domain concepts by relations like is_part_of, requires, and suggested order, for the link with LO. Learner models are created and maintained in learning system. These models contain, mainly, the learner cognitive state and preferences (knowledge level, cognitive and perceptive abilities, relations with the actors of the learning process etc.).

In the system a learning unit will be composed from a selected set of goals of the training (key concepts that the learner must learn) and from a learning path (a sequence of LOs that will be used for a learner in order to reach the goals). Once established these elements, begins the complete cycle of the learning-training process. The ontology of the project management course contains 201 concepts and 3 types of relationship between concepts.

The following table presents an sample of the concepts of ontology, in connection with ICB competence elements ([3]):

ICB competence elements	**The learning system concepts**	**ICB competence elements**	**The learning system concepts**
Project management success	INT, PRJ, SCS, DSC, FSP, SUC	Assertiveness	CO4
Interested parties	MSP, MSE, PIN, RMS, ACO, QAD, AAN, SRP	Relaxation	CO5
Project requirements and objectives	ENT, ASI, STO, RST, OBV, OOB, DOB, OBP, OSA, NOB	Openness	CO6
Risk & opportunities	MRO, IER, ACA, MOC, PMM	Creativity	CO7
Quality	MCP, PCP, PPR, PCR, PCM, ASC, ADP, CON, COA	Results orientation	CO8
Project organisations	SOP, ORG, OPR, RPR, PRP, MGP, CMP, STP, ASO, CER, FDP, CAM, MEP, COL, EPR	Efficiency	CO9

The following table describes the type of the relationship between concepts:

ID	Relationship type	Symbol
1	Has-Part	─▷
2	Is-required-by	──►
3	Suggested-Order	‒ ‒ ►

The figure no. 5 presents an overall view of the ontology. The figure no. 6 presents detailed views of the following parts of the ontology: project (figure 6a), project management (figure 6 b) and project oriented organizations (figure 6 c).

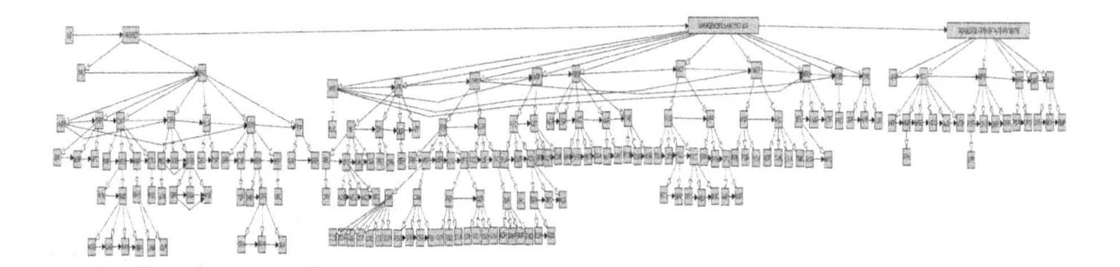

Fig. 5. The ontology of the project management course – a general view

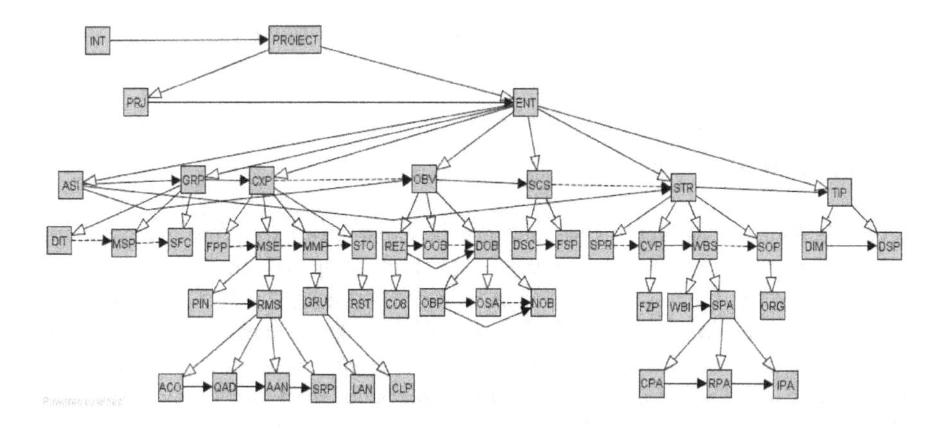

a) The detailed view of the *Project* part

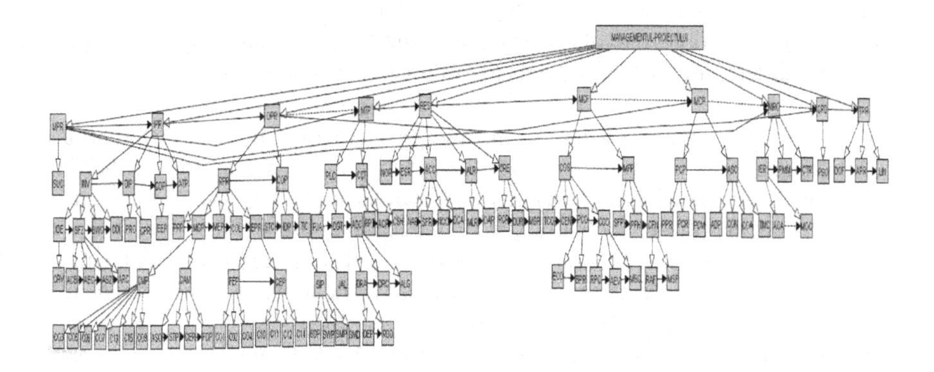

b) The detailed view of the *Project management* part

Fig. 6. The ontology of the project management course – detailed views

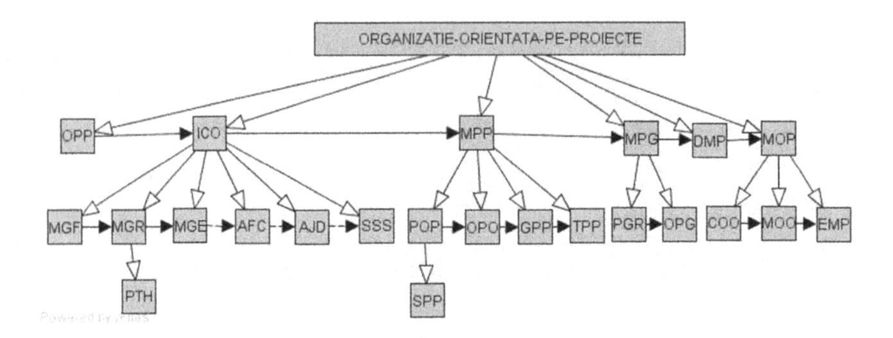

c) The detailed view of the *Project oriented organizations* part

Fig. 6. (*continued*)

6 Conclusions

To be a knowledge worker, "a person must be capable of exploiting the knowledge and expertise produced by other people, be able to develop his or her core competence on a continual basis, know how to operate in networks, master ICT, and be able to build such a space where he or she can co-operate with others" ([7]). Competence development has become an increasingly important issue today. Competence development is achieved through formal/informal education and training and using different learning technologies. The e-learning systems are widely accepted and used for competences development. In lifelong learning, one of the main requirements for the e-learning systems is the content personalization, because each person/user has a specific target in the competence development process.

Besides the main objective of content personalization, the ontology based approach adopted in our learning system presents the following advantages:

- A unitary interpretation of the content structure by the different users categories or software agencies – with major advantages for all the actors that participated in the creation and maintenance of a complex educational content and those who participate in the learning process
- The explicit specification of the domain – It will permit a actualization of concepts that refer to the knowledge area, without major modifications in the realized e-learning tools/programs
- Facility of reusing the respective knowledge domain – with economical advantages by saving the costs of (re)writing the course supports for different learning forms
- The separation of the knowledge domain from the operational knowledge – it will permit reusing the created tools with small modifications and for other knowledge domain than project leading.

References

1. Castells, M., Pekka, H.: The Information Society and the Welfare State: The Finnish Model. Oxford University Press, Oxford (2002)
2. Garcia, A.C.B., Kunz, J., Ekstrom, M., Kiviniemi, A.: Building a Project Ontology with Extreme Collaboration and VD&C, CIFE Technical Report #152, Stanford University (November 2003)
3. International Project Management Association, International Competence Baseline – ICB, V 3.0 (2006)
4. Kanellopoulos, D., Kotsiantis, S., Pintelas, P.: Ontology-based learning applications: a development methodology. In: Proceedings of the 24th IASTED International Multi-Conference Software Engineering, Innsbruck, Austria, February 14-16 (2006)
5. Lytras, M.D., Carroll, J.M., Damiani, E., Tennyson, R.D. (eds.): WSKS 2008. LNCS (LNAI), vol. 5288. Springer, Heidelberg (2008)
6. Markku, M.: Creating Favourable Conditions for Knowledge Society through Knowledge Management, eGovernance and eLearning. In: FIG Workshop on e-Governance, Knowledge Management and e-Learning in Budapest, Hungary, April 27-29 (2006)
7. Teekaput, P., Waiwanijchakij, P.: eLearning and Knowledge Management, Symptoms of a Reality. In: Third International Conference on eLearning for Knowledge-Based Society, August 3-4, Bangkok, Thailand (2006), http://www.ijcim.th.org/v14nSP1/pdf/p27.1-6-fin-36.pdf
8. Vargas-Vera, M., Lytras, M.D.: Personalized learning using ontologies and semantic web technologies. In: Lytras, M.D., Carroll, J.M., Damiani, E., Tennyson, R.D. (eds.) WSKS 2008. LNCS (LNAI), vol. 5288, pp. 177–186. Springer, Heidelberg (2008)
9. Liu, X., El Saddik, A., Georganas, N.D.: An implementable architecture of an e-learning system. In: Electrical and Computer Engineering, IEEE CCECE (2003)
10. ***Advancement of the knowledge society. Comparing Europe, the US and Japan, European Foundation for the Improvement of Living and Working (2004), http://www.eurofound.eu.int

Virtual Cultural Tour Personalization by Means of an Adaptive E-Learning System: A Case Study

Carla Limongelli[1], Filippo Sciarrone[2], Marco Temperini[3], and Giulia Vaste[1]

[1] Dept. Computer Science and Automation, Roma Tre Un.
Via della Vasca Navale, 79 00146 Rome, Italy
{limongel,vaste}@dia.uniroma3.it
[2] Open Informatica s.r.l. - E-learning Division
Via dei Castelli Romani, 12A - 00040 Pomezia, Italy
f.sciarrone@openinformatica.org
[3] Dept. Computer and Systems Science, SAPIENZA Un.
Via Ariosto, 25 00184 Rome, Italy
marte@dis.uniroma1.it

Abstract. Visiting a real or virtual museum or an archaeological site can be a hard task, especially in case of large sites provided with many works of art or ancient ruins. For this reason most historical sites provide guided tours, to improve visitors satisfaction and interest. In this work we explore the use of an e-learning environment, called LECOMPS5, to provide museums or other cultural sites with the capability of automatically planning personalized tours, according to visitors needs and interests. LECOMPS5 allows a domain expert, through a suitable GUI, to build a pool of learning components concerning a given site. Then the system, by means of an embedded planner, generates a personalized tour through the works of art, on the basis of the visitor's artistic interests and needs. We propose a first application of this system to an ancient archaeological site called *Lucus Feroniae*, showing how an e-learning platform can be successfully used for guiding visitors as well.

1 Introduction

Personalized learning and personalized tours are two very important objectives that are not so far apart. In both cases, people are requested to navigate in a knowledge domain, composed either of virtual Learning Objects, as it happens for distance learning, or of real works of art, as it happens when visiting a museum. Artificial Intelligence provides many techniques to deal with the problem of performing personalization and adaptation through automatic systems and with Cultural Heritage domains in general [1,2,3]. The PEACH project [4], concerns with the development of a novel integrated framework for museum visits. In [5], the adaptation aspects of a mobile museum guide are addressed, investigating the relationships between personality traits, such as the emotional state, and the attitudes towards some basic dimensions of adaptivity, such as *cognitive*

M.D. Lytras et al. (Eds.): WSKS 2009, LNAI 5736, pp. 40–49, 2009.

state and *learning styles.* The CHIP project [6] proposes an ontology based approach for bridging the vocabulary gap between domain experts and end users and provides personalization of museum tours based on user's explicit feedbacks.

In this work we propose a multifaceted system for accompanying visitors and enjoy an augmented overall visiting experience. We explore the use of an adaptive e-learning platform for the automated generation of personalized tours in virtual Cultural Heritage domains. This environment is based on a user model represented by the current user's cognitive state and learning styles, according to the Felder and Silverman model [7]. In particular, we use the LECOMPS5 e-learning system [8,9,10] which, through an embedded Pdk planner [11], is able to build automatically for each learner a learning path, personalized so to reach a stated *Target Knowledge* basing on learner's individual *Starting Knowledge* [9]. Mutuating the experiences done in the educational context, we extended the system to operate in the Cultural Heritage domain, so to allow a domain expert, by means of a suitable Graphical User Interface, to build a pool of *Learning Components*, each one representing (instead of usual didactic-oriented content) relevant information about resources such as drawings, pictures, works of art in general, and also ancient Roman ruins, as it is in the present paper.

The main idea of our work is that a tour in a museum or in an archaeological site can be dealt with through an e-learning environment, where the works of art are part of the concepts taught through the Learning Objects, and a personal learning path is a sequence of Learning Objects that mirrors an actual tour in the cultural site. We propose a case study based on a real archaeological site called *Lucus Feroniae*, an ancient site close to Rome. The interests and personal traits of the visitor, intended either as a single person, or as a whole group, are modeled by the system through an initial questionnaire. Then the system can generate a personalized tour tailored on such interests. Here we present a first application of this approach, generating different tours for different visitors.

The paper is structured as follows: Sec.2 illustrates the characteristics of the LECOMPS5 system, together with the analogies between personalized learning and personalized tours; Sec.3 presents our *Lucus Feroniae* case study; then, Sec.4 shows an example of tour personalization, and conclusions are drawn in Sec.5.

2 The System

In the following, after a brief description of the LECOMPS5 e-learning platform, we show the analogies between the generation of personalized e-learning courses and the personalized tour generation in cultural sites.

2.1 Lecomps5

LECOMPS5 [9] is a web-based e-learning environment supporting functionalities for teachers, students, and administrators, capable to generate personalized and adaptive courses on the basis of the students' starting knowledge on the domain of interest, and on the basis of the student's learning styles. A personalized course, related to a given subject matter, is characterized by the *Target*

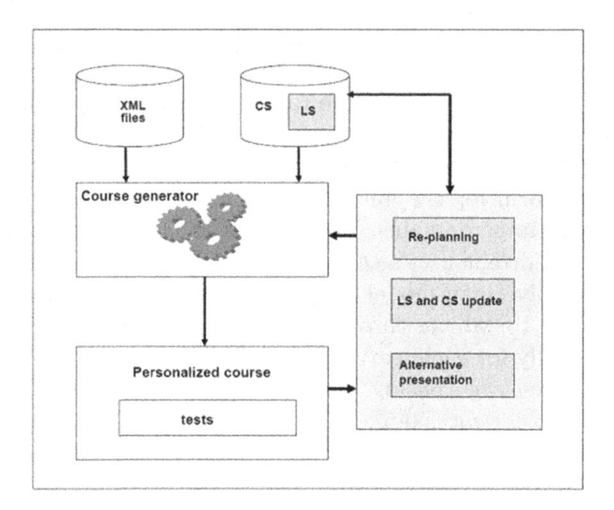

Fig. 1. The LECOMPS5 architecture

Knowledge (*TK*), that is the knowledge to be acquired by the student through the course, and by the *Starting Knowledge* (*SK*), that is the student's knowledge about the topic before taking the course. The knowledge is represented by atomic elements, called *Knowledge Items* (*KI*). A course is composed by a set of *Learning Components* (*LC*), i.e., Learning Objects enriched with the specification of the *Required Knowledge* (*RK*, prerequisites) and the *Acquired Knowledge* (*AK*), related to the learner's study of the component's learning content (both expressed as sets of *KI*); a value for the effort needed by the learner to study the material contained in the *LC* is also specified. The effective acquisition of the *AK* of a given *LC* can be evaluated through questions, included in the *LC* and related to the concepts explained there. All the *LC*s related to a given subject matter are collected together into a pool, that is a sort of knowledge database. The teacher defines prerequisite relationships among *LC*s. This task is made easy by the graphic visualization of such relationships.

LECOMPS5 configures the personalized course for a given student on the basis of her *SK*, measured by a pre-test, of her *TK*, pre-defined by the teacher, and on the basis of *LC*s, as arranged by the teacher in the graph. LECOMPS5 selects the *LC*s such that the *AK* of all such selected components, together with the *SK*, covers the *TK*. The automated configuration of the course is performed by means of the Pdk planner, described in the following. Fig. 1 shows the overall architecture of the system. The grey boxes represent the functional modules of the system directly connected to the course generation. In particular, the *Alternative presentation* module is used to propose alternative learning materials to students, according to her current learning styles.

2.2 Tour/Course Generation

In order to put the e-learning environment on trial, for generating personalized tours, we have to highlight some similarities and differences we think do exist between didactic courses and cultural tours. The main differences are in that visitors are not evaluated during their tours, while students are evaluated during their learning process and the relationships among works of art are not necessary prerequisite relationships. Then, the path of a cultural tour is actually a sequence of *stations*, i.e., steps, each one modeled by one of the Learning Components in the sequence. In Fig. 2 the home page of the virtual tour is presented: the user can launch several tasks such as for example *Lezioni* (lessons) and *Test Finale* (final test), starting her tour when she wants.

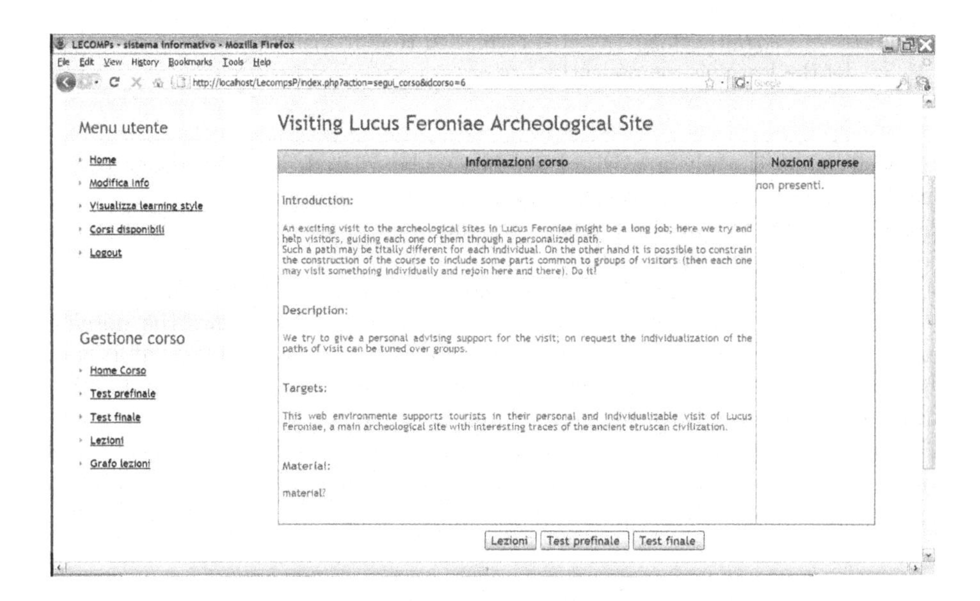

Fig. 2. The LECOMPS5 Home page

Besides these differences, cultural sites tours can be thought as learning problems as well, as shown in Tab. 1.

The domain expert defines a set of relationships among the learning components, that is the works of art describing the cultural site. Tour generation corresponds to plan a sequence of actions leading to a personalized goal, that is, to visit all the places and works of art, on the basis of the visitors interests as indicated in the initial questionnaire. When a visitor executes an action of the plan, she is offered the description, according to her user model, of the part of the site (or the work of art) she is looking at. In order to perform tour generation we need to use a tool that allows to specify the requirements of domain experts and visitors in an easy way. To this aim, we focus our attention on logic-based planners, which can exploit some important functionalities such as: domain validation, redundant actions detection or control knowledge specification, i.e., additional information that can enrich the planning domain (given as mere list of actions with their preconditions and effects) and guide the plan synthesis. For instance, once a pool is arranged, the domain expert might want

Table 1. LECOMPS5: analogies between personalized learning and personalized tours

	e-learning domain	Cultural Heritage domain
SK	starting knowledge	interests and prior knowledge
TK	target knowledge	the set of works of art to be visited
KI	knowledge item	the work of art
RK	required knowledge	the work of art that should have been visited
AK	acquired knowledge	the work of art visited during a step of the tour (acquisition of the step)
LC	learning component	a model of a step in the tour
effort	cognitive load of the learning component	estimated time to visit the work of art

to specify that if a given work of art has to be visited, it is necessary to see another one before, or that a visitor prefers to see only works of art with a given theme, or that a visitor is already expert and she wants to know only elaborations about a given work, or that a visitor does not know anything about the place she is visiting, or she is a child, and the explanation has to be as simple and direct as possible. What is needed is a language that allows the domain expert to specify such kind of *control knowledge*. The Pdk planner conforms to the "planning as satisfiability" paradigm, and the logic used to encode planning problems is the propositional Linear Time Logic (LTL) [12]. The related planning language PDDL-K [11] guides the domain expert for the specification of control knowledge. Pdk accepts PDDL-K as input language, parses the problem description into its LTL representation and reduces planning to model search.

3 The Case Study: Lucus Feroniae

In this Section we show a brief description of the archaeological site we chose for our case study. Lucus Feroniae stands on a travertine platform located in *Capena*, a little old town close to Rome. It has very ancient origins, as ancient as the origins of worship of the *Feroniae Goddess*, a testimony of an italic cult like those discovered in sanctuaries of *Trebula Mutuesca*, *Terracina*, and *Amiterno*.

The shrine is located at the 18*th* Mile of *Via Tiberina*, at *Scorano*, and the exact location was identified only in 1953, when Prince Victor Maximus, owner of the *Scorano Castle*, and surrounding lands, signaled to the Southern Etruria Superintendency the outcropping, during some works, of the archaeological findings. Fig.3 shows snapshots of the ancient *Via Tiberina*, where the site is located, and of the *Amphitheater*.

When accessing the archaeological site, we immediately meet a crossroad between the old *Via Tiberina* and the road to join the sanctuary to *Capena*: the *Capenate Road* where we can see the remains of an ancient gate. This crossroad was a very important road junction and in this place were found the *Cippi Miliari*, dated to the third century B.C., which is the dating of the most recent restoration of these roads. Continuing along the *Via Tiberina*, we immediately

Fig. 3. A Section of the ancient Via Tiberina and the Amphitheatre

notice on our right some not very large environments, which have been identified as meeting and refreshment points, perhaps *Tabernae*. After we meet a rectangular square with an East-West orientation where there is still a part of the ancient pavement made with rectangular slabs of limestone. Another interesting characteristic is the *Amphitheater* of which the load-bearing structures remained. It has a very unique form: it is almost circular, but, although very small, it presents all the characteristic aspects of a true *Amphitheater* with its doors still very well preserved, with the *Vomitoria*, that is, exits for the public, and service environments below the stairs. Finally, the south side is less preserved and recently, precisely in this area, came to light some structures certainly of the Republican Roman era. On the North there is the purely religious area, the focal point of the ancient political life and administration of worship in the city.

4 Examples of Personalized Tours Generation

In this section we show two examples of personalized tours generation, based on the user interests. Let us suppose that a visitor does not want to see all the works of art at the site, but she is interested in a special "theme track", e.g.: she wants to see all the epigraphies, or all the tombs or all the statues and so on.

Table 2. The two different generated tour

Visitor A epigraphies	**Visitor B Statues, fountains marbles and columns**
Entrance to the Archeological Site	Entrance to the Archeological Site
Amphitheatrum	Tabernae
Forum	Basilica
Augusteum	Forum
	Augusteum

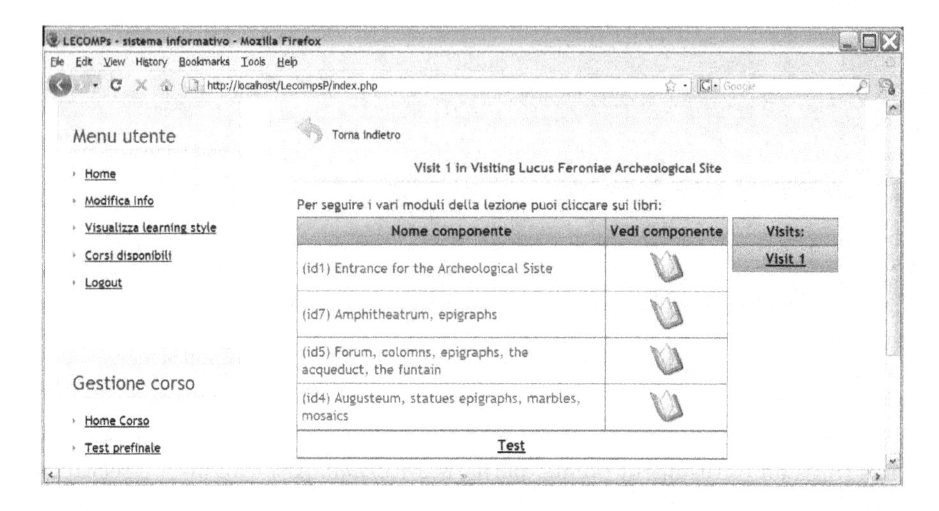

Fig. 4. The LECOMPS5 window with the proposed tour for the first visitor

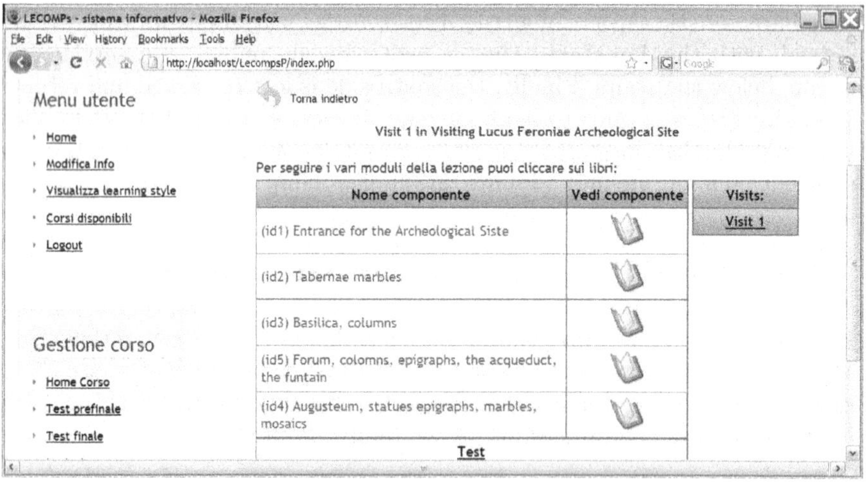

Fig. 5. The LECOMPS5 window with the proposed tour for the second visitor

In order to plan a desirable tour for a given visitor, LECOMPS5, through its GUI, allows the domain expert to configure the location of the works of art and the preferences of the visitors in an easy and expressive way.

In our case study, there are two visitors, with different preferences. The first one, with specific historical interests, wants to examine epigraphies, while the second one wants to see more in general fountains, statues, marbles and columns. The personalized tours, as generated by the system, are shown in Tab. 2, while Fig.4 and Fig.5 illustrate directly the tours proposed by the LECOMPS5 system.

These generated tours are compatible with the preferences of the two different visitors. In fact, in the domain description, the *LC* present the following characteristics:

- Amphitheatrum: epigraphies;
- Tabernae: marbles;
- Termae: ceramics and mosaics;
- Forum: columns, epigraphies, fountains and aqueduct;
- Augusteum: statues, epigraphies, marbles and mosaics;
- Basilica: columns.

From the previous Tab. 2, we can see that none of the two visitors have to visit the *Terme*, since they are not interested neither in ceramics nor in mosaics.

5 Conclusions and Future Work

In this paper we presented an application of the LECOMPS5 e-learning environment, to the automatic generation of personalized tours. The main idea, here, is that a learning environment can be seen as a suitable tool for describing cultural sites and their artifacts, and arrange such descriptions in different sequences and with different levels of depth, according to the attitudes and interests (or, in other words, *needs*) shown by the end-users/visitors. We showed that there are many analogies between personalized learning and personalized tours, and that such analogies allow to apply basically the same environment to both problems.

We think that, besides the availability of personalization, there are several advantages offered by the described framework. One is in the user-friendly environment used by the domain expert to arrange the description of the site elements. In addition, the architecture of the system (with the PDDL-K language embedded in LECOMPS5) can allow for several other aspects of personalization, such as the offering of an explanation level whose content depends on the user's background, a characteristic which is already managed by the system in the e-learning courses. Moreover, being the system based on the management of electronically-delivered resources, it is apparent that it can be used to cover both the *on-site* ("physical") approach to a visit and the *virtual* one, while all the possible interleaving between such approaches are supported (one visitor could move physically only through a portion of the path, and see the rest virtually). This is relevant once we think that a visiting path is composed by a sequence of steps, and its design might be based on the topological relations between close steps, as well as on the logical/cultural relations between steps. If a path has to be followed entirely on site, the topological constraints might prevail on the others; if we allow for interleaved - on site and virtual - steps we can reach a good trade-off between cultural consistence and physical fruition of the path.

With respect to the on-site approach, we are working on an efficient use of the planner to manage topological constraints for path finding; an application of GPS technology might easily allow to support visitors during their personalized site tour. About virtual visits, we are working on a case study related to a big museum, that usually a visitor cannot visit in a single session: the visitor is supposed to select a desirable time-span to spend in the museum, a preferred author, an historical period, an artistic current, and the system configures an initial visit plan. The visitor is anyway free to follow or not the system suggestions and, in case she looks to be interested in other artifacts, the system can re-plan the visit on the base of her decision.

Presently there are efforts to apply e-learning techniques into other fields of interest ([13]). And in fact an aspect of the present work concerns also the integration, into the area of management of cultural and touristic knowledge, of concepts and techhniques that are well established in the field of personalized e-learning. We think that, by applying typical tools of personalized e-learning, we can obtain "good quality" visiting tours, meaning that they are interesting, culturally appropriate, and physically affordable with respect to the personal needs of the individual.

It is useful to conclude with some remarks about the significance of this work with respect to the general area of the *Knowledge Society.* The system we have been describing is aimed at supporting the construction of visiting tours and their possibly partial virtual fruition. A first application scenario of the system is in the possibility to define visiting paths according to the personal needs and attitudes of the individual. There are several aspects of a visit that can be personalized: a site could be quite large and one could prefer to have a limited (incomplete) visit of it; a person could prefer to follow the path on foot, or virtually, or to proceed partly physically and partly virtually; satisfying such needs would make *affordable* for a person paths that otherwise could not be followed; in either cases, the path constructed by the system would be sound, from the point of view of the sequential relation among the various steps in the path. Another application scenario for such a tool is in the area of support to impaired persons, which is of great relevance with respect to the democratic developments and the care for rights into the knowledge society, and is of rising economic importance [14].

References

1. Bordoni, L., Pasqualini, L., Sciarrone, F.: Chem: A system for the automatic analysis of e-mails in the restoration and conservation domain. In: LREC 2004: Language and Resource Evaluation Conference, Lisboa (2004)
2. Colagrossi, A., Sciarrone, F., Seccaroni, C.: A methodology for automating the classification of works of art using neural networks. Leonardo, Journal of the International Society for the Arts Sciences and Technology 36(1), 69 (2003)
3. Gentili, G., Micarelli, A., Sciarrone, F.: Infoweb: An adaptive information filtering system for the cultural heritage domain. Applied Artificial Intelligence 17(8-9), 715–744 (2003)
4. Stock, O., Zancanaro, M., Busetta, P., Callaway, C.B., Krüger, A., Kruppa, M., Kuflik, T., Not, E., Rocchi, C.: Adaptive, intelligent presentation of information for the museum visitor in PEACH. User Model. User-Adapt. Interact 17(3), 257–304 (2007)

5. Goren-Bar, D., Graziola, I., Pianesi, F., Zancanaro, M.: The influence of personality factors on visitor attitudes towards adaptivity dimensions for mobile museum guides. User Model. User-Adapt. Interact 16(1), 31–62 (2006)
6. Wang, Y., Aroyo, L., Stash, N., Rutledge, L.: Interactive user modeling for personalized access to museum collections: The rijksmuseum case study. In: Conati, C., McCoy, K.F., Paliouras, G. (eds.) UM 2007. LNCS (LNAI), vol. 4511, pp. 385–389. Springer, Heidelberg (2007)
7. Felder, R.M., Silverman, L.K.: Learning and teaching styles in engineering education. Engineering Education 78(7), 674 (1988)
8. Limongelli, C., Sterbini, A., Temperini, M.: Automated course configuration based on automated planning: framework and first experiments. In: Chiazzese, G., et al. (eds.) Methods and Technologies for Learning, Proc. ICMTL 2005, WIT Press (2005)
9. Limongelli, C., Sciarrone, F., Vaste, G., Temperini, M.: Lecomps5: A web-based learning system for course personalization and adaptation. In: IADIS International Conference e-Learning 2008, Amsterdam, The Netherlands (2008)
10. Limongelli, C., Sciarrone, F., Temperini, M., Vaste, G.: Lecomps5: a framework for the automatic building of personalized learning sequences. In: Lytras, M.D., Carroll, J.M., Damiani, E., Tennyson, R.D. (eds.) WSKS 2008. LNCS (LNAI), vol. 5288, pp. 296–303. Springer, Heidelberg (2008)
11. Mayer, M.C., Limongelli, C., Orlandini, A., Poggioni, V.: Linear temporal logic as an executable semantics for planning languages. J. of Logic, Lang. and Inf. 1(16), 63–89 (2007)
12. Wolper, P.: The tableau method for temporal logic: an overview. Journal of Logique et Analyse 28, 119–152 (1985)
13. Sammour, G., Schreurs, J., Al-Zoubi, A., Vanhoof, K.: The role of knowledge management and e-learning in professional development. Int. Journal of Knowledge and Learning 4(5), 465–477 (2008)
14. Wu, Y.-C.J., Cheng, M.J.: Accessible tourism for the disabled: Long tail theory. In: Lytras, M.D., Carroll, J.M., Damiani, E., Tennyson, R.D. (eds.) WSKS 2008. LNCS (LNAI), vol. 5288, pp. 296–303. Springer, Heidelberg (2008)

An Integrated Model to Monitor and Evaluate Web 2.0 Project-Based Learning

Francesca Grippa, Giustina Secundo, and Giuseppina Passiante

e-Business Management, Scuola Superiore ISUFI – University of Salento, Via per Monteroni s.n. 73100 Lecce Italy
{francesca.grippa,giusy.secundo,
giuseppina.passiante}@ebms.unile.it

Abstract. This paper presents an integrated model to monitor and evaluate web 2.0 project-based learning through its application in a real higher education environment. Creating communities in the classroom has been the traditional method of engaging students. The web 2.0 technologies are changing the way learning communities interact. We designed a Social Computing Environment to support a community of students coming from Morocco, Tunisia, Egypt and Jordan involved in an International Master's Program. This study suggests that Web 2.0 tools can be integrated in a model to assess learning effectiveness in terms of Learners' Satisfaction, Knowledge Creation and Learning Performance. Preliminary results suggest that the Social Computing Environment helped students to better collaborate with peers, tutors and company's supervisors, and also provided an environment where the learning assessment phase could be much easier to perform.

Keywords: Web 2.0 tools, project-based learning, learning monitoring and evaluation, knowledge creation, social computing environment.

1 Introduction

The increasing adoption of high-speed connections, mobile communication technologies, and real-time collaboration practices have the potential to make learning communities more effective. With the emergence of the Internet - and Web 2.0 technologies in particular - students are able to utilize innovative tools to gain experience through experimentation and action.

Some of the Web 2.0 applications like wikis and blogs (or weblogs) have experienced a rapid growth in recent years. These tools are considered to be highly beneficial applications for supporting distributed learning communities. Web 2.0 technologies increase and accelerate learners' ability to work in group for project purpose. Project-based learning (PBL) is an approach for classroom activity that emphasizes learning activities that are long-term, interdisciplinary and student-centered. This approach is generally less structured than traditional, teacher-led classroom activities; in a project-based class, students often must organize their own work and manage their own time. Within the project-based learning framework students collaborate and work together to answer real problems. Web 2.0 technologies allow supervisors to monitor knowledge

M.D. Lytras et al. (Eds.): WSKS 2009, LNAI 5736, pp. 50–59, 2009.

acquisition and knowledge sharing processes in all the project phases, thus improving collaborative learning within and across the community [1]. Despite these assumptions there is still a lack of empirical research to better understand how learning methods can be improved using these applications [2], [3], [4].

In this paper we address the following question: *how to monitor and evaluate Web 2.0 Project based Learning*? As an attempt to provide some evidence to this question, in this paper we propose the application of a Social Computing Environment (SCE), in which web 2.0 technologies are applied to support a project-based learning approach. We present the main functionalities of the SCE and we report how students responded to it.

After a brief introduction on the benefits of Web 2.0 technologies in a PBL perspective, we discuss how a learning community can be nurtured more effectively when members use a social computing environment for their interaction with peers and tutors. We describe the main functionalities and tools implemented in the SCE and we use the International Master's Program in e-Business Management (IMeBM) as a research setting to test its effectiveness. Finally, some discussions and implications will be presented to understand how web 2.0 can support PBL based on the empirical evidence.

2 Theoretical Background: Web 2.0 for Project-Based Learning

With the emergence of a new generation of web-based technologies students have the possibility to use innovative tools to build new competencies in experimentation and action. Web 2.0 tools can be applied to any classroom but have been found to be very helpful in the online learning environment by engaging learners in activities using the web as a resource [5], [6], [7].

Web 2.0 technologies support knowledge sharing processes that allows active learning within and across a community [8]. Web 2.0 may lead to a shift from a traditional teacher-centered perspective to a dynamic learner-centred approach. This transition represents a profound change in the higher education sector. Web 2.0 technologies such as blogs, wikis, podcasts, and RSS feeds have been defined as "social software", given their capacities for collaborative content development [9]. These technologies place learners at the center of online activities, enabling new methods for content co-creation, collaboration, and consumption, creating new ways of interacting with web-based applications. The challenging task for today's educational institutions is to find the most suitable way to integrate these tools in the classroom experience and collaboration in distance learning. Wikis and blogs have the potential to enrich classroom activities through up-to-date content development and active discussion that help learners achieve better scores and acquire new skills (e.g. collaborate in virtual team and time management).

Empirical research indicates how web 2.0 solutions in learning environments have the potential to support a collaborative learning process, that results in a positive interdependence of group members, future face-to-face meetings, individual accountability, and appropriate use of collaborative skills [2], [3]. As shown by Johnson and Johnson [10], cooperative teams achieve higher levels of performance and retain information longer than learners working individually. A social constructivist perspective has been also used to explain the success of Web 2.0 applications, especially wikis, in making the learning process more effective [11], [12]. Project-based Learning is a pedagogical methodology that emphasizes the collaborative dimension of learning. It has roots in constructivism, as it engages students in authentic student-centred tasks to enhance learning. This method has an important role in the context of educational environments as the project is seen as a way to reach education goals.

One of the Web 2.0 applications showing high potential in supporting the development of the project-based method is the wiki [2]. Wikis give the opportunity to meet virtually at students' convenience and work on projects together in a synchronous or asynchronous way [13]. Because all comments and ideas are consolidated on a single webpage, a wiki creates a clearer visual representation of team direction than do individual e-mail messages [14].

Fountain [15] suggests several applications of a wiki in project integration work, including managing a long-term design process, problem solving, permitting constructive critique of pedagogical projects, allowing commentaries/critiques on project integration work, and cross class/course projects. Schaffert et al. [2] suggest ways in which wikis can be useful in project knowledge management, including brainstorming and cross-fertilization of ideas, coordination of activities, scheduling of meetings, and serving as a sort of notepad for common information items. Chen et al. [16] discuss the benefits of wikis in their design engineering group project, including the possibility for students to collect, organize, and share photos, videos and presentations. A pedagogical challenge common in project-based courses is that students see what they have produced but they do not see what they have learned [16]. Wikis seems to have the potential to alleviate this problem.

3 Research Design

In our research we assume that the new generation of web applications may represent a valid support in the creation of new learning environments, where blogs, wikis, RSS feeds and folksonomy promise to support cooperative learning and Project-based Learning [17]. Our study aims to provide an empirical contribution to understand how web 2.0 technologies and project-based learning approaches can be integrated to support learning communities. To this purpose, we observed a community of Master's students in their use of a web 2.0 platform and we administered a final survey to collect data on their feedback, opinion, level of satisfaction.

The research setting for this experimentation is the IMeBM program, designed and managed by the e-Business Management Section of Scuola Superiore ISUFI- University of Salento (Italy). To support collaboration and knowledge creation within and across the IMeBM community, we designed and applied a Social Computing Environment in which wiki, blog, RSS and folksonomies were integrated. This technological platform has been developed using Drupal as the most appropriate Content Management System (CMS) for the development of the wiki-learning system. The software architecture is based on the following applications: Apache web server; PHP release 5.2.4; MySQL database; Java Virtual Machine, Ver. 1.5.

3.1 Research Methodology

In this paper we would like to provide preliminary evidence to the following research question: *how to monitor and evaluate Web 2.0 Project based Learning*?

At this purpose, we propose the experimentation of a Social Computing Environment built as a collaborative platform where Web 2.0 technologies are extensively used to support a learning community. In particular, we will present the case of a community created by the International Master's Program in e-Business Management. We will describe how learners are supported in their interaction with mentors and peers, posting comments, evaluating and rating peers' deliverables, providing constructive feedback in order to improve individual learning effectiveness and other teams' performance.

We conducted this study in two phases. Phase 1 was the design, development and testing of the PBL application; since this phase has been already addressed, it is not the objective of this work to further describe it. This paper explores the preliminary results of phase 2, in which we analyzed the integration of web 2.0 within PBL approaches in a learning community.

We observed the IMeBM community from March 2008 till October 2008. During this period, 23 learners (34% coming from Morocco, 48% coming from Jordan, 17% coming from Tunisia, 1% coming from Egypt) were involved in the IMeBM education activities and research projects. According to the PBL methodology, researchers and executive members of partner organizations actively participated in the IMeBM activities, projects and classroom phases.

3.2 The Integrated Model for Monitoring and Evaluating Web 2.0 PBL

In this paragraph we propose the integrated model that we have designed to monitor and evaluate learners' satisfaction, knowledge creation and learning performance in a community using web 2.0 tools to collaborate. To assess Learners' Satisfaction we used monthly questionnaires through which students evaluated aspects like mentors' effectiveness and quality of materials. As for the Learning Performance variable, we collected data on students' outcomes at both individual and team level over a period of 8 months using a Likert scale (1=high, 5=low). The criteria used for assessing students were based on their motivation, involvement in classroom, contribution to the project deliverables and results to structured exams. As for the dimension Knowledge Creation we evaluated a wiki created by students on the topic "Internet Business Management". The criteria for the assessment were:

- Content consistency: degree of logical organization of content and absence of contradictions in the interlinked materials.
- Level of depth: degree of clear description of the topics.
- Accuracy: the ability to prepare a complete overview of the topic, reporting all the available information and providing the correct list of references.
- Interdisciplinary nature: the extent of connection and integration between several disciplines.
- Collective contribution: the ability to create a consistent wiki that is function of different perspectives and contributions.

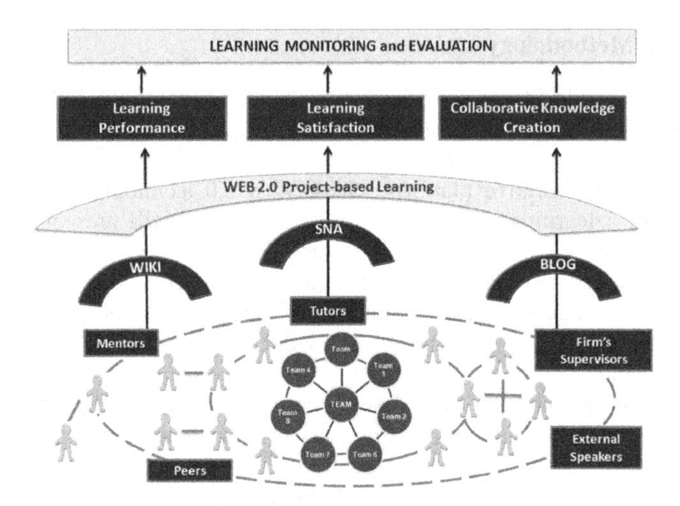

Fig. 1. The Integrated Model

The assessment of the ability to create knowledge in a collective manner has been conducted by tutors and mentors who assigned a value from 1 (low) to 5 (high) to the teamwork and the individual contribution. Figure 1 presents the main processes, functionalities and actors supported by the Social Computing Environment.

4 Monitoring and Evaluating Web 2.0 PBL: Preliminary Evidence

In the following sections, we will understand how PBL - supported by Web 2.0 - impacts on learning effectiveness, presenting the application of a Social Computing Environment in the ongoing 2008 edition of the International Master in e-Business Management (IMeBM). We observed how the SCE was adopted by this learning community and we present the preliminary results of the experimentation. The Master's Program is organized in mobility between Italy and Southern Mediterranean Countries and is offered on annual basis, full–time.

4.1 The Web 2.0 Project-Based Approach in Practice: The IMeBM Case

As stated by Bridges and Hallinger [18], in project-based approaches, most of the learning occurs within the context of small groups rather than lectures. The pedagogical goal is to foster problem solving skills by exposing learners to real-life dilemmas: they are helped to develop reasoning skills in a wide range of settings, to benefit from solving a real problem and to feel a sense of accomplishment useful to improve motivation that together with the learning need are the fundamental levers for an effective self-organized learning.

The technological and functional properties of the Social Computing Environment have been developed to implement the PBL pedagogical strategy where learning effectiveness was maximized.

Three main roles have been identified as core members of the community:

- Administrator: technical system supervisor who manages users and roles.
- Superpeer: domain expert who stimulates the creative discussion and collaborative problem solving inside the community.
- Peer: key "owner" of the learning process.

Table 1 describes how to translate processes and actions related to Project-based Learning into web 2.0 functionalities. For each of the five processes we associated a set of actions required to perform those processes and the required roles. Then, for each action we describe how web 2.0 functionalities have been used in the SCE to support them.

Table 1. PBL processes and related Web 2.0 functionalities

PBL Processes and Roles	**WEB 2.0 System Functionalities**
PROJECT CREATION Administrator creates learners' profile and learners create personal page	Learners navigate their profile, update personal information, interests, skills, login information.
PROJECT PLANNING Superpeers submit an assignment as project or problem, provide just in time feedback, and provide contents or topics as guidelines for a specific solutions to problems	-The Project is introduced through audio blogging, podcasting, posts on mentor's blog. -Posting assignments: to facilitate the collaboration within the team, learners are asked to create their own team blog. -RSS feeds are used as tool to support tutors/mentors in tracking assignments' submission.
PROJECT IMPLEMENTATION Superpeers store community knowledge, Peers develop Knowledge maps and systematically find the suitable information	The technological platform helps peers create a semantic map to formalize the collective knowledge on a problem. In the platform, to each assignment is attributed a title, a description, a deadline, and additional notes.
PROJECT COMPLETION Peers debrief after attending seminars, systematize knowledge into deliverables, papers, diagrams, project results, access wikis pages and on-line repository of files.	Project Team deliverables can be Semantic maps, SWOT Analysis matrix, Open-ended questions to extract lessons-learned, Cross-case application to other cases/industries.
PROJECT DISCUSSION Peers share comments on other teams' work, and superpeers provide feedback on peers' assignment	Peers are required to share their comments within the collaborative space, the mentors' blog or the team's blog. They are asked to rate each of their group members and other groups' performance by posting on the superpeer's blog.

Among the processes enabled by the Social Computing Environment, we also included the application of methods and tools of Social Network Analysis (SNA) to investigate the deep patterns of interaction among peers and superpeers through quantitative methods. The role of social analyst has been designed in the SCE to analyse reports and provide suggestions on the basis of the emerging relational trends.

5 Discussion of Preliminary Results

The SCE platform is being used to enable collaborative learning processes among the participants through the creation of a virtual learning community. Superpeers introduce business topics during classroom activities, and illustrate real life experiences under different points of view. After this face-to-face period, superpeer and peers start sharing ideas, knowledge resources within the learning space.

Masters' participants collaborate to exchange knowledge, solve problems, complete assignments, and develop competencies in a peer-based, non-hierarchical and community-oriented learning environment. Traditional psychological obstacles and distances among experts/professors and learners fall down, resulting in a more efficient hybridizing processes of collective experience. The IMeBM community has been mentored and monitored by three superpeers, who were asked to ensure a continuous "virtual presence" in the platform so to feed interaction, answer questions, and give suggestions. After 8 months, we collected data through web surveys and by tracking the utilization of the platform:

- 52 participants created a personal profile including detailed professional information and target competencies;
- more than 15 business case studies have been added by superpeers ;
- more than 15 different types of assignments have been submitted by superpeers;
- a community wiki section has been populated and about 9 wiki pages have reported more than 350 reads;
- comments, messages and knowledge resources have been tagged by peers and superpeers resulting in a folksonomy made of about 40 keywords.

The questionnaire we sent to students to gather data on their attitude and satisfaction for the SCE has shown preliminary results that indicate way of improving the platform towards more customization of the web 2.0 tools. Response rate was 78% (18 respondents out of 23). Figure 2 presents an example of wiki page created within the SCE.

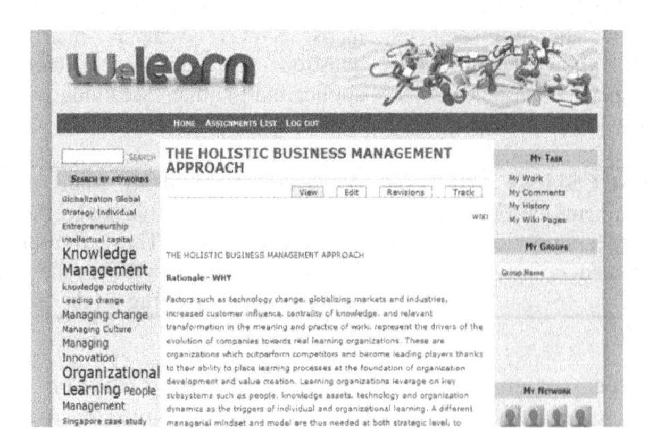

Fig. 2. An example of collaborative knowledge creation: a wiki page

The majority of respondents (70%) had no technical problem in downloading multimedia files available on the SCE platform. A student reported: "*It all depends on the Internet availability and on my own time, because social life is not through the technology [...], I like to discuss face-to-face rather than to post comment in blogs or create a wiki*".

Figures 3a and 3b illustrate the friendship and problem solving networks built through a web-survey after eight month of classroom and project team activities. Each node is an actor (peer and superpeer) and the length of the tie indicates how strongly a student reported to rely on others for solving a problem or how strongly he/she reported others as being friend. It is interesting to notice how more densely connected the friendship network appears, compared to the problem solving network, where dispersed clusters of actors are more evident.

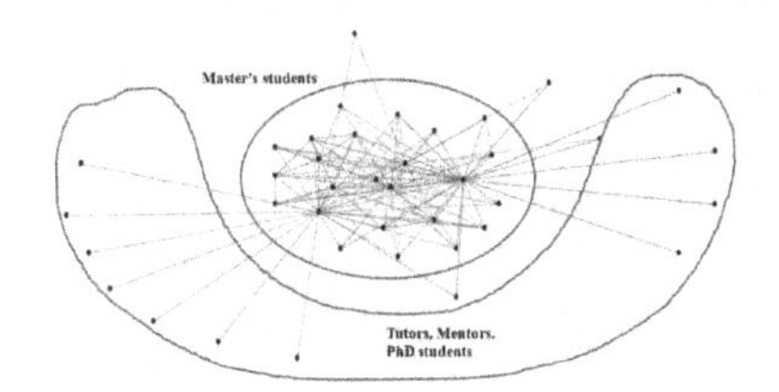

Fig. 3a. Friendship Network

Fig. 3b. Problem Solving Network

Almost all respondents (83%) agreed that the use of an inquiry and project-based approach helped them to retain more contextual knowledge and establish more fruitful contact with other community's members. 68% of students found very useful to use blogs and wikis to collaborate with peers, and many of them considered very important the discussion of real business cases (72%). As reported by a student, blogs and wikis are useful tools, and they should be created "*by us and by others, as it is better to engage seniors and experts in the process of knowledge transfer. Their role and experience might help control and correct the information created by us*".

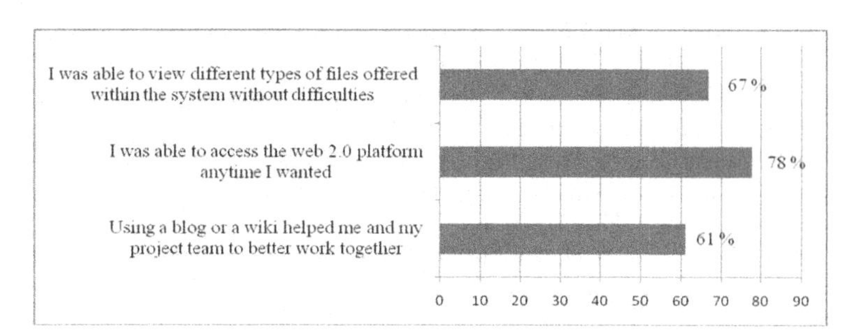

Fig. 4. Percentage of students who agreed on the benefits of using web 2.0

As Member 3 commented: "*The use of new case studies and the possibility to work on real projects, helped me learn how to apply knowledge to real situations. It allowed me to connect the business management concepts with my previous working experience. As for the SCE platform, I used it very often to create wiki pages for my various projects*".

6 Conclusions and Future Research

In this paper we propose a Web 2.0 based platform supporting a Project-based Learning approach, providing preliminary evidence from a case study. We observed and then surveyed a learning community in order to understand their satisfaction and utilization of the social computing platform.

Following the feedback provided by students in the survey and the observation of the degree of utilization of the SCE, we conclude that the use of Web 2.0 technologies in education has a high potential in supporting the acquisition of learners' competencies and skills. Nonetheless, it requires an important effort from the mentors called to change their mindset and behaviour. The active participation of a "community energizer" might help to change in fieri the tools following users' feedback.

The main benefits perceived in using the SCE platform are connected to:

- the possibility to collectively create a knowledge base of business case studies and other learning materials
- the absence of hierarchical relations among participants that makes it easier to share ideas and files
- the availability of many web 2.0 tools that can help support collaboration.

The SCE helped students to better collaborate with peers, tutors and mentors, and provided an environment where the learning assessment phase was much easier to perform. The social analyst role supported in monitoring the community evolution in the virtual environment, looking at individual communication patterns and group dynamics.

Future research might assess data stored in the SCE platform to monitor the community evolution and correlate indicators of communication and contribution with learning outcomes.

References

1. O'Reilly, T.: What is Web 2.0: Design Patterns and Business Models for the next generation of software. O'Reilly, Sebastopol (2005)
2. Schaffert, S., Bischof, D., Buerger, T., Gruber, A., Hilzensauer, W., Schaffert, S.: Learning with semantic wikis. In: Proceedings of the First Workshop on Semantic Wikis–>From Wiki To Semantics, Budva, Montenegro, pp. 109–123 (retrieved, November 2006)
3. Parker, K.R., Chao, J.T.: Wiki as a Teaching Tool. Interdisciplinary Journal of Knowledge and Learning Objects 3, 57–72 (2007)
4. Dawnes, S.: E-learning 2.0, eLearn Magazine, Association for Computing Machinery, Inc. (2005), http://elearnmag.org/subpage.cfm?section=articles&article=29-1 (retrieved November 9, 2005)
5. Chapman, C., Ramondt, L., Smiley, G.: Strong community, deep learning: Exploring the link. Innovations in Education & Teaching International 42(3), 217–230 (2005)
6. Corallo, A., De Maggio, M., Grippa, F., Passiante, G.: Evolving Mechanisms of Virtual Learning Communities. In: Lessons Learned from a case in Higher Education. LNCS, pp. 303–313. Springer, Heidelberg (2008)
7. Reiser, R., Dempsey, J.V.: Trends and issues in instructional design and technology, 2nd edn. Prentice-Hall, Inc., Upper Saddle River (2007)

8. Delich, P.: Pedagogical and interface modifications: What instructors change after teaching online, Published doctoral dissertation, Pepperdine University, Malibu, CA (2006)
9. Tapscott, D., Williams, A.D.: Wikinomics. How mass collaboration changes everything, Portfolio 1st edn. (2006)
10. Johnson, D.W., Johnson, R.T.: Cooperation and the use of technology. Handbook of research for educational communications and technology 1, 1017–1044 (1996)
11. Miers, J.: BELTS or Braces? Technology School of the Future (2004), http://www.tsof.edu.au/research/Reports04/miers.asp (retrieved November 2006)
12. Higgs, B., McCarthy, M.: Active learning – from lecture theatre to field-work. In: O'Neill, G., Moore, S., McMullin, B. (eds.) Emerging issues in the practice of university learning and teaching, Dublin, All Ireland Society for Higher Education (AISHE), pp. 37–44 (2005)
13. Byron, M.: Teaching with Tiki. Teaching Philosophy 28(2), 108–113 (2005)
14. Naish, R.: Can wikis be useful for learning? e.learning Age(2006), http://www.qiconcepts.co.uk/pdf/Can%20Wikis%20be%20useful%20for%20learning.pdf (retrieved, November 2006)
15. Odamtten, T., Millard, J.: Learning from others within the landscape of "transitional economies" and the challenge in ICT development for African countries. AI & Society 3(1), 51–60 (2009)
16. Chen, H.L., Cannon, D., Gabrio, J., Leifer, L., Toye, G., Bailey, T.: Using wikis and weblogs to support reflective learning in an introductory engineering design course. In: Proceedings of the 2005 American Society for Engineering Education Annual Conference & Exposition, Portland, Oregon, June 12-15 (2005)
17. Blumenfeld, P.C., Soloway, E., Marx, R.W., Krajcik, J.S., Guzdial, M., Palincsar, A.: 'Motivating Project-Based Learning: Sustaining the Doing, Supporting the Learning'. Educational Psychologist 26(3&4), 369–398 (1991)
18. Bridges, E.M., Hallinger, P.: Problem based learning for administrators, Eugene, Ore.: ERIC Clearinghouse on Education Management (1992)

Linking Semantic Web and Web 2.0 for Learning Resources Management

Adeline Leblanc and Marie-Hélène Abel

HEUDIASYC CNRS UMR 6599,
Université de Technologie de Compiègne
BP 20529, 60205 Compiègne CEDEX, France
{adeline.leblanc,marie-helene.abel}@utc.fr

Abstract. The growth and development of information and communication technologies implies new working forms and new learning forms. In such a context (1) the learning social process is increased, (2) methods to manage the numerous learning resources at disposal on the web are required. In this paper we present why and how we associated knowledge management methods, Semantic Web and Web 2.0 to develop a collaborative learning environment in the framework of the approach MEMORAe.

Keywords: Collaborative Learning Environment, Organizational Learning, Semantic Web, Web 2.0.

1 Introduction

Due to the growth and development of information and communication technologies, information takes a more and more important place in our society which can be qualified of information society. This context implies new working forms and new learning forms that constitute a part integral of the industrial challenges. Becoming a Learning Organization is a way for an organization to stay competitive. Such an organization is an organization in which work is anchored in an organizational culture which allows and encourages the training at various levels: individual, group and organization. Each organization actor is a kind of continuous learner. He has to reach the good resource at the right moment. That necessitates: a) This resource is stored, well described and well indexed; b) And/or facilitators to engage learners into social process (interaction, conversation).

Within the MEMORAe approach we are interested in these new learning forms. We consider that they are connected to the knowledge management practices and we developed a learning environment based on the concept of learning organizational memory. This environment is dedicated to support an organizational learning and consist in a web platform using semantic annotations and web 2.0 technologies.

In the following, we present why e-learning is a social process and how web 2.0 technologies can facilitate this social process. Then, we bring closer the e-learning area with the knowledge management area in an organizational learning context. We justify the Semantic Web/Web 2.0 association to facilitate the learning social process. Finally we present the approach MEMORAe.

M.D. Lytras et al. (Eds.): WSKS 2009, LNAI 5736, pp. 60–69, 2009.

2 e-Learning and Social Process

Most of the early web-based courses were designed to complement classical teaching methods for dissemination of courses materials. The online learning environments were used as tools for pedagogical material (learning object) delivery in which the students' role was passive: no exploitation of the web communication potential. We know that learning is an active process where learners build their knowledge and understanding [1] [2]. However, according to [3], we have not to forget learning is also a social process which proceeds through conversation [4] and interaction [5].

E-learning has the potential to put into practice this social process. However, to exploit this potential, e-learning requires facilitators to engage learners into interaction [6].

Web 2.0 technologies seem to be a good candidate to build e-learning environments taking into account a social process. With Web 2.0 technologies, users are readers as well as contributors. Thus Web 2.0 technologies offer to users distributed collaboration facilities [7]. They seem to be a good way to produce facilitators in order to engage learners into interaction. These distributed collaboration facilities start to have a significant impact on e-learning. They allow a distributed control and coordinated actions between learners.

The most well-known web 2.0 applications are wikis and blogs. For few years wikis have been used in educational institutions as tools that promote sharing and collaborative creation of Web contents. Wikis are crucially different from blogs, which are also used in educational context, in that users can modify any entry, even material posted by others. Let's note that, although wikis are a tool for creating contents, they serve at the same time collaborative skills learning. Thus, wikis offer to students a collaborative environment where they learn how to work with others, how to create a community and how to operate in our society that we can qualify of cognitive society. In such a society the creation of knowledge and information is increasingly becoming a group effort [8].

Thus, the Web 2.0 key components are the facility of using creation tools and the collaboration and social interactions they offer.

Meanwhile, according to [9]:

"Web 2.0 is an attitude not a technology. It's about enabling and encouraging participation through open applications and services. By open I mean *technically open* with appropriate APIs but also, more importantly, *socially open*, with rights granted to use the content in new and exciting contexts."

Finally, the Web 2.0 success comes from the technology/attitude association. Following this way, Downes coined the term e-learning 2.0 which results from the combination of web 2.0 technologies and the collaboration/social interactions they offer in the context of learning application [10].

3 Organizational Learning Process

The role of organizational learning for company's survival and performance has been described in works [11][12]. Organizational learning is the process by which organizations learn. Problems solving in collaboration becomes the main activity producing value in companies. The new working forms generate more and more informal situations of learning. The discussions around problems and\or projects, various meetings, etc. produce situations in which members learn and develop knowledge and particular competences. That's why an organizational learning process can be finally seen as a collective capability based on experiential and cognitive processes and involving knowledge acquisition, knowledge sharing and knowledge utilization [13].

Thus a learning organization is an organization in which processes are imbedded in the organizational culture that allows and encourages learning at the individual, group and organizational level [14]. A learning organization must be skilled at creating, acquiring, and transferring knowledge, and at modifying its behaviour to reflect knew knowledge and insights [15]. According to [16], a learning organization is a firm that purposefully constructs structures and strategies so as to enhance and maximize organizational learning.

An organization cannot learn without continuous learning by its members. Individual learning is not organizational learning until it is converted into organizational learning. The conversion process can take place through individual and organizational memory [17]. The results of individual learning are captured in individuals' memory. And, individual learning becomes organizational learning only when individual memory becomes part of organizational memory.

Finally, organizational learning seldom occurs without access to organizational knowledge. In contrast to individual knowledge, organizational knowledge must be communicable, consensual, and integrated [18]. According to [17], being communicable means the knowledge must be explicitly represented in an easily distributed and understandable form. The consensus requirement stipulates that organizational knowledge is considered valid and useful by all members. Integrated knowledge is the requirement of a consistent, accessible, well-maintained organizational memory.

Thus companies try to take into account in their development policy: a) The 'formal' training in the framework of e-learning projects; b) The 'informal' training through the recognition of knowledge exchanges networks; c) The formalization of knowledge capitalization procedures coming from the theory of the Knowledge Management (KM).

Knowledge Management comprises a range of practices used by organizations to identify, create, represent, and distribute knowledge for reuse, awareness and learning[1].

So, KM systems and e-learning systems serve the same objectives: facilitate the development of competences and learning in organizations [19]. They are moreover complementary in an Organizational Learning context. E-learning systems are used as support by learners so that they can develop their knowledge. They offer to them structured educational contents and opportunities of intercommunication about specific subjects. The KM systems offer possibilities of access to knowledge by means of contents management systems. These kinds of systems aim at managing all the contents of a company.

In spite of this context, several studies showed that the connections between e-learning and KM are not operationnalized [20]. According to Ras [21], this is due to various barriers on conceptual or technical level. We can mention:

---

[1] http://en.wikipedia.org/wiki/ Knowledge_management.

- Problems on conceptual level: for example [22] propose an environment of workplace composed of three spaces: a working space, a knowledge space and a learning space. The main problem concerns the connections between these spaces. Each space has its own structure which reflects the mental model of its users.
- Problems on a technical level: the spaces of work, knowledge or learning are implemented on different systems. Each system possesses its own structure of contents.

Associating Web 2.0 and Semantic Web approaches could offer a platform that overcomes these barriers.

4 Linking Web Semantic and Web 2.0

Web 2.0 technologies enable non-specialist users to contribute to the Web. This leads to new requirements of Networked Information Retrieval [13]. Indeed, in the web 1.0 context, only the web masters created content and thus used professional data in order to enable any users to retrieve it by way of different methods such as key words-based retrieval, metadata-based retrieval... In the Web 2.0 context, due to the lack of semantic relation among contents, massive information produced by users can't be processed. This information generally consists in microcontents creation. Microcontents come from the various contributions of users such as images, collected bookmarks, queries and answers of forum and so on. Unfortunately, much of these contributions are currently confined into private space or published in formats that hinder its reuse.

In the context of Semantic Web, data on the web are published in machine-readable format using shared ontologies to give them a formal semantic, and inter-linked on a massive scale [23]. Thus data can be retrieval easily. Publishing data using languages dedicated to the Semantic Web (RDF[2], OWL[3] or Topic Maps[4]), has different advantages: (a) Makes data retrieval by using a standard query language (SPARQL[5], TMQL[6]); (b) Facilitates the integration of data from different resources; (c) Allows the creation of machine-readable links between data resources. However, create Semantic Web applications necessitates specialist skills and it is a brake to their growth.

Web 2.0 and the Semantic Web have been previously considered as independent. Each approach is supported by its own community. Due to their strengths and their weakness, we believe, like several colleagues, that these two visions are complementary:

"The Semantic Web can learn from Web 2.0's focus on community and interactivity, while Web 2.0 can draw from the Semantic Web's rich technical infrastructure for exchanging information across application boundaries." [24].

According to the same authors, this is possible by providing simple, well-structured Web forms through which users can add comments, information to a web site without requiring any knowledge of the underlying technologies or principles. By

[2] http://www.w3.org/RDF/
[3] http://www.w3.org/2004/OWL/
[4] http://www.topicmaps.org/
[5] http://www.w3.org/TR/rdf-sparql-query/
[6] http://www.isotopicmaps.org/tmql/

following this approach, an environment should enable users to create content that is immediately usable on the Semantic Web. Users are guided and place their contribution in such a way that semantic annotations are automatically associated.

Semantic wikis are an illustration of this complementarily. They try to combine the strengths of Semantic Web and Wiki technologies. The wikipedia definition[7] specifies that a semantic wiki is a wiki that has an underlying model of the knowledge described in its pages.

5 The Approach MEMORAe

Our aim, within the approach MEMORAe, is to operationalize connections between e-learning and knowledge management in an organizational learning context. We chose to associate: a) knowledge engineering and educational engineering; b) Semantic Web and Web 2.0 technologies to model and build a collaborative learning environment.

In the e-learning side, we decided to follow the resources modelling approach based on the learning objects paradigm;

In the knowledge management, we chose to adapt the concept of Organizational Memory. Dieng define such a concept as an "explicit, disembodied, persistent representation of knowledge and information in an organization, in order to facilitate its access and reuse by members of the organization, for their tasks" [25]. Extending this definition, we propose the concept of Learning Organizational Memory for which users' task is learning.

In the Semantic Web side, we decided to structure and organize our memory content by way of ontologies.

In order to facilitate exchanges and social processes we decided to use Web 2.0 technologies and to model and organize micro resources created by way of ontologies.

The approach MEMORAe is organized around two projects. The project MEMORAe deals mainly with semantic aspects. The project MEMORAe2.0 is an extension of the project MEMORAe and deals mainly with collaborative aspects.

In order to assess our approach, we chose to build Learning Organizational Memory for academics organization: a course on algorithms and programming at the Compiègne University of Technology (France) and a course on applied mathematics at the University of Picardy (France).

5.1 The Project MEMORAe

Within the project MEMORAe [26], we were interested in the knowledge capitalization in the context of organizations. To that end, we developed the environment E-MEMORAe as support for Learning Organizational Memory. In such a system resources are indexed to knowledge organized by means of ontologies: domain and application. The domain ontology, defines concepts shared by any organization; the application ontology defines concepts dedicated to a specific organization. In our context, the domain ontology defines concepts shared by any training (Course,

[7] http://en.wikipedia.org/wiki/Semantic_wiki

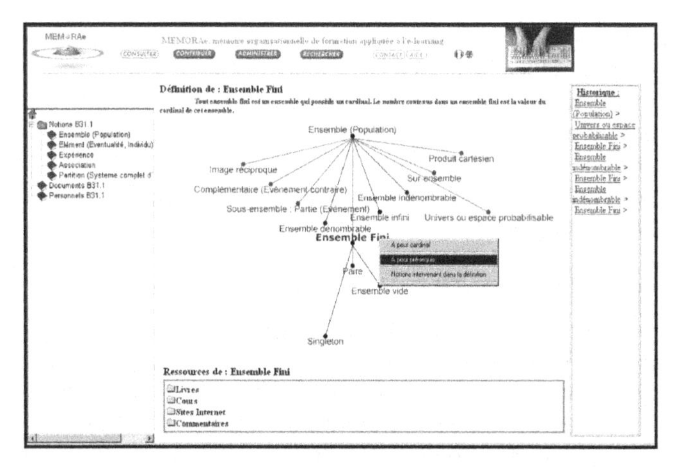

Fig. 1. E-MEMORAe navigation interface (in French)

Exercise, Professor, etc.); The application onologies define concepts dedicated to a specific training (For the applied mathematics training: Set, SubSet, etc.). Using these ontologies, actors can acquire knowledge by doing different tasks (solving problems or exercises, reading examples, definitions, reports, etc.). We used Topic Maps[8] as a representation formalism facilitating navigation and access to the resources. Such formalism has a XTM[9] specification that consists in an XML syntax allowing its expression and its interchange. It also has its own query language TMQL[10].

The ontology structure is also used to navigate among the concepts as in a roadmap. The user has to access to the resources which are appropriate for him.

E-MEMORAe aims at helping the users of the memory to acquire organization knowledge. To this end, users have to navigate through the application ontology that is related to the organisation, and to access to the indexed resources thanks to this ontology.

The general principle is to propose to the user, at each step, either precise information on what he is searching for, or graphically displayed links that allow him to continue its navigation through the memory.

To be more precise, the user interface (cf. Figure 1) proposes:

- Entry points (left of the screen): they enable users to start their navigation with a given concept. An entry point provides a direct access to a concept of the memory and consequently to the part of the memory dedicated to it. They were chosen by the head of the course who considers them as essential.
- Resources (bottom of the screen) related to the current concept: they are ordered by type (books, course notes, sites, examples, etc.). Starting from a map concept or an entry point, the user can directly access to associated resources. Descriptions of these resources help him to choose among them.

---

[8] http://www.topicmaps.org/
[9] http://www.topicmaps.org/xtm/
[10] http://www.isotopicmaps.org/tmql/spec.html

- A short definition of the current concept: it enables users to get a preview of the concept and enables them to decide if they have to work it or not.
- A history of the navigation: it enables the user to remind and to be aware of the path he followed before. Of course, he can get back to a previously studied notion if he wants to.
- Least but not last, the part of the ontology describing the current resource is displayed at the centre of the screen.

5.2 The Project MEMORAe2.0

Within the project MEMORAe2.0 [27] we are interested in developing memory collaborative functionalities and social processes. To that end, we take into account different levels of memory and different ways to facilitate exchanges between the organizational actors. In such an environment, we distinguish knowledge and resources of: a) the whole organization; b) a group of individuals in the organization – the organization is constituted of different groups of individuals even if it can be seen as a group itself; and c) an individual.

To that end, we modelled different level of memories. In order to facilitate and to capitalize exchanges between organization members we added the concept of exchange resources (currently, we tested the forum concept). These models extended the MEMORAe domain ontology. The idea is to capitalize exchanges concerning any concepts of the organization ontologies.

For example, when actors need to know who works on a project, they have to access to the information relative to the project itself. A way to do this is to navigate through a concept map based on an ontology defining the organization knowledge. According to their access rights, they can visualize space/memory resources. In case of exchange resources (forum), they can exchange ideas or information (externalization of tacit knowledge).

In order to put into practice this modeling we developed a new environment called E-MEMORAe2.0[11] (cf. Figure 2). It re-uses general principles of E-MEMORAe and

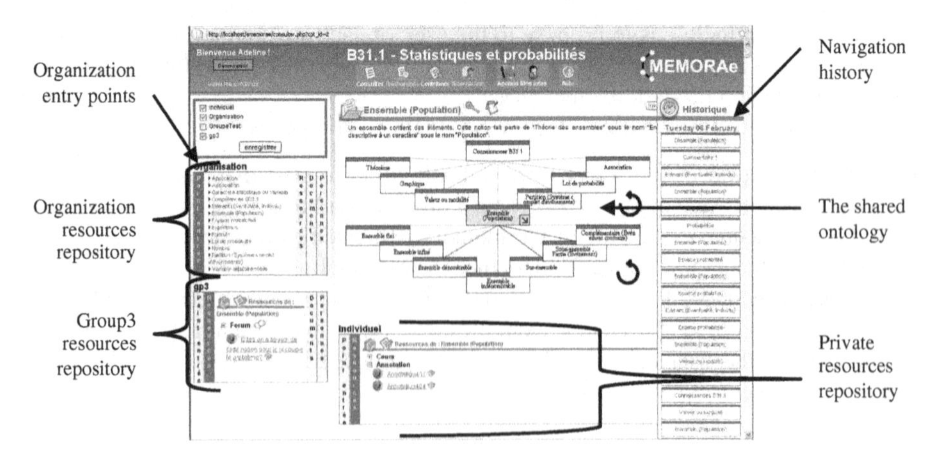

Fig. 2. E-MEMORAe2.0 navigation interface (in French)

[11] http://www.hds.utc.fr/~ememorae/Site-MEMORAe2.0/

gives the possibility of learners to have a private space and participate to share spaces according to their rights. All these spaces (memories) share the same ontologies but store different resources and different entry points. In order to facilitate the resource transfers, users can visualize different spaces/memories content at the same time and make a drag and drop to store a resource from a specific memory to another one.

In order to facilitate users' exchanges, we modeled a forum as a set of microcontents or micro-resources. We decided to manage these micro-resources like any resources in the memory. Thus we defined the concept of MEMORAe-Forum which is an exchange resource and is linked to one micro-resource of question type and 0 or n resources of answer type. Each micro-resource is indexed by a concept of the application ontology and has an author, a date of contribution.

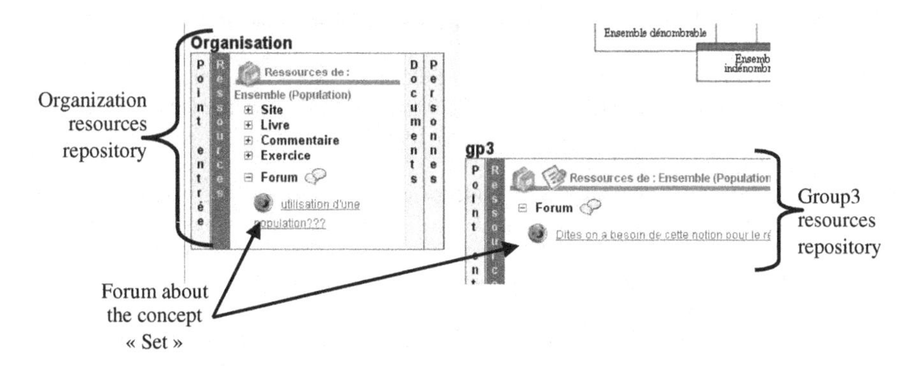

Fig. 3. Forum Access (in French)

In such a way, each group memory has its own forum organized around the shared ontology. All the forum contributions are distributed in the resource space among the other resources. Users don't access to the forum itself but to the memory resource space and then select resources of MEMORAe-Forum type to participate to the forum. So there is not explicit forum where users can visualize all contributions of members but it will possible to build this presentation form with the functionality export we plan to develop.

6 Conclusion

In this paper we presented links we made between knowledge management, e-learning, semantic web, and web 2.0 technologies to build a collaborative environment in the framework of the approach MEMORAe. We implemented a semantic forum in such a way that users contribute to produce resources semantically described without specialist skills. Seeing queries and answers as resources should offer facilitators to engage learners into interaction: they see these resources when they access to the resources part and then can be questioned. Even it is a semantic collaborative environment; E-MEMORAe2.0 is not a semantic wiki. It is a memory where it is possible to organize any resources or micro-resources in different work spaces (individual, group, organization) around shared ontologies. Users can easily transfer resources from one space to another one; they can contribute to exchange microresources (semantic forum). We plan to model semantic chat, semantic blogs and semantic e-mail in the same way we modelled semantic forum. All these micro-resources will be capitalized and accessed like any resources in the memory (course, web site, exercise, etc.). In the case of blogs we will be in the context of blog farm.

E-MEMORAe and E-MEMORAe2.0 evaluations gave us good results [26] [27]. Learners used their different memories and tested forums. Currently our environment is used by academics. We have contact with industrials in order to evaluate such an environment to foster learning and innovation in their organization.

References

1. Wittrock, M.C.: Learning as a generative process. Educational Psychologist 11, 87–95 (1974)
2. Papert, S., Harel, I.: Constructionism. Ablex Publishing, Norwood (1991)
3. Ebner, M., Holzinger, A., Maurer, H.: Web 2.0 technology: Future interfaces for technology enhanced learning? In: Stephanidis, C. (ed.) HCI 2007. LNCS, vol. 4556, pp. 559–568. Springer, Heidelberg (2007)
4. Motschnig-Pitrick, R., Holzinger, A.: Student-Centered Teaching Meets New Media: Concepts and Case Study. IEEE Journal of Educational Technology & Society 5(4), 160–172 (2002)
5. Preece, J., Sharp, H., Rogers, Y.: Interaction Design: Beyond Human-Computer Interaction. Wiley, New-York (2002)
6. Sargeant, J., Curran, V., Allen, M., Jarvis-Selinger, S., Ho, K.: Facilitating interpersonal interaction and learning online: Linking theory and practice. Journal of Continuing Education in the Health Professions 26(2), 128–136 (2006)
7. O'Reilly, T.: What is Web 2.0 – design patterns and business models for the next generation of software (2005), http://www.oreillynet.com/lpt/a/6228
8. Richardson, W.: Blogs, wikis, podcasts, and other powerful web tools for classrooms. Corwin Press, Thousand Oaks (2006)
9. Davis, I.: Talis, Web 2.0 and All That (2005), http://iandavis.com/blog/2005/07/talis-web-20-and-all-that?year=2005&monthnum=07&name=talis-web-20-and-all-that
10. Downes, S.: E-learning2.0 (2008), http://www.elearnmag.org/subpage.cfm?section=articles&article=29-1
11. Agyris, C., Schön, D.A.: Organizational learning II: Theory, method, and practice. Addison-Wesley, Reading (1996)
12. Senge, P.M.: The fifth discipline. Doubleday Publishing, New York (1990)
13. Zhang, Z., Tang, J.: Information Retrieval in Web 2.0. In: Wang, W. (ed.) IFIP International Federation for Information Processing. Integration and Innovation Orient to E-Society, vol. 251, pp. 663–670. Springer, Boston (2007)
14. Sunassee, N., Haumant, V.: Organisational Learning versus the Learning Organisation. In: Proceedings of South African Institute of Computer Scientists Information Technologists 2004, SACSIT, pp. 264–268 (2004)
15. Garvin, D.: Building a learning organization. Business Credit 1, 19–28 (1996)
16. Dodgson, M.: Organizational Learning: A Review of Some Literatures. Organizational Studies 14(3), 375–394 (1993)

17. Chen, J., Ted, E., Zhang, R., Zhang, Y.: Systems requirements for organizational learning. Communication of the ACM 46(12), 73–78 (2003)
18. Duncan, R., Weiss, A.: Organizational learning: Implications for organizational design. In: Staw, B. (ed.) Research in Organizational Behavior, pp. 75–123. JAI Press, Greenwich (1979)
19. Schmidt, A.: Bridging the Gap between Knowledge Mabagement and E-Learning with Context-Aware Corporate Learning. LNCS, pp. 203–213. Springer, Heidelberg (2005), http://herakles.fzi.de/aschmidt/Schmidt_LOKMOL05_Extended.pdf
20. Efimova, L., Swaak, J.: KM and E-learning: toward an integral approach? In: Proceedings of KMSS 2002, EKMF, pp. 63–69 (2002)
21. Ras, E., Memmel, M., Weibelzahl, S.: Integration of E-Learning and Knowledge Management – Barriers, Solutions and Future Issues. In: Althoff, K.-D., Dengel, A., Bergmann, R., Nick, M., Roth-Berghofer, T. (eds.) (2005)
22. Ley, T., Lindstaedt, S.N., Albert, D.: Supporting Competency Development in Informal Workplace Learning. In: Althoff, K.-D., Dengel, A.R., Bergmann, R., Nick, M., Roth-Berghofer, T.R. (eds.) WM 2005. LNCS (LNAI), vol. 3782, pp. 189–202. Springer, Heidelberg (2005)
23. Shadbolt, N.R., Hall, W., Berners-Lee, T.: The semantic Web revisited. IEEE Intelligent System 21, 96–101 (2006)
24. Ankolekar, A., Krötzsch, M., Tran, T., Vrandecic, D.: The two cultures: mashing up Web 2.0 and the Semantic Web. In: Proceedings of the 16th International Conference on the World Wide Web, WWW 2007, Banff, Canada, pp. 825–834 (2007)
25. Dieng, R., Corby, O., Giboin, A., Ribière, M.: Methods and tools for corporate knowledge management. In: Proceedings of the 11th workshop on Knowledge Acquisition, Modeling and Management (KAW 1998), Banff, Canada, pp. 17–23 (1998)
26. Abel, M.-H., Benayache, A., Lenne, D., Moulin, C.: E-MEMORAe: a content-oriented environment for e-learning. In: Pierre, S. (ed.) E-learning networked environments and Architectures: A Knowledge processing perspective. Springer Book Series: Advanced Information and Knowledge Processing, pp. 186–205 (2006)
27. Leblanc, A., Abel, M.-H.: Using Organizational Memory and Forum in an Organizational Learning Context. In: Proceedings of the Second International Conference on Digital Information Management ICDIM 2007, pp. 266–271. Springer, Heidelberg (2007)

Quantum Modeling of Social Networks
The Q.NET Project

Cristian Bisconti, Angelo Corallo, Marco De Maggio, Francesca Grippa, and Salvatore Totaro

eBusiness Management Section, Scuola Superiore ISUFI - University of Salento
via per Monteroni, sn 73100 Lecce, Italy
{cristian.bisconti,angelo.corallo,marco.demaggio,francesca.grippa, salvatore.totaro}@ebms.unile.it

Abstract. This research aims at the application of the models extracted from the many-body quantum mechanics to describe social dynamics. It is intended to draw macroscopic characteristics of communities starting from the analysis of microscopic interactions with respect to the node model. In the aim to experiment the validity of the proposed mathematical model, the Q.NET project is intended to define an open-source platform able to model nodes and interactions of a network, to simulate its behaviour starting from specific defined models. Q.NET project will allow to visualize the macroscopic results emerging during the analysis and simulation phases through a digital representation of the social network.

Keywords: Quantum Physics, Social computing, Social Networks.

1 Introduction

In recent years new organizational forms are emerging in response to new environmental forces that call for new organizational and managerial capabilities. Organizational communities, interpreted as catalytic networks, are becoming the governance model suitable to build a value-creating organization, representing a viable adaptation to an unstable environment [3,6]. The theoretical framework used in this project to describe the nature and evolution of communities is known as *complexity science*. According to this approach organizational communities are viewed as *complex adaptive systems* (CAS): they co-evolve with the environment because of the self-organizing behavior of the agents determining a fitness landscape of market opportunities and competitive dynamics. A system is complex when equations that describe its progress over time cannot be solved analytically [11]. Understanding complex systems is a challenge faced by different scientific disciplines, from neuroscience and ecology to linguistics and geography.

A number of tools have been developed in recent years to analyse complex systems. Amaral and Ottino [2] identify three types of tools belonging to areas well known by physicists and mathematicians: Social Network theory, Quantum

M.D. Lytras et al. (Eds.): WSKS 2009, LNAI 5736, pp. 70–77, 2009.

Mechanics, Statistical Physics. A number of researchers have shed light on some topological aspects of many kinds of social and natural networks [1,10]. As a result, we know that the topology of a network is a predictable property of some types of networks that affects their overall dynamic behaviour and explains social processes.

Social Network Analysis (SNA) and Dynamic Network Analysis (DNA) represent the most adopted methodological approaches to the study of organizational networks during the past years [15,8,12,4]. Social Network Analysis proposes methods and tools to investigate the patterning of relations between social actors [15]. It studies organizational communities, providing a visual representation and relying on the topological properties of the networks so measuring the characteristics of the network object of the analysis.

The main limitation of SNA is to be mainly a quantitative social science method, ignoring sometimes the importance of qualitative issues to explain phenomena. Its unit of analysis is not the single actor with its attributes, but the relations between actors, defined identifying the pair of actors and the properties of the relation among them. By focusing mainly on the relations, SNA might underestimate many organizational elements which could influence the ability of an organization to reach its goals.

The empirical work on network information advantage is still "content agnostic" [7]. As stated by [5], SNA globally considered is a framework to investigate the information structure of groups, the structural aspect of correlation, disregarding the content of relationships, and the nodes' properties. Paying attention only to the structural facets of community interactions is like considering all the ties as indistinguishable and homogeneous.

Dynamic Network Analysis uses data coming from the SNA to perform an evolutionary study of the organizational networks, to predict possible network transformation over time. During the last years a trend emerged in the application of the statistical physics to several interdisciplinary fields like biology, information technology, and social sciences, and physicists showed a growing interest for modeling systems also far from their traditional context. In particular, in social phenomena the basic elements are not particles but individuals, and each of them interacts with a limited amount of others, generally close, and negligible if compared to the number of people within the system.

In the light of the considerations above it seems plausible the model of social systems built as a many-body system based on a quantum structure. We will try for the first time to describe a model of the agent with his own identity regardless of the role and the interactions of the communities he belongs to. We consider also very important the relationship between the state of the agent and the interactions typical of the community in which he lives. The goal of agent-based modeling is to identify he higher (macro) level of the social system emerging from the characteristics of its lower (micro) level.

The Artificial Life community has been the first in developing agent-based models [9,13,14], but since then agent-based simulations have become an important tool in other scientific fields and in particular in the study of social systems.

In this context it is worth mentioning the concept of Brownian agent which generalizes that of Brownian particle from statistical mechanics. A Brownian agent is an active particle which possesses internal states, can store energy and information and interacts with other agents through the environment. Again the emphasis is on the parsimony in the agent definition as well as on the interactions, rather than on the autonomous actions. Agents interact either directly or in an indirect way through the external environment, which provides a feedback about the activities of the other agents. Direct interactions are typically local in time and ruled by the underlying topology of the interaction network. Populations can be homogeneous (i.e., all agents being identical) or heterogeneous. Differently from physical systems, the interactions are usually asymmetrical since the role of the interacting agents can be different both for the actions performed and for the rules to change their internal states. Agent-based simulations cover now a central role in modeling complex systems and a huge literature has been developed about the internal structure of the agents, their activities and the multi-agent features.

2 Objectives of the Research Project

Q.NET is a basic research project conceived by the eBMS Section of Scuola Superiore ISUFI, University of Salento (Italy). It aims at the application of the models extracted from the many-body quantum mechanics to describe social dynamics. It is intended to draw macroscopic characteristics of the network starting from the analysis of microscopic interactions basic respect to the node model. In the aim to experiment the validity of the proposed mathematic model, the Q.NET project is intended to define an open-source platform able to model nodes and interactions, to simulate the network behaviour starting from specific defined models, and to visualize the macroscopic results emerging during the analysis and simulation phases through a Digital representation of the social network.

The research presented in the Q.NET project will start with the study of methodologies and instruments from the inferential statistics that showed their usefulness for the comprehension of the level of reliability of the observed interactions, mirroring the real behaviour of the population under investigation, being not only the result of a casual event.

A discussion about the limitations and potentialities of different inferential statistical models (e.g. Bayesian Theory) will represent the theoretical basis to suggest new models based on quantum physic principles to model the complex space of the actors and of the interactions among them. Quantum physics is plenty of complex models of interaction between states described by vectors in representative spaces, resulting in well defined collective states. Representative space means the space where the actor (the body of the system) is defined. The actor definition is provided through the definition of his state, that could be expressed as distribution function named probability strength and related to the real probability distribution of the actor. This probability strength is a vector of the representative space defined as Hilbert space infinite dimension. Based on this consideration, the Q.NET Project intends to answer to the following question: "Which models from the quantum physics are suitable to model the behaviour and the evolution of organizational communities?". To answer to this question, a framework will be proposed based on three dimensions:

- Modeling language and primitives able to describe social phenomena based on the individual characteristics and on the interactions among individuals;
- Physic and mathematical methods applied in the field of organizational communities analysis;
- Set of Information Technology Tools represented by:
  - Applications for the node and network modeling;
  - Applications for the network behaviour simulation;
  - Applications for real and simulated variables analytic visualization.

The Q.NET project aims at satisfying the need of intrdisciplinarity that is the basis of all the studies about complex systems. The framework of analysis and simulation that is proposed here tries to identifies physical and mathematical sciences, information technologies and artificial intelligence, statistics and sociology. This integration represents a further expected result of the research that is intended to be conducted in the Q.NET Project.

3 Description of the Q.NET Project

This project will be organized in six phases:

- Recognition of the state of the Art
- Formulation of modelling primitives and language able to describe the social phenomena based on the individual characteristics and interactions.
- Formulation of a mathematical model to represent the organizational dynamics
- Formulation of simulation tools based on agent-based modeling
- Development and implementation of the simulation tools based on agent-based modelling
- Framework validation to cases of organizational communities

3.1 State of the Art

Based on the most recent contributions in literature, we will define a taxonomy based on dimensions like: community purpose, size, degree of physical proximity, leadership and membership, degree of internal diversity, life cycle, sponsorship and degree of institutionalization. We will focus on the relation between actors' behavior and the socio-technical environment in which actors are involved. A description of the methodologies and the tools of inferential statistics will be provided with reference to relational data analysis. In addition to the studies on statistical modeling of social behavior, we will take into consideration scientific contributions related to their physical modeling. In particular, we will focus on the efforts done in the field of Econophysics, a new science that studies both the behavior of social actors (individuals, groups, organizations), considered as "particles" of a physical system, and the effect of their interactions on the system.

3.2 Formulation of a Language and Modeling Primitives to Describe the Social Phenomena on the Basis of Individual Characteristics and Interactions between Individuals

We will describe actor's behavior and the social characteristics that will provide input for defining both the logical structure and the abstract language for modeling actors and interactions.

In order to create the quantum mathematical model, it is important to define initially which is the space of configurations where the node has to be modeled, that is to define the properties of the node to which precise functions of distribution will be associated. These functions will act both on ordinary spaces and on spaces with suitably defined characteristics. This is done in analogy to the description of particles of a quantum system by a wave function that depends both on the coordinates of the ordinary space, and the coordinates of the spin area. The models and structures that will shape the organizational community will be described by a formal language that can express at the same time the actors' characteristics and those of the entire network, as well as the management of the simulation and related results.

3.3 Formulation of the Mathematical Framework for Building a Quantum Model of the Agent and of the Interactions between Actors

This part of the project will address the formulation of the mathematical framework for modeling the actor and the possible interactions with other actors. Following the paradigm of quantum physics, the process will consist of building an algebraic abstract structure of operators able to describe the characteristics of the actor. Once identified the algebraic space on which to operate, we will define the possible expressions of the interactions involving actors. This phase has an abstract and mathematical nature, so to consider the whole set of possible properties that need to be represented. After this phase we will formulate the model itself, based on the mathematical framework developed earlier. The stages of formulation of the model will follow the typical paradigms of microscopic description of many-body systems. The actor-agent model that we want to create will be based on the specification of its location along different dimensions, which can be defined in a more or less interdependent way. This opens the way to a vector representation of the state of an agent and a possible tensorial representation of the interactions between agents, and leads to a vision in which the bonds in which the community is structured emerge from such interactions.

3.4 Design and Implementation of Simulation Tools Based on Agent-Based Modeling

The logical structure for describing the participants and the interactions in algebraic structures will be represented in a digital way, using an object-oriented language (known or to be developed), capable of representing actors, communities and simulations in the most efficient, flexible and complete manner. This work will also determine the pattern of interaction between actors, and therefore, how the networks will be defined in the simulation environment, in addition to the events to be managed during the temporal evolution of simulation and the interactions made by the graphical interface. The digitalized actors and communities will be simulated in their evolution. This will involve the design of a simulation engine based on parallel processing, that can provide the simulation of large communities. A set of primitives will be developed able to statistically set the parameters of certain actors within the community. The digital model will also be able to access the output variables for the visual representation of the community's situation, both in the total and in the partial way. To test the network is therefore natural to define the model using graphical tools, by simulating it from a statistical or deterministic configuration, and then by observing the results.

The Project of the graphic interface will consist of the implementation of a framework that will implement all the specifications relating to the handling of digital models in a graphical environment. In this framework three applications will be implemented:

- *Q.NET-Creator*, with which actors and community will be shaped.
- *Q.NET-Simulator*, through which, by interacting with the engine it will be possible to set, start and manage the simulation of the community.
- *Q.NET-Reporter*, which will interact with the engine, and will be able to produce the report of the simulation and analysis of the simulated community.

3.5 Validation of the Framework on Cases of Organizational Communities

In this phase we will identify the organizational communities on which the validity of the framework will be tested. These communities will be described in terms of differences in nature, community objective, size, degree of physical proximity, leadership and membership, level of internal diversity, life cycle, sponsorship and degree of institutionalization. Then collected data will be suitable to be processed with the system of analysis and simulation. They include, but are not limited to, information flows (e.g. exchanges of email, instant messaging, formal and informal meeting, e-meeting, conference call) and profiling of users (e.g. personal characteristics, professional experience, skills, roles, tenure) . The methodology for data collection will include the administration of questionnaires, email tracking, recording of data archives, which will be translated into a format suitable for computation by the given system of analysis. The computation of the data collected will provide a representation of the community useful as a yardstick for the configuration of the community. Objective of the task is to start from the lessons learned by applying the model to real organizational communities and extract possible scenarios in which the framework can be useful for new organizational communities. Managerial recommendations on how to interpret the predictions that emerge from the framework.

4 Expected Results

The expected results of Q.NET project are so defined:

- Definition of the space of configurations and creation of the space of operators that afterward will be specified according to their functional form for each particular property that is intended to be described.
- Development of a formal language to describe at the same time the actors' and whole network's characteristics.
- Development of an actor-actor interaction model able to describe possible communication channels on which an organizational network can be built.
- Definition of a model of organizational network based on microscopic variables able to describe collective characteristics of the system.

Q.NET Framework will consist of classes and libraries able to analyze of social networks applying the proposed quantum models. The modules and applications designed will represent the key instruments applied by scientific communities involved in social network analysis to develop and validate digital models of social phenomena with whom the behaviors of organizational communities will be analyzed and predicted. Then the works obtained will be suitable to be organized and maintained in a simple and effective way. These modules and the applications aim to be interoperable, scalable and to support many simulations at the same time.

To face these requirements the modules of the framework that are intended to be developed will be designed to work on an atomic scale on systems different from an architectural point of view and devoted to support different storage systems located in different places and communicated through TCP/IP networks.

5 Conclusions

The application of the quantum model and of the simulation model to the behaviour of communities will support the social network analysis within the description of possible evolutionary scenarios of organizational communities based on the comparison between real and simulated models. The project will provide managerial results related to:

- Definition of guidelines for the application of the analysis and simulation framework to organizational communities different in objectives, size, culture, level of institutionalization.
- Development of digital models to be applied to new contexts able to provide analysis and forecasts about the evolution of their organizational properties. Such models will allow to suggest modalities to improve and optimize resources, information flows, competences. The focus of the re-organization could vary according to the choice of the organizational variable to be processed by the analysis and simulation model.

Possible scenarios where the system described in this paper might be applied are:

- Monitoring the technical and social knowledge exchange within and across communities of innovation.
- Defining key performance indicators integrated with network metrics, so to link the rate of information flow to the organizational and individual performance.
- Monitoring the evolution of virtual learning communities.

References

1. Albert, R., Barabàsi, A.-L.: Statistical Mechanics of Complex Networks. Rev. of Mod. Phys. 74, 47 (2002)
2. Amaral, L.A.N., Ottino, J.M.: Complex networks - Augmenting the framework for the study of complex systems. The European Physical Journal 38, 147–162 (2004)
3. Clippinger III, J.H.: The biology of business. Jossey-Bass Publishers, San Francisco (1999)
4. Gloor, P., Grippa, F., Kidane, Y.H., Marmier, P., Von Arb, C.: Location matters - measuring the efficiency of business social networking. International Journal of Foresight and Innovation Policy 3/4, 230–245 (2008)
5. Goodwin, J., Emirbayer, M.: Network Analysis, Culture, and the problem of Agency. American Journal of Sociology (1999)
6. Grippa, F., Di Giovanni, A., Passiante, G.: Open Business Innovation Leadership. In: Romano (ed.) Fostering Innovation through Value-Creating Communities, The Emergence of the Stakeholder University, pp. 109–169. Palgrave Macmillan, UK (2009)
7. Hansen, M.: The search-transfer problem: the role of weak ties in sharing knowledge across organization subunits. Administrative Science Quarterly 44(1), 82–111 (1999)
8. Carley, K.M.: Smart Agents and Organizations of the Future. In: Lievrouw, L., Livingstone, S. (eds.) The Handbook of New Media, ch. 12, pp. 206–220. Sage, Thousand Oaks (2002)
9. Maes, P.: The agent network architecture (ANA). SIGART Bulletin 2(4), 115–120 (1991)
10. Newman, M.E.J.: The structure and function of complex networks. SIAM Review 45, 167–256 (2003)
11. Pavard, B., Dugdale, J.: The contribution of complexity theory to the study of sociotechnical cooperative systems. In: Third International Conference on Complex Systems, Nashua, NH, May 21-26 (2000), `http://www-svcict.fr/cotcos/pjs/` (Retrieved, October 2005)
12. Chandra, S.: Tests of Evolution in Social Networks mimeo. Graduate School of Public and International Affairs, University of Pittsburgh (2008)
13. Steels, L.: When are robots intelligent autonomous agents. Robotics and Autonomous Systems (1995)
14. Varela, F.J., Bourgine, P.: Toward a Practice of Autonomous Systems (1992)
15. Wasserman, S., Faust, K.: Social Network Analysis. Cambridge University Press, Cambridge (1996)

FlexSPMF: A Framework for Modelling and Learning Flexibility in Software Processes

Ricardo Martinho[1], João Varajão[2], and Dulce Domingos[3]

[1] School of Technology and Management, Polytechnic Institute of Leiria, Portugal
rmartin@estg.ipleiria.pt

[2] Department of Engineering, University of Trás-os-Montes e Alto Douro, Portugal
jvarajao@utad.pt

[3] Department of Informatics, Faculty of Sciences, University of Lisboa, Portugal
dulce@di.fc.ul.pt

Abstract. Software processes are dynamic entities that are often changed and evolved by skillful knowledge workers such as software development team members. Consequently, flexibility is one of the most important features within software process representations and related tools. However, in the everyday practice, team members do not wish for total flexibility. They rather prefer to learn about and follow previously defined advices on which, where and how they can change/adapt process representations. In this paper we present FlexSPMF: a framework for modelling controlled flexibility in software processes. It comprises three main contributions: 1) identifying a core set of flexibility concepts; 2) extending a Process Modelling Language (PML)'s metamodel with these concepts; and 3) providing modelling resources to this extended PML. This enables process engineers to define and publish software process models with additional (textual/graphical) flexibility information. Other team members can then visualise and learn about this information, and change processes accordingly.

1 Introduction

Software process modelling involves eliciting and capturing informal process descriptions, and converting them into process models. A process model is expressed by using a suitable Process Modelling Language (PML). A type of Process-Aware Information Systems (PAISs) called Process-centred Software Engineering Environments (PSEEs) supports the modelling, instantiation, execution, monitoring and management of software process models and instances.

Software processes are commonly held as dynamic entities that must evolve in order to cope with changes occurred in: the real-world software project (due to changing requirements or unforeseen project-specific circumstances); the software development organization; the market; and in the methodologies used to produce software [1]. Therefore, it should be possible to quickly implement new processes, to enable on-the-fly adaptations of running ones, to defer decisions regarding the exact process logic to runtime, and to evolve implemented processes over time.

M.D. Lytras et al. (Eds.): WSKS 2009, LNAI 5736, pp. 78–87, 2009.

Consequently, process flexibility has been identified as one of the most important features that both PMLs and PSEEs should support [2].

However, allowing for total process flexibility questions the usefulness of models themselves as guidance to a software development team work plan. Unlimited and unclassified flexibility hampers seriously the learning and reuse of information on changes made in past processes that can be useful in the future. Moreover, in the everyday business practice, most people do not want to have much flexibility, but would like to follow very simple rules to complete their tasks, making as little decisions as possible [3,4].

To corroborate this, case studies on flexibility in software processes (see, e.g., [5]) make evidence on the need of having (senior) process participants expressing and controlling the amount of changes that other process participants are allowed to make in the software process. This *controlled flexibility* can be defined as *the ability to express, by means of a PML,* which, where *and* how *certain parts of a software process should change, while keeping other parts stable* [4]. This faces strong requirements on PML comprehensibility, both because some users initially lack experience with modelling, and because evolution and learning will cause more frequent updates to the models.

We present in this paper the Flexibility Software Process Modelling Framework (FlexSPMF). It comprises the three following contributions:

1. *Identification* - consists in defining and systematising concepts that will be used to express this controlled flexibility in software processes;
2. *Metamodelling* - maps those concepts onto an existing PML's metamodel. Here, we adopt the Software & Systems Process Engineering Metamodel (SPEM) [6], and extend it by adding a flexibility sub-metamodel;
3. *Modelling* - concerns to the PML representation and PSEE tool support for modelling controlled flexibility. From our new metamodel structure, we derived a UML profile called FlexUML, which enhances UML Activity Diagrams (ADs) as the PML with a set of flexibility-related stereotypes and tagged values.

FlexSPMF provides process engineers the power of modelling controlled flexibility as guidance to software development team changes within software processes.

The rest of the paper is organised as follows: the next section provides an overview of FlexSPMF. Then, sections 3, 4 and 5 describe the details of each of the aforementioned contributions. Section 6 refers most prominent related work and section 7 concludes the paper.

2 Framework Overview

FlexSPMF fits our *identification*, *metamodelling* and *modelling* contributions within the software process model lifecycle, as depicted in Fig. 1. Here we can observe two main (human) roles: the process engineer and the software development team. The former contributes to both identification and metamodelling activities of the lifecycle, and is responsible for modelling software processes with

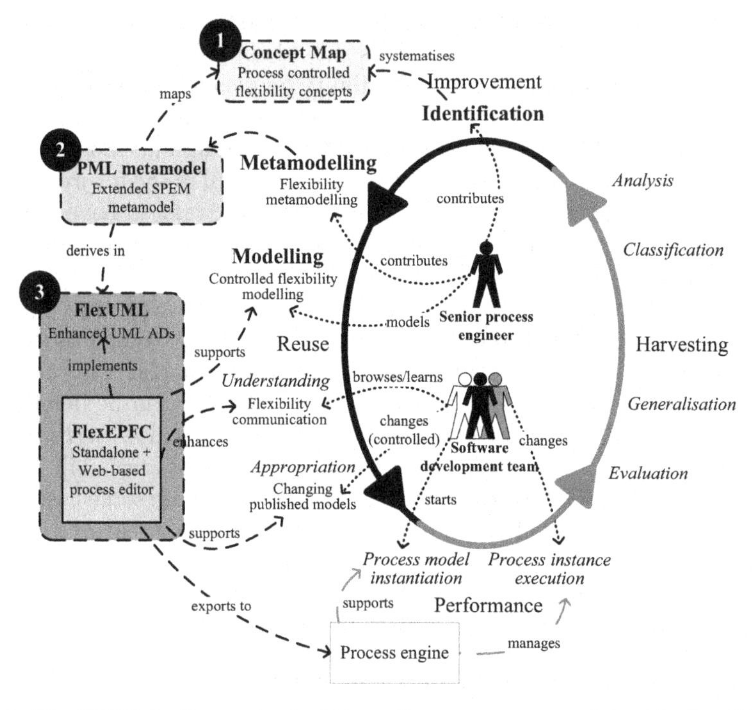

Fig. 1. FlexSPMF in the context of the software process model evolution lifecycle

controlled flexibility. Then, the software development team can browse and learn those models, including flexibility information on their process elements. Hence, they can also change those elements by following advice on previously configured flexibility options.

To better illustrate the framework's purpose, let us consider the example of an organisation that has a process engineer who is responsible for the general modelling and management of software processes and associated models. For a particular process model, s/he requires as mandatory the production of test cases associated with a `Test solution` activity.

On the other hand, s/he also wishes to allow the software testing team to perform a detailed modelling of this activity, which is not completely specified until process model instantiation (e.g., when starting up a software project). S/he does this on the assumption that the testing team members have better skills for the task, and that there is often the need for model adaptation to certain project management constraints (like time, cost or human resources).

Therefore, s/he requires PML resources to express within the process model *which*, *where* and *how* the `Test solution` activity can or cannot be changed. To accomplish this, s/he firstly needs to identify and systematise a delimited set of process controlled flexibility concepts which all other team members will understand and share in the language. Figure 1's first solution (number 1) refers to a Concept Map (Cmap) [7]. Cmaps are used as an informal ontology-based approach to clarify and improve learning about concepts and relationships of a certain knowledge domain (see, e.g., [8]). In this case, the domain is the modelling of controlled flexibility within software process representations.

Solution number 2 refers to *metamodelling*, i.e., the mapping of the concepts onto a MetaObject Facility (MOF)-compliant metamodel. Here, we adopted SPEM, which serves as a reference metamodel for many Model-Driven Development (MDD) software processes, such as the Rational Unified Process (RUP), Open Unified Process (OpenUP), eXtreme Programming (XP) and Scrum. We extended it by adding our sub-metamodel, which comprises a set of class elements that addresses particularly the flexibility concern.

Solution number 3 refers to *modelling*, more precisely to the FlexUML profile. It introduces UML stereotypes to enhance SPEM ADs as a flexibility-aware PML. It also comprises its implementation in the FlexEPFC PSEE: a customised version of the original IBM's Eclipse Process Framework Composer (EPFC[1]). We added the possibility for a process engineer to express, in a first modelling step, controlled flexibility within a process model. After this, the process engineer can use FlexEPFC to publish the process model to an automatically generated web application. Besides providing an efficient way for the software development team to learn and browse process models, it also includes a web process editor. This allows the team to change the published models, according to the controlled flexibility previously modelled by the process engineer.

The next sections describe in more detail each FlexSPMF contribution.

3 Identification

Figure 2 illustrates the main Cmap for the process controlled flexibility domain, picturing its core concepts and relationships (please consult [9] for more detailed information on these). Briefly, it states that a *software process* is a combination of *elements* represented by a process *model*.

Software process modelling *elements* include: 1) *functional* elements (e.g. *phases*, *activities* and *steps*); 2) *behavioural* elements (e.g. control flow nodes such as *fork*, *join*, *decision* and *merge* nodes, as also iterative/parallel region elements); 3) *organisational* elements; and 4) informational elements (e.g. *data*, *artifact*, *work product* (intermediate and end) and *object* elements).

A process *model* depends on its *metamodel*, which establishes the structure of concepts, relationships and constraints that can be used in defining a process model. The metamodel usually defines a Process Modelling Language (PML), where all process modelling elements are specified. The Software & Systems Process Engineering Metamodel (SPEM) [6] is an example of a software process metamodel which defines UML ADs as the core PML.

Process models are then created as instances of the metamodel. They represent an arrangement of process elements in more or less specified sequences of activities, resources and resulting work products that will provide guidance for specific project workplans. Process models are templates which represent

[1] http://www.eclipse.org/epf

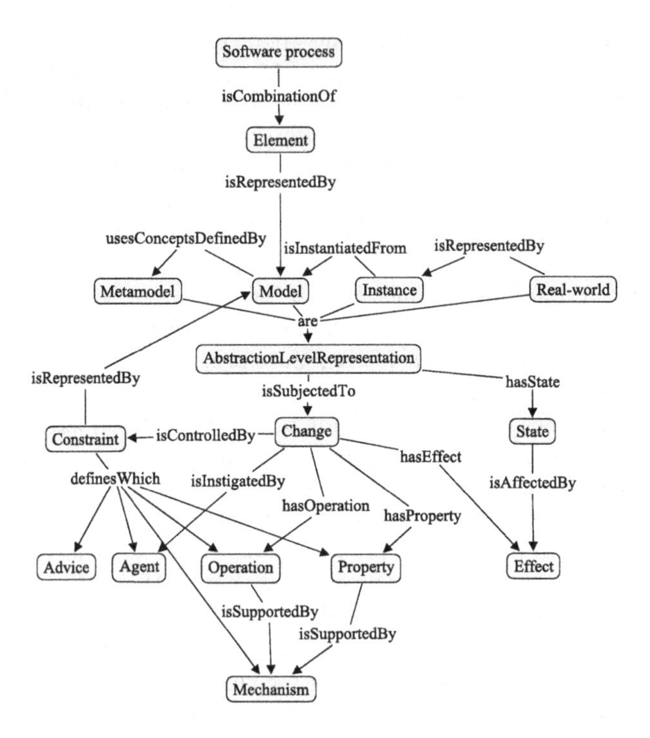

Fig. 2. Core Concept Map for sharing knowledge on process controlled flexibility

reusable process knowledge. Examples of software process models include the process structures comprised by well known software methods like the Unified Process [10].

Applying a process model for a specific software project is called process *instantiation*. An *instance* follows the model and provides specific data for each distinct project, such as activities' duration, (human) resource assignments, cost estimations and monitoring/control data updates. Multiple process instances may share the same process model.

On the contrary, *real-world* processes have a 1:1 multiplicity to process instances, as they reflect the activities, resources and products that are actually performed, used and produced by humans or tools. It describes what is really happening, and process participants may retrieve feedback which is used to update process running instances.

Metamodel, model, instance and real-world are distinct but correlated *abstraction levels* of process element *representations*. These representations are subjected to *changes*, which in turn have *effects* that can affect their *states*.

A *change* is characterised by *properties* and *operations*. Performing *change operations* includes creating, updating and deleting process elements, as well as moving them or realising element- and representation-specific operations such as *undo*, *skip* or *redo* an `Activity` in a process running *instance*. Actually, change operations are the actions that will change the state of the process elements.

Properties of change are not dependent on a process element's type, but characterise multiple and general dimensions of a change. Possible implementations of properties of change commonly referred in literature include [11,12]: 1) *extent* (incremental or revolutionary); 2) *duration* (temporary or permanent); 3) *swiftness* (immediate or deferred propagation between models and running instances) and; 4) *anticipation* (planned or ad-hoc changes). Both operations and properties are supported by corresponding *mechanisms*. For example, executing an *add* operation on a process model implies the use of a software tool that, besides supporting process editing features, also provides verification of conformance, consistency or compliance rules associated with that operation.

Changes are instigated (put into action) by *agents* of change. Agents of change need to have the ability to set a mechanism of change into motion. In the software process context, the agent of change is responsible for triggering mechanisms of change that will result in an effect of change endured by one or more process element representations. Agents of change may be software components that automatically change process element representations under some criteria, or humans such as software process engineers, project managers, analysts, designers, programmers and testers, that need to change/adjust software processes.

A change can be controlled by *constraints*, which are also represented by *models*. These contain *advice* on operations, properties, mechanisms and/or agents that should be considered when changing a certain process element representation. Advice on a change can be a *value-* or *text-*based attribute (e.g., *60%* or *recommended*) , or any other combination of values that best fit the process element representation to which the advice is associated.

For example, a constraint of change may impose that *a* `Test solution` *activity instance representation cannot be skipped.* The modelling of this constraint can be made using a three component tuple of the form *(abstraction level, operation, advice)*, with the values *(instance, skip, denied)*.

4 Metamodelling

We use a UML class metamodel structure that reflects the above concepts, and that can be used to extend the existing metamodel of a PML. Here, we provide a concrete example by using SPEM and UML ADs as the metamodel and corresponding PML to be extended. Figure 3 pictures our derived FlexSPMF metamodel.

The structure is based on the decorator design pattern from Gamma et al. [13]. This is the leftmost and gray-shaded structure which inherits from a common abstract `SPEMProcessModellingElement`. The decorator pattern allows the attachment of additional responsibilities to an object dynamically. In our context, when designing a process model, a process engineer can pick any process element (such as an `Activity`) and decorate it with one or more change decorators, represented by `ChangeDecorator`'s specialised classes. These derive from the aforementioned concepts of change operation, property, mechanism, advice and abstraction level.

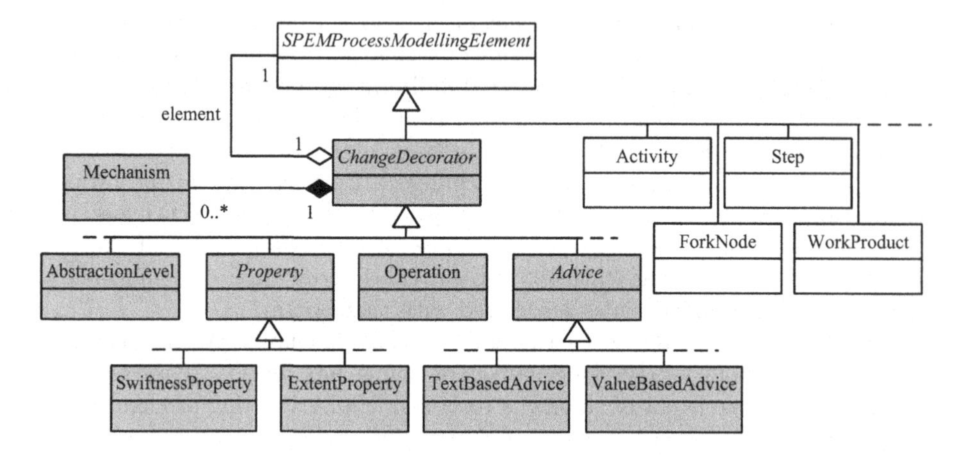

Fig. 3. FlexSPMF metamodel structure as an extension to the SPEM metamodel

When applied to a process element, a decorator represents one of the components that form a *constraint of change*. Taking from our former example, the tuple *(abstraction level, operation, advice)* maps to a combination of three decorators, namely: `AbstractionLevel` (*instance*), `Operation` (*skip*) and `TextBasedAdvice` (*denied*), which are applied to instances of the `Test solution` activity.

Mechanisms can be associated with any decorator, providing implementation details according to a process element's type.

The next section describes the way a process engineer can use the extended SPEM ADs and FlexEPFC to model controlled flexibility in software processes.

5 Modelling

For the *modelling* contribution we developed FlexUML: a UML profile for ADs (see [14] for further details). Each `ChangeDecorator` concrete specialisation class of Fig. 3's metamodel corresponds to a FlexUML stereotype with its own tagged values. Figure 4 presents two applications of FlexUML, noted by the names of the stereotypes between «guillemets» above the process elements' graphical notations and callouts with their respective tagged values (according to [15]).

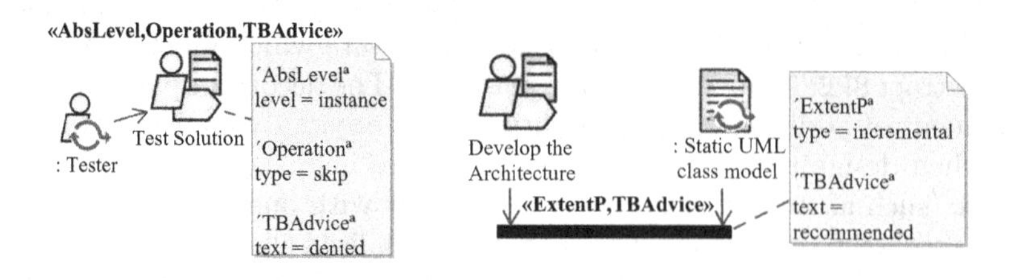

Fig. 4. `Activity` (left) and `Join node` (right) FlexUML stereotype applications

The profile applications are made to an activity (`Test solution`) and to a join node. The former includes the «`AbsLevel,Operation,TBAdvice`» stereotypes, which values advise to not skip the activity's instances. The latter has the «`ExtentP,TBAdvice`» stereotypes that advise as *recommended* an incremental change.

Once a process engineer finishes applying the flexibility stereotypes, s/he can use FlexEPFC to generate a web-based published version of the process model. The original auto-generation from EPFC does not allow any changes to this published version. To fulfil our requirement of also enabling the team to change the models, we embedded the process editor also in the web-based application. This way, besides browsing, the team can also learn about the advised flexibility, and perform changes accordingly.

Figure 5 is a snapshot of this customised web-based process editor which we called WebFlexEPFC (see [16] for more details). The snapshot shows the editing/changing of an OpenUP `Elaboration phase` join node, but according to

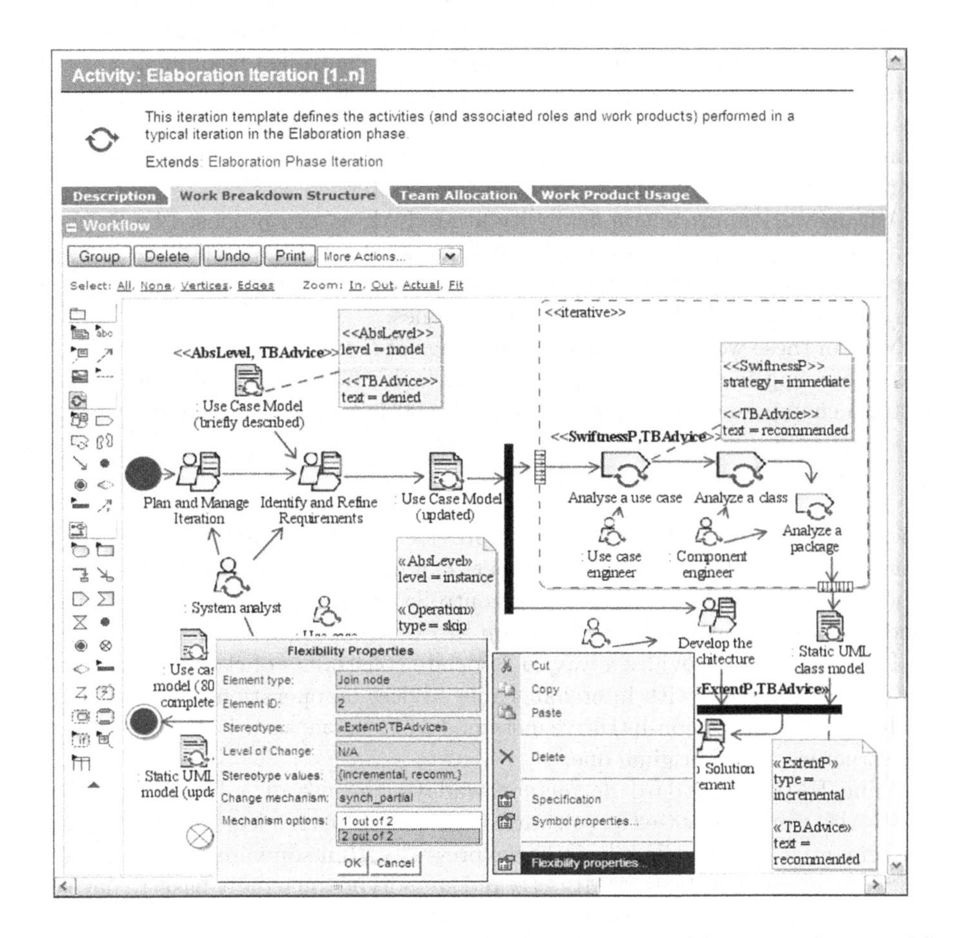

Fig. 5. WebFlexEPFC: changing a join node of an OpenUP Elaboration phase model

the previously modelled constraint of change (noted by the «ExtentP,TBAdvice» stereotype applications).

6 Related Work

Recent process-aware flexibility taxonomies include the one by Regev et al. in [11]. We use some of the concepts referred here, including *abstraction level*, *subject* and *property of change*. These and other change concepts are also present in the taxonomy proposed by Schonenberg et al. in [12], under a four type-based flexibility classification: *design*, *deviation*, *underspecification* and *change*. *Constraints of change* in process models are analysed by Wörzberger et al. in [17]. They propose modelling tool artifacts to support correctness, compliance and consistency constraint modelling and checking.

Although there is no shared classification of concepts among these works, they have the same intent of FlexSPMF *identification* contribution. However, our concept map focuses mainly on the definition of a core set of concepts, delegating on the decorator-based metamodel structure the possibility of combining them interchangeably, to form adjustable categories on controlled flexibility.

In [5], Cass and Osterweil advocate that, in spite of software design requiring a lot of creativity and insight, some process rigidity seems necessary. They conclude that combining flexibility rules with process guidance helps designers to produce better software designs, and to produce them faster.

A framework approach on flexibility-aware PAIS is also proposed by Reichert et al. in [2]. The ADEPT2 flexible PAIS is able to adapt process instances to changes occurred in real-world processes, on a quest to support most of the workflow patterns, including the exception ones.

None of these works approach the challenge of modelling controlled flexibility, i.e., expressing in process models how much flexibility is allowed, in order to restrain changes on declarative (starting as 100% flexible) software processes.

7 Conclusions

The main objective of FlexSPMF is to provide process engineers with means to control the flexibility that team members can enjoy when guided by a software process. Our contributions include a controlled flexibility concept map, metamodel and modelling language based on SPEM UML ADs.

The metamodel provides a way to associate constraints of change by decorating process elements with interchangeable advices on operations and properties of change. It is also a non-intrusive metamodel extension, since it does not modify the structure of the original one.

Being UML a standard *de facto* nowadays, the use of a PML other than UML ADs would increase the software development team's learning curve about the semantics of controlled flexibility expressed within software process models. Software teams already use UML ADs and stereotypes on a daily basis to develop software. They also recognise them as the PML used for modelling well known SPEM-based processes like RUP, XP, OpenUP and Scrum.

References

1. Cugola, G.: Tolerating Deviations in Process Support Systems via Flexible Enactment of Process Models. IEEE Transactions on Software Engineering 24(11), 982–1001 (1998)
2. Reichert, M., Rinderle-Ma, S., Dadam, P.: Flexibility in process-aware information systems. LNCS Transactions on Petri Nets and Other Models of Concurrency (ToPNoC) 2, 115–135 (2009)
3. Bider, I.: Masking Flexibility Behind Rigidity: Notes on How Much Flexibility People are Willing to Cope With. In: Pastor, Ó., Falcão e Cunha, J. (eds.) CAiSE 2005. LNCS, vol. 3520, pp. 7–8. Springer, Heidelberg (2005)
4. Borch, S.E., Stefansen, C.: On Controlled Flexibility. In: Proc. of the 7th Workshop on Business Process Modeling, Development and Support (BPMDS), pp. 121–126 (2006)
5. Cass, A.G., Osterweil, L.J.: Process Support to Help Novices Design Software Faster and Better. In: Proc. of the 20th IEEE/ACM Intl. Conference on Automated Software Engineering (ASE), pp. 295–299 (2005)
6. OMG: Software Process Engineering Metamodel Specification, v2.0. Technical report, Object Management Group (2007)
7. Novak, J.D., Cañas, A.J.: The theory underlying concept maps and how to construct and use them. Technical report, IHMC CmapTools, 2006-01 Rev 2008-01, Florida Institute for Human and Machine Cognition (2008)
8. Razmerita, L., Lytras, M.D.: Ontology-based user modelling personalization: Analyzing the requirements of a semantic learning portal. In: Lytras, M.D., Carroll, J.M., Damiani, E., Tennyson, R.D. (eds.) WSKS 2008. LNCS (LNAI), vol. 5288, pp. 354–363. Springer, Heidelberg (2008)
9. Martinho, R., Domingos, D., Varajão, J.: On a concept map for the modelling of controlled flexibility in software processes. Technical report TR-2009-12. Dep. de Informática, Faculdade de Ciências da Universidade de Lisboa (May 2009)
10. Jacobson, I., Booch, G., Rumbaugh, J.: The Unified Software Development Process. Addison-Wesley Longman Publishing Co., Inc, Boston (1999)
11. Regev, G., Soffer, P., Schmidt, R.: Taxonomy of Flexibility in Business Processes. In: Input to the 7th Workshop on Business Process Modeling, Development and Support (BPMDS 2006) (June 2006), `http://lamswww.epfl.ch/conference/bpmds06/taxbpflex`
12. Schonenberg, H., Mans, R., Russell, N., Mulyar, N., van der Aalst, W.M.P.: Towards a taxonomy of process flexibility. In: CAiSE 2008, pp. 81–84 (2008)
13. Gamma, E., Helm, R., Johnson, R., Vlissides, J.: Design Patterns: Elements of Reusable Object-Oriented Software. Addison-Wesley, Reading (1995)
14. Martinho, R., Domingos, D., Varajão, J.: FlexUML: A UML Profile for Flexible Process Modelling. In: Proc. of the 19th Intl. Conference of Software Engineering and Knowledge Engineering (SEKE), pp. 215–220 (2007)
15. OMG: Unified Modeling Language: Superstructure, version 2.0. Technical report, Object Management Group (2005)
16. Martinho, R., Varajão, J., Domingos, D.: A two-step approach for modelling flexibility in software processes. In: Proc. of the 23rd IEEE/ACM Intl. Conference on Automated Software Engineering, ASE (2008)
17. Wörzberger, R., Kurpick, T., Heer, T.: On correctness, compliance, and consistency of process models. In: Proc. of the 17th IEEE Intl. Workshops on Enabling Technologies: Infrastructures for Collaborative Enterprises, WETICE 2008 (2008)

Learning Objects Quality: Moodle HEODAR Implementation

Carlos Muñoz, Miguel Ángel Conde, and Francisco J. García Peñalvo

Factultad de Ciencias, University of Salamanca, Plaza de los caídos S/N, 37008 Salamanca, Spain
{carlosmm,mconde,fgarcia}@usal.es

Abstract. One of the most important aspects of a "continuously in change" society is to improve everything everywhere. In order to obtain the best products, they should be periodically evaluated and reengineering. So the evaluation task and, of course, the adequate results interpretation, can make all the difference between competitors. eLearning is similar to this products. Different issues can be evaluated to make learning process getting better and better, such as tutors, platform software and contents. In this last issue, it can be included the minimum knowledge unit: the Learning Object (LO). There exists different models and methods for LO evaluation. What is pretended with this work is to choose one model and implement a singularly tool, in order to automatically evaluate this Learning Objects and produce a set of information, that can be used to improve those Learning Objects. In this case it is implemented the evaluation model called HEODAR in the University of Salamanca framework.

Keywords: Learning Object, evaluation, Moodle, module, agent, JADE, quality.

1 Introduction

The processes automation that supports the work of experts in certain subjects or that even allows for the release of them and their dedication to other tasks is a constant that has been occurring on an ongoing basis in different spectrum of our society.

This automation has already reached the level of education encompassing all its modalities and many of its processes (empty classroom control, students attendance, rating records and so forth) covering planes like Web content evaluation [1] and multimedia evaluation [2] which leads to the conclusion that eLearning is not out of reach.

More precisely, this article focuses on the study of the current state of the processes associated with the Learning Objects evaluation and on the implementation of tools which allow these processes automation and extract the advantages, previously mentioned, applied to OnLine learning field.

Thus, the evaluation can be considered as the last step of a quality Learning Objects management process [3], being identified as a cyclical process in which periodic content evaluation by students and experts, reverses in a continuous improvement of Learning Objects stored in appropriate repositories.

M.D. Lytras et al. (Eds.): WSKS 2009, LNAI 5736, pp. 88–97, 2009.

The reasons for this study stems from the relevance that Learning Objects are acquiring as portable minimum information units between Learning Management Systems (LMS). There should be methods to evaluate the efficiency and pedagogical quality of these objects for their reuse in different contexts. If an object has reached an optimal adaptation level for a learning activity it should be used in such activities and not others that may provide a more tangential value. The only way to determine these Learning Objects characteristics is through the quality evaluation provided for the students and through real experiences in a platform or a set of platforms.

Obviously, these processes evaluation enhancement, using software tools that automates the entire process or part of it, will represent a clear advantage which will make possible to focus the effort on the main objective: to improve the content, and not on the extraction of evaluations results.

Throughout this paper it will be discussed how the evaluation tool is implemented. Firstly it will be considered the Learning Objects evaluation evolution and also the evolution of existing software tools to perform this work. After that it will be described the tool planning and analysis steps. Finally it will be talked about following stages in tool development, ending with a list of conclusions.

2 Learning Objects Evaluation and Tools

The evaluation task of any kind of entity, component, object or concept has been occurring over time. However, the definition spirit of that term has been maintaining during that time, as long as it is a technique used to enhance the improve of any of them.

In eLearning scope, there are many factors that can be evaluated: tutors, students, platforms, documents and so on. Of course, since Learning Objects emergence, these can be added to the list as a feasible element to be improved, analyzing the results obtained by the application of appropriate evaluation techniques.

From this point of view, several studies have been focused on the acquisition process models and evaluation models that produce the best results and allow Learning Objects improvement in order to achieve greater satisfaction of all actors (students, tutors and, of course, content authors).

It can be find processes like CLOE (Co-operative Learning Object Exchange) which is based on the content review by content, instructional design and other aspects experts, and other processes that rely on systems or repositories with a social and technological fundament, in which evaluators use software tools that guide the processes evaluation like DLNET (Digital Library Network for Engineering and Technology), MELROT (Multimedia Educational Resource for Learning and Online Teaching) [4] and eLera [5] through an specific evaluation tool called LORI (Learning Object Review Instrument) [6] which propose the result visualization using ratios, extracted in an automatic form from the actors registered evaluations [7].

However, in this field of Learning Objects quality evaluation tools, there is not tools which perform this tasks integrated into the most widely used LMS, which contains the majority of such Learning Objects, beyond institutional repositories like the previously mentioned.

That is why it is set out the design and development of a software tool that implements an evaluation model. The selected one is the result of Erla Mariela Morales thesis called HEODAR (*Herramienta de Evaluación de Objetos de Aprendizaje*) which principles [8] where studied and considered as appropriated for being used in an institutional environment: the University of Salamanca, experimentally integrated in a LMS platform with expansion possibility to other existing platforms.

3 Initial Planning

Among the variety of LMS platforms existing nowadays, it must be taking into account the selection of the optimal platform for an initial integration of HEODAR tool. Due to the recent implantation at the University of Salamanca of a new LMS platform, which is used by teachers and students, it is resolved to implement the tool to woks under such LMS: Moodle. One of the biggest motivations is the possibility of making impact, effectiveness, utilization and usability studies about the model and also about the designed tool in a practice and guarantied form. It is consider getting these results during the first months after the integration tool process. The results will be used for review and correction.

In addition to these factors, this LMS brings another advantages set over other existing LMS, beyond its use in an academic environment and its use in real formative experiences:

1. Moodle is an Open Source software system.
2. Moodle is implemented and maintained by an international community integrated for more than 600.000 members (Januaty 2009 information, http://moodle.org/stats)
3. Moodle has a base of about 50.000 installed servers around the world (199 with more than 10.000 registered users) in whitch there is more than 27 millions of inscribed students.
4. Moodle is translated to 75 languages [9].

Moodle is used as the primary LMS in most of Spaniard universities. In Spain, Moodle community is really consolidated, and annual congresses are celebrated with great participation grades, especially of every level public and private education centers agents and of company's directives. Nowadays, Moodle is the most popular eLearning environment in Spaniard educative centers and every day the number of interested companies increased. The versatility and easy to use philosophy, an impeccable community attention and an original business model are the succeed keys [10].

4 Form Study

Once determinate the platform in which the development and tests are going to be focused on, without undermining its possible expansion to other platforms when the functionality has been thoroughly tested, there is a discussion about the best integration form of HEODAR tool into Moodle.

From a platform point of view, there are several ways to accomplish the utilities integration with Moodle: the incorporation to the Moodle core (as a resource or an activity) or the development of independent tools linked with the platform using data communication techniques.

From a HEODAR point of view, and taking into account its collaborative characteristics [11], it is not considered the design and development of the tool as an external complement, but the definition and development as a functional element inside the platform that can be deployed on it. Obviously, and knowing the HEODAR characteristics, in which appears a clear relation with tutors and content authors, it is finally decided a tool functional an architectonical definition as a Moodle activity module.

5 Architectonical Definition and Data Sources

Once the decision regarding the environment and the initial design and implementation form is reached, it starts the study of the different ways to research and define the module architecture.

While the Moodle module definition and schema sets a group of parameters that must be considered in order to obtain a correct integration and working of the module, there are still several aspects to be evaluated. The HEODAR tool succeeds rely on the correct selection of such aspects.

The first one focuses on the data model. We worked on the correct definition of this model towards meeting the needs the evaluation model sets out, but also thinking about how to satisfy other factors suggested by the model authors and that, probably, will constitute a first evaluation model evolution. Such factors should be translated into the data model and are referred to the following characteristics:

- The possibility, not actually considered, of modifying weights that will affect to the calculation process of the evaluation score. These weights are associated to the different evaluation model classifiers. At this moment, in the original evaluation model, every classifier has the same weight.
- To provide intelligent capabilities to the tool, through software agents results analysis. These agents interpret and transmit the process results to the stakeholders (mainly content authors) without the need for them to access the platform and get the information.

This second factor has been considered as really interesting by some content authors, while it represents a tangible improvement on the way in which these authors receive feedback about the quality of its contents. With this information they may be able to improve them.

Note that the content authors do not usually publish them (especially in a postgraduate and continuing education fields) on one platform (even on homogeneous platforms). In that way they may find it useful the existence of an element (agent) that extracts the valuable information about their content, and send it to them automatically indicating the origin information platform.

Thus, with the gathered information a definition of the architecture model is made. This model can be observed in Fig. 1.

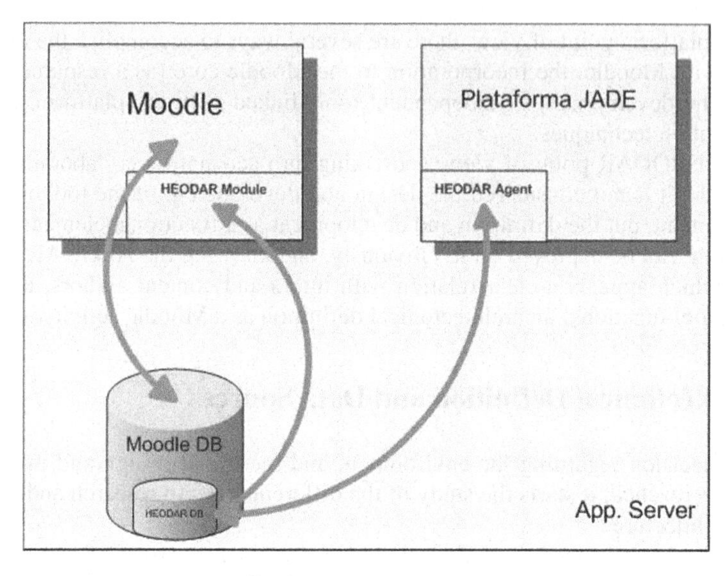

Fig. 1. Architecture model

6 Development Phases

In order to implement the two main blocks that constitute the proposed architectural model, the tool itself and the agent system, the project is divided into two phases:

- **First phase:** it is constituted for the definition and construction of a stable data model and the implementation and implantation of HEODAR tool as a Moodle module.
- **Second phase:** it concentrates the effort in the design and implementation of an agent system which provide the module with the intelligent behavior previously defined and clearly useful for tutors and content authors.

Both phases are integrated into an incremental construction model, which allows putting into operation the first phase resultant product and, after certain time, integrating the second phase resultant product, without modifying the working process of the initial module or the stored evaluation result.

Actually the first phase is carried looking for being incorporated into the Moodle platform of the University of Salamanca (Studium) and perform rigorous tests and impact studies.

7 The Developed Module

The functional and information characteristics of the first phase implemented module are adapted to the requested necessities includes in the theoretical evaluation model definition. All that information was evaluated in order to define the data model in which the module relies on. This data model can be seen in Fig. 2.

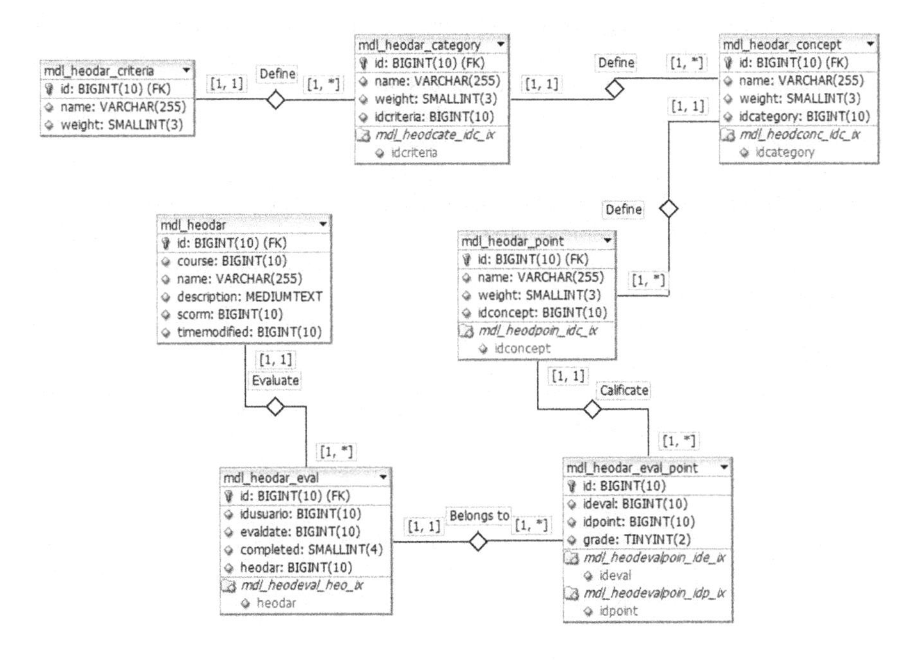

Fig. 2. Module data model

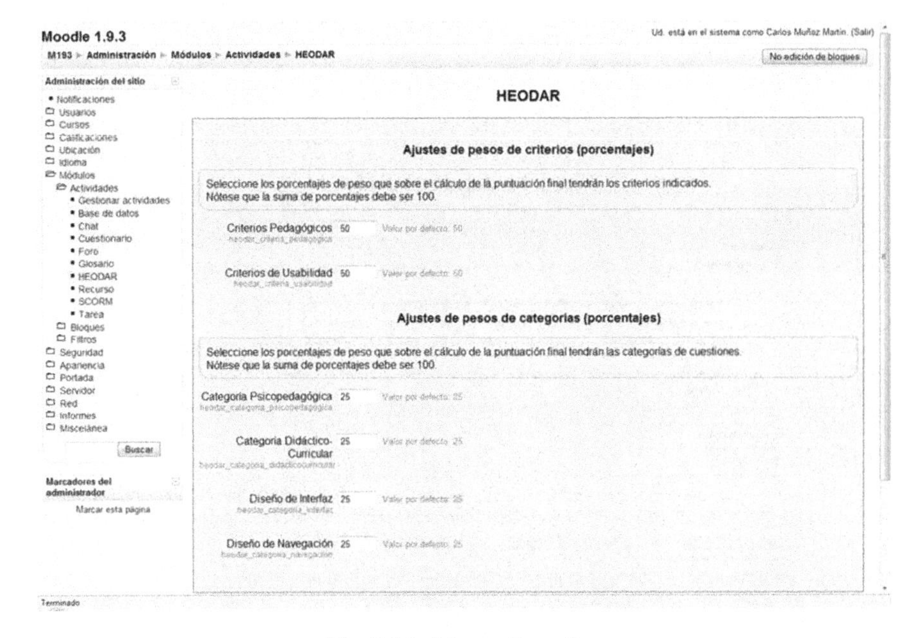

Fig. 3. Module configuration

It is notable that the incorporation of evaluation questions into the data model responds to an issue indicated in establishing the foundations of the tool itself [3]. This issue is the possibility that in the future, factors and parameters measured by each of the questions can vary in the future due to the own experience. Whit this model the question and classification modification is possible.

This is obviously reflected in the score calculation process implementation for a particular Learning Object evaluation through the evaluation model questions. This calculation process must be implemented considering the number of questions of each category and another factor: the score of a concept depends on the weights of each sub concept it involves.

Thus, it is possible to define, using the module configuration, the influence weights as seen in Fig. 3.

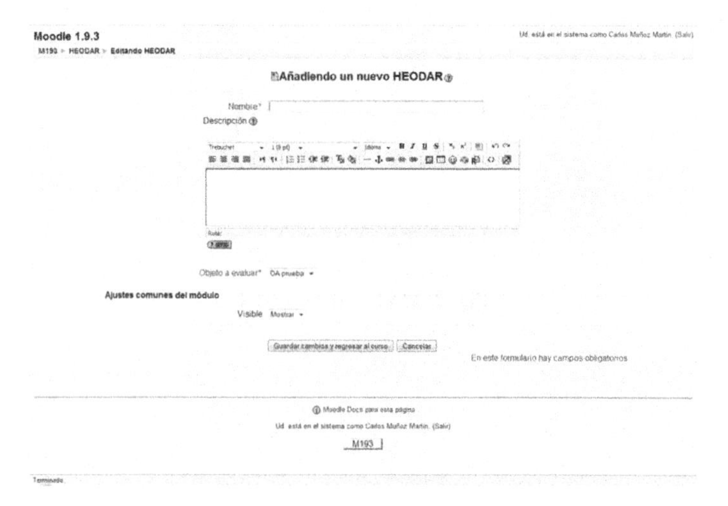

Fig. 4. Module instantiation

Fig. 5. Evaluation form

With regard to the configuration of the activity in which this tool is integrated, only minimum information is requested to the creation of an evaluation activity. This information is a name to name the activity in the course, a description and one of the course learning objects. These learning objects are obtained from available platform data for the course in which the activity is being configured. An example could be shown in Fig. 4.

With regard to evaluation display, this is represented as a test to each of the participants of the course. The factors are classified and presented according to the data definition, making easy the completion of each of the factors, which evaluation levels are specified in the theoretical definition of the tool. Fig 5 shows an example of such representation.

8 The Second Development Phase

Once the module resulting from the development of the first phase is integrated, following steps should be, the recovery and analysis of the first data released from the module use in Studium, and the beginning of the development of the second stage.

In an initial evaluation made about the different techniques that allow the implementation of considered behavior, the best option is working over a JADE agents platform. This platform will be the base of the implementation and integration of a set of agents which autonomously do the following tasks: 1) Periodic Retrieval of information gathered through the tests. 2) Processing and transmission of information analyzed and transformed to the contents authors.

Obviously, there is a system configuration that will allow agents to identify those environments (platforms) on which the use of this system is possible. Also they were allowed to access to the appropriate stored data, obtained through the module evaluation.

Another aspect is the different contexts in which this module is thought to be used. Initially it will be tested in an academic context, at the University of Salamanca in a Moodle platform that is a part of the Online campus Studium.

The second step will be to test the tool in an enterprise context, more specifically in Clay Formación, so we will be able to compare the results in both context, and finally the tool is pretended to be installed in a foundation model linked to a rural environment.

With all this information a results review will be made and the conclusions will be applied for the tool improvement.

9 Integration with JADE Agents

There are several attempts to integrate an LMS platform with a system of agents in order to provide a degree of intelligence to this system. Considering Moodle LMS, some initiatives and integration trials has been made, among which, Forums will be enhanced. This integration was the addition of some new parameter to the forum and which allows the definition of the communication way between agents and the platform. Those parameters include the option to enable or disable the use of information by the system of agents in order to prevented external queries. Another important parameter is the frequency with which the agents carry out the tasks entrusted to it.

All these parameters can be exported to the system of agents focused on a JADE platform. This is because there are evidences of the potential for integration and operation of that exportation.

The result, once developed, will be added to the module already produced and put into operation in the platform Studium of the University of Salamanca.

10 Conclusions

Considering evaluation of learning objects and its related processes, HEODAR implementation could be seen as an example that changes this preceding way. This is because we can evaluate learning elements in a different context. This context is directly where teachers and students use them, in a Learning Management System.

The integration as a Moodle module has resulted in a success due to the possibility of it integration in the platform for Online Teaching of the University of Salamanca. Since that integration, results will begin to be extracted from the data generated by the assessments of the actors involved on that platform. This information will be used for the improvement of learning objects stored in repositories of that university.

This represents an important advantage in comparison with those which have not yet integrated the developed module, this is because it allows continuous improvement of content, which is expected to increase satisfaction among students. This fact will be confirmed by the better grades granted by evaluators to all that reevaluated objects.

Also, future work, focusing mainly on intelligent systems for communication with the authors of content, will make a substantial improvement in the implementation of HEODAR which level was measured by comparative studies between the results obtained in the initial stage and those obtained after the integration of the second phase of development.

The quality of learning is conditioned to, among other factors, the quality of content and the quality of learning objects, such as basic units of knowledge. Also implementations and evaluation techniques must be considered, as a way to improve learning quality and thus the satisfaction of students and tutors. This is the reason because the evaluation and its application in real contexts is so important to improve learning processes.

References

1. Marquès, P.: Criterios de calidad para los espacios web de interés educativo, `http://decroly.org/escola/formacio/criterios_%20calidad_web_educ.pdf`
2. Gibbs, W., Graves, P.R., Bermas, R.S.: Evaluation guidelines for multimedia courseware. Journal of Research on Technology in Education 34(1), 2–17 (2001)
3. Morales, E., García, F., Barrón, A., Berlanga, A., López, C.: Propuesta de Evaluación de Objetos de Aprendizaje. II Simposio Pluridisciplinar sobre Diseño, Evaluación y Descripción de Contenidos Educativos, SPEDECE (2005)
4. MERLOT (2009), `http://www.merlot.org/`

5. eLera. Simon Fraser University, Surrey, British Columbia, Canada (2009), http://www.elera.net
6. Nesbit, J.C., Li, J.: Web-based tools for learning object evaluation. In: International Conference on Education and Information Systems: Technologies and Applications, Orlando, Florida (2004)
7. Nesbit, J., Belfer, K., Leacock, T.: Learning Object Review Instrument (LORI) User Manual E-Learning Research and Assessment Network (2003)
8. Morales, E.: Gestión del Conocimiento en Sistemas eLearning, basado en Objetos de Aprendizaje, Cualitativa y Pedagógicamente definidos. University of Salamanca, Salamanca, Spain (2008)
9. Cole, J., Foster, H.: Using Moodle, 2nd edn. O'Really, Sebastopol (2007)
10. Molist, M.: Moodle llena la geografía educativa española de campus virtuales. Diario el Pais (2008), http://www.elpais.com/articulo/portada/Moodle/llena/geografia/educativa/espanola/campus/virtuales/elpeputec/20081204elpcibpor_1/Tes
11. Morales, E., Gomez, D.A., Garcia, F.J.: HEODAR: Herramienta para la Evaluación de Objetos Didácticos de Aprendizaje Reutilizables. In: SIIE 2008, Salamanca, Spain (2008)
12. Scutelnicua, A., Linb, F., Kinshukb, L.T., Grafd, S., McGrealb, R.: Integrating JADE Agents into Moodle. In: Proceedings of the International Workshop on Intelligent and Adaptive Web-based Educational Systems (IAWES), Hiroshima, Japan, pp. 215–220 (2007)

Web 2.0 Applications, Collaboration and Cognitive Processes in Case-Based Foreign Language Learning

Margrethe Mondahl, Jonas Rasmussen, and Liana Razmerita

Copenhagen Business School,
Department of International Language Studies and Computational Linguistics
mm.isv@cbs.dk, jonas0707@gmail.com, lr.isv@cbs.dk

Abstract. Web 2.0 has created new possibilities for students to engage, interact and collaborate in learning tasks that enhance learning processes and the overall learning experience. The challenge for educators is to design and integrate a new set of tools based on specific didactic principles associated with a specific domain of learning. This article discusses experiences and challenges of using a Web 2.0-based collaborative learning environment in case-based teaching of foreign languages. The cognitive processes underlying learning and in particular foreign language learning are assumed to be facilitated by means of collaborative Web 2.0 applications. Based on the notions of social constructivism and social constructionism, we argue that foreign language learning is an individual as well as collaborative process and cognitive processes underlying learning and in particular foreign language learning are facilitated by means of collaborative Web 2.0 tools. Preliminary experiments have indicated that collaborative learning processes that are embedded in an e-learning platform are supportive and conducive to successful problem-solving which leads to successful adult foreign language learning.

Keywords: Web 2.0, collaborative learning, foreign language learning, learning, case-based teaching, social learning, cognitive processes.

1 Introduction

The central idea of social-constructivism is that knowledge construction is a social process that occurs through connectivity and collaboration with others. Constructivism postulates that human knowledge is constructed and that the learner builds new knowledge based on the foundations of previous learning. In a constructivism approach to learning, the learner is no longer a simple passive receiver of knowledge; (s)he is stimulated to play an important role in constructing her/his knowledge. Learning processes may also take place through complex interactions such as games, conversations, and collaborations with colleagues and friends. In this context social learning may be defined as a more ludic form of learning [1].

In terms of social-constructionism, we would argue that the learning process also occurs in communities that constantly interact with the individual's constructions in the internal learning process [2]. As a result, foreign language learning occurs as part

M.D. Lytras et al. (Eds.): WSKS 2009, LNAI 5736, pp. 98–107, 2009.

of a social interplay, which is influenced by the culture and communicative understandings that surround the individual learner. Moreover, by using communicative web-based tools, learners are prompted to describe the learning process and take in feedback, which may support learning processes and facilitate foreign language learning. Within the learning platform, this is implemented in the form of a learning log.

Knowledge creating learning processes are key to the development of the knowledge society [3] and needs to be supported by new learning platforms that are facilitative of collaborative and constructive learning. The project outlined in this article focuses on an interplay between foreign language learning and Web 2.0 applications integrated in a collaborative learning platform. Returning to social- constructionist and –constructivist considerations, we would argue that new learning and teaching strategies may be designed using Web 2.0 tools [4, 5]. Social software including wikis, blogs, user created videos, podcasting have created new possibilities for students to engage, interact and collaborate in learning tasks that enhance learning processes and the overall learning experience. However, in order to design new learning platforms that enhance the learning experience, educators must plan and conceptualize the pedagogical principles, the associated tools and the strategy that enable them to test their assumptions according to specific learning objectives. The growing popularity of Web 2.0 suggests that students' learning might be accomplished under circumstances different from those currently used in universities. Furthermore recent studies suggests that the digital generation of students learn differently from the previous generation and they are dependent on the Web for accessing information and interacting with the others [6]. Even though Web 2.0 applications are promising for use in the educational setting, more considerations and evaluation studies are needed in order for "pedagogy 2.0" to be established [7]. Web 2.0 social software may be approached from different perspectives: as a new social media tool, a facilitator of new forms of interaction and knowledge sharing [8], enabler of personal information and knowledge management tools and new didactic tools that facilitate interaction and social processes.

In this article we investigate the possible use of a Web 2.0 collaborative platform for foreign language learning in a case-based setting. We discuss why case-based learning is particularly relevant for adult, foreign language learning at university level and how students' learning logs may support reflection and processes that facilitate deep-learning. Preliminary studies have emphasized the need for a collaborative platform that supports and enhances foreign language learning processes which are inherently social and individual at the same time. A next step would involve the design of a dynamic semantic-enhanced learning environment that accommodate personalization and enable users to set their own personal agenda for learning [9].

The article is structured in 5 sections. The following section describes the theoretical and didactic principles that underlie learning and in particular foreign language learning. Section 3 introduces the case-based foreign language learning setting and highlights the importance of collaboration and the associated social processes for foreign language learning. Section 4 provides an overview of the pilot study and discusses the preliminary results, while section 5 provides a summary and a future outlook towards new experiments and challenges for an enhanced version of the platform.

2 Learning, Cognition and Foreign Language Learning

Efficient and *successful learning strategies* are crucial to educational success and lifelong, adult learning may successfully take its starting point in the learner's understanding of what it means to learn.

However, what is rested in the concept of learning, and how may it be defined in terms of foreign language learning? First of all, learning a foreign language resembles learning how to solve all other sorts of problems ranging from learning how to drive a car to studying astronomy. One definition of learning is "*the process which leads to the creation of new knowledge thus changing the learner's behaviour and his or her understanding of the surrounding world*"[10]. Learning is thereby individual and process oriented in contrast to instruction, which focuses on subject matter and aims at disseminating information. Contrary to child learning, conscious *cognitive strategies* are key to adult learning. Additionally, understanding why and how for instance languages work and hearing or reading explanations may be used to monitor processes that are useful shortcuts to taking in new knowledge [11].

Learning comprises reflection on one's own learning processes – a form of meta cognition – where the ability to stop and think about one's own learning process becomes central and adds to personal development. Additionally, it facilitates new insights and thereby raises cognitive awareness [12]. For this reason, learning should be viewed as a life-long process where assimilation and accommodation processes substitute rote learning and remembering of facts [10]. However, this process is dependent on individual learning styles "*the way in which each individual learner begins to concentrate on, process, absorb, and retain new and difficult information* [13]". From this perspective learning is an individual matter and each learner has his or her own method of acquiring knowledge.

Motivation is another important factor in learning including foreign language learning – if you cannot see the raison d'être of learning something new, you probably will not bother to pay attention or take in new knowledge. In other words, new intake about a foreign language requires that your *affective filter* is low and that you are willing to incorporate new and sometimes conflicting information in order to move ahead. In terms of looking at motivation in educational settings, the aspect of time is also an essential element. During the complex processes of foreign language learning, "*motivation does not remain constant, but is associated with a dynamically changing and evolving mental process, characterised by constant (re)appraisal and balancing of the various internal and external influences that the individual is exposed to*" [14]: 617. As a result, most learners experience a fluctuation of their enthusiasm and commitment during a learning process.

From a foreign language learning perspective, this is not always consistent with the exposure of the formal classroom or other formal, predesigned learning platforms, as the process is individual, characterised by individual learning styles and based on the needs and capabilities of the individual learner. This, however, does not mean that there are not many patterns of similarity by which foreign language learning may be organised collectively, but individualised learning platforms may serve the purpose better since individual learning patterns may be taken into account. In contrast to this, it is part and parcel of language learning that it takes place in collaboration with others – you cannot learn a language without hearing and reading what others produce.

Learning takes place in interaction and this must be facilitated together with individualisation in order for language learning to be as successful as possible.

Foreign language learning rests on the learner's ability to identify where the problems lie and on his/her ability to address areas of lack of competence. Again, similar to all other learning processes, foreign language learning is a construction process where previous knowledge is used as building blocks and where matches and mismatches with previous knowledge are brought into play. Acknowledging this means that learning a foreign language may be addressed in similar ways as other forms of learning. However, foreign language learning processes do require a very high degree of practice as well, for which reason learners should be allowed to experience the *'flow'*[1] that motivates and creates new impulses. As highlighted by Rasmussen [15], "*We as actors are situated within a framework that contains a past, present and a future – i.e. our temporal standpoint moves and writes a part of the history, and creates a culture in which learning occurs*". Following this, flow should be viewed as a condition in which people are so absorbed in a specific task that they forget all about time and place. Csikszentmihalyi [16] defines flow as, "*a deep and uniquely human motivation to excel, exceed, and triumph over limitation*".

The next section discusses a learning platform that facilitates collective as well as individual learning and foreign language learning at the same time.

3 Collaborative Learning and Web 2.0

Web 2.0 based applications include online chat forums, wikis, blogs, social networking sites that make knowledge sharing easy and unobtrusive for the individual. This type of tools facilitates communication, sharing information and online socialization. Web 2.0 social applications may be a facilitator for the exchange of items of interest such as: bookmarks, business contacts, music, videos, photos, articles, views etc. Using Web 2.0, users may easily express or share their opinions, 'think by writing', seek others' opinions and feedback and be connected with the others.

Web 2.0 applications facilitates social processes, communication, online interaction and eventually enable social learning where emphasis is on collaboration, debate, critique, peer review. Contextual collaboration seamlessly integrates content sharing, communication channels and collaboration tools into a unified user experience that enables new levels of productivity [17].

Ultimately, personal knowledge management becomes possible and thus individualization together with collaboration, whenever this is called for, becomes a motivating factor that enhances knowledge acquisition, deep learning and student performance. Additionally, it enables the learner to optimize his/her management of knowledge, as (s)he is able to reflect upon his/her knowledge during the creative process. Finally, it is particularly interesting in terms of foreign language learning, as the acquisition of effective problem-solving, self-directed learning and team skills is probably more important than the content learned [18]:631.

---

[1] Hermansen defines flow as the pre-occupied time in which activities in the present absorb the past and present in an imploding the present of vitality (2005: 42).

3.1 Case-Based Language Learning

Traditionally, cases have been used to highlight and discuss decision making processes, to address problem solving procedures and to address issues in leadership and management. Teaching with cases is a useful tool if the goal is to experience problem solving in simulations that resemble real life situations and with case content based on real life events or on ongoing developments. However, if we accept that language learning is a construction process, where new knowledge is added through experiencing successes or the opposite in communication situations, and if we accept the notion that personal involvement and motivation are key elements in all learning, then case-based language learning is an obvious possibility and challenge for the language learner and the learning platform designer.

Research has shown that if students work with language problems in an electronic case environment, they become more motivated for collaboration, resulting in successful planning of communication [19]. In more traditional learning environments where case work is limited to the simulation scenario and where no collaborative services are offered early on, process-oriented information sharing and learning are very limited. These findings together with a very clear focus on the language elements through case-based teaching within a Web 2.0 enabled e-learning platform suggest that learning may be efficient if the students' attention is focused on communication oriented problem solving in a collaborative environment.

Cases are not new to the language learning classroom, but cases have not been used to learn languages, rather to discuss cultural, business related and political issues in intercultural business courses. In these case-based teaching classrooms, learning how to solve managerial problems or solve issues in Human Resources departments may be valuable assets to understanding differences and problems in for instance global business, but language issues per se are not part of the package.

Foreign language learning with cases means that focus in the case is on decoding messages, constructing and producing new texts and on successful communication and dissemination of information. For these elements to be featured in the case, we need to take problem solving and the establishment of new language related knowledge seriously. We have to address issues in linguistics, pragmatics, discourse and culture as well as strategies that will assist the learner in understanding and producing the best texts possible – either in written or spoken formats.

Having established that language learning, decoding and production are collaborative processes where meaning is often negotiated by interlocutors, we need to look at how these tie in with case-based teaching. First of all, we need to move attention away from problem-solving related to company processes and decision making and over to problem-solving related to communication and language related problems. How is information processed if the purpose of that information processing is for instance the production of a press release for a global business? What are the issues at stake if we need to address new target audiences, to produce texts that differ in genre, if we need to draw on intercultural knowledge sharing? Then the case addresses problem solving in relation to language and interculturality, information processing, and informed reflection addresses successful dissemination of information and communication. The problem-solving involved is as complex and differentiated as that related to solving management problems and the processes that learners need to master are of a similar nature. The language case deals with analysis and evaluation of a communication situation where focus is on professional communication across cultures and languages. Therefore, the decisions made are product-oriented and relate to cultural and language related parameters; decision-making focuses on communication strategies, on target audience characteristics and adaptation to cultural parameters. Based on these assumptions, the language case is one of the answers to adult, foreign language learning but it requires setting up a collaborative case-based learning platform.

As stated above, the collaborative nature of language learning means that case-based work should also facilitate online exchange of information, information sharing and information management possibilities.

3.2 From Blogs to Learning Logs

Traditionally, blogs are textual but they vary widely in their content. They can be devoted to politics, sharing opinions, news, or technical issues. Using a blog, students can demonstrate critical thinking, take creative risks and make sophisticated use of language [4].

In this context, blogs are used to reinforce learning processes and create a forum for students to reflect on what and how they learn. In order to know more about how to make learning more efficient, *learning logs* that track progress, obstacles, successes and lack of the same are very useful tools both for the learner and for the lecturer.

Learning logs do not require special training – reflecting on your own processes is sufficient, especially as this increases awareness of processes and meta-knowledge of the processes experienced. The purpose of the learning log is twofold: giving the learner an insight into his/her own processes and own problem solving strategies, difficulties overcome and new challenges that must be met and giving researchers and lecturers insights and new data on learners' processing of a foreign language in a multifaceted process that involves both foreign language acquisition and the intake of non-language related information. This information is of the utmost importance for the learner who is given access to own learning style characteristics, reflection on own cognitive processes and to the researcher/lecturer who is provided with valuable insights into what works and what does not work for the individual learner and for a group of learners.

The learning log may facilitate reflection and enhance deep-learning by aiming to track student cognitive processing similar to think-aloud protocols and retrospection of data. The learning log thus becomes an incentive to students to reflect on their learning as well as tool to measure the success of the learning process. Qualitative as well as quantitative analyses have been and are currently being carried out, but as stated in the section 2, learning demands an individual effort which the learner may monitor, gain meta-knowledge about and use in his or her personal development. Affective filters, such as the need to learn, may influence learning. Accordingly, motivation may be high or low as well as successes and failures to incorporate new knowledge that change a little bit of you through a process of acquisition and may be influencing the pattern of learning experienced.

4 Pilot Study

In order to test some of the central hypotheses sparked by the assumption that adult foreign language learning will be facilitated and individualised through a collaborative case-based, electronic learning platform, a pilot study was set up in the spring of 2007. The case presented was developed to match the concept of a *language case* and thus reflected the language case setup outlined above. The study focused on two student groups evaluating learning processes, reflection and the quality of the portfolio-based assignments. The first group was presented with the case-based portfolio assignment in a Joomla[2]-based electronic platform and content management system, whereas the second group received the material as a paper-based case. The material included a press release, a translation assignment and a short report. The research questions focused on four aspects: motivation, student information processing, student use of methods and models introduced in the case and the final results of the students' work.

In regard to the information processing and the work processes involved, it was assumed that the electronic language learning platform would lead the students to reflect more on their processes, share information early on in their assignment work, and facilitate deep-learning through focusing on the processing of information rather than on the end result. The research assumptions were that if students were encouraged to collaborate, to share information, to use peer review opportunities, their learning outcome would be deeper, they would take in new information that would be internalised and accessible for later use as it was individualised through problem-solving and not via instruction.

Methodologically, the study was organised as follows: the case and the research questionnaires were designed by the research group; a teacher who was not a member of the research group taught both groups of students. Each group of students comprised approximately 30 students of English at the Copenhagen Business School bachelor programme in English and international business communication. The teacher and the students were asked to evaluate the work processes, and their performance during and after four weeks of case-based learning work. Additionally, the teacher was asked to write a log book for each group and she was interviewed by an external consultancy to ensure non-partisanship and objectivity. The two groups' assignments were evaluated by the teacher and anonymously by a research group member.

The results of the qualitative interviews and the questionnaires were that in regard to the motivation factor, both teacher and students were more motivated in the electronic learning platform than in the traditional case-based learning work because it was a novel and innovative educational initiative, where the students behaved "as usual" with a decrease in participation, a decrease in the number of assignments handed in and a high level of focus on the final product rather than on learning processes that would lead to the final product.

The results of the pilot study show that the students were more successful in regard to solving assignment problems that were of a discoursal or pragmatic nature than the control group, whereas the control group was more successful at the linguistic level, ie solving problems of syntax and morphology. This has led the research group to conclude that – based on the pilot study – there seems to be a relationship between students' ability to solve more complex foreign language problems related to genre, target group adaptation, script and situation adequacy if their knowledge sharing during assignment work is facilitated. The pilot study did not set up ideal conditions for

[2] http://www.joomla.org/

knowledge sharing; however, the students responded favourably to the possibility. The study does not provide information on the long-term effects of the knowledge taken in – it is not clear whether the knowledge acquired has become declarative and thus ready for re-use in new contexts.

In regard to methods and models used and discussed by the students, opinion differ between the teacher and the research group on the one side and the students on the other side. The teacher and the research group can trace students' successful use of models and methods in both groups, but different models and methods are applied depending on the language learning platform selected. The electronic platform invited methods that were collaborative whereas the paper-based case work seemed to be less collaborative and more result-oriented. The students, on the other hand, did not reflect on these differences and did not see themselves as more or less able to draw on methods and models that might assist them in their problem solving.

On the question of learning outcomes, the students, who had worked with the electronic platform, were of the opinion that the role of the teacher as facilitator and sparring partner was very positive. They also stated that the electronic learning platform was more realistic than traditional case work, especially the "publication aspect", which meant that the students' final products were made available to all students in a shared forum. These real-life elements contributed to students' motivation and eagerness to perform well, but did not mean that the number of portfolio-assignments increased. The students were happy to read what other students had produced, but did not after all feel the inclination to process the more time-consuming and challenging assignments of the case.

5 Discussion and Outlook

The article has introduced case-based foreign language learning that is facilitated by a Web 2.0 enabled collaborative platform that supports social learning as well as individual learning. Based on the above theoretical considerations and the pilot study, the following challenges may be identified: the setting up of a new and improved foreign language learning platform that allows students to interact online, collaborate and share knowledge in all phases of the process that leads to the production of a foreign language text. Furthermore, a learning log must be incorporated to capture cognitive development associated learning processes and reinforce reflection as well as the students' experience in regard to the completion of case-based teaching assignments. Preliminary results have indicated that motivation and collaboration are influencing the quality of students' work, that the electronic platform is a facilitator for collaboration and for knowledge sharing. The electronic platform invited collaborative methods whereas the paper-based case work seemed to be less collaborative and more result-oriented.

The assumption is that the case-based learning platform integrating Web 2.0 tools supports deep-learning of foreign languages and the intake of new words and knowledge that will be turned into new, re-usable knowledge, as students are made aware of their own processing and successful roads to intake that will remain in their knowledge base. This means that reflection that supports and reinforces learning must be part of the case-based learning setup. The learning log is one answer to this – but not just a haphazard description of what the students did and did not do; this part of the learning process needs awareness-raising elements that will provide new insights to the individual student and enable the student to understand and benefit from the strong elements of his or her learning style. Here, reflection questions may enhance students' learning log use and self-understanding.

The design of appropriate learning logs is a key element in further development of the learning platform together with a more social, user-friendly interface which allows students to collaborate, share information, experiences and connect through synchronous and asynchronous communication services. In the new version of the platform, the students will be offered the possibility to present critique and comment on the other student's work, be able to collect and share references and materials that are relevant for their portfolio assignments. We expect that using the platform the students will acquire communicative, critical and collaborative skills that are useful both for scholarly and professional contexts.

References

[1] Razmerita, L., Gouarderes, G., Comte, E.: Ontology-based User Modeling and e-Portfolio Grid Learning Services. Applied Artificial Intelligence Journal 19, 905–931 (2005)

[2] Illeris, K.: How We Learn-Learning and Non-learning in School and Beyond. Routlege, New York (2007)

[3] Naeve, A., Yli-Luoma, P., Kravcik, M., Lytras, M.D.: A modelling approach to study learning processes with a focus on knowledge creation. International J. Technology Enhanced Learning 1, 1–34 (2008)

[4] Duffy, P.: Engaging the YouTube Google-Eyed Generation: Strategies for Using Web 2.0 in Teaching and Learning. In: European Conference on ELearning, ECEL 2007, Copenhagen, Denmark, pp. 173–182 (2007)

[5] Doolan, M.: Using Web 2.0 Technologies to Engage With and Support the net Generation of Learners. In: Proceedings of the 6 Conference on eLearning, ECEL, Copenhagen, Denmark, pp. 159–172 (2007)

[6] Benson, V., Avery, B.: Embedding Web 2.0 Strategies in Learning and Teaching. In: Miltiadis Lytras, P.O.D.P. (ed.) Web 2.0: The Business Model. Springer Science and Business Media, USA (2008)

[7] Benson, V.: Is the Digital Generation Ready for Web 2.0-Based Learning? In: Proceedings of The Open Knowledge Society: A Computer Science and Information Systems Manifesto: First World Summit on the Knowledge Society, Athens, Greece, September 24-26 (2008)

[8] Kirchner, K., Razmerita, L., Sudzina, F.: New Forms of Interaction and Knowledge Sharing on Web 2.0. In: Miltiadis Lytras, E.D., De Pablo, P.O. (eds.) Web2.0: The Business Model, pp. 21–37. Springer Science and Business Media, USA (2008)

[9] Razmerita, L., Lytras, M.D.: Ontology-Based User Modelling Personalization: Analyzing the Requirements of a Semantic Learning Portal. In: Lytras, M.D., Carroll, J.M., Damiani, E., Tennyson, R.D. (eds.) WSKS 2008. LNCS (LNAI), vol. 5288, pp. 354–363. Springer, Heidelberg (2008)

[10] Lauridsen, O.: Learning Styles in ICT based and ICT supported learning: a Foundling? In: Emerging Technologies in Teaching Languages and Cultures, Language on the Edge: Implications for Teaching Foreign Languages and Cultures, vol. 4, pp. 133–145 (2004)

[11] Mondahl, M., Jensen, A.K.: Lexical Search Strategies in Translation. Meta 41, 97–114 (1993)
[12] Hermansen, M.: Relearning. Danish University of Education Press and CBS Press, Kobenhaven (2005)
[13] Dunn, R., Dunn, K.: The complete guide to the learning-styles inservice system. Allyn & Bacon, Needham Heights (1999)
[14] Dörnyei, Z., Skehan, P.: Individual differences in second language learning. In: The handbook of second language acquisition, pp. 589–630 (2003)
[15] Rasmussen, J.: Mesterlære og den almene paedagogik. In: Nielsen, K.K. (ed.) Maesterlære: Laering som social praksis. Hans Reitzel, Kobenhaven (1999)
[16] Csikszentmihalyi, M.: Flow: The psychology of optimal experience, Harper, UK (1991)
[17] Geyer, W., Silva Filho, R.S., Brownholtz, B., Redmiles, D.F.: The Trade-offs of Blending Synchronous and Asynchronous Communication Services to Support Contextual Collaboration. Journal of Universal Computer Science 14, 4–26 (2008)
[18] Barrows, S.H.: The Essentials of Problem-Based Learning. Journal of Dental Education 63, 630–632 (1998)
[19] Ingstad, L., Mondahl, M.: The electronic language case. Copenhagen Business School, Copenhagen (2009)

A Filtering and Recommender System Prototype for Scholarly Users of Digital Libraries

José M. Morales-del-Castillo[1], Eduardo Peis[1], and Enrique Herrera-Viedma[2]

[1] Dpt. Of Library and Informaction Science, University of Granada,
Colegio Máximo de Cartuja s/n (18071), Granada, Spain
[2] Dpt. Of Computer Science and A.I., University of Granada, c/ Periodista Daniel Saucedo Aranda s/n (18071), Granada, Spain
{josemdc,epeis}@ugr.es, viedma@decsai.ugr.es

Abstract. The research and scholarly community of users in a digital library presents several characteristics that make necessary the development of new services capable of satisfying their specific information needs. In this paper we present a filtering and recommender system prototype that applies two approaches to recommendations in order to provide users valuable information about resources and researchers pertaining to domains that completely (or partially) fit that of interest of the user. As an outlook, we briefly enumerate its main features and elements, and present an operational example, which illustrates the overall system performance. Additionally, the outcomes of a simple evaluation of the prototype are shown.

Keywords: Filtering and Recommender Systems, Digital Libraries, Scholarly Community, Semantic Web technologies, Fuzzy Linguistic Modeling.

1 Introduction

One of the key aims of the so-called Information Society is to facilitate the interconnection and communication of sparse groups of people, which can collaborate with each other by exchanging on-line information from distributed sources [1]. In specific contexts, such as in the research and scholarly domain, where many times work is developed relaying on team-based research [5], finding colleagues and associates to build collaborative relationships has become a crucial matter. Actually, this is one of the pillars of the conduct of research and production of scholarship [17]. Nevertheless, this task can be specially difficult when the research activity implies opening new multidisciplinary lines of investigation, since it is hard to know *what's hot* and *who's in* in a certain domain out of that of our specialization (even if both areas are related or close to each other).

Due to this, scholarly libraries in general and digital scholarly libraries in particular (which are considered by the research and scholarly community as main nodes to access scientific information) must provide their users new services and tools to ease such kind of tasks.

In this paper we present a filtering and recommender system prototype for digital libraries that serves this community of users. The system makes available different

M.D. Lytras et al. (Eds.): WSKS 2009, LNAI 5736, pp. 108–117, 2009.

recommender approaches in order to provide users valuable information about resources and researchers pertaining to knowledge domains that completely (or partially) fit that of interest of the user. In such a way, users are able to discover implicit social networks where is possible to find colleagues to form a workgroup (even a multidisciplinary one).

The paper is structured as follows. In section 2 we briefly present the main features and elements of the prototype. An operational example of the performance of the prototype is shown in section 3, and the outcomes of an experiment to evaluate the system are presented in section 4. Finally, in section 5 some conclusions are pointed out.

2 Overview of the Prototype

The system here proposed is based on a previous multi-agent model defined by Herrera-Viedma et al. [14], which has been improved by the addition of new functionalities and services. In a nutshell, our prototype eases users the access to the information they required by recommending the latest (or more interesting) resources acquired by the digital library, which are represented and characterised by a set of hyperlink lists called *feeds* or *channels* that can be defined using vocabularies such as RSS 1.0 (*RDF Site Summary*) [3]. The system is developed by applying different fuzzy linguistic modeling approaches (both ordinal [24] and 2-tuple based fuzzy linguistic modeling [13]) and Semantic Web technologies [4]. While fuzzy linguistic modelling [24] supplies a set of approximate techniques to deal with qualitative aspects of problems, defining sets of linguistic labels arranged on a total order scale with odd cardinality, Semantic Web technologies allow making Web resources semantically accessible to software agents [11]. In such a way, is possible to improve *user-agent* and *agent-agent* interaction, and settle a semantic framework where software agents can process and exchange information.

Based on this infrastructure, the system is able to filter and recommend resources from two different approaches that are explained in the following section.

2.1 Recommender Approaches

Traditionally, filtering and recommender systems have been classified into two categories [18]: systems that provide recommendations about a specific resource according to the opinions given about that resource by different experts with a profile similar to that of the active user (known as collaborative recommender systems) and systems that generate recommendations according to the similarity of a resource with other resources assessed by the active user (i.e. content-based recommender systems).

In both of them, the likeness can be measured using different similarity functions [19][22] which are usually interpreted in a *linear way* (i.e. the higher the similarity value is, the more relevant it is to generate a recommendation). This is what we call *monodisciplinary* approach, since it lets users deepen into their knowledge in a specific area.

Nevertheless, as discussed above, it's quite common (and almost a need) for many researchers the need to keep the track of new developments and advances in other fields related to their specialization domain. In this way, it is possible for them to widen their research scope, open new research lines and create multidisciplinary workgroups.

In such circumstance, users require to get recommendations about resources whose topics are related to (but not exactly fit) their preferences, but without modifying their starting preferences at all. In this case it makes sense considering as relevant an interval of *mid-range* similarity values instead of those close to one (i.e. both extremely similar and dissimilar similarity values are discarded). Therefore, it would be necessary defining some kind of center function [23] that enable constraint the range of similarity values we are going to consider as relevant. In our model, the interpretation of similarity is defined by a Gaussian function μ as the following:

$$\mu\left(Sim\left(p_i, r_j\right)\right) = e^{\left(Sim\left(p_i,\ r_j\right) - k\right)^2}$$

where $Sim\ (p_i, r_j)$ is the similarity measure among the resources p_i and p_j, and k represents the center value around which similarity is relevant to generate a recommendation (in this case k=0'5). This is what we call *multidisciplinary* approach.

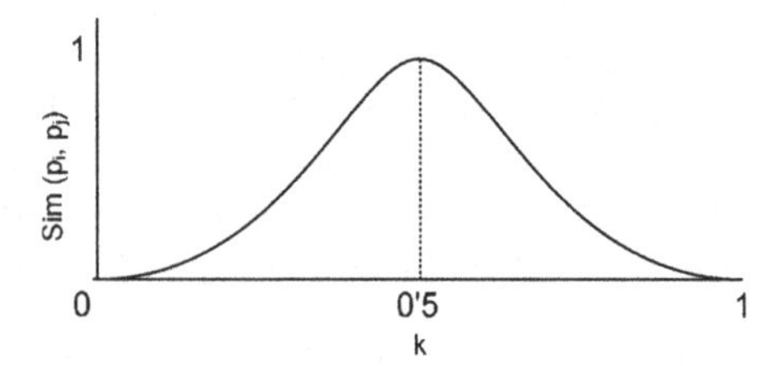

Fig. 1. Gaussian center function

2.2 Architecture and Modules

To carry out the filtering and recommendation process we have defined 3 software agents (interface, task and information agents) that are distributed in a 5 level hierarchical architecture:

- *Level 1. User level*: In this level users interact with the system by defining their preferences, providing feedback to the system, etc.
- *Level 2. Interface level*: This is the level defined to allow interface agent developing its activity as a mediator between users and the task agent. It is also capable to carry out simple filtering operations on behalf of the user.
- *Level 3. Task level*: In this level is where the task agent (normally one per interface agent) carries out the main load of operations performed in the system such as the generation of information alerts or the management of profiles and RSS feeds.
- *Level 4. Information agents level*: Here is where several information agents can access the system's repositories, thus playing the role of mediators between information sources and the task agent.
- *Level 5. Resources level*: In this level are included all information sources the system can access such as a full-text documents repository and a set of resources described using RDF-based vocabularies [2] (RSS feeds containing the items featured

by the digital library, a user profile repository and a thesaurus that describes the specialization domain of the library).

The underlying semantics of the different elements that make up the system (i.e. their characteristics and the semantic relations defined among them) are defined through several interoperable web ontologies [9][10] described using the OWL vocabulary [15].

In the prototype there are also defined 3 main activity modules:

- *Information push module*: This module is responsible for generating and managing the information alerts to be provided to users. The similarity between user profiles and resources is measured according to the hierarchical lineal operator defined by Oldakowsky and Byzer [16], which takes into account the position of the concepts to be matched in a taxonomic tree. Once defined this similarity value, the relevance of resources or profiles is calculated according to do the concept of *semantic overlap*. This concept tries to ease the problem of measuring similarity using taxonomic operators, since all the concepts in a taxonomy are related in a certain degree and, therefore, the similarity between two of them would never reach 0 (i.e. we could find relevance values higher than 1 that can hardly be normalized). The underlying idea in this concept is determining areas of maximum semantic intersection between the concepts in the taxonomy. To obtain the relevance of profiles to other profiles we define the following function:

$$Sim\left(P_i,P_j\right)=\frac{\sum_{k=1}^{MIN\,(N,M)} H_k\left(Sim\left(\alpha_i,\delta_j\right)\right)\left(\frac{\omega_i+\omega j}{2}\right)}{MAX\,(N,M)}$$

where $H_k(\mathrm{Sim}\ (\alpha_i,\ \delta_j))$ is a function that extracts the k maximum similarities defined between the preferences of $P_i = \{\alpha_1, \ldots, \alpha_N\}$ and $P_j=\{\delta_1, \ldots, \delta_M\}$, and ω_i, ω_j are the corresponding associated weights to α_i and δ_j. When matching profiles $P_i = \{\alpha_1, \ldots, \alpha_N\}$ and items $R_j = \{\beta_1, \ldots, \beta_M\}$, since subjects are not weighted, we will take into account only the weights associated to preferences so the function in this case is slightly different:

$$Sim\left(P_i,R_j\right)=\frac{\sum_{k=1}^{MIN\,(N,M)} H_k\left(Sim\left(\alpha_i,\beta_j\right)\right)\omega_i}{MAX\,(N,M)}$$

- *Feedback or user profiles updating module*: In this module, the updating of user profiles is carried out according to users' assessments about the set of resources recommended by the system. This updating process consists in recalculating the weight associated to each preference in a profile and adding new entries to the recommendations log stored in every profile. We have defined a matching function, which rewards those preference values that are present in resources positively assessed by users and penalized them, on the contrary, when this assessment is negative. Let $e_j \in S'$ be the satisfaction degree provided by the user, and $\omega^{j}_{il} \in S$ the weight of property i (in this case i=«Preference») with value l. Then, we define the following updating function g: $S' \mathrm{x}\ S \rightarrow S$:

$$g(e_j, \omega_{li}^j) = \begin{cases} s_{Min(a+\beta, T)} & if \ s_a \leq s_b \\ s_{Max(0, a-\beta)} & if \ s_a > s_b \end{cases}$$

$$s_a, s_b \in S \mid a, b \in H = \{0, \ldots T\}$$

where, (i) $s_{a=} \omega'^{j}_{li}$; (ii) $s_{b=} e_j$; (iii) a and b are the indexes of the linguistic labels which value ranges from 0 to T (being T the number of labels of the set S minus one), and (iv) β is a bonus value which rewards or penalize the weights of the preferences. It is defined as β=*round*(2|b-a|/T) where *round* is the typical round function.

- *Collaborative recommendation module*: The aim of this module is generating recommendations about a specific resource in base to the assessments provided by different experts with a profile similar to that of the active user. The different recommendations (expressed through linguistic labels) are aggregated using the *Linguistic Ordered Weighted Averaging* (LOWA) operator [12], which is capable to combine linguistic information. It also allows users to explicitly know the identity and institutional affiliation data of these experts in order to contact them for any scholarly purpose. This feature of the system implies a total commitment between the digital library and its users since their altruistic collaboration can only be achieved by granting that their data will exclusively be used for contacting other researchers subscribed to the library. Therefore, becomes a critical issue defining privacy policies to protect those individuals that prefer to be *invisible* for the rest of users. Nevertheless, we have to point out that this functionality is still in development and has not been implemented yet.

3 Operational Example

To clarify the performance of the system here we show an operational example. Let's start defining a set of premises:

- A generic user that wants to obtain *monodisciplinar* recommendations from the system, with a profile P where preferences α_1, α_2 (N=2) and their associated weights ω_1, ω_2 are defined.
- An item R of the RSS feed of the system represented by the subjects β_1, β_2, β_3 (M=3).

First of all the system proceeds to calculate the similarity between the resources in the RSS feed and the profile of the active user applying the taxonomic linear operator defined in [16]. Let α_1 be the concept "*Control instruments*" with a depth of 2 in the thesaurus of the system and β_2 the concept "*Record group classification*" with a depth of 3 (being 6 the maximum depth of the thesaurus). As the common parent (*ccp*) of both concepts is "*Archival Science*" (which depth is 0 by default), the distance between them is $d(\alpha_1, \beta_2) = 0.83$, and its associate similarity $Sim(\alpha_1, \beta_2) = 0.17$.

In the next step, the relevance of the item R to the profile P is calculated. Let the importance value for the preference α_1 be the linguistic label "*Very high*" (i.e. $\omega_1=0.83$) and for α_2 the label "*Medium*" (i.e. $\omega_2= 0.5$). Besides, if the number of preferences and subjects is respectively N=2 and M=3, then the 3 maximum similarities are chosen to calculate the relevance value (in this case, let's suppose Sim (α_1, β_3)=0.88, Sim (α_2, β_1)=0.84, and Sim (α_2, β_2)=0.93). The resulting relevance value is *Rel (P, R)*= 0.54 so, as the relevance threshold has been fixed in *k*=0.50, the resource R is selected to be retrieved.

Then, applying the 2-tuple based fuzzy linguistic modeling approach [13], relevance is displayed as a linguistic label extracted from the linguistic variable "*Relevance level*" together with a numeric value (also called "*symbolic translation*"). Therefore, for the relevance value *Rel (P, R)*= 0.54, the outcome is "*Medium* + 0.04". This implies that "*Medium*" is the closest linguistic label for that relevance value, and that the corresponding numeric value for this label has been exceeded in 0.04.

The following step consists in searching profiles (similar to the profile of the active user) with recommendations about the resource R in order to generate a collaborative recommendation. Supposed two users that have respectively assessed the resource R with the linguistic labels "*High*" and "*Medium*" (which have been extracted from the linguistic variable "*Level of satisfaction*"), when applying the LOWA operator [12] the resulting aggregated label is the following: k= MIN{6,3 + round (0.4*(4-3))}=3. Then l_k= "*Medium*".

As the non-weighted average similarity of the preference α_1 (with a value of 0.80) is lower than that of α_2 (with a value of 0.88), this last preference value will be the chosen to be updated. Let's see an example of the updating process.

Supposed the user assesses the resource R (which has satisfied his information needs) defining a satisfaction level with the linguistic label e_j="*Very High*" (where $e_j \in S'$= *{null, very low, low, medium, high, very high, total}*). In this case, the associated weight to α_2 is $\omega^j_{(Preference,\ \alpha 2)}$= "*Medium*"(where $\omega^j_{li} \in S$ = *{null, very low, low, medium, high, very high, total}*). Considering that $s_a \leq s_b$, whose index values are *a*=3 and *b*=5, and *T*=6, we have that β=*1*, so the new associated weight for α_2 is increased in a factor of one $(\omega^j_{(Preference,\ \alpha 2)})$'= *g* (*Very high, Medium*) = "*High*".

If the user decides to get multidisciplinary recommendations the process is carried out in a slightly different manner. Let R and R' be the set of retrieved resources with relevance values *Rel(P, R)*=0.57 and *Rel(P, R')*=0.83 respectively the system recalculates both relevance values according to the centering function: μ(Rel (P, R))=1.005; μ(Rel (P, R'))=1.110. Then, the system re-arranges the retrieved items and considers as more relevant the values which are closer to one (in this case, R is more relevant than R').

4 Prototype Evaluation

We have set up an experiment to evaluate the content-based module of the prototype in terms of precision [6] and recall [7] (since the collaborative recommendation module is not fully implemented yet and suffers from *cold start problem* [21]). These two measures (together with the F1 measure [20] are normally used in filtering and recommender systems to assess the quality of the set of retrieved resources.

To carry out the evaluation and according to users' information needs, the set of items recommended by the system have been classified into four basic categories:

relevant suggested items (Nrs), relevant non-suggested items (Nrn), irrelevant suggested items (Nis) and irrelevant non-suggested items (Nin). We have also defined other categories to represent the sum of selected items (Ns), non-selected items (Nn), relevant items (Nr), irrelevant items (Ni), and the whole set of items (N).

Based on to these categories we have defined in our experiment precision, recall and F1 as follows:

Precision: Ratio of selected relevant items to selected items, i.e., the probability of a selected item to be relevant.

$$P= Nrs/Ns$$

Recall: Ratio of selected relevant items to relevant items, i.e., the probability of a relevant item to be selected.

$$R= Nrs/Nr$$

F1: Combination metric that equals both the weights of precision and recall.

$$F1=(2*P*R)/(P+R)$$

The goal of the experiment is to test the performance of our prototype in the generation of accurate and relevant content-based recommendations for the users of the system, exclusively considering the mono-disciplinary search. To do so, we have asked a random sample of twelve researchers in the field of Library and Information Science that develop their activity at the University of Granada to evaluate the results provided by the prototype.

One of the premises of the experiment is that at least one of the topics defined for a relevant resource and one of the experts' preferences must be constraint to the same sub-domain of the thesaurus. In such a way, we can leverage a better terminological control on subjects and preferences and extrapolate the output data to the whole thesaurus. In this case, the sub-domain selected is "*Archival science*", which is composed of 96 different concepts. We also require two more elements:

- an RSS feed containing 30 items extracted from the E-LIS open access repository [8], from which only 10 of them are semantically relevant (i.e. with at least one subject pertaining to the selected sub-domain).
- a set of profiles with at least one preference pertaining to the targeted sub-area.

The prototype is set to recommend up to 10 resources and then users are asked to assess the results by explicitly stating which of the recommended items they consider are relevant. The results of the experiment are shown in table 1:

Table 1. Experimental data

	User1	User2	User3	User4	User5	User6	User7	User8	User9	User10	User11	User12
Nrs	6	5	3	6	4	5	5	4	6	3	7	6
Nrn	2	3	2	1	2	3	2	2	2	2	1	2
Nis	4	5	7	4	6	5	5	6	2	7	3	4
Nr	8	8	5	7	6	8	7	6	8	5	8	8
Ns	10	10	10	10	10	10	10	10	10	10	10	10

Precision, recall and F1 for each user are shown in table 2 (in percentage) and represented in the graph in figure 2. The average outcomes reveal a quite good performance of the prototype.

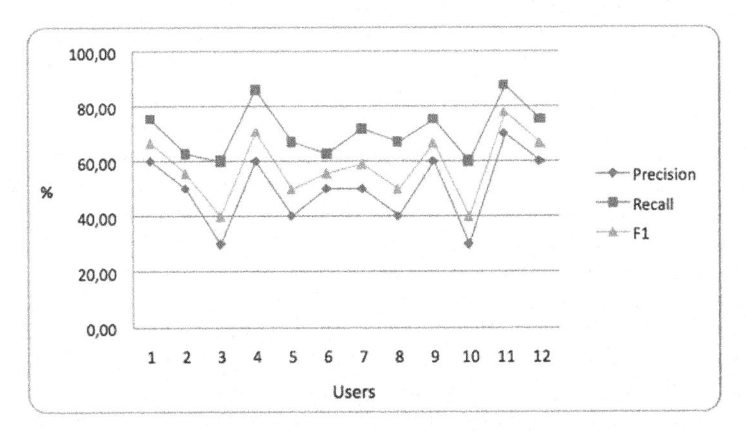

Fig. 2. Precision, recall and F1

Table 2. Detailed experimental outcomes

%	User1	User2	User3	User4	User5	User6	User7	User8	User9	User10	User11	User12	**Aver.**
P	60.00	50.00	30.00	60.00	40.00	50.00	50.00	40.00	60.00	30.00	70.00	60.00	50.00
R	75.00	62.50	60.00	85.71	66.67	62.50	71.43	66.67	75.00	60.00	87.50	75.00	70.66
F1	66.67	55.56	40.00	70.59	50.00	55.56	58.82	50.00	66.67	40.00	77.78	66.67	58.19

4 Conclusions

In this paper we have presented a multi-agent filtering and recommender system prototype for digital libraries designed to be used by the scholarly community, that provides an integrated solution to minimize the problem of accessing relevant information in vast document repositories and finding new research colleagues.

The prototype combines Semantic Web technologies and fuzzy linguistic modeling techniques to improve communication processes and user-system interaction.

The system is able to generate both *monodisciplinary* recommendations (to deepen into users' specialization area) and *multidisciplinary* recommendations, which allow users eliciting resources whose topics are tangentially related to their preferences.

The prototype makes possible for researchers to uncover implicit social networks, which relate them with other researchers from different domains, thus easing the task of forming multidisciplinary working groups. Nevertheless, this implies that the system should apply privacy policies to protect those individuals that prefer to be *invisible* for the rest of users.

The system has been evaluated and experimental results show that the model is reasonably effective in terms of precision and recall, although further detailed evaluations may be necessary.

Acknowledgements

This work has been supported by FEDER funds in the National Spanish Projects TIN2007-61079, PET2007_0460 and FOMENTO-90/07.

References

1. Angehrn, A.A., Maxwell, K., Luccini, A.M., Rajola, F.: Designing collaborative learning and innovation systems for education professionals. In: Lytras, M.D., Carroll, J.M., Damiani, E., Tennyson, R.D. (eds.) WSKS 2008. LNCS (LNAI), vol. 5288, pp. 167–176. Springer, Heidelberg (2008)
2. Beckett, D. (ed.): RDF/XML Syntax Specification (2004), http://www.w3.org/TR/rdf-syntax-grammar/
3. Beged-Dov, G., Bricley, D., Dornfest, R., Davis, I., Dodds, L., Eisenzopf, J., Galbraith, D., Guha, R.V., MacLeod, M., Miller, E., Swartz, A., van der Vlist, E. (eds.): RDF Site Summary (RSS) 1.0 (2001), http://web.resource.org/rss/1.0/spec
4. Berners-Lee, T., Hendler, J., Lassila, O.: The Semantic Web: A new form of Web content that is meaningful to computers will unleash a revolution of new possibilities. The Scientific American (May 2001), http://www.sciam.com/article.cfm?id=the-semantic-web
5. Borgman, C.L.: Scholarship in the digital age: Information, infrastructure, and the Internet. MIT Press, Cambridge (2007)
6. Cao, Y., Li, Y.: An intelligent fuzzy-based recommendation system for consumer electronic products. Expert Systems with Applications 33(1), 230–240 (2007)
7. Cleverdon, C.W., Mills, J., Keen, E.M.: Factors Determining the Performance of Indexing Systems, Test Results, vol. 2. ASLIB Cranfield Project (1966)
8. E-LIS: Homepage (2009), http://eprints.rclis.org/
9. Gruber, T.R.: Toward principles for the design of ontologies used for knowledge sharing. International Journal of Human-Computer Studies 43(5-6), 907–928 (1995)
10. Guarino, N.: Formal ontology and information systems. In: Guarino, N. (ed.) Formal Ontology in Information Systems, pp. 3–17. IOS Press, Amsterdam (1998)
11. Hendler, J.: Agents and the Semantic Web. IEEE Intelligent Systems, 30–37 (March-April 2001)
12. Herrera, F., Herrera-Viedma, E., Verdegay, J.L.: Direct Approach Processes in Group Decision Making using Linguistic OWA operators. Fuzzy Sets and Systems 79(2), 175–190 (1996)
13. Herrera, F., Martinez, L.: A 2-tuple fuzzy linguistic representation model for computing with words. IEEE Transactions on Fuzzy Systems 8(6), 746–752 (2000)
14. Herrera-Viedma, E., Peis, E., Morales-del-Castillo, J.M., Anaya, K.: Improvement of Web-based service Information Systems using Fuzzy linguistic techniques and Semantic Web technologies. In: Liu, J., Ruan, D., Zhang, G. (eds.) E-Service intelligence: methodologies, technologies and applications, pp. 647–666. Springer, Heidelberg (2007)
15. McGuinness, D.L., van Harmelen, F. (eds.): OWL Web Ontology Language Overview (2004), http://www.w3.org/TR/2004/REC-owl-features-20040210/
16. Oldakowsky, R., Byzer, C.: SemMF: A framework for calculating semantic similarity of objects represented as RDF graphs (2005), http://www.corporate-semantic-web.de/pub/SemMF_ISWC2005.pdf

17. Palmer, C.L., Teffeau, L.C., Pirmann, C.M.: Scholarly information practices in the online environment: Themes from the literature and implications for library service development. Report commissioned by OCLC Research (2009), http://www.oclc.org/programs/publications/reports/2009-02.pdf
18. Popescul, A., Ungar, L.H., Pennock, D.M., Lawrence, S.: Probabilistic models for unified-collaborative and content-based recommendation in sparse-data environments. In: Breese, J.S., Koller, D. (eds.) Proc. of the 17th Conference on Uncertainty in Artificial Intelligence (UAI), pp. 437–444. Morgan Kaufmann, San Francisco (2001)
19. Salton, G.: The Smart retrieval system–experiments. In: Salton, G. (ed.) Automatic document processing. Prentice-Hall, Englewood Cliffs (1971)
20. Sarwar, B., Karypis, G., Konstan, J., Riedl, J.: Analysis of recommendation algorithms for e-commerce. In: Jhingran, A., Mason, J.M., Tygar, D. (eds.) Proc. of ACM E-Commerce 2000 conference, pp. 158–167. ACM, New York (2000)
21. Schein, A.I., Popescul, A., Ungar, L.H.: Methods and metrics for cold-start recommendations. In: Jarvelin, K., Beaulieu, M., Baeza-Yates, R., Myaeng, S.H. (eds.) Proc. of the 25'th Annual International ACM SIGIR Conference on Research and Developmentin Information Retrieval (SIGIR 2002), pp. 253–260. ACM Press, New York (2002)
22. van Rijsbergen, C.J.: Information Retrieval. Butterworths (1979)
23. Yager, R.R.: Centered OWA operators. Soft Computing 11(7), 632–639 (2007)
24. Zadeh, L.A.: The concept of a linguistic variable and its applications to approximate reasoning. Information Sciences 8(1), 199–249; 8(2) 301–357; 9(3) 43–80 (1975)

Towards the Extraction of Intelligence about Competitor from the Web

Jie Zhao[1,2] and Peiquan Jin[3]

[1] School of Industry and Administration,
Anhui University, 230039, Hefei, China
[2] School of Management, University of Science and Technology of China, Hefei, China
[3] Department of Computer Science and Technology,
University of Science and Technology of China, Hefei, China
zjpq@mail.hf.ah.cn

Abstract. In this paper we present a system framework for the extraction of intelligence about competitor from the Web. With the surprising increasing of the data volume in the Web, how to get useful intelligence about competitor has been an interesting issue. Previous study shows that most people prefer to look up information by competitor. We first analyze the requirements on the extraction of competitor intelligence from the Web and define three types of intelligence for competitor. And then a system framework to extract competitor intelligence from the Web is described. We discuss the three key issues of the system in detail, which are the profile intelligence extraction, the events intelligence extraction, and the relations intelligence extraction. Some new techniques to deal with those issues are introduced in the paper.

Keywords: competitive intelligence; competitor; Web; intelligence extraction.

1 Introduction

Nowadays, it is no doubt that Web has brought a lot of impacts both on personal life and business filed. In the business field, with more and more information can be searched in the Web, enterprises begin to pay much attention to the utilization of the Web, among which the most interested issue is to extract some competitive intelligence about competitor. As a recent survey reported [1], most people prefer to look up information by competitor. It is obvious that enterprises can receive many benefits and even enhance the competitive power if they can obtain lots of intelligence about their competitor from the Web [2].

Unfortunately, currently people are limited in the ways of acquiring the intelligence about competitor. The commonly-used way is to use a search engine to collect information about competitor. Hence, people will get a large set of Web pages, because current search engines usually support a text-based searching mechanism. For example, if you search the information about a corporation “Microsoft” in Google, you may get a result containing more than six millions of Web pages. It is hard for one to manually get intelligence from so a big data set. Recently, some people introduced the text mining techniques into the intelligence acquiring process [3]. However, while they are capable of filtering the non-related text blocks in a Web

M.D. Lytras et al. (Eds.): WSKS 2009, LNAI 5736, pp. 118–127, 2009.

page, it divides a Web page into a set of text blocks. This eventually brings more information processing work in order to produce the competitive intelligence.

This paper mainly concentrates on the issue of extracting intelligence about competitor from the Web. We present a framework for competitor extraction from the Web, and the key issues are discussed. The main contributions of the article are summarized as follows:

(1) The requirements of extracting competitor intelligence from the Web are analyzed. This is to give an answer to the questions: "What intelligence about competitor can people extract from the Web?" and "What intelligence about competitor do they want?"

(2) A system framework to automatically extract competitor intelligence from the Web is presented. The main issues are discussed in the paper. We also introduce some new techniques to treat the critical issues in the system.

The following of the paper is structured as follows. In Section 2 we discuss the related work. Section 3 discusses the intelligence requirement on extracting competitor intelligence from the Web. Section 4 gives the discussion about the framework for competitor extraction from the Web. And conclusions and future work are in the Section 5.

2 Related Work

Competitive intelligence refers to the process that gathering, analyzing and delivering the information about the competition environment as well as the capabilities and intensions of the competitors, and then transforming them into intelligence [4]. Competitive intelligence is acquired, produced and transmitted through the competitive intelligence systems (CIS).

Traditionally, people usually utilize some publications to acquire competitive intelligence, such as news paper, magazines, or other industry reports. With the rapid development of the Web, people can search any information in a real-time way, thus it has become an important way to obtain competitive intelligence from the Web [2].

The detailed procedure of producing competitive intelligence from the Web can be described as follows. For example, suppose the company wants to get the competitive intelligence about one of its competitors, namely, the company C, they will first search the information about the company C through some search engines, e.g. Google, typically using some keywords like "C Company". Then the experts analyze the gathered Web pages to make out a report about the company C. In this paper, we call this type of intelligence acquiring "Web-page-based competitive intelligence acquiring". The disadvantages of the Web-page-based way are obvious. Since the search engine will usually return a huge amount of Web pages, e.g. when you search in Google using the keywords "Microsoft Office 2008" you will get billions of Web pages, it is ultimately not feasible for experts to analyze all the searching results and produce valuable competitive intelligence.

Recently, researchers introduced the Web text mining approach into the CIS. The Web text mining aims at finding implicit knowledge from a huge amount of text data [3]. It depends on some fundamental technologies, including the computing linguistics, statistical analysis, machine learning, and information retrieval. So far, re-searchers have proposed some approaches to processing Web pages, such as extracting text from Web pages [5] and detecting changes of Web pages [6]. According to the text-mining-based approaches, the noisy data in Web pages can be eliminated, and a set of text blocks are obtained and even clustered in some rules. However, since a Web page typically contains a lot of text blocks, this method will consequently produce a large number of text blocks which is much more than the number of Web pages. Besides, if the text blocks are clustered under specific rules, the information about competitors and competition environment will spread among different clusters and bring too much work for information analysis.

Competitive intelligence serves for companies and people, so in order to make the competitive intelligence systems more effective, first we should study what competitive intelligence companies need. As a survey indicated [1], most people prefer to look up information by competitor. When we further ask one more question: "What is the competitive intelligence about the competitors?", most companies will give out the answer: "We want to know everything about our competitors, their history, products, employees, managers, and so on." Are these information only Web pages? The answer is definitely "no". Web pages are only the media that contain the needed in-formation, but note they are NOT competitive intelligence. The CIS is expected to produce competitive intelligence about competitors or competition environment from a large set of Web pages, but not just deliver the Web pages or the text blocks in them. This means we should transfer the Web-page-based viewpoint into an entity-based viewpoint. In other words, the CIS should deliver competitive intelligence about the entities such as the competitors (or sub-entities such as the products of a specific competitor), rather than just deliver the Web pages that surly contain the basic information.

3 Requirements on Extracting Competitor from the Web

Before we design a software system which is able to automatically generating competitive intelligence about competitor from the Web, the first issue we should focus on is to clarify the requirements of competitor extraction. The survey in [1] indicates that most people want to know the product information, news, and announcements about the competitors in the same market. Based on a systematic view, we give out the following description of the requirements on competitor extraction from the Web (see Table 1).

Table 1. Intelligence requirements on extracting competitor from the Web

Type	Description
Profile Intelligence	This type of intelligence refers to the basic information about competitor, e.g. company name, telephone number, address, products set, managers' names, etc.
Events Intelligence	This type of intelligence refers to the news usually co-rellated with time and location, e.g. establishment of the company, release of new products, staff reduction, Being listed stock, etc.
Business Relations Intelligence	This type of intelligence refers to the business relations between competitor and other companies, e.g. suppliers of the company, investors, customers served, etc.

3.1 Profile Intelligence

The profile intelligence is the general information about competitor. Many websites such as Wikipedia [7] provides some general information about companies, such as names, employee counts, managers' names, etc. Fig.1 shows the extracted general information of the Lenovo Corporation.

lenovo联想

Type	Public company (Red chip)
Founded	1984
Headquarters	Beijing, China Morrisville, North Carolina, United States Hong Kong (registered)
Key people	Liu Chuanzhi, Chairman Yang Yuanqing, CEO[1]
Industry	Computing
Products	Personal computers, laptops, imaging, mobile phones
Revenue	▲$16.4 billion USD (2008)
Employees	Approx.23,000 (2008)
Website	www.lenovo.com

Fig. 1. Example of profile intelligence extracted from the Web[1]

3.2 Events Intelligence

Events about competitor usually refer to the news about it. Many websites provide news which is updated frequently. Through the events expressed in the news, people are able to know the recent development of the competitors. Typical events are the establishment of the competitor, the listed-in-stock of the competitor, the progress of some specific project, etc. Fig.2 shows some recent events about Oracle.

April 20, 2009 *Anaheim*	Oracle at Defense Information Systems Agency (DISA) Customer Conference 2009 Register Show Details Email to a Friend
April 20, 2009 *San Francisco*	Oracle at RSA Conference 2009 Register Show Details Email to a Friend
April 21, 2009 *Grapevine*	Oracle at Maintenance Repair & Overhaul (MRO) Annual Conference 2009 Register Show Details Email to a Friend
April 22, 2009 *San Francisco*	Oracle at High-Tech Forecasting & Planning Summit Register Show Details Email to a Friend

Fig. 2. Example of events in the Web[2]

[1] Data source: http://en.wikipedia.org/wiki/Lenovo
[2] Data source: http://events.oracle.com/

3.3 Business Relations Intelligence

Compared with profile and events, the business relations are usually more implicit. This is because most companies do not want that the competitors know their suppliers or customers. However, this type of competitive intelligence may be more useful than others. For example, if you know exactly the suppliers of your competitor, you may have some countermeasures to control those suppliers so as to leave the competitor in a passive situation. To obtain the business relations about competitor, we must perform an intelligent analysis on the contents of Web pages. For example, from the Web page shown in Fig.3, we get to know that Mayfield Fund is an investor of the Consorte Media Corporation.

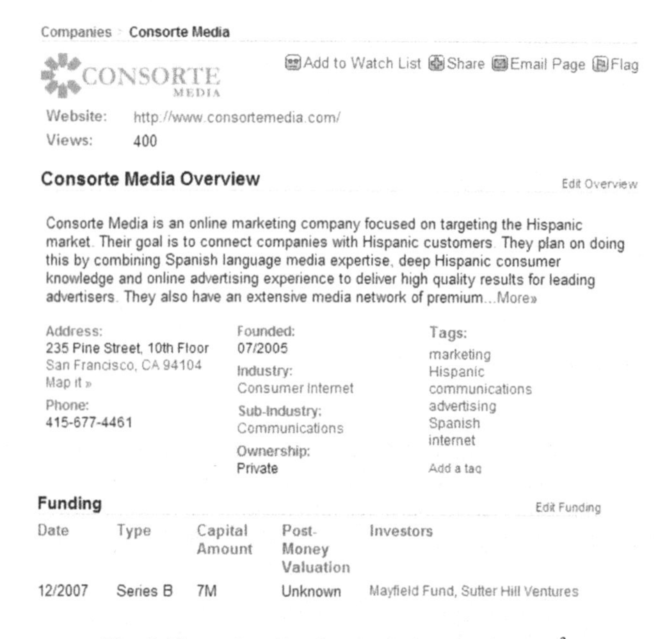

Fig. 3. Example of business relations in the Web[3]

4 A Framework for Competitor Extraction from the Web

The architecture of competitor intelligence extraction from the Web is shown in Fig.4. Each of the three components shown in the figure contains some critical issues. Experts (or users) run the three components and obtain the profile, events, and business relations respectively and finally generate the specification of competitor intelligence. The details about the three components are described in the following text.

[3] Data source: http://www.tradevibes.com/ company/ profile/ consorte-media

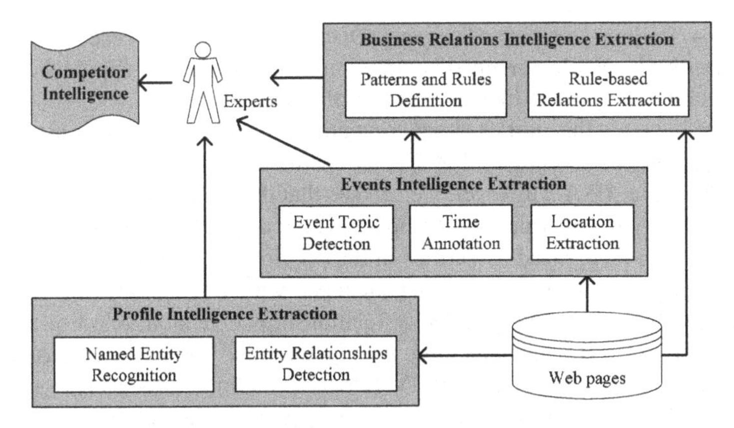

Fig. 4. Architecture of competitor intelligence extraction from the Web

4.1 Profile Intelligence Extraction

For the extraction of profile intelligence about competitor, we use a two-step approach: named-entity recognition and entity relationships detection.

Named-entity recognition is a hot research field in Web information extraction and retrieval [8]. It was first introduced as a sub-task in the MUC (Message Understanding Conference) conference [9]. Its main task is to recognize and classify the specific names and meaningful numeric words from the given texts. Typical named entities are company names, person names, addresses, times, etc. Most of the previous research in this field focused on three types of named entities: time entities, number entities, and organization entities [10]. According to the context of competitor intelligence extraction, several types of named-entities are needed to be studied. However, different methods are also required for different named-entities. For example, we use a hierarchy method to recognize the addresses from Web pages, i.e. "China

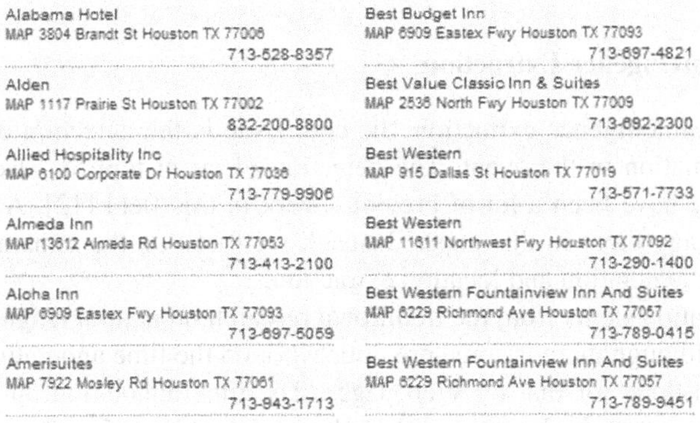

Alabama Hotel
MAP 3804 Brandt St Houston TX 77006
713-528-8357

Alden
MAP 1117 Prairie St Houston TX 77002
832-200-8800

Allied Hospitality Inc
MAP 6100 Corporate Dr Houston TX 77036
713-779-9906

Almeda Inn
MAP 13612 Almeda Rd Houston TX 77053
713-413-2100

Aloha Inn
MAP 6909 Eastex Fwy Houston TX 77093
713-697-5059

Amerisuites
MAP 7922 Mosley Rd Houston TX 77061
713-943-1713

Best Budget Inn
MAP 6909 Eastex Fwy Houston TX 77093
713-697-4821

Best Value Classic Inn & Suites
MAP 2536 North Fwy Houston TX 77009
713-692-2300

Best Western
MAP 915 Dallas St Houston TX 77019
713-571-7733

Best Western
MAP 11611 Northwest Fwy Houston TX 77092
713-290-1400

Best Western Fountainview Inn And Suites
MAP 6229 Richmond Ave Houston TX 77057
713-789-0415

Best Western Fountainview Inn And Suites
MAP 6229 Richmond Ave Houston TX 77057
713-789-9451

Fig. 5. The entity relationships detection issue in the profile intelligence extraction[4]

[4] Data source: http://www.yellowpagecity.com/ US/TX/ Houston/ Hotels

[*country*] → Beijing [*city*] → Chaoyang District [*district*] → Peace Road [*street*] No.128 [*number*]", while for email extraction we use another approaches such as pattern matching, i.e. strings like "[*strings*]@[*strings*].[*strings*]".

While we have got the entities about competitor, the next step is to construct the relationships among those entities. Fig.5 is an example. This figure contains some hotels in the Houston, USA. However, we can see that there are several companies listed in the Web page, each of which has company name, address, and a telephone number. Note those are all recognized as entities in the previous step. And now we must clarify which address or telephone number is matched with which company. This is what we do in the second step, i.e. the entity relationships detection. We use a distance-based method to create the matching among entities. The distance-based idea is straightforward. We first create the DOM (Document Object Model) [11] structure of the processed Web page, and for a given entity we group the entities that have minimal distance to it with this entity. Correlated entities are usually located in the same cell of a table (see Fig.6), or in the same paragraph, and those entities are very close in the DOM graph.

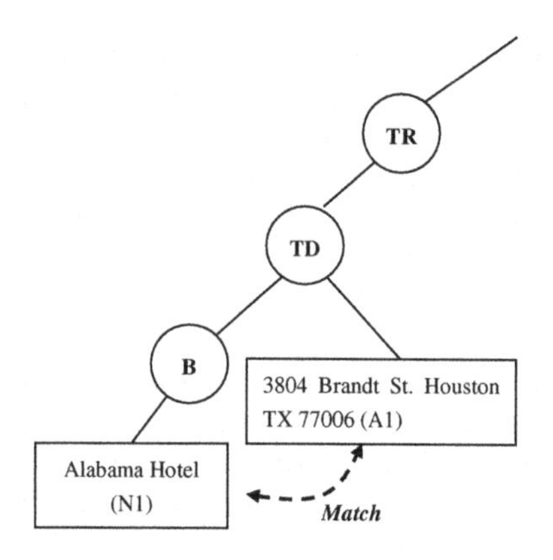

Fig. 6. The distance-based approach to find entity relationships

4.2 Events Intelligence Extraction

For the events intelligence extraction, the core issue is the detection of the time and location information in the event. The detection of event topic is also an important issue, but there have been a lot of previous work in this field [12]. And we will not mention them any more in this paper. Instead, we focus on the remained two issues, which are time annotation and location extraction.

Time annotation stems from the traditional research on natural language processing (NLP) [13]. Although there is some previous work on the time annotation on text, rare work has been done for that on Web pages. The time annotation on text is usually based on the two standards: TIMEX2 [14] and TimeML [15]. The most important difference between Web page and text is that a Web page has some tags. So in this

```
Algorithm Web_Time_Extraction (page w)
Input: w: a Web page
Output: S: a set of time periods.
  Begin
    Transform the Web page into a DOM graph G
    For each node g in G Do
       Detect time elements from g based on TIMEX2
       Format each time element
       Add formatted times into S
    End For
    Return S
  End
End Web_Time_Extraction
```

paper, we first eliminate the tags in a Web page, and then apply the traditional time annotation approaches to obtain the time information in the Web page. The detailed algorithm *Web_Time_Extraction* is shown in the above.

For the location detection in a Web page, we conduct a pattern-based approach to detect location information in a Web page. First, we construct a hierarchy dictionary to store the hierarchy location names according to their geographical relationships, e.g. "[*country*]→[*city*] → [*district*] → [*street*] → [*number*]". We have found that there are common patterns in the expression of locations and addresses. Thus a set of patterns are defined and then used to detect locations from Web pages.

4.3 Business Relations Extraction

Business relations are very important for companies. Generally, there are several types of business relations. The ACE (Automatic Content Extraction) has defined six types of relations in English texts [16], which are listed in Table II. However, those relations are not defined for competitor intelligence. The only interested types in ACE are the Person-Social relation and ORG-Affiliation relation. But these relations are too rough for business relations intelligence extraction. We classify the business relations into two types: Inner-ORG relations and Inter-ORG relations. The Inner-ORG (ORG is the abbreviation of the word "organization") relations refer to the business relations between a company and its components, e.g. company-manager, company-employee, and so on. According to our discussion in the Section "*Profile Intelligence Extraction*", those relations are all captured by the entity relationships matching procedure. The reason is that all the related entities in the Inner-ORG relations are all the entities about one single company. The Inter-ORG relations are relations among different companies. Examples of the Inter-ORG relations are company-investor, company-supplier, company-partner, etc.

Table 2. The relations defined by ACE [16]

Type	Subtypes
Phisical	Located, Near
Part-Whole	Geographical, Subsidiary
Personal-Social	Business, Family, Lasting-Personal
ORG-Affiliation	Employment, Ownership, Founder, Student-Alum, Sports-Affiliation, Investor-Shareholder, Membership
Agent-Artifact	User-Owner-Inventor-Manufacturer
Gen-Affiliation	Citizen-Resident-Religion-Ethnicity

5 Conclusions and Future Work

As competitive intelligence plays more and more important role in the development of enterprises, how to acquire competitor intelligence from the Web has been one of the focuses in most companies. In this paper, we briefly introduce the requirements on the extraction of the intelligence about competitor from the Web, based on which a system framework to extract competitor intelligence from the Web is presented and some of its key issues are discussed. Currently we are working on the algorithms mentioned in the article. Our future work will concentrate on the implementation of the system proposed in the paper and conduct experiments to demonstrate the performance of the algorithms and the whole system.

Acknowledgments. This work is supported by the National Natural Science Foundation of China (No. 70803001 and 60776801), and the Science Research Fund of MOE-Microsoft Key Laboratory of Multimedia Computing and Communication (No. 06120804).

References

1. LaMar, J.: Competitive Intelligence Survey Report (2007), `http://joshlamar.com/documents/CITSurveyReport.pdf`
2. Thompson, S., Wing, C.Y.: Assessing the Impact of Using the Internet for Competitive Intelligence. Information & Management 39(1), 67–83 (2001)
3. Mikroyannidis, A., Theodoulidis, B., Persidis, A.: PARMENIDES: Towards Business Intelligence Discovery from Web Data. In: Proc. of IEEE/WIC/ACM International Conference on Web Intelligence (WI 2006), pp. 1057–1060 (2006)
4. Kahaner, L.: Competitive Intelligence. Simon & Schuster, New York (1996)
5. Hotho, A., Nürnberger, A., Paass, G.: A Brief Survey of Text Mining. LDV Forum (LDVF) 20(1), 19–62 (2005)
6. Khoury, I., El-Mawas, R.M., El-Rawas, O., et al.: An Efficient Web Page Change Detection System Based on an Optimized Hungarian Algorithm. IEEE Transaction on Knowledge Data Engineering (TKDE) 19(5), 599–613 (2007)
7. `http://www.wikipedia.org`

8. Whitelaw, C., Kehlenbeck, A., Petrovic, N., et al.: Web-scale Named Entity Recognition. In: Proc. of CIKM 2008, pp. 123–132 (2008)
9. Sundheim, M.: Named Entity Task Definition-Version 2.1. In: Proc. of the Sixth Message Understanding Conference, pp. 319–332 (1995)
10. Khalid, M., Jijkoun, V., de Rijke, M.: The impact of named entity normalization on information retrieval for question answering. In: Macdonald, C., Ounis, I., Plachouras, V., Ruthven, I., White, R.W. (eds.) ECIR 2008. LNCS, vol. 4956, pp. 705–710. Springer, Heidelberg (2008)
11. http://www.w3.org/DOM/
12. Sun, B., Mitra, P., Giles, C.L., et al.: Topic segmentation with shared topic detection and alignment of multiple documents. In: Proc. of SIGIR, pp. 199–206 (2007)
13. Wong, K.F., Xia, Y., Li, W., et al.: An Overview of Temporal Information Extraction. Int. Journal of Computer Processing of Oriental Languages 18(2), 137–152 (2005)
14. TIMEX2, http://timex2.mitre.org/
15. TimeML, http://www.timeml.org/site/index.html
16. ACE (Automatic Content Extraction) English Annotation Guidelines for Relations, Version 6.2, Linguistic Data Consortium (2008), http://www.ldc.upenn.edu/Projects/ACE/

Modeling Learning Technology Interaction Using SOPHIE: Main Mappings and Example Usage Scenarios

Sinuhé Arroyo and Miguel-Ángel Sicilia

Information Engineering Research Unit
Computer Science Dept., University of Alcalá
Ctra. Barcelona km. 33.6 – 28871 Alcalá de Henares (Madrid), Spain
sinuhe.arroyo@uah.es, msicilia@uah.es

Abstract. Learning technology is nowadays subject to intensive standardization efforts that have lead to different specifications defining the functions and information formats provided by different components. However, differences persist and in some cases, the proliferation of proposed standards and protocols leads to increased heterogeneity. Further, the terminologies used to describe resources and activities are diverse, resulting in semantic interoperability problems and conversations between elements that have still not been addressed. The SOPHIE model for semantically describing choreographies has the potential to mediate between these heterogeneous behaviors and representations. This paper sketches the main elements of a mapping of service-based learning technology components based on OKI OSIDs, and provides some examples of mediation based on SOPHIE.

Keywords. Choreographies, ontologies, learning technology, OKI, OSID, SOPHIE.

1 Introduction

Learning technology is in a continuous process of increased standardization and different consortia push to produce specifications that enable higher levels of interoperability (Devedžić, Jovanović and Gašević, 2007). There are many different types of leaning technology components that are addressed in proposed standards, including digital asset repositories, different kinds of contents, learner model servers, etc. The widespread use of Web 2.0 applications has pushed forward the need for component interoperability, considering easier integration, and eventually capabilities of aggregation into mash-ups (Severance, Hardin and Whyte, 2008).

Service-orientation emphasizes a view of interoperability based on clearly defined interfaces that can be accessed by means of Internet messages (Dagger et al., 2007). In the area of component interoperability, the Open Knowledge Initiative (OKI[1]) has resulted in a comprehensive reference architecture and normalized interface definitions aimed at enhancing the *plugability* of learning technology systems, currently

[1] http://www.okiproject.org/

M.D. Lytras et al. (Eds.): WSKS 2009, LNAI 5736, pp. 128–136, 2009.

considered in broader frameworks as the IMS *Abstract Framework*[2] and ELF[3]. The OKI software architecture applies the concepts of separation, hiding, and layering towards the goal of interoperability and easy integration in order to pull the common elements out of a given problem, leaving the remaining portions more tractable. The OKI initiative has grown and evolved to become an important learning technology integration framework and it is being implemented on top of relevant systems such as Moodle and Sakai. Architectural components in OKI are represented basically by *Open Service Interface Definitions* (OSIDs) specifying common functions that are typical of different learning technology components, e.g. the `Repository` OSID in version 2.0 provides methods as `getAssetsBySearch()` to search resources inside digital content repositories.

Specifications as OKI do not address the heterogeneous semantics of service interfaces and the different behaviors of the components they give access to, as they assume their common interfaces or interconnection "buses" will eventually be used as adaptors. However, the reality of learning technology (both commercial and open source) reveals slow adoption of proposed standards (Jayal. and Shepperd, 2007) and the co-existence of heterogeneous implementations, and even of different proposed standards for the same purpose. This is the case of IMS DRI[4], SQI[5] and OKI interfaces for repositories, which specify similar functionality but with significant syntactic and semantic divergences. Also, proposed standards evolve continuously, causing a degree of transitory heterogeneity.

Further, learning technology scenarios comprising more than a single request/ response pattern have still not been subject to a systematic specification effort, as it has occurred for example in the field of electronic business with models as OAGIS[6]. The `Workflow` OSID provides a means for coordinating and managing workflow based on some predetermined logic, among one or more actors, acting as a specific orchestration service with capabilities similar to BPEL4WS, but this is different to the non-centralized interoperation that takes place in service choreographies.

Rius, Sicilia and García-Barriocanal (2008a, 2008b) have recently approached the automation of scenarios between learning technology components as a problem of defining message interchange patterns that can be described by means of a common ontology. In that direction, the SOPHIE model (Arroyo and Sicilia, 2008) has addressed the syntactic and semantic heterogeneity of service choreographies, providing a core ontology that can be combined with domain ontologies and ontology mappings to mediate between heterogeneous systems collaborating in decoupled interaction. The SOPHIE model is prepared for extension to any domain, and this paper outlines the main elements that are required to use the SOPHIE model in the context of learning technology interactions. The entities and relations in OKI are taken as a base model that reflects a degree of consensus on the service interfaces of several learning technology components. From that base ontology, several kinds of syntactic, behavioral and semantic mismatches between component interfaces can be mediated by a

---

[2] http://www.imsglobal.org/af/
[3] http://www.elframework.org
[4] http://www.imsglobal.org/digitalrepositories/
[5] Simple Query Interface Specification. http://www.prolearn-project.org/lori/, Version 1.0 Beta, 2004-04-13, 2005.
[6] http://www.oagi.org/

SOPHIE-based middleware, including reconciling non-compatible message exchange patterns (Arroyo, Sicilia and López-Cobo, 2008).

The rest of this paper is structured as follows. Section 2 briefly describes the main elements required for the application of the SOPHIE model to the domain of learning technology. Then, Section 3 provides example usage scenarios, illustrating some of the benefits of mediator architectures in learning technology. Finally, conclusions and outlook are provided in Section 4.

2 Specializing SOPHIE for the Domain of Learning Technology

SOPHIE defines a conceptual framework, which in a first cut is divided into syntactic and semantic models. The syntactic model details the syntax of the framework as three different complementary models, namely: structural, behavioral and operational (Arroyo, et al., 2007). It provides the means to establish the compatibility among the structure and behavior of messages, by using a semantic description of message exchange patterns (MEPs), which delineate the skeleton of the message exchanges. For a complete reference of SOPHIE see (Arroyo and Sicilia, 2008). In what follows, the main elements required to adapt SOPHIE to service-base learning technologies are sketched, as a first step to a complete description of such application.

2.1 Domain Ontologies

Domain ontologies are used (or usually reused, as they are often previously available) in SOPHIE to represent specific party information that details the meaning of the concepts used in messages, relating them to their meaning in a particular application domain. In this case, a translation of OKI version 2 interfaces into ontology form provides a base ontology reusing the effort of the project. This ontology provides concepts for each of the OSIDs, e.g. `Repository`, `Agent` or `CourseManagement`.

OKI version 2 describes the messages that can be *received* by a given entity. Then, each of the methods in the interfaces is associated to entity types that provide them via a `providesMethod` property. For example, the `EntityType SchedulingManager` provides the `Method` called `GetScheduleItemsForAgents`, which models the corresponding method in the OSID interface of the same name. Parameters are modeled also through a `ParameterType`. Some of these parameters are in turn defined through OKI interfaces. For example, in this method:

```
getScheduleItemsForAgents(long start, long end,
             Type status, Id[] agents): ScheduleItemIterator
```

The `Id` type is an OSID representing the identification of agents (individuals or groups). In addition to OKI OSID definitions, any other arbitrary domain ontology can be used, be them describing element interfaces or not. The OKI ontology described describes only component types. The actual running instances of those processes are modeled separately and they would be used for runtime processing inside a SOPHIE implementation.

2.2 Choreography Ontologies

For each given integration scenario, choreography ontologies need to be developed to specify the behavior and semantics of each party. The first step is developing for each party to be integrated a choreography ontology importing the SOPHIE base ontology and any needed domain ontology as the OKI OSID one sketched above. Considering the definitions above related to modeling entity types, methods and parameter types, a correspondence with the SOPHIE core ontology can be summarized as follows:

Interface elements	SOPHIE core ontology	Description
`EntityType`	`PartyType`	Entity types are the parties in the message interchange, so entity types are subsumed by SOPHIE parties.
`Method`	`MessageType` (separating request and response)	Method invocation represents a message in which data is transferred. The typical method invocations can be considered as a pair of messages (request/response[7]).
`ParameterType`	`Document-Type/ElementType`	Each parameter can be modeled as a document, or as Elements. Alternatively, a single document (with some default name as `DParam`) can be used to model all the elements in the method invocation.

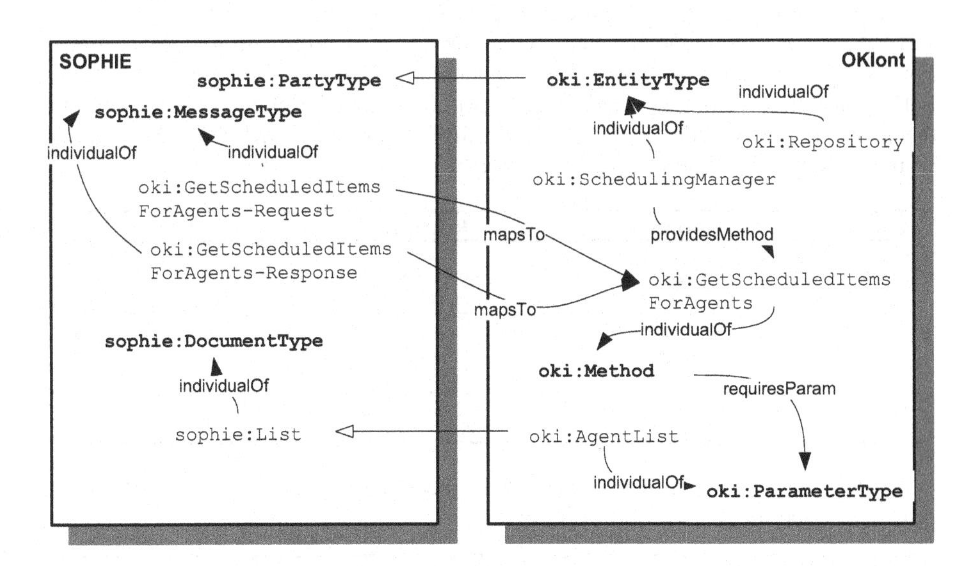

Fig. 1. Example mappings of entity, method and message types between SOPHIE and OKI

Once the basic mapping is done, the resulting ontology importing the domain and the core SOPHIE ontology can be used to further refine the mappings. Figure 1 depicts an example mapping of a single method.

Once the type structure mappings are established, individual components will be modeled as instances of the above, e.g. the class `RepositoryEntity` models concrete repository interfaces (including endpoints for invocation).

Document types and element types in SOPHIE are mapped to parameters in the OKI ontology. In this case, the semantics of the data types require considering each possible type separately, e.g. lists of OSID references are a kind of list with iteration semantics.

3 Example Usage Scenario

This section sketches some example usage scenarios of SOPHIE in the context of learning technology.

3.1 Matching Repository Queries

One of the search interfaces in OSID version 2 repositories is realized by the following operation:

```
getAssetsBySearch(Serializable criteria,
                  org.osid.shared.Type searchType,
                  org.osid.shared.Properties searchProperties)
```

The only requirement of the searching criteria (`searchCriteria`) is to be `Serializable`, i.e., this parameter can contain any type of information as Java objects. For example, it can be a simple string containing keywords or some query terms. Let's consider a learning management system *L* that includes client logic for invoking the above If we consider a stateless repository R implementing SQI, the ontology would include the following basic definitions:

Concept	Elements	Relations
EntityType	SQISource SQITarget	 providesMethod synchronousQuery providesMethod setQueryLanguage
Method	synchronousQuery setQueryLanguage[8] setResultsFormat	 hasParameter queryLanguageId hasParameter resultsFormat
ParameterType	queryLanguageId resultsFormat	

The structural mapping in SOPHIE between the OKI-based client and the SQI-based provider can be summarized in the following:

L (client)	R (provider)
MGetAssetsBySearch.DParam.searchType	MSetQueryLanguage.DParam.queryLanguageId
MGetAssetsBySearch.DParam.criteria	MSynchronousQuery.DParam.queryStmt
MGetAssetsBySearch.DParam.searchType	MSetResultsFormat.DParam.resultsFormat

The `searchType` in OKI is intended to provide all the information regarding the kind of query language and results expected, so that it is mapped to both of the initial configuration tasks, namely, setting the query language and the results format.

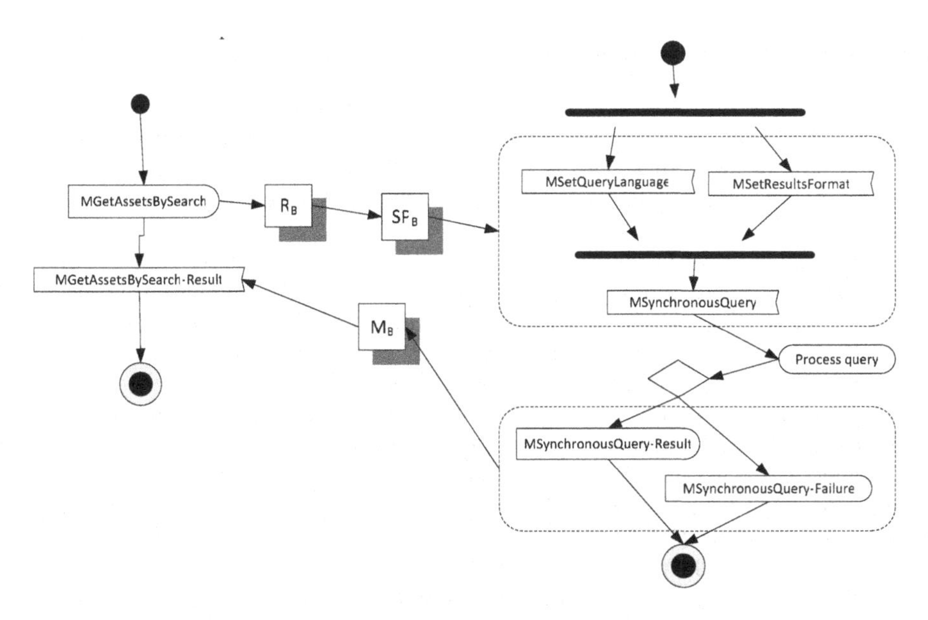

Fig. 2. Example mapping of SQI and OKI message passing

With respect to the mapping of Message Exchange Patterns, the following Figure summarizes the main operational mismatches that have to be resolved.

A *refiner box* R_B is used for the mapping of the input messages as described above. Then, a *split box* SP_B is used to transform the single invocation in L to the three messages required. A finite state machine (FSM) would be used in SOPHIE to define the sequencing of the three resulting messages (not showed in this paper). A *merge box* M_B is then used to combine the two possible responses in one.

3.2 Mapping Classification Systems

The mapping between classification systems can be done by referring to a mapping ontology. Following the above example, the refiner box R_B can be extended to include mappings for the `MGetAssetsBySearch.DParam.criteria` before it is transformed. The ontology mapping capabilities currently defined in SOPHIE are based on term equivalence. For example, mappings between biological ontologies are used in the OBO Foundry[7]. An example of such a mapping between the Gene Ontology (GO) and the PFAM database of protein families is the following:

```
id: GO:0000166
xref: Pfam:PF00133 "tRNA-synt_1"
```

Then, if L is using the GO but *R* uses PFAM for indexing resources, the specification of the *refiner box* R_B can be extended to map both ontologies according to the mappings collection above. It should be noted that this mappings do not necessarily represent logical equivalence, as they rely in external services that serve as black boxes for the mapping process.

[7] http://www.obofoundry.org

3.3 Integrating Heterogeneous Tools

The example described above represent a rather simple mapping adapting disparate interface definitions. In other cases, choreographies come from running workflows at one or several of the parties. This is the case of IMS LD execution engines as CopperCore[8] for which service interfaces for specific purposes have been devised (Vogten et al., 2006).

Combining an IMS LD engine with an OKI `Scheduling` service so that some concrete schedule items (e.g. those concerning new student assignments) result in scheduled items in notifications for new activities in the IMS LD run is an example of an integration of elements serving disparate purposes, combining for example scheduling services with the activity-oriented workflow of IMS LD.

Notifications in IMS LD are mechanisms to trigger new activities, based on an event during the learning process. In this case, the OKI method `createScheduleItem` in the interface `SchedulingManager` will be invoked by a dispatcher (e.g. the LMS represented as an OSID instance of `CourseManagementManager`), and synchronized with an ongoing IMS LD.

This case requires the mapping of the OKI ontology with an integrated architecture as defined in the CopperCore Service Integration (CCSI) described in (Vogten et al., 2006). In this case, a `Dispatcher` will be driving the scenario, and the CCSI method[9] will be used as the main message in the side of the IMS LD engine:

```
CopperCoreAdapter.notify(String userId, int runId,
                                   String notificationXml) : void
```

The following Table summarizes some of the main mappings in the message structure.

OKI SchedulingManager	CopperCore Integration Service
MCreateScheduleItem.DParam.start MCreateScheduleItem.DParam.end	(1)<<send MGetActivitityTree>> (2)<<modify activity tree>> (3)MNotify.DParam.notificationXml.learning-activity-ref
MCreateScheduleItem.DParam.displayName	MNotify.DParam.notificationXml.subject
MCreateScheduleItem.DParam.agents	MNotify.DParam.userId
<<No mapping, known only by the dispatcher>>	MNotify.DParam.runId
MCreateSchedule-Item.DParam.masterIdentifier	<<No mapping, known only by the dispatcher>>

Note that service-specific identifiers as `runId` and `masterIdentifier` are in this kind managed and correlated by the intermediate dispatcher that is driving the integration, serving in this case as an orchestration element.

In the case that `DParam.agents` contains more than one reference, several `MNotify` messages are required, which will be achieved by using an *add box* of a special kind, so that an iteration on the same message structure is implemented.

The mapping of `start` and `end` of the schedule items requires the creation of an additional learning-activity at the IMS LD side, which would require using the

[8] http://coppercore.sourceforge.net/
[9] Taken from the Java source code of the adapter.

message `MGetActivityTree` and then changing in (which can not be done directly through the `CopperCoreAdapter`), and finally sending the notification that the new activity has been added. In this case, a *merge box* and (merging start and end) and then an *add box* (as there are additional messages) can be used to transform the messages.

The above integration bridges two existing service interfaces through an intermediate dispatcher, by doing a small number of data mappings and resolving some basic mismatches between the message exchange patterns that are implicitly prescribed by the design of the adapter interfaces. These can be defined declaratively inside a SOPHIE implementation, avoiding the development of adapters per each pair of possible interface matchs.

4 Conclusions and Outlook

Learning technology comprises a diversity of components that provide heterogeneous architectures and interfaces to external systems. Standardization efforts in that domain are also diverse, resulting in a variety of situations regarding the services exposed by concrete implementations. Further, the conversations between these services have still not been included in proposed standards except for a few cases tied to specific components and initiatives. The interoperability at the level of choreography is thus still unrealized, which calls for devising mediating frameworks that are able to overcome mismatches at the syntactic, behavioral and semantic levels. The SOPHIE model provides a framework for that task, based on a core choreography ontology that can be mapped to the ontologies used by different technology components. We have described here the main elements resulting from the adaptation of SOPHIE to the learning technology domain, using the OKI specifications as a core domain ontology that provides the broader view of learning technology. Examples have been sketched illustrating the kind of mapping and specifications required to bridge some common learning technology services that rely in disparate proposed standards.

Future work should develop a complete mapping of the set of relevant learning technology specifications to the SOPHIE model. Also, further work in an implementation of the adapted SOPHIE model will be addressed. OKI is currently in the process of transitioning to version 3, which includes several important innovations, including a concept of orchestration used to coordinate related OSIDs that might be relevant for the approach used to map learning technology in SOPHIE. Orchestration is also mentioned in the ELF framework.

Acknowledgements

This work has been partially supported by the Spanish Ministry of Industry (Ministerio de Industria, Comercio y Turismo) as part of the activities of project SUMA: e-learning multimodal y adaptativo (TSI-020301-2008-9).

This work reuses outcomes of LUISA, an EU funded project (IST- FP6 - 027149).

References

Arroyo, S., Duke, A., López-Cobo, J.M., Sicilia, M.A.: A model-driven choreography conceptual framework. Computer Standards & Interfaces 29(3), 325–334 (2007)

Arroyo, S., Sicilia, M.A.: SOPHIE: Use case and evaluation. Information and Software Technology 50(12), 1266–1280 (2008)

Arroyo, S., Sicilia, M.A., López-Cobo, J.M.: Patterns of message interchange in decoupled hypermedia systems. Journal of Network and Computer Applications 31(2), 75–92 (2008)

Dagger, D., O'Connor, A., Lawless, S., Walsh, E., Wade, V.P.: Service-Oriented E-Learning Platforms: From Monolithic Systems to Flexible Services. IEEE Internet Computing 11(3), 28–35 (2007)

Devedžić, V., Jovanović, J., Gašević, D.: The pragmatics of current e-learning standards. IEEE Internet Computing 11(3), 16–24 (2007)

Jayal, A., Shepperd, M.J.: An evaluation of e-learning standards. In: Proceedings of the 5th International Conference on E-Governance, Hyderabad, India, December 28-30 (2007)

Rius, A., Sicília, M.A., García-Barriocanal, E.: An ontology to automate learning scenarios? An approach to its knowledge domain. Interdisciplinary Journal of Knowledge and Learning Objects 4, 151–165 (2008)

Rius, A., Sicília, M.A., García-Barriocanal, E.: Towards automated specifications of scenarios in enhanced learning technology. Int. J. of Web-Based Learning and Teaching Technologies 3(1), 68–78 (2008)

Severance, C., Hardin, J., Whyte, A.: The coming functionality mash-up in Personal Learning Environments. Interactive Learning Environments 16(1), 1744–5191 (2008)

Vogten, H., Martens, H., Nadolski, R., Tattersall, C., van Rosmalen, P., Koper, R.: CopperCore Service Integration - Integrating IMS Learning Design and IMS Question and Test Interoperability. In: Proc. of the Sixth IEEE International Conference on Advanced Learning Technologies (ICALT 2006), pp. 378–382 (2006)

SeMatching: Using Semantics to Perform Pair Matching in Mentoring Processes

Ricardo Colomo-Palacios, Juan Miguel Gómez-Berbís,
Ángel García-Crespo, and Myriam Mencke

Computer Science Department, Universidad Carlos III de Madrid
Av. Universidad 30, Leganés, 28911, Madrid, Spain
{ricardo-colomo,juanmiguel.gomez,angel.garcia,
myriam.mencke}@uc3m.es

Abstract. The importance of the human factor in 21st century organizations means that the competent development of professionals has become a key aspect. In this environment, mentoring has emerged as a common and efficient practice for the development of knowledge workers. Following the surge of concepts such as eMentoring, advancements of the Internet and its evolution towards a Semantic Web, such developments present novel opportunities for the improvement of the different characteristics of mentoring. Basing itself on such advancements, this paper presents SeMatching, a semantics-based platform which utilizes different personal and professional information to carry out pair matching of mentors and mentees.

Keywords: Semantic Web, Social Software, Mentoring, e-Mentoring, Pair Matching.

1 Introduction

The dramatic spread of the Internet throughout all levels of society has substantially transformed forms of communication, entertainment and acquisition of knowledge. A constantly increasing number of people encounter responses to their questions on the Internet on a daily basis, and have adapted it as a new form of communication. These novel forms of social behavior have had the result that user preferences and behavior can be easily obtained, which combined with a user's professional data, represent an opportunity for knowledge management initiatives. In order to fully exploit such initiatives, it is necessary to rely on technological tools which enable the organization and exploitation of data for a determined objective.

Regarding the concept of mentoring, since the end of the 1970s, mentoring has become a business practice for staff development which has attained general application. This practice has also been influenced by the boom of the Internet, and the immediate consequence of this has been the appearance of e-Mentoring. The focus of the current paper finds itself at the union of mentoring with the new capabilities of the Internet, and has been named SeMatching. SeMatching has been conceived as a tool based on the Semantic Web (SW) to facilitate pair matching in mentoring. The tool can be used in both traditional mentoring environments, as well as for e-Mentoring.

M.D. Lytras et al. (Eds.): WSKS 2009, LNAI 5736, pp. 137–146, 2009.

2 eMentoring: A New Social Tool, an Ancient Way of Career Development

The concept of mentoring dates back to the earliest stages of human civilization [1]. The origins of the mentoring term can be traced back to the history of Ancient Greece. In Homer's masterpiece, "The Odyssey", Ulysses, king of Ithaca, delegates his house and the education of his son, Telemachus to Mentor, when he leaves for the Troy War (traditionally dated 1193 BC-1183 BC). However, different authors [2] claim that, despite the term having its origin in Ancient Greece, the concept stems from methods and techniques of three Chinese kings, Yao, Shun and Yu between approximately 2333 and 2177 BC. Therefore, despite the importance of the classical Greek etymology, the Chinese origin is earlier than the Greek one.

Apart from its origin, current literature stemming from a number of disciplines (Management, Social Psychology, Sociology...) has provided a significant number of studies about mentoring from the late seventies to the 20th century. As a consequence of the interest raised by the topic and its broad application to business environments, multiple definitions of the term have been coined. Hence [3] have undertaken a re-conceptualization of the term from an in-depth study of existing literature definitions. Mentoring has thus been defined as an improvement process concerning a number of aspects related to a professional career, but also with the global improvement of the individual, which requires a senior advisor and a junior protégé. The relationship established implies benefits for both sides involved. The protégé obviously achieves a remarkable improvement in his professional career, promotion-wise [4], [5], a higher income [4], [6] and more satisfaction and social acceptance in the working environment [7]. On the other hand, mentors benefit from high-speed promotions, reputation and personal satisfaction [8], [9], [5]. Finally, organizations consequently gain a higher motivation from employees, more working stability and the improvement of leadership and development skills in its core [8], [10], [11], being able to rely on employees with more adaptation skills, ready to face a decision making process with more guarantees [12], develop social capital in broader social networks [13] and finally, support knowledge transfer across projects [14].

As discussed in [15], there are three main factors, codenamed as "demographic" that might influence the productivity of the mentoring relationship: firstly, the duration, and secondly the type (formal or informal) and, lastly, the demographic composition of the relationship (in terms of gender and race, mostly, the latter quite variable and more relevant in inter-cultural societies such as in the USA). The first two variables are interconnected, it has been proved that informal mentoring relationships take more time and outperform formal relationships in terms of professional development [7]. Concerning demographic compositions, different features of the binomial structure also affect the final outcome of the process, both sides being of the same race and gender being the most productive relationships [15].

Due to the capabilities of technology of setting up new communication means and paradigms among people, the envisagement of electronic communication as a means for mentoring relationships was immediate. E-Mentoring refers to the process of using electronic means as the primary channel of communication between mentors and protégés [16]. The key distinction between electronic mentoring and traditional mentoring (t-Mentoring) is reflected in the face-to-face time between mentors and protégés. The communication means used by both sides is absolutely different in the two mentoring types. While traditional mentoring uses face-to-face relationships, e-Mentoring, which also harnesses face-to-face relationships, particularly at the beginning of the relationship, is principally based on email, chat, instant messaging and several other Internet applications. In [17], authors point out that e-Mentoring communication can take place synchronously (for example, electronic chat, instant messaging) or asynchronously (for example, email, message boards).

As previously outlined, e-Mentoring is a type of mentoring, based totally or partially on electronic communication. Due to the large amount of electronic communication instruments, a remarkable number of different names for the e-Mentoring concept have been provided. According to Perren [18], e-Mentoring can be seen to encompass a range of terms: computer-mediated mentoring, tele-Mentoring, e-mail mentoring, Internet mentoring, online mentoring and virtual mentoring.

Although Evans and Volery [19] suggested from their survey of experts that e-Mentoring is "second-best" and should only be seen as a supplement to face-to-face mentoring, there are many other studies in which e-Mentoring is considered as a valid vehicle to overcome some of the barriers posed by t-Mentoring. E-Mentoring provides flexibility and easy access, which is highly beneficial to those who may face barriers to being mentored, because of their gender, ethnicity, disability or geographical location [20]. Hamilton and Scandura [16] also analyze the advantages of e-Mentoring compared with t-Mentoring. These advantages are classified into three groups:

- Organizational structure
- Individual & interpersonal factors
- Flexible / alternative work arrangements

Firstly, regarding the organizational structure, e-Mentoring eliminates geographical barriers characteristic of face-to-face interactions, smoothes status differences within the organization and increases the pool of available mentors. Secondly, concerning individual and interpersonal factors, the absence of face-to-face interactions decreases and minimizes gender or ethnical issues impact by increasing the effectiveness of mentors with a lack of social skills. Finally, regarding flexible or alternative work arrangements, the ability and actual capability of performing asynchronous communications implies the elimination of temporal barriers or caveats. In addition, communication is not geographically bound.

Other studies provide further arguments which support the work of Hamilton and Scandura [16]. Hence, following this trend, Warren and Headlam-Wells [21] observed that t-Mentoring, typically operating in large organizations, tends to cast the mentee as a passive recipient of structured formal provision of mentoring. In contrast, the use of the Web as a communication means provides improved access for the mentor and the mentee, and as stated by [16], creates a larger pool of potential mentors and mentees [22]. In addition, as with other e-learning programmes, a major advantage of an e-Mentoring system is its cost effectiveness [23], [24]. There are high start-up costs, but once established, the operational costs are relatively low. Costs related to travel or time away from the job and costs of updating learning resources can be reduced. Lastly, a record of the ''discussion'' usually exists for later reflection and learning [24]. To summarize, it is possible to state, as discussed by Clutterbuck and Cox [25], that e-Mentoring will be able to overcome many of the problems characteristic of t-Mentoring.

Nevertheless, not all e-Mentoring features are win-win for the mentor-mentee relationship. Eby, McManus, Simon and Russell [26] conducted empirical work in this area that further examined the dark side of mentoring, and developed a useful taxonomy of negative experiences from the mentee perspective using qualitative data. These authors make important distinctions between what is considered negative and how that might be different from the perspective of the mentee and the mentor. A later study conducted by Ensher, Heun and Blanchard [27] uses the findings of [26] to identify five major challenges in e-Mentoring: (1) likelihood of miscommunication, (2) slower development of the relationship online than in face-to-face mentoring, (3) the relationship requires competency in written communication and technical skills, (4) computer malfunctions, and (5) issues of privacy and confidentiality.

Pair matching criteria is one of the most important issues in both e-Mentoring and t-Mentoring. This concept is based on the assignment of a mentor to a mentee depending of a number of parameters defined with the purpose of harnessing the mentorship. Such parameters might include values, gender coincidences, related professional experience, and so on. Despite it being a key issue of concern for mentoring, crucial according to [24], there is little research concerning the matching of pairs [23]. Cohen and Light [28] argue that matching solely on the basis of mentees needs and mentors' skills may not be enough to ensure successful matches, and suggest that personality factors may also be significant. Indeed, successful mentoring relationships are often reported as those where mentees felt they shared their mentors' personal values [23].

With the exception of a small number of mentoring programmes which use a formal matching system, most tend to use a 'hand-sift' method, whereby mentees are matched with a mentor who suits their needs [29]. However, if a number of people of remarkable size is faced, this type of selection can unfortunately not be applied. Therefore, [29] suggest a set of eleven criteria that allow the automation of the process, together with application criteria:

Table 1. Matching criteria proposed by [29] for automatic pair matching

Criterion	Explanation
Age	Mentee matched with older mentor.
Number of years work experience	Mentee matched with mentor with more work experience.
Level of qualification	Mentee matched with mentor with higher qualification level.
Marital status	Mentee matched with mentor with same marital status.
Children	Mentee matched with mentor in a similar situation to themselves (having/had children).
Dependent care	Mentee matched with mentor in a similar situation.
Life/career history	Identify similarities in life/career experiences, e.g. having experienced barriers to progression.
Personal skills	Mentee matched with mentor who could help them develop the personal skills they need to improve.
Professional skills	Mentee matched with mentor who could help them develop the professional skills they need to improve.
Vocational sector	Mentee matched with mentor who worked/had worked in a similar occupational area.
Personal values	Mentee matched with mentor who shared similar core values.

3 SeMatching: New Tool, Ancient Needs

In today's organizations, the capabilities of Information Communication Technology (ICT) have transformed not only forms of communication, but also the forms of personal development in professional environments. In particular, the arrival of the SW represents a revolution in the forms of accessing and storage of information. The SW term was coined by [30] to describe the evolution from a document-based web towards a new paradigm that includes data and information for computers to manipulate. The SW provides a complementary vision as a knowledge management environment [31] that, in many cases has expanded and replaced previous knowledge management archetypes [32]. In this new scenario, SW technology has been identified as a factor which can be exploited in the environment of mentoring. According to [33], "Mentors can be semantically selected by matching profiles".

The application of semantics in relation to the management of human capital in organizations is not a new concept. One of the research areas which holds the longest tradition relates to the analysis of competencies. According to McClelland [34], competence concerns the relation between humans and work tasks: rather than knowledge and skills themselves, competence involves the knowledge and skills required to perform a specific job or task in an efficient way. More recently, HR-XML defined competency as a specific, identifiable, definable, and measurable knowledge, skill, ability and/or other deployment-related characteristic (for example, attitude, behavior, physical ability) which a human resource may possess and which is necessary for, or fundamental to, the performance of an activity within a specific business context. Various initiatives which propose the use of this new technology have been seen as a result of the popularity of the competency concept and the growth of the SW. The technology has been applied for the training of work teams [35], filling competency gaps in organizations [36], analyzing competency gaps [37], knowledge management for software projects [38], knowledge sharing and reuse [39], assist the learning process [40] or assist work assignment [41] to cite some of the most recent initiatives. Zülch and Becker [42] expressed the need for a fixed terminology of competence-related concepts, and Schmidt and Kunzmann [43] pointed out that Ontology-based approaches are the solution for the crucial trade-off in competency modeling needs. Taking those two conclusions into account, all of the works mentioned use ontologies as a tool. Thus ontologies are the appropriate formalisms to represent competency concepts. Similarly, in the current work, ontologies can be used to represent pair matching concepts. The theory which supports the use of ontologies is a formal theory within which not only definitions but also a supporting framework of axioms is included [44].

However, neither the competency ontologies used, nor some of the standardization efforts on modeling competencies can cover all of the characteristics which pair matching spans. An adaptation of the available competency ontologies is required. The LUISA Project (http://luisa.atosorigin.es) addresses the development of a reference semantic architecture for the major challenges in the search, interchange and delivery of learning materials. A deliverable of this project is the Generic Competence Ontology (GCO).Using this specification as a base which is adapted for mentoring pair matching issues, many of the criterions can be directly included just by using concepts from the ontology (person, competency, attitude, skill, job position...), while others must be updated (Vocational sector, Personal values…). As a result of this, a modified version of GCO adapted to mentoring scenario is used here. The figure below demonstrates the architecture of Sematching:

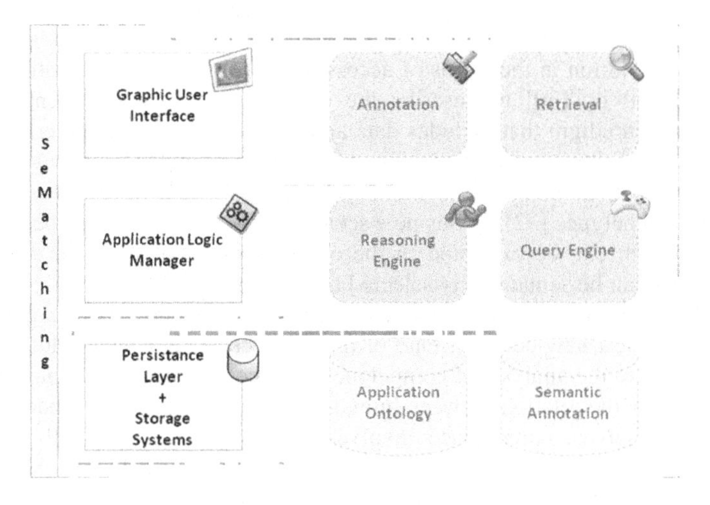

Fig. 1. Sematching Architecture

The current section details the different components of the architecture without specifically focusing on the software layer where they belong. This is not necessary, since the three functionalities are well defined and have a commonly shared and used pattern:

- Annotation GUI: This component interacts with the user by providing a set of graphical elements to annotate the resources by means of semantic annotations based on the ontology used.
- Retrieval GUI: It offers a semantic annotation retrieval functionality for the user, based on both the Reasoning Engine and the Query Engine. In the former, retrieval is envisaged as location of a subset of concepts by means of Description Logics subsumption. In the latter, the retrieval is provided by SPARQL definitions to find, manage and query RDF triples following a particular criteria.
- Reasoning Engine: This component derives facts from a knowledge base, reasoning about the information with the ultimate purpose of formulating new conclusions. In the SeMatching framework, it consists of an OWL Description Logics based reasoner, such as the Renamed ABox and Concept Expression Reasoner (RACER). It uses subsumption to find sets and subsets of annotations based on logical constraints.
- Query Engine: The Query Engine component uses the SPARQL RDF query language to make queries into the storage systems of the back end layer. The semantics of the query are defined not by a precise rendering of a formal syntax, but by an interpretation of the most suitable results of the query. This is because SeMatching stores mostly RDF triples or OWL DL ontologies, which also present an RDF syntax.

- Semantic Annotation and Application Ontology Repositories: These two components are semantic data store systems that enable ontology persistence, querying performed by the Business Logic layer components and offer a higher abstraction layer to enable fast storage and retrieval of large amounts of OWL DL ontologies, together with their RDF syntax. This ensures that the architecture has a small footprint effect and a lightweight approach. An example of such systems could be the OpenRDF Sesame RDF Storage system, or the Yet Another RDF Storage System (YARS), which deal with data and legacy integration.

The SeMatching architecture is a self-contained, loosely coupled open architecture which allows using a wide range of software technologies for its implementation. The added value of SeMatching will be, taking into account its SW orientation, the inclusion of really "soft" aspects like values in the ontology model. Many of the concepts that are related to pair matching (soft and hard skills, career history, qualification...) are used in other semantic efforts e.g. [38], [41], however the inclusion of values (which according to some authors are also present in competences, but as a part of it) in this scenario is new. In the other hand, in comparision with traditional pair matching process, the use of ontologies can provide a common vocabulary and the undeniable matching that is associated with this approach.

Many organizations are facing vitualization, outsourcing and offshoring. In this new scenario, mentoring, and, in particular, eMentoring can be a way to preserve organizational culture in complex work layouts. Thus, SeMatching can be seen as enabling technology for these organizations in which both knowledge and workers are scattered. Those organizations, like any others, need to preserve their culture by performing a good mentoring process based on new pair matching processes.

4 Conclusions and Future Work

Taking into account the progressive virtualization of organizations, SeMatching represents a latent opportunity for organizations interested in the development of their intellectual capital. SeMatching represents an innovative and technologically advanced initiative, which takes advantage of the capacities that the SW provides.

The current work proposes two types of initiatives which should be explored in future research. In the first place, the integration of the functionalities provided by Web 2.0 in the development of the profiles of the mentors and mentees. Populating the ontologies based on available social information represents a research opportunity which has been previously developed by the authors, and this data is considered here as an invaluable source for the creation of user preference profiles. Furthermore, these preference profiles may be very useful data sets for the addition of personal preferences in the pair matching process. In the second place, it is aimed to test the platform developed empirically by evaluating the capabilities of SeMatching from a qualitative viewpoint. The researchers envisage testing the platform using a set of tests carried out by experts who validate the platform from a qualitative perspective. From a quantitative perspective, authors are working in the prototype of SeMatching in order to perform a pilot study in an Spanish university.

References

1. Kammeyer-Mueller, J.D., Judge, T.A.: A quantitative review of mentoring research: Test of a model. Journal of Vocational Behavior 72, 269–283 (2008)
2. Huang, C.A., Lynch, J.: Mentoring: The Tao of Giving and Receiving Wisdom. Harper Collins, San Francisco (1995)
3. Friday, E., Friday, S.S., Green, A.L.: A reconceptualization of mentoring and sponsoring. Management Decision 42(5), 628–644 (2004)
4. Dreher, G.F., Ash, R.A.: A comparative study of mentoring among men and women in managerial, professional, and technological positions. Journal of Applied Psychology 75, 539–546 (1990)
5. Scandura, T., Tejeda, M., Werther, W., Lankau, M.: Perspectives on Mentoring. Leadership & Organization Development Journal 17(3), 50–58 (1996)
6. Whitely, W.T., Dougherty, T.W., Dreher, G.F.: Relationship of career mentoring and socioeconomic origin to managers' and professionals' early career progress. Academy of Management Journal 34, 331–351 (1991)
7. Chao, G.T., Walz, P.M., Gardner, P.D.: Formal and informal mentorships: A comparison on mentoring functions and contrast with nonmentored counterparts. Personnel Psychology 45, 619–636 (1992)
8. Hunt, D.M., Michael, C.: Mentorship: a career training and development tool. Academy of Management Review 8(3), 475–485 (1983)
9. Zey, M.G.: The Mentor Connection. Irwin, Homewood (1984)
10. Viator, R.E., Scandura, T.A.: A study of mentor-protégé relationships in large public accounting firms. Accounting Horizon 5(3), 20–30 (1991)
11. Levesque, L.L., O'Neill, R.M., Nelson, T., Dumas, C.: Sex differences in the perceived importance of mentoring functions. Career Development International 10(6), 429–443 (2005)
12. Ragins, B., Scandura, T.: Burden or blessing? Expected costs and benefits of being a mentor. Journal of Organizational Behavior 20(4), 493–510 (1999)
13. Hezlett, S.A., Gibson, S.K.: Linking Mentoring and Social Capital: Implications for Career and Organization Development. Advances in Developing Human Resources 9(3), 384–411 (2007)
14. Landaeta, R.E., Kotnour, T.G.: Formal mentoring: a human resource management practice that supports knowledge transfer across projects. International Journal of Learning and Intellectual Capital 5(3/4), 455–475 (2008)
15. Nielson, T.R., Eisenbach, R.J.: Not All Relationships are Created Equal: Critical Factors of High-Quality Mentoring Relationships. The International Journal of Mentoring and Coaching 1(1) (2003)
16. Hamilton, B.A., Scandura, T.A.: E-Mentoring: implications for organizational learning and development in a wired world. Organizational Dynamics 31, 388–402 (2003)
17. Smith-Jentsch, K.A., Scielzo, S.A., Yarbrough, C.S., Rosopa, P.J.: A comparison of face-to-face and electronic peer-mentoring: Interactions with mentor gender. Journal of Vocational Behavior 72, 193–206 (2008)
18. Perren, L.: The role of e-Mentoring in entrepreneurial education & support: a meta-review of academic literature. Education & Training 45(8/9), 517–525 (2003)
19. Evans, D., Volery, T.: Online business development services for entrepreneurs: an exploratory study. Entrepreneurship and Regional Development 13(4), 333–350 (2001)
20. Bierema, L., Hill, J.: Virtual mentoring and HRD. Advances in Developing Human Resources 7(4), 556–568 (2005)

21. Warren, L., Headlam-Wells, J.: Mentoring women entrepreneurs: A better approach. Organisations and People 9(2), 11–17 (2002)
22. Packard, B.W.: Web-based mentoring: Challenges traditional models to increase women's access. Mentoring and Tutoring 11(1), 53–65 (2003)
23. Headlam-Wells, J., Gosland, J., Craig, L.: There's magic in the web: e-Mentoring for women's career development. Career Development International 10(6/7), 444–459 (2005)
24. Hunt, K.: E-Mentoring: solving the issue of mentoring across distances. Development and learning in organizations 19(5), 7–10 (2005)
25. Clutterbuck, D., Cox, T.: Mentoring by wire. Training Journal 35–39 (November 2005)
26. Eby, L.T., McManus, S.E., Simon, S.A., Russell, J.E.: The Protégés perspective regarding negative mentoring experiences: the development of a taxonomy. Journal of Vocational Behavior 57, 1–21 (2000)
27. Ensher, E.A., Heun, C., Blanchard, A.: Online mentoring and computer-mediated communication: New directions in research. Journal of Vocational Behavior 63(2), 264–288 (2003)
28. Cohen, K.J., Light, J.C.: Use of electronic communication to develop mentor-protégé relationships between adolescent and adult AAC users: pilot study. Augmentative and Alternative Communication 16, 227–238 (2000)
29. Headlam-Wells, J., Gosland, J., Craig, L.: Beyond the organisation: The design and management of E-Mentoring systems. International Journal of Information Management 26, 372–385 (2006)
30. Berners-Lee, T., Hendler, J., Lassila, O.: The Semantic Web. Scientific American 284(5), 35–40 (2001)
31. Warren, P.: Knowledge Management and the Semantic Web: From Scenario to Technology. IEEE Intelligent Systems 21(1), 53–59 (2006)
32. Davies, J., Lytras, M., Sheth, A.P.: Semantic-Web-Based Knowledge Management. IEEE Internet Computing 11(5), 14–16 (2007)
33. Sicilia, M.A., Lytras, M.D.: The semantic learning organization. The learning organization 12(5), 402–410 (2005)
34. McClelland, D.C.: Testing for competence rather than for 'intelligence'. American Psychologist 28, 1–14 (1973)
35. Gómez-Berbís, J.M., Colomo-Palacios, R., García Crespo, A., Ruiz-Mezcua, B.: ProLink: A Semantics-based Social Network for Software Project. International Journal of Information Technology and Management 7(4), 392–404 (2008)
36. Colomo-Palacios, R., Ruano-Mayoral, M., Gómez-Berbís, J.M., García Crespo, A.: Semantic Competence Pull: A Semantics-Based Architecture for Filling Competency Gaps in Organizations. In: García, R. (ed.) Semantic Web for Business: Cases and Applications. IGI Global (2008)
37. De Coi, J.L., Herder, E., Koesling, A., Lofi, C., Olmedilla, D., Papapetrou, O., Sibershi, W.: A model for competence gap analysis. In: Proceedings of 3rd International Conference on Web Information Systems and Technologies (WEBIST), Barcelona, Spain (2007)
38. Colomo-Palacios, R., Gómez-Berbís, J.M., García-Crespo, A., Puebla Sánchez, I.: Social Global Repository: using semantics and social web in software projects. International Journal of Knowledge and Learning 4(5), 452–464 (2008)
39. Lanzenberger, M., Sampson, J., Rester, M., Naudet, Y., Latour, T.: Visual ontology alignment for knowledge sharing and reuse. Journal of Knowledge Management 12(6), 102–120 (2008)

40. Naeve, A., Sicilia, M.A., Lytras, M.D.: Learning processes and processing learning: from organizational needs to learning designs. Journal of Knowledge Management 12(6), 5–14 (2008)
41. Macris, A., Papadimitriou, E., Vassilacopoulos, G.: An ontology-based competency model for workflow activity assignment policies. Journal of Knowledge Management 12(6), 72–88 (2008)
42. Zülch, G., Becker, M.: Computer-supported competence management: Evolution of industrial processes as life cycles of organizations. Computers in Industry 58(2), 143–150 (2007)
43. Schmidt, A., Kunzmann, C.: Sustainable Competency-Oriented Human Resource Development with Ontology-Based Competency Catalogs. In: Cunningham, M., Cunningham, P. (eds.) eChallenges (2007)
44. Smith, B.: Ontology. An Introduction. In: Floridi, L. (ed.) Blackwell Guide to the Philosophy of Computing and Information, pp. 155–166. Blackwell, Oxford (2003)

A Mediating Algorithm for Multicriteria Collaborative Knowledge Classification

Imène Brigui-Chtioui[1] and Inès Saad[2,3]

[1] GRIISG – Institut Supérieur de Gestion, 147 Avenue Victor Hugo 75116 Paris, France
imene.brigui-chtioui@isg.fr
[2] MIS, University of Picardie Jules Vernes, 33 Rue Saint Leu, 80039 Amiens, France
[3] Amiens School of Management, 18 Place Saint-Michel, 80000 Amiens, France
ines.saad@u-picardie.fr

Abstract. In this paper we present a mediating algorithm to provide a conflict resolution in the knowledge management system K-DSS which is a decision support system for identifying crucial knowledge. In the knowledge base of K-DSS, inconstancies can appear because of conflicts between decision makers evolved in the process of identification of crucial knowledge. Our objective is to solve conflicts by the mean of argumentative agents representing decision makers. Multiagent theory is suitable to our context due to the autonomous character of agents able to represent faithfully each human actor evolved in the process of identification of crucial knowledge.

Keywords: Knowledge Management, multiagent systems, conflict resolution, Knowledge classification.

1 Introduction

Capitalizing on all the company's knowledge requires an important human and financial investments. To optimize the capitalizing operation, one should focalize on only the so called "crucial knowledge", that is, the most valuable knowledge. This permits particularly to save time and money.

In practice, decision makers use tacit and explicit knowledge available in various forms in the organization to select, from a set of options, the alternative(s) that better response(s) to the organization objectives. The main objective of capitalizing is to extract tacit knowledge. Thus, companies should invest in engineering methods and tools [5] in order to preserve the knowledge.

In our case study, the goal is to propose a method to identify and qualify crucial knowledge in order to justify a situation where knowledge capitalization is advisable. The method is supported by a K-DSS [10][11] which is a decision support system. The aim of this paper is to improve K-DSS. Mainly, to cope with inconsistency in decision rules we shall use an argumentative approach.

In this work, we aim to solve conflicts in the crucial knowledge classification. During this process, decision makers can have contradictory opinions that lead to inconsistencies in the shared knowledge base.

M.D. Lytras et al. (Eds.): WSKS 2009, LNAI 5736, pp. 147–155, 2009.

This article is structured in the following way. Section 2 gives an overview on the related works. Section 3 presents the knowledge management system K-DSS, followed by the multiagent system details in section 4. The experimentations and results are presented in section 5. Section 6 summarizes our contribution.

2 Related Works

Only few theoretical works are available in literature to identify crucial knowledge. We think that the method proposed by [8] enables to study the area and to clarify the needs in knowledge required to deal with pertinent problems through the modeling and analysis of sensitive processes in the company. This approach involves all the actors participating in the area of the study.

In addition, the method proposed by Tseng and Huang [14] propose to compute the average score of each attribute of the knowledge as a function of the evaluations provided by each analyst. Then, the analyst evaluates the importance of knowledge in respect to each problem. Finally, the average global is computed for each analyst. One limitation of this method is that the scales used are quantitative. However, due to the imprecise nature of the knowledge, qualitative scales are preferred.

In a shared decision making, conflicts can appear mainly because of opinion divergence. Many theoretical works treat the issue of argumentation advantages in several domains like negotiation [9] and conflict resolution [13].

Many works treat the problem of conflict resolution by having recourse to argumentation [6][12] and especially in the field of knowledge management [3][4].

3 The Knowledge Decision Support System: K-DSS

Figure 1 describes the functional architecture of K-DSS. Two phases may be distinguished in this figure. The first phase is relative to the construction of the preference model. The preference model is represented in terms of decision rules. The second phase concerns the classification of potential crucial knowledge by using the rules collectively identified by all the decision makers in the first phase. In this paper, attention is especially devoted to present the phase 1 "Construction of the preference model". This phase consists in identifying, from the ones proposed, an algorithm for computing the contribution degrees. The selection is collectively established by all the decision makers with the help of the analyst. Whatever the selected algorithm, it uses the matrices Knowledge-Process (K-P), Process-pRoject (P-R) and pRoject-Objective (R-O) extracted from the database to compute the contribution degree of each piece of knowledge into each objective.

Once these matrices are generated, the contribution degrees are first stored in a decision table and then introduced in the database. The decision table contains also the evaluation of the knowledge concerning the vulnerability and use duration criteria extracted from the database. These evaluations are collectively defined and introduced by the analyst in the database. The analyst should introduce in the decision table, and for each decision maker k, the decisions concerning the assignment of knowledge of reference into the classes Cl1 and Cl2. Two decision classes have been defined: Cl1: "not crucial knowledge" and Cl2: "crucial knowledge".

The decision table contains, in addition to the columns relative to vulnerability and those relative to contribution degree and use duration criteria, as many columns as decision makers. It consists in determining, with the help of each decision-maker, assignments of a set of knowledge items "Reference crucial knowledge" in the following decision classes: Cl1 and Cl2.

Once the decision table is generated, it will be used as the input of the induction algorithm selected by the decision makers (DOMLEM or Explore) [7]. This algorithm permits to generate the list of the initial decision rules for each decision maker *d*. During this step, opinion conflicts can appear concerning knowledge assignments. These conflicts lead to inconsistencies in the shared knowledge base.

Then the next step consists in modifying sample assignments or evaluations with the concerned decision-maker, when inconsistencies are detected in the decision rules base. Finally, the last step consists in determining decision rules that are collectively accepted that will be used latter by JESS for the classification phase.

In the first version of our system K-DSS [10], the procedure used to solve the inconsistency in the decision rules is performed by every decision maker evolved in the process of the evaluation of the knowledge assisted by the analyst.

In the following section, we present the multiagent system aiming to reach agreements on the knowledge assignment by the mean of argumentation.

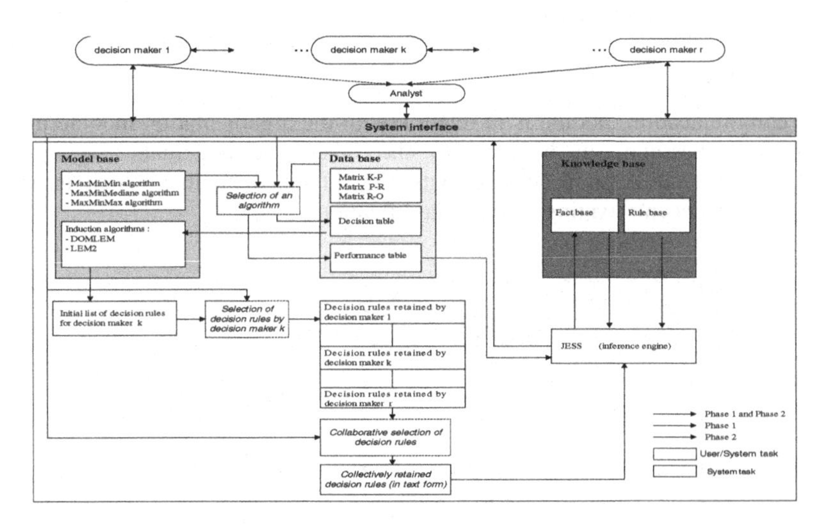

Fig. 1. Functional architecture of K-DSS [10]

4 The Multiagent System

Multiagent theory is particularly suitable to our context due to the autonomous character of agents able to represent faithfully each human actor taking part in the classification process. In the other hand, an automated process is appropriate to our context because of the large amount of knowledge to analyze, the large number of decision maker involved in the assignments of knowledge and hard delay constraints.

Our multiagent system [2] is made up of a set of autonomous behavior-based agents that act on behalf of their beliefs (Figure 2). The organization contains two types of agents:

a. **the mediator agent *m*** that is responsible for the knowledge base management. Its goal is to resolve conflicts in order to have a consistent knowledge base. Its role consists on detecting conflicts, putting in touch decision maker agents source of the conflict and prompting them to reach an agreement. If an agreement can not take place, the mediator agent is destined for making an objective decision due to its meta-rules. Note that only the mediator agent is entitled to modify the collective knowledge base. The meta-rule notion will be detailed in section 4.4.

b. **the decision maker agents a_i** that are responsible for the knowledge classification on the basis of its beliefs. Each decision maker agent represents a human decision maker and manages an individual rule base allowing him the classification and the argumentation.

Agents involved in the knowledge classification process have the same goal: sharing a consistent knowledge base. Decision maker agents are made up of 3 interdependent modules:

- *Communication module* allowing the message exchange between agents;
- *Inference module* responsible for inferring rules from the individual rule base and deducing classification for each knowledge;
- An *argumentation module* which is able to construct arguments that enhance a given classification.

The communication module is in relation with the argumentation module in order to construct messages to be sent to the other decision maker agents. The argumentation module is in relation with the inference module which is able to generate arguments motivating a given classification.

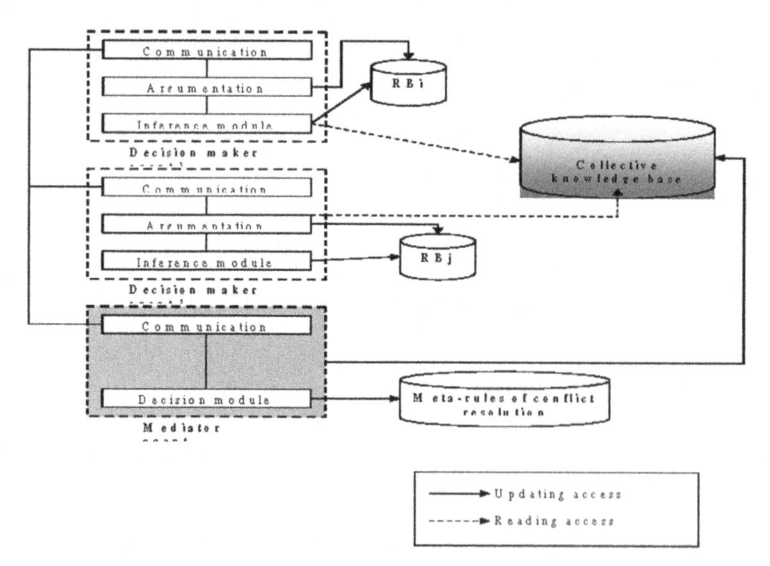

Fig. 2. Multiagent architecture. This shows a figure consisting of modules communicating in order to classify shared knowledge.

4.1 Preliminaries

- We denote by a_1, a_2, ... a_n decision maker agents involved in the knowledge classification process;
- We denote by k_1, k_2,...k_n knowledge to classify;
- We denote by K, the collective knowledge base;
- We denote by α, β, γ,...knowledge classification;
- We denote by $\prec$, *the preference relation between classifications*

Definition 1. Classification. A classification α is represented by a triplet $< a_i , k, c >$ where a_i represents the decision maker agent, k denotes the classified knowledge and c the decision class of classification.

Definition 2. Conflict. A conflict is detected iif: $\exists\ \alpha <a_i , k, c >$ *and* $\beta < a_j, k, c' >$ / c $\neq$ c'.

Definition 3. Consistency. It exists Consistency iif $\forall\ \alpha <a_i, k, c >$ *and* $\beta < a_j, k, c' >$, *+Conflict.*

Definition 4 Argument. An argument is represented by a pair $< \alpha, R_\alpha >$ where α denotes a classification and R_α the set of rules establishing α.

4.2 Communication Protocol

The communication protocol [1] specifies actions agents are authorized to take during the classification process. The argumentation process is initiated by the mediator agent if a conflict is detected (cf. Definition 2). A *call_for_arguments* message is sent by the mediator agent to the two agents in conflict which are asked to reach an agreement.

After receiving this call, agents start the argumentation process. This process can be viewed as an exchange of *justify* messages finished by an *accept* or a *reject* message.

An *accept* message stays that an agreement is reached. On the other hand, a *reject* message implies that the mediator agent should come to an objective decision based on its meta-rules. The mediator agent algorithm is detailed in the next section.

4.3 Mediator Agent Algorithm

The mediator agent m is responsible for solving conflicts between classifications on the basis of its meta-rules. Figure 3 shows the state graph of the mediator agent. When a conflict is detected, the mediator agent sends a *Call_for_arguments* message to the concerned agents and stays idle. At the end of the argumentative process, decision maker agents inform mediator agent about their decision. If the mediator agent receives an *Accept*, the process is complete and the classification appointed is established. On the other hand, if a *Reject* message is received, the mediator agent should make a decision based on its meta-rules.

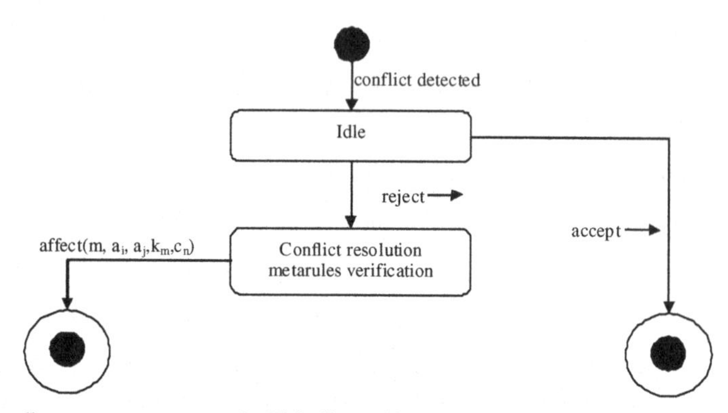

Fig. 3. The mediator agent state graph. This figure illustrates the mediator agent algorithm for the decision makers conflict resolution.

4.4 Meta-rules

Table 1 represents the knowledge classification criteria and their associated rules. A meta-rule consists on giving a weight ω_i to each criterion *i*. We choose the additive linear function as aggregation model. A classification α is then evaluated by a utility function U_α:

$$U_\alpha = \sum \omega_i U_i(x_\alpha{}^i)$$

U_i the scoring function that reduces all criteria on the same scale [0, 100].
x_α^i the value of the classification α on the criterion *i*.

Table 1. Classification criteria and associated rules

Criterion	*Domain*	*Description*	*Associated rule*
$NAg(\alpha)$	[0,N]	The number of agents establishing α	If $NAg(\alpha) < NAg(\beta)$ Then $\alpha \prec \beta$
$\gamma(A(\alpha))$	[0,100]	The approximation quality of the agent establishing α.	If $\gamma(A(\alpha)) < \gamma(A(\beta))$ Then $\alpha \prec \beta$
$\vert R_\alpha \vert$	[0,∞]	The number of rules conducting to establish α	If $R_\alpha < R_\beta$ Then $\alpha \prec \beta$
$\partial(R_\alpha)$	[0,100]	The average of the rules strength in R_α.	If $\partial(R_\alpha) < \partial(R_\beta)$ Then $\alpha \prec \beta$

4.5 Illustrative Example

In order to show the use of the proposed approach, this section presents an illustrative example. This example concerns 3 agents that negotiate in order to classify 3 knowledge in the shared knowledge base. The example data are detailed in Table 2.

Table 2. Knowledge classification scenario (3 agents and 3 knowledge)

Agent	Approximation quality	Agent classifications	$\vert R\alpha\vert$	$\partial(R_\alpha)$
a_1	75	$\alpha_1 <a_1, k_1, c_1>$ $\alpha_2 <a_1, k_2, c_1>$ $\alpha_3 <a_1, k_3, c_2>$	$\vert R\ \alpha 1\vert=2$ $\vert R\ \alpha 2\vert=5$ $\vert R\ \alpha 3\vert=8$	$\partial(R_{\alpha 1})=10$ $\partial(R_{\alpha 2})=20$ $\partial(R_{\alpha 3})=40$
a_2	60	$\beta_1 <a_2, k_1, c_1>$ $\beta_2 <a_2, k_2, c_2>$ $\beta_3 <a_2, k_3, c_2>$	$\vert R\ \beta 1\vert=5$ $\vert R\ \beta 2\vert=3$ $\vert R\ \beta 3\vert=11$	$\partial(R_{\beta 1})=11$ $\partial(R_{\beta 2})=10$ $\partial(R_{\beta 3})=15$
a_3	35	$\gamma_1 <a_3, k_1, c_1>$ $\gamma_2 <a_3, k_2, c_2>$ $\gamma_3 <a_3, k_3, c_2>$	$\vert R\ \gamma 1\vert=2$ $\vert R\ \gamma 2\vert=12$ $\vert R\ \gamma 3\vert=4$	$\partial(R_{\gamma 1})=12$ $\partial(R_{\gamma 2})=3$ $\partial(R_{\gamma 3})=25$

The conflict that appears between decision makers concerns K_2 which is classified in Cl1 by a_1 and in Cl2 by Ag_2 and Ag_3. After a dialogue between the concerned agents and a reject message, the mediator agent applies the meta-rule presented in Table 3 in order to make an objective multicriteria decision about the classification of K_2.

Table 3. The mediator agent meta-rule

	Approx. Quality	$NAg\ (\alpha)$	$/R_{\alpha l}$	$\partial\ (R_\alpha)$
Weight	0.3	0.1	0	0.6

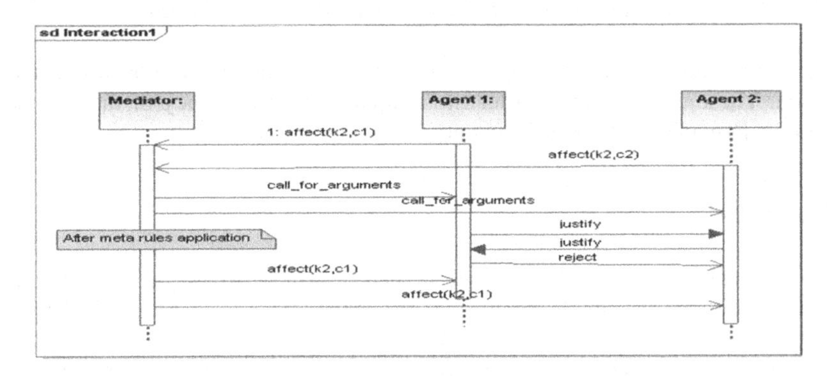

Fig. 4. Sequence diagram. This diagram shows a communication between the mediator agent and two decision makers about the K2 classification. As shown in this example, agent1 wants to classify K2 in Cl1 and agent2 wants to classify the same knowledge in Cl2. After an exchange of *justify* messages, a deal can't be reached. Then, by applying his meta-rule (table4), the mediator agent classifies K2 in Cl1 and informs the concerned agents by sending an *affect* message.

In order to compare the agents' classifications, we apply the multicriteria decision model on the basis of the weights assigned by the mediator agent as follows:

$U(\alpha 2) = (0.3*75) + (0.1*1) + (0*5) + (0.6*20) = 34.6$

$U(\beta 2) = (0.3*60) + (0.1*2) + (0*3) + (0.6*10) = 24.2$

$U(\gamma 2) = (0.3*35) + (0.1*2) + (0*12) + (0.6*3) = 12.5$

$U(\alpha 2) > U(\beta 2) > U(\gamma 2)$ ➔ $\alpha 2$ is the winner classification and k_2 is classified in Cl1.

The scenario of K2 classification is illustrated in the sequence diagram of the Figure 4.

5 Conclusion

This paper details the issue of identification of crucial knowledge and proposes a mediating algorithm in order to provide a conflict resolution in the context of crucial knowledge classification. The aim of the proposed multiagent system is to manage conflicts between decision makers by argumentation and to lead to a consistent shared knowledge base.

We propose in this work the detail of the conducted experiments in order to appreciate the advantages of using an argumentative approach in our study context. Experiments show two main points. First, the argumentative approach leads to a decreasing number of conflicts between decision maker agents. Second, an argumentative approach is less sensitive to an increasing number of reference knowledge.

Future work will include *attacks* relations between arguments in order to improve the argumentation framework. Take as an example 2 arguments *Arg1 : P1 -> C1* and *Arg2 : P2 -> C2*. For example, we maintain that Arg2 *attacks* Arg1 *iff* (C2 -> C1) ∨ (C2 -> P1).

Finally, we aim to experiment the proposed argumentative approach considering real data. These experiments can lead to compare the consistency of the knowledge base in the human and in the argumentative multiagent context.

References

1. Brigui-Chtioui, I., Saad, I.: Solving conflicts in crucial knowledge classification: a multiagent approach. In: Proceedings of the 2007 IADIS International Conference on Applied Computing, Salamanca, pp. 282–288 (2007)
2. Brigui-Chtioui, I., Saad, I.: Solving Conflicts in Knowledge Management System: A Multiagent Approach. In: Proceedings of the 12th International Symposium on the Management of Industrial and Corporate Knowledge (ISMICK 2008), Nitero, Rio de Janeiro, 3-5 November 2008, pp. 150–161 (2008)
3. Chesñevar, C.A., Maguitman, P., Simari, G.: Argument-Based User Support Systems using Defeasible Logic Programming. In: Maglogiannis, I., Karpouzis, K., Bramer, M. (eds.) Artificial Intelligence Applications & Innovation. IFIP International Federation for Information Processing, vol. 204, pp. 61–69. Springer, Boston (2006)

4. Chesñevar, C.A., Brena, R., Aguirre, J.: Solving Power and Trust Conflicts through Argumentation in Agent-mediated Knowledge Distribution. International Journal of Knowledge-based and Intelligent Engineering Systems (KES), special issue on agent-based Knowledge Management (2006)
5. Dieng, R., Corby, O., Giboin, A., Rybière, M.: Methods and tools for corporate knowledge management. Technical report, INRIA, ACACIA project (1999)
6. Elvang-Goransson, M., Krause, P., Fox, J.: Dialectic reasoning with inconsistent information. In: Heckerman, D., Mamdani, E. (eds.) UAI 1993: Proceedings of the Ninth annual Conference on Uncertainty in Artificial Intelligence, The Catholic University of America, Providence, Wa Artificial Intelligence, vol. 53(2-3), pp. 125–157 (1992)
7. Greco, S., Matarazzo, B., Slowinski, R.: Rough sets theory for multicriteria decision analysis. European Journal of Operational Research 129, 1–47 (2001)
8. Grundstein, M.: From capitalizing on Company Knowledge to Knowledge Management. In: Morey, D., Maybury, M., Thuraisingham, B. (eds.) Knowledge Management, Classic and Contemporary Works, ch. 12, pp. 261–287. The MIT Press, Massachusetts
9. Kraus, S., Sycara, K., Evanchik: Argumentation in Negotiation: A Formal Model and Implementation. Artificial Intelligence 104(1-2), 1–69 (1998)
10. Saad, I.C., Chakhar, S.: A decision support for identifying crucial knowledge requiring capitalizing operation. European Journal of operational research 195, 889–904 (2009)
11. Saad, I., Rosenthal-Sabroux, C., Grundstein, M.: Improving the Decision Making Process in The Design project by Capitalizing on Company's Crucial Knowledge. Group Decision and Negotiation 14, 131–145 (2005)
12. Simari, G.L., Loui, R.P.: A mathematical treatment of defeasible reasoning and its implementation. Artificial Intelligence 53(2-3), 125–157 (1992)
13. Sycara, K.: Arguments of Persuasion in Labour Mediation. In: Proceedings of the Ninth International Joint Conference on Artificial Intelligence, pp. 294–296 (1985)
14. Tseng, B., Huang, C.: Capitalizing on Knowledge: A Novel Approach to Crucial Knowledge Determination. IEEE Transactions on Systems, Man, and Cybernetics Part A: Systems and Humans 35, 919–931 (2005)

Knowledge Modeling for Educational Games

Miroslav Minović[1], Miloš Milovanović[1], Dusan Starcevic[1], and Mlađan Jovanović[2]

[1] Faculty of Organizational Sciences, Laboratory for Multimedia Communications, University of Belgrade
{mminovic,milovanovicm,starcev}@fon.rs
[2] Faculty of Electrical Engineering, University of Belgrade
mladjan@rcub.bg.ac.rs

Abstract. Use of educational games during teaching process does not represent a new topic. However question "How to address the knowledge in educational games?" is still open. The purpose of this paper is to propose a model that will attempt to establish the balance between knowledge integration into game on one side, and its reusability on other. Our model driven approach is relying on use of Learning Objects (LO) as constructing pieces of knowledge resources which are specialized for educational game design purpose. Presented models contribute to methodology of educational games development in a way that they embrace principles of learning and knowledge management early in design process. We demonstrated applicability of our models in design case study, where we developed educational game editor where educator can easily define new educational game utilizing existing knowledge, assessment and multimedia from repository.

Keywords: Knowledge modeling, Educational game, MDA, Metamodels.

1 Introduction

There is a promising role of digital games in education process. Traditional forms of teaching and passing knowledge lose their strength daily, due to development of technology and different motivational factors for the upcoming generations. Digital game-based learning is a novel approach applying at universities' courses and lifelong learning. In search for new role of universities in changing context of education, gaming is becoming a new form of interactive content, worth of exploration [1]. Features of games that could be applied to address the increasing demand for high quality education are already identified as [2]: clear goals, lessons that can be practiced repeatedly until mastered, monitoring learner progress and adjusting instruction to learner level of mastery, closing the gap between what is learned and its use, motivation that encourages time on task, personalization of learning, and infinite patience.

Use of educational games during teaching process does not represent a new topic. Since the need for new forms of education has been recognized by the researchers in this area, new problem arose. The main issue in this area of research is "How to address the knowledge in educational games?" At this moment, development of educational games includes knowledge integration during game development process. This

M.D. Lytras et al. (Eds.): WSKS 2009, LNAI 5736, pp. 156–165, 2009.

approach establishes strong coupling of game context and integrated knowledge which further disables reuse of that particular knowledge. In order to increase knowledge reuse, there is a need for a certain level of separation from the game. That extraction, on the other hand, can lead to poor integration with game context, which disrupts the flow of the game. Finding the right balance is essential regarding this matter.

Finding the right way to model the knowledge for use in educational games presents an important issue. Video games teach players certain skills and knowledge [3,4]. The problem arises when there is a need to teach specific matter such as subject curriculum at universities etc. Integrating that kind of knowledge, while still making game interesting and playable presents a big challenge. The purpose of this paper is to propose a model that will attempt to establish the balance between knowledge integration on one side, and its reusability on other. Our model driven approach is relying on use of Learning Objects (LO) as constructing pieces of knowledge resources which are specialized for educational game design purpose. Learning objects represent a small, reusable pieces of content relevant for learning (for example, an online exercise; a coherent set of introductory readings on a specific topic; or an assessment test) [5]. Reusability of LO represents using LO in different courses, by different teachers and learners [6]. In this case LOs can be reused in different educational games as well as other eLearning forms, online classes, tests etc.

The paper is structured as follows. In part 2, we give a brief discussion about this area of research and survey of research result regarding this matter. Next, we give a description of the proposed approach inspired by model-driven development that represents a basis of this work. Detailed description of our metamodel is a subject of part 4 of this paper. An example of application of the described metamodel, is described in part 5. Part 6 gives a conclusion and issues in respect to our future work.

2 Games in Education

The essence of e-learning lies in knowledge management [7,8]. The ever-increasing importance of knowledge in our contemporary society calls for a shift in thinking about innovation in e-learning.

In the context of e-learning, ontologies serve as a means of achieving semantic precision between a domain of learning material and the learner's prior knowledge and learning goals [9]. Ontologies bridge the semantic gap between humans and machines and, consequently, they facilitate the establishment of the semantic web and build the basis for the exchange and re-use of contents that reaches across people and applications [10].

It is important to be able to separate content from expression within a LO in order to be able to clearly distinguish two important types of questions: those dealing with the meaning that has to be conveyed by the LO, and those dealing with how meaning is to be expressed [11].

On the other side, main purpose of educational games is to teach and pass knowledge. That is why a majority of educational games is focused mainly on knowledge. Different skills and knowledge can be taught differently. Some games are using well-known, popular environment and set of rules, adapted for purposes of education - for example, the educational game based on "Who wants to be a millionaire?" quiz [12]. It uses all elements of the TV show, but questions are chosen by the teacher.

Some games are developed with certain subject matter in mind, like games for teaching electromagnetism called Supercharged! [13] or a fantasy adventure game for teaching the basic concepts of programming [14]. In some cases, the modification of popular games (game modding) was used for teaching computer science, mathematics, physics and aesthetics [15]. Game design can be used to achieve similar goals - developing problem solving skills and teamwork [16].

Regardless of the rapid growth of this research field, knowledge modeling for the purpose of educational games is still in its initial phases. While there are many examples of practical work in knowledge modeling and knowledge management, there is very little practical work done in the field of knowledge modeling and integration with educational games. While there are numerous efforts that games can be applied to learning, relatively few attempts can be found where principles of learning and knowledge management were explicitly followed a priori in design [9]. Cognitive modeling and assessment tools have to be incorporated in to educational games, giving insight into learning outcomes and enabling their evaluation. The challenge with these games is also that they are very costly to develop, as they must compete with commercial video games in terms of quality of graphics, challenges, and game play [2].

3 Proposed Approach

Our approach is inspired by the model-driven development, where software development's primary focus and products are models rather than computer programs. In this

Table 1. Mapping Educational Games Concepts to the OMG's MDA Levels

OMG MDA Level	Educational Game Metamodeling Architecture	Description
M3 – Meta-metamodel	The Meta Object Facilities (MOF)	The MOF is an OMG standard that defines a common, abstract language for the specification of metamodels. MOF is a meta-metamodel – the model of the metamodel, sometimes also called an ontology
M2 – Meta models	The Educational Game Metamodel (EGM)	The Educational Game Metamodel provides a common and standardized language about phenomena from various domains relevant to the design of educative games. It is called a metamodel as it is the abstraction of platform specific models.
M1 – Models	Platform-specific Shemas (XHTML, SAPI, SWIXml Schemas...)	Platform specific models of educational game content.
M0 – Objects, data	Content data (XHTML, SAPI, SWIXml files...)	Instances of platform specific models.

way, it is possible to use concepts that are much less bound to underlying technology and are much closer to the problem domain [17].

Table 1 gives an overview of educational game development through MDA levels. It uses a platform-independent base model (PIM), and one or more platform-specific models (PSM), each describing how the base model is implemented on a different platform [18]. In this way, the PIM is unaffected by the specifics of different implementation technologies, and it is not necessary to repeat the process of modeling an application or content each time a new technology or presentation format comes along. The views on game content from different levels of abstraction can be derived by *model transformations*. In MDA, platform-independent models are initially expressed in a platform independent modeling language, and are later translated to platform-specific models by mapping the PIMs to some implementation platform using formal rules. The transformation of the content models can be specified by a set of rules defined in terms of the corresponding higher level metamodels. The transformation engine itself may be built on any suitable technology such as XSLT tools. Our approach is based on standard technologies such as the Unified Modeling Language (UML) and XML, which are familiar to many software practitioners and are well supported by tools. Therefore, it is not necessary to develop complex solutions from scratch, and it is possible to reuse existing model-driven solutions and experiences from other domains. In our work we rely on existing UML modeling tools, XML parsers and software frameworks, developing only code that extends, customizes, and connects those components according to common and standardized language defined in the Educational Game Metamodel.

4 Educational Game Metamodel

Defining educational game models requires a vocabulary of modeling primitives. Therefore, our metamodel describes basic educational game concepts. Figure 1 shows a simplified educational game metamodel.

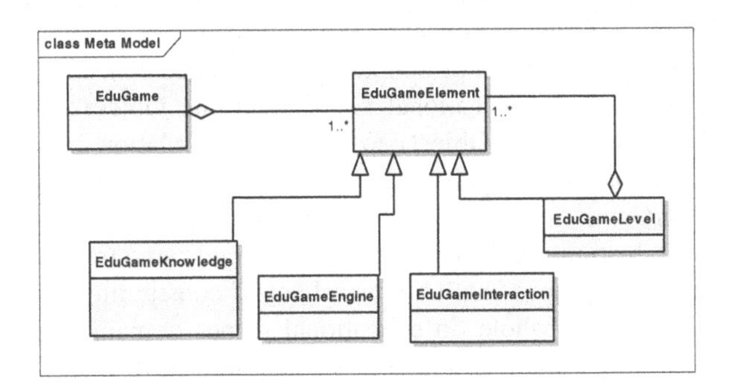

Fig. 1. Educational game basic concepts

The metamodel's main concept is *EduGameElement* which is used as a basis for defining other concepts of educational game. *EduGameKnowledge* defines educational content that aims to convey to players in learning process. There is a need for expertise in science area for managing complexities of the underlying knowledge.

Educational content needs some form of presentation to the user, therefore we introduce concept of *EduGameEngine*. It describes the mechanism used to present knowledge, which, for example, might be the learning tool to generate and answer questions that guide learner through the exploration and discovery of the required science area. *GameInteraction* concept describes communication between player and game. This concept describes interaction at high level of abstraction regardless of specific manifestations. In broad outline, interaction is established using multiple channels of communications, and concept is derived from our existing metamodel of multimodal human-computer interaction [19]. The overall goal is to convey knowledge in interactions rather than static data. *EduGameLevel* comprises previous modeling primitives in order to provide inherent mechanism for game progress as well as creating a sense of achievement. It also allows creating games at multiple scales of knowledge and skills.

4.1 Knowledge Metamodel for Educational Game

In this paper, we will focus on knowledge modeling for educational games. Further development of knowledge metamodel for educational game should provide a good basis for modeling domain knowledge and integration with the game. Model should enable manageable learning path, through the game, as well as knowledge reusability. In addition, model must provide the ability of knowledge assessment and integrating that assessment with the game.

In order to structure domain knowledge, we introduce basic metamodel shown on the Figure 2. *EduGameKnowledge* consists of *EduGame KnowledgeCategory*, which defines hierarchy of knowledge, and contains zero or more domain models.

Domain model represents a specific knowledge area and consists of Domain Concepts, which are self-related. Domain Concept represents a specific unit of knowledge that constitutes a building block of mentioned knowledge area. Relation between Domain Concepts has two important aspects. If concept relates to other concept, than correlation attribute will have value between zero and one (one for exactly the same concept and zero for non-related concepts). Second important relation is prerequisite, which signifies concepts that must be adopted before related concept.

We define *EduGameLO* (Educational Game Learning Object) and *EduGameAO* (Educational Game Assesment Object) to introduce a relation between game and knowledge. One *EduGameLO/EduGameAO* is related with one or more Domain Concepts. Finally, *EduGame Scene* consists of zero or more *EduGameLO* and zero or more *EduGameAO*. Inspiration for using the name *Scene* came from the field of movies. As in movies, scene represents an integral set of constituting parts that are presented to the viewers as a whole. In educational games domain, scene represents a composition of learning materials and assessments, that have a specific educational purpose, but its presentation depends on many different factors. Adequate interaction with learner required from us to develop another part of our framework, targeting Game Interaction [20].

Major benefit of our metamodel is that construction of educational games is driven by "learning scenario", which actually defines domain concepts and learning path that learner should adopt. Less experienced educator can construct educational game, simply relying on domain model developed by experts in specific knowledge area, utilizing already established relations between concepts for given domain model.

On the other side, established relation enables us to create transformations on lower MDA levels, in order to automatically generate *EduGame Scene*. Educator can define "learning scenario" (or use existing one from repository), and educational game will be generated (by use of game template from repository). *EduGame Engine* can generate adequate *EduGame Scene*, by choosing Learning and Assesment Objects, and pass it to *EduGameInteraction* for presentation to learner.

Next model (Fig. 2) gives in more detail specification of Learning and Assessment Objects. Although learning and assessment is often overlapping, in our model we distinguish between Learning Object and Assessment Object. In this way, we can achieve separated management of learning and assessment paths, as well as easier manipulation by computers. Specific nature of educational games leads us to separation of LO and AO. In order to keep learners motivation high, game should provide a sense of achievement. Learner should be provided with a challenge adequate to his current knowledge state, which is established by use of AOs. Game platforms enable implementation of assessment objects which are different than in classical eLearning (for example mini-game inside the game which will verify acquired knowledge or skills). Finally, for learner this separation does not have to be so clear, since advanced educational games should mix these two concepts and blur the separation line between them. Further, *EduGameLO* can be Simple or Complex type. Simple LO can be: *TextEduGameLO*, *PictureEduGameLO*, *VideoEduGameLO* or *AudioEduGameLO*. *ComplexEduGameLO* represents any combination of Simple Learning Objects as well as combination with other Complex Learning Objects. *EduGameAO* has same specialization, as *Simple* and *Complex EduGameAO*.

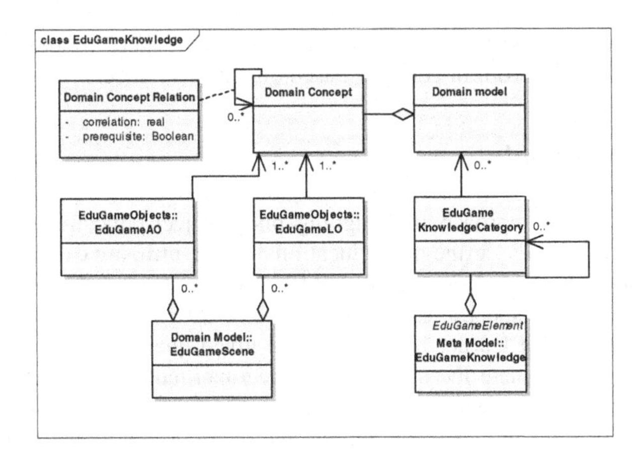

Fig. 2. Educational game knowledge metamodel

We decided to continue specializing simple AO into *Question, Simulation, Puzzle* and *Mini-game*. *Question* covers all standard question types for knowledge assessment (like multiple-choice or free-text question). *Simulation* represents a specific kind of assessment, where learner has some kind of mini-model for manipulation and has to use it in order to solve a given problem. *Simulation* is particularly convenient for skill assessment. *Puzzle* refers to a group of tasks described as logical assignments or logical hurdles. *Mini-game* is assessment object in a form of a game. Important characteristic of every assessment object is to provide *EvaluationPoints* value (for example, at implementation level this can be valued between 0 and 100) in order to enable verification of knowledge. *Complex EduGameAO* can be aggregation of Simple EduGameAO (mini-game for example).

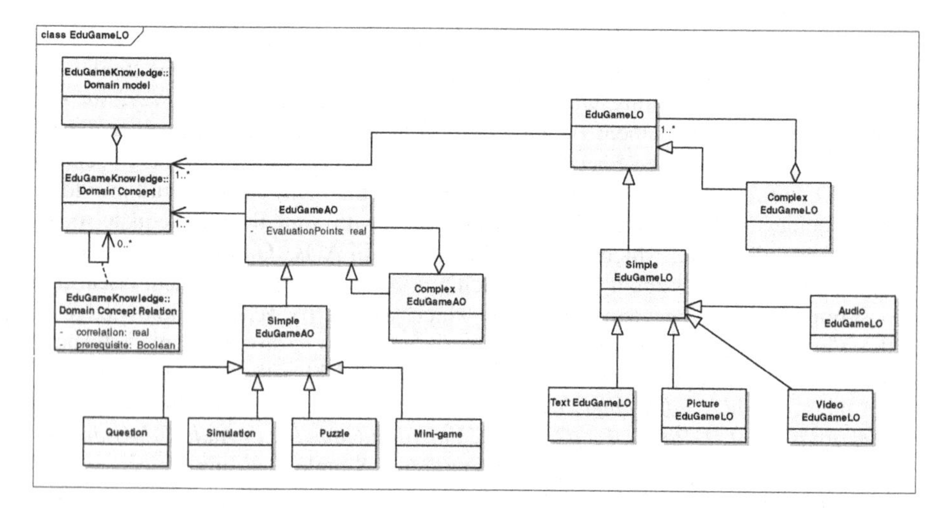

Fig. 3. Educational game Learning Object and Assessment Object metamodel

In next section we will present our design case study, where we developed educational game editor as a proof of concept, based on our metamodels.

5 Design Case Study

We have applied our approach in designing pilot project educational game editor, where educator can easily define new educational game, utilizing existing knowledge, assessment and multimedia from repository. For now, we provided support for adventure game type. For this specific game type, we use a 2D game environment.

The main idea is that learner has to solve all given quests inside adventure game world. In order to complete the quest, player (learner) must pass all assessment objects successfully. First, educator defines a map, which presents a game world environment. Upon that, he defines game regions and relations among them, and proceeds with detail definition of each region, where he can use existing learning objects or create new one for specific domain concept. Assessment objects can belong to a region, too. Also every *region* contains Non Playing Characters (NPC) that act as enemies, partners and support characters to provide challenges, offer assistance and support the storyline. After the creation of regions, we can make different avatars, assign them different abilities, and make quests and assignments for future players.

The result of the process described is a uniquely structured XML document [21], which EduGameEngine uses for interpretation and presentation to learner.

Figure 5. shows how user see interpreted adventure game, implemented as Java Applet. By advancing through the game, and by solving the quests, the player gains knowledge and learns new concepts.

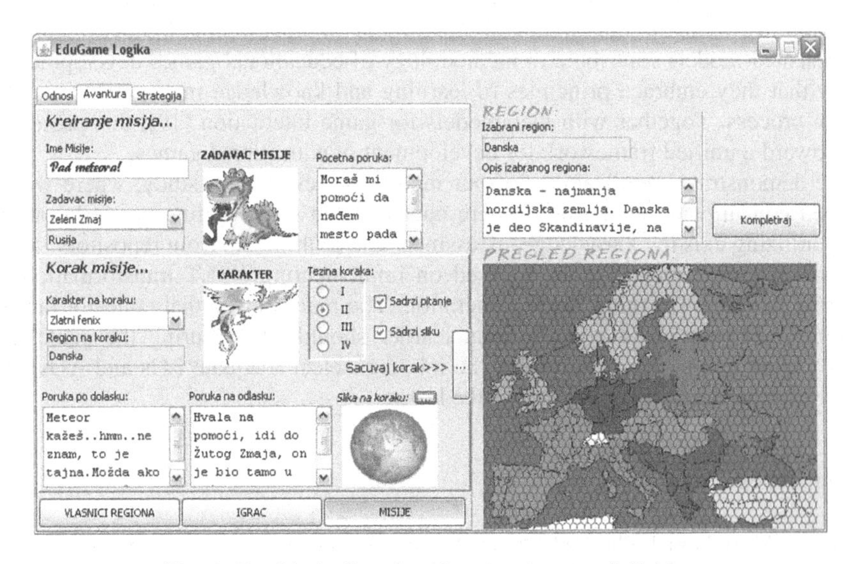

Fig. 4. Graphical editor for educational games definition

Fig. 5. Web client game interface

6 Conclusion

This paper describes our research in knowledge modeling for educational games. Extending our previous work on Educational Game Metamodel [22], this article provides further details of this approach and introduces a knowledge metamodel that enriches learning by creating platform for seamless integration between knowledge and game.

Presented models contribute to methodology of educational games development in a way that they embrace principles of learning and knowledge management early in design process. Together with metamodels for game interaction [19], this presents a step toward a unified framework for development of educational games.

We demonstrated applicability of our models in design case study, where we developed educational game editor where educator can easily define new educational game utilizing existing knowledge, assessment and multimedia from repository.

Further development will be focused on implementing XSLT transformations in order to automate Learning Scene generation. The main idea is that educator defines "learning scenario" with concepts that learner should adopt during the game, and game engine will perform the rest of the job, e.g. select adequate LOs and AOs, produce Learning Scene and present it to the learner.

References

1. Pivec, M.: Editorial: Play and learn: potentials of game-based learning. British Journal of Educational Technology 38(3), 387–393 (2007)
2. Federation of American Scientists, R&D Challenges in Games for Learning. Tech. Rep., Washington D.C. (2006), `http://www.fas.org/learningfederation`
3. De Aguilera, M., Mendiz, A.: Video games and Education. Computers in entertainment 1(1) (2003)
4. Estallo, J.A.: Los videojugos. Juicios e prejuicios. Planeta, Barcelona (1995)
5. Jovanović, J., Gašević, D., Brooks, C., Devedžić, V., Hatala, M., Eap, T., Richards, G.: Using Semantic Web Technologies to Analyze Learning Content. IEEE Internet Computing 11(5), 45–53 (2007)
6. Gašević, D., Jovanović, J., Devedžić, V.: Enhancing Learning Object Content on the Semantic Web. In: Proceedings of the IEEE International Conference on Advanced Learning Technologies ICALT 2004, Joensuu, Finland (2004)
7. Ronchetti, M., Saini, P.: Knowledge management in an e-learning system. In: Proceedings of the 4th IEEE International Conference on Advanced Learning Technologies ICALT 2004, Joensuu, Finland, pp. 365–369 (2004)
8. Kostas, M., Psarras, J., Papastefanatos, S.: Knowledge and information management in elearning environments: The user agent architecture. Inf. Manage. Comput. Security 10(4), 165–170 (2002)
9. Kickmeier-Rust, M.D., Albert, D.: The ELEKTRA ontology model: A learner-centered approach to resource description. In: Leung, H., Li, F., Lau, R., Li, Q. (eds.) ICWL 2007. LNCS, vol. 4823, pp. 78–89. Springer, Heidelberg (2008)
10. Antoniou, G., van Harmelen, F.: Web ontology language: OWL. In: Staab, S., Studer, R. (eds.) Handbook on ontologies, pp. 67–92. Springer, Heidelberg (2004)

11. Brajnik, G.: Modeling Content and Expression of Learning Objects in Multimodal Learning Management Systems. In: Proceedings of Universal Access in Human-Computer Interaction, pp. 501–510. Springer, Heidelberg (2007)
12. Reinhardt, G., Cook, L.: Is This a Game or a Learning Moment? Decision Sciences Journal of Innovative Education 4(2), 301–304 (2006)
13. Squire, K.B.: Electromagnetism supercharged! Learning physics with digital simulation games. In: Proceedings of the 2004 International Conference of the Learning Sciences, Santa Monica (2004)
14. Moser, R.: A fantasy adventure game as a learning environment: Why learning to program is so difficult and what can be done about. In: Proceedings of the ITiCSE 1997, Uppsala, Sweden (1997)
15. El-Nasr, M., Smith, B.: Learning through game modding. ACM Computers in entertainment 4(1), article 3B (2006)
16. Steiner, B., Kaplan, N., Moulthrop, S.: When play works: Turning game-playing into learning. In: Proceedings of IDC 2006, Tampere, Finland (2006)
17. Selic., B.: The pragmatics of model-driven development. IEEE Software, pp. 19–25 (2003)
18. Bezivin, J.: From Object Composition to Model Transformation with the MDA. In: Proceedings of TOOLS. IEEE CS Press, Santa Barbara (2001)
19. Obrenovic, Z., Starcevic, D.: Modeling multimodal Human-Computer interaction. IEEE Computer 37(9), 62–69 (2004)
20. Minović, M., Milovanović, M., Jovanović, M., Starčević, D.: Model Driven Development of User Interfaces for Educational Games. Accepted for presentation at Human System Interaction conference, Catania, Italy, May 21-23 (2009)
21. Minović, M., Milovanović, M., Lazović, M., Starčević, D.: XML Application For Educative Games. In: Proceedings of European Conference on Games Based Learning ECGBL 2008, Barcelona, Spain (2008)
22. Jovanovic, M., Starcevic, D., Stavljanin, V., Minovic, M.: Educational Games Design Issues: Motivation and Multimodal Interaction. In: Lytras, M.D., Carroll, J.M., Damiani, E., Tennyson, R.D. (eds.) WSKS 2008. LNCS (LNAI), vol. 5288, pp. 215–224. Springer, Heidelberg (2008)

SISCOVET: Control System of Transport Vehicle Drivers Using GPS Location and Identification through the Electronic ID Card

Luis de-Marcos, José-Ramón Hilera, José-Antonio Gutiérrez, and Salvador Otón

Computer Science Department, University of Alcalá
Ctra Barcelona km 33.6, Alcalá de Henares, Madrid, Spain
{luis.demarcos,jose.hilera,jantonio.gutierrez,
salvador.oton}@uah.com

Abstract. This paper presents new methodologies in the analysis of the driving and rest times demanded by the Spanish Ministry of Public Works to all transport vehicle drivers through a control system using GPS location and identification through the electronic ID card. It is proposed the creation of a control system for the vehicle driver through GPS technology. Furthermore, technologies will be used for the data transport between the device that will be included in the vehicle and a server containing the characteristics of each route traveled by different drivers. The proposed control system will use as access control the electronic ID card in order to avoid any sort of vulnerabilities related to phishing (identity thefts). It aims to study the technology used by the electronic ID card. So, the main technological innovation is the use of the newborn electronic ID card alongside with the most advanced wireless technologies in the field, integrating both concepts in the same electronic device.

Keywords: DNI-e (e-IDs), GPS, wireless technologies, Wi-Fi, bluetooth, transport.

1 Introduction

The recent introduction of the electronic ID card in Spain allows a new kind of interaction between people and electronic media, faster and easier than the traditional one. This kind of interaction is similar to that used with credit cards, in which users introduce the card into a reader, type their personal identification number (PIN), and if all data are right they can begin to operate.

Thus, one of the main aims of the electronic ID card is that users interact in this way, but without limiting the area of tasks to a single field.

Security is one of the most interesting fields. In this field they can be developed systems that taking into account the universal the universal nature of the electronic ID card avoid the need to generate a means of identification (ie card) for each system.

Control systems of transport vehicle drivers nowadays are the classic analog tachograph, which displays the driving and rest times by means of an approved paper

M.D. Lytras et al. (Eds.): WSKS 2009, LNAI 5736, pp. 166–175, 2009.

disc. The ministry is demanding a new device; the digital tachograph which seems like a car radio. It is used a digital card for the identification of the driver.

This project proposes the creation of a control system of the vehicle driver through a GPS technology. This technology aims to:

- Locate the transport vehicle to track the trajectory. As a result of this tracking the following parameters will be obtained:
  - Measurement of the average speed and instant speed, controlling at every time the speeding.
  - Measurement of the distance traveled by a transport vehicle.
  - Control of driving times by detecting the vehicle movements.
  - Control of the driving times by not detecting the vehicle movements.

Moreover, technologies will be used for the data transport between the device included in the vehicle and a server containing all the characteristics of each route followed by each driver. These technologies are:

- Wi-Fi [1]: it will be used in order to transmit all the measured parameters during the route to a server that will calculate the possible infractions.
- GPRS [2]: it will be used in order to transmit all the measured parameters during the route to a server that will calculate the possible infractions, if the driver does not have Wi-Fi connection at the end of the route.
- Bluetooth [3]: it will be used in order to transmit all the measured parameters during the route to a device located in a short distance.

The proposed control system will use as access control the electronic ID card in order to avoid any sort of vulnerabilities related to phishing (identity thefts). It aims to study the technology used by the electronic ID card.

With the appearance of the new electronic ID card there are new possibilities related to the access control and control systems of transport vehicle drivers. The attempt of manipulating the current control systems causes the search for new less vulnerable systems; therefore the aim is to identify the transport vehicle driver through the electronic ID card, allowing a practical and safe identification.

The electronic ID card also allows to unify all identification systems into a single system. This idea is trying to be developed in order to make it possible the identification in all European Union countries customs by means of the ID card, replacing the passport.

The control systems of transport vehicle drivers are limited at the moment. As an example, until a few months ago the famous analog tachograph was still used. It allowed an easy manipulation by the drivers, issue that attempts to be remedied by using the electronic ID card.

However, since January 2008 the Spanish Ministry of Public Works has imposed the new digital tachograph, which detects possible manipulations of the performed measurements. Therefore, it requires a digital identification card specific for the tachograph to operate.

This project will provide the possibility of identifying the transport vehicle driver by means of a common document as it will be the electronic ID card. It will avoid any sort of manipulations, since the measurements of the required parameters to control the driving and rest times of the driver will be carried out by GPS location.

The system is composed of:

- A **control center**, with database management and connectivity to the tachograph placed in each vehicle.
- **Collection of devices** required to perform the measurements of each vehicle and communicate with the rest of the system, comprising:
  - **Tachometer:** device to measure the shaft turn speed, usually the engine turn speed in revolutions per minute (RPM).
  - **Electronic ID card reader:** reader device or the e-ID, which allows the driver to be authenticated and then the whole system will start to operate.
  - **Tachograph:** electronic device that records several events generated by a vehicle during its driving. The recorded events usually are: speed (average and maximum), RPM, mileage, sudden braking and accelerations, idle time (the vehicle is stopped with the engine running), among others. These data can be collected by a computer and stored in a database or printed as graphs for further analysis.
- **Multimedia mobile device** for the traffic policeman, easy-to-use, intuitive and with connection to the corresponding vehicle through Bluetooth technology.

2 Main Aims

The main aim of this project is to find new methodologies to analyse the driving and rest times demanded by the Spanish Ministry of Public Works, for all transport vehicle drivers through a control system using GPS location and identification by the electronic ID card. The system will select the most suitable transmission technology for each case automatically, ie, not user intervention is required. Among the used technologies there have been included GPS, GPRS, WIFI, BLUETOOTH and ZIGBEE4. The idea is to switch automatically from one technology to another depending on factors such as coverage, cost or transmission speed.

As a specific aim it is the development of the information system required to carry out the tracking, control and analysis tasks of the different routes traveled by each one of the drivers. They will be developed specifically:

- **Tracking system and data collection.**
  - Based on GPS for the location and wireless technologies (GPRS / UMTS / WiFi / ZigBee and Bluetooth) for the communication.
  - Data storage in the database.
- **Data obtaining system by the agent.**
  - Quantitative data analysis according to repports obtained from the information stored in the tachograph.

Thus, the different scenarios that may occur and the different possibilities they show are exhibit in the following images:

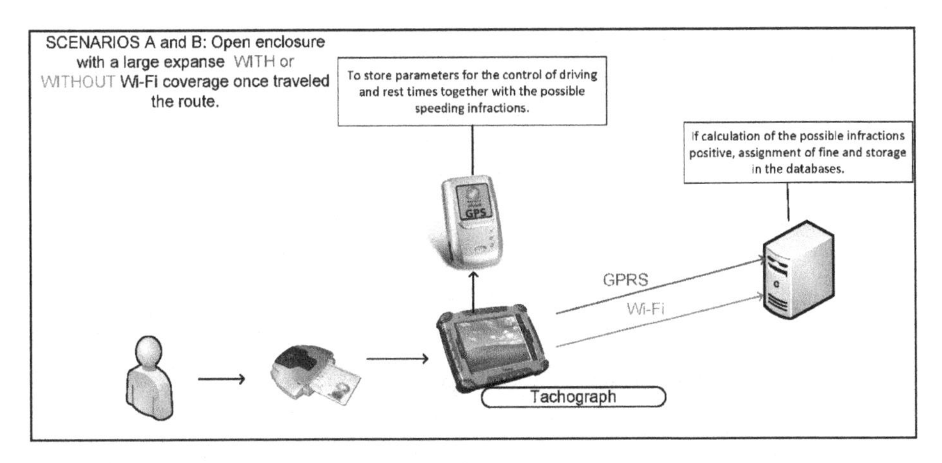
SCENARIOS A and B: Open enclosure with a large expanse WITH or WITHOUT Wi-Fi coverage once traveled the route.
To store parameters for the control of driving and rest times together with the possible speeding infractions.
If calculation of the possible infractions positive, assignment of fine and storage in the databases.
GPS
GPRS
Wi-Fi
Tachograph

Fig. 1. SISCOVET Operation for Checking Completed Routes

The following picture shows a scenario situated in an open enclosure with the possibility of transmitting data through Bluetooth, in which it is expected to check infractions of previous routes:

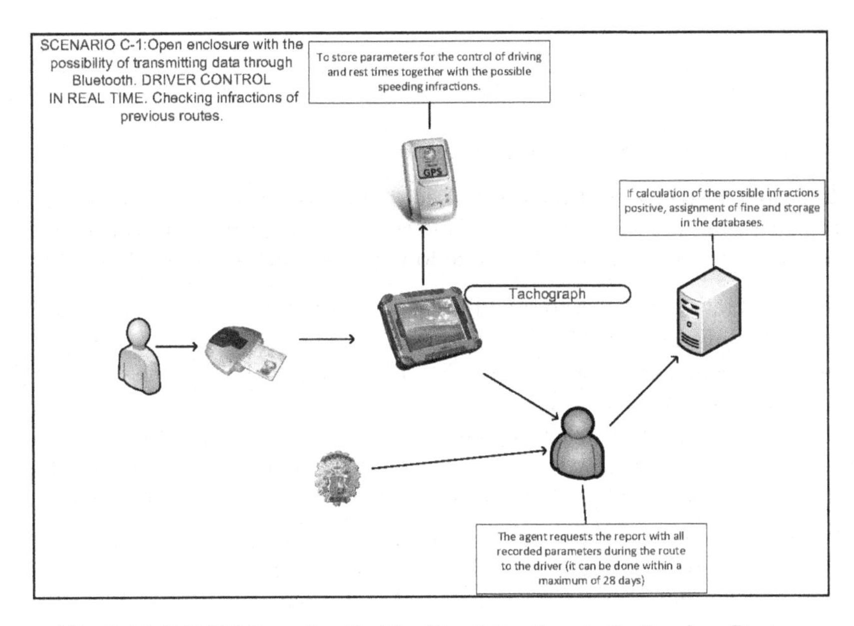
SCENARIO C-1:Open enclosure with the possibility of transmitting data through Bluetooth. DRIVER CONTROL IN REAL TIME. Checking infractions of previous routes.
To store parameters for the control of driving and rest times together with the possible speeding infractions.
GPS
If calculation of the possible infractions positive, assignment of fine and storage in the databases.
Tachograph
The agent requests the report with all recorded parameters during the route to the driver (it can be done within a maximum of 28 days)

Fig. 2. SISCOVET Operation for Checking Infractions in the Previous Routes

The following picture shows a scenario situated in an open enclosure with the possibility of transmitting data through Bluetooth, in which it is expected to check infractions of the current route:

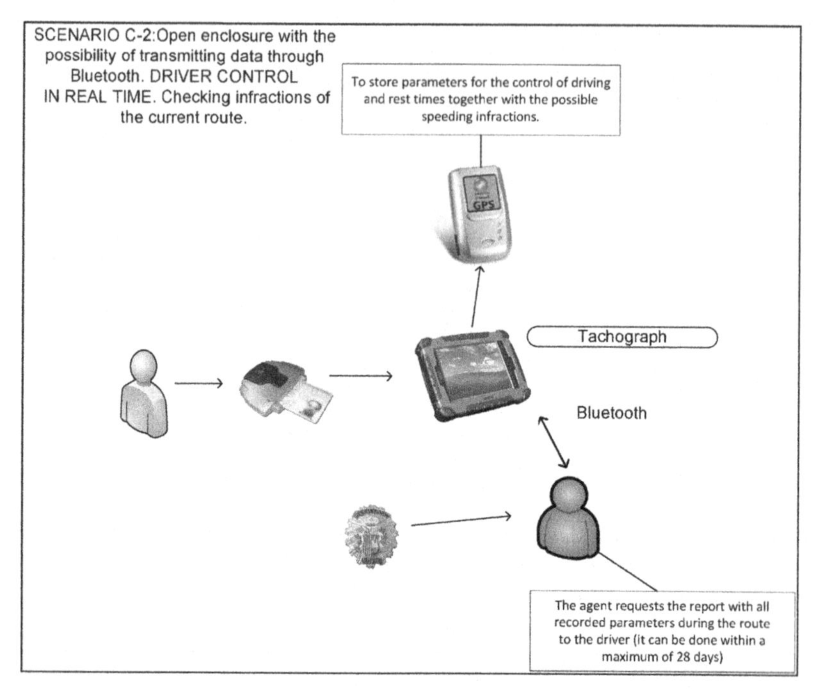
SCENARIO C-2:Open enclosure with the possibility of transmitting data through Bluetooth. DRIVER CONTROL IN REAL TIME. Checking infractions of the current route.
To store parameters for the control of driving and rest times together with the possible speeding infractions.
GPS
Tachograph
Bluetooth
The agent requests the report with all recorded parameters during the route to the driver (it can be done within a maximum of 28 days)

Fig. 3. SISCOVET Operation for Checking Infractions in the Current Route

The system will comprise the following elements:

Tachograph device: used by the driver to communicate with the server and send collected data of the route. It will be able to communicate with the mobile device and the server. Through a user interface the following possibilities will be offered:

- Generation of reports from different periods of activity.
- Transmission of the stored information to the server.
- Connection by Bluetooth with the mobile device and transmission of the report requested by the agent.

The storage capacity of the tachograph will be up to 28 days of information and the generated reports will be able to contain the information of those 28 days. When sending the information to the server it will be used any of the possible technologies.

Mobile device: through which the search of the driver infractions will be provided to the traffic policeman. By activating the Bluetooth device, the license plate number will be sent, the infractions will be obtained and a report will be displayed in the screen.

Server: where the data from the tachograph are stored and processed. It comprises an object-relational database management system PostgreSQL which will manage the data sent by the tachograph.

The SISCOVET "Control system of Transport Vehicle Drivers using GPS location and Identification through the electronic ID card" system must fulfil three basic functions:

- **To locate the driver and store information** sent by him about the route being traveled.
- **To allow the collection of these recorded data by the agent.** If a traffic policeman stops a transport vehicle, once the required identification is done, he/she will be able to obtain reports and data stored in that moment in the tachograph of the vehicle. This way, he/she will carry out as many investigations, warnings… as needed.

It is important to think of them separately, since the requirements of each one of these systems are quite different. The location system must have the same reliability than the telecommunications network, and be compatible with the telecommunications standards and network equipments from different manufacturers.

3 Description of the Project

Here is described the SISCOVET system specifying its scope, technological environment and main users.

3.1 Determining the Scope of the System

This project stems from the advisability of developing a specialized computer system that allows centralizing data from a GPS, a digital tachograph, and the electronic ID card, to perform a better tracking of a transport vehicle.

The system comprises:

- A control center, with database management and connectivity to the rest of devices.
- Mobile device with Bluetooth connectivity for the traffic policemen.
- A GPS device to obtain the location.
- A digital tachometer, integrated in the vehicle.
- An electronic ID reader.
- Web server and database.

This system is conceived to improve the present tachometer system. So, thanks to a GPS device it will be possible to know the route traveled by a vehicle or the exact point where it is. The same applies to the electronic ID card; it will provide more data than the provided by the present driver card.

Likewise, the system controls more exhaustively that the vehicle driver complies the traffic regulations established for the sector.

The main characteristics of the system are:

- Establishment of a system to detect the vehicle location using GPS technologies in order to create a route of travel.
- Establishment of a connection with mobile devices that request information about the stored routes in the device, through Bluetooth.
- Data storage in the device for further verification by a traffic policeman.
- Sending of data to a server to store them and subsequent consults.
- From the digital tachograph, reading the driver data and extraction of speeds and rest times.

The equipment can use different wireless communication technologies; it will choose one over another depending on the coverage of each place relating to the cost it entails:

- Preference of using WiFi in regards to GPRS because of the cost of the telephone operators, and the data transmission speed.
- Bluetooth connection in connections with mobile devices used to control infractions.

It has been adopted an object-oriented approach (O.O), in order to make the most of its advantages:

1. Greater structuration when programming.
2. Compaction of data (Encapsulation).
3. Easy to reuse code (Easy maintenance and expansion)
4. Better presentation of the problem.

3.2 Identification of the Technological Environment

In the following subsections it is carried out a high-level definition of the technological environment required to meet the needs of information, specifying the possible constraints and restrictions. To do so the technological environment has been taken into account.

3.2.1 Area of Action

The area of action will be focused on all Spanish geography, specifically in those vehicles devoted to transport.

3.2.2 Control Center

The control center is the server to which all data from the tachographs will be sent.

This control center will have a database that will store all information sent from the tachographs. Also it will have a program capable of calculating infractions and generating reports with them.

Furthermore the control center will have a web server that, through a page, will offer information about the stored data.

3.2.3 Multimedia Mobile Devices

The application developed for these devices uses a programming language that allows the execution in embedded devices, such as PDAs or Tablet PCs. It offers a Bluetooth connectivity, but also taking into account Wi-Fi and GPRS connectivity.

This application will be used by traffic policemen; they will establish through it a Bluetooth connection with the vehicle tachograph to obtain the route reports.

This application will be able to ask for reports to the tachograph, both in the current route and in previous ones. The application will receive this reports (in XML format) to show it in the screen for the traffic policeman's requirements.

3.2.4 Point Capture Device

Module to make GPS positions regularly at a predetermined set interval of time, to store in the tachograph the coordinates the vehicle has traveled.

The mobile point capture device will be activated once the tachograph is activated, sending at that moment the current location; and it won't be deactivated until the tachograph is switched off, sending then the current location again.

Thus, it is possible to control the whole route of a vehicle, no matter whether it is in motion or stopped.

3.2.5 Personal Data Reader Device

This device will be the electronic ID card reader.

Once the system is started, the tachograph will ask the ID card to be introduced in the reader device in order to register the driver. If no electronic ID card is introduced in the reader, the tachograph will emit a warning failure.

By reading the electronic ID cart it is possible to authenticate that the driver is really driving the vehicle.

The system stores the ID number, surnames and name of the person.

The law establishes that the ID card cannot be compulsorily trapped in any device or place. So the ID card is not required to be introduced in the reader during the whole route. When starting up it will be indispensable to introduce the ID card in the reader, but it can be removed, if desired, once the system indicates it.

3.2.6 Device for Calculating Speed and Working Times

This device corresponds to the current digital tachograph, included by law in all transport vehicles. This device nowadays interacts with the driver in the following ways:

- When starting up the vehicle, it asks for the insertion of the driving card by the driver.

Likewise, this device will control automatically:

- State of the vehicle (stopped or in motion) to establish the worker's state.
  - If the vehicle is stopped, the device will establish the worker's state "working".
  - If the vehicle is in motion, the device will establish the worker's state "driving".
- Instantaneous speed. By means of sensors placed on the vehicle the device will instantly know the speed it has. This way it controls if the vehicle exceeds the established limits, emitting a warning in that case.
- Rest times. According to the working time there is a rest time. This device will control that these times are not exceeded, emitting a warning in that case.

These data will be stored (together with the data related to the driver identification) to, among other things, generate reports that may be subsequently requested.

3.3 Identification of Participants and End Users

Here are identified the participants and end users both in the procurement of requirements and the validation of the different products and final acceptance of the system.

Given the importance the collaboration among users has in the process of obtaining requirements, it is worth determining who is going to participate in the work sessions, specifying the functions and assigning responsibilities.

3.3.1 Driver

This is the user who drives the transport vehicle, that is, the person from which all data is collected during the route; the protagonist of all measurements and operations of the system. So, the tasks of this user are:

- He/she will be responsible for the mobile device during its use.
- He/she will use all user functions of the application.
- He/she will provide his/her personal and registering data in the first use of the application, in order to be registered in the system, through the introduction of the electronic ID card in the reader and the driving card in the digital tachograph.
- He/she will be responsible for the maintenance of the device during its use.
- He/she will interact with the digital tachograph to introduce working and rest times as appropriate. The driving times will be automatically detected by the system. The working, rest or available times have to be introduced by the driver.
- He/she will activate the Bluetooth connection for the device to communicate with an agent's mobile device.
- He/she will take the driving card out, indicating the end of an activity.
- He/she will drive the vehicle travelling the corresponding routes.

3.3.2 Administrator

This is the user in charge of the complete system. The administrator must know how they work each of the applications comprising the final system. Thus, the operations focused on being performed by him/her are:

- Responsible for managing and maintaining the application in the server.
- Responsible for maintaining the accuracy and integrity of the database.
- Responsible for detecting errors of malfunctions of the system and communicate them to the maintenance team.
- He/she will add the different devices that the tachograph deals with, such as the e-ID or GPS reader.
- He/she will add the users of devices, ie the drivers.

3.3.3 Traffic Policeman

This is the user who may ask for certain data to the vehicle driver with intent to control its work. The characteristics and operations at his/her scope are:

- Responsible for the road safety.
- In case of inspection of a vehicle he/she will control through his/her mobile device that it has not made any infractions, and carrying out the corresponding sanctions if required.

4 Conclusions

A project represents an effort to achieve a specific aim through a particular set of interrelated activities and the efficient use of resources. Among the specific aims there are those related to the final functionality, aspect, characteristics, etc. and those related to make it within the established period.

Once finished, it is important to value some aspects about the work carried out, such as:

- **Scope:** the project has been developed thinking of a national scope - Spain. Thus, the scope of an application like this is quite important, due to the data registered by the Ministry of Public Works, reporting about the flow of goods [4]. The information about transport by road with Spanish vehicles has been obtained from the Permanent Survey of Goods Transport by Road, a periodic publication of the Ministry of Public Works. This survey reveals that most of the records correspond to transport operations within the country. The operations carried out between Spain and other countries or out of Spain by Spanish carriers are much limited. Hence the usefulness this project can provide to the territory for which it is developed.
- **Possible improvements:** the developed application is a prototype that can be executed in a computer. Therefore, the main improvement would be the implementation of the complete project, fulfilling the main functionality; being executed inside a vehicle. Still, the developed prototype presents all functionalities required to get an appropriate execution and offer all the characteristics.

Acknowledgements

This research is co-funded by: (1) the University of Alcalá FPI research staff education program, (2) the Spanish Ministry of Industry, Tourism and Commerce PROFIT program (grants FIT-020100-2008-23, TSI-020302-2008-11).

References

1. Mathew, S.: Gast. 802.11 Wireless Networks. O'Really Media, Sebastopol (2005)
2. Seurre, E., Savelli, P., Pietri, P.-J.: GPRS for Mobile Internet. Artech House Publishers, Norwood (2003)
3. Huan, A.S., Rudolph, L.: Bluetooth Essential for Programmers. Cambridge University Press, Cambridge (2007)
4. Ministerio de Fomento de España.: Resultados de la Encuesta Permanente de Transportes de Mercancías por Carretera (EPTMC), http://www.fomento.es/MFOM/LANG_CASTELLANO/INFORMACION_MFOM/INFORMACION_ESTADISTICA/Publicaciones/transporte_mercancias_carretera/parte1-2007.htm (Accessed, 23 March 2009)

Visual Modeling of Competence Development in Cooperative Learning Settings

Kathrin Figl[1], Michael Derntl[2], and Sonja Kabicher[2]

[1] Institute for Information Systems and New Media, WU Vienna University of Economics and Business, Augasse 2-6, 1010 Vienna, Austria
[2] Research Lab for Educational Technologies, University of Vienna, Rathausstrasse 19/9, 1010 Vienna, Austria
kathrin.figl@wu-wien.ac.at, michael.derntl@univie.ac.at, sonja.kabicher@univie.ac.at

Abstract. This paper proposes a novel approach to modeling cooperative learning sequences and the promotion of team and social competences in blended learning courses. In a case study – a course on Project Management for Computer Science students – the instructional design including individual and cooperative learning situations was modeled. Specific emphasis was put on visualizing the hypothesized development of team competences during the design stage. These design models were subsequently compared to evaluation results obtained during the course. The results show that visual modeling of planned competence promotion enables more focused design, implementation and evaluation of collaborative learning scenarios.

1 Introduction

Almost every job posting today includes the requirement that the applicant must be able to work in teams, communicate effectively in interdisciplinary work groups, and collaborate with coworkers both in face-to-face and distant settings. While most higher education curricula still have a strong focus on promotion of technical skills and factual knowledge, more recent developments such as the Bologna Process [2] accommodate the new qualification requirements in the knowledge society and promote a shift of focus towards developing *competences* of students based on a desired target competence profile for graduates. Competence is defined as the ability to *use* knowledge, skills and personal, social and/or methodological abilities [3]. One set of highly job-relevant competences are those related to teamwork, and one effective way of facilitating the development of those teamwork competences is provision of collaborative learning environments. However, introduction of effective collaborative learning, as will be argued in the following section, is a demanding and non-trivial endeavor. Collaborative learning may introduce a considerable amount of uncertainty, it requires more flexibility, and it generally follows a non-linear flow of events and activities; therefore, planning for those settings is particularly difficult.

To overcome these obstacles, we adopt a visual-language approach [4] to designing for collaborative learning with particular emphasis on planning for the promotion of

M.D. Lytras et al. (Eds.): WSKS 2009, LNAI 5736, pp. 176–185, 2009.

team competences. We build upon an existing UML-based design language [5] and extend it with visual icons that allow the explicit representation of (a) the social setting of an activity and (b) the hypothesized promotion of team competences within a learning activity. The distinction of different types of team competence supported by learning activities is adopted from [1], who differentiate teamwork knowledge, skills and attitudes.

The paper is structured as follows. In Section 2 we introduce relevant background work on cooperative learning and team competences. In Section 3 we propose the visual extensions needed to include cooperation and team competence promotion in the design models. In Section 4 we present a case study in which we compare the hypothesized competence promotion during course design with the actual perceived competence shifts as rated by students after the course. The final section concludes the paper and gives an outlook on further work.

2 Cooperative Learning and Team Competences

Cooperative learning is defined as "the instructional use of small groups so that students work together to maximize their own and each other's learning" [6, p. 3], or as "an educational approach in which the learning environment is structured so that students work together towards a common learning goal" [7, p. 119]. Computer supported cooperative learning (CSCL) [8] [9] is a combination of group-based learning and its supporting technology. Cooperative learning goes beyond putting students in groups and giving them assignments [10, p. 45]; it should be regarded as an overall goal in education, since this kind of learning meets the demands of the modern knowledge society better than teacher-centered lectures [11, p. 631]. Cooperative learning is proven to be capable of producing various positive effects, for instance less dropouts – especially at the beginning of the studies – because it can contribute to students' sense of belonging to colleagues and feeling of security [12]. A meta-study on the effects of cooperative small-group learning on science, mathematics, engineering, and technology students found significant positive effects on achievement, persistence and attitudes (towards subject matter, self esteem and motivation) [13]. Another meta-analysis on cooperative learning methods showed that students showed higher academic achievements – with respect to grades, quality of products such as reports – than competitive and individualistic efforts [14].

In the context of this paper we are particularly interested in the fact that cooperative learning can promote social competences like communicating effectively and managing conflicts more effectively than individual learning [15]. Working in teams usually allows students to realize the benefits of teamwork [16], even though some studies showed that students prefer individual work because their individual effort is recognized higher than in teamwork, which is likely if students did not receive proper training before teamwork [17]. Negative experiences with teamwork, especially with "social loafing," can undermine students' attitudes towards working in teams [16]. If teamwork is not well managed, negative team experiences might discourage students and create negative attitudes towards teamwork in class [17]. In addition, these negative experiences with teamwork not only have negative effect on students' attitudes toward team participation, but may also contribute to poor team performance on the job and should therefore be avoided [18]. It is evident that cooperative learning calls for developing and demonstrating a certain set of skills and attitudes that we subsume under the term "team competence."

Parry [19, p. 60] defines competences as "a cluster of related knowledge, skills and attitudes that affects a major part of one's job (a role or responsibility), that correlates with performance on the job, that can be measured against well-accepted standards, and that can be improved via training and development." Team competences are a main factor for team performance at work or in a learning environment; on the individual level they are the characteristics a team member has to possess to successfully engage in teamwork [20]. They are team-generic, held by individuals and can be transported to other teams. According to Cannon-Bower et al. [1] there are three important types of team competences: *knowledge competences*, *attitude competences* and *skill competences*. Knowledge competences include, for instance, knowing about proper behavior in teamwork, roles in a team or the team's goals. With respect to attitude competences, positive attitudes towards teamwork are important for effective teamwork. Skill competences represent the learned capacity to interact with other team members and include group decision-making skills, adaptability/flexibility skills, interpersonal relations skills and communication skills [1]. Collaborative learning is one "natural" method to prepare students for working in teams and for promoting team competences.

3 Visual Modeling of Learning Activities

The discussion in Section 2 leads us to the conclusion that incorporating teams in courses or other educational offerings is definitely a complex task. To support instructors and learning designers in this task, we propose a method to improve the planning of teamwork and associated facilitation and promotion of team competences by using visual design models.

There are a plethora of visual instructional design languages [4] and tools already on the market. The use of visual models can help instructors and learning designers in visual thinking – i.e., planning and reflecting the design of cooperative learning and the promotion of teamwork competences through drawing visual representations of learning activities and environments. To model cooperative learning and the promotion of team competences, we used the coUML visual design language [5], which basically extends UML activity diagrams with symbols to visualize the mode of presence (web/distant, face-to-face, blended) of learning activities. In its basic notation, coUML does not include means for modeling (a) hypothesized promotion of competences and (b) social setting – individual vs. cooperative – of learning activities. To enable visualization of those complementary information assets, we extended the coUML notation with additional visual elements (icons), which are depicted in Table 1. Attaching these icons to learning activities in the course design models allows planning and hypothesizing about the intended use of teamwork and promotion of team competences. For an example of a course design model employing this extended notation see Fig. 1 in the following case study section.

Table 1. Additional UML activity diagram elements

Icon	Description
	Attached to activity symbols that model course participants cooperating in teams.
	Attached to activity symbols that model activities executed by individual participants.
K S A	Attached to activities to model the level of competency, which is addressed by the activity. The team competency "pyramid" shows three levels: knowledge (K), skills (S), and attitudes (A). The competency pyramid was inspired by the team competencies model proposed by Cannon-Bower et al. [1], who introduced this competency distinction. Note that light fill color on a level indicates moderate promotion of this competency level, dark fill color is used to indicate strong promotion.
Note: the following two icons are part of the coUML notation and only included in this table for the purpose of clarity.	
Web activity	Light fill color is used to model online / web-based activities.
♂	Attached to an activity to indicate that it includes several sub-activities, which are modeled in detail in a separate diagram.

4 Case Study: Project Management Course

4.1 Course Description

The Project Management course is a combined lecture and lab course and is part of the Computer Science curriculum at the Faculty of Computer Science, University of Vienna. The overarching learning goal of course is that students know project management methods and techniques, and that they are able to apply particular methods and techniques of planning and controlling IT projects after finishing the course. During the course, students were familiarized with specific criteria of IT projects, learned to use project management tools (in particular MS Project®) and to work efficiently in teams. Subject-specific topics covered in the course were network plans, cost and time estimation of projects, project metrics, quality assurance and program management. Students created and planned projects in small teams, developed project deliverables and used MS Project as a tool for planning and controlling. In the face-to-face units of the course, solutions were presented and discussed, practical exercises were done, and teamwork issues were reflected.

Pedagogical Elements. In order to experience both a scientific viewpoint on project management as well as its practical application, the course was designed as an interactive cooperative learning scenario. In the course units, students received compact subject-matter inputs and experimented with project management tools during workshops and practical exercises. One of the main concerns was to design the course as interactive as possible and to offer opportunities for project-based teamwork. Particularly the practical (e.g. MS Project, time and effort estimation, peer reviews), game-based (e.g. online self test), and communication facilitation activities (e.g.

brainstorming on students' expectations) in the course enabled students to actively deal with project management. During workshops, teams performed tasks that referred to theoretical inputs or their team project. Student teams elaborated various project management topics which they presented in the face-to-face units.

Assessment of Students' Performance. Students were assessed according to the achievement of the following issues:

- *Individual project planning task*: Each student planned a small IT project using MS Project. The projects included about 25 activities, some milestones and resource planning. The aim was to ensure that students acquired basic skills in working with the software before they began working in their team.
- *Team project*: Students formed teams and chose a software project or an organizational project as their work context. The team had to work through a number of project phases such as project definition und vision; work breakdown structure, activity list, milestone plan, Gantt chart, risk analysis, etc.
- *Active participation*: Participation in the course units, in online self tests, in the online assessment, as well as in the final face-to-face assessment meeting were additional evaluation criteria.

4.2 Visual Course Design

Fig. 1 gives an overall design model of the Project Management course. It includes instructor activities and student activities amended with visual icons (introduced in Table 1) to emphasize the promotion of team competences in particular activities. The model includes various activities that are designed to promote, among other competences, students' teamwork competences. These activities include:

- *Team projects and team presentations*: typically involved the elaboration of a topic/milestone and its brief presentation in class via a PowerPoint presentation; projects were partly performed outside of class, partly in workshops in class and uploaded in a virtual space.
- *Constellation*: Constellation was used in the context of courses as an "icebreaker" and socializing game. In the back of the lecture room the participants were asked to choose their physical location according to their answers to questions asked by the instructor (e.g. length of study, distance to university, and experience with project management). This vivid exercise breaks the ice and favors peer exchange.
- *Helium stick*: The "helium stick" exercise is a typical team building exercise where participants stand in two lines in front of each other. A two-meter long stick is placed on their fingers by the instructor and they need to lower the stick in a way that each student's hands support the stick in each instant of its movement. Interestingly, the stick will initially move up rather than down until the people manage to coordinate their movements. This playful exercise helps to developing team-based strategies and is accompanied by collective discussions and reflections.
- *Team project management training*: Training included topics like roles in a project team as for example project manager, time and effort estimation of work packages, reallocating tasks in case of resource conflicts or due to unexpected events as staff shortfalls and handling of the tool MS Project.

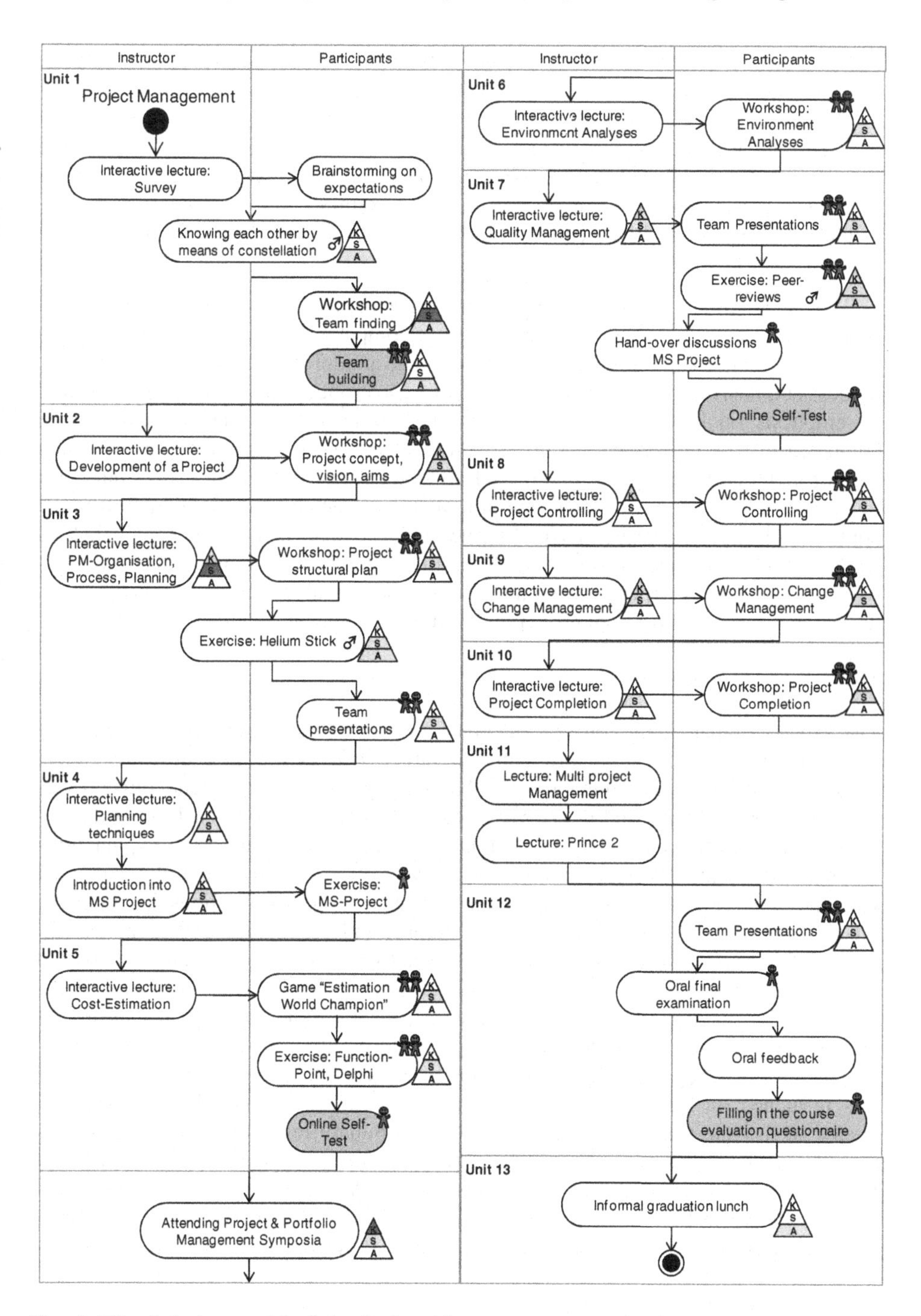

Fig. 1. Visual design model of the Project Management course (activity graph starts in left column and continues in right column)

- *Reflection*: Teams were asked to share their team experience in class (part of the final oral assessment) and online, for instance as part of the self evaluation in the online course evaluation questionnaire.
- *Peer reviews*: Peer reviews were employed both face-to-face and online. In the early stages of assigned projects, students reviewed their peers' project documents using a paper-based peer review form. They examined the documents and completed a review checklist. This was followed up by face-to-face a discussion with the partner team.

The model of the Project Management course includes 25 activities that are hypothesized to have a positive influence on team competences (i.e. those activities with a team competence pyramid icon attached). The model evidently displays that the highest emphasis is put on skill competences (2 × strong + 19 × moderate), followed by emphasis on knowledge competences (1 × strong + 6 × moderate) and attitude competences (6 × moderate). In comparison to design models of other courses (which are not included in this paper due to lack of space), the Web Engineering course for instance had lower emphasis on promotion of team competences, while the Soft Skills course had higher emphasis. In Web Engineering, the focus is more on technical skills using web technologies, while in Soft Skills the focus is almost completely on teamwork and related interpersonal competences. The actual effect on competences in these courses as rated by students is presented in the next sub-section.

4.3 Course Evaluation

The Project Management course (and other courses of the study) was evaluated by means of a post-hoc questionnaire including rating-scale as well as open-ended items. As depicted in Fig. 2, students indicated that in the Project Management course it was easier to work in teams and to establish positive interactions with each other than in other courses of the study. In an additional open-ended item they gave explanations for their judgments: "Very open atmosphere with a lot of talking and closer contact with the lecturers" — "It was very easy for me to establish relationships with the others because I was very interested in the subject and because the course matter was very well presented (one simply *had* to be active)" — "After completion of the theoretical part of the course we iteratively put it into practice, which promoted learning" — "By means of the relative free organization of teaching, by means of teamwork, by means of the pleasant atmosphere during the courses."

To find out whether students would judge Project Management to have more or less influence on team competences than other courses, additional items were included in the questionnaire. According to a Chi-Square test, students rated the perceived changes of their team knowledge, skill and attitude competences to be different for the courses considered (i.e., Soft Skills, Project Management and Web Engineering). As shown in Fig. 3, students rated the perceived effect of the Soft Skills course on these competences as higher than the effect of Project Management and Web Engineering, respectively. This is congruent with the distribution of emphasis on team competences in the visual design models. In accordance with the visual model of the learning activities included in Project Management and their influence on team competences, the course's effect on attitude competences was rated lower than the effect on knowledge and the top-rated skill competences.

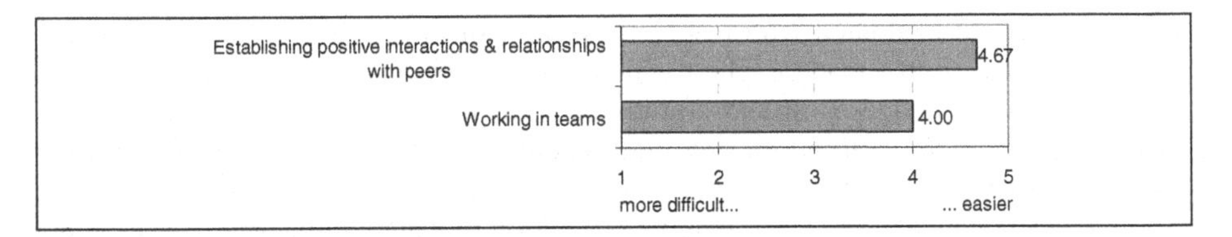

Fig. 2. Teamwork in Project Management in comparison to other courses of the Computer Science study (scale: 1 = more difficult … 5 = easier; n = 15)

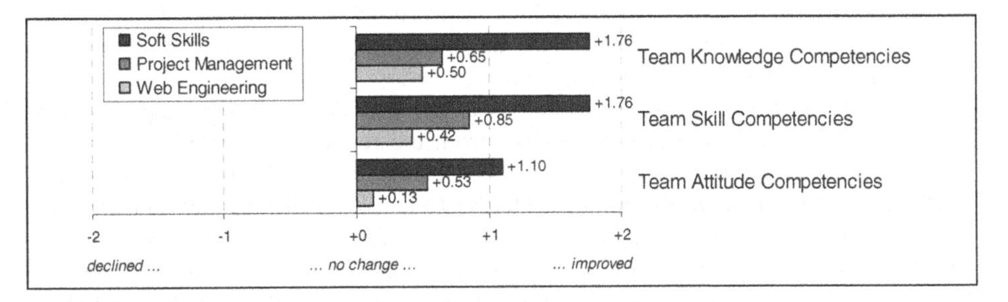

Fig. 3. Promotion of Team Competences in Project Management (n=20)

Additionally we asked students to assess if the Project Management course enhanced specific team-related skills (31 items) on a 5-point scale. Results showed that group decision making was fostered most strongly, followed by communication and flexibility competences and interpersonal relation competences were fostered least. Data reveals that students rated the overall influence on team competences highest in the Soft Skills course; however, when asked about specific team skill competences like gathering and sharing information or reallocating tasks, students rated the Project Management course to have more effect. In particular, the Project Management course had positive effects on competences like identifying possible alternatives or considering different ways of doing things. These positive evaluation results could be traced back to modeled learning activities concerning planning techniques, project environment analysis and project controlling, which requested students to deal with several eventualities that could happen in and to a project team.

5 Summary and Outlook

In discussions on key qualities for job qualifications, team competences are frequently mentioned as important generic competences; correspondingly, the promotion of team competences is more and more demanded in study programs. Since the inclusion of team competences and cooperative learning is a tough challenge for teachers and designers, this paper proposed a novel approach to modeling cooperative learning and team competence promotion, which is intended to support the courses design process.

UML activity models were extended with simple visual icons to enable the modeling of hypotheses about how courses would promote team competences. Additionally the paper presented a case study, in which a complete course on Project Management was modeled and evaluated with respect to promotion of team competences. Visual modeling of the course made explicit which course elements were designed for promoting team knowledge, skill and attitude competences. Empirical quantitative and qualitative evaluation results reflected the hypothesized and modeled positive influence of the course on team skill competences. The perceived changes in competences as rated by students demonstrated that the promotion of team competences at different levels (knowledge, skills, and attitudes) as planned in the visual design model was achieved during runtime. Survey data showed that the course was successful in fostering students' team skill competences like identifying possible alternatives or considering different ways of doing things in a project team. Additionally the case study illustrated the usefulness of employing visual models for identifying, planning and evaluating important learning activities.

Therefore, several possible contributions to the knowledge society, especially in the area of learning and education emerge from the research presented. Firstly, the process of modeling and testing hypotheses on competence and knowledge build-up supported by visual models can be transferred to a variety of teaching and training settings. There are several face-to-face as well as technology-enhanced learning scenarios besides the presented one, in which the concept can be used – for instance, in higher education as well as staff development. Secondly, the concept is not limited to the promotion of team competences, but could be generalized to address additional competences, since most learning activities include the levels of knowledge, skills and attitudes. While we acknowledge that planning for competence promotion can never immediately lead to actual effects, we point to the importance of thoughtful integration and planning of learning activities in today's knowledge society to support the creation, sharing and use of knowledge.

Acknowledgements. This research was supported by the University of Vienna through the projects "Technology-Enhanced Learning" (SP395001) and "E-Learning Umsetzung" (SP395002).

References

1. Cannon-Bowers, J.A., Tannenbaum, S.I., Salas, E., Volpe, C.E.: Defining competencies and establishing team training requirements. In: Guzzo, R., Salas, E. (eds.) Team effectiveness and decision making in organizations, pp. 333–380. Jossey Bas, San Francisco (1995)
2. European Commission: The Bologna process (2007), http://ec.europa.eu/education/policies/educ/bologna/bologna_en.html
3. European Commission: The European Qualifications Framework for Lifelong Learning (2008), http://ec.europa.eu/education/policies/educ/eqf/eqf08_en.pdf
4. Botturi, L., Stubbs, T.: Handbook of Visual Languages for Instructional Design. Information Science Reference, Hershey (2007)
5. Derntl, M., Motschnig-Pitrik, R.: coUML – A Visual Language for Modeling Cooperative Environments. In: Botturi, L., Stubbs, T. (eds.) Handbook of Visual Languages for Instructional Design: Theories and Practices, pp. 155–184. Information Science Reference, Hershey (2007)

6. Johnson, D., Johnson, R., Smith, K.A.: Cooperative Learning: Increasing College Faculty Instructional Productivity. The George Washington University, School of Education and Human Development, Washington, D.C. (1991)
7. Prichard, J.S., Bizo, L.A., Stratford, R.J.: The Educational Impact of Team-Skills Training: Preparing Students to Work in Groups. British Journal of Educational Psychology 76, 119–140 (2006)
8. Beatty, K., Nunan, D.: Computer-mediated collaborative learning. System 32, 165–183 (2004)
9. Strijbos, J.W., Martens, R.L., Jochems, W.M.G.: Designing for interaction: Six steps to designing computer-supported group-based learning. Computers & Education 42, 403–424 (2004)
10. Fellers, J.W.: Teaching teamwork: exploring the use of cooperative learning teams in information systems education. The DATA BASE for Advances in Information Systems 27, 44–60 (1996)
11. Reinmann-Rothmeier, G., Mandl, H.: Unterrichten und Lernumgebungen gestalten. In: Krapp, A., Weidenmann, B. (eds.) Pädagogische Psychologie, Beltz Verlag, Weinheim (2001)
12. Seymour, E., Hewitt, N.M.: Talking About Leaving: Why Undergraduates Leave the Sciences. Westview Press (1997)
13. Springer, L., Stanne, M., Donovan, S.: Effects of Small-Group Learning on Undergraduates in Science, Mathematics, Engineering, and Technology: A Meta-Analysis. Review of Educational Research 69, 21–51 (1999)
14. Johnson, D., Johnson, R., Stanne, M.B.: Cooperative Learning Methods: A Meta-Analysis (2000)
15. Johnson, D., Johnson, R.: Cooperative Learning And Social Interdependence Theory (1998)
16. Ruiz, U.B.C., Adams, S.G.: A conceptual framework for designing team training in engineering classrooms. In: Proc. American Society for Engineering Education Annual Conference & Exposition, Salt Lake City, UT (2005)
17. Ulloa, B.C.R., Adams, S.G.: Enhancing teaming skills in engineering management students through the use of the Effective Team Player – Training Program (ETP-TP). In: Proc. American Society for Engineering Education Annual Conference & Exposition (2004)
18. Buckenmyer, J.A.: Using Teams for Class Activities: Making Course/Classroom Teams Work. Journal of Education for Business, 98–107 (2000)
19. Parry, S.B.: Just What is a Competency? (And Why Should You Care?). Training 35, 58–64 (1998)
20. Baker, D.P., Horvarth, L., Campion, M.A., Offermann, L., Salas, E.: The ALL Teamwork Framework. In: Murray, T.S., Clemont, Y., Binkley, M. (eds.) International Adult Literacy Survey, Measuring Adult Literacy and Life Skills: New Frameworks for Assessment, vol. 13, pp. 229–272. Ministry of Industry, Ottawa (2005)

LAGUNTXO: A Rule-Based Intelligent Tutoring System Oriented to People with Intellectual Disabilities

Angel Conde[1], Karmele López de Ipiña[1], Mikel Larrañaga[1], Nestor Garay-Vitoria[1], Eloy Irigoyen[1], Aitzol Ezeiza[1], and Jokin Rubio[2]

[1] University of the Basque Country
[2] LEIA CDT

Abstract. In order to face the problems that people with disabilities find in their integration into working environments, one of the key issues is the implementation of solutions offered by new technologies by using what experts call "Support Technologies". The development of Intelligent Tutoring Systems (ITS) based on mobile platforms offers new perspectives for better integration of people with disabilities. The LAGUNTXO System aims to achieve the performance of human tutors, going a step beyond classical tutoring systems which perform organizational tasks. Due to the wide diversity related to people with disabilities, an intelligent structure that may achieve a convenient tutoring system configuration for each case has been incorporated. With an appropriate design of the structure and architecture of this task handler, it is very easy to operate by stakeholders. An automaton-based mechanism has been performed to technologically adapt the large amount of possibilities related to the interaction between people with disabilities, the task that is going to be made autonomously by these people, and the mobile system elements. In this paper, LAGUNTXO architecture, operational ways, and several use cases are presented.

Keywords: People with intellectual and physical disabilities, intelligent tutoring system.

1 Introduction

Integrating people with disabilities into working and social environments is one of the main issues in applying ITC into the assistive field. Particularly, it is necessary to pay special attention to the integration problem of people with intellectual disabilities.

This project's origin lies in a request of GUREAK ARABA S.L. (GRUPO GUREAK), a company that works in the integration of people with disabilities. GUREAK ARABA, S.L. considered that a computer aided support system may help to grow the autonomy, both in social and working environments, of users with intellectual disabilities.

Computer aided systems have been successfully applied in many fields [1]. Intelligent Tutoring Systems (ITSs) are computer-based instructional systems with models of instructional content that specify *what* to teach, and teaching strategies that specify *how* to teach [2, 3]. The ITS monitors the learner performance to determine the student's mastery on certain topics or tasks and how to satisfy his/her requirements by selecting the most appropriate pedagogical strategy and content to be taught.

M.D. Lytras et al. (Eds.): WSKS 2009, LNAI 5736, pp. 186–195, 2009.

Intelligent Tutoring Systems working into mobile platforms are an appropriate response to one of the main problems of people with disabilities: their integration into social and working environments. These devices are designed in order to reach the user adaptation and to obtain an interaction that compensates personal disabilities, for increasing the performance, individual autonomy, working capability, personal security and a healthy environment in workplaces [4].

Initially, the Intelligent Tutoring System will have to cope with several features:

- To allow tutoring every task of people with disabilities, giving more autonomy in working environments.
- To have a multimodal Task Management System for data integration from different sources (speech, images, videos, and text) associated with each personalized profile.
- To be integrated into a mobile platform, i. e. a mobile telephone or PDA.
- To contain a multimedia interface that has to be friendly, reliable, flexible, and ergonomically adapted.
- To integrate a human emotional predictive management in order to prevent risk, emergency and blockage situations that can damage these people and interfere with their integration into working and social environments.
- To be entirely configurable by stakeholders without technological knowledge in order to enable an easy and flexible access.
- To show the capability of exporting the system to other collectives, i. e. the elderly.

Ethical issues also have to be taken into account [5] while developing Assistive Technology, in the particular case of people with intellectual disabilities. The pitfall of generalization must be avoided. Therefore, it is of great importance not typifying each person with an intellectual disability with general labels.

Due to broad diversity of people with intellectual disabilities, we have incorporated an intelligent structure that may achieve an appropriate tutoring system configuration for each particular case. This implies a personal study and a related profile to each person, made by human tutors, caregivers or relatives. All these items lead to design a system with a configuration profile easily accessible to the stakeholder.

At the moment, the project has been carried out by a multidisciplinary research group with researchers from different fields such as Computing, Psychology, Medicine, and Engineering. These studies have also caused several works with social environment associations and companies devote to the industrial integration of people with disabilities. The level of success achieved within the project life is described in detail in the next sections. Section 3 describes in detail the architecture of LAGUNTXO SYSTEM. Finally, some conclusions and future work are shown in section 4.

2 Intelligent Tutoring Systems

Intelligent Tutoring Systems apply Artificial Intelligent techniques and methodology to the development of computer based learning systems in order to construct adaptive systems [2]. An ITS is based on the education as a process of cooperation between tutor and student. In general, the process is guided by the tutor who must analyse the behaviour, the mastery level and the satisfaction of the student. Tutor has to determine and apply the more appropriate teaching strategies at every moment [6]. These strategies must answer several questions to ensure that the learning process is successfully carried out [7]: what to explain, what level of detail is necessary, when and how to interrupt student, and how to detect and to correct errors. The four basic components that classically are identified in a ITS are the Domain Module, the Pedagogic Module, the Student Model and the Dialog Module [2, 3]:

- Domain Module: It contains the knowledge to be taught, and it fits pedagogical principles in order to facilitate the work to the Pedagogic Module. In this work, the domain module contains the information that guides people to perform the task they have to do on their environment.
- Pedagogic Module: This module determines the content or tasks to be assigned to the student, as well as the pedagogic strategy to be applied based on the information of the Domain Module and the Student Model.
- Student Model: It represents the belief of the system about the student's mastery during the instruction process. Besides, it includes information about his/her preferences, performance, motivation, and so on. It is used to observe and evaluate the learning progress of the student.
- Dialog Module: It provides the communication interface between the system and the user.

3 Architecture of LAGUNTXO System

The LAGUNTXO system has been developed in order to facilitate the integration of people with intellectual disabilities into their working and social environments. It has to achieve the performance of human tutors, going a further step than those classical tutoring systems [3, 7, 8] by dealing not only with the management of the tasks to be performed but also with the broad diversity of people with cognitive disabilities. A Task Management System (TMS) has been developed in order to achieve this goal.

In order to improve the integration of people with disabilities using computer supported systems, the characteristics of each person have to be considered in order to determine not only Domain Module of the tutoring system but also the device where it will be installed. The kind of disability may impose some constraints on the type of tasks and some devices might be more accurate for some users than others. This implies a personal study and a related profile for every person, made by human tutors, caregivers or relatives. However, the diversity of people with disabilities is so broad that tools for lightening this work are needed. LAGUNTXO provides an assisting tool that allows any stakeholder (tutors, caregivers and relatives) to configure the ITS in two dimensions considering the characteristics of the operational task and the diversity of the disabilities. In this sense, an automaton-based mechanism has been performed to technologically adapt the large amount of possibilities related to the interaction between people with disabilities, the task that is going to be made autonomously by these people, and the system elements. This mechanism is designed in a general way for providing some characteristics such as portability for people with different disabilities, as well as solutions for other communities, e. g. the elderly. The tasks that have been considered so far are related to working environments, like labelling products in stores and cleaning surfaces in complex buildings, but at the moment tasks devoted to independent living in tutored housed are being developed.

It is possible to configure different devices that are involved with different interfaces, for instance keyboards, touch screens, audio devices, and any combination of them. Furthermore, some devices for working in outside environments are considered.

Moreover, an emotional module to increase the reliability and tutor scope has been included. The emotional module analyses several non intrusive biomedical signals, for instance: heart rhythm, skin perspiration and relative movements. This module identifies emotional changes of those persons that are being tutored. By means of this identification, the critical blockage states will be detected. In this way, it will be possible to perform direct interventions for solving these eventualities.

For testing the emotional module, a new experiment set has been performed. This is made with a standard biometric testing system that obtains several biological signals. The experiment set consists of several changing environmental situations and the research of those tested signals through intelligent machine learning techniques.

Hence, the designed LAGUNTXO prototype has been structured in four main subsystems that will be described in next sections: the Task Management System (TMS), the Intelligent Tutoring System in Mobile Platform (ITS-MP), the Intelligent Dialog System (IDS), and the Human Emotions Analysis System (HEAS).

3.1 Task Management System (TMS)

The Task Management System (TMS) has been developed to overcome the lack of suitable tools that deal properly with the broad diversity of people with disabilities when defining the tasks to perform. In this work, a system that provides the configuration possibility has been created; it handles any possible case in separate profiles due to the diversity. The TMS is composed by three modules where the information is divided in the following parts:

- The tasks to carry out: Taking into account the work features, the specified characteristics of workers, the subtask divisions, etc.
- Ergonomic characteristics: Defining more specifically characteristics that can be used for increasing the reliability of the jobs perform by people with disabilities. It is relevant to introduce information about these people's interaction with the different devices of Intelligent Tutoring.
- Users' personal information: In order to know how workers can manage in different environments, it is necessary to introduce their personal profile. In this way, it is possible to prevent accidents and emotional blockage situations for avoiding personal and physical damage. The purpose of this information must be helping and attending these people with disabilities in the integration into social and working environments, not to control them. Thus, it has to be carefully stored and used.

In the TMS (Figure 1) this information is organized in several databases which can be continuously updated by users, tutors, caregivers and relatives. These databases are in a server in order to provide access from any remote stations for performing each device configuration. The particular profile configuration is loaded into the intelligent tutoring devices of any user. Encryption of databases and transmissions are ensured to prevent personal data misuse.

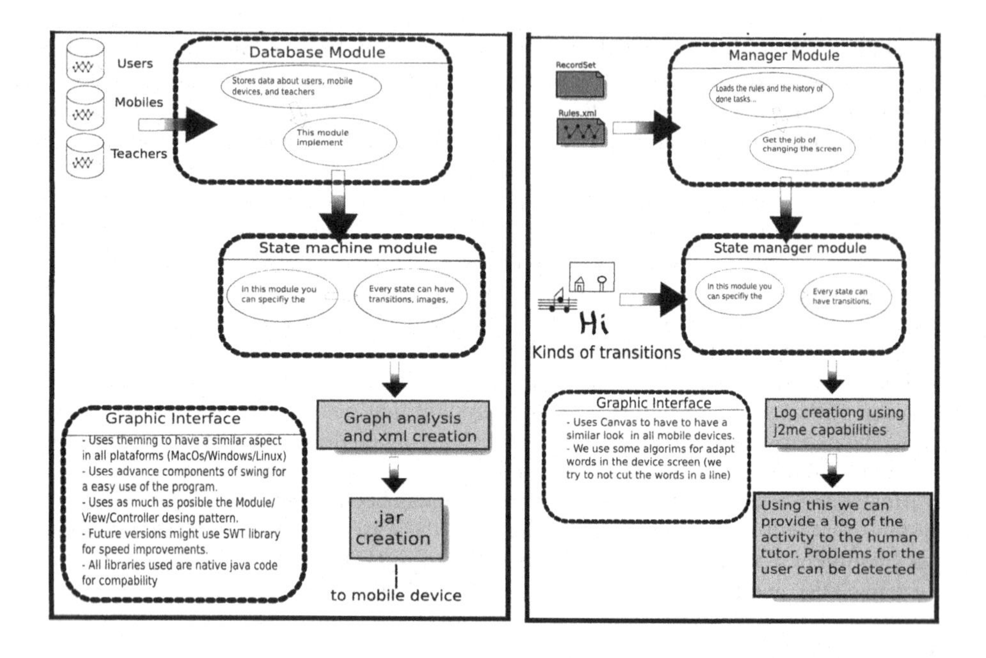

Fig. 1. Structure of the TMS

Fig. 2. Structure of the ITS-MP

Moreover, the TMS has been designed to allow a comfortable and simple configuration, giving to users an easy way to build the profile that will be loaded into ITS-MP (Figure 2). In this sense, TMS has inside an automaton-based mechanism supporting several functions. First, the automaton handles the communication with tutors, caregivers and relatives in order to allow better understanding of its functionality. Also, it organizes in a correct way the information supplied by users. Finally, it generates the characteristics map of all Intelligent Tutoring configurable devices which will be activated. Figure 3 shows an edition screen to configure automaton states. These states are organized in several levels which are connected by conditional transitions. Each state represents a different subtask to perform in order to solve the entire tasks. Depending on users disabilities, tasks profiles, and handled mobile platforms, appropriate states and transitions will be charged on those platforms. Furthermore, the appropriate media type (image, sound, etc.) will be used considering both, the user profile and the device features.

First data set to introduce into the database have been obtained by a previous study about the real situation of people with disabilities at different working environments. That information will be completed by human tutors, caregivers and relatives while observing how attended people develop different jobs with several mobile platforms. This study is carried out in several workshops of some social organizations, respecting all familiar and individual privacy rights, considering ethical questions, as well as observing the legislation under these circumstances.

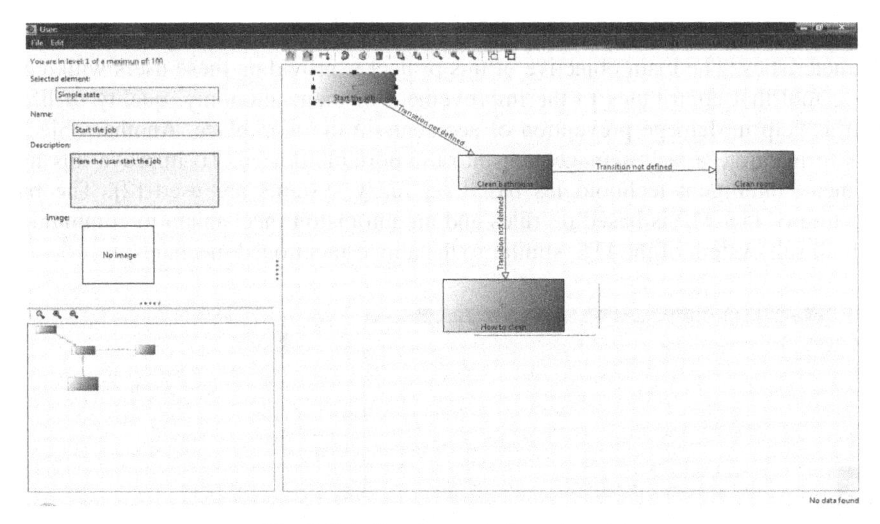

Fig. 3. Interface to create the automata

In this way, the performance of these workers and their integration process to the working and daily life would be enhanced. Besides, designed system is registered like a health product through a clinical research plan according to the current legislation.

Taking into account that the stored information covers a large diversity of cases frequently changing and it is necessary the adaptation to new technologies, the solution implemented allows adding new information to databases at any time. This strategy achieves intelligent tutoring with better assistance. TMS manages information about people with disabilities and existing mobile platforms. For every device, information that might be used in order to determine if it is appropriate for a particular user or characteristic is provided. Meanwhile the people list contains their personal information, as well as the personal involved tasks.

Database structure is composed by several states. These states have items as images, videos, texts, etc., configuring the skeleton of the task. There exist two states:

- Simple state: With a single feasible task, but not abstract description.
- Complex state: With a set of steps that has to be defined into the automaton-based mechanism. In order to adapt the feature of the task to one person, it will be necessary to define several particular items or steps.

For interconnecting states, different transitions have been created. Each transition has an associated condition for moving from one state to another. Initially, the number of transitions is unknown. This is the reason why new transitions have to be created by stakeholders. The program is user-friendly, reliable, usefulness, agreeable, with a clear interface to be used by people with low computing knowledge. These interfaces are presented on tables or menus, depending on the data handled (Figure 4).

3.2 Intelligent Tutoring System in Mobile Platform

Looking at ITS-MP in Figure 2, based on the characteristics of the people who will use these devices, it is absolutely necessary to design an interface that shows the following features: friendly, comfortable, flexible and ergonomically adapted to their characteristics. The main objective of this project is providing these users with a cognitive tool that contributes to the improvement of their autonomy, quality of life as well as help in damage prevention of accidents in the workplace. Another objective tries to integrate a task management into the portable device. To improve this management, intelligent technologies based on fuzzy systems are used [2]. The basic structure of ITS-MP is based on rules and an automaton mechanism to communicate with all sub-system of the ITS, similar to the above-mentioned mechanism.

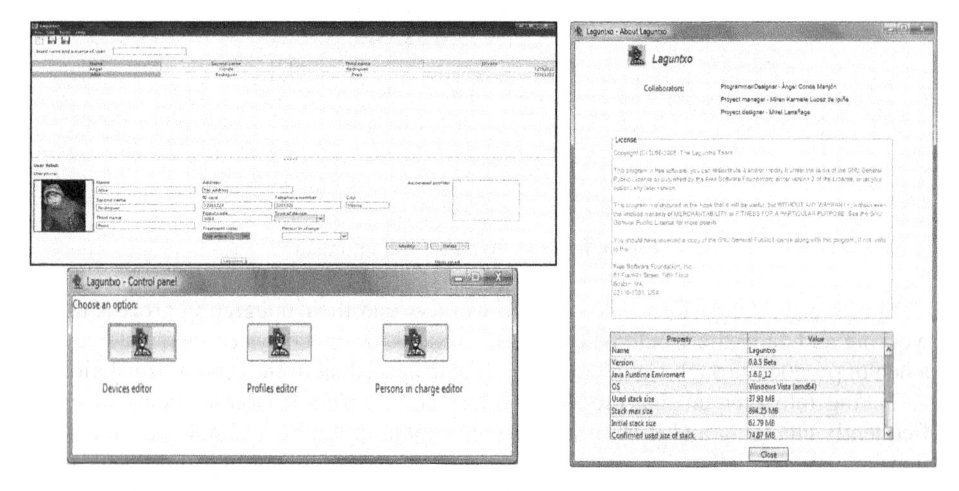

Fig. 4. Several interfaces for configuring the Task Management System

An interface example can be seen at Figure 5. It shows several screens where users can select the next subtask step by audio-video messages.

3.3 Intelligent Dialog System (IDS)

Linguist engineering and intelligent tools have been included in these systems in order to increase the reliability when they are used for tutoring people with disabilities, especially with intellectual disabilities [8]. Due to the integration and adaptation of these devices it is necessary to made bigger efforts in finding out adequate solutions. In the development of appropriate tools, the ergonomic directives as well as the specific needs of these persons are fundamental. For instance, people with intellectual disabilities have several physical and psychological common characteristics that have to be considered, such as heavy and fine mobility altered, smaller capacity to stay out, difficulty to anticipate or to understand consequences of their conduct, better visual perception and retention than auditory, longer response times, and difficulty in understanding instructions given in sequential form.

The IDS interacts with other sub-systems through the automaton mechanism. It uses the information of the automaton states and the user profile to present the information the user needs to perform the assigned task or subtask (Fig. 3).

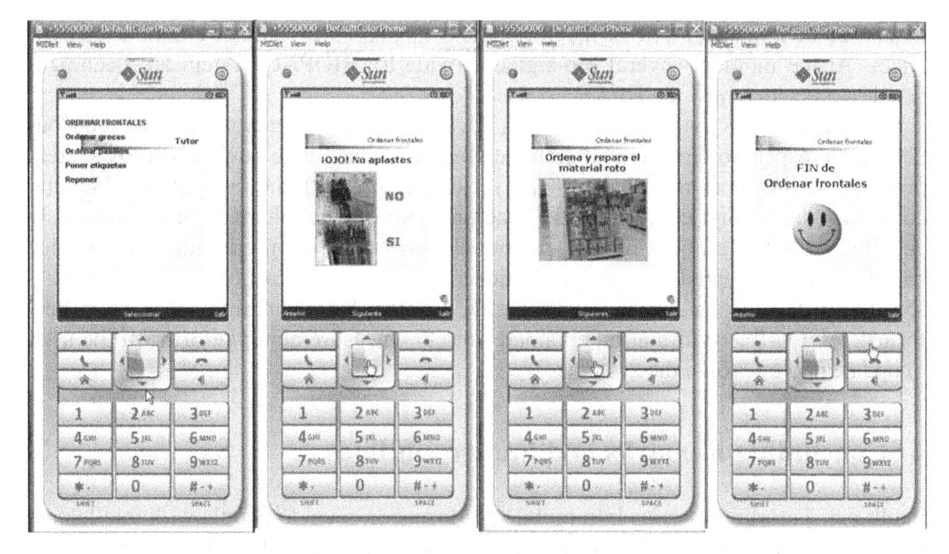

Fig. 5. ITS-MP interfaces

3.4 Human Emotion Analysis System (HEAS)

Human emotions appear as response to this changing and partially unpredictable world where any intelligent system (natural or artificial) needs the emotions for surviving due to limited abilities and multiple causes [9]. Emotions are composed by similar components to the cultural, subjective, physiologic, and behaviour components that express the personal perception with respect to the mental and body status, and the way for interacting with the environment [10].

Emotions are mechanisms that allow a description of the universe to the mind when there is not a symbolic representation. An artificial emotional system that generates and processes different emotions based on physiological reactions and predictable experiences (emotional memory, social emotion) can improve automatic systems where is included if interact with the environment.

One of the first objectives of this work is to develop a convenient measuring system for identifying non visible human emotions, by mean of human behaviour emulation based on an automatic emotional learning. Emotion per se, it is an interdisciplinary topic which can be studied in Philosophy, Neuroscience, Computational Intelligence, Machine Learning, and Robotics fields [11]. At the same time, the crucial question of data and possible databases to be used in emotion research has to be addressed, like the ontology described in [12].

In this work a different measuring system has been developed incorporating both individual artificial emotional patterns (emotional data base of human emotional patterns) and emotional memories (data bases of human experiences). LAGUNTXO platform will be more accurately managed by including human emotion analysis. This system also gives information about emotional state transitions, in order to prevent potential blockage situations.

The devices intended for the data capture of emotional states information will measure heart rhythm, body temperature, movements, facial expressions and blood pressure. The new human emotions model will rest on emotional human patterns, databases of human emotions memories, and databases of human emotions experiences. At the moment several bio-signal provide by BIOPAC system are been analysed in order to adjust the HEAS.

The final HEAS system will contain several technical innovations and contributions with respect to the classical architectures so far used. The new measuring system proposed in this work is building by a hybrid structure with the integration of several components. The first component is a machine learning module, trained by previously acquired knowledge about human emotional answer. Also, an emotional knowledge based system is created. Simple perception information is measured by non-intrusive sensors to develop an Emotional Predictive Control based on simulating brain performances. On-line information obtained from the measurement platform will be used to update and to evolve the system.

4 Concluding Remarks and Future Outlines

Intelligent Tutoring Systems into mobile platforms oriented to people with intellectual disabilities have been developed. Due to the wide casuistic of the problem a friendly, comfortable, flexible and ergonomically adapted system has been designed. User-centred approach has been suited when developing all modules of the system.
A prototype called LAGUNTXO has been developed. It works as an active distributed support system, and allows compensating user disabilities through task programming, facilitating suspension and resume of tasks, offering help in blockage situations and reminding key points or steps of a task. It also facilitates tasks by offering to stakeholders an easy and innovative tool to be used specially in training processes.

Pilot tests indicate a high level of satisfaction of both the users and the stakeholders. On the one hand, it is increasing the users' autonomy, improving their training and the quality of their work. On the other hand, stakeholders have a new and easily configurable tool that reduces the time they have to dedicate in training. Especially in the working environment, the better quality of service and the cost reduction in hours can be an opportunity to advance in professional integration of disabled people collective. In any case, in order to get more significant results new tests will be made in the new situations that are being created at the moment.

Despite everybody could benefit of the results of research on assistive technology, accessibility and intelligent environments, participation of people with disabilities in these research experiments is fundamental, since these persons have many great difficulties of adaptation, and they are very sensible to bad technological design.

Moreover, the job made so far has offered an opportunity to establish a solid link of communication between the social world of disabled people, caregivers and trainers, and the technological world of researchers. This link is leading to improve the knowledge and to overcome the gap between these two often separated worlds.

In these sense, this work is already being done to fulfil the new challenges that the project collaboration is generating. Complementary tools are being integrated in the system to improve its usefulness and to include the system within an ambient intelligence architecture. These tools are based on pattern matching (images and speech), human emotional feeling analysis, contextual information deployment (GPS, networking), and Artificial Intelligent (AI) techniques, giving the system the capacity of dynamic adaptation to the learning process.

References

1. Sammour, G., et al.: The role of knowledge management and e-learning in professional development. International Journal of Knowledge and Learning 4(5), 465–477 (2008)
2. Wenger, E.: Artificial Intelligence and Tutoring Systems. Morgan Kaufmann, Los Altos (1987)
3. Yazdani, M.: Intelligent tutoring systems: an overview. In: Lawler, C.R., Yazdani, M. (eds.) Learning environments & tutoring systems, pp. 182–201. ablex, norwood (1987)
4. Kenny, C., Pahl, C.: Personalised correction, feedback and guidance in an automated tutoring system for skills training. International Journal of Knowledge and Learning 4(1), 75–92 (2008)
5. Ezeiza, A., et al.: Ethical Issues on the Design of Assistive Technology for people with mental disabilities. In: International Conference on Ethics and Human Values in Engineering, pp. 75–84 (2008)
6. Nowak, E.: The role of anthropometry in design of work and life environments of the disabled population. Department of ergonomics research. Institute of industrial design Press, Poland (1999)
7. Chia-fen, C.: A study on job placement for handicapped workers using job analysis data, Department of industrial management. National Taiwan University of Science and Technology Press, Taipei (2002)
8. García, J., et al.: Intelligent Tutoring System to Integrate people with Down Syndrome into work environments. In: International Conference on Education. IADAT. Innovation, Technology and Research on Education, pp. 120–123 (2006) ISBN: 84-933971-9-9
9. Cañamero, L.: Emotion understanding: From the perspective of autonomous robots research. Neural Networks 18, 445–455 (2005)
10. Cowie, R., Douglas-Cowie, E., Cox, C.: Beyond emotion archetypes: Databases for emotion modelling using neural networks. Neural Networks 18, 371–388 (2005)
11. Taylor, J., et al.: Emotion and brain: Understanding emotions and modelling their recognition. Neural Networks 18, 313–316 (2005)
12. López, J.M., Gil, R., García, R., Cearreta, I., Garay, N.: Towards an ontology for describing emotions. In: Lytras, M.D., Carroll, J.M., Damiani, E., Tennyson, R.D. (eds.) WSKS 2008. LNCS (LNAI), vol. 5288, pp. 96–104. Springer, Heidelberg (2008)

Extending the REA Ontology for the Evaluation of Process Warehousing Approaches

Khurram Shahzad

Department of Computer and Systems Science (DSV),
Royal Institute of Technology (KTH) / Stockholm University, Sweden
mks@dsv.su.se

Abstract. A data warehouse developed for analysis of business processes is called Process Warehouse (PW). The design of process warehouse is seldom evaluated, since a generic PW model that can be used as a benchmark for the evaluation is missing. Therefore, in this paper, we extend the REA ontology and use it for developing a generic PW model that can further be used as criteria for evaluation of PW design. Moreover, the generic PW model is used to evaluate various process warehousing approaches, collected through a comprehensive survey.

Keywords: Business Process Management, Process Analysis, Business Process Monitoring, REA Ontology, Process Warehouse.

1 Introduction

Process Analysis (PA) is used as a tool that helps in the identification of bottlenecks, in making operating decisions and to improve business processes [1, 2]. PA is used for monitoring, optimization and it works as a milestone in continuous process improvement [3]. In order to analyze business processes the data collected from the execution of processes is used [2]. For some years, data warehouse (DW) technology has been used for business analysis. A DW that can be used for the analysis of business processes is called Process Warehouse (PW) [4].

The REA (Resource, event, agent) ontology has been in existence since 1980's [5]. It is used to model information about the agents who trigger an event and the resource transferred or consumed during the event [6, 7]. Also, it has been recognized as an established way of modeling business processes, so, it should be possible to develop a benchmark (based on REA ontology) for analysis of business processes.

A number of efforts [8, 9, 10] have been made to design process warehouses, but evaluation and comparison of PW designing approaches is missing. This is due to the absence of a generic PW model that can be used as a benchmark for the evaluation and comparison of these approaches.

The purpose of this study is to extend the REA ontology and use it for developing a generic PW model that can further be used as criteria for the evaluation of PW design. The extensions to REA are required because it can't support multi-level representation of its concepts in its current form, therefore, we extend REA by developing

M.D. Lytras et al. (Eds.): WSKS 2009, LNAI 5736, pp. 196–206, 2009.

hierarchies of its core concepts (resource, event, agent). From the extended REA a generic PW-model is developed, and thereby used as a benchmark for evaluation of PW approaches.

The rest of the paper is organized as follows. Section 2 introduces the extended REA ontology. Section 3 contains a generic process warehouse model. In section 4, results of the evaluation of process warehousing approaches are given. A discussion about evaluation, conclusions and future research directions is given in section 5.

2 Extending the REA Ontology (E-REA)

The REA (Resources, Events, Agents) ontology was originally proposed by McCarthy [5]. It was a domain specific framework that has been used for designing values transfer between actors and it can be tracked back to double-booking keeping in accounting systems [11]. Subsequently, REA has been enhanced to be used in various fields like e-commerce [12]. The REA core concepts are defined as [13]:

Resource, Event and Agent are the core concepts of REA ontology, as shown in fig.1. Resources are the entities that are of value for a business, e.g. book, table, data, information etc. A resource can be either converted to another resource or it can be transferred (exchanged) to another agent. For example, flour (a resource) can be converted to bread (a resource) and it can also be transferred from one to another actor. However, there are some resources that are neither convertible nor transferable between agents [7].

Events are the business transactions that take place at a certain time e.g. booking, renting, validating or invoicing, etc. As an outcome of an event, a resource is either transferred from one agent to another or converted from one to another form depending upon the nature of the event e.g. the nature of baking (an event) is conversion and the nature of selling (an event) is exchange.

Agents are the entities that are custodial (have control) of a resource e.g. a person, company, department or a business unit etc. Agents are the entities that trigger an event as an outcome of which values are transferred. For example, a customer (an agent) purchases bread from a store (an agent).

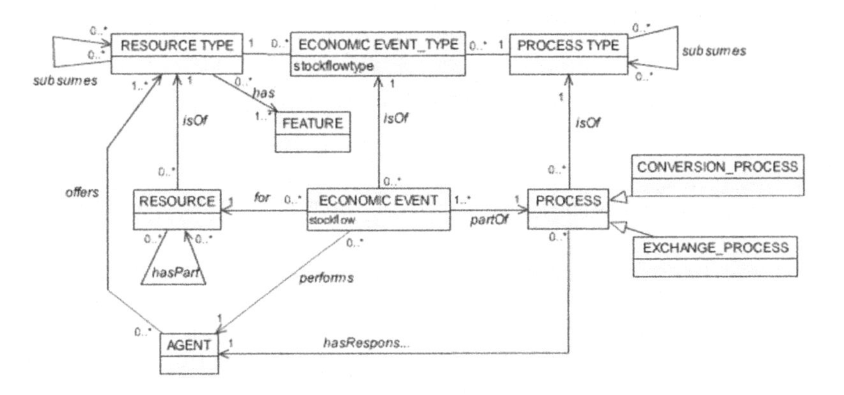

Fig. 1. REA Ontology Core Concepts [7]

Motivation for using REA ontology: REA ontology has been extended to form a basis for enterprise information systems architectures [11], and it is applicable in a number of domains, for instance e-commerce frameworks [12], relational databases [14], model driven design of software applications [15] and in teaching accounting information systems [5]. It is also used for modeling business processes in order to represent transfer/conversion of resources between actors at high level [6]. Unlike other approaches, REA not only reveals why a business process occurs but also proves traceability of business transactions [7]. Therefore, due to wider applicability of REA ontology it is included in this study.

Motivation for extending REA: Dimensional schemas of a process warehouse (PW) support analysis of business processes from high level to low level of details. This multilevel analysis can be done by drill-up and drill-down operations (slicing and dicing operations) on dimensions of a PW, where each level of detail is represented by a granularity level. On the other hand, REA is capable of representing business processes at a single level of detail, so, information about different levels of details cannot be captured by the generic REA Ontology presented in fig. 1.

A process in REA only represents changes in value of the resources and it does not address control flow aspects [14]. Moreover, each business process has its input, output and a number of activities that are performed in a specific order [17], which can not be represented by the REA ontology. These limitations of REA ontology motivate the need for extending core REA concepts.

The Extended REA (E-REA): In this section, based on literature, we present extensions to the core concepts of the REA ontology.

Resources can be of two types, traditional and information resources [18]. a) Traditional resources, these are the entities that are of value for a company. Traditional resources can be of different nature some of them are concrete and transferable (like book) and others are psychological and cannot be directly transferred (like pleasure, status) [7]. These are called tangible and non-tangible resources respectively. b) Information resources, it is the data related to an activity produced during realization of a business process. The information is related to the input resources required for the process, the resources consumed during the process and the output generated by the process, as shown in figure 2(a). It is there, in order to represent the specific input and output and data related to a process.

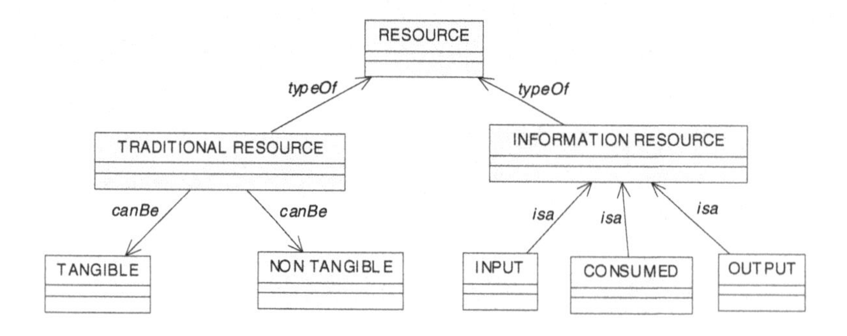

Fig. 2. (a) Resource hierarchy

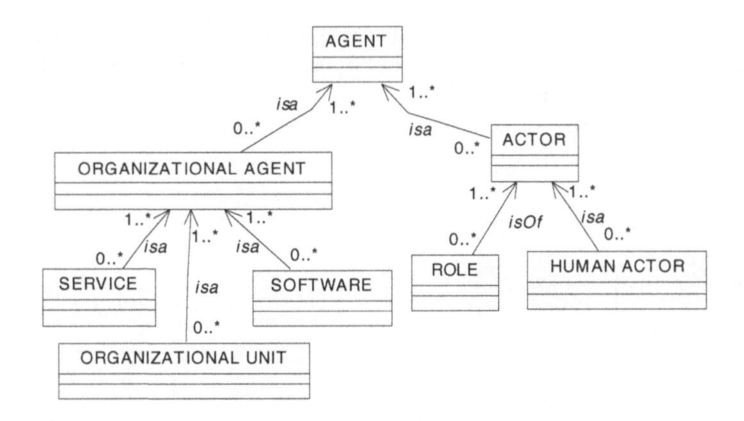

Fig. 2. (**b**) Agent hierarchy

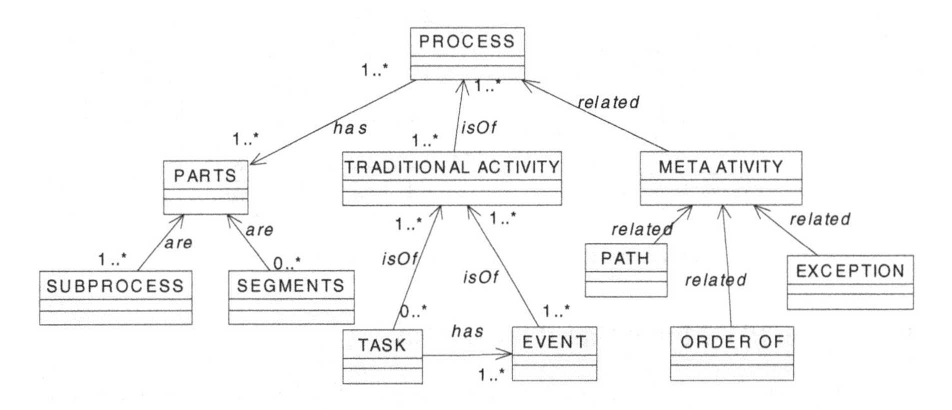

Fig. 2. (**c**) Process hierarchy

Agent has between divided into two major categories, a) actor, is the individual who participate in a process. It could be human actor or the role assigned to human actor. b) Organizational agent, is the part (subset) of organization which has participated in a process. An organizational agent could be, a unit of organization, a service provided by external organization or software that is working as an agent [18], as shown in figure 2(b). An activity (event) can be performed internally by an organization or with the help of a service from outside the organization. Therefore, service is considered as an external organizational agent, which can perform an activity for an organization.

According to [17], a process consists of a number of activities that are performed in an order. Also, a process may have subprocesses [19] and it can be divided into smaller segments [20]. Our process hierarchy consists of parts of a process, activities of process and meta-activity of a process. Meta activity contains detailed meta information (like order of activities, different paths or exceptions) related to activities.

We extend the REA ontology (presented in figure 1) by adapting the changes presented in figure 2 (a, b, c). On putting all the concepts together, the Extended REA

(E-REA) is shown in figure 3. It is notable that, a) the concepts from figure 1 that are not related to the extensions are skipped in figure 3, to reduce the representation complexity, b) the core concepts of REA and their relationships are represented by dotted boxes in order to clarify the extensions, c) strong borders represent the root of the hierarchies in figure 2 (a, b, c).

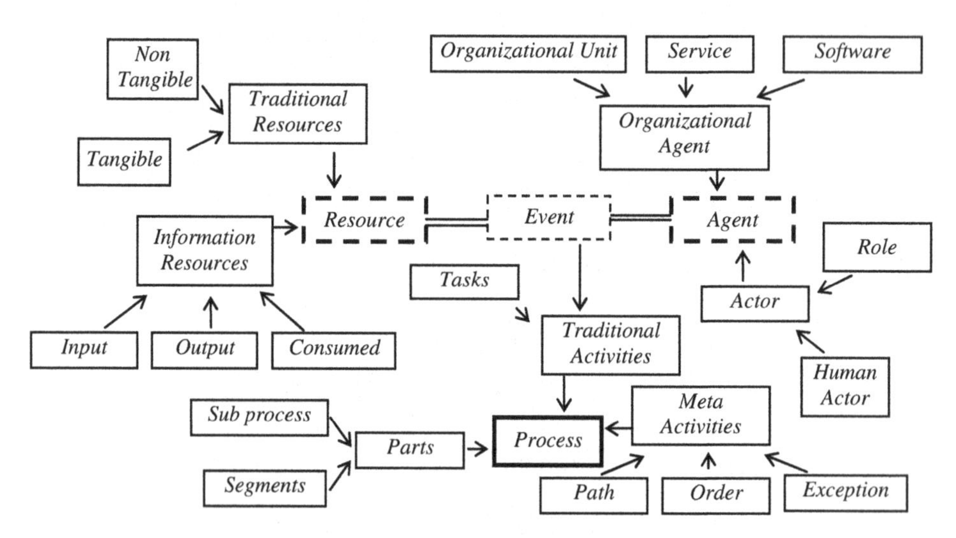

Fig. 3. Extended REA (related information only)

3 Generic PW Model Based on E-REA

Dimensional model (DM) has emerged as an alternative to the relational data model, optimized for data analysis. DM has measures (to be analyzed) surrounded by different dimensions (known as perspectives of analysis) [21]. For designing data warehouse, DM is used because of its two-fold benefits [28], on one hand, it is close to the way of thinking of data analyzers and therefore it helps users in understanding data. On the other hand, it allows designers to predict user intentions. Similarly, a process warehouse that is used for analysis of processes has a DM that facilities analysis of processes from various perspectives.

It has been established that process analysis can effectively improve efficiency of execution of process, reduce cost and increase productivity [1, 2, 4]. Here, we derive the dimensions that can be used for multi level analysis and optimization of resources, agents and processes from the Extended REA (E-REA). For each dimension, we discuss three things, how optimization can be done, what questions are answered by the dimension and how multilevel analysis is supported.

Resource: Process analysis can be used for optimal utilization of resources [4]. This can be done by analyzing affect of presence of resource, disruption caused by the resource, demands on resources and resource allocation etc. The resource dimension answers several questions like what are different types of resource, categories of resources, input, output and consumed resources etc. For multilevel analysis of resources, the resource hierarchy given in fig 2 (a) is used in the resource dimension.

Process: Process analysis can effectively be used for optimal execution of a process and its activities [4]. This can be done by analyzing activities, bottlenecks, throughputs and accessibility of actors etc. This dimension answers several questions like which sub processes constitute a process, how many segments of a process and there are and the activities that form a process. For multilevel analysis of process, the process hierarchy given in fig 2 (c) is used in the resource dimension.

Agent: Process analysis can be used for optimal utilization of agents [4]. This can be done by the analysis of an actor's performance, task assignment, workload prediction and avoiding unnecessary hiring. This dimension answers several questions like, what are the agents, how many tasks are assigned and completed, when agents are available and how they are treated. For multilevel analysis of process, the process hierarchy given in fig 2 (b) is used in the resource dimension.

Time: This is one of the most important dimensions of DM and in various studies it has been emphasized as a compulsory part of warehouse definition [21]. In a process warehouse, the time dimension can be used for the comparison of recent execution of a process with previous process executions. The standard time hierarchy is used for this purpose.

By putting all these dimensions and hierarchies together, we have developed a generic model for process warehouses (shown in fig 4). The dotted boxes represent the elements that are common between the generic meta-model and REA. It is important to observe that the hierarchies are borrowed from the extended REA model but due to lack of space complete hierarchies are not presented.

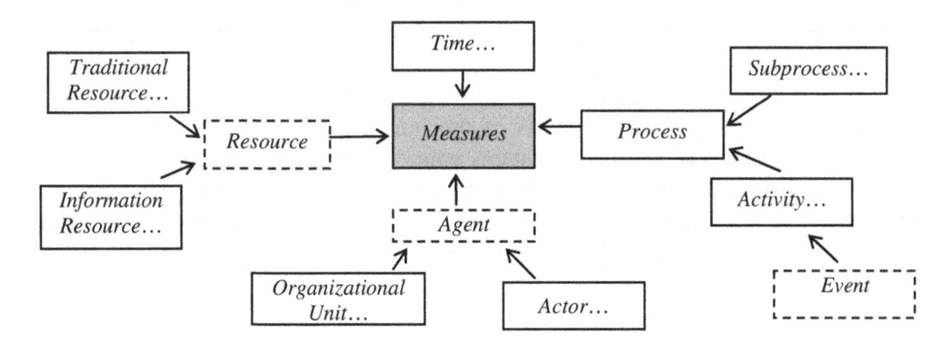

Fig. 4. A Generic Process Warehouse Model

4 Evaluation of PW Approaches Using the Generic PW Model

Using the generic PW model we evaluate nine approaches of designing process warehouses. In this section, we evaluate whether the important analysis aspects (or dimensions given in fig.4) are supported by the approach or not.

The approaches are selected by a comprehensive survey of major digital libraries of known publishers. For searching through these repositories we used several keywords and phrases related to process warehousing, dimensional modeling and business process analysis. We also searched through all the proceedings of major conference related to databases and information systems. After this survey we only selected the approaches that present multidimensional schema of process warehouse.

The approaches under investigation are: a) Goal-driven design of a DW based (GD) [8], b) DW technology for Surgical workflow (WS) [9], c) multidimensional data model (MD) [22], d) generic warehouse for business process data (GW) [10], e) Performance warehouse (PW) [23], f) goal-oriented requirement for DW design (RD) [24], g) real time process data store (PD) [25], h) business oriented development of DW structure (PS) [26], i) DW for audit trail analysis (DT) [27].

The evaluation results are represented on a scale of Yes/No. The values is a) Yes, if the dimension is available, b) No, if the dimension is not available. Also we add a sign (-) beside yes if the aspect is discussed in the paper but it is not modeled as a separate dimension. Table 1-4 shows the results of the evaluation of nine approaches.

Table 1. Resource Dimension

	GD	WS	MD	GW	PW	RD	PD	PS	DT
Traditional Resource	Yes	Yes	Yes	No	No	Yes	Yes	Yes	No
Tangible	No	Yes	Yes	No	No	No	Yes	Yes	No
Non-Tangible	Yes	Yes	Yes	No	No	Yes	Yes	Yes	No
Information Resource	Yes	Yes	Yes	No	No	Yes	Yes	Yes	No
Input	Yes	Yes	Yes	-Yes	No	Yes	No	Yes	No
Output	Yes	No	Yes	No	No	No	Yes	No	No
Consumed	No	No	No	No	No	No	No	No	No

In table 1 the results of the evaluation of resource dimension are presented. There is no approach that provides analysis of the information resources consumed in a process. However, GW approach contains some discussions on inputs of processes but inputs are not available as dimensions. By using GW, PW and DT approaches it is not possible to analyze traditional resources. Moreover, several approaches are incapable of supporting output analysis of processes.

Table 2. Agent Dimension / participant

	GD	WS	MD	GW	PW	RD	PD	PS	DT
Actor	Yes	Yes	Yes	Yes	Yes	Yes	Yes	Yes	Yes
Human Actor	Yes	No	No	No	Yes	Yes	Yes	Yes	Yes
Role	Yes	Yes	Yes	Yes	Yes	Yes	Yes	Yes	Yes
Organizational Agent	Yes	Yes	Yes	Yes	Yes	No	No	No	Yes
Service	No	No	No	No	No	No	No	No	No
Software	No	Yes	Yes	-Yes	No	No	No	No	No
Organizational Unit	Yes	Yes	Yes	No	Yes	No	No	No	Yes

Analysis of Actors is supported by all the approaches. However, organizational agents' analysis are usually missing. The exceptions are that, organizational unit analysis is supported by some approaches (GD, WS, MD, PW, DT). Some discussion about software is available in GW approach but it is not available in the form of a separate dimension. Moreover, service analysis is also missing in all the approaches. The detailed evaluation results of the approaches are given in table 2.

Table 3. Process Dimension

	GD	WS	MD	GW	PW	RD	PD	PS	DT
Parts of Process	No	No	Yes	Yes	Yes	No	Yes	No	No
Sub Process	No	No	Yes	-Yes	-Yes	No	-Yes	No	No
Segment	No	No	No	No	No	No	No	No	No
Traditional Activity	Yes	Yes	Yes	Yes	Yes	Yes	Yes	-Yes	Yes
Event	No	Yes	Yes	No	No	Yes	Yes	-Yes	Yes
Task	Yes	Yes	Yes	Yes	Yes	No	Yes	No	Yes
Meta Activity	No	Yes	Yes	Yes	Yes	No	Yes	No	Yes
Path	No	No	-Yes	No	No	No	Yes	No	Yes
Exception	No	No	No	No	Yes	No	No	No	No
Order	No	No	-Yes	-Yes	No	No	No	No	-Yes

More than half of approaches (GD, WS, RD, PS, DT) don't support analysis of part of process. However, there are some approaches in which sub process is discussed but not available as an independent dimension. Segments can not be analyzed by using any approach. Task analysis is supported by most of the approaches but the information related to the meta activity is not available in most of the approaches. Exceptions cannot be analyzed, different parts can not be followed by most of the approaches and execution order is not available as independent dimension although they have been discussed. The detailed evaluation of approaches is given in table 3.

Table 4. Time Dimension

	GD	WS	MD	GW	PW	RD	PD	PS	DT
Time	Yes	Yes	Yes	Yes	Yes	Yes	Yes	Yes	Yes

Time dimension is a key part of data warehouse design and the time dimension is available in all the approaches. It is used for analysis and comparison of various metrics with overtime changes.

5 Discussion and Conclusion

Efforts have been made to design a process warehouse, but evaluation and comparison of process warehouse approaches is missing. This is due to the absence of a generic PW model that can be used as a benchmark for the evaluation and comparison of these approaches. Therefore, in this study REA ontology is extended and used for developing a generic PW model. A brief discussion on evaluation of process warehousing approaches is as follows:

Analysis of traditional resources is supported by some approaches (like WS, MD, PD and PS). In contrast to it, information resources are not fully supported by any approach. However, partial support is available in all the approaches except PW and DT. Analysis of actors is supported by most of the approaches (GD, PW, RD, PD, PS, DT), whereas complete analysis of organizational agents cannot be done by any approach. This is because services acquired by the organization cannot be analyzed.

Multilevel analysis of a process and its activities is partially supported by all the approaches because analysis of parts of process and segments of process cannot be done due to the absence of their granularity levels in approaches like (GD, WS, PS, DT). Activity analysis is supported by many approaches but meta information about activities is not available in any approach. Time dimension is available in all process analysis approaches.

In our study, we present a generic PW model and based on the evaluation of process warehousing approaches we conclude that: a) design of PW model can be evaluated by using the generic PW model, b) time dimensions is not ignored by any approach, c) no process warehousing approach supports complete analysis of business processes.

There are some limitations of the study, a) during the evaluation of PW approaches some dimensions are found that are not covered by the generic PW model. This is becausPe REA is high level therefore the generic PW model that is developed based on REA is not complete. b) The scope of the generic PW model is limited to dimensions and measures are neither discussed nor evaluated and compared in this study.

Future research aims to develop a goal-oriented approach for collecting requirements of a process warehouse design and a method for designing process warehouse that supports complete business process analysis.

References

1. Grigori, D., Casati, F., Castellanos, M., Dayal, U., Sayal, M., Shan, M.C.: Business Process Intelligence. Computer in Industry 53(3), 321–343 (2004)
2. Aalst, W.M.P., Dongen, B.F., Gunther, C.W., Mans, R.S., Medeiros, A.K., Rozinat, A., Rubin, V., Song, M., Verbeek, H.M.W., Weijters, A.J.M.M.: ProM 4.0: Comprehensive Support for Real Process Analysis. In: Kleijn, J., Yakovlev, A. (eds.) ICATPN 2007. LNCS, vol. 4546, pp. 484–494. Springer, Heidelberg (2007)
3. `http://oeas.ucf.edu/process_analysis/what_is_pa.htm` (last accessed, 13 March 2009)
4. Junginger, S., Kabel, E.: Business Process Analysis. In: eBusiness in Healthcare, pp. 57–77. Springer, London (2007)
5. McCarthy, W.E.: The REA Accounting Model: A generalized framework for accounting systems in a shared data environment. The Accounting Review LVII (3), 554–578 (1982)
6. Dunn, C.L., McCarthy, W.E.: Conceptual Models of Economic Exchange Phenomena: History's 3rd wave of Accounting Systems. In: Proceedings of the 6th World Congress of Accounting Historians, pp. 133–164 (1992)

7. Johannesson, P., Andersson, B., Bergholtz, M., Weigand, H.: Enterprise Modelling for Value Based Services Analysis. In: Proceedings of the 1st International IFIP WG 8.1 Working Conference on Practices of Enterprise Modeling. LNBIP, vol. 15, pp. 153–167. Springer, Heidelberg (2008)
8. Niedrite, L., Solodovnikova, D.: Goal-Driven Design of a data warehouse based business process analysis system. In: Proceedings of the 6th WSEAS International Conference on Artificial Intelligence, Knowledge Engineering and Database, vol. (6), pp. 243–249 (2007)
9. Neumuth, T., Mansmann, S., Scholl, M.H., Burgert, O.: Data warehouse technology for surgical workflow analysis. In: 21st IEEE International Symposium on Computer-based Medical Systems, pp. 230–235 (2008)
10. Casati, F., Castellanos, M., Dayal, U., Salazar, N.: A generic data warehousing business process data. In: Proceedings of the 33rd International Conference on Very Large Databases (VLDB 2007), pp. 1128–1137 (2007)
11. Geerts, G.L., McCarthy, W.E.: The Ontological Foundation of REA Enterprise Information Systems. Paper presented at the Annual Meeting of the American Accounting Association. Philadelphia, PA (2000)
12. UN/CEFACT Modelling methodology, http://www.unece.org/cefact/umm/UMM_userguide_220606.pdf
13. REA Technology, http://reatechnology.com/what-is-rea.html (last access, 13 March 2009)
14. Geerts, G.L., McCarthy, W.E.: Using Object Templates from the REA Accounting Model to Engineer Business Processes and Tasks. The Review of Business Information Systems 5(4), 89–108 (2001)
15. Hruby, P.: Ontology-Based Domain-Driven Design. In: Proceedings of the Best Practices for Model Driven Software Development workshop, in conjunction with OOPSLA 2005, San Diego, USA (2005)
16. Jarke, M., List, T., Koller, J.: The Challenge of Process Data Warehousing. In: Proceedings of the 26th International Conference on Very Large Data Bases (VLDB 2000), Cairo, pp. 473–483 (2000)
17. The Business Process Model- UML Tutorial, http://www.sparxsystems.com/downloads/whitepapers/The_Business_Process_Model.pdf (last accessed, 13 March 2009)
18. List, B., Korherr, B.: An evaluation of conceptual business process modelling languages, In: Proceedings of ACM Symposium on Applied Computing (SAC 2006), Dijon, pp. 1532–1539 (2006)
19. MIT repository, http://process.mit.edu/Activity.asp?ID=1700 (last accessed, 13 March 2009)
20. Sampaio, P., He, Y.: Process design and implementation for customer segmentation e-services. In: Proceedings of the IEEE International conference on e-technology, e-commerce and e-service, pp. 228–234 (2005)
21. Kimball, R.: The Data Warehouse Toolkit: The Complete Guide to Dimensional Modeling, 2nd edn. John & Wiley, Chichester (2000)
22. Mansmann, S., Neumuth, T., Scholl, M.H.: Multidimensional data modeling for business process analysis. In: Parent, C., Schewe, K.-D., Storey, V.C., Thalheim, B. (eds.) ER 2007. LNCS, vol. 4801, pp. 23–38. Springer, Heidelberg (2007)
23. Kueng, P., Wettstein, T., List, B.: A holistic process performance analysis through a performance data warehouse. In: 7th Americas conference on information systems (AMCIS 2001), pp. 349–356 (2001)

24. Giorgini, P., Rizzi, S., Garzetti, M.: Goal-oriented requirement analysis for data warehouse design. In: Proceedings of 8th ACM International Workshop on Data warehousing and OLAP (DOLAP 2005), pp. 47–56 (2005)
25. Schiefer, J., List, B., Bruckner, R.M.: Process data store: A real-time data store for monitoring business processes. In: Mařík, V., Štěpánková, O., Retschitzegger, W. (eds.) DEXA 2003. LNCS, vol. 2736, pp. 760–770. Springer, Heidelberg (2003)
26. Bohnlein, M., Ende, A.U.: Business Process Oriented development of data warehouse structures. In: Jung, R., Winter, R. (eds.) Proceedings of Data Warehousing, pp. 3–21. Springer Physica (2000)
27. Pau, K.C., Si, Y.W., Dumas, M.: Data warehouse model for audit trial analysis in workflows. In: Proceedings of IEEE International Conference on e-Business Engineering, ICEBE 2007 (2007)
28. Rizzi, S.: Conceptual Modeling Solutions for the Data Warehouse. In: Wang, J. (ed.) Data Warehousing and Mining: Concepts, Methodologies, Tools and Applications, IGI Global Publishers (2007)

The Social Impact on Web Design in Online Social Communities

Maimunah Ali and Habin Lee

Brunel Business School, Brunel University
Maimunah.Ali@brunel.ac.uk, Habin.lee@brunel.ac.uk

Abstract. Social influence and behaviour of online communities and groups research had gained momentum with the popularity of web as a medium of social interactions. These studies on the other hand had focused on communication factors and had largely ignored the influence of social ties on user interface experience. Since user interface experience depends on user interface design, social influence on quality design would be valuable information in enhancing user quality experience in information technology. Taking into consideration that past studies had examined the impacts of culture and social separately, this research intends to interpret the effects of social and cultural influence on web design. With results from observation of sample blogs show that there are indications of social influence on web design, further exploration into the issue will benefit future research in information systems.

Keywords: weblogs, web design, social impact, Malaysia.

1 Introduction

Social interactions exists in every facet of life, in groups large and small, traditional or non-traditional settings. Although studies into the online social behaviour exist, the influence of elements and nature of social interactions on design behaviour are not known in the literature of information systems. Studies had focused on the explicit behaviour of online communities' members ranging from buying decision to offline interaction behaviour (Dholakia et al. 2004; Bagozzi et al., 2007) leaving the implicit behaviour of design preferences an area of potential research interest. Interpreting the design behaviour on the web should be focused on the design choice, which is the explicit representation of individual design preferences (Kryssanov, Tamaki and Kitamura, 2001).

For the past decades, web design preferences had been defined along the traits of culture. Authors like Marcus and Gould (2001), Cook and Finlayson (2005) and Singh et.al (2008) argued on the necessity of design preferences adhering to specific cultural variations and determinants. However, there arise question whether should culture be the main decisive factor in design preferences if the web sites in focus are personal blogs. With the increase number of online social communities stimulated by the rise in blogs, the obvious impact would be increased online social interactions

M.D. Lytras et al. (Eds.): WSKS 2009, LNAI 5736, pp. 207–217, 2009.

among bloggers that lead to the diffussion of ideas, knowledge, experience as well as best practices, fashion and trends on the web. The increase in online social interactions give rise to two issues:

Does increased social interactions among bloggers affect personal weblog design?
Does social interaction make bigger impact to weblog design than culture?

The issues will be addressed in three stages namely the operationalisation of identified variables, the collection and the analysis of data from the study. The initial process will involve the operalisation of variables into quantifiable indicators. This covers three identified items namely the network affliation, the network age and design elements. The network affliation involves searching for network of blogs with commonality in members based on demographic data. The blogs are then sorted according to years of existence to determine the network age. Design elements are coded into different segments based on selected indicators namely the author's profile, blog profile, information design, navigation design and visual design.

The implication of the findings is twofold. First, it enriches the literature in information systems by providing insights into how online social interaction is related with the usage of the online systems. The paper serves as pioneer literature on the influence of social interactions on blogs design. Second, it allows web designers and online marketers greater understanding of how people influence is interwoven into design by giving information on how online community users can be detained. Findings would illustrate the depth of understanding of the issue and achieve more desirable solutions for design practices of the web. The paper will be broken down into 5 sections. Section 1 will provide brief introduction on the issue followed by section 2 which will illustrate review on the theoretical foundation of understanding culture and social aspects on the web. Section 3 touches on the research hypothesis while section 4 draws the proposed methodology of the research. Brief analysis on the initial observation of weblogs will be presented in section 5.

2 Literature Review

Interest in the information systems literature research particularly to the impact of cultural differences on the usage and development of ICT had focused on a wide variety of issues, ranging from the general management issues, information systems and infrastructure, to the issue of cultural transfer and culture in system analysis and design (Burn et al., 1993; Cummings and Guynes, 1994; Sackmary and Scalia, 1999; Marcus and Gould, 2001; Robins and Stylianou, 2002; Park, 2004; Marcus and Alexander, 2007; Singh, 2008). Burn et al. (1993) for example, used the dimensions of culture to study the relationship between information systems and culture with focus on the top management issues in Hong Kong. Cummings and Guynes (1994) on the other hand, studied the variation of culture among staff at the headquarters and subsidiaries of Multinationals Corporation. Singh et al. (2008) however had conducted a study of design preferences of a distinct culture within a nation, focusing on the Hispanic ethnic group in the US. The state-of-the-art in defining design preferences had been focused on the cultural impact on design.

There are also researchers that proposed design elements in accordance to cultural dimensions without referring to any particular culture or nations. Marcus and Gould (2001), Ross (2001), Simon (2001), Cook and Finlayson (2005), Marcus (2006), Wurtz (2006), Wan Mohd Isa et al. (2007) provided cultural indicators and cultural markers for design variations that appeal to specific cultural dimensions. Depending on the cultural environment of user and the context of use, studies on cultural indicators on design variations had been bias toward culture and commercial web sites with regard to web design. That assumption may not hold under the social infrastructure of the web.

No research so far had revealed the influence of social interactions on design choices of blogs. Literatures on online communities' social factors and behaviours started in the 1990s with researchers like Keisler (1984), Galletta et al. (1995), Kraut et al. (1998), Marc (1998), McKenna and Bargh (1999), Compeau et al. (1999) propagated the ideas that social interactions on the web is an area of research that has vast potential to be explored and examined. Currently, virtual communities had been perceived as an interest group that interacts online to achieve personal or shared goals of their members. Recent studies on online networking focus more on the influence of social interactions have particular actions and behaviours of members through understanding the nature and the role of the influence. Postmes et al. (2001) for example, argued that in groups that used computer-mediated communication, the existence of group norms is significant and influential. Dholakia, Bagozzi and Pearo (2004) examined the effects of group participation of online communities from the perspective of marketing on two different online communities i.e. network and small-group-based. In a study of online interaction of Facebook, Ellison, Steinfield and Lampe (2007) found that the use of Facebook as a medium of interaction had a strong association with social capital and encourage psychological well-being.

Several mechanisms and approaches had been utilised in the studies of social influence on web communities and behaviour. Most of these studies examined and investigated individual intentions, motives and participation of web users to a particular online behaviour, focusing on the nature and roles of social influence. Interestingly, the issue of social influence mechanisms on web design had not been touched by any researchers as of date. Social determinants of web design changes as a result of social interactions on weblogs is not known in the literature of information systems.

3 Research Model and Hypotheses

The research model is shown in Figure 1.

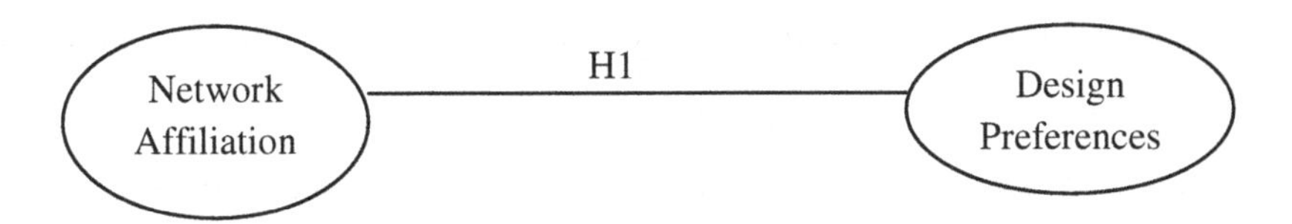

Fig. 1. Socal Interactions Model on Design Preferences

Dholakia et al. (2004) found that decision-making in online membership is a direct function of social influence and indirect function (through social influence) of value perceptions. Given that personal weblogs are social groups, social interactions may induce changes in design based on users experience and usage and as frequent interactions among the same individuals result in greater knowledge and interpersonal relationships, ideas and knowledge are exchange frequently among regular groups of friends. It is hypothesise that:

H1: Network affiliation affects the design preferences of webloggers

Researchers find that online memberships are playing bigger and wider roles in various aspects of members' life from friendship, learning, giving advice and opinion, purchasing and consuming products and obtaining services (Bagozzi and Dholakia, 2002). Indeed, Bagozzi *et al.* (2002) suggests groups that are formed through identification are very influential in shaping and changing members' opinion, preferences and actions.It is hypothesise that:

H1.1: Blogs in one network will share similar design elements.

Social influence is exerts differently in different groups membership (Bagozzi and Lee, 2002). There exists different group behaviours among different group membership. Online social grouping tend to share similar sense of belonging, values and preferences among members of the same online community. The hypothesis is as follows:

H1.2: There will be significant difference in design preferences among heterogenous networks of blogs

Figure 2 shows the proposed model between blog age and design preferences. Blog age in term of years of existence is propose to be the dominant factor to determine design elements preferences between culture and social influence. To evaluate whether social interactions or culture has a bigger impact on personal weblog design, it is hypothesise that:

H2: Age of blogs will determine the level of impacts between culture and social interaction.

H2.1: New blog will be more affected by culture

H2.2: Social interaction plays a bigger role in the older blogs

Therefore, for personal blogs, maintaining interpersonal connectivity and social interactions are coherent attributes that emphasise the types of influence available in online communities. Bagozzi et al. (2005) established that in online high-interactivity groups, social influence effects incorporated into the values and goals of decision makers that are shared with members of their group. It is hypothesise that:

H3: The stronger the social interactions, the bigger the social influence on design preferences.

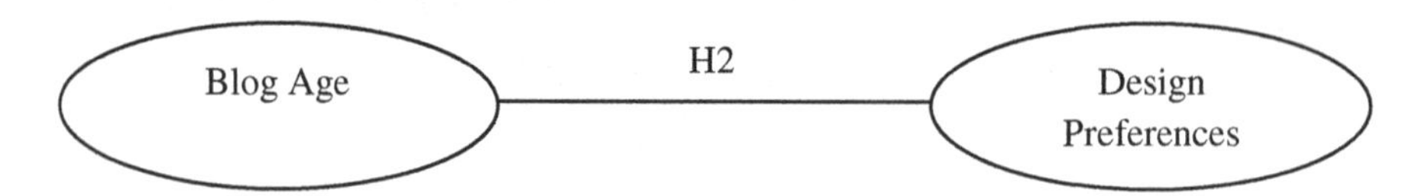

Fig. 2. The Proposed Model between Blog Age and Design Preferences

4 Research Method

The issues will be addressed in three stages namely:

(i) The operationalisation of identified variables;
(ii) The collection of data and;
(iii) The analysis of data from the study.

The initial stage will involve the operalisation of variables into quantifiable indicators. This covers three identified items namely the network affliation, the network age and design elements. The network affliation involves searching for network of blogs with commonality in members based on demographic data. The blogs are then sorted according to years of existence to determine the network age. Design elements are coded into different segments based on selected indicators namely the author's profile, blog profile, information design, navigation design and visual design. Based on literature, online indicators are identified as follows.

Table 1. Indicators of Weblogs Design Component

Component of Design	Indicators	Measurement
Author's Profile	• Name • Age • Gender • Location • Occupation • Interest • Education	• Full-name disclosure • Age disclosure (scale 1 – 6) • Gender differentiation • Regional location (scale of 1 – 6) • Types of employer (scale 1 – 5) • Stated / Not stated • Levels of education (scale 1 – 5)
Blogs Features	• Purpose • Advertisement • Chat Title • Archives • Recent Comments • Search Engine • Blog Survey • Credits/Awards • Statistics	• Blogging reason (scale 1 – 4) • Existence of advertisement (Yes/No) • Types of journal (scale 1 – 6) • Yes / No • Exist / Absence • Exist / Absence • Survey provider (scale 1 – 4) • Yes / No • Exist / Absence
Navigation	• Navigation System • Site Registration • Security Provision • Visitor Counter • Navigational Links • Blogrolls	• Customise / Contextual • Required / Not required • Exist / Absence • Exist / Absence • Control / Supportive • Customise / None
Content	• Information Sorting • Accessibility • Organisation of Information • Types of Information • Contain • Focus • Symbols • Audio • Colour • Emphasis	• Hierarchical / Non-hierarchical • Restricted / No barriers • Priority / Equal importance • By task / By modular • Personal achievement / Group • Youth and action / age and experience • Materialism / Family / None • Exist / Absence • Exist / Plain • Relationship / Rules
Visual	• Symbols • Picture • Animation • Avatar • Artwork	• Rules / No symbol • Personal / Group • Exist / Absence • Exist / Absence • Design / Task / None

The data collection constitutes the second stage of the study method. The research will employ technique from content analysis method to examine and predict the relationships between culture and social on weblog design. Each weblog in the sample will be examined based on design elements to determine evidence of homogeneity characteristics. Observation of each weblog will be done over a period of time tracing changes in design. Frequency counts will be used to detect similar design elements.

The last stage involves data analysis from the study. Analysis of variance (ANNOVA) will be used to examine social explorations of each weblog involving the use of Chi-square (χ^2) analysis and an F-test statistics. The χ^2 analysis and F-test statistics are useful to determine blogs variation and heterogeneity. In addition, qualitative analysis will be conducted on the network of blogs to examine social and cultural characteristics based on demographic information and cultural background.

5 Weblog Sampling and Design Components

The selection of networks of blogs will be conducted randomly. Eah network of blogs will consists of one main blogger and 50 co-bloggers. To ensure the interdependency of each network, it is ascertain that there is no overlapping of member bloggers between networks. Figure 3 shows the independency of each network of blogs.

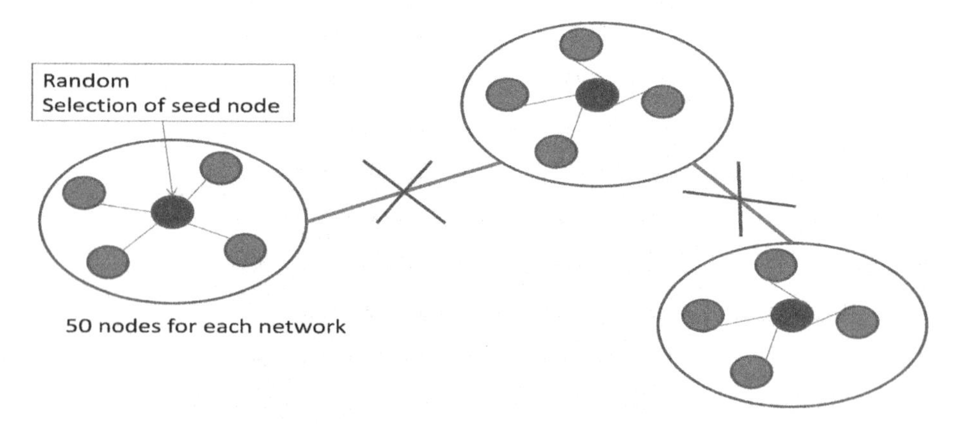

Fig. 3. Weblogs sampling and independency

The design components chosen are author's profile, blog features, navigation design, content design and visual design. Author's profile is a section where an individual blogger state his or her personal details in terms of name, age, gender, where the blogger is located, blogger's occupation and level of education. The degree of exposure i.e. the extent to which a blogger will reveal oneself will be at the discretion of the blogger.

The layout which the blogger chooses as the framework will determine the features of the intended blog. These features are visible on either side of the layout page and most of the features serve as auxiliary functions providing additional information to the bloggers and visitors regarding blogs' traffic movement, number of visitors, blog recognition and survey. Design characteristics that form part of the blog features are blog purpose, existence of advertisement, statement of chat title, archives, recent comments, search engine, blog survey, credits or awards and statistics.

The three other design components namely the navigation design, the content design and the visual design constitute the core aspects of the blogs. Navigation design refers to navigational functions that help blogger access different sections of the blogs (Cyr, 2008). The design feature could either aid in or hinder blogger from getting and searching the required information on the blogs. The content or information design on the other hand, constitutes design elements that state either the accurate or misguided information about the blog. It could be in the form of icon, structure of the information, the focus and the emphasis of the blog. Nonetheless, the visual design elements deals with the outlook of the blog in terms of colour, photograph, shapes, animation, artwork and symbol. The aesthetic aspect of this type of design elements tend to appeal to the emotional side of the blogger by capturing the beauty characteristics of the web (Cyr, 2008).

6 Initial Observation and Analysis

An initial study had been undertaken with the aim of examining indicators of social influence on personal blogs. A total of 150 personal blogs was chosen as sample in the initial study. The selection of blogs for the purpose of initial observation is taken from three Malaysian network blogs that had been observed for a period of three months from December 2008. Each network consists of one main blogger and 50 cobloggers categorise as 'friends' or members of the same network. Steps are taken to ensure that there is no overlapping of members from one network to the other by comparing and sorting out each member of the two networks to avoid biasness. An initial analysis was done on both network to determine indicators of social influence on design features.

The initial results and analysis of the networks are shown in Table 2. Column 1, 2, 3 and 4 show the number of similar occurrences with regard to the design elements in network 1 – 4. Looking at the number of similar occurrence, blogs in networks 1, 2 and 3 show that there exist similarity in design elements among blogs in each network. For example, 50 bloggers in network 1 allow comments in their blogs while 34 bloggers in network 2 permit the same function and 47 bloggers in network 1 use statistics counter to capture the traffic movement of his or her personal blogs and 24 bloggers do so in network 2. This indicates that member bloggers in a network do share similar design elements across blogs thus supporting the hypothesis that blogs in a network share similar design elements.

Table 2. Initial Results and Analysis of Three Independent Network Blogs

Component of Design	Network 1 (n=51)	Network 2 (n=51)	Network 3 (n=51)	Network 4 (n=51)
Author's Profile				
• Name	15	42	46	39
• Age	50	22	34	35
• Gender	50	44	50	51
• Location	42	18	37	38
• Occupation	48	25	35	37
• Interest	48	26	14	23
• Education	3	9	25	20
Blogs Features				
• Purpose	44	38	50	51
• Advertisement	46	24	42	24
• Chat Title	50	28	47	45
• Archives	48	44	46	44
• Recent Comment	50	34	44	49
• Search Engine	2	20	44	39
• Blog Survey	1	26	27	13
• Credits/Awards	0	3	5	0
• Statistics	47	24	29	27
Navigation				
• Navigation System	47	27	47	2
• Site Registration	0	1	0	36
• Security Provision	0	1	0	22
• Hit Counter	48	24	29	21
• Navigational Links	50	27	47	28
• Blogrolls	0	4	49	39
Content				
• Information Sorting	0	0	0	0
• Accessibility	0	1	0	18
• Organisation	0	0	0	0
• Types	0	49	51	0
• Contain	50	30	47	28
• Focus	50	44	51	39
• Symbols	0	0	28	6
• Audio	0	1	2	0
• Colours	4	8	1	6
• Emphasis	50	30	47	45
Visual				
• Symbols	0	0	0	6
• Picture	50	38	46	51
• Animation	0	1	31	6
• Tracker	0	0	15	40
• Chatbox	0	4	43	2

The analysis involved the four network of blogs and a control group Network 0 which contains blogs that are chosen randomly. Each group are tested for homogeneity of variance within group and heterogeneity between groups using Analysis of Variance (ANOVA). The results are shown in Table 3.

Table 3. Analysis of Variance

Network	F	Sig	χ^2
0	3.847	0.049	
1	0.858	0.521	
2	0.196	0.938	59.063
3	0.538	0.709	
4	1.601	0.213	

There was a significant similarities of design preferences in network 1 to 4, $\rho \geq 0.05$. The variances are not significantly difference within groups. Therefore, accept H1.1: Blogs in one network will share similar design elements.

The test for heterogeneity between networks is shown by the value of χ^2. The χ^2 value of 59.063 is significant at 0.05% critical value. Accept H1.2: There will be significance difference in design preferences among heterogenous networks of blogs.

7 Discussion and Conclusion

The aim of the pilot study was to survey indications of social influence on design elements of personal blogs among members of the same online network. Initial findings found that design elements of blogs in a same network do share similar design preferences shown by the number of similarity occurrence of design preferences across blogs in the same network. The sense of sharing among member bloggers in a virtual community is in line with findings of Bagozzi (2007), Dholakia et al. (2004), Postmes (2001) that members of virtual communities tend to share sense of belonging, values and preferences among each other. The blogger-to-blogger communication is influential in shaping opinions and behaviours (Bagozzi et al., 2002) such that member bloggers in a same network have influence on design elements preferences in blogs. Indeed, Bagozzi et al. (2002) suggests groups that are formed through identification are very influential in shaping and changing members' opinion, preferences and actions. This coincides with members identified with certain network in term of blogging and membersip.

So far, the initial observation had been conducted on weblogs networks in one country under the assumption that the weblogs are influenced by the same culture. The objective is to minimise the influence of culture while finding indications of social influence on blogs. To explore the co-existence of culture on design preferences of networks of blogs, a comparative study between Malaysia and another country would provide enlightenment on design preferences in networks of blogs under different cultural values. This would incorporate another dimension on design preferences in network of blogs under the influence of culture. It would be interesting to study the level of influence social and cultural have on design preferences of blogs and compare them between countries.

References

1. Bagozzi, R.P., Lee, K.H.: Multiple routes for social influence: The role of compliance, internalization and social identity. Social Psychology Quarterly 65(3), 226–247 (2002)
2. Bagozzi, R.P., Dholakia, U.M.: Intentional social action in virtual communitie. Journal of Interactive Marketing 16(2), 2–21 (2002)
3. Bagozzi, R.P., Dholakia, U.M., Pearo, L.K.: Antecedents and consequences of online social interactions. Media Psychology 9, 77–114 (2007)
4. Bergami, M., Bagozzi, R.P.: Self-categorization, affective commitment and group self-esteem as distinct aspects of social identity in the organization. British Journal of Social Psychology 39, 555–577 (2000)
5. Burn, J., Saxena, K.B.C., Ma, L., Cheung, H.K.: Critical issues in IS management in Hong Kong: A cultural comparison. Journal of Global Information Management 1(4), 28–37 (1993)
6. Compeau, D., Higgins, C.A., Huff, S.: Social cognitive theory and individual reactions to computing technology: A longitudinal study. MIS Quarterly 23(2), 145–158 (1999)
7. Cook, J., Finlayson, M.: The impact of cultural diversity on web site design. Advanced Management Journal 70(3), 15–23 (2005)
8. Cummings, M.L., Guynes, J.L.: Information systems activities in transnational corporation: A comparison of US and non-US subsidiaries. Journal of Global Information Management 2(1), 12–27 (1994)
9. Cyr, D.: Modeling Web Site Design Across Cultures: Relationships to Trust, Satisfaction, and E-Loyalty. Journal of Management Information Systems 24(4), 47–72 (2008)
10. Dholakia, U.M., Bagozzi, R.P., Pearo, L.K.: A social influence model of consumer participation in network and small group-based virtual communities. International Journal of Research in Marketing 21, 241–263 (2004)
11. Ellison, N.E., Steinfield, C., Lampe, C.: The benefits of Facebook Friends: Social capital and college students' use of online social network sites. Journal of Computer-Mediated Communication 12, 1143–1168 (2007)
12. Evers, V., Day, D.: The role of culture in interface acceptance, pp. 260–267. Chapman and Hall Ltd, Boca Raton (1997)
13. Fink, D., Laupase, R.: Perceptions of web site design characteristics: A Malaysian/Australian comparison. Internet Research: Electronic Networking Applications and Policy 10(1), 44–55 (2000)
14. Fraternali, P., Tisi, M.: Identifying cultural markers for web application design targeted to a multi-cultural audience. In: Proceedings of Eighth International Conference on Web Engineering, pp. 231–239. IEEE, Los Alamitos (2008)
15. Galletta, D.F., Ahuja, M., Hartman, A., Teo, T., Peace, G.A.: Social influence and end-user training. Communication ACM 38(7), 70–79 (1995)
16. Girgensohn, A., Lee, A.: Making web sites be places for social interactions. In: Proceedings of the 2002 ACM Conference on computer supported cooperative work, pp. 136–145 (2002)
17. Kraut, R., Patterson, M., Lundmark, V., Keisler, S., Mukopadhyay, T., Scherlis, W.: Internet paradox: A social technology that reduces social involvement and psychological well-being? American Psychologist 53, 1017–1031 (1998)
18. Keisler, S., Siegel, J., McGuire, T.: Social psychological aspects of computer-mediated communication. American Psychologist 39, 1123–1134 (1984)
19. Klobas, J.E., Clyde, L.A.: Social influence and internet use. Library Management 22(2), 61–68 (2001)

20. Kryssanov, V.V., Tamaki, H., Kitamura, S.: Understanding design fundamentals: How synthesis and analysis drive creativity, resulting in emergence. Artificial Intelligence in Engineering 15, 329–342 (2001)
21. Marc, B.: The politics of technology: On bringing social theory into technological design. Science, Technology and Human Values 23(4), 456–490 (1998)
22. Marcus, A., Gould, E.W.: Cultural dimensions and global web design: What? So what? Now What? In: Proceedings of the sixth conference on human factors and the web, Texas (2001), http://www.amanda.com/resources/hfweb2000/AMA_CultDim.pdf (Retrieved December 19, 2008)
23. Marcus, A.: Culture: Wanted? Alive or dead? Journal of Usability Studies 2(1), 62–63 (2006)
24. Marcus, A., Alexander, C.: User validation of cultural dimensions of a website design. In: Aykin, N. (ed.) HCII 2007. LNCS, vol. 4560, pp. 160–167. Springer, Heidelberg (2007)
25. Mason, W.A., Conrey, F.R., Smith, E.R.: Situating social influence process. Dynamic, multi directional flows of influence within social networks Personality and Social Psychology Review 11, 279–300 (2007)
26. McKenna, K.Y.A., Bargh, J.A.: Causes and consequences of social interaction on the internet. Media Psychology I, 249–269 (1999)
27. Ning Shen, k., Khalifa, M.: Exploring multi-dimensional conceptualization of social presence in the context of online communities. In: Jacko, J.A. (ed.) HCI 2007. LNCS, vol. 4553, pp. 999–1008. Springer, Heidelberg (2007)
28. Park, M.H.: A Study of a Cultural Relativism in Web Interface Design. International Journal of Diversity in Organisations, Communities and Nations 5(2), 157–162 (2004)
29. Postmes, T., Spears, R., Sakhel, K., de Groot, D.: Social influence in computer-mediated communication: The effects of anonymity on group behavior. Personality and Social Psychology Bulletin 27, 1243–1254 (2001)
30. Postmes, T., Spears, R., Lea, M.: The formation of group norms in computer-mediated communication. Journal of Human Communication Research 26(3), 341–371 (2000)
31. Powell, T.A.: Web design: The complete reference. Osborne/McGrawHill, USA (2000)
32. Robbins, S.S., Stylianou, A.C.: A study of cultural differences in global corporate web sites. Journal of Computer Information Systems, 3–9 (winter 2002)
33. Simon, S.J.: The impact of culture and gender on websites: An empirical study. Databases for Advances in Information Systems 32(1), 18–37 (2001)
34. Singh, N., Zhao, H., Hu, X.: Cultural adaptation on the web: A study of American companies' domestic and Chinese websites. International Journal of Global Information Management 11(3), 63–80 (2003)
35. Singh, N., Baack, D.W., Pereira, A., Baack, D.: Culturally customizing websites for US Hispanic online customers. Journal of Advertising Research, 224–233 (June 2008)
36. Sun, H.L.: Building a culturally-competent corporate web site: An exploratory study of cultural markers in multilingual web design. In: Proceedings of the 19th Annual International Conference on Computer Documentation, pp. 95–102. ACM Press, New York (2001)
37. Terry, D.J., Hogg, M.A., White, K.M.: The theory of planned behaviour: Self-identity, social identity and group norms. British Journal of Social Psychology 38, 225–244 (1999)
38. Wan Mohd Isa, W.A.R., Md Noor, N.L., Mehad, S.: Incorporating the cultural dimensions into the theoretical framework of website information architecture. In: Aykin, N. (ed.) HCII 2007. LNCS, vol. 4559, pp. 212–221. Springer, Heidelberg (2007)
39. Wurtz, E.: Intercultural communication on web sites: A cross-cultural analysis of web sites from high-context cultures and low-context cultures. Journal of Computer-Mediated Communication 11, 274–299 (2006)

Semantic-Based Tool to Support Assessment and Planning in Early Care Settings

Ruben Miguez, Juan M. Santos, and Luis Anido

Department of Telematics Engineering, University of Vigo
Vigo, Spain
{rmiguez,jsgago,lanido}@det.uvigo.es

Abstract. In recent years, governments and educational institutions have increased their interest in enhancing and expanding pre-school programs and services. In order to achieve a high-quality education, governments have developed several national frameworks. A careful activity planning and the provision of a personalized learning experience are key factors of these programs. In this paper, we present the main design guidelines of an ICT-based tool that supports the tracking and assessment of a child's development, as well as the recommendation and planning of new learning activities. We also propose a semantic description of the domain based on e-learning standards and specifications. Use of semantic technologies enhances aspects as tools interoperability, processes automation as well as the relevance and precision of the recommendations made by the system. Besides, this tool creates a virtual common framework that encourages collaboration between families and practitioners and it supports parents' involvement in early care settings.

Keywords: early childhood education, assessment, semantic web, personalization, framework.

1 Introduction

Modern Knowledge Societies are involved in a deep transformation process concerning the education for children. For historical reasons, the care and the upbringing of children have been traditionally developed separately [1]. However, these aspects are currently addressed by the governments from a unique perspective in order to meet the demands of citizenship. Recent studies [2] show the benefits of a quality early education in the short and long term. This fact, together with factors related to policies of equality and social inclusion, has turned the early childhood education into a key area within of educational policies of the governments.

Families and schools are co-responsible for the education of children, and they together should act as a team, through a set of attitudes, expectations and working methods, in the planning of appropriate educational activities. The teacher training and the involvement of the families in the children activities are key factors towards a quality education [3][4]. The observation, recording and discussion of the progress made in both home and school are key elements in planning and adapting the learning experiences to the degree of development and interests of a particular child.

M.D. Lytras et al. (Eds.): WSKS 2009, LNAI 5736, pp. 218–227, 2009.

The use of Information and Communication Technologies (ICT) can play a crucial role in these processes [5]. One of the most common uses of ICT in this area is to serve as support and guidance of parents and educators. Governments, educational institutions and user communities have launched several web portals which provide a wide range of services including discussion forums, blogs, educational multimedia repositories, policy and legislation information services, etc. Furthermore, a limited collection of ICT technologies are being used in early childhood settings to strengthen many aspects of childhood educational practices. Although there is a clear under-implementation compared to higher educational levels [6], several institutions and schools currently apply these technologies at the children schools. Some research projects consider the use of devices such as interactive whiteboards and smart tables in kindergartens to foster cooperative work. However, TV programs and other popular media products as DVDs are the most broadly used devices in these settings. Other projects such as KidSmart [7] allow children to use specifically designed computers (taking into account aspects such as ergonomics, ease of use and durability in design). Specifically designed tools as digital toys, social robots or videogames are also commonly used in childhood schools. Children collaborate, play, interact and experiment with these devices to enhanced their motor, language and social skills [8]. New ICT-based tools will be added progressively to the child's education and they will play a key role at child homes and classrooms of tomorrow.

Given the wide variety of devices that can be used by the child, it is necessary the development of systems that allow the gathering, processing and evaluation of the interaction of users with those devices. That information can be used to automatically track the progress of the child and offer new experiences and activities appropriate to their profile, interests, abilities and needs. This paper discusses the main issues of such a system, aimed at parents and educators to facilitate the assessment, monitoring and planning of learning activities. Based on the child's development state, its capacity and predefined learning objectives, the system is able to offer automated recommendations for new educational experiences to accomplish. The analysis of the gathered information also allows for early detection of possible developmental disorders as well as providing a set of activities designed to improve children's skills in particular underdeveloped areas.

This proposal is described throughout this paper. Section 2 gathers a set of particular issues in the area of early childhood education that must be taken into account in the development of such a system. Section 3 briefly describes the main objectives of the proposed framework. The general Reference Model of the system is discussed on Section 4 and Section 5 gives an overview of the identified supporting Semantic Model that allows the construction of personalized services. Section 6 outlines the Reference Architecture and, finally, Section 7 concludes and summarizes the paper.

2 Distinctive Features of Early Childhood Education

Currently, there is a wide range of early childhood educational programs promoted by governments and education departments of various countries. Despite the obvious differences in approaches and methodologies, the idea underlying them all is the same: to create an educational community in which learning is seen as a social experience, primarily interactive and experimental, which serves as a prelude to compulsory education. In this sense, the fundamental objective is to fulfill the principles of Delors [9]: "Learning to be, learning to do, learning to know and learning to live together and to live with others".

Any approach in this area should take into consideration the unique characteristics that make a child's education a special educational field:

- Early care settings are characterized by a fragmented educational scenario [1]. A broad range of centres, nursemaids, community groups and institutions provide educational services from birth to six years old. Consequently, the quality of this education is very heterogeneous. It depends highly on the qualification grade of educators, available resources (material, environment, adult-child ratio) and parents' involvement [3]. Providing the area with mechanisms to support the creation of a common framework and to facilitate the formation and participation of the actors is one of the fundamental priorities of governmental policies in scope.
- Childhood is a time of rapid growth and personal development due to the natural curiosity of children. Practitioners need to provide daily opportunities that encourage children to practice new developed skills and to plan for experiences through which they will acquire confidence and competence. In this way, personalized learning is of paramount importance to enhance children's skill development and motivation.
- Children have difficulties to maintaining their focus on one activity for long periods of time. Video and multimedia resources that combine education and entertainment (*edutainment*) have wide acceptance among children [10].
- The literature identifies a common set of core development areas for the children [11]: i) psychomotor development, ii) emotional, personal and social development; iii) cognitive development; iv) health and welfare; v) creativity and communication capacity and vi) knowledge of the environment. Progress in these areas can be observed both in the classroom and at home, so it is especially relevant in this field the home-school communication and the involvement of families in the activities of their children.
- The daily analysis of progress is a critical factor in child education. Appropriate evaluation allows defining educational activities adapted to the needs and interests of the child, facilitating the development of their skills and promoting a common framework for discussion and meeting between families and educators [12]. The evaluation process consists of a series of steps starting from the observation, recording and assessment of activities undertaken by the child. Based on the study and discussion of the results, it is possible to extract conclusions to planning the activities in the classroom or at home.

3 An ICT-Based Tool to Support Planning and Evaluation

Our final aim is to develop a distributed system that facilitates the automated tracking and registry of children's activities, the evaluation of their progresses and the planning of new learning activities taken into account the specifically needs of each child. As these tasks take place at home and at school settings, the developed applications must give support to both environments. Therefore, the final system contributes to strengthen the home-school link. To achieve this general goal, a set of partial objectives have been identified:

- Definition of a standard-based electronic curriculum. It takes into account the specific needs and most relevant issues in this age range.
- Development of a mechanism aimed at supporting the formal specification of educational scenarios and children's profiles potentially susceptible of learning underdevelopment.
- Definition of an architectonical framework that facilitates interoperability among different devices and services.
- Modeling of a system that supports personalized services and processes automation such as, for example, recommender and planning procedures.
- Development of a system aimed to allow families and practitioners to add and manage observations, audiovisual resources and reports about the learning activities carried out by the children.
- Development of a software agent that monitors and analyzes children's profiles in order to detect and identify underdevelopment situations.
- Development of a collaborative system to support management, edition and ranking of the learning activities by the early care community.

4 The Reference Model

A Reference Model is an abstract framework that identifies the most relevant relationships among the entities in a particular environment. In order to build this model we use as information sources: a) the study of the domain; b) interviews with specialists and practitioners; c) government frameworks and educational policies and d) the previous experiences of the authors in the domain [13]. After a process of discussion and reflection we have identified the following actors and relationships: *children* (from birth to six years old), *families* (parents, grandparents, siblings, etc.), *practitioners* (responsible of children's education) and *specialists* in the early childhood field as pediatricians or speech therapists.

The system acts as a link among the actors that may use a broadly set of on-line services. The system offers a virtual environment, shared by the users, composed by the logical entities described below (Figure 1):

- *Register* - It stores the results of the activities carried out by the children using different electronic devices.
- *Profile Manager* - It allows to query and edit observations in the child's profile (e.g. comments, achieved skills, photographs, videos of activities, and so forth).
- *Activity Manager* - It facilitates the management of the learning activities and their collaborative edition.
- *Reporter* - It is used by practitioners, families and specialists to generate reports of different issues of a child's profile.
- *Notifier* - It analyses and monitors stored profiles in order to detect those children susceptible of being in risk of underdevelopment.

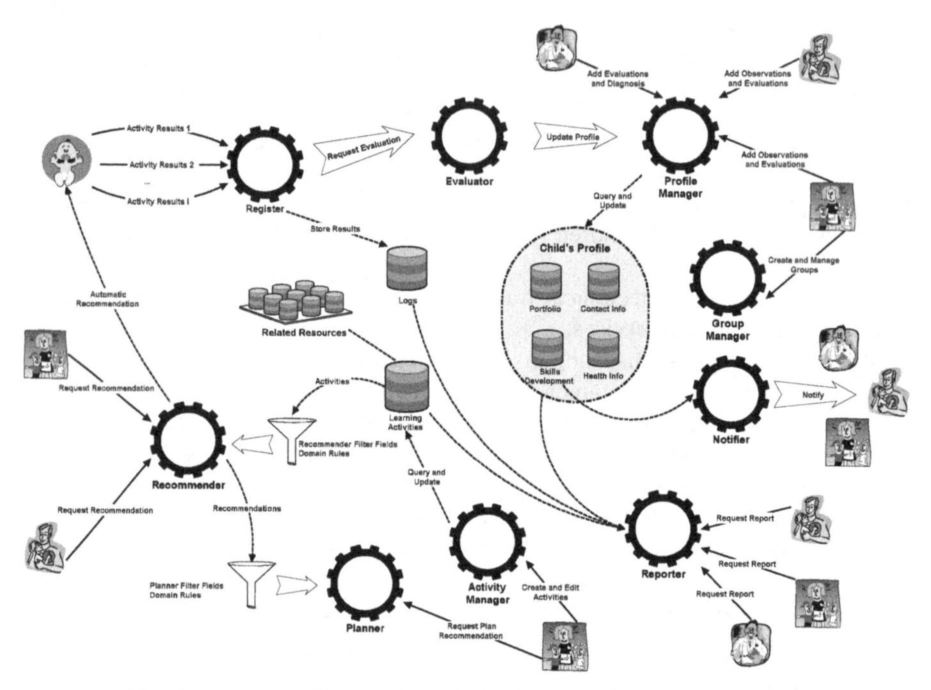

Fig. 1. Assessment System: Identified Relationships and Logic Entities

- *Recommender* - It offers personalized recommendations of new learning activities to a child. The profiles stored by the system are used in order to improve the relevance of the recommendations.
- *Planner* - It offers a set of learning paths that satisfy some predefined constraints as the children's and practitioners' agenda, the requirements of the activities, or a group of learning goals to achieve.
- *Group Manager* - It supports the creation and definition of groups of individuals. The recommender and planner systems consider these individuals as a whole, so they can make recommendations, for example, taking into account the profiles of all the children belonging to the same classroom.

5 The Semantic Model

Providing a formalized semantic description of the entities and relationships involved in the domain is of paramount importance to our proposal. In this case, we formalize the knowledge using a collection of OWL-DL ontologies [14].The construction of these ontologies was guided by the activities proposed in *Methontology* [15], a mature methodological process in the Knowledge Engineering area. As the basis for the definition of the ontology concepts we have used several specifications and standards of the e-learning domain as well as some terms commonly used by national departments of education. In this way, we encourage the keep of the knowledge already defined and agreed upon the domain. The identified concepts were grouped in three main categories:

- *Learning Competences*: The formalized description of a child's competences is a key point in any assessment system. We use as the basis for this ontology the assessment scale developed by the Early Years Foundation Stage (EYFS) [11]. It is divided in 13 different knowledge areas. Each one considers different evolution stages numbered from 1 to 9 (higher numbers correspond to higher competency levels). As the modeling information schema we use the specification IMS-RDCEO [16]. It describes a competence in terms of: *Identifier*, *Title*, *Description* and *Definition*. We needed to extend this schema in order to represent the information defined by the EYFS assessment scale. We have added the elements: *Scale Value*, *Competence Covered*, *Knowledge Topic*, *Recommended Age*, *and Related Resource*.
- *Child Profile*: We consider a child's profile composed by 4 basic elements: i) contact information, ii) health records, iii) the portfolio and iv) progress development. As the basis for this modeling we use the IMS-LIP and IMS-ePortfolio specifications [16]. In some cases, we have extended and adapted these recommendations in order to use them in the early care setting. For example, the *QCL* category identified by IMS-LIP does not have any use in childhood education. Our learner description ontology is composed of: *Identification*, *Learning Objective*, *Interest*, *Competency*, *Affiliation*, *Accessibility*, *Relationship*, *Product*, *and Health Record*.
- *Learning Activities*: The Dublin Core and LOM metadata schemas are used as a basis for this ontology development [16]. Particularly, we have identified the elements: *Identifier*, *Title*, *Abstract*, *Description*, *Background*, *Topic*, *Creator*, *Duration*, *Recommended Age*, *Requisite*, *Environment*, *Type*, *Adult Support*, *Learning Goal* and *Related Resource* (see Figure 3). Adult guide and support is a key feature in the domain. So *Adult Support* field indicates if the activity can be, or no, carried out by a child on their own. Besides, we define different categories of learning activities (excursions, songs, games, etc.) through the *Type* element, and other issues as the environment, learning goals, duration or the intended audience are also defined.

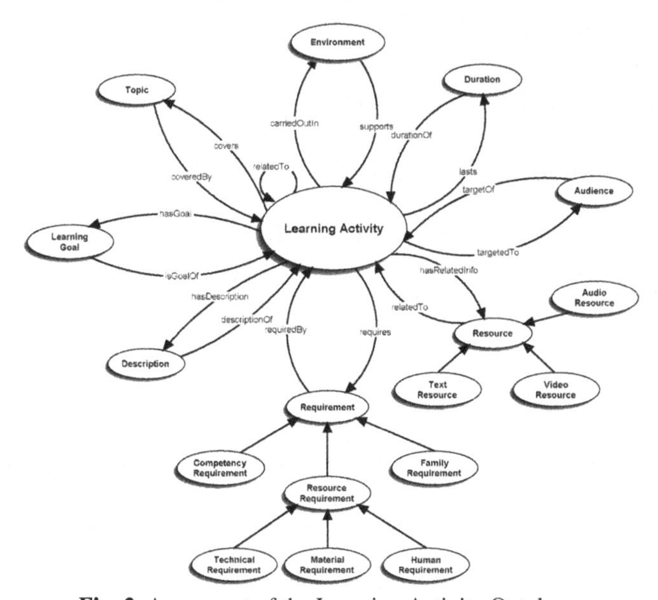

Fig. 2. An excerpt of the Learning Activity Ontology

Inference engines, using a set of pre-defined logic rules and the stored information are able to extract and add new knowledge to the system.

Some of these rules can be expressed using the syntax provided by the OWL-DL language.These rules are used when the properties and concepts in the ontology are defined. As an example we show the definition of the property *covers* as transitive:

```
<owl:TransitiveProperty rdf:ID="covers">
            <rdfs:domain rdf:resource="#Competence"/>
            <rdfs:range  rdf:resource="#Competence"/>
</owl:TransitiveProperty>
```

This definition allows the inference system to reason automatically that: "If a competence A covers all the features in a competence B, and B covers a competence C, then A also covers C".

Sometimes the expressivity capabilities of OWL-DL are not enough and we need to use Horn-like rules. In our particular case these rules are expressed using the semantic rule language SWRL [17]. Using these rules we can complete this reasoning. "If a child has mastered the competence A, then the system can automatically infer that he/she also masters the competences B and C because A covers B and A covers C". So:

```
Child(?x) • Competence(?y) • isAbleTo(?x,?y) • covers(?y,?c) •
isAbleTo(?x,?c)
```

Sometimes this additional expressivity is not enough and we need to use SWRL language extensions (*built-ins*) or particular solutions offered by the inference engines. If these methods are used to infer new knowledge, non-monoticity may be a consequence (i.e. the addition of new knowledge can invalidate a previously inferred statement). Below we present an example of this type of rules. The rule allows the system to detect a linguistic underdevelopment in a four years old child:

```
Child(?x) • hasAge(?x,?age) • swrlb:greaterThan(?age,?3) •
swrlb:maxCompetenceValue(?value,?x,"Language") •
swrlb:lessThan(?value,?3) • LanguageUnderdevelopment(?x)
```

6 Architectonical Framework

We consider the Reference Architecture as a decomposition of the Reference Model in a collection of software pieces and data flows that implements its functionality. Next sections briefly describe the Reference Architecture defined in our proposal.

6.1 Layer Structure

We have developed a SOA architecture following the design guidelines provided by existing e-learning abstract architectural systems such as IMS Abstract Framework and the IEEE LTSA. We have hierarchically grouped the services into three different tiers where each one relies on functionality provided by services at bottom layers. Over this set of service layers we have defined an additional one, the Agent Layer, that group those pieces of software that copes with the most complex system needs. Issues covered by each layer are briefly described below (Figure 3):

- *Infrastructure Services Layer*: It provides end-to-end transaction and communications functionalities.
- *Common Services Layer*: It provides a broad range of cross-domain functionalities (e.g. authorization, format conversion, group management and so forth).
- *Domain Services Layer*: It provides the domain specific functionalities. It gathers features as profile management or tracking.
- *Agent Layer*: It provides high-level functionalities to developers (e.g. Planner or Recommendation systems)

6.2 Agent and Service Modeling

From the study of the domain and making use of the main results of international projects as the e-Framework [18] and the OKI [19], we identified a set of services that fulfills the system needs (Figure 3). For each one, we formally define its functionality by means of an *OSID* (Open Service Interface Definition) document. Whenever it is possible we reuse those specifications already developed by the OKI project such as the *Assessment* and *Authentication* OSIDs. Finally, identified services are grouped into logical layers and implemented and deployed using web services technologies.

In section 4 we showed a collection of logical entities that capture the main system functionalities. For each one we have developed a software agent responsible for implement its features. The identified agents are: *Register*, *Evaluation Agent*, *Profile Manager*, *Activity Manager*, *Reporter*, *Notifier*, *Recommender*, *Planner* and *Group*

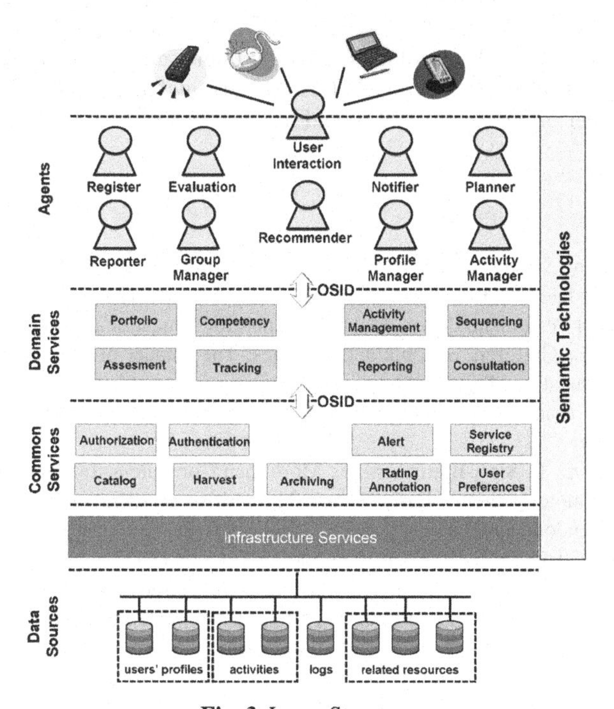

Fig. 3. Layer Structure

Manager. Besides, the proposal considers an additional element, the *User Agent* responsible for facilitating the user-system interaction using different access devices.

Finally, and taking as a reference the architecture defined by the SIF project [20], we have defined a central element, the *Core Communication System* (CCS). It is responsible for verifying, managing and monitoring the system data flows, as well as the authentication, authorization and services orchestration procedures.

7 Conclusions

In order to improve the quality of the young children's education, the tracking of the child's progress and the careful planning of the learning activities are key features. In this paper we have shown the main design guidelines of an ICT-based system that gives support to the tracking, assessment and planning processes. This framework creates a cooperative virtual environment where families and practitioners of different institutions can participate. Using the system they can look up relevant information about the upbringing of the children, establish discussions about a child's progression or even ask for a personalized learning activity recommendation.

Semantic technologies support the provision of personalized learning activities and facilitate the management of the children's profiles. Automated semantic-agents can be developed to analyze the stored children's profiles. If some under-development in any knowledge area is detected, the system automatically notifies parents and practitioners of this situation and recommends a set of activities to correct it.

Since 2006, the authors have collaborated in different projects to introduce ICT in the early care settings [13]. The initial prototype is being tested in a real environment: the "Rede Galega de Escolas Infantís", a public childhood schools network in Galicia, northwest of Spain. Currently, this network includes +120 kindergartens and a growing number of families (+4,500) and teachers (+800).

References

1. OECD: Starting Strong II: Early Childhood Education and Care. Technical Report (2006)
2. Wylie, C., Hodgen, E., Ferral, H., Thompson, J.: Contributions of Early Childhood Education to Age-14 Perfomance. New Zealand Council for Educational Research (2006)
3. Department for Children, Schools and Families: The Impact of Parental Involvement on Children's Education. Technical Report (2008)
4. Learning and Teaching Scotland. A Curriculum Framework for Children 3 to 5. Technical Report (1999)
5. Siraj-Blatchford, I., Siraj-Blatchford, J.: More than Computers: Information and Communication Technology in the Early Years. The British Association for Early Childhood Education, London (2003)
6. O'Hara, M.: ICT in the Early Years. Continuum, London (2004)
7. Lee, L., O'Rourke, M.: Information and Communication Technologies: Transforming Views of Literacies in Early Childhood Settings. Early Years 26(1), 49–62 (2006)
8. Plowman, L., Stephen, C.: Technologies and Learning in Pre-School Education. American Educational Research Association Conf. (2006)
9. Delors, J., et al.: Learning: The Treasure Within. UNESCO (1998)

10. Walldén, S., Soronen, A.: Edutainment. From Television and Computers to Digital Television. University of Tampere Hypermedia Laboratory (2008)
11. Qualifications and Curriculum Authority: Early Years Foundation Stage. Profile Handbook (2008)
12. Department for Children, Schools and Families: Practice Guidance for the Early Years Foundation Stage. Technical Report (2008)
13. Anido, L., Míguez, R., Santos, J.M.: Computers and Advanced Technology in Early Childhood Education. CATE 2008 (2008)
14. McGuinness, D.L., van Harmelen, F.: OWL Web Ontology Language Overview. W3C Recommendation (2004)
15. Fernández, M., Gómez, A., Sierra, A.: Building a Chemical Ontology Using Methontology and the Ontology Design Environment. Intelligent Systems 14, 37–45 (1999)
16. Anido, L., Rodríguez, J., Caeiro, M., Santos, J.: Observing standards for web-based learning from the web. In: Laganá, A., Gavrilova, M.L., Kumar, V., Mun, Y., Tan, C.J.K., Gervasi, O. (eds.) ICCSA 2004. LNCS, vol. 3044, pp. 922–931. Springer, Heidelberg (2004)
17. Horrocks, I., Patel-Schneider, P.F., Boley, H., Tabet, S., Grosof, B., Dean, M.: SWRL: A Semantic Web Rule Language Combining OWL and RuleML. W3C Member Submission (2004)
18. E-Framework, `http://www.e-framework.org`
19. OKI, `http://www.okiproject.org/`
20. SIFA: SIF. Implementation Specification 2.2 (2008)

Open Source Resources and Web 2.0 Potentialities for a New Democratic Approach in Programming Practices

Lucia Tilio, Viviana Lanza, Francesco Scorza, and Beniamino Murgante

Laboratory of Urban and Territorial Systems, University of Basilicata, Via dell'Ateneo Lucano 10, 85100, Potenza, Italy
lucia.ilio@unibas.it, viviana.lanza@unibas.it,
francesco.scorza@unibas.it, beniamino.murgante@unibas.it

Abstract. This paper reports about an experience concerning the implementation of a WEBGIS, a BLOG and an ontology. WEBGIS and BLOG allow to promote a spread of spatial data knowledge, to consult planning documents in the Internet, increasing transparency level of programming choices and involving different stakeholders participation. The ontology is intended as an in progress powerful tool, increasing knowledge rationality. Ontologies can help the community by defining and explicating a shared language and strengthening the efficacy of direct interactions. The case study has been applied to Marmo Platano–Melandro PIT (Territorial Integrated Projects), an area with high potentialities in the North-Western part of Basilicata Region (Italy), responsible for the accomplishment of POR (Regional Operative Program) Basilicata 2000-2006 and for the elaboration of a common and shared strategy to manage an integrated program of interventions for local development.

Keywords: Democracy, e-democracy, e-participation, open source, WEBGIS, ICT, BLOG, OGC standards, ontologies, programming assessment.

1 Introduction

The debate about electronic democracy has become very interesting during the last ten years. The development of information and communication technologies is finding a dynamic and responsive environment, especially at the local level, where municipal governments are playing a proactive role in achieving new forms of interaction between institutions and citizens. At present the scientific debate aims to investigate the way in which new technologies are changing the relationship between government and citizens. The growing importance of communication aspects in the contemporary planning has been conferring a more and more determining role in planning on electronic world [1]. Using ICT in democratic processes, in fact, makes possible, for public administrations, to involve citizens and all stakeholders in decision-making processes.

Nowadays, despite this situation, great part of programming documents proposes a bottom-up approach, facilitated by new tools [2], considering municipalities as a maximum level of shared decision, ignoring citizen's ideas, opinions and imagination

M.D. Lytras et al. (Eds.): WSKS 2009, LNAI 5736, pp. 228–237, 2009.

which might improve their quality. WEBSITE, WEBGIS and BLOG (the first two as information tools and the last one as interaction tools) are the instruments increasing participation level.

The case study has been applied to Marmo Platano–Melandro PIT (Territorial Integrated Project), an area with high potentialities in the North-Western part of Basilicata Region (Italy), including fifteen municipalities and two consortiums of communes in mountain areas. It is a local organization responsible for the implementation of POR (Regional Operative Program) Basilicata 2000-2006 and for the elaboration of a common and shared strategy to manage an integrated program of interventions for local development. The attempt of a new governance, based on cohesion and cooperation among local authorities, is the way to improve efficacy and effectiveness into policies.

This project research has been characterized by several and different objectives, like sharing of territory knowledge, innovating programming procedures, using open source tools, introducing the possibility of assessing, evaluating, measuring efficacy and effectiveness of policies. Moreover, a new way to define a common language and to strengthen the efficacy of direct interactions, through ontologies was introduced, and a participation approach for the next programming period 2007-2013 was set up.

Here we define what do we mean for programming practices in the democratic approach. We intend democracy and e-democracy paradigms according to several aspects: first, it is important to figure out which tools communicate and spread information about strategies, policies, decisions, expected outcomes, etc.. Simply implementing a WEBSITE was not enough to explicate democracy meaning; therefore we achieved it by implementing a WEBGIS, giving spatial dimension to programming interventions and allowing everyone knowledge. Then, every common citizen (and not only) should participate and be involved in programming activities giving feedback to local organizations. The best way we found was establishing a BLOG, as a tool to allow a simple, immediate and effective dialogue between all stakeholders in the territorial context. Democracy means also guaranteeing the coherence among all programming and planning tools. We achieved this purpose by implementing a common language framework referred to programming context, which cannot be misunderstood. Finally, we highlight that making transparent all programming processes is the main character of democracy, especially concerning efficacy and effectiveness evaluation phases.

In this work we describe components and functions of an e-democracy platform developed by the PIT Marmo Platano - Melandro under the scientific direction of the Laboratory of Territorial and Regional Systems Engineering (LISUT) of the University of Basilicata. It is an operative prototype, implemented few months ago, which still has not produced relevant data for the assessment of its efficacy, but it represents an integrated platform with high potentialities for promoting participation in programming development at a local scale.

2 Spread Tools for Programming Practices: WEBGIS Using OGC Standards

The growing diffusion of the Internet allows the transition from "geographic information systems" to "spatial data infrastructures" (SDI), in order to facilitate data dissemination and to reduce data duplication [3]. A Spatial Data Infrastructure means technologies, policies, standards, and human resources necessary to acquire, process, store, distribute, and improve utilization of geospatial data [4]. In order to implement a Spatial Data Infrastructure, a strong cooperation among local authorities is necessary. Marmo Platano – Melandro PIT played an aggregation role among municipalities in building dialogue and cooperation and encouraging synergy. PIT also played a central role in developing this local spatial data infrastructure, stimulating GIS culture, completely absent in this area. PIT produced main data for the infrastructure and defined information contents and spatial data infrastructure model. After the implementation, a local authority staff has been trained for everyday system update. It may seem that this approach does not properly follow INSPIRE directive [5], [6], [7] but in this phase it is the only way to realize a SDI, because municipalities are too small with scarce resources to develop their own information systems.

WEBGIS is one of the most interesting aspects of PIT information system (http://www.pitmpm.basilicata.it/PIT/map.phtml).

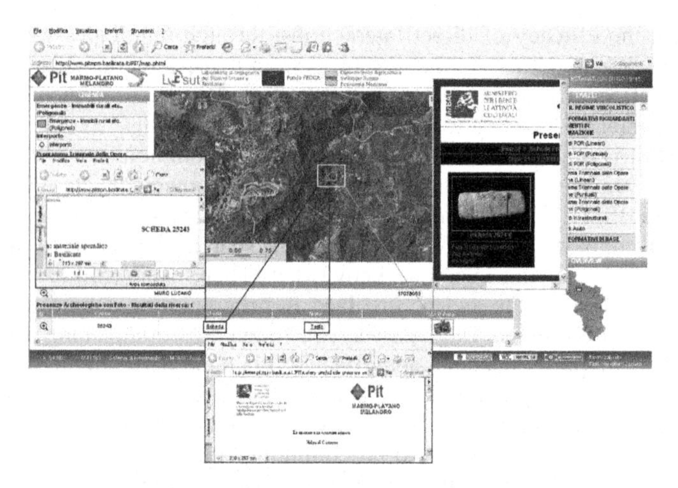

Fig. 1. Marmo Platano – Melandro PIT WEBGIS

This tool pursues many objectives, from the promotion of a spread of spatial data knowledge to the increasing involvement of different stakeholders: it allows citizens, local entrepreneurships, practitioners and employed of local authorities to access geographic databases surfing on the net and to check for spatial funds redistribution in a transparent way.

WEBGIS is based on a *client-server* architecture accessing rules via internet or intranet in order to navigate, update and maintain data. The whole architecture has been developed on an Open Source platform, according to Open GIS Consortium specifications. The operative system adopted is *Debian GNU / Linux*, and the most common applications of GFOSS (Geospatial Free and Open Source Software) have been used; for each informative level, standard ISO 19115 metadata are also made available, following the Metadata National Repertory (CNIPA).

It has been implemented a Web Map Service (WMS), accessible on URL www.pitmpm.basilicata.it/cgi-bin/wms_pit, allowing each user to add PIT data to its own data and to work with his GIS software.

3 The Way to Collect and Take Citizen's Opinions into Account: The PIT 2.0

Over recent years, the Internet has become a popular medium for carrying out all kinds of commercial, social and governmental activities. Presumably it has become a part of society quicker than any other new technology and it is now considered as a new democratizing tool, supposedly bringing people closer together and allowing them to participate in civilized society [8].

During the last decade, all governments and local organizations have been more and more using the Internet and ICT e-government tools in order to give more opportunities for citizen participation and therefore for enhancing information and service delivery to citizens. It is necessary to consider that citizen participation needs constant communication, using new tools in order to facilitate a bottom-up participation process [2]. This approach offers the advantage to stimulate citizen involvement in the choice of design alternatives in programming processes, overcoming time and space constraints. It is possible to separate two main tools: information tools, representing a low participation level and providing a "one-way" participation, and interaction tools, providing a "two-way" participation, including citizen's opinions in process, and considering them as process actors through a mutual exchange of comments, questions, discussion channels, etc. [9].

A typical information tool is the WEBSITE (www.pitmpm.it); although it is a very popular communication tool among governments and institutions, and despite the fact that Marmo Platano – Melandro PIT WEBSITE is very complete in content and information, it represents a low participation level. In order to achieve a high participation level, we activate an interaction tool, the WEBGIS.

An increasing number of local authorities applies the use of Information and Communication Technologies (ICT) to increase the participation of stakeholders in democratic processes, so that the number of e-gov projects and relevant tools is rapidly growing [10]. WEBSITE and WEBGIS are not enough for public participation; in this case information flow goes only from PIT to citizens. It is important to create a sort of virtual space where people can discuss, compare and exchange information, suggesting ideas to public administrations. The support of web 2.0 and ICT technologies aims to capitalize collective intelligence in programming processes. Interesting aspects are related, on one hand, to the creation of a real local organization network, in order to apply transparency, participation to choices, equity, redistribution principles and, on the other hand, to the application of ICT new tools to promote citizen participation in community activities.

Theories about communicative planning have forcefully emphasized how language and modes of communication play a key role in shaping planning practices, public dialogues, policy making, and collaboration processes [11], and today the most popular and effective tools for exchanging opinions and collecting information is the BLOG.

The BLOG (fig. 2) designed for the Marmo Platano - Melandro PIT derives from the need to ensure citizens information and interaction with the institutions on government policies.

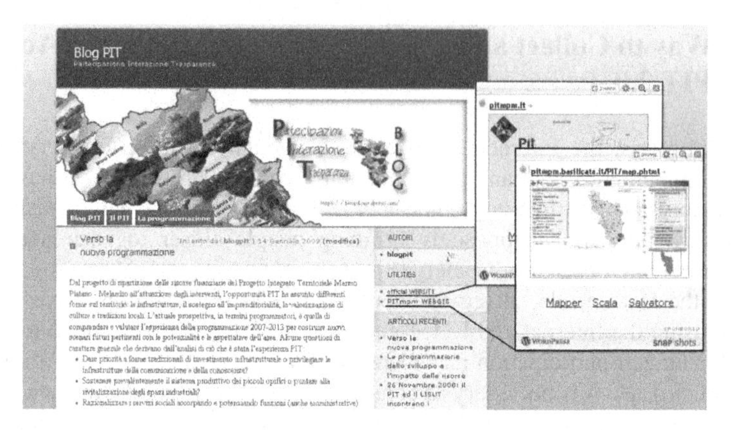

Fig. 2. Marmo Platano – Melandro PIT BLOG (http://blogpit.wordpress.com)

The BLOG represents a resource for local development, community life and identity; it promotes collaborative relationships and constant citizen involvement in public decisions, overcoming typical participation constraints and giving more emphasis to the role of citizens. At present, we cannot give any interesting results concerning the use of the BLOG because it has been recently made available, but we hope that it might guarantee a real citizenship involvement into the next programming phase (2007-2013 period).

The PIT 2.0 appears ideal for the activation of participatory practices among citizens because they are increasingly familiar to them. Having "familiar" tools will greatly increase potential "participation" [12]. Therefore, e-government produces a transformation from face-oriented or file-oriented services to a digital platform, determining increased effectiveness, improved public information diffusion and enhanced equity opportunities for citizens [13].

More innovative or democratic forms of participation in development planning (voluntarily entered into by local planning authorities) are still relatively rare. But where they do exist, they can provide an effective mean through which planners can fully engage with the communities they serve and generate more informed discourses on planning policy matters [14]. This has happened with the PIT experience: it is rare but possible that in a small context, like the PIT one, new participation forms in development programming can draw up policies and strategies and then "advertise" them to the public to assure transparency and accessibility.

4 Ontologies in Planning and Programming

During the last years, agencies with planning competences have remarkably increased; consequently local authorities are overregulated by a huge number of planning documents. Administrative functions related to territory government are attributed to elective institutions (e.g. municipalities, Regions, etc.). In the same way, sector-based institutions (e.g. monuments and fine arts bureau, national and regional parks, etc.) manage territory by prescriptive plans, more effective than administrative ones.

In a lot of cases a plan by an institution cannot be coherent with a plan by another institution, and some actions admitted by a plan can be forbidden by another. The problem of integration among different plans occurs. A way to tackle this problem is the use of ontologies, basing the integration of geographic information primarily on its meaning [15],[16]. Overcoming the traditional philosophical definition of ontology as the "discipline dealing with theories of being", we will use the slightly different notion (proposed, among others, by Grüber): a specific ontology *as a model* can be defined as "the explicit specification of an abstract, simplified view of a world we desire to represent" [17].

In order to be useful, the ontology has to be shared. In an international community of users, the first difficulty comes from languages, but a similar problem comes out when we match together different programs or plans adopted by different bodies. Ontology can help the community to define and explicit a common language and to strengthen the efficacy of direct interactions [18].

The development of an ontology might be quite different depending on the level of user's involvement [19]. In the present case, the ontological approach has been developed by a limited group of experts (managing the research project) and then imposed to the community members through the tools developed (WEBSITE, WEBGIS, BLOG).

Ontology design is a crucial step in the process of applying ontologies to e-democracy tools and processes. Special attention should be paid to the structural elements of the ontology: domain (or 'scope' of the ontology), concepts ('classes'), hierarchy, attributes of concepts, restriction and relations between concepts, instances. The definition of such elements represents the 'ontology design'.

The domain is the abstraction of the reality we want to represent. In the study case it is composed by physical elements, relations among them, value systems, program actions, social issues, policy goals. The first issue, in order to improve process rationality, is to circumscribe the domain. Ceravolo and Damiani [20] propose some questions to be answered in this phase:

- Q1: Which is the portion of real world we want to describe through the ontology?
- Q2: Which are the answers we expect from our ontology?

In our case we add two other questions concerning the geographic dimension:

- Q3: Which is the spatial dimension of the domain (in other words: "where does the ontology work")?
- Q4: Is the domain open or close?

Our objective is to represent the whole of plans and programs acting on the PIT area (Q1). These policy tools refer to several institutional levels (from the European level to the municipality one) and they affect several sectors of intervention. Many integrations but also several contradictions are evident.

Answering Q3 question might appear to be a consequence of the administrative border of the study area (the PIT area). This choice is an element of strong simplification of the reality and so it implies errors in gathered evaluations. A way to control such errors is to consider the domain as open in space, time and objects (Q4).

The second methodological question (Q2) is probably the key of the ontology design. What do we expect from our work? In a synthetic view, we implemented e-democracy tools to improve participation in local development processes. This ontological representation aims to obtain an improvement of rationality in policy making. This could be possible if contradictions and conflicts among different planning tools are removed or at least reduced. The activity (considered as a bottom-up and participated approach) leading to such an ambitious objective is the evaluation intended as comprehensive and context based evaluation [21].

The result of this approach leads us to define five main classes of ontology domain for our application:

1. Plan, defined as "Written account of intended future course of action (scheme), aims at achieving specific goal(s) or objective(s) within a specific timeframe. It explains in a detailed way what, when, how, and by whom needs to be done and often it includes best case, expected case, and worst case scenarios"[22].
2. Project, defined as "Planned set of interrelated tasks to be executed over a fixed period and within certain costs and other limitations"[22].
3. Policy, defined as "A specific statement of principle or of guiding actions that implies clear commitment but is not mandatory. A general direction that a governmental agency sets to follow, in order to meet its goals and objectives before undertaking an action program"[23].
4. Tools, defined as "Financial or normative instruments for policy implementation" (our definition).
5. Actors, defined as "Groups of private, public, no-profit bodies involved in development process" (our definition).

5 Efficacy and Effectiveness into Programming: Assessment Opportunities

The main effort of this work is to provide a widely shared view of programming scenarios in the area of Marmo Platano – Melandro PIT. Such scenarios merge together structural characteristics of the context with the wider framework assisted by European Structural Funds. WEBGIS facilitates the involvement of local government agencies and citizens in the evaluation of the heterogeneous impacts of public expenses.

Within the complex system of functions, relations and procedures for programming and managing EU funds, a strategic need is to definitively link intervention strategies and actions to the local context of implementation [21]. The context becomes the reference term for programming and evaluating public investments. The realized WEBGIS is a tool which gives shots of the territorial condition through a complex data system. It also allows us to go over the static representation, since it provides dynamic perspectives for updating data.

The local development programming during 2000-2006 period followed the traditional procedures with a strong control by Regional Authority in Basilicata region. The involvement of territories (bottom-up approach) has been developed through the experimental tool of Territorial Integrated Projects (PIT). This experimental approach produced heterogeneous results in the different Objective One Regions in Europe and especially in Italy [24]. The Basilicata Region case is considered as a Best Practice for the governance model implemented and in terms of expense capacity of the brand new territorial bodies (the PITs). An open question is to understand the real perception of the development policy by the final beneficiaries (territories, local communities, productive system, etc.). Considering the study case, citizens of Savoia di Lucania municipality are conscious that about € 17.000,00 of public (mainly EU) funds have been spent for each of them in local development interventions. Tools implemented in this work allow to improve the consciousness of local communities concerning the public effort to support local development, especially giving precise information about "how" and "where" the public intervention took place. The long term impact of such a process is to get an active participation of local communities to decisional processes, in order to obtain more transparent and shared decisions. The final evaluation of the impact of public expenses in the area is far to be achieved, since a lot information is incomplete and the temporal horizon of data collection should be wider. Anyway, this framework allows us to anticipate some future challenges for the territories of the Marmo Platano - Melandro PIT and also to remark some first-level assessment of the informative tool developed. The innovation promoted by Marmo Platano - Melandro PIT started a new governance model alternative to the traditional one [25], as it resulted more effective in the use of public resources.

6 Discussion and Conclusion

The effectiveness of the e-democracy process depends on the participation level considered, both in a qualitative and a quantitative way. It means that a massive participation in terms of number of stakeholders is crucial and at the same time that this participation has to be informed and rational. Stakeholders have to be conscious of the contents of the debate (often technical contents) and they also have to respect a procedure ensuring the rationality of the process itself. A way to ensure the effective use of e-participation tools such as WEBSITE, WEBGIS and BLOG is to build an agreement on semantic interpretation of reality. Such effort depends on the programmer, who should match technical knowledge with "popular" view of reality, in order to facilitate the interaction among stakeholders, administrative bodies and technicians. The ontological approach can give relevant results. Through the application of a well structured ontology to the participative process, it is possible to ensure more transparent evaluation for programming and managing functions, peculiar to the administrative body.

Many opportunities to spread e-democracy come from tools implemented by Marmo Platano – Melandro PIT. This ensured better information and greater opportunities for citizens to be involved in planning process. Benefits of such an effort will come in the future programming period (especially the 2007-2013 EU convergence policy). In our opinion it is important to emphasize the potential of these tools in order to achieve a high level of interaction G2C [26] and the commitment of a local organization, to ensure the improvement of the following principles: participation, interaction and transparency. According to a study accomplished by Lanza and Prosperi (2009), the project described in this work is a part of "Collaborative E-Governance". The implemented project appears like a world of real e-participation focused on virtual and digital tools, belonging to Open Source and Free Web-ware instruments [12]. The application described can lead Public Administrations, stakeholders, technicians and, in a general view, citizens towards a renewed approach in local development planning. Participation and interaction between components of local communities may produce more true interpretations of the context.

Acknowledgements. This study has been supported by the Marmo Platano - Melandro PIT, Project "Banca studi, progetti ed immagini per la tutela e valorizzazione del patrimonio rurale". Authors are grateful to PIT Manager Ing. Gaetano Schiavone and project participants involved in data collection and implementation: Ing. Francesco Lasala, Ing. Gabriele Nolè, Ing. Giuseppe Zaccagnino.

References

1. Hajer, M.: The need to zoom out: understanding planning processes in a post-corporatist society (1997); copy available from M Hajer, `http://www.postmaster@wrr.nl`
2. Knapp, S., Coors, V.: The use of eParticipation systems in public participation: the VEPs example. In: Coors, V., et al. (eds.) Urban and Regional Data Management, pp. 93–104. Taylor and Francis, London (2008)
3. Nerbert, D.D.: Developing Spatial Data Infrastructure: The SDI Cookbook, Version 2.0. `http://www.gsdi.org/docs2004/Cookbook/cookbookV2.0.pdf` (2004)
4. Executive Order 12906 Coordinating Geographic Data Acquisition and Access: The National Spatial Data Infrastructure (1994)
5. Directive 2007/2/EC of the European Parliament and of the Council of 14 March 2007 establishing an Infrastructure for Spatial Information in the European Community (2007)
6. Annoni, A.: La nuova iniziativa della commissione europea per lo sviluppo di una infrastruttura di dati spaziale europea (INSPIRE). 6° Conferenza Nazionale ASITA – GEOMATICA per l'ambiente, il territorio e il patrimonio culturale, Perugia (2002)
7. Drafting Teams (Specification, Data, Network Services, Metadata), INSPIRE Technical Architecture–Overview, Drafting Teams (2007), `http://inspire.jrc.ec.europa.eu/`
8. Woolgar, S. (ed.): Virtual Society? - technology, cyberbole, reality. Oxford University Press, Oxford (2002)
9. Evans-Cowley, J., Conroy, M.M.: The growth of e-government in municipal planning. Journal of Urban Technology 13(1), 81–107 (2006)
10. Tambouris, E., Liotas, N., Tarabanis, K.: A Framework for Assessing eParticipation Projects and Tools. In: Proceedings of the 40th Hawaii International Conference on System Sciences (2007)
11. Pløger, J.: Public participation and the art of governance. Environment and Planning B: Planning and Design 28, 219–241 (2001)
12. Lanza, V., Prosperi, D.: Collaborative E-Governance: Describing and Pre-Calibrating the Digital Milieu in Urban and Regional Planning. Taylor and Francis, London (2009)
13. Conroy, M.M., Evans-Cowley, J.: E-participation in planning: an analysis of cities adopting on-line citizen participation tools. Environment and Planning C: Government and Policy 24, 371–384 (2006)
14. Tewdwr-Jones, M., Thomas, H.: Collaborative action in local plan-making: planners' perceptions of "planning through debate". Environment and Planning B: Planning and Design 25, 127–144 (1998)
15. Fonseca, F., Egenhofer, M., Davis, C., Borges, K.: Ontologies and Knowledge Sharing in Urban GIS. Computer, Environment and Urban Systems 24(3), 232–251 (2000)

16. Laurini, R., Murgante, B.: Interoperabilità semantica e geometrica nelle basi di dati geografiche nella pianificazione urbana. In: Murgante, B. (ed.) L'informazione geografica a supporto della pianificazione territoriale, pp. 229–244. FrancoAngeli, Milano (2008)
17. Gruber, T.R.: Toward principles for the design of ontologies used for knowledge sharing. Int. J. Hum, Comput. Stud. 43(5/6), 907–928 (1995)
18. Damiani, et al.: KIWI: A Framework for Enabling Semantic Knowledge Management. In: Zilli, et al. (eds.) Semantic Knowledge Management: An Ontology-Based Framework. Information science reference. Hershey, New York (2009)
19. Corallo, et al.: Enhancing communities of practice: an ontological approach. In: 11th International Conference on Industrial Engineering and Engineering Management, Shenyang, China (2005)
20. Ceravolo, P., Damiani, E.: Introduction to Ontology Engineering. In: Zilli, et al. (eds.) Semantic Knowledge Management: An Ontology-Based Framework. Information science reference. Hershey, New York (2008)
21. Las Casas, G., Scorza, F.: Comprehensive evaluation and Context based approach for the future of Regional Operative Programming in Europe. In: Proceedings of 48th European Regional Science Association Congress 2008, Liverpool, UK (2008)
22. Business Dictionary, http://www.businessdictionary.com/ (retrieved)
23. The California general plan glossary, http://www.cproundtable.org/cprwww/docs/glossary.html (retrieved)
24. Moccia, D.F.: Resistenze alla pianificazione strategica: una analisi trans-culturale della ricezione ed uso della pianificazione strategica nella pianificazione integrata italiana. In: Archibugi, F., Saturnino, A. (eds.) Pianificazione Strategica e governabilità ambientale, Alinea, Firenze (2004)
25. Scorza, F. (ed.): Contributi alla innovazione degli strumenti per lo sviluppo locale. ERMES Edizioni, Potenza (2008)
26. Prosperi, D.C.: PPGIS: Separating the Concepts and Finding the Nexuses. Proceedings, Urban Data Management Symposium (2004)

University and Primary Schools Cooperation for Small Robots Programming

G. Barbara Demo[1], Simonetta Siega[2], and M. Stella De Michele[3]

[1] Computer Science Department, University of Turin, c.so Svizzera 185, 10149 Turin, Italy
barbara@di.unito.it
[2] Istituto Comprensivo "Fogazzaro" di Baveno (VB), via Brera 12, 28831 Baveno - Verbania, Italy
simo.si@alice.it
[3] I Circolo Didattico Settimo Torinese, v. Buonarroti 8, 10036 Settimo Torinese, Italy
primocircolo@comune.settimo-torinese.to.it

Abstract. In July 2007, in the Italian northwest region named Piedmont, a number of teachers and school headmasters created a School-Net for k-12 "Educational use of robotics". The School-Net aims at promoting Papert's constructionism in a cooperative environment and at setting up a model of small robots programming activities integrated in standard curricula covered in k-12 school years. The project is based on the cooperation between the School-Net and the Computer Science Department of the Turin University for providing technical competences with mini-languages, designing and implementing program development environments pupils oriented and maintaining a community of practice supporting teachers during their activities with robots. Here we concentrate on primary school activities where educational aspects concerned by using small robots fill a long list with, of course, mathematics but also education to affectivity, creativity, communication, geography and others. Experiences from the project are here described.

Keywords: cross-disciplinary activities, inquiry based teaching and learning techniques, pupil centered teaching, programming mini-languages.

1 Introduction

In July 2007 a group of Italian primary and secondary schools headmasters signed an agreement called "Net for the Educational use of robotics" aiming at carrying out mutual interest activities using small robots in their schools. This agreement involved schools scattered in Piedmont, an Italian northwest region. The First Teaching District of Beinasco (Turin), with its headmaster V. Termini, was chosen as the Net leader Institute and S. Siega as the pedagogical responsible. The net also had the cooperation of G. Marcianó, leading the *Robotica Laboratory* of the Regional Institute for Researches in Education (IRRE), and of G. B. Demo from the Department of Computer Science of the University of Turin.

M.D. Lytras et al. (Eds.): WSKS 2009, LNAI 5736, pp. 238–247, 2009.

The School-Net aimed at promoting Papert's constructionism in a cooperation environment for setting up a model of small robots programming experiences in support to standard curricula covered during k-12 school years [1]. All educators members of the net had already been involved in ICT projects different in time and in kind of activities. In particular, most of them had been cooperating with G. Marcianó in his *Robotica Laboratory* activities promoted by Piedmont IRRE, an Institute that was going to change its functions in summer 2008. Thus the idea of organizing several schools in a Net partly had administrative and financial purposes, yet and most importantly, had pedagogical purposes originated primarily from teachers working in the field. They selected a net of schools organization in order to gather experiences from quite different institutions and to create both a shared pedagogical environment and a common professional guidance. This conceptual change in school organization was felt very important particularly in a situation where schools are struck by repeated changes. The shared environment likely provides better stability.

In their previous activities, educational researchers grouped in the Net already showed the same professional conviction of educating schoolchildren by always connecting the current technology challenges to their common roots as for pedagogy [2] and for didactics [3]. This mingling between tradition and innovation has given rise to a project for an original education methodology where technology is used in order to offer children the pleasure to learn every subject "beyond the pencil and the book" [1]. In minutes of a meeting of the Piedmont School-Net Technical Group we read that the Net aims at "developing, documenting, evaluating and disseminating k-12 educational activities with small robots that must be concrete, feasible and strongly affecting the daily curriculum of students following Marciano's idea of robotics as a learning environment" [4]. Teachers also wanted an experience exposing pupils to the method during several years of their education. Thus a k-12 project was decided where robots should be used with continuity rather than in occasional laboratories hours. Though also some junior and senior secondary schools are involved, most up to now School-Net experiences concern kindergarten and primary schools, likely because primary school teachers are most prone to cross-disciplinary activities and for innovative methods of teaching standard subjects are considered more successful if applied from the very beginning of children school life.

As we said above, several members had already been involved in activities connected with small robot programming before the Net was set up. To give an idea of these early experiences, in Section 2 S. Siega sketches activities in a fourth grade class in Baveno primary school during year 2003/2004 when a single Lego RCX robot was used. These can be considered first Net experiences because S.Siega currently is the Piedmont School-Net pedagogical responsible. Sections 3 and 4 concern recent activities. In Section 3 M. S. De Michele describes her 2007/2008 experiences in a second grade class with the Bee-Bot, by the TTS-group, programmable by pushing buttons on its back. Several teachers active in the School-Net have used the Bee-Bot. For sake of space, here only the De Michele's activity is sketched that we consider interesting for she was novice to programmable robots. Her experience can be useful to teachers thinking of approaching robotics with their first grades pupils and can inspire confidence that good results are achievable when pupils and teachers learn together. Section 4 is a short overview of recent activities where students write programs. From about the beginning, schools of the Net have used different types of robots and programming languages. Among programming languages used to program the RCX Lego robot, Siega and her schoolchildren in 2004 began to use the NQC (Not Quite C) textual language, proposed by D. Baum [5]. Most pupils found using iconic languages less clear than using the textual NQC particularly when icons have to be connected in a behavior description. On their side, teachers found that using the same textual format allows interesting mutual influence exchanges between linguistic competences pupils are collecting in their native language and those from conceiving and writing robot programs [6], [7]. Thus when G. Marcianó began to think of a children oriented robot programming language thus easier to be used in primary schools, he defined the textual language NQCBaby, a Logo-like programming mini-language [8]. NQCBaby is sketched in Section 5 where also a short description is given of the software tools developed around it for a better use by pupils and teachers. Future directions of the Piedmont School-Net work are given in the conclusive Section 6.

2 An Historical Perspective, from 2003/2004

As we wrote in the Introduction, schoolteacher S. Siega is the current pedagogical coordinator of the network of Piedmont schools involved in the educational use of robotics. Since 2003 she began to program one RCX Lego Mindstorm in a fourth grade primary class, after having worked with her pupils using Microworld software and the Logo language. The pupils criticized both the RCX manual, which proposed poorly varied models, and the robot programming language, that was found to be not user-friendly enough. Pupils also stated that the "robot" concept should not only apply to an object built using Lego bricks, but to any programmable, autonomous and mobile object. Due to this observation the awareness arose that by using different kits a larger number of children, belonging to different ranges of age, could be involved in robot activities. This is the important result that the School-Net today can be proud of having achieved.

After the 2003/2004 single-class experience, G. Marcianò proposed the project "Educational use of Robotics" for the three school years 2004-2007. Three schools agreed to his plan: Siega's Istituto Comprensivo of Baveno, the Direzione didattica of Tortona and the Istituto tecnico of Novara. The latter is a senior secondary school. The project made possible to continue studying and, above all, experimenting with primary school pupils the belief that robotics in school should be regarded as a topic pertaining not as much to the "new technologies" area, rather to the "new possible teaching methods" in a school-laboratory, i.e. a school environment where to "learn how to learn".

First experiences were often randomly initiated but shortly consolidated by student's responses greatly positive. Thus scientific measures of possible recognition and validation of educational applications have been proposed and documented [9]. Meanwhile the NQCBaby language was developed as a new instrument specifically designed for an educational use of robots in the school.

After three years, the correct evolution of the IRRE project was the creation of the network of Piedmonts schools to which this paper refers, because of the spreading of good practices produced in nearby schools. The network shares in its work the realization of what S. Papert wrote: "The child programs the computer and, in doing so, both acquires a sense of mastery over a piece of the most modern and powerful technology and establishes an intimate contact with some of the deepest ideas from science, from mathematics, and from the art of intellectual model building. ... Programming a computer means nothing more or less than communicating to it in a language that it and the human user both "understand". And learning languages is one of the things children do best", from the Introduction of [1].

The use of different languages enables students to communicate with different robots. If a student likes better to use icons, she/he may use them rather than a textual language: what matter is the concept of programming. Pupils enter commands to a robot and check if it performs what they want. The immediate feedback allows understanding if we have done a good job OR IF we made an ERROR. In this case we can correct and change the action of the robot immediately!

Practicing a method of learning by doing is a peculiarity of the our network of schools. This allows pupils understand what they are doing rather than learn mostly by heart what to do. "If today a student in school learns something the content is not the most important, yet the methodology of learning, that can be applied again in the future"[1].

3 Current First Programming Activities Using the Bee-Bot

The Bee-Bot, produced by the TTS group, is a big bee that can be programmed by pushing buttons on its back for moving forward, backward, turning left, right, starting to move or deleting previous commands. As we wrote, several teachers in our School-Net have carried out activities with Bee-Bot. Here we recall fragments from the report that M. Stella De Michele wrote to document the activities she, novel to robots, has carried out with her second grade schoolchildren during last 2007/2008 school-year. M. Stella is specialized in teaching humanities but in 2007 promptly agreed to become responsible for the robots experiences in her school and to use the Bee-Bot with her 7 years old pupils in their second grade in order to begin learning with them how to program small robots and how to use them for standard curriculum teaching.

"I consider necessary that schools face the technology surrounding pupils. I used for some years computers with my classes but I was curious of using an object that can move around the way you can teach him either by writing a description of a path or giving the description by pushing buttons as in the Bee-Bot case.

Our story with robots began when pupils found one Bee-Bot on our classroom windowsill. We began trying to understand why this bee, different from those we are used to, was there. Possibly she got lost because of the pollution and entered in our classroom to rest. The bee was greeted, given a nickname (Maya) and pupils introduced themselves to her. By pushing the buttons on its back children discovered that they could teach the Bee-Bot how to move on the floor (i.e. in a two dimensional space): going straight or turning left or right exactly a **quarter of a cake** (in second grade pupils have not yet dealt with angles and their measures). We discovered the bee could **stroll around** the classroom by pushing more buttons one after the other in a sequence and pushing the **go** button. When pupil asked whether we could make the bee go from one child to another, i. e. from a starting point to an end point, some

[1] [15], page. 3.

pupil observed that buttons could not be pushed randomly as they had been doing when they wanted the bee go strolling on the floor.

Making the Bee-Bot go from one child to another requires children to take decisions: first we must decide where to go from where, i.e. design a path connecting two points. Different children may suggest different paths. We shall consider some of them and for each path we decide which buttons to push and how many times. Then we verify if the Bee-Bot moves as we want. If it does not, we have not been good teachers: we taught our bee wrong, we have to change its behavior, i.e. the sequence of buttons pushed. If we want to teach from the beginning the Bee-Bot a new behavior we have to take some time *planning* what we want the Bee-Bot to do.

We have to "be precise" and discover how far the bee moves each step and so on. Thus we introduced the **concept of measure:** if Maya has moved for a while how can we say how far she went? How do we measure the space covered? First we used several non-conventional tools then we choose the ruler because it is a common tool and gives a number for the quantity of space covered for each step. To determine how far the bee goes with a given number of pushes one child suggests to sum (one step plus the previous ones) another to multiply (we count all the steps times the space covered by each step). Thus teacher recalls that both are right because we define multiplication using the sum and some child shouts: <Teacher, is this robotics or math? >. Children used their exercise book for drawing the paths each child liked better. At this point introducing the Cartesian plane turned out natural, also suggested by some pupil."

After one-year experience we are not proposing here a generalization. The activities described are an excerpt from a class robot activity journal that we will compare and discuss with other colleagues from the School-Net having first grades primary school experiences. Though we have not yet performed a specific evaluation of pupils' achievements, we can compare BeeBot pupils' competences with those of other second grade pupils seen in over twenty years of teaching experience. We notice that, by using a Bee-Bot, more numerous first grades pupils develop skills for

- logical thinking and counting
- solving topological problems (what is right and left and what are right and left of something outside me,
- accessing problem-solving education
- getting used to an *inquiry-based learning (and teaching) technique* even in activities, as those described above, perceived as near to mathematics. This is an uncommon experience in first grades [10].

Besides, we perceive that pupils have a playful approach to robotics and begin to understand what programming a robot is. For our young pupils we plan an evaluation session adapted from the one described by Kurebayashi for older students [11].

It is important to point out that above activities naturally involved several educational aspects other than those concerning mathematics, more obvious. For example, we considered different reasons why the bee entered our classroom. The pollution possibility was found acceptable and children all together wrote the "Bee-Bot Story". Moreover, different forms of pollution, causes, consequences and remedies were discussed: thus some Environment preserving Education has been covered. Pupils introduced themselves to the bee, gave their welcome holding it in their hands, gave it a name, involved it in their school life showing regard for the new "thing": this is

Education to affectivity and to diversity. For every robot experience we had a common discussion time followed by a self–activity where each child wrote down few lines on what we had done. Children learned by doing activities with a concrete object and teachers learned with them.

4 Programming Languages Currently Used in Primary Schools

The current methodology of the Piedmont School-Net includes the use of 4 different kinds of robot kits, with different peculiarities and functions that allow different kinds of learning: the Bee-Bot, the Scribbler by Parallax, RCX and NXT by Lego. Children can use five programming languages, according to their skills but also in line with the robot kit that is being used. Also, by means of a long-lasting cooperation with G. B. Demo of the University of Turin, a compiler for the NQCBaby language is available with a user interface very simple to use and friendly for children who have immediately accepted it. Pupils describe the desired robot behaviors in NQCBaby programs that are translated in NQC [5]. Thus schoolchildren maintain a competent use of the language primitives and are enabled to learn.

After four years of experiments, enrichments and modifications of the methodology, the schools involved in the Piedmont School-Net project may claim that the educational use of robotics, in favorable circumstances, allows kids to attain powerful skills for their cognitive development. Schools with longer time experience have been able to observe that students involved in robotics activities for six school years, i.e. from their primary school second grade to the junior secondary third grade, are able to solve meaningful problems and write related programs with robots equipped with sensors and actuators.

Since last year it was possible to experiment both in the kindergarten and in the first years of the primary school the *Bee-Bot*, the bee-shaped robot, a programmable machine that involves children in the use of the first computer procedures, as it was said in the previous section. After the Bee-Bot, it is possible to work with the *Scribbler*, the blue turtle (also called "the messy robot") that aims at simulating what the children program in Logo with Microworlds.

In the following years of the primary school, the Lego Mindstorms bricks allow to use various languages (iconic and textual) and to implement paths with several types of sensors and to find different meaningful solutions to given problems. To conclude, in junior secondary school activities using the most recent Lego NXT robot, complex and refined in its components, meets different needs of teenagers, without forgetting the application of the Piedmont School-Net methodology aiming at students' cognitive development rather than at promoting coding skills. During 2007/2008, in Baveno School, four different types of small robots have been used programmed with six different languages depending on pupils' grades and previous experiences. These numbers show the growth of experiences as for robot use in that school during about five years from first activities. Students educated through Robotics activities in schools of the network, end up having a concept of technology not so much as of a black box rather as of a world they can control because they understand it.

5 Cooperation with the University of Turin

The cooperation of the Department of Informatica of the University of Turin in the School-Net project began in autumn 2006. An integrated development environment (IDE) and compilers of the programming language NQCBaby into the NQC language for the RCX robot and into the NXC language for the NXT are now available to schools, developed by students of the University of Turin [10]. Our current work is toward providing a unique textual language, children oriented, to be used for programming all different robot types. This unique language is based on NQCBaby: thus it is a textual language mother-tongue-based and, according to the Logo philosophy, with primitives coming from children language. Indeed our approach is to make children use easier languages rather than building tools to make easier the existing languages difficult for pupils such as "wood icons" for the iconic programming language proposed in [12]. The pedagogical methodology integrated in the IDE already available to schools provides a gradual introduction of pupils to NQCBaby with language enrichments from children at beginning-to-write level that use NQCBaby0 to NQCBaby6 level, usually for junior secondary schools. NQCBaby0 is the kernel of the language: it is the textual form of the button commands on the Bee-Bot back.

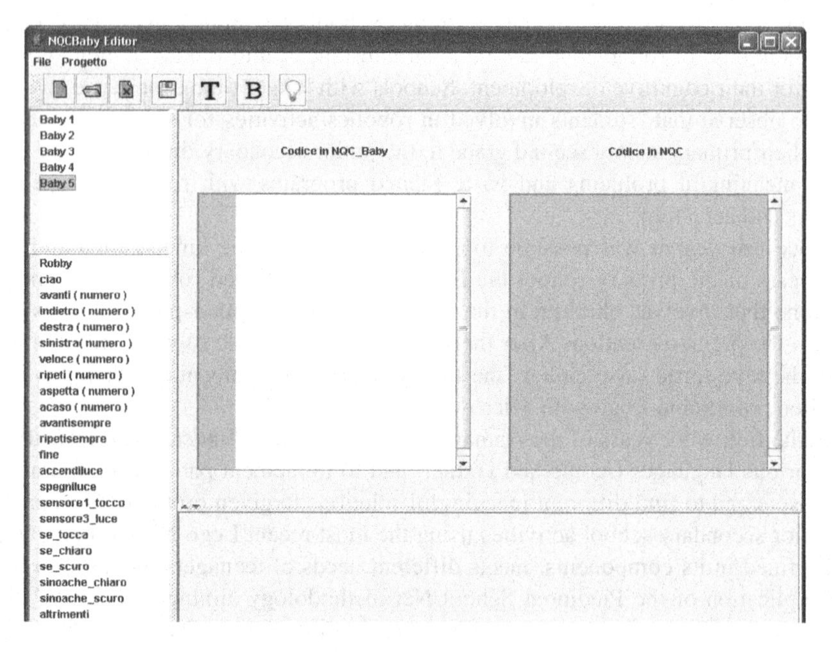

Fig. 1. The interface of the integrated development environment

Children write their NQCBaby programs using the Integrated Development Environment (IDE) interface shown in Figure 1. The "white board" central to the window is where children write their NQCBaby code. On top left side, we have the toolbar where the button T is used for translating the NQCBaby code. Errors are reported on the bottom with code line. Language levels are written on top of the left column indicated as Baby1, Baby2 and so on. Each next level encapsulates the previous ones and deals with either a different robot needing/allowing new primitives or new hardware components, sensors or actuators. Ordered introductions of new components, for example sensors, and related primitives for using them in robot-programmed behaviors shall comply the advances of schoolchildren logical and linguistic abilities [7], [10]. So doing robot-programming fits learning achievements children accomplish and becomes an original tool contributing to strengthening standard linguistic and logical curricula advances. The language grows with children, with their school education and with what they can/want to do with their different robots.

According to the mini-languages approach, NQCBaby is not a complete language because our purpose is not making children become good programmers but rather giving them the opportunity to use concrete robots for doing **concrete programming** [14].

When the RCX robot is used, NQCBaby is translated into NQC [Demo 2008]. When an NXT robot is used, NQCBaby is translated into the NXC (Not eXactly C) language. Actually the last extensions of the language NQCBaby providing primitives best fitting the NXC language, is called NXCJunior and NXC is the target of the translation. The following Progr-1 and Progr-2 are two examples of programs written by, respectively, schoolchildren 10 years old and 12 years old. The first one specifies a robot strolling around: it might be a program where pupils check primitives of the language without a specific goal: the left column is the NQCBaby English version for sake of comprehension, right column is the same code translated into the NQC language.

Progr-1:

```
Hi Robbi
   speed(3)
   forward(100)
   speed(7)
   backward(100)
   repeat(3)
      right(90)
      left(90)
   end-repeat
   repeat(2)
      backward(10)
      forward(20)
   end-repeat
thanks-bye.
```

```
task main()
{  SetPower(OUT_A+OUT_C,3);
   OnFwd(OUT_A+OUT_C); Wait(100);
   SetPower(OUT_A+OUT_C,7);
   OnRev(OUT_A+OUT_C); Wait(100);
   repeat(3)
   {  OnFwd(OUT_A);OnRev(OUT_C);
      Wait(90);
      OnFwd(OUT_C);OnRev(OUT_A);
      Wait(90);
      Off(OUT_A+OUT_C);
   }
   repeat(2)
   {  OnRev(OUT_A+OUT_C);Wait(10);
      OnFwd(OUT_A+OUT_C);Wait(20);
      Off(OUT_A+OUT_C);
   }
   Off(OUT_A+OUT_C);
}
```

In Progr-2 we find the function **`flip-coin`** that in both the NQC and NXC languages corresponds to a call of the function **`random`**. The program describes the behavior of a robot that goes forward for a while then chooses to turn left or right depending on the result of flipping a coin. Thus the robot is randomly going left or right. The NXC version of the program is shown on the right column.

Progr-2:

```
Hi Susi
   repeat-always
      speed(75)
      forward(500)
      if (flip-coin = heads)
         right(75)
      else   // it's cross
         left(75)
      end-if
   end-repeat
thanks-bye.
```

```
task main()
{  while(true)
   {  OnFwd(OUT_AC, 75);
      Wait(500);
      if (Random() >= 0)
      {  OnRev(OUT_C, 75); }
      else
      {  OnRev(OUT_A, 75); }
      Wait(360);
   }
}
```

Comparing the NBCBaby in the left columns of the above programs and the target code versions in the right columns here shown, we have examples of what we mean by saying that NQCBaby is a programming language children oriented rather than robot oriented.

6 Conclusions

Experiences here described began with one teacher and a small number of pupils. As for primary schools, nowadays the project counts 100 teachers in 17 different schools for about 1000 schoolchildren from 5-6 years old to 13 years old. With these already large numbers of pupils having done robot experiences here described for some years now, future activities will concern evaluating competences acquired by these students. Moreover, teachers of the Piedmont School-Net will continue developing the methodology but also using it as an every day teaching tool in several disciplines, which is one of the peculiar aim of the project. An effort is also toward extending the number of junior secondary schools involved in order to follow the students that have programmed robots in primary school grades as they progress in their education life. The homogeneousness and the common support of the pedagogical method while carrying out robot activities, tough the geographical distribution and the different types of schools involved, is another peculiar aspect of our project.

Exhibitions of the robot programming activities are requested by an increasing number of schools. We can satisfy a very small number of these requests and each for very short time though pupils, with their remarks, show that their minds are well prepared to this kind of project as it will involve more classes (among pupils' interesting remarks it is worth mention the one given by a seven years child. We were at the end of less than two hours of work with an NXT programmed using our programming language in our IDE and he said: <<My robot and yours have the same eyes, though mine is more beautiful, but mine can do three different things only while we can make this one do what we want>>).

Besides all the cross-disciplinary innovative activities that students will experience with robot programming, other important results specifically concern digital literacy. Indeed pupils learn how to write in a formal language, what an integrated development environment tool is and how to use the one we implemented specifically for this project. They learn what a translator is, its syntax error finding action and can have a look at the different translations for the different robots. We can say that their digital competences are to those of pupils only using any Office suite or similar, as the musical technique of piano players is to the one of stereo players, following the *Pianos Not Stereos* paper by M. Resnick, Bruckman and Martin [14].

Future investigations concern how pupils used to solve problems with programmable robots in primary schools, can develop inquiry-based learning techniques. Some hints have been given here in Section 3. This would be quite a positive change with respect to learning and teaching techniques we often observe around us where mathematics is an exercise of memory particularly in primary school but also in secondary where the chance to face creative problems is quite reduced: as an example, Euclidean geometry is almost disappearing.

Acknowledgments. Many thanks to all the members of the Piedmont School-Net project, to the schoolchildren who robot programmed with us, and to the University of Turin students who designed and implemented software tools here sketched.

References

1. Papert, S.: Mindstorms: Children, Computers, and Powerful Ideas. Basic Books, New York (1980)
2. Vygotskij, L.S. (ed.): Thought and language. Universitaria G. Barbèra, Florence (1934)
3. Vegetti, M.S.: Man psychology: for a science of a socio-historic education. In: Liverta Sempio, O., Vygotskij, Piaget, Bruner (eds.) Concepts for the Development, Raffaello Cortina, Milan (1998)
4. Marcianò, G.: Robotics as a learning environment. In: Didamatica Conference 2007, pp. 22–30 (2007)
5. Baum, D.: NQC language, `http://bricxcc.sourceforge.net/nqc`
6. Marcianò, G.: Robotics languages for schools. In: Didamatica Conference 2006, pp. 185–197 (2006)
7. Demo, G.B.: Marcianó, G., Contributing to the Development of Linguistic and Logical Abilities through Robotics. In: 11th European Logo Conference, Comenious University Press, Bratislava, p. 46 (2007), `http://www.eurologo2007.org/proceedings`
8. Brusilovsky, P., Calabrese, E., Hvorecky, J., Kouchnirenko, A., Miller, P.: Mini-languages: A Way to Learn Programming Principles. Education and Information Technologies 2(1), 65–83 (1997)
9. Marcianò, G., Siega, S.: Informatics as a language. In: Proc. Didamatica 2005, Potenza (2005)
10. Demo, G.B.: Programming Robots in Primary Schools Deserves a Renewed Attention. In: Proc. First World Summit Knowledge Society, Athens, September 24-28 (2008)
11. Kurebayashi, S., Kanemune, S., Kamada, T., Kuno, Y.: The Effect of Learning Programming with Autonomous Robots for Elementary School Students. In: 11th European Logo Conference, p. 46. Comenious University Press, Bratislava (2007), `http://www.eurologo2007.org/proceedings`
12. Horn, M.S., Jacob, R.J.K.: Tangible programming in the classroom with Tern. In: CHI 2007 Conference On Human factors in computing systems, 965-1970, San Jose (2007)
13. Demo, G.B., Marcianó, G., Siega, S.: Concrete Programming using Small Robots in Primary Schools. In: 8th IEEE International Conference on Advanced Learning Technologies, pp. 301–302. IEEE Press, New York (2008)
14. Resnick, M., Bruckman, A., Martin, F.: Pianos Not Stereos: Creating Computational Con-struction Kits. J. Interactions 3, 6 (1996)
15. Kopciowsky, J.: The learning process according to Feuerstein. La Scuola, Brescia (2002)

Improving Access to Services through Intelligent Contents Adaptation: The SAPI Framework

Nicola Capuano[1,2], Gaetano Rocco De Maio[3], Pierluigi Ritrovato[1,2], and Giacomo Scibelli[4]

[1] Research Center on Software Technologies, University of Sannio, Italy
[2] Dept. of Information Engineering and Applied Mathematics, University of Salerno, Italy
{capuano,ritrovato}@crmpa.unisa.it
[3] Information Technology Services S.p.a., Italy
gaetano.demaio@its.na.it
[4] Poste Italiane S.p.A., Italy
Scibel19@posteitaliane.it

Abstract. Nowadays, every public or private company has to provide the access to their services through Internet. Unfortunately, the access channels and devices increase both in numbers and heterogeneity. We started few years ago with a PC, wired connected to Internet, moved to wireless access through mobile phone and looking, in the next future, at wearable devices. If we also would like to take into account the user's preferences and his possible handicap we easily realise that combinations increase exponentially making impossible to adapt everything. The paper presents a rule based framework allowing to automatically adapt contents and services according to device capability, communication channel, user preferences and access context.

Keywords: Hypermedia Adaptive Systems; Context aware adaptive systems; Rule based adaptation systems.

1 Introduction to SAPI Adaptation Framework

In order to satisfy the growing request by the most people concerning with improvement and access facilitation to contents and services provided by the PosteItaliane S.p.A[1] the SAPI (Sistema Automatico Per Ipovedenti – Partially Sighted Automatic System) Project was proposed. SAPI aims to develop a framework for providing blind and half-blind users with intelligent services and contents, turning one's attention to adapting services and interfaces to *Situation*. In SAPI Situation is intended for the user-context and includes users' needs and features, environment, device and network.

SAPI belongs to the family of MultiChannel and MultiModal Adaptive Information System context-aware, so it has also implications for accessibility. SAPI pays great attention to the user interface design [1] [2] in order to satisfy advanced

[1] Poste Italiane is biggest mail service provider in Italy with more than 14.000 postal office, 200 sorting hubs, about 5000 ATM and about 50.000 POS on the country.

M.D. Lytras et al. (Eds.): WSKS 2009, LNAI 5736, pp. 248–258, 2009.

accessibility and adaptation requirements. In particular SAPI adapts its user interfaces in *batch* and at *runtime* (*on-the-fly*). In batch it uses evolution algorithms, while runtime, every time user-context changes it adapts executing the right adaptation rule. Since in SAPI a user interface is a collection of *Entities*, adaptation rules act on them and since entities are provided with semantic description, adaptation rules are described through an abstract and generic rule description language.

The flexibility and the efficiency of the system have been tested over practical scenarios of use. In case of need, the rules will be modified without modifying the application which uses them. Moreover using a rule language semantically interoperable and expressive allows an easy use and management of rules. So it will be possible to deal with the rule discovery using an innovative approach and, within the composition of adaptation rules, it will be possible to solve many problems about rule conflicts and find the best adaptation path. In fact, in SAPI, the description of a rule allows also to build adaptation paths in order to satisfy a more complex adaptation goal. The remainder of the paper is organized as follows. In section 2 there is a briefly description of the adaptation approach in SAPI. In section 3 we give some definitions and then illustrate the adaptation model which allows to define the new rule description language. Section 4 analyse the state of the art comparing current approaches with SAPI ones. Finally, in section 5 we describe how the adaptation engine works and illustrate some results.

2 Presentation Adaptation in SAPI

SAPI provides its users with many services each of them has a business logic, a workflow of activities, and a presentation logic that presents the results coming from the business logic in a specific user interface. Each service is able to adapt itself adapting its business and presentation logic. While the business logic is only adapted in batch, presentation logic is both adapted in batch and on the fly. Since the presentation logic is the composition of several user interfaces which, as indicated above, are collections of entities, adapting presentation logic means adapting entities. The entity is the atomic elements of SAPI and constitutes the basic element to compose a service. The Entity is a digital object (button, inputbox, hyperlink, etc.) or a media resource (text, audio, video, image, etc.), provided with a semantic description. It identifies a content which can have different versions and features but the same semantics. For each Entity there are two semantic sections: the first, called declarative section, contains metadata and relations between entities used for describing the content and related use; the second, called rule section, contains information about rules that allows an external actor to adapt on the fly the content to the specific situation. The semantic description is written in OWL.

Figure 1 shows the model of an Entity. An entity can be atomic or composed. A composed entity groups many atomic or composed entities and has: a navigation model that indicates what are the previous and next entity; an interaction model that indicates the logic order in which the entity can be presented to the users; a presentation model that indicates the layout of the presentation. Finally every entity has one or more adaptation rules that can be applied on it in order to satisfy user needs. Every rule is implemented and stored in the rule repository.

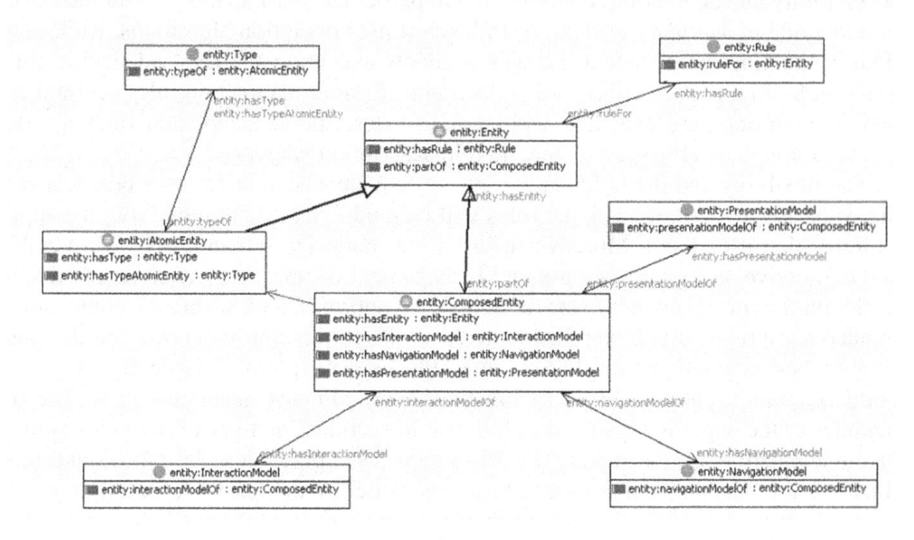

Fig. 1. Entity Model

Figure 2 shows the main SAPI's framework components involved in the adaptation. Every entity is stored in the "Entity Repository" which is accessible from the "Service Provider" (SP), a module responsible for the management and execution of the business and presentation logic. Execution of business logic consists in executing one o more activities, each of them refers at least one or more entities as results or as container to put results.

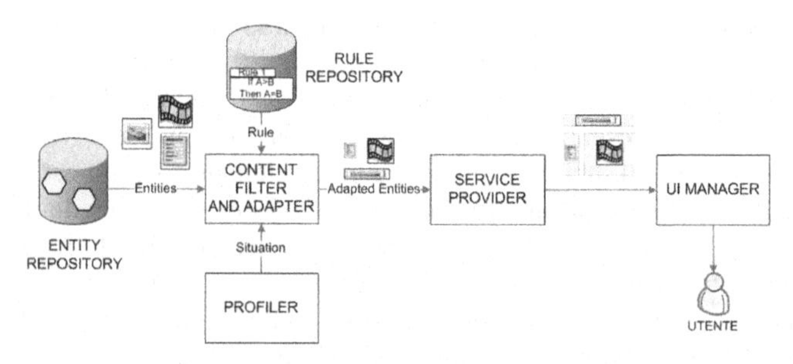

Fig. 2. Adaptation Architecture of SAPI

Once the entity is selected and set, before it could reach the SP have to be sent to the "Content Filterer and Adapter" (CFA) to adapt it to the specific situation. CFA is a rule engine that fires the right rule in order to adapt the entity to the situation that is sent from the "Profiler" which gets information about the user, the environment, the network and the device, querying an ontology and using physical sensors.

When Content and Adapter Filterer receives the entity, it analyses entity's description, decides which of its rules must be fired and selects them from the rule repository. Finally the entity will be sent to the UI manager which constructs the interface and presents it to the user.

3 The Adaptation Model

Before defining the adaptation model is necessary to fix some concepts. In particular, the concept of profile and stereotype, which contains the information the CFA needs to adapt entities, and the concept of adaptation and adaptation rule. Let's start with some definitions.

3.1 Core Definition

Definition – Profile: Given $C = \{c_{1,\ldots,}c_n\}$, a set of characteristic, and Dc_i, the domain of values which i-th characteristic can assume; an instance of profile P is every n-tupla î belonging to the set $P = Dc_1 X \ldots X Dc_n$.

In SAPI we have defined the following characteristics:

- C_U are the characteristics of user: preferences, behaviours, abilities, etc.;
- C_E are the characteristics of environment: brightness, noise, etc.;
- C_D are the characteristics of device: CPU, OS, screen size, color depth, etc.;
- C_N are the characteristics of network channel: delay, bandwidth, etc..

Definition – Stereotype of a Profile: The stereotype of a profile $P = Dc_1 X \ldots X Dc_n$ is a profile $S = Sc_1 X \ldots X Sc_n$ which satisfies the next conditions:

- has the same characteristics of P;
- Sc_i is a subset of $Dc_{i,}$ $i = 1,\ldots, n$;
- is consistent;
- identifies a class of instances of profile very similar among themselves;
- is dissimilar from other stereotypes.

Definition – Adaptation: The adaptation of an entity E in the state S_0 of a given Situation S is every sequence of conditioned actions such as, every time S changes, is able to lead E in a state compatible with S in a deterministic way. Because it will be useful in a short time, once given the definition of adaptation, is possible to give the definition of event too.

Definition – Event: An event is every change of state of the situation S represented by an instance of profile.

Definition – Adaptation rule: An adaptation rule is one of the conditioned action executed in order to adapt one or more entities. According to the literature on Active Database in SAPI the adaptation rule follows the Event Condition Action (ECA) formalism.

Definition – Adaptation model: An adaptation model M_a is a set of adaptation rules $R_a = \{r_1, \ldots, r_n\}$ whose *termination* and confluence are guaranteed.

Since it is an unsolvable problem, Termination is often guaranteed defining a sufficient number of constraints on the properties of the rules which constitute the adaptation model [3].

3.2 The SAPI Adaptation Rule Model

In order to specify an abstract and generic model to describe adaptation rule, we started from results of two research areas: *Adaptive Hypermedia Systems* (AHS) and *Active Database*. Brusilovsky in [11] classified adaptation methods and techniques basing on main concepts which mark an adaptation rule: *Adaptation Goals* (**why**); *Something* (**what**); *Situation* (**to what**); *Methods and Techniques* (**how**).

Starting from these concepts, it is possible to define an abstract model which allows to describe a generic adaptation rule, i.e. a rule on which basis it is possible to adapt *something* (of the *Universe*) on a given *situation* (of the *Universe*) and/or on changes of a given situation which the rule is sensitive (*situation-awareness*). In SAPI *something* refers to the Entity.

Before starting to work out the model it is right to do a distinction between absolute transformation ability and relative transformation ability of an Entity.

Absolute transformation ability: the absolute transformation ability of an Entity is the set of Entity transformation methods which are situation-independent. These methods ignore the *situation* concept, so they are application-independent .

Relative transformation ability: the relative transformation ability of an Entity is the set of Entity transformation methods which are able to generate a new version of the Entity as congenial as possible to a given situation. Differently from the previous case, these methods must have consciousness about the concept of situation and its coding. So they are application-dependent. Figure 3 shows an example of relative transformation ability.

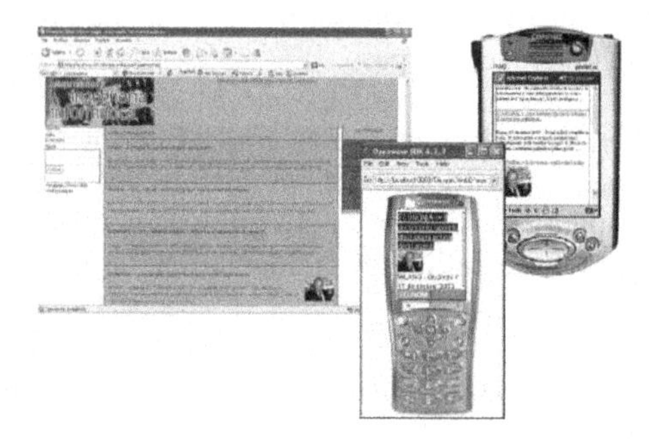

Fig. 3. Example of relative adaptation ability to three different situations

For the *activation* of an adaptation rule we assume that an adaptation rule can be fired, if and only if, some *conditions* (*Pre-Conditions*) are satisfied, in the following cases (*event/triggers*): on explicit request made by an actor (*Adaptor*); whenever an Event Detector finds a *situation* change (*sensor*); as a *propagation* of a rule which is able to trigger an internal event.

As a direct consequence of above, a formal description of adaptation rules comes down naturally. We show this description through the UML class diagram represented in Figure 4. For the sake of simplicity, all of possible triggers of an adaptation rule are considered sub-classes of the EVENT class.

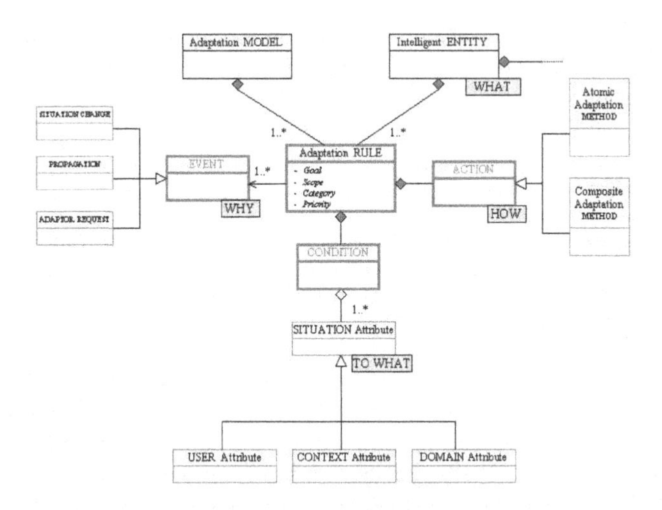

Fig. 4. SAPI Adaptation Model

Definition – Horn Clause: In mathematical logic, a Horn clause is a clause (a disjunction of literals) with at most one positive literal.

A Horn clause is a rule if and only if it has one, and only one, positive literal and at least a negative literal so that: $\sim A_1 \vee \sim A_2 \ldots \vee \sim A_k \vee C$ which is equivalent to $(A_1 \wedge A_2 \ldots \wedge A_k) \Rightarrow C$, i.e. an implication as *if Antecedent then Consequent* with one or more conditions and just only one conclusion.

So an *Adaptation Rule* Ar of one *Entity* En on a given *Situation* S, fireable when the event Ev occurs, in which the *implication* is an *Horn Rule*, is defined as follows:

$$\text{Ar: on Ev, Cd} \Rightarrow \text{Ad (En, S)} \tag{1}$$

where:

- Cd is a condition consisting of a union of atoms as $C_i \gamma C_j$ or $C_i \gamma K$ where:
  - K= costant,
  - γ = comparative operator
  - C_i and C_j characteristic of S;
- Ad (En, S) = adaptation of En to situation S.

Therefore the intuitive meaning of an adaptation rule is: when an event Ev occurs on a given situation S, if the condition Cd is verified then do the adaptation of the Entity En on situation S. Basing on the (1) it is possible to observe that an adaptation rule is a specific ECA Rule (reaction rule) which paradigm, in his most generic form, can be considered as the next:

On Event *if* Condition *then do* Action(s)

For example:

On *Event: = < Brightness_ Sensor_Value <= Low_Threshold >*

If *((C_D = ($d_1, d_2, \ldots,$ PDA $, \ldots, d_k$)) **AND** (C_E = ($e_1, e_2, \ldots,$ NIGHT, MOBILE $, \ldots, e_l$)) **AND** (C_N = ($e_1, e_2, \ldots, e_l$)) **AND** (C_U = ($u_1, u_2, \ldots,$ AGE > 60 $, \ldots, u_n$)))*
Then *(Adaptation Method = A_{FE})*

Where A_{FE} uses *font enhancement* as adaptation technique. This adaptation rule says that when a man at least 60 year old uses a PDA at night the system must increment the font dimension so it will be easier readable.

4 Related Work

In the following we shortly present the state of the art on two relevant research topics: Adaptive Systems and rule descriptions.

4.1 Adaptive Systems

Adaptation of content and user interfaces in order to achieve accessibility is a great challenge that the research community have been studying for year, although the laws on accessibility are very recent. In [8] a tool for the automatic generation of adaptive Web sites is presented. FAWIS relies on two basic notions: the profile, used to model a variety of contexts and their characteristics, and the configuration which describes how to build the various levels of a Web application (content, navigation and presentation). These are respectively similar to SAPI profile and SAPI entity, which is richer thanks to its semantic description. The automatic adaptation of content delivery is achieved by means of an original notion of production rule that allows to specify declaratively how to build a configuration satisfying the adaptation requirements for a given profile. The module responsible of this work is the Context Manager. It selects the adaptation rules from a repository and, avoiding conflicting among them, generates the configuration that describes the final adaptation. This is a solution a little bit different from SAPI. SAPI repository of adaptation rules contains the implementation of rules but not the description that is part of the entity.

[9][10] propose an ontology-based approach to adaptation. The work in [10] defines a user model (an XML file) that consists of concepts and attributes and uses adaptation rules in order to perform updates of this model. The adaptation rules are production rules and define how the user model is updated. For example, when the user accesses a page (or an object included in a page) the rules associated with the access attribute are triggered. The approach proposed in [9] tracks user interactions and tries to present the content chosen by the user. This work is done taking in account the relationship existing among this content that are represented by entity with a semantic description. Since the adaptation is carried out using SWRL, [9] uses production rules too. SAPI uses the approach proposed in [10] only to guarantee the evolution of the user profile and implements the approach proposed in [9] through the interaction model of each entity. In addition SAPI doesn't use production rules but ECA rules and instead of presenting every time a different content, changes its presentation (for example summarizing it or changing to another version) without changing the entity and its description.

4.2 Rule Description Languages

Rules are generally fragment of knowledge *self-contained* which are usually used in a form of reasoning. They can specify, for example: static or dynamic integrity constraints of a system (*integrity rules*); derivations of new concepts (*derivation rules*); a reactive behaviour of system in reply to particular events (*reaction rules*).

Furthermore, rules can be described with three abstract level:

1. *Computation Independent Business Domain Level* (CIM in MDA of OMG) – rules are described in a declarative way using an informal or visual language;
2. *Platform Independent Operational Design Level* (PIM in MDA of OMG) – rules are described using a formal language or a computational paradigm which is easy translatable in instruction that a software system can execute.
3. *Platform Specific Implementation Level* (PSM in MDA of OMG) – rules are described using a language of a specific execution environment such as Oracle 11g, Jess 7.1, JBoss Rules (aka Drools), XSB 3.1 Prolog, o Microsoft Outlook Rule Wizard.

In PIM can be included *Rule Markup Languages* which main goal is to allow publication, deploying, reuse and rule interchange on the Web and distributed system. Rule markup languages also allow to describe rule in a declarative way and as modular unit self-contained and interchangeable between different systems and tools.
Among the main rule languages, actually available or developing, we mention RuleML[4], SWRL[5], R2ML[6].

RuleML Initiative started in 2000 to define a markup language able to support different kind of rules and different semantics. Even though it is an XML-based language and allows to describe all kind of rules, the latest version RuleML 0.91 (2006) hasn't yet a syntax for integrity rule and reaction rule. In order to get around this was proposed Reaction RuleML[7] which defines new constructs within separated modules which are added to the RuleML family as additional layers.

Semantic Web Rule Language (SWRL) is based on a combination of the OWL-DL and OWL-Lite sublanguages of the OWL Web Ontology Language with the Unary/Binary Datalog RuleML sublanguages of the Rule Markup Language. Its rules are of the form of an implication between an antecedent (body) and consequent (head): whenever the conditions specified in the antecedent hold, then the conditions specified in the consequent must also hold. Rules must be safe, i.e. variables that occur in the antecedent of a rule may occur in the consequent.

The REWERSE II Rule Markup Language (R2ML) allows interchanging rules between different systems and tools. It integrates the Object Constraint Language (OCL), a standard used in information systems engineering and software engineering, SWRL and RuleML and includes four rule categories: derivation rules, production rules, integrity rules and ECA/reaction rules.

Even though these languages allow to describe rules, to execute them is necessary an execution environment. For the experimentation in SAPI, we used JBoss Rules (aka DROOLS), an open source framework that provides standards-based business rules engine and business rules management system (BRMS) for easy business policy access, change, and management. It allows to encode business rules in the IT application infrastructure and supports a variety of language and decision table inputs, making it easy to quickly modify the business policies. It also simplify applications by separating business policy or rules logic from process, infrastructure, and presentation logic. This modularity enables to develop, deploy, modify and manage a business process' rules without modifying code nor rebuilding the application. Using JBoss Rules it was possible to translate the rule adaptation model in a set of executable rules. The code fragment below shows the code used to translate the previous adaptation rule which says that when a man at least 60 year old uses a PDA at night the system must increment the font dimension so it will be easier readable.

```
package sapi.adapter.rules
import sapi.adapter.ws.EntityAdapter;
import sapi.KnowledgeManager.bean.Entity;
import sapi.KnowledgeManager.bean.DeviceContext;
import sapi.KnowledgeManager.bean.EnvironmentContext;
import sapi.KnowledgeManager.bean.NetworkContext;
import sapi.KnowledgeManager.bean.UserContext;
rule "font_enhancement"
    salience 10
    no-loop true
    agenda-group "content_style"
    dialect "java"
    when
      $ent: Entity();
      $DC: DeviceContext();
      $EC: EnvironmentContext();
      $NC: NetworkContext();
      $UC: UserContext();
      (DeviceContext(device == "LowBrightness") and
      EnvironmentContext(environment ==
  "LowBrightness") and
       EnvironmentContext(ubiquity == true) and
       UserContext (age >= 60))
then
      $ent = EntityAdapter.fontEnhancement($ent);
end
```

This rule is stored in a file and if someone want to change it, it is not necessary to change the application or re-compile it.

5 Adaptation Examples

In the following we present some adaptation experiments carried out using the SAPI framework prototype.Keeping in mind the components presented in Figure 2, SAPI's components communication is based on web service, thus when SP needs contents, it invokes the CFA web Service which implements the adaptation rule engine. In this case SP act as the adaptor. The CFA invocation means the situation is changed so it is fired the event which causes the adaptation. CFA is a module that receives an entity and analyzing its description is able to understand which the rules it can fire are. After, it takes the rules described in the entity from the repository. Obviously not all the rules will be fired, it depends on the situation. Thus CFA needs to receive the profile too. Once CFA has the rules and the profile, matching them will be able to understand which rule it must fire in order to return an entity adapted to SP. According to the previous section each rule consists of two parts: a condition and an action. The condition is expressed in the *when* clause as a Boolean expression using attribute of concepts which model the profile. The action is expressed in the *then* clause and consists of a function that has the same name mapped on the entity description. It is implemented in java and stored in the rule repository. When the condition is false, CFA usually doesn't do anything, but optionally there can also be a second action to perform. This solution allows us the separate rule description and rule implementation. This entails that if someone wants to change the conditions of a rule it hasn't to compile or write back the rule implementation. The actions of a triggered rule temporarily update some attributes of some concepts of the entity model. After that, entities are transformed in a Java object and returned to SP that will insert the entities in the right layout in order to allow the UI Manager to present the contents to the users.

Figure 5 and Figure 6 show two examples of adaptation.

The first is an example of a simple adaptation. The left part of Figure 5 shows the raw entities without adaptation. If a user half-blind request the same content SAPI recognise the user (for concrete adaptation even for the simple access the user is provided with an RFID tag that allow SAPI to identify the profile stereotype), retrieve the profile, analyse the context and adapts all the entities resizing each trying to use the whole page. The second example shows a more complex adaptation. A user is using a mobile phone and it isn't experienced in this service. These information are in the profile and CFA applies three rules. The first one adapts the content to the new device and since it can contain less entities it is forced to split the page in more than one. The second rule adapts the content for a user not experienced. The third is a resizing of the entities. In this case the dimension is decreased. In this example CFA fired three adaptation rules. Every entity could be adapted through more than one function. The order in which they are fired is established through a priority that helps avoiding loops. Drools allows to indicate priorities in the *salience* clause.

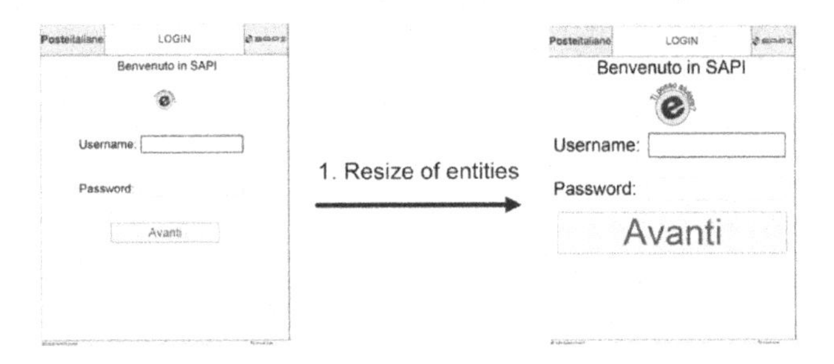

Fig. 5. Entities resize

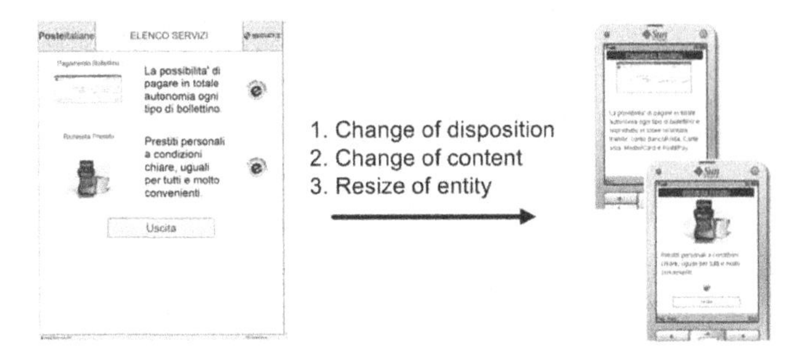

Fig. 6. Contents and layout adaptation

References

1. Troiano, L., Birtolo, C., Armenise, R., Cirillo, G.: Optimization of menu layout by means of genetic algorithms. In: van Hemert, J., Cotta, C. (eds.) EvoCOP 2008. LNCS, vol. 4972, pp. 242–253. Springer, Heidelberg (2008)
2. Troiano, L., Birtolo, C., Miranda, M.: Adapting palettes to color vision deficiencies by genetic algorithm. In: GECCO 2008: Proceedings of the 10th annual conference on Genetic and evolutionary computation, Atlanta, Georgia, USA, 12-16 July 2008, pp. 1065–1072 (2008)
3. Wu, H., De Bra, P.: Sufficient Conditions for Well-behaved Adaptive Hypermedia Systems (2001)
4. RuleML Iniziative, http://www.ruleml.org
5. SWRL, http://www.w3.org/Submission/SWRL/
6. Wagner, G., Damásio, C.V., Lukichev, S.: First-Version Rule Markup Languages (2005)
7. Paschke, A., Kozlenkov, A., Boley, H.: A Homogenous Reaction Rule Language for Complex Event Processing. In: 2nd International Workshop on Event Drive Architecture and Event Processing Systems (EDA-PS 2007), Vienna, Austria (2007)
8. De virgilio, R., Torlone, R.: FAWIS: A Tool for the Automatic Generation of Adaptive Web Sites, MAIS (2006)
9. Tran, T., Cimiano, P., Ankolekar, A.: Rules for an Ontology-based Approach to Adaptation. In: First International Workshop on Semantic Media Adaptation and Personalization (2006)
10. De Bra, P., Aerts, A., Berden, B., de Lange, B., Rousseau, B., Santic, T., Smits, D., Stash, N.: Aha! The adaptive hypermedia architecture. In: Proceedings of the ACM Hypertext Conference, Nottingham, UK (2003)
11. Brusilovsky, P.: Methods and Techniques of Adaptive Hypermedia (1996)

Semantic Based Access over XML Data*

Nikos Bikakis, Nektarios Gioldasis, Chrisa Tsinaraki, and Stavros Christodoulakis

Technical University of Crete, Department of Electronic and Computer Engineering
Laboratory of Distributed Multimedia Information Systems & Applications (TUC/MUSIC)
University Campus, 73100, Kounoupidiana Chania, Greece
{nbikakis,nektarios,chrisa,stavros}@ced.tuc.gr

Abstract. The need for semantic processing of information and services has lead to the introduction of tools for the description and management of knowledge within organizations, such as RDF, OWL, and SPARQL. However, semantic applications may have to access data from diverse sources across the network. Thus, SPARQL queries may have to be submitted and evaluated against existing XML or relational databases, and the results transferred back to be assembled for further processing. In this paper we describe the *SPARQL2XQuery* framework, which translates the SPARQL queries to semantically equivalent XQuery queries for accessing XML databases from the Semantic Web environment.

Keywords: Semantic Web, XML Data, Information Integration, Interoperability, Query Translation, SPARQL, XQuery, SPARQL2XQuery.

1 Introduction

XML has been extremely successful for information exchange in the Web. Over the years XML was established as a tool for describing the content of diverse structured or unstructured resources in a flexible manner. The information transferred with XML documents across the internet lead to needs of systematic management of the XML documents in organizations. XML Schema and XQuery [7] were developed to give the users database management functionality analogous to the Relational Model and SQL. In the Web application environment the XML Schema acts also as a wrapper to relational content that may coexist in the databases.

The need for semantic information processing in the Web on the other hand has lead to the development of a different set of standards including OWL, RDF and SPARQL[0]. Semantic Web application developers expect to utilize SPARQL for accessing RDF data. However, information across the network may be managed by databases that are based on other data models such as XML Schema or the Relational model. Converting all the data that exist in the XML databases into Semantic Web data is unrealistic due to the different data models used (and enforced by different standardization bodies), the management requirements (including updates), the difficulties in enforcing the original data semantics, ownership issues, and the large volumes of data involved.

* An extended version of this paper is available at [20].

M.D. Lytras et al. (Eds.): WSKS 2009, LNAI 5736, pp. 259–267, 2009.

In this paper we propose an environment where Semantic Web users write their queries in SPARQL, and appropriate interoperability software undertakes the responsibility to translate the SPARQL queries into semantically equivalent XQuery queries in order to access XML databases across the net. The results come back as RDF (N3 or XML/RDF) or XML [1] data. This environment accepts as input a set of mappings between an OWL ontology and an XML Schema. We support a set of language level correspondences (rules) for mappings between RDFS/OWL and XML Schema. Based on these mappings our framework is able to translate SPARQL queries into semantically equivalent XQuery expressions as well as to convert XML Data in the RDF format. Our approach provides an important component of any Semantic Web middleware, which enables transparent access to existing XML databases.

The framework has been smoothly integrated with the *XS2OWL* framework [15], thus achieving not only the automatic generation of mappings between XML Schemas and OWL ontologies, but also the transformation of XML documents in RDF format.

The design objectives for the development of the *SPARQL2XQuery* framework have been the following: a) Capability of translating every query compliant to the SPARQL grammar b) Strict compliance with the SPARQL semantics, c) Independence from query engines and working environments for XQuery, d) Production of the simplest possible XQuery expressions, e) Construction of XQuery expressions so that their correspondence to SPARQL can be easily understood, f) Construction of XQuery expressions that produce results that do not need any further processing, and g) In combination with the previous objectives, construction of the most efficient XQuery expressions possible.

The rest of the paper is organized as follows: In Section 2 the related work is presented. The mappings used for the translation as well as their encoding are described in Section 3. Section 4 describes the query translation process. An example presented at Section 5. The transformation of the query results described at Section 6. The paper concludes in section 7.

2 Related Work

Various attempts have been made in the literature to address the issue of accessing XML data from within Semantic Web Environments [2, 3, 6, 8, 9, 10, 11, 15, 16, 17, 18]. More relevant to our work are those that use SPARQL as a manipulation language. To this end, the *SAWSDL Working Group* [8] uses XSLT to convert XML data into RDF and a combination of SPARQL and XSLT for the inverse. Other approaches [9, 10, 11] combine Semantic Web and XML technologies to provide a bridge between XML and RDF environments. *XSPARQL* [11] combines SPARQL and XQuery in order to achieve Lifting and Lowering. In the Lifting scenario (which is relevant to our work), *XSPARQL* uses XQuery expressions to access XML data and SPARQL Construct queries for converting the accessed data into RDF. The main drawback of these approaches is that there is no automatic way to express an XML retrieval query in SPARQL. Instead, the user must be aware of the XML Schema and create his/her information retrieval query accordingly (XQuery or XSLT). In our work, the user is not expected to know the underlying XML Schema; (s)he expresses his/her query only in SPARQL in terms of the knowledge that (s)he is aware of, and (s)he is able to retrieve data that exist in XML databases. The aforementioned attempts, as well as others [12, 13, 14] that try to bridge relational databases with the Semantic Web using SPARQL, show that the issue of accessing legacy data sources from within Semantic Web environments is a valuable and challenging one.

3 Mapping OWL to XML Schema

The framework described here allows XML encoded data to be accessed from Semantic Web applications that are aware of some ontology encoded in OWL. To do that, appropriate mappings between the OWL ontology (*O*) and the XML Schema (*XS*) should exist. These mappings may be produced either automatically, based on our previous work in the *XS2OWL* framework [15], or manually through some mapping process carried out by a domain expert. However, the definition of mappings between OWL ontologies and XML Schemas is not the subject of this paper. Thus, we do not focus on the semantic correctness of the defined mappings. We neither consider what the mapping process is, nor how these mappings have been produced.

Such a mapping process has to be guided from language level correspondences. That is, the valid correspondences between the OWL and XML Schema language constructs have to be defined in advance. The language level correspondences that have been adopted in this paper are well-accepted in a wide range of data integration approaches [2, 3, 6, 15, 16, 17]. In particular, we support mappings that obey the following language level correspondence rules: *O* Class corresponds to *XS* Complex Type, *O* DataType Property corresponds to *XS* Simple Element or Attribute, and *O* Object Property corresponds to *XS* Complex Element.

Then, at the schema level, mappings between concrete domain conceptualizations have to be defined (e.g. the *employee* class is mapped to the *worker* complex type) either manually, or automatically, following the correspondences established at the language level.

At the schema level mappings a mapping relationship between *O* and an *XS* is a binary association representing a semantic association among them. It is possible that for a single ontology construct more than one mapping relationships are defined. That is, a single source ontology construct can be mapped to more than one target XML Schema elements (1:n mapping) and vice versa, while more complex mapping relationships can be supported.

3.1 Encoding of the Schema Level Mappings

Since we want to translate SPARQL queries into semantically equivalent XQuery expressions that can be evaluated over XML data following a given (mapped) schema, we are interested in XML data representations. As a consequence, based on schema level mappings for each mapped ontology class or property, we store a set of XPath expressions (*"XPath set"* for the rest of this paper) that address all the corresponding instances (XML nodes) in the XML data level. In particular, based on the schema level mappings, we construct:

- A **Class XPath Set** X_C for each mapped class C, containing all the possible XPaths of the complex types to which the class C has been mapped to.

- A **Property XPath Set** X_{Pr} for each mapped property *Pr*, containing all the possible XPaths of the elements or/and attributes to which *Pr* has been mapped.

Example 1: Encodings of Mappings

Fig. 1 shows the mappings between an OWL Ontology and an XML Schema.

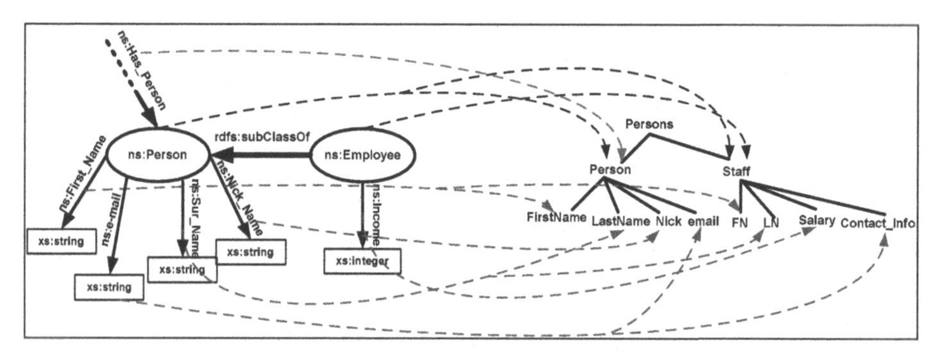

Fig. 1. Mappings Between OWL & XML

To better explain the defined mappings, Fig. 1 shows the structure that the XML documents (which follow this schema) will have. The encoding of these mappings in our framework is shown in Fig. 2.

Classes:	**DataType Properties:**
$X_{ns:Person}$={/Persons/Person, /Persons/Staff}	$X_{ns:First\_Name}$={/Persons/Person/FirstName, /Persons/Staff/FN}
$X_{ns:Employee}$={/Persons/Staff}	$X_{ns:Sur\_Name}$={/Persons/Person/LastName, /Persons/Staff/LN}
	$X_{ns:Nick\_Name}$={/Persons/Person/Nick }
Object Properties:	$X_{ns:e-mail}$ ={/Persons/Person/email, /Persons/Staff/Contact_Info}
$X_{ns:Has\_Person}$={/Persons/Person }	$X_{ns:Income}$={/Persons/Staff/Salary}

Fig. 2. Mappings Encoding

4 Query Translation Process

In this section we present in brief the entire translation process using a UML activity diagram Fig. 3 shows the entire process which starts taking as input the given SPARQL query and the defined mappings between the ontology and the XML Schema (encoded as described in the previous sections). The query translation process comprises the activities outlined in the following paragraphs.

4.1 SPARQL Graph Pattern Normalization

The *SPARQL Graph Pattern Normalization* activity re-writes the Graph-Pattern (*GP*) of the SPARQL query in an equivalent normal form based on equivalence rules. The SPARQL *GP* normalization is based on the *GP* expression equivalences proved in [4] and re-writing techniques. In particular, each *GP* can be transformed in a sequence *P1*

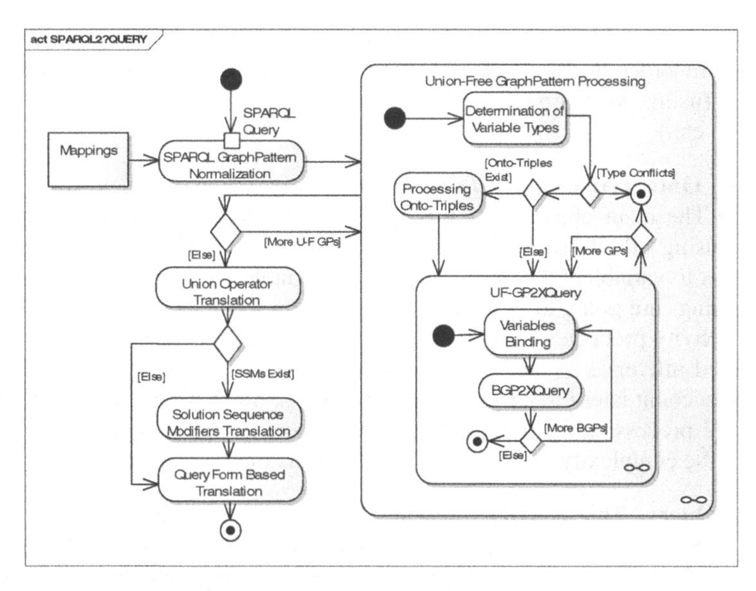

Fig. 3. Overview of the SPARQL Translation Process

UNION P2 UNION P3 UNION...UNION Pn, where *Pi* (*1≤i≤n*) is a Union-Free *GP* (i.e. *GPs* that do not contain Union operators) [4]. This makes the *GP* translation process simpler and more efficient, since it decomposes the entire query pattern into sub-patterns that can be processed independently of each other.

4.2 Union-Free Graph Pattern (UF-GP) Processing

The *UF-GP Processing* translates the constituent *UF-GPs* into semantically equivalent XQuery expressions. The *UF-GP Processing* activity is a composite one, with various sub-activities. This is actually the step that most of the "real work" is done since at this step most of the translation process takes place. The *UF-GP Processing* activity is decomposed in the following sub-activities:

Determination of Variable Types. This activity examines the type of each variable referenced in each UF-GP in order to determine the form of the results and, consequently, the syntax of the Return clause in XQuery. Moreover, variable types are used by the *"Processing Onto-Triples"* and *"Variables Bindings"* activities. Finally, this activity performs consistency checking in variable usage in order to detect any possible conflict (e.g. the same variable name is used in the definitions of variables of different types in the same *UF-GP*). In such a case, the *UF-GP* is not going to be translated, because it is not possible to be matched with any RDF dataset.

We define the following variable types: The *Class Instance Variable Type* (*CIVT*), The *Literal Variable Type* (*LVT*), The *Unknown Variable Type* (*UVT*), The *Data Type Predicate Variable Type* (*DTPVT*), The *Object Predicate Variable Type* (*OPVT*), The *Unknown Predicate Variable Type* (*UPVT*).

The form of the results depends on the variable types and they are structured in such a way that allows their transformation to RDF syntax. The transformation can be done by processing the information regarding the form of the results and the input mappings. In order to allow the construction of result forms, appropriate XQuery functions (using standard XQuery expressions) have been implemented (like *func:CIVT,* etc.).

Processing Onto-Triples. *Onto-Triples* actually refer to the ontology structure and/or semantics. The main objective of this activity is to process onto-triples against the ontology (using SPARQL) and based on this analysis to bind (i.e. assigning the relevant XPaths to variables) the correct XPaths to variables contained in the onto-triples. These bindings are going to be used in the next steps as input *Variable Binding* activity. This activity processes *Onto-Triples* using standard SPARQL in order to perform any required inference so that any schema-level query semantics to be analyzed and taken into account later on in the translation process. Since we are using SPARQL for *Onto-Triple* processing against the ontology, we can process any given *Onto-Triple* regardless the complexity of its matching against the ontology graph.

UF-GP2XQuery. This activity translates the *UF-GP* into semantically equivalent XQuery expressions. The concept of a *GP*, and thus the concept of *UF-GF*, is defined recursively. The *BGP2XQuery* activity translates the basic components of a *GP* (i.e. Basic Graph Patterns-*BGPs* which are sequences of triple patterns and filters) into semantically equivalent XQuery expressions. To do that a variables binding step is needed. Finally, *BGPs* in the context of a *GP* have to be properly associated. That is, to apply the SPARQL operators among them using XQuery expressions and functions. These operators are: *OPT*, *AND*, and *FILTER* and are implemented using standard XQuery expressions without any ad hoc processing.

- **Variables Binding.** In the translation process the term "variable bindings" is used to describe the assignment of the correct XPaths to the variables referenced in a given Basic Graph Pattern (*BGP*), thus enabling the translation of *BGP* to XQuery expressions. In this activity, *Onto-Triples* are not taken into account since their processing has taken place in the previous step and their bindings are used as input in this activity. The same holds for Filters, since they don't affect the binding process (more details can be found at [19]).
- **BGP2XQuery.** This activity translates the *BGPs* to semantically equivalent XQuery expressions based on the *BGP2XQuery* algorithm. The algorithm manipulates a sequence of triple patterns and filters (i.e. a *BGP*) and translates them into XQuery expressions, thus allowing the evaluation of a *BGP* on a set of XML data. The algorithm takes as input the mappings between the ontology and the XML schema, the *BGP*, the determined variable types, as well as the variable bindings and generates XQuery expressions (more details can be found at [19]).

4.3 Union Operator Translation

This activity translates the *UNION* operator that appears among *UF-GPs* in a *GP*, by using the *Let* and *Return* XQuery clauses in order to return the union of the solution sequence produced by the *UF-GPs* to which the Union operator applies.

4.4 Solution Sequence Modifiers Translation

This activity translates the SPARQL solution sequence modifiers. Solution Modifiers are applied on a solution sequence in order to create another, user desired, sequence. The modifiers supported by SPARQL are *Distinct*, Order By, *Reduced*, *Limit* and *Offset*.

For the implementation of the *Distinct* and *Reduced* modifiers, our software generates XQuery functions (in standard XQuery syntax) (*func:DISTINCT*, *func:REDUCED*) according to the number and the names of the variables for which the duplicate elimination is to be performed. Regarding the rest of the solution sequence modifiers, the next table shows the XQuery expressions and built-in functions that are used for their translation in XQuery (the XQuery variable *$Results* has been bound to the solution sequence produced by XQuery expressions, and *N, M* are positive integers).

Table 1. Translation of Solutions Sequence Modifiers

SPARQL Modifier	*XQuery Translation*
LIMIT N	**return**($Results[**position**( )<= N])
OFFSET N	**return**($Results[**position**( )> N])
LIMIT N & OFFSET M	**return**($Results[**position**( )> M **and position**( )<= N+M])
ORDER BY DESC(?x) ASC(?y)	**for** $res **in** $Results **order by** $res/x **descending empty least**, $res/y **empty least** **return** $res

4.5 Query Forms Based Translation

SPARQL has four forms of queries (*Select, Ask, Construct* and *Describe*). According to the query form, the structure of the final result is different. The query translation is heavily dependent on the query form. In particular, after the translation of any solution modifier is done, the generated XQuery is enhanced with appropriate expressions in order to achieve the desired structure of the results (e.g. to construct an RDF graph, or a result set) according to query form.

5 Example

We demonstrate in this example the use of the described framework in order to allow a SPARQL query to be evaluated in XML Data (based on Example 1). Fig. 4 shows how a given SPARQL query is translated by our framework into a semantically equivalent XQuery query.

Consider the query: *"For Person subclasses, return their instances, their last name and their income, the first name of which is "Nick", the last name begins with "B", and they have an e-mail address. The (existence of) income is optional. The query will return at most 20(LIMIT 20) solutions ordered by last name value at descending order and skipping the first 5 solutions (OFFSET 5)"*.

SPARQL Query:

```
PREFIX ns:  <http://example.com/ns#>
PREFIX rdfs: <http://www.w3.org/2000/01/rdf-schema#>
PREFIX rdf:  <http://www.w3.org/1999/02/22-rdf-syntax-ns#>

SELECT ?empl ?lname ?inc
WHERE{ { ?emplCl rdfs:subClassOf ns:Person .
         ?empl    rdf:type        ? emplCl .
         ?empl    ns:First_Name   "Nick" .        BGP_1
         ?empl    ns:Sur_Name     ?lname .
         ?empl    ns:email        ?email .
         FILTER regex( ?lname, "^B")    }
       OPTIONAL{ ?empl ns:Income ?inc. }          BGP_2
     }
ORDER BY DESC(?lname)
LIMIT 20 OFFSET 5
```

Translated XQuery Query :

```
declare namespace func = "http://www.music.tuc.gr/funcs";
let $doc := collection("http://www.music.tuc.gr/...")
let $Modified_Results :=(
  let $Results :=(
    let $BGP_1:=(
      for $empl  in  $doc/Persons/Staff[./FN= "Nick"]
      for $lname in  $empl/LN
      let $email :=  $empl/Contact_Info
      where( exists($email) and matches($lname, "^B" ) )
      return(<Result> <empl>{func:CIVT($empl)}</empl>,
                      <lname>{ string($lname)}</lname> </Result>)
    )
    let $BGP_2:=(
      for $empl in $doc/Persons/Staff
      for $inc in $empl/Salary
      return(<Result> <empl>{func:CIVT($empl)}</empl>,
                      <inc>{ string($inc)}</inc> </Result>)
    )
    return ( func:OPTIONAL($BGP_1, $BGP_2) )
  )
  return (let $Ordered_Results:=(
                        for $iter in $Results
                        order by $iter/lname descending empty least
                        return($iter) )
          return ($Ordered_Results[position( )>5 and position( )<=25]))
)
return (<Results>{$Modified_Results }</Results>)
```

Fig. 4. SPARQL Query Translation Example

6 Transformation of the Query Results

An important issue in the entire approach is the structure of the returned results. In our work and for the *Ask* and *Select* query forms we encode the returned results according to the SPARQL Query Result XML Format [1], which is a W3C recommendation. Moreover the values returned with the results, can be easily transformed into RDF (N3 or RDF/XML) syntax by processing the information of the results and the input mappings.

7 Conclusions

We have presented an environment that allows the evaluation of SPARQL queries over XML data which are stored in XML databases and accessed with the XQuery language. The environment assumes that a set of mappings between the OWL ontology and the XML Schema exists. The mappings obey certain well accepted language correspondences.

The *SPARQL2XQuery* framework has been implemented as a prototype software service using Java related technologies (Java 2SE, Axis2, and Jena) on top of the Berkeley DB XML. The service can be configured with the appropriate mappings (between an ontology and an XML Schema) and translates the input SPARQL queries into XQuery queries that are answered over the XML Database.

This work is part of as more generic framework that we are pursuing which aims to providing algorithms, proofs and middleware for the transparent access from the Semantic Web environment to federated heterogeneous databases across the web.

References

1. Beckett, D. (ed.): SPARQL Query Results XML Format. W3C Recommendation (January 15, 2008), http://www.w3.org/TR/rdf-sparql-XMLres/
2. Bohring, H., Auer, S.: Mapping XML to OWL Ontologies. Leipziger Informatik-Tage, 147–156 (2005)
3. Lehti, P., Fankhauser, P.: XML Data Integration with OWL: Experiences & Challenges. In: Proceedings of the International Symposium on Applications and the Internet (2004)
4. Pérez, J., Arenas, M., Gutierrez, C.: Semantics and Complexity of SPARQL. In: Cruz, I., Decker, S., Allemang, D., Preist, C., Schwabe, D., Mika, P., Uschold, M., Aroyo, L.M. (eds.) ISWC 2006. LNCS, vol. 4273, pp. 30–43. Springer, Heidelberg (2006)
5. Prud'hommeaux, E., Seaborne, A. (eds.): SPARQL Query Language for RDF. W3C Recommendation (January 15, 2008), http://www.w3.org/TR/rdf-sparql-query/
6. Rodrigues, T., Rosa, P., Cardoso, J.: Mapping XML to Exiting OWL ontologies. In: International Conference WWW/Internet 2006, Murcia, Spain, October 5-8 (2006)
7. Siméon, J., Chamberlin, D. (eds.): XQuery 1.0: an XML Query Language. W3C Recommendation (January 23, 2007), http://www.w3.org/TR/xquery/
8. Farrell, J., Lausen, H.: Semantic Annotations for WSDL and XML Schema. W3C Recommendation, W3C (August 2007), http://www.w3.org/TR/sawsdl/
9. Groppe, S., Groppe, J., Linnemann, V., Kukulenz, D., Hoeller, N., Reinke, C.: Embedding SPARQL into XQuery/XSLT. In: SAC 2008, pp. 2271–2278 (2008)
10. Droop, M., Flarer, M., Groppe, J., Groppe, S., Linnemann, V., Pinggera, J., Santner, F., Schier, M., Schoepf, F., Staffler, H., Zugal, S.: Embedding XPATH Queries into SPARQL Queries. In: Proc. of the 10th International Conference on Enterprise Information Systems, ICEIS 2008 (2008)
11. Akhtar, W., Kopecký, J., Krennwallner, T., Polleres, A.: XSPARQL: Traveling between the XML and RDF worlds – and avoiding the XSLT pilgrimage. In: Bechhofer, S., Hauswirth, M., Hoffmann, J., Koubarakis, M. (eds.) ESWC 2008. LNCS, vol. 5021, pp. 432–447. Springer, Heidelberg (2008)
12. Bizer, C., Cyganiak, R.: D2R Server, http://www4.wiwiss.fu-berlin.de/bizer/d2r-server/index.html
13. OpenLink Software: Virtuoso Universal Server, http://virtuoso.openlinksw.com/
14. CCNT Lab. Zhejiang Univ. China: Dart Grid, http://ccnt.zju.edu.cn/projects/dartgrid/
15. Tsinaraki, C., Christodoulakis, S.: Interoperability of XML Schema Applications with OWL Domain Knowledge and Semantic Web Tools. In: Proc. of the ODBASE 2007 (2007)
16. Cruz, I.R., Xiao, H., Hsu, F.: An Ontology-based Framework for XML Semantic Integration. In: Database Engineering and Applications Symposium (2004)
17. Christophides, V., Karvounarakis, G., Koffina, I., Kokkinidis, G., Magkanaraki, A., Plexousakis, D., Serfiotis, G., Tannen, V.: The ICS-FORTH SWIM: A Powerful Semantic Web Integration Middleware. In: Proceedings of the First International Workshop on Semantic Web and Databases 2003 (SWDB 2003), pp. 381–393 (2003)
18. Amann, B., Beeri, C., Fundulaki, I., Scholl, M.: Querying XML Sources Using an Ontology-Based Mediator. In: Meersman, R., Tari, Z., et al. (eds.) CoopIS 2002, DOA 2002, and ODBASE 2002. LNCS, vol. 2519, pp. 429–448. Springer, Heidelberg (2002)
19. Bikakis, N., Gioldasis, N., Tsinaraki, C., Christodoulakis, S.: Querying XML Data with SPARQL. In: Proceeding of the 20th International Conference on Database and Expert Systems Applications (DEXA 2009) (2009)
20. Bikakis, N., Gioldasis, N., Tsinaraki, C., Christodoulakis, S.: The SPARQL2XQuery Framework. Technical Report, http://www.music.tuc.gr/reports/SPARQL2XQUERY.PDF

Building a Collaborative Semantic-Aware Framework for Search

Antonella Carbonaro

Department of Computer Science, University of Bologna,
Mura Anteo Zamboni, 7, Bologna, Italy
antonella.carbonaro@unibo.it

Abstract. The paper presents an ontological approach for enabling personalized searching framework facilitating the user access to desired contents. Through the ontologies the system will express key entities and relationships describing resources in a formal machine-processable representation. An ontology-based knowledge representation could be used for content analysis and concept recognition, for reasoning processes and for enabling user-friendly and intelligent content retrieval.

Keywords: Ontology, Semantic Web Applications, collaboration in information searching, personalized searching framework.

1 Introduction

The Web is increasingly becoming important than ever, moving toward a social place and producing new applications with surprising regularity: there has been a shift from just existing on the Web to participating on the Web. Community applications and online social networks have become very popular recently, both in personal/social and professional/organizational domains [1]. Most of these collaborative applications provide common features such as content creation and sharing, content-based tools for discussions, user-to-user connections and networks of users sharing common interest, reflecting today's Web 2.0 rich Internet application-development methodologies.

The Semantic Web offers a generic infrastructure for interchange, integration and creative reuse of structured data, which can help to cross some of the boundaries that Web 2.0 is facing. Currently, Web 2.0 offers poor query possibilities apart from searching by keywords or tags. There has been a great deal of interest in the development of semantic-based systems to facilitate knowledge representation and extraction and content integration [2], [3]. Semantic-based approach to retrieving relevant material can be useful to address issues like trying to determine the type or the quality of the information suggested from a personalized environment. In this context, standard keyword search has a very limited effectiveness. For example, it cannot filter for the type of information, the level of information or the quality of information.

By exploiting each other's achievements the Semantic Web and Web 2.0 together have a better opportunity to realize the full potential of the web [4].

Potentially, one of the biggest application areas of social networks might be personalized searching framework (e.g., [5],[6]). Whereas today's search engines provide

M.D. Lytras et al. (Eds.): WSKS 2009, LNAI 5736, pp. 268–275, 2009.

largely anonymous information, new framework might highlight or recommend web pages created by recognized or familiar individuals. The integration of search engines and social networks can lead to more effective information seeking [7].

Additionally, we can consider semantic information representation as an important step towards a wide efficient manipulation and retrieval of information [8], [9], [10]. In the digital library community a flat list of attribute/value pairs is often assumed to be available. In the Semantic Web community, annotations are often assumed to be an instance of an ontology. Through the ontologies the system will express key entities and relationships describing resources in a formal machine-processable representation. An ontology-based knowledge representation could be used for content analysis and object recognition, for reasoning processes and for enabling user-friendly and intelligent multimedia content search and retrieval.

In this work we explore the possibilities of synchronous, semantic-based collaboration for search tasks. We describe a search system wherein searchers collaborate intentionally with each other in small, focused search groups. Developed framework (SWS2 – Semantic Web Search 2.0 - project) goes beyond implementation of ad hoc user interface. It also identifies information that one group member searches and uses it in realtime to improve the effectiveness of all group members while allowing semantic coverage of the involved domain. The semantic approach is exploited introducing an ontology space covering domain knowledge and resource models based on word sense representation.

There are many scenarios in which small groups of users collaborate on Web search tasks to find information, such as school students or colleagues jointly writing a report or a research, or arranging joint travel. Although most search tools are designed for individual use, some collaborative search tools have recently been developed to support such collaborative search task [11]. These tools tend to offer two classes of support: i) awareness features (e.g., sharing and browsing of group members' query histories, and/or comments on results and on web pages rating), ii) division of labor features (e.g., to manually split result lists among group members, and/or algorithmic techniques for modifying group members' search results based on others' actions) [12]. Collaborative search tools are relatively novel and thus not widely available.

2 Personalized Searching Framework

One of the areas in which information retrieval is likely to see great interest in the future is synchronous collaborative search. This concerns the common scenario where two or more people working together on some shared task, initiate a search activity to satisfy some shared information need. Conventionally, this need is satisfied by independent and uncoordinated searching on one or more search engines, leading to inefficiency, redundancy and repetition as searchers separately encounter, access and possibly re-examine the same documents. Information searching can be more effective as a collaboration than as a solitary activity taking advantage of breadth of experience to improve the quality of results obtained by the users [13]. Community-based recommendation systems [14], [15] or user interfaces that allow multiple people to compose queries [12] or examine search results [16] represent various forms of collaboration in search.

Traditional approaches to personalization include both content-based and user-based techniques. If, on one hand, a content-based approach allows to define and maintain an accurate user profile (for example, the user may provides the system with a list of keywords reflecting him/her initial interests and the profiles could be stored in form of weighted keyword vectors and updated on the basis of explicit relevance feedback), which is particularly valuable whenever a user encounters new content, on the other hand it has the limitation of concerning only the significant features describing the content of an item. Differently, in a user-based approach, resources are processed according to the rating of other users of the system with similar interests. Since there is no analysis of the item content, these information management techniques can deal with any kind of item, being not just limited to textual content. In such a way, users can receive items with content that is different from that one received in the past. On the other hand, since a user-based technique works well if several users evaluate each one of them, new items cannot be handled until some users have taken the time to evaluate them and new users cannot receive references until the system has acquired some information about the new user in order to make personalized predictions. These limitations often refer to as the sparsity and start-up problems. By adopting a hybrid approach, a personalization system is able to effectively filter relevant resources from a wide heterogeneous environment like the Web, taking advantage of common interests of the users and also maintaining the benefits provided by content analysis. A hybrid approach maintains another drawback: the difficulty to capture semantic knowledge of the application domain, i.e. concepts, relationships among different concepts, inherent properties associated with the concepts, axioms or other rules, etc [17].

In this context, standard keyword search is of very limited effectiveness. For example, it does not allow users and the system to search, handle or read concepts of interest, and it doesn't consider synonymy and hyponymy that could reveal hidden similarities potentially leading to better retrieval. The advantages of a concept-based document and user representations can be summarized as follows: (i) ambiguous terms inside a resource are disambiguated, allowing their correct interpretation and, consequently, a better precision in the user model construction (e.g., if a user is interested in computer science resources, a document containing the word 'bank' as it is meant in the financial context could not be relevant); (ii) synonymous words belonging to the same meaning can contribute to the resource model definition (for example, both 'mouse' and 'display' brings evidences for computer science documents, improving the coverage of the document retrieval); (iii) synonymous words belonging to the same meaning can contribute to the user model matching, which is required in recommendation process (for example, if two users have the same interests, but these are expressed using different terms, they will considered overlapping); (iv) finally, classification, recommendation and sharing phases take advantage of the word senses in order to classify, retrieve and suggest documents with high semantic relevance with respect to the user and resource models.

For example, the system could support Computer Science last-year students during their activities in courseware like Bio Computing, Internet Programming or Machine Learning. In fact, for these kinds of courses it is necessary an active involvement of the student in the acquisition of the didactical material that should integrate the lecture notes specified and released by the teacher. Basically, the level of integration depends both on the student's prior knowledge in that particular subject and on the comprehension level he wants to acquire. Furthermore, for the mentioned courses, it is necessary to continuously update the acquired knowledge by integrating recent information available from any remote digital library.

2.1 Use Case Analysis

A first level of system analysis can be achieved through its functional requirements. Such functional requirements are described by the interaction between users and the systems itself. Therefore, users may be interested in semantic-based search or collaborative semantic-based search.

We define an interaction between users as a collaborative search session managed by the system using specialized components: in particular, the system should cover both user manager and sessions between users manager roles.

2.2 System Modules

In the following we list the components able to handle user data:

i) User Interface Controller: it coordinates the information flow between interface control and other system components and allows to perform data presentation for the GUI visualization.
ii) Semantic searcher: it implements semantic-based searches extracting concepts related to introduced keywords using a thesaurus and searching in the underling ontology corresponding documents.
iii) Interest coupler: it performs intersection between user interest matching relevant terms extracted from semantic searcher.
iv) User Manager: it deals with user. For example through the User Manager, it is possible to register new users or to search for their information. Moreover, it is able to associate mail boxes to user to enhance communication.
v) Session Manager: it manages collaborative search sessions allowing user insertion and search terms shared between users. It allows to maintain consistency between session views and creates message boxes for the specific session whose content is available to all the participants.

2.3 Data Analysis

i) OWL

The ontology developed to test implemented framework maintains relation between courses, lessons, teachers and course material. Ontology is a representation model in a given domain that can be used for the purposes of information integration, retrieval and exchange. The ontology usage is widely spread in not only the artificial intelligent and knowledge representation communities, but most of information technology areas. In particular, ontology has become common in the Semantic Web community in order to share, reuse and process domain information between humane and machine. Most importantly, it enables formal analysis of domain knowledge, for example, context reasoning becomes possible by explicitly defining context ontology.

There are several possible approaches in developing a concept hierarchy. For example, a top-down development process starts with the definition of the most general concepts in the domain and subsequent specialization of the concepts, while a bottom-up development process starts with the definition of the most specific classes, the leaves of the hierarchy, with subsequent grouping of these classes into more general concepts. The hybrid development consists in a combination of the top-down and bottom-up processes. Due to our personal view of the domain we took the combination approach. Once we have defined the classes and the class hierarchy we described the internal structure of concepts defining the properties of classes. Over the evolving ontology we perform diagnostics to determine the conformance to common ontology-modeling practices and to check for logical correctness of the ontology.

ii) User Data

It maintains data of the users handled by the system.

iii) Session Data

It maintains data corresponding to collaborative search sessions.

2.4 Developed System Interaction

The developed system proposes three different interaction between the users.

i) Search interaction

This interaction starts when a user performs a search proposing one or more keywords. The Semantic searcher module returns a list containing relevant documents and recommends terms for the possible following searches. Therefore, the User Interface Controller is able to find similar user with similar interest in performed searches using Interest Coupler module.

ii) Collaborative search session interaction

A user can decide to contact another user, proposed by the system similar to his interests, to start collaborative search session. The request produces an Invitation message in the message box of the target user. Concurrently, a listening permanent loop allows to User Interface Controller to advise target user. In the case of positive response, the User Interface Controller creates a new collaborative search session and a Session Join request is sent.

iii) Interaction during a collaborative search session

The user could modify the list of search terms adding or removing some keyword. The request, managed by the User Interface Controller, is forwarded to Session Manager that updates search terms, replacing term list and requiring GUI updates. The same interaction can be used to implement a session chat, allowing more collaboration value to the system.

2.5 System GUI

The search home page is showed in Figure 1. Box A allows to the user to insert his nickname to use during SIG sessions dynamically showed in box B.

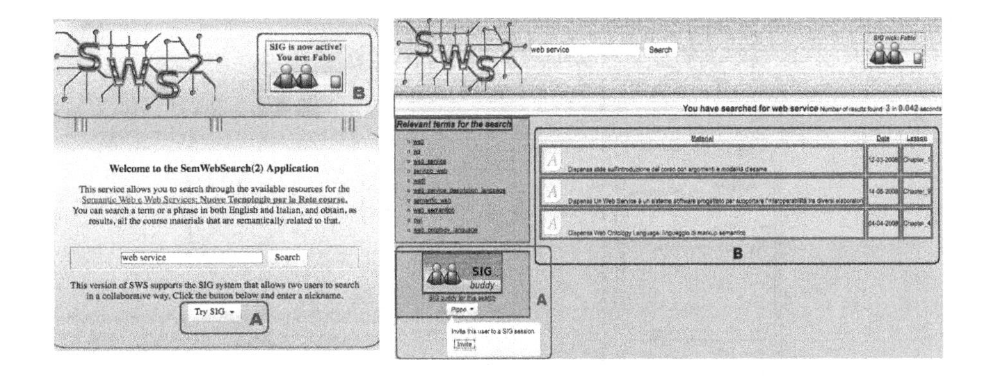

Fig. 1. SWS2 home page

Fig. 2. Collaborative search session

If a user participate to collaborative search sessions, the system proposes in his search result page a new box containing similar users (Figure 2, box A). This button also allows to send Invitation message to target user; a background function verify the presence of new Invitation messages and, eventually, notify them to the user.

Figure 3 box A shows communication facilities offered to system users, while Figure 3 box B shows terms actually used to search session. Using components showed in box B1 the user may add search terms, while using the component showed in box B2 the user may remove session search terms. The button showed in box B3 is twofold: on one hand, it allows users to accept the lists of terms created by the system. On the other, through such button, it is possible to perform the described semantic searches.

3 Considerations

Golovchinsky et al. [7] distinguish among the various forms of computer-supported collaboration for information seeking, classifying such systems along four dimensions: intent, depth of mediation, concurrency, and location.

The intent could be explicit or implicit. In our framework two or more people set out to find information on a topic based on a declared understanding of the information need, which might evolve over time. So, our framework implements explicit information seeking scenarios.

The depth of mediation is the level at which collaboration occurs in the system. Our system implements algorithm mediation at the search engine level explicitly consider ongoing collaboration and coordinate users activities during the search session.

People can collaborate synchronously or asynchronously. In our system the collaboration is synchronous involving the ability of people to influence each other in real time.

Finally, collaboration may be co-located (same place at the same time) or, as in our framework, distributed, increasing opportunities for collaboration but decreasing the fidelity of possible communications.

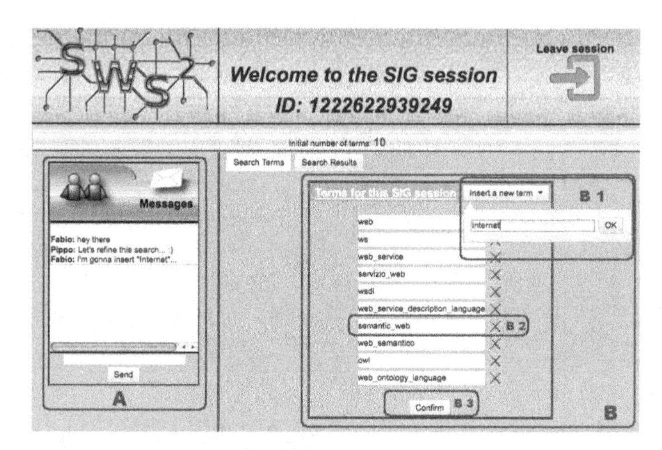

Fig. 3. Collaborative search terms specification

An important step in the searching process is the examination of the results retrieved. In order to test developed framework we have collected over 50 different documents concerning actual domain. We have extracted several concepts used during the annotation phase and performed tests to verify searching functionalities. It is currently difficult to replicate or make objective comparisons in personalized retrieval researches, so to evaluate search results we have considered the order used by the framework to present retrieved results. During this step, the searcher browses through the results to make judgments about their relevance and to extract information from those found to be relevant. Because information is costly (in terms of time) to download, displays of result lists should be optimized to make the process of browsing more effective. We have also evaluated the effect that the proposed framework has on collaboration and exploration effectiveness. Using implemented tools, searchers found relevant documents more efficiently and effectively than when working individually and they found relevant documents that otherwise went undiscovered.

The work described in this paper represents some initial steps in exploring semantic-based search retrieval collaboration within a focused team of searchers. It could be considered as one possible instance of a more general concept. While the initial results are encouraging, much remains to be explored. For example, most of the current research on sensemaking has been at the individual level, with little understanding of how sensemaking occurs in collaborative search tools.

References

1. Kolbitsch, J., Maurer, H.: The Transformation of the Web: How Emerging Communities Shape the Information we Consume. Journal of Universal Computer Science 12(2), 187–213 (2006)
2. Henze, N., Dolog, P., Nejdl, W.: Reasoning and Ontologies for Personalized E-Learning in the Semantic Web. Educational Technology & Society 7(4), 82–97 (2004)
3. Bighini, C., Carbonaro, A.: InLinx: Intelligent Agents for Personalized Classification, Sharing and Recommendation, International Journal of Computational Intelligence. International Computational Intelligence Society 2(1) (2004)

4. Bojars, U., Breslin, J.G., Finn, A., Decker, S.: Using the Semantic Web for linking and reusing data across Web 2.0 communities. Web Semantics: Science, Services and Agents on the World Wide Web 6, 21–28 (2008)
5. Pickens, J., Golovchinsky, G., Shah, C., Qvarfordt, P., Back, M.: Algorithmic Mediation for Collaborative Exploratory Search. To appear in Proceedings of SIGIR
6. Freyne, J., Smyth, B.: Collaborative Search: Deployment Experiences. In: The 24th SGAI International Conference on Innovative Techniques and Applications of Artificial Intelligence, Cambridge, UK, pp. 121–134 (2004)
7. Golovchinsky, G., Qvarfordt, P., Pickens, J.: Collaborative Information Seeking. Computer 42(3), 47–51 (2009)
8. Calic, J., Campbell, N., Dasiopoulou, S., Kompatsiaris, Y.: A Survey on Multimodal Video Representation for Semantic Retrieval. In: The Third International Conference on Computer as a tool. IEEE, Los Alamitos (2005)
9. Carbonaro, A.: Defining Personalized Learning Views of Relevant Learning Objects in a Collaborative Bookmark Management System. In: Ma, Z. (ed.) Web-based Intelligent ELearning Systems: Technologies and Applications, pp. 139–155. Information Science Publishing, Hershey (2006)
10. Bloehdorn, S., Petridis, K., Simou, N., Tzouvaras, V., Avrithis, Y., Handschuh, S., Kompatsiaris, Y., Staab, S., Strintzis, M.G.: Knowledge Representation for Semantic Multimedia Content Analysis and Reasoning. In: Proceedings of the European Workshop on the Integration of Knowledge, Semantics and Digital Media Technology (2004)
11. Paul, S.A., Morris, M.R.: CoSense: enhancing sensemaking for collaborative web search by. In: CHI 2009: Proceedings of the 27th international conference on Human factors in computing systems, pp. 1771–1780 (2009)
12. Morris, M.R., Horvitz, E.: Searchtogether, an interface for collaborative web search. In: Proceedings of UIST, pp. 3–12 (2007)
13. Baeza-Yates, R., Pino, J.A.: A first step to formally evaluate collaborative work. In: GROUP 1997: Proc. ACM SIGGROUP Conference on Supporting Group Work, pp. 56–60 (1997)
14. Carbonaro, A.: Defining personalized learning views of relevant learning objects in a collaborative bookmark management system. In: Web-Based Intelligent e-Learning Systems: Technologies and Applications. Idea Group Inc., USA (2005)
15. Smyth, B., Balfe, E., Boydell, O., Bradley, K., Briggs, P., Coyle, M., Freyne, J.: A live-user evaluation of collaborative web search. In: Proceedings of IJCAI, Ediburgh, Scotland, pp. 1419–1424 (2005)
16. Smeaton, F., Lee, H., Foley, C., McGivney, S., Gurrin, C.: Collaborative video searching on a table. In: Multimedia Content Analysis, Management, and Retrieval, San Jose, CA (2006)
17. Carbonaro, A., Ferrini, R.: Considering semantic abilities to improve a Web-Based Distance Learning System. In: ACM International Workshop on Combining Intelligent and Adaptive Hypermedia Methods/Techniques in Web-based Education Systems (2005)

A Fuzzy Data Warehouse Approach for Web Analytics

Daniel Fasel and Darius Zumstein

Information Systems Research Group, Department of Informatics, University of Fribourg, Boulevard de Pérolles 90, 1700 Fribourg, Switzerland
{Daniel.Fasel,Darius.Zumstein}@unifr.ch

Abstract. The analysis of web data and metrics became an important task of e-business to control and optimize the website, its usage and online marketing. Firstly, this paper shows the use of web analytics, different web metrics of Google Analytics and other Key Performance Indicators (KPIs) of e-business.

Secondly, this paper proposes a fuzzy data warehouse approach to improve web analytics. The fuzzy logic approach allows a more precise classification and segmentation of web metrics and the use of linguistic variables and terms. In addition, the fuzzy data warehouse model discusses the creation of fuzzy multidimensional classification spaces using dicing operations and shows the potential of fuzzy slices, dices and aggregations compared to sharp ones. The added value of web analytics, web usage mining and the fuzzy logic approach for the information and knowledge society are also discussed.

Keywords: Fuzzy logic, fuzzy classification, data warehouse, web analytics, Google Analytics, web usage mining, web metrics, electronic business.

1 Introduction

Since the development of the World Wide Web 20 years ago, the internet presence of companies became a crucial instrument of information management, communication and electronic business. With the growing importance of the web, the *monitoring* and *optimization* of a website and online marketing has become a central task. Therefore, *web analytics* gains in importance for both business practice and academic research. It helps to understand the traffic on the website, the behaviour of visitors and customers.

Today, many companies are using *web analytics software* (e.g. Google Analytics, Yahoo, WebTrends, Nedstad, Omniture, SAS) to collect, store and analyse web data. They provide dashboards and reports with important web metrics to the responsible persons. Like web analytics software, a *data warehouse* is an often used information system. It reports integrated, unchangeable, time-related and summarized information to the management for analysis and decision making purposes.

So far, the *classification* of metrics or facts in data warehouses has always been done in a sharp manner. This paper proposes to classify web metrics fuzzily within a data warehouse. After an introduction in web analytics in *section 2*, *section 3* explains the fuzzy classification approach and a fuzzy data warehouse model. *Section 4* shows the use for the knowledge society and *section 5* gives a conclusion and an outlook.

M.D. Lytras et al. (Eds.): WSKS 2009, LNAI 5736, pp. 276–285, 2009.

2 Web Analytics

According to the Web Analytics Association [1], *web analytics* is the measurement, collection, analysis and reporting of Internet data for the purposes of understanding and optimizing web usage. Weischedel et al. [2] define web analytics as the monitoring and reporting of web site usage so that enterprises can better understand the complex interactions between web site visitor actions and web site offers, as well as leverage insight to optimise the site for increased customer loyalty and sales.

However, the *use of web analytics* is manifolds (see Table 1) and is not restricted to the optimization of the website and web usage (for an overview see [3, 4, 5, 6, 7]). By analyzing the *log files* (server-site data collection) or using *page tagging* (client-site data collection with web analytics software), the traffic on the website and the visitors' behaviour can be observed exactly with different *web metrics* listed in Table 2. The number of *page views*, *visits* and *visitors* are often discussed standard metrics. However, as Key Performance Indicators (KPIs) are considered: *stickiness, frequency, depth* and *length of visit*, *conversion rate(s)* and other e-business metrics like the *order rate* or *online revenue* (compare Fig. 1).

Table 1. Use of web analytics

Web analytics is useful for the ongoing ***optimization*** *of …*
▪ *website quality* (navigation/structure, content, design, functionality, usability)
▪ *online marketing* (awareness, image, campaigns, banner/keyword advertising)
▪ *online CRM* (customer acquisition and retention)
▪ *individual marketing* (personalized recommendations, mass customization)
▪ *segmentation* of the traffic and visitors
▪ *internal processes* and *communication* (contacts, interactions, relations)
▪ *search engine optimization* (visibility, Google ranking, PageRank, reach)
▪ *traffic* (page views, visits, visitor)
▪ *e-business profitability* (efficiency and effectiveness of the web presence)

Table 2. Definitions of web metrics (also measured in Google Analytics [8])

Web Metric	*Definition*
Page Views	The number of page views (page impressions) of a web page accessed by a human visitor (without crawlers/spiders/robots)
Visits	A sequence of page views of a unique visitor without interruption (of usually 30 minutes)
Visitors	The number of unique visitors (users) on a website
Pages/Visits	The Ø number of page views during a visit for all visitors
Time on Site	The Ø length of time for all visitors spent on the website
Stickiness	The capability of a web page to keep a visitor on the website
Bounce Rate	The percentage of single page view visits
Frequency	The number of visits, a visitor made on the site (= *loyalty*)
Recency	The number of days, since a visitor last visit on the site
Length of Visit	The time of a visit, the visitors stay on the site (in seconds)
Depth of Visit	The number of pages, the visitors visited during one visit
Conversion rate	The percentage of visitors who converts to customers

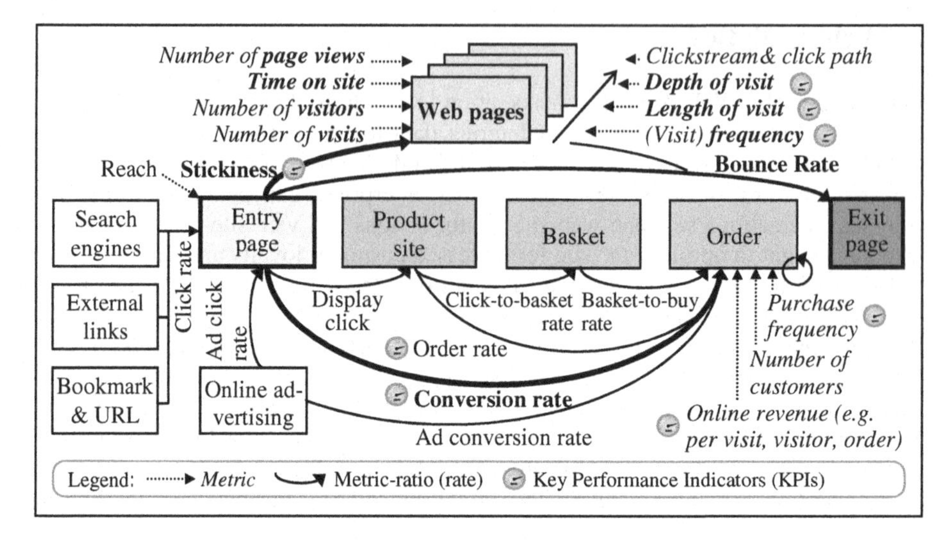

Fig. 1. Relations between web and e-business metrics

3 Using a Fuzzy Data Warehouse for Web Analytics

3.1 The Fuzzy Classification Approach

The theory of fuzzy sets or fuzzy logic goes back to Lofti A. Zadeh in 1965 [9]. It takes the subjectivity, imprecision, uncertainty and vagueness of human thinking and language into account, and expresses it with mathematical functions.

A fuzzy set can be defined formally as follow [10, 11]: if X is a set, then the *fuzzy set* A in X is defined in (1) as

$$A = \{(x, \mu_A(x))\} \tag{1}$$

where $x \in X$, $\mu_A : X \rightarrow [0, 1]$ is the *membership function* of A and $\mu_A(x) \in [0, 1]$ is the *membership degree* of x in A.

For example, in a sharp set (Fig. 2a), the *terms* “few”, “medium” or “many” page views of a visitor can be either true (1) or false (0). A value of 1 of the membership function μ (Y-axis in Fig. 2a) means that the number of page views (on the X-axis) is corresponding to one set. A value of 0 indicates that a given number of page views do not belong to one of the sets. The number of page views of one visitor per month are defined as “few” between 0 and 32, 33 to 65 page views are “medium” and more than 66 are classified as “many” page views.

However, to classify the page views – or any other web metric – sharply, is *problematic* as following example shows. If visitor 1 has 65 page views, he is classified in the “medium” class, visitor 2 with 70 has “many” page views. Although the two visitors have visited nearly the *same* number of pages (visitor 2 visited only 5 pages more), they are assigned to two *different* sets, or classes respectively.

By defining *fuzzy sets* (Fig. 2b), represented by the membership functions, there are continuous transitions between the terms “few”, “medium” and “many”. Fuzzily,

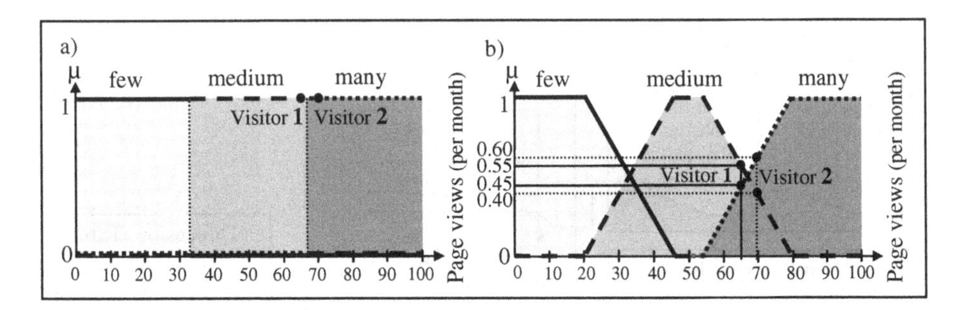

Fig. 2. Sharp (a) and fuzzy (b) classification of the web metric page views

visitor 1 page views are classified both as "medium" (0.55 resp. 55%) *and* "many" (0.45 resp. 55%). Also visitor 2 belongs partly to two classes (60% to "many" and 40% to "medium") *at the same time*. Obviously, the fuzzy logic approach allows a more *precise* and *fair* classification, and the risk of misclassifications can be reduced.

3.2 Building the Fuzzy Data Warehouse

In the example above, a simple, one-dimensional classification is considered. In contrast, data warehouses are *multidimensional*. Furthermore, a *data warehouse* is as a subject-oriented, integrated, time-varying, non-volatile collection of data in support of the management's decision-making process [12].

Multidimensional models as data warehouses usually consist of two components: dimensions and facts. The *dimensions* are the qualitative part of the model and are used for presenting as well as navigating the data in a data warehouse. They serve as the primary source of query constraints, groupings and report labels [13].

The *facts*, usually performance indicators of a company, are the quantitative component of the data warehouse and stored in a certain granularity (see [13, 14]). Following [13], dimension schemes contain a partial ordered set of *category attributes*

$$(\{D_1, \ldots, D_n, Top_D ; \rightarrow\}) \tag{2}$$

where $\rightarrow$ is a functional dependency and Top_D is a generic top level element in a way that all the attributes determine Top_D like $(\forall i(1 \leq n) : D_i \rightarrow Top_D)$.
D_i is the base level attribute that defines the *lowest level of granularity* of a dimension

$$(\exists i(1 \leq i \leq n)\forall j(1 \leq j \leq n, i \neq j) : D_i \rightarrow D_j). \tag{3}$$

Every category attribute can have several dimensional attributes that describe functionally the dimension. A flexible approach to integrate fuzzy concepts in the dimensions structure of a data warehouse is to use *meta fuzzy tables.*

A category attribute that has to be handled fuzzily is extended with two meta tables. The first is containing a semantic *description* about a fuzzy concept and the second meta table contains *membership degrees* of each dimensional attribute of a category attribute.

Fig. 3 shows the schema of the table of the *fact* "page views" and the tables of the three *dimensions* "time", "visitor" and "web page". The main advantage of including

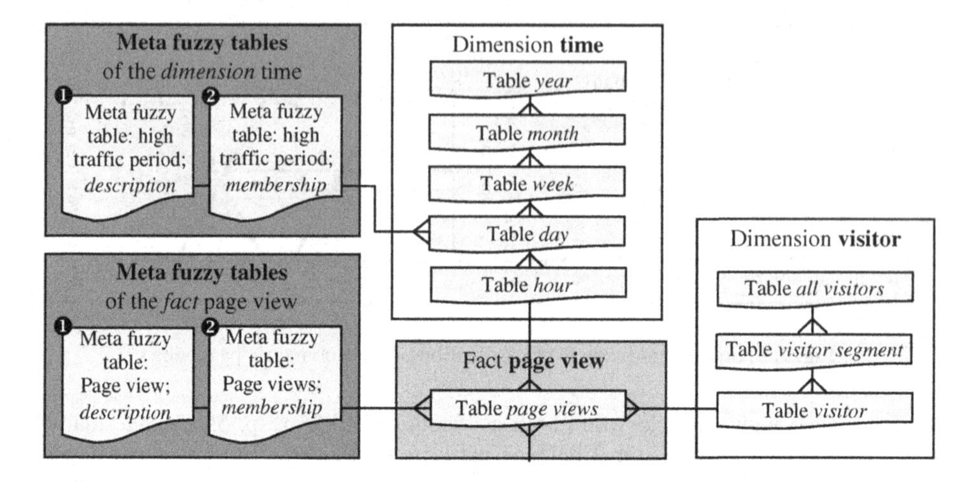

Fig. 3. Schema of a fuzzy data warehouse for web analytics

fuzziness over *meta fuzzy tables* in dimensions is the reduction of the complexity to maintain the summarizability as explained in [15, 16]. This is more flexible than approaches that build the fuzziness directly in the dimensions as proposed by [17, 18]. Using meta fuzzy tables fulfills the additional requirements of the architectural rework of data warehousing Inmon et al. describes as DW 2.0 [12].

As considered in Fig. 2, for the fact "page views" linguistic terms like "few", "medium" or "high" can be defined (or any other words or word combination used by the IT or marketing department). Each single value of the fact "page view" is then embedded in the fuzzy concept over its membership degrees μ_{few}, μ_{medium} and μ_{many}.

Instead of integrating the fuzziness directly in the fact table, meta fuzzy tables provide the same flexibility for the facts as discussed for the dimensions. If a fact is aggregated over dimensions, the granularity of the fact changes. The transition of the granularity is defined within an aggregation function in the data warehouse. A *count operation* is used to aggregate the page views over the dimension time. Consequently, the monthly page views are the total of page views of all days in the month.

To aggregate the fuzzy concept, the membership degrees of the Monthly Page Views (MPV) are calculated as the arithmetic average of the Daily Page Views (DPV)

$$\mu_{few}(MPV) = \sum_{i=1}^{n} \frac{\mu_{few}(DPV_i)}{DPV_i} \tag{4}$$

where n is the amount of days in the month.

3.3 Definition of Fuzzy Constructs

One big advantage of the fuzzy data warehouse approach and the use of linguistic terms is that *new constructs* can be defined. For instance, a web analyst could define the fuzzy concepts "*high traffic period*", "*attractive web pages*" (which generate high value for e-business) or "*high visitor value*" (see following Section 3.4).

Considering the definition of time constructs, the fuzzy logic is especially suited. For example, it becomes not suddenly *evening* at 6 pm, or *night* at 10 pm, but human

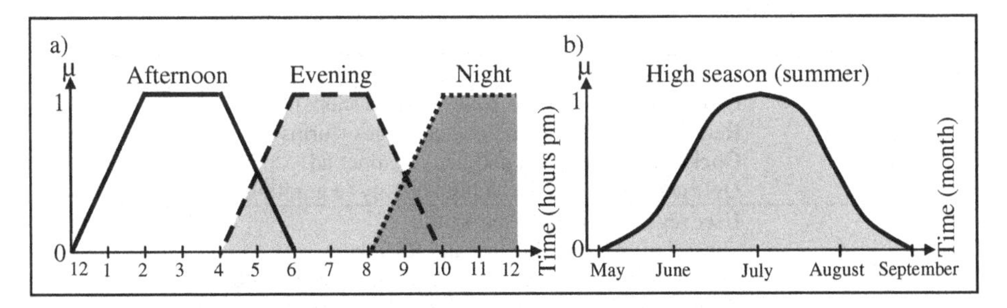

Fig. 4. Fuzzy time constructs: afternoon, evening and night (a), and high season (b)

beings perceive a fluent transition between afternoon, evening and night (see Fig. 4a). Different seasons do not start and end abruptly too, and neither do seasonal variations, like a *high season* in summer (in Fig. 4b) or winter tourism. Such *fuzzy time concepts* are interesting for web analytics, since they allow new types of analysis. For example, the web manager can query all web pages or visitors with many page views in high season or in the evening. With a sharp classification, all page views between 6 pm and 8 pm are displayed only (but arbitrarily not these at 5.59 or 8.01!). With a fuzzy classification, already page views after 4 and until 10 pm are considered at a certain membership degree, or percentage (e.g. 50% afternoon and 50% evening at 5 pm).

In Fig. 2 and 4, one dimensional fuzzy classifications are shown. If two or *more dimensions* or constructs are considered simultaneously (for instance: all web pages with many page views and a low bounce rate of loyal visitors in the evening), the operations slicing and dicing are used in data warehousing.

3.4 Fuzzy Slicing and Dicing

One strength of data warehouses is the possibility to provide *user-* or *context-relevant* information using *slicing* (i.e. the restriction of a dimension to a fix value) and *dicing* (i.e. the focus on a partition of the data cube). Slicing, dicing, drill-down and roll-up (that means dis-/aggregation) based on fuzzy logic enables the definition of extended dimensional concepts. For example, with the *dicing* operation it is possible to create multidimensional *fuzzy segments*. Since *segmentation* (e.g. page views segmented by time on site and new visitors) is a crucial task of web analytics, the fuzzy approach provides new insights and knowledge into website traffic and visitor behaviour.

A promising segment from the e-business point of view is for example visitors with high *visitor value*. Visitor value is defined here as the attractiveness of a visitor for the website. Like customer value or customer equity, visitor value can be defined and measured differently, depending on the strategy or goals of a website.

In the following example, website users may have *high visitor value*, if they visit a high number of "attractive" web pages. A web page in turn is attractive, if it generates high monetary or non-monetary value for the company (Table 3). Consequently, *page attractiveness* is considered here from a company's point of view. The combination of the fuzzy classification of the fact page views (in Fig. 2b) and the dimensions time and web page (whose attractiveness is classified fuzzily and described by meta fuzzy tables) defines a *fuzzy multidimensional classification space* in Figure 5.

Table 3. Example of a categorisation of page attractiveness

Attractiveness	*Examples of web pages*
High	Orders and transactions pages (online shops) Request, submission and contact pages (forms) Clicks on paid content and links (banner ad) User generated content of high quality (e.g. ratings, reviews)
Medium	User registration and user profile Subscriptions to newsletter and RSS feeds Recommendation pages (e.g. "send this page" ,"tell a friend") Downloads or printouts of files and documents
Low	Information pages Homepage and entry pages Exit pages

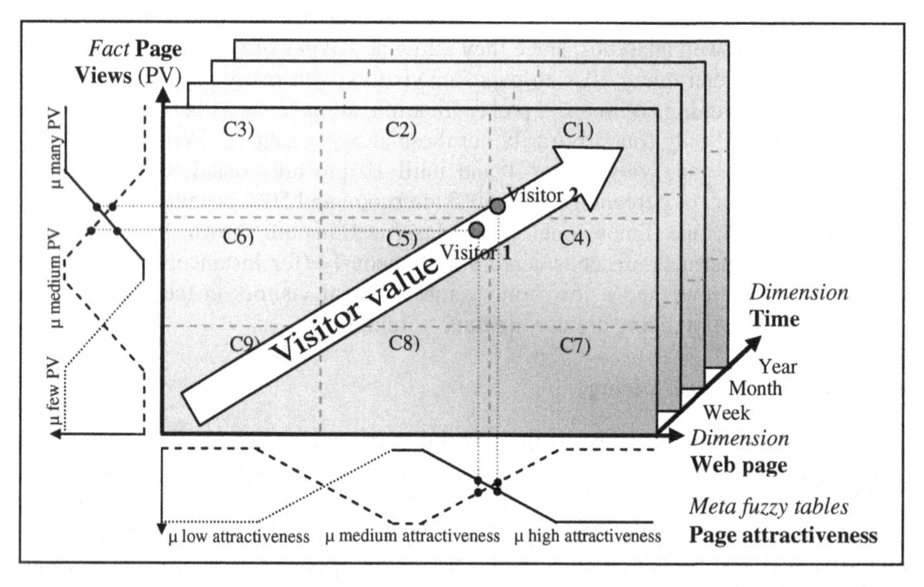

Fig. 5. *Visitor value* is defined by the fact page views and the dimension web page, whereby the visitor value is high if a visitor has viewed many attractive web pages (in Table 3). That means that class C1, which can be considered as a *fuzzy segment*, has the highest visitor value and C9 the lowest. Note that the *page attractiveness* is defined in the meta fuzzy tables as well as its membership degrees. In the fuzzy data warehouse model, an operation over several dimensions (here: web page and time) is called *fuzzy dicing*. Visitor value can be dynamically analysed and aggregated over time from day to week, month and year as defined in formula 4.

4 Use of Fuzzy Web Analytics for the Knowledge Society

4.1 Acquiring Knowledge about and from Website Visitors

By analyzing the *sources* of the traffic (see Fig. 6), often visited web pages, users' interactions, their *search requests* on the website, and the *keywords* in referred search engines, web analytics and fuzzy data warehousing provide valuable insights about the *characteristics* and the *information demand* of visitors. It is possible to constantly gain knowledge about the motivation and behaviour of users. This enables an improvement of *website quality*, for instance by providing *visitor-oriented* content, high information quality and a *user-friendly* navigation or structure. Web analytics can therefore support the detection of problems and failures of a websites' *usability* (e.g. by analyzing the bounce rates). This approach complements the methodology to evaluate and measure the usability proposed by [19] at the WSKS 2008.

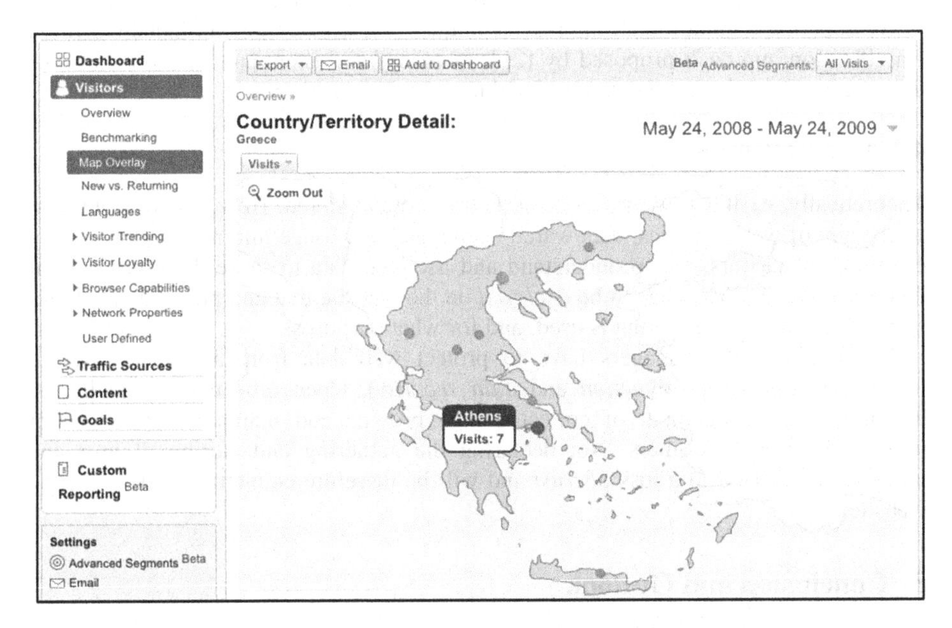

Fig. 6. With the *map overlay* of Google Analytics [8], analysts know from which continent, countries and cities visitors come from. In addition, it can be analyzed which language they speak and which web pages they visited: when, how often, how many and how long.

4.2 Web Analytics as a Measurement System of Knowledge Demand

Websites provide knowledge and digital content by web publishers, or in the Web 2.0 also User Generated Content (UCG) by contributors. Knowledge workers and web managers need to know, which information and knowledge offers has been accessed, to align this offer to the knowledge demand. Web analytics is thus an information and *knowledge measurement and management system* to analyze and survey the demand of knowledge. Measuring how knowledge is consulted provides additional value.

4.3 Creating Knowledge Using Web Usage Mining

Web Analytics as a knowledge measurement and management tool not only generates information or knowledge about and from visitors and online customers. Additional knowledge derived from *patterns* can also be created for website quality improvement using web usage mining. *Web usage mining* is a part of web mining, beside web structure and content mining [20, 21]. It refers to discovery and analysis of patterns in click-streams and associated data collected or generated as a result of user interactions with web resources on one or more websites [22]. A common web usage mining task is the analysis of *associations* between visited pages [23].

From an e-marketing point of view, for instance, it is interesting to know which web pages visitors viewed, *before* they visited the online shop or another attractive web page of Table 3. Web usage mining can also be used for individual marketing to provide *personalized content* or *recommendations* and to build web ontologies.

By applying common data mining techniques [see e.g. 24], or an inductive fuzzy classification approach proposed by [25], meaningful clusters and relations between different web metrics can be detected and used for web analytics.

4.4 Critical Remarks

Theoretically, each action of users can be tracked and stored. However, it should *not* be the aim of web analytics, data warehousing and web usage mining, to x-ray, tail or even control visitors, but to understand and use web data to serve the users in their own interests. It always has to be *declared* on the website that and how web analytics software or web usage mining is used, and for what purposes.

In addition, data managers have to protect web data from attacks, abuses or loss (keywords: *data protection* and *data security*). Open information and honest handling and management of collected data provide and maintain confidence of users and online customers. Not declaring the gathering and storing of user and usage data will lead to a loss of *trust* and will be therefore counterproductive for the website.

5 Conclusion and Outlook

In our information and knowledge society, websites create value both for their visitors and operators. Web analytics provide insights and knowledge about added value, the traffic on the website and about the behaviour of visitors or customers on web pages, in order to control and optimize website success.

However, sharp classifications of web metrics are often inadequate, as academic research shows. Therefore, this paper proposes a fuzzy data warehouse to overcome the limitations of sharp data handling in web analytics and information systems.

A fuzzy data warehouse is a powerful instrument to analyze web metrics and other performance measures in different granularity and from various dimensions, using slicing or dicing operations. The fuzzy data warehouse model allows an extension of classical performance indictors in a way that descriptive environmental information can be directly integrated into the querying processes using linguistic variables.

The research center *Fuzzy MarketWing Methods* (www.FMsquare.org) applies the fuzzy classification to data base technologies and marketing. It provides several *open source prototypes* and is *open for collaboration* with researchers and practitioners.

References

1. Web Analytics Association, http://www.webanalyticsassociation.org/aboutus
2. Weischedel, B., Matear, S., Deans, K.: The Use of eMetrics in Strategic Marketing Decisions. Int. Journal of Internet Marketing and Advertising 2, 109–125 (2005)
3. Sterne, J.: Web Metrics. Wiley, New York (2002)
4. Peterson, E.: Web Site Measurement Hacks. O'Reilly, New York (2005)
5. Kaushik, A.: Web Analytics – An Hour a Day. Wiley, New York (2007)
6. Clifton, B.: Advanced Web Metrics with Google Analytics. Wiley, New York (2008)
7. Hassler, M.: Web Analytics. MIT Press, Heidelberg (2008)
8. Google Analytics, http://www.google.com/analytics
9. Zadeh, L.A.: Fuzzy Sets. Information and Control 8, 338–353 (1965)
10. Zimmermann, H.-J.: Fuzzy Set Theory and its Applications. Kluwer, London (1992)
11. Werro, N.: Fuzzy Classification of Online Customers, Dissertation, University of Fribourg (2008), http://ethesis.unifr.ch/theses/downloads.php?file=WerroN.pdf
12. Inmon, W.H., Strauss, D., Neushloss, G.: DW 2.0 – The Architecture for the Next Generation of Data Warehousing. Morgan Kaufmann, Burlington (2008)
13. Lehner, W.: Datenbanktechnologien für Data-Warehouse-Systeme – Konzepte und Methoden. dpunkt, Heidelberg (2003)
14. Kimball, R., Ross, M.: The Data Warehouse Toolkit. Wiley, New York (2002)
15. Lenz, H.J., Shoshani, A.: Summarizability in OLAP and Statistical Data Bases. In: Proceedings of the Ninth International Conference on Scientific and Statistical Database Management, pp. 132–143. IEEE Computer Society, Washington (1997)
16. Schepperle, H., Merkel, A., Haag, A.: Erhalt von Imperfektion in einem Data Warehouse. In: Bauer, A., Böhnlein, M., Herden, O., Lehner, W. (eds.) Internationales Symposium: Data-Warehouse-Systeme und Knowledge-Discovery, pp. 33–42. Shaker, Aachen (2004)
17. Delgado, M., Molina, C., Sanchez, D., Vila, A., Rodriguez-Ariza, L.: A fuzzy multidimensional model for supporting imprecision in OLAP. In: IEEE International Conference on Fuzzy Systems, July 25-29, vol. 3, pp. 1331–1336 (2004)
18. Pérez, D., Somodevilla, M., Pineda, I.: Fuzzy Spatial Data Warehouse: A Multidimensional Model. In: Proceedings of the Eight Mexican International Conference on Current Trends in Computer Science, Morelia, September 24-28, 2007, pp. 3–9 (2007)
19. Alva, M.E., Martínez, A.B., Gayo, J.E.L., del Carmen Suárez, M., Cueva, J.M., Sagástegui, H.: Proposal of a Tool of Support to the Evaluation of User in Educative Web Sites. In: Lytras, M.D., Carroll, J.M., Damiani, E., Tennyson, R.D. (eds.) WSKS 2008. LNCS (LNAI), vol. 5288, pp. 149–157. Springer, Heidelberg (2008)
20. Liu, B.: Web Data Mining. Springer, New York (2007)
21. Markow, Z., Larose, D.: Data Mining the Web. Wiley, New York (2007)
22. Mobasher, B.: Web Usage Mining. In: Liu 2007, pp. 449–483 (2007)
23. Escobar-Jeria, V.H., Martín-Bautista, M.J., Sánchez, D., Vila, M.: Web Usage Mining Via Fuzzy Logic Techniques. In: Melin, P., Castillo, O., Aguilar, L.T., Kacprzyk, J., Pedrycz, W. (eds.) IFSA 2007. LNCS (LNAI), vol. 4529, pp. 243–252. Springer, Heidelberg (2007)
24. Berry, M., Linoff, G.: Mastering Data Mining. In: The Art and Science of Customer Relationship Management. Wiley, New York (2000)
25. Zumstein, D., Kaufmann, M.: A Fuzzy Web Analytics Model for Web Mining. In: Proc. of the IADIS European Conference on Data Mining, Algarve, Portugal, June 18-20 (2009)

Interoperability for LMS: The Missing Piece to Become the Common Place for Elearning Innovation

Marc Alier Forment[1], María José Casañ Guerrero[1], Miguel Ángel Conde González[2], Francisco José García Peñalvo[2], and Charles Severance[3]

[1] Universitat Politècnica de Catalunya, c/Jordi Girona Salgado 1-3, 08034 Barcelona, Spain
{malier,mjcasany}@lsi.upc.edu
[2] Universidad de Salamanca, Plaza de los caídos S/N. 37008, Salamanca, Spain
{mconde,fgarcia}@usal.es
[3] University of Michigan
{csev}@umich.edu

Abstract. This paper speculates about the future of LMSs considering the upcoming new learning applications and technologies, and the different attitudes of learners and teachers, given their technological background described using the digital natives and immigrants metaphor. Interoperability is not just a nice to have feature, but a must have features for LMS if these systems are going to be the common place where the ICT empowered learning innovation happens. After analyzing some standards and initiatives related to interoperability on LMS, the authors present an overview of the architecture for interoperability they propose. This architecture is being implemented for the well known Open Source LMS Moodle.

Keywords: Elearning, Web 2.0, Mobile Learning, Interoperability, LMS.

1 Introduction

1.1 The LMS, the Dinosaur and the Meteor

ELearning has experienced an extraordinary growth over the last years, learning paradigms; technological solutions, methods and pedagogical approaches have been developed, discarded and adopted. We have reached a point in time when most of learning institutions have adopted the use of Learning Management System (LMS) software, either from commercial vendors or Free Open Source Communities. LMS have reached the balance to meet the structure and (traditional) ways of schools, universities and other educational institutions.

As Dr Charles Severance [1] (founder of the Open Source LMS Sakai and currently working for IMS) states: LMS "*are all mature enough that the majority of faculties and student users are generally satisfied regardless of which system chosen*". It seems that LMS systems have achieved a stability and maturity, a Golden Age of LMS; LMS adopt each other's features and are slowly beginning to look like clones of one another. Severance wonders if this stable ecosystem of LMS is waiting to

M.D. Lytras et al. (Eds.): WSKS 2009, LNAI 5736, pp. 286–295, 2009.

become extinct by a meteor strike. We expect that meteor strike will be a set of disruptive learning innovation practices that do not fit inside the bounds of the current crop of LMS (Moodle, Sakai, Angel, Blackboard) [1].

The disruptive changes may strike from one of several directions. (1) The new generations of mobile devices have escaped from the fences that telecommunications operators have been erecting for so long, and becoming platforms opened to software developers ready to create great mobile learning applications. The same thing applies for the next generations of game consoles and all sorts of gadgets that surround us in our digital life, ready to be part of our learning processes. (2) The so-called Web 2.0 which is not really a new technology, is the way that people have decided to relate to their peers through technological means: an enhanced way to relate to others, produce, share and consume information. Relations among students, production and consumption of information are the basic tools of education. Thus Web 2.0, and other extensions such as Virtual Worlds [2], are influencing already the way we teach and learn. And (3) finally there is learning content and the producers of learning content.

Nowadays there are mostly two kinds of educational content: The content owned and distributed commercially buy publishers, and the Open Educational Resources such as the Open Courseware initiatives from MIT (http://ocw.mit.edu), UOC (http://ocw.uoc.edu) and others. According to Severance "There is a significant un-*solved problem when moving course information between Learning Management Systems and Open Educational Resources. There is a similar tension moving course content between commercial publishers and LMSs. [...] The meteor strike will happen when the owners and holders of content tire of the current situation and decide to take the initiative and simply change the rules of the game.*" [1].

The LMS can be a tool to spread learning innovation. Because most of teachers and learners use them at some level or sometimes because its mandatory. Anyway the LMS is a common ground for teachers and learners. We think that the right approach to spread innovation and transform learning will actually come from within the LMS itself.

To avoid extinction or becoming a barrier for the spreading and adoption of learning innovations or contents, LMS need to evolve and adapt to what is coming next. But, what kind of evolution are we talking about? Is it about features? No. It is about flexibility and interoperability.

1.2 The Common Ground for Digital Natives and Digital Immigrants

LMS are on the right track to meet the needs of the learning institutions [3]: integration with back office systems, library repositories, academic portfolios, semantic syllabus etc.

But, learning does not happen in the institution "management" of learning, it happens among students and teachers using whatever technology and resources they find and use in their learning.

Teachers usually like the web based LMS, sometimes they even love them!. The Open Source LMS Moodle (http://moodle.org) is a vivid example of this. Moodle is created and maintained by a huge community (more than half a million members worldwide) of developers and users (most of them teachers). Teachers, developers and institutions share experiences and enthusiasm about using Moodle in *Moodlemoot* conferences (http://moodlemoot.com http://moodlemoot.net) that gather hundreds of people in many countries every year. The Universitat Politecnica de Catalunya decided on 2004 to use Moodle as its corporate LMS (http://atenea.upc.edu) and found out that several teachers had already started their own Moodle installations, even some of them where active members of the early Moodle community. As other universities have reported on *Moodlemoot* conferences this is a common fact. Even the teachers who do not do learning innovation using ICT, after a painful process of adoption, consider a web based environment to be useful to get information about their students, distribute some power point files and grade the students.

Students usually do not have problems using ICT tools in their learning. Some students fall in the category of **digital natives**, introduced by Prensky [4], as opposed to the rest of students and most of the teachers who are **digital immigrants**.

Digital natives are those users who were born in the digital age and who are permanent skillfully users of new technologies. Digital natives feel attracted to new technologies. They satisfy their needs in many ways. They quickly gain expertise in the use of every kind of devices; they are at ease in any situation that implies their use, they look for new services and they use each new service to reach their targets.

According to Presnky, when digital natives involve in learning processes tend to:

1. Learn without paying attention; introduce learning elements, resources such as games.
2. Learn from other contexts, mobile devices, game consoles, digital television, etc.
3. Use of multimedia tools in learning such as contents trough podcasts
4. Include social software in learning process, integration of tools such as Facebook in learning platforms.

On the other hand digital immigrants were not born immersed in a ICT-saturated environment, and had to adapt themselves to new technologies in their lives and their learning (and teaching) processes. What do digital immigrants want?

1. A guided learning process and not necessarily shared or parallel.
2. Integration of tools they usually use in the new LMS contexts.
3. Ease the integration of back office tasks with the LMS.
4. Possibility to control from a single environment regardless of the activity in which the device is carried out.

ICT empowered learning innovation is likely going to come from the side of the digital natives users. While the digital immigrants learners will have to make their best to keep the pace, digital immigrant teachers might prefer to stay in the environments they know and feel comfortable.

This could create a big gap between teachers and learners, leading to a scenario where students might feel that they can learn better on their own way, using Open Educational Resources, Web 2.0 technologies and other sources of information.

But, right now the LMS are the common ground for digital natives and immigrants in the community of learners and teachers. If LMS can evolve to be able to embrace new kinds of practices, applications and technologies then, the LMSs can be a very valuable tool for integration and diffusion of new practices where all users can find their place.

1.3 The Missing Piece

The missing piece in today LMSs is interoperability. Using interoperability methods, LMS will be able to easily integrate and share information and services with other applications, platforms and even other LMS. The rest of this paper is about this matter. We will discuss about interoperability, we will review the standards being developed and adopted and some related projects.

2 Interoperability in Learning ICT Environments

2.1 Introduction

New generations of learning applications are rapidly emerging. For example, game-based learning has a huge potential in the learning process of children, adolescents and even grownups (for example Big Brain Academy), and has been an important field of research since late 1970s [5]. More recent studies [6], [7] explore the potential of using game consoles and other portable devices such as Nintendo DS or PlayStation Portable for education purposes since children spend so much time with these devices. Game players can learn to do things such as driving a car, but deeper inside, they learn thinks such as take information from many sources and make decisions quickly, deduce the games rules rather than being told, create strategies, overcome obstacles or learn to collaborate with others though the Network.

Other learning applications use portable technology such as digital cameras, mobile phones, MP4 players, or GPS devices to enhance the learning process. These applications are often called mobile learning (m-learning) applications. Although mobile learning is in its infancy, there are many experiences using mobile technology [8].

Blogging, wikis, podcast, screen-cast, contents from YouTube, Google Maps, pictures in Flickr, and social interaction in Facebook or Twitter, are common sources of information used by students while they learn or work in their assignments. The consumers of these applications are the digital natives, children who have lived all their lives with technology.

Right now we can find lots of learning applications, like the ones described in the previously, living outside the LMS ecosystems (Mobile applications in particular). Teachers willing to innovate are using applications and technologies not supported by their institution LMS, and by doing so they are taking their students outside the virtual campus. Thus the students need to go to several different sites (using different usernames and passwords) in a scrambled learning environment. This may cause confusion and frustration to students.

We need to allow the use of these new kinds of learning technologies and applications inside the LMS. To incorporate the new generation of learning applications inside the LMS there is a need for interoperability between systems.

2.2 The SOA Approach to Interoperability

Interoperability is defined by IEEE as “the ability of two or more systems, or components to exchange information and to use the information that has been exchanged” [6]. The IEEE definition for interoperability is 16 years old, and nowadays software systems can do more things together than just exchange information, for example share functionality. The Open Knowledge Initiative (OKI) offers a new definition for interoperability: "the measure of ease of integration between two systems or software components to achieve a functional goal. A highly interoperable integration is one that can easily achieved by the individual who requires the result". According to this definition, interoperability is about making the integration as simple and cost effective as technologically possible [9].

One new approach to reach interoperability between different systems is the Service Oriented Architecture (SOA). The Service Oriented Architecture (SOA) is a software engineering approach that provides a separation between the interface of a service, and its underlying implementation. For consumer applications of services, it does not matter how services are implemented, how contents are stored and how they are structured. In the SOA approach consumer applications can interoperate across the widest set of service providers (implementations), and providers can easily be swapped on-the-fly without modification to application code.

SOA preserves the investment in software development as underlying technologies and mechanisms evolve and allow enterprises to incorporate externally developed application software without the cost of a porting effort to achieve interoperability with an existing computing infrastructure.

For the previous reasons the authors believe that a SOA architecture can be used to provide interoperability between LMS and new generation of learning applications. Using a SOA architecture to integrate new generation of learning applications into the Web-based LMS, we will get the following benefits:

1. The students could use the new generation of learning applications in the learning process.
2. The teachers could use the Web-based LMS they are used to, to create new tasks for their students. These new tasks may be done using an external application from the LMS interface.
3. The use of external applications from the LMS would be done from a consistent software interface that would not create confusion to students and teachers.

2.3 Interoperability Specifications

2.3.1 The Open Knowledge Initiative

The Open Knowledge Initiative (OKI) was born in 2003 with the purpose of creating a standard architecture of common services across software systems that need to share, such as Authentication, Authorization, Logging [9]. The OKI project has developed and published a suite of interfaces know as Open Service Interface Definitions (OSIDs) whose design has been informed by a broad architectural view. The OSIDs specifications provide interoperability among applications across a varied base of underlying and changing technologies. The OSIDs define important components of a SOA as they provide general software contracts between service consumers and service providers. The OSIDs enable choice of end-user tools by providing plug-in interoperability. OSIDs are software contracts only and therefore are compatible with most other technologies and specifications, such a SOAP, WSDL. They can be used with existing technology, open source or vended solutions.

Each OSID describes a logical service. They separate program logic from underlying technology using software interfaces. These interfaces represent a contract between a software consumer and a software provider. The separation between the software consumer and provider is done at the application level to separate consumers from specific protocols. This enables applications to be constructed independently from any particular service environment, and eases integration. For example, services such as authentication are common functions required by many systems. Usually each application has built their own unique solution to this specific function. As a result the authentication functions are implemented in many ways and this results in information being maintained in different places and adding barriers to reuse. The OKI approach separates the authentication function from the rest of the system and provides a central authentication service for all the applications.

OKI describes with OSIDs the basic services already available in e-learning platforms. Among others, these basic services used by many e-learning platforms are described in the following OKI OSIDs:

- The authentication OSID is used to register a new user or to know if the user is connected to the system. This is a basic service in any software system.
- The authorization OSID is used to know if a user has rights to access a service or function. This service is necessary in any system using roles.
- The logging OSID is used to capture usage information. It is useful to know how the system is working for system diagnostics and performance.
- The internationalization OSID is used to change the language of the application or add new languages.
- The configuration OSID is used to change configuration parameters.

Thus using the OKI OSIDs has the following advantages:

- Ease to develop software. The organization only has to concentrate in the part of the problem where they can add value. There is no need to redo common functions among the systems.
- Common service factoring. OKI provides a general service factory so that services can be reused.
- Reduce integration cost. The current cost of integration is so high that prevents new solutions from being easily adopted. OSIDs are a neutral open interface that provides well-understood integration points. This way there is no need to build a dependency on a particular vendor.

Software usable across a wider range of environments, because OKI is a SOA architecture. But OKI still has a long way to go before becomes a de facto standard of interoperability. So far about 75 projects have implemented the OSIDs and given feedback to the OKI community process.

2.3.2 The IMS Global Learning Consortium Initiatives for Interoperability in Learning Systems

The IMS Global Learning Consortium is also working since 2005 in standards towards interoperability and integration of learning services and systems. The IMS Abstract Framework is set of (abstract) specifications to build a generic e-learning framework, which might be able to interoperate with other systems following the IMS AF specifications. IMS AF describes a e-learning system as the set of services that need to be offered. IMS AF is a standard that can be complemented by the OKI OSIDs because OKI provides more specific information about the semantics of the services, how to use them and in what kind of situations they could be used. IMS also defines the IMS Learning Technologies for Interoperability. While IMS AF and OKI work on the exchange of information and services, IMS Simple LTI developed under supervision of Dr. Charles Severance, focuses on the process on how a remote tool is installed on a web based learning system [10].

The OSIDs tells us how to exchange information between the LMS and an external learning application, but how will the teacher and the student reach the application form the LMS? These kinds of proxy bindings are described by the IMS LTI 1.0 and 2.0 standards. Another initiative from IMS towards interoperability is the IMS Learning Tools for Interoperability (IMS LTI). The basic idea of IMS LTI is that the LMS has a proxy tool that provides an endpoint for an externally hosted tool and makes it appear if the externally hosted tool is running within the LMS. In a sense this is kind of like a smart tool that can host lots of different content.The proxy tool provides the externally hosted with information about the individual, course, tool placement, and role within the course. In a sense the Proxy Tool allows a single-sign-on behind the scenes using Web services and allows an externally hosted tool to support many different LMS's with a single instance of the tool.

The IMS LTI 2.0 architecture focuses on the launch phase of the LMS-to-tool interaction. The launch accomplishes several things in a single Web service call:

- Establish the identity of the user (effectively like a single sign-on).
- Provide directory information (First Name, Last Name, and E-Mail address) for the user.
- Indicate the role of the current user whether the user is an Administrator, Instructor, or Student.
- Provide information about the current course that the Proxy tool is being executed from such as Course ID and Course Title.
- Provide a unique key for the particular placement of the Proxy Tool.
- Securely provide proof of the shared secret.
- Hints as to display size.
- An optional URL of a resource, which is stored in the LMS – which is being provided to the external tool as part of a launch

2.4 SOA Initiatives

There have been several initiatives to adapt SOA services for LMS and the integration of external applications into the LMS [11], [12], [13].

But, the previous initiatives have the following limitations:

- A defined application domain. Not all LMS services are provided, only those which are useful to a specific application domain.
- Unidirectional Interoperability. Architectures work only in one direction, which is, provide information from the LMS to other applications or integrate it with other tools. It not possible to provide information from external applications to the LMS.
- Interoperability Specifications. Definition of a service structure that does not use specifications for interoperability.

3 Adding SOA Interoperability to LMS: An Architectural Proposal

LMS are web-based applications. Their purpose is to present information in web pages, and most of them have been designed and developed when the current trends of web design (XHTML 1.0, CSS, AJAX) had just been proposed and therefore are not supported by the existing browsers. So, the internal design of most LMS is intended to do just that: serve web pages; not to provide services to be consumed by third applications or to consume external services and provide them as if they were their own services.

The authors have been working on way to adapt an LMS software to provide services to be consumed by third applications or to consume external services, and provide external services as if they were LMS own services. Out of this work we have defined a SOA architecture that tries to integrate the existing external educational applications with existing LMS. This architecture can also be used to extend the LMS resources and activities to other applications, such as mobile-learning applications [14].

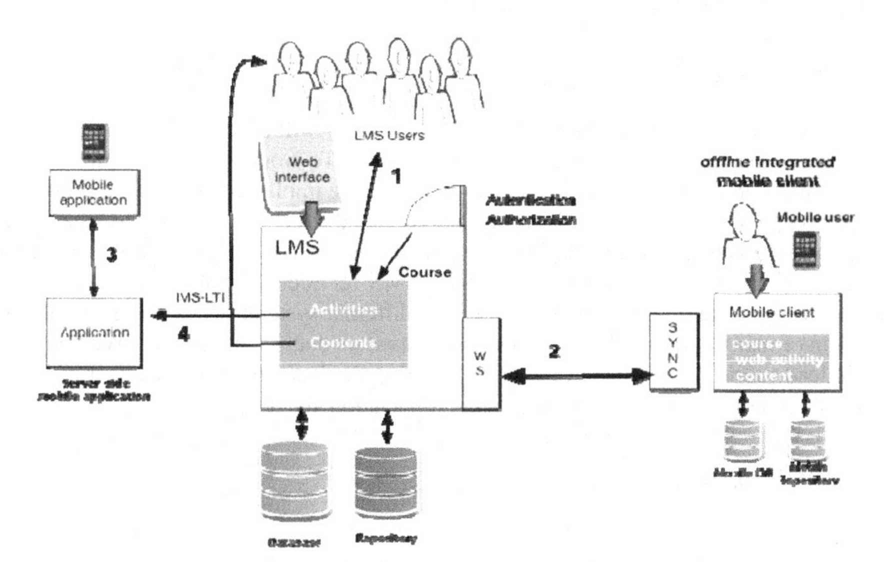

Fig. 1. Architecture to integrate LMS with external mobile applications

The architecture has the following elements:

1. If needed, a refactored version of the LMS core providing a clear layer with the basic services of the LMS.
2. A web-service layer that provides the basic services (such as authentication, authorization or course enrollment). These services are used by external applications using a web-service layer. This web-service layer is based on the OKI OSIDs. While layer 1 is specific to each LMS, the second layer just provides services defined by the OSID's and behind this layer any LMS could be working. This layer provides a framework for application developers that allows to integrate their work with any LMS, not just one.

3. A web-service consumer on IMS LTI 2.0 to integrate (launch) external educational applications within the LMS.

The proposed architecture presents important advantages such as:

1. LMS independence. The webservice layer defines the necessary services to access contents and activities in any LMS, because it is based on the OKI OSIDs. The only thing to do is the implementation of this WS layer on a specific LMS.
2. Mobile device or other platforms independence. Any kind of mobile device (or other devices) can be used as a client application (i.e, cell phones, tablet PC, One Laptop per Child). It is possible to have different implementations in different clients.
3. External application independence. The architecture is independent from the external applications that need to be integrated with the LMS.

This architecture will come in Moodle 2.0, due to the collaboration between the group of UPC and Moodle.org and Moodle.com.

4 Conclusions

The work being developed with the Moodle community will become a widespread reference implementation of interoperable LMS platform. Te work for the Moodle community is based on several standards and can lead to more implementations and compatible developments. The authors are also working on related open sourced projects oriented to implement mobile applications plug-gable to this interfaces build on Moodle. These projects use the SOA architecture, to expand LMS services to many scenarios (such as mobile learning) and allows the integration of different system, overcoming an inherent problem of many current e-learning standards based on data exchange [15]. We hope this might lead to the development of Mobile Learning applications strongly integrated with the LMS and thus with a greater potential of application and adoption.

Acknowledgment

Thanks also to Dr. Xavier Franch and our researchers colleagues from the GESSI research group in UPC. This work has been supported by the Spanish project Education TIN2007-64753. Thanks also to the GRupo de investigación en InterAcción y eLearning members from the University of Salamanca for their support. This work is partialy supported by the KEOPS research project (TSI2005-00960) from the Ministerio de Educación y Ciencia, and by the project SA056A07 from the Consejería de Educación de la Junta de Castilla y León.

References

1. Severance, C.: A Meteor Strike May Cause LMS Systems to become Extinct!. Dr Chuck Severance Weblog, http://www.dr-chuck.com/csev-blog/000606.html
2. Alier, M., Senges, M.: Virtual worlds as environment for learning comunities. In: Lytras, M.D., Tennyson, R., Ordonez de Pablos, P. (eds.) Knowledge Networks: The Social Software Perspective (2009)

3. Rengarajan, R.: LCMS and LMS: Taking advantage of tight integration. Click 2 Learn (2000), http://www.e-learn.cz/soubory/lcms_and_lms.pdf
4. Prensky, M.: Digital natives, digital immigrants. On the Horizon 9 (5), 1–6 (2001), http://www.marcprensky.com/writing/
5. Aguilera de, M., Mendiz, A.: Video games and education (education in the face of a "parallel school"). Computers in Entertainment (CIE) 1, 1–10 (2003)
6. Prensky, M.: Digital Game-Based Learning. McGraw-Hill, New York (2001)
7. Prensky, M.: Mark Prensky writing, http://www.marcprensky.com/writing/default.asp
8. Brown-Martin, G.: It's the learning, stupid! In: Arnedillo, I., Isaias, P. (eds.) IADIS International Conference Mobile Learning, pp. XVII-XIX. IADIS PRESS (2008)
9. Merriman, J.: Redefining interoprability. The Open Knowledge Initiative (OKI), http://www.okiproject.org/view/html/node/2916
10. IMS Global Learning Consortium. IMS Tools Interoperability Guidelines, http://www.imsglobal.org/ti/index.html
11. Kurz, S., Podwyszynski, M., Schwab, A.: A Dynamically Extensible, Service-Based Infrastructure for Mobile Applications. In: Song, I.-Y., Piattini, M., Chen, Y.-P.P., Hartmann, S., Grandi, F., Trujillo, J., Opdahl, A.L., Ferri, F., Grifoni, P., Caschera, M.C., Rolland, C., Woo, C., Salinesi, C., Zimányi, E., Claramunt, C., Frasincar, F., Houben, G.-J., Thiran, P. (eds.) ER Workshops 2008. LNCS, vol. 5232, pp. 155–164. Springer, Heidelberg (2008)
12. LUISA. Learning Content Management System Using Innovative Semantic Web Services Architecture, http://luisa.atosorigin.es
13. Pätzold, S., Rathmayer, S. and Graf, S.: Proposal for the Design and Implementation of a Modern System Architecture and integration infrastructure in context of e-learning and exchange of relevant data. In: ILearning Forum 2008. European Institute For E-Learning, pp. 82–90 (2008).
14. Casañ, M., Conde, M., Alier, M.: García,F.: Back and Forth: From the LMS to the Mobile Device. In: IADIS International Conference Mobile Learning, pp. 114–120. IADIS Press, Barcelona (2009)
15. Vossen, G., Westerkamp, P.: Why service-orientation could make e-learning standards obsolete. Int. J. Technology Enhanced Learning 1(1/2), 85–97 (2008)

Combining Service Models, Semantic and Web 2.0 Technologies to Create a Rich Citizen Experience*

Maria Kopouki Papathanasiou[1,2], Nikolaos Loutas[1,2], Vassilios Peristeras[2], and Konstantinos Tarampanis[1]

[1] Information Systems Lab, University of Macedonia, Thessaloniki, Greece
{mkoppap,nlout,kat}@uom.gr
[2] National University of Ireland, Galway, Digital Enterprise Research Institute
{firstname.lastname}@deri.org

Abstract. e-Government portals suffer from various shortcomings, such as the lack of formal service models for documenting services, the complexity of the user interfaces, the badly structured content etc. In this work we start from a mature model for modelling public services and employ semantic and Web 2.0 technologies in order to develop a citizen-centric e-Government portal. Content managers are provided with a mechanism for adding structured service descriptions. A rich user experience has been created for all the users of the portal. The user interface can be personalized, visual browsing mechanisms and efficient search mechanisms have been implemented on top of the underlying service repository. Structured service descriptions, which are semantically annotated using the RDFa mechanism, are made available to the users. In addition, the users can annotate the service descriptions using tags similar to what is currently done for popular Web 2.0 portals and the portal also logs the users' behaviour when using the portal. Hence, user-defined service metadata are created, which then support service and tag recommendation mechanisms.

Keywords: Web 2.0, e-Government, social, semantic, annotation, tag, GEA.

1 Introduction

During the last years, governments have made significant efforts for improving both their internal processes and the services they provide to citizens and businesses. This led to several successful e-Government applications (i.e. see www.epractice.eu). One of the most popular tools that were used by governments in order to modernize their services and make them accessible is e-Government portals, i.e. [1].

Nevertheless, most of these efforts did not succeed and according to a Gartner study [2] "most e-Government strategies have not achieved their intended objectives and have failed to trigger sustainable government transformation that will ultimately lead to greater efficiency and citizen-centricity".

* This work is supported in part by the SemanticGov (www.semantic-gov.org) and the Ecospace (www.ip-ecospace.org) FP6 projects.

M.D. Lytras et al. (Eds.): WSKS 2009, LNAI 5736, pp. 296–305, 2009.

The problems and the obstacles that prevented e-Government efforts from taking up are both organizational as well as technical. As governments are complex organizations with numerous diverse activities, we have to scope our discussion to e-Government service portals and examine the problems that they face.

First the lack of mature public service models hinders domain as well as technology experts from grasping a deep understanding of public services and tackling with their inherent complexity [3]. Having a coherent model to document services is necessary if these services are to be offered electronically. Such a model contributes towards the homogenization of public services descriptions and facilitates the sharing and reuse of these descriptions. Governmental portals will have services described in a common way, thus enabling interoperability and cross-portal querying. The model facilitates the discovery of public services which are available online supporting complex queries and advanced search options.

Technical and design issues, such as the lack of metadata or the complex user interfaces and the stovepipe information system implementations, also hinder e-Government efforts. It is interesting that only 14% of the citizens who use governmental portals were always able to find the information they were seeking [4]. Currently, public services are categorized in e-Government portals following a yellow page directory approach, where hierarchical category trees are used in order to organize the services' space.

In this work we are investigating how Semantic Web and Web 2.0 technologies can be combined with a mature model for public services, thus allowing governments to improve the quality of service provision and create what is currently known as a rich user experience. In particular, this work aims at the development of an e-Government portal which will:

- Support the documentation of services following a standard public service model and make available structured service descriptions that are both human as well as machine understandable.
- Allow citizens to have an active role, contrary to what was the common practice until now, and let them describe the services in their own terms, thus bridging the service discovery gap [5].
- Allow citizens to search for public services using user friendly, efficient and easy to use mechanisms, thus facilitating the service discovery problem.
- Improve the citizen's experience in the portal by means of personalization and recommendation mechanisms.

The rest of the paper is organized as follows. Section 2 discusses related efforts. Section 3 presents the architecture of the portal and the different types of users supported by the portal. Details on the functionalities of the portal are given in section 4. The evaluation of the portal is discussed in section 5. Section 6 concludes the paper and discusses possible research directions.

2 Related Work

Our portal tries to improve existing e-Government portals. We refer here to two national e-Government portals, which have been awarded as best practices.

KEP (http://www.kep.gov.gr) is the Greek Government's point of reference for its citizens. It makes available online descriptions of public services and makes it possible to submit electronic applications for some of these. KEP was given a Good Practice award by EU in 2007. Nevertheless, the content of the portal is not homogenized; the descriptions are not always complete and are usually hard to understand. Finding something in the portal can be quite a challenge as information about services is structured as a yellow-page directory.

Citizens Information (http://www.citizensinformation.ie/categories) is an Irish e-Government portal that provides public service information for Ireland. In 2007 Citizens Information won a World Summit Award. Services are organized either from generic ones to more specific or based on life events.

Our work is influenced by Web 2.0 platforms and tries, where applicable, to transfer best practices to the eGovernment domain. Table 1 shows the main functionalities of three major Web 2.0 portals, namely YouTube, Flickr, and Del.icio.us, and indicates whether these have been implemented in our portal.

It is worth mentioning, that unlike other Web 2.0 platforms, where all users are allowed, if not encouraged to add content, in our case the users add metadata to describe the content, but the content itself, namely the service descriptions, is added only by the service providers. At a later step, though, these service descriptions are enhanced by user-defined metadata.

Table 1. Web 2.0 portals and the GEA eService Tool

Web 2.0 Platforms	Users' Profile	Upload Content	Add Comments	Rate Content	Keyword Search	Tag Content	Recommend Tags	Browse Using Tag Clouds	Access Related Content
YouTube	×	×	×	×	×	×			×
Flickr	×	×	×		×	×		×	
Del.icio.us	×				×	×	×	×	
GEA eService Portal	×	×			×	×	×	×	×

3 Architecture of the GEA eService Portal

In order to address our objectives we have designed and implemented an e-Government service portal which capitalizes on Semantic Web and Web 2.0 technologies and makes available public service descriptions based on the Government Enterprise Architecture Public Service Model [6], [7]. Thus, we call this portal "GEA eService Tool". The conceptual architecture of the system consists of the following three layers:

The users interact with the portal through the *user interface layer* so as to use the desired functionalities. As explained later in detail four different types of users have been identified and the user interface is adapted according to these types.

The *application layer* implements the business logic of the portal and provides its main functionalities. It is modular and comprises of set of components, which inter-operate in order to implement the various functionalities of the portal, which are described in detail in section 4.

Finally, the content of the portal, e.g. service descriptions, tags etc, is stored in the underlying *repository*.

The portal has been developed using Drupal (http://drupal.org/), which was selected as it provides good support for Web 2.0 and Semantic Web technologies and has a very active developers' community.

The "GEA eService Portal" distinguishes the users into the following four categories and makes available different functionalities to each type of users:

a. *Anonymous Users.* They are the citizens that visit the portal without having registered. They can only browse and read the content of the portal.
b. *Authenticated Users.* They are the citizens that have registered in the portal. Authenticated users can take advantage of the additional functionalities, such as annotating the portal's content or getting recommendations about services.
c. *Content Managers.* They are the civil servants that are responsible for managing the content of the portal. They add and update the service descriptions that are made available by the portal. For that purpose, they use a data entry wizard which is based on GEA Public Service Model.
d. *System Administrators.* They are the administrators of the portal and their main task is its technical maintenance, i.e. they ensure the faultless operation of the system, they add/remove functionalities etc. They can also monitor the users' behavior and have access to statistical information with respect to the portals usage (e.g. page hits), the pages accessed more frequently, the most active users, the pages that have been recently visited etc.

The four user types led us to a role-based grouping of the portals functionalities. Nevertheless, the focus of this work is primarily on showing how semantic and Web 2.0 technologies enhance service provision and search and offer a rich user (in our case citizen) experience. Thus, we will limit our discussion only to those functionalities provided to citizens (either anonymous or authenticated users) which are based either on semantic or on Web 2.0 technologies.

4 Functionalities Provided to the Citizens

This section discusses the functionalities that are provided either to anonymous or authenticated users or to both of them.

4.1 Searching for Services and Finding Service Information

The portal goes beyond simple keyword search and supports advanced search functionalities. Thus, both anonymous and authenticated users can use multiple search criteria and combine them with logical operators such as AND, OR etc. They can also use the guided search, which is an inherent Drupal functionality tailored to our needs, where the search results can be stepwise refined using different attributes of the service descriptions. When a service is found, then the service description is presented either as a structured online report, which follows the GEA Public Service Model, or visually as shown in Fig. 1 [8]. In both cases the citizens access information such as the service description, the cost, the inputs and outputs, the service provider etc.

The structured online reports produced by the portal make use of the power of semantics. In fact we have followed the popular slogan "a little semantics goes a long way". Therefore, all the reports are implemented as XHMTL pages which are then annotated using the RDFa mechanism. RDFa allows embedding RDF triples into an XHTML document. Thus, every XHTML report contains a set of RDF triples each of which has as subject the URL at which the service is invoked and as object either a URI or a URL to a resource, depending on the predicate [9]. The predicates of the triples come from two different service models, SAREST [9] and an implementation of the GEA Public Service Model in OWL.

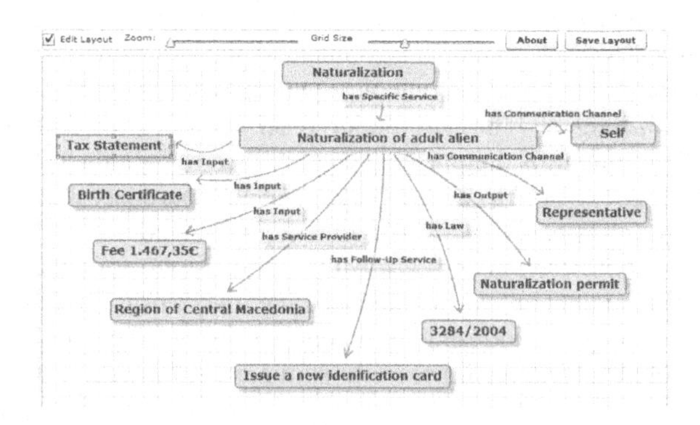

Fig. 1. The visual representation of the "Naturalization of adult alien" service

```
<div xmlns:sarest="http://knoesis.wright.edu/srl/sarest"
      xmlns:islab="http://islab.uom.gr">
<div about="http://islab.uom.gr/Naturalization">The Greek Naturalization public
service is provided by the <div rel="islab:serviceprovider"
  resource="http://islab.uom.gr/gea.owl#Prefecture"/>          Regional
Authorities and belongs to the <div rel="sarest:domain-rel"
  resource="
http://islab.uom.gr/gea.owl#SubDomainCommunityAndSocialServices"/>
Community and Social Services domain </div>
```

Fig. 2. Semantically annotated service description for the Greek Naturalization public service

SA-REST could not meet our needs when it came to annotating domain specific information such as input documents, service providers etc. Domain specific semantics were required in that case and this is why the implementation of the GEA Public Service Model in OWL was used. This combination allowed us to semantically annotate all important attributes of a public service, such as input/ output documents, the service providers, the tags, the place from which the public service is invoked etc. An example of a semantically annotated service description for the Greek Naturalization public service is presented in Fig. 2. The annotation process is transparent to the user. The semantically annotated reports are then interoperable and can be crawled and used by Semantic Web search engines.

4.2 Social Annotation of Services

One of the main principles on which our portal is built is that the users are not passive information recipients, like in typical e-Government portals, but they participate actively and express their view on the services offered through the portal.

This practically means that the service descriptions that the content managers add in the portal are enhanced with additional user-defined metadata, which come from the users either directly or indirectly. This implements what is defined as the Social Contract in [10]. In order to facilitate the direct input of user-defined metadata, a tagging mechanism has been implemented. Thus, the authenticated users can describe any service using a comma-separated list of tags. A snapshot of this is presented in Fig. 3. The recommended tags and the tag cloud of the service are also shown.

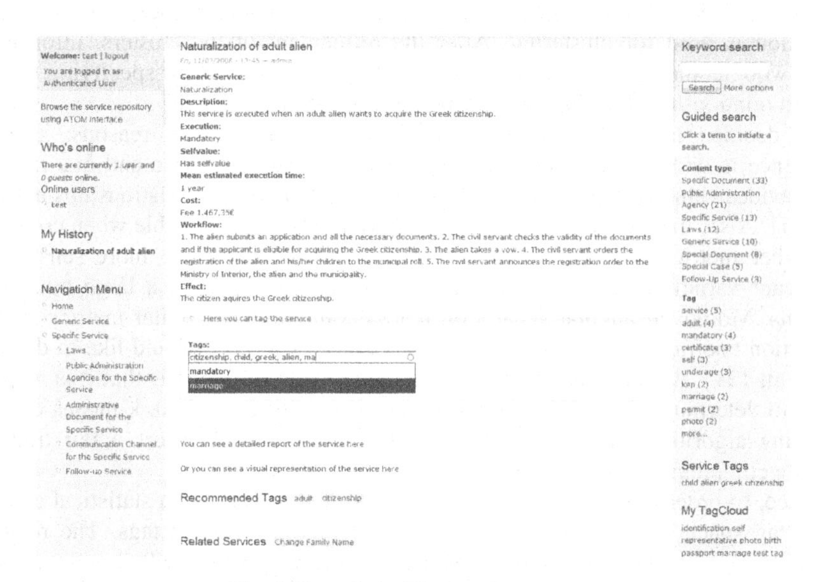

Fig. 3. Snapshot of the tagging process

In order to support the tagging process a tag recommendation mechanism has been implemented, which suggests to the user tags that:

a. have already been used in order to annotate a service. This reduces lexical variations (e.g. plurals, capital letters, abbreviations etc.), spelling errors etc.
b. are popular for this service, i.e. the tags that have the highest frequency.
c. belong to the same tag cluster with the popular ones.

Many researchers suggest tag recommendation mechanisms in order to enhance tagging systems. The tagging services of popular Web 2.0 platforms are discussed in [11]. Some of these approaches, e.g. [12] and [13], use the idea of "virtual" users and reputation scores. Thus, they recommend meaningful tags based on the content and the context of the resources to be annotated, the users' reputation and their preferences. Others, such as [14] and [15] use graphs for recommending tags. The tag space is represented as a graph and algorithms such as [16] are used to cluster the graph.

Tags that belong to the same cluster are candidate for recommendation. In [17], [18], they convert the folksonomy into an undirected graph and then apply FolkRank in order to generate recommendations. In a similar line of work, in [19] folksonomies are represented as hypergraphs with ternary edges where each edge indicates that an actor associated a certain concept with a particular instance. After applying network analysis techniques, clusters that contain semantically related tags are created.

We have implemented the approach proposed by [20], where they try to find the semantics behind the tag space in social tagging systems. Their approach consists of three steps: preprocessing, clustering and concept/relation identification. Firstly, the tag space is cleaned up by filtering out nontrivial tags, grouping together morphologically similar tags and not taking into consideration the infrequent ones or those which stand in isolation. Then, statistical analysis is used so as to define groups of possibly related tags. The clustering methodology is based on co-occurrence and the angular separation is used for clustering. After the refinement of the clusters, information from Swoogle and external data (Wikipedia, Google) are used for specifying the discovered relationships. We have not implemented this last step yet.

We decided to implement this approach for a number of reasons. The co-occurrence is widely used while creating clusters of the tag space and is a measure that provides valuable input for algorithms to extract taxonomic relationships between tags [21]. Also, the angular separation is believed to be more suitable when compared with other metrics such as Euclidian and Manhattan, which are more sensitive to significant variations in a few elements than little variations in a large number of elements. Moreover, angular separation is less complex than similar metrics such as correlation coefficient and is the measure to choose when one would like to discover synonyms [21]. Another strong point of this clustering technique is that it is not necessary to determine a priori the total number of the clusters, unlike k-means or other clustering algorithms. Lastly, we believe that this approach offers a fair trade-off between complexity and quality of results.

Hence, in order to create tag-clusters, we first had to perform a statistical analysis of the tag space so as to determine clusters of possibly related tags. The relations between tags were detected based on their co-occurrence. This resulted in the creation of a co-occurrence matrix, where every column/row is a vector which corresponds to one of the tags. We calculated the similarity between two tags using the angular separation. Afterwards, we determined a threshold so as to filter out pairs of tags that are not highly similar to each other. Every pair of similar tags initiates a cluster, which can be extended by a tag that is computed to be similar with the rest of the tags in the cluster. This process is repeated for all the tags of the dataset. Whenever the tag space is fully checked and the cluster cannot be further extended, a new similar pair initiates another cluster and the algorithm starts over again till all the similar pairs are used. Two clusters may be (almost) identical although the initial similar pair is different. Thus, the three smoothing heuristics proposed by [20] are used so as to minimize the number of the constructed clusters.

Before concluding this section, it should be reminded that the portal collects user metadata indirectly as well. This is done by monitoring the user's behavior and identifying usage patterns, for example service description *b* was visited after service description *a* or those that view service descriptions *a* and *b* also view service description *c*. These metadata are also used for enhancing the expressivity of the service description. In order to achieve this, during every session, the system stores information that refers to the visited pages, the browsing order and a timestamp of each visit. Thus, information such as the date, the exact time while the visitor access a page and the referrer (through which page the user accessed the current) is kept.

4.3 Browsing the Service Repository Using Tag Clouds

The users can browse the service repository by clicking on the tags of the tag clouds. Each time the selected tag acts as a filter and filters out the services that are not annotated with it. The portal supports three different types of tag clouds:

Overall Tag Cloud, which is generated from all the tags that have been added from every user (anonymous or authenticated) to every service. The popularity of a tag, shown by its font size, indicates the number of services that have been annotated with it.

Service Tag Cloud, which consists of all the tags that have been used by all authenticated users to annotate a service. It appears whenever a user (anonymous or authenticated), accesses a service description. Here the popularity of a tag indicates its usage frequency for the specific service.

Personal Tag Cloud, which includes all the tags, that an authenticated user has selected in order to annotate the services of his/her interest. In this case, the popularity of a tag indicates its importance for that specific user.

4.4 Recommendation of Related Services

The portal uses recommendation mechanisms in order to suggest related services based on the similarity of the Service Tag Clouds. Services are considered to be related when: (a) they belong to the same life event; (b) they have common characteristics (i.e. input documents, service provider, tag etc); (c) they are executed sequentially (one after the other). In order to compute the similarity between services the algorithm proposed in [20] is used again. This time the process is based on a matrix where the columns are vectors that correspond to Service Tag Clouds. Afterwards, the angular separation is applied to these vectors and thus the similarity between services is computed. If two vectors are similar, which means that the corresponding Service Tag Clouds are also similar, then the services are related [5].

5 Evaluation

In order to evaluate the prototype of the GEA eService Tool, a set of evaluation activities was organized once a stable version was released. The main objective of the evaluation was to assess the overall quality of the portal's user interface and functionalities in terms of usability, efficiency and user satisfaction. For that reason, an evaluation workshop was organized. Approximately 20 citizens of different ages, sexes and backgrounds participated in the workshop. Questionnaires were prepared and were handed out to the participants, who were encouraged to write their opinions after using the portal. The questionnaires included a set of 10 questions which had to be answered by assigning a weight starting from 1 (not at all) to 7 (very much).

The results of the evaluation were very encouraging for our work. For example, almost 89% of the citizens liked very much the idea of implementing such a portal for public services. Approximately 80% of the participants were very satisfied with the browsing the service repository using tag clouds functionality. It is worth mentioning that more than 90% of the participants really liked the fact that they could tag the public services that they use and seemed to comprehend its added value. Finally, 40% of the participants needed some help in order to use the portal and another 50% of them encountered some problems/faults while using it.

6 Conclusions and Future Work

In the era of the social Web, e-Government poses new challenges with respect to improving citizens' experience and transforming citizens from passive information recipients to active information contributors.

In order to address these requirements, we adopted new trends, technologies and principles that Web 2.0 and the Semantic Web propose and developed a citizen-centric e-Government portal. Citizen-centricity in our work is primarily supported by the fact that citizens can actively participate in the portal and describe the services in their own terms. The visual browsing of the service repository, the advanced search functionalities, the tag and service recommendation mechanisms and the detailed service descriptions enhance the citizens' experience when using the portal. The evaluation results suggest that our effort is in the right direction.

As part of our future work, we aim at dealing with the problems of synonyms and lack of consistency that all tagging systems face. In fact, [20] offer a solution for that which remains to be incorporated in our work. For example, what if a citizen annotates a service with the word marriage while another uses the word wedding or what if s/he uses dates instead of date.

Another possible research direction is to develop recommendation mechanisms that will be based on the information about the user's behavior or on their profiles or even on a combination of both. For instance, if we know that after viewing the "Issuing a marital status certificate" service the majority of users accesses the "Issuing a family allowance" service, then the system can suggest the second service as a follow-up of the first one. In another example, the portal could recommend services based on the marital status that the citizen specified when registering, i.e. if you are married you are eligible for the marriage allowance.

Concluding, it is worth mentioning that our portal is used in a pilot study and it will provide the infrastructure for documenting more than 100 public services provided by the Cypriot Ministry of Finance. A running prototype of the portal is available at http://195.251.218.39/cyprus/.

References

[1] Drigas, A.S., Koukianakis, L.G., Papagerasimou, Y.V.: An E-Government Web Portal. WSEAS Transactions on Environment and Development 1(1), 150–154 (2005)

[2] DiMaio, A.: Web 2.0 in Government: a blessing and a curse, Gartner (2007)

[3] Tambouris, E.: Introducing the need for a Domain Model in Public Service Provision (PSP) eGovernment Systems. In: 3rd International Conference on Data Information Management, London, UK, November 13-16 (2008)

[4] Reddick, C.G.: Citizen interaction with e-government: From the streets to servers? Government Information Quarterly 22, 38–57 (2005)
[5] Fernandez, A., Hayes, C., Loutas, N., Peristeras, V., Polleres, A., Tarabanis, K.: Closing the Service Discovery Gap by Collaborative Tagging and Clustering Techniques. In: ISCW 2007, Workshop on Service Discovery and Resource Retrieval in the Semantic Web (2008)
[6] Peristeras, V., Tarabanis, K.: The Governance Architecture Framework and Models. In: Saha, P. (ed.) Advances in Government Enterprise Architecture, Hershey, PA, IGI Global Information Science Reference (2008)
[7] Peristeras, V., Tarabanis, K.: The Governance Enterprise Architecture (GEA) Object Model. In: Wimmer, M.A. (ed.) KMGov 2004. LNCS (LNAI), vol. 3035, pp. 101–110. Springer, Heidelberg (2004)
[8] Basca, C., Corlosquet, S., Cyganiak, R., Fernandez, S., Schndl, T.: Neologism: Easy Vocabulary Publishing. In: Workshop on Scripting for the Semantic Web at ESWC, Tenerife, Spain (2008)
[9] Sheth, A.P., Gomadam, K., Lathem, J.: SA-REST: Semantically Interoperable and Easier-to-Use Services and Mashups. IEEE Internet Computing 11(6), 91–94 (2007)
[10] Loutas, N., Peristeras, V., Tarabanis, K.: Extending Service Models to Include Social Metadata. In: WebSci 2009: Society On-Line, Athens, Greece (2009)
[11] Derntl, M., Hampel, T., Motschnig-Pitrik, R., Pitner, T.: Inclusive Social Tagging: A Paradigm for Tagging-Services in the Knowledge Society. In: Lytras, M.D., Carroll, J.M., Damiani, E., Tennyson, R.D. (eds.) WSKS 2008. LNCS (LNAI), vol. 5288, pp. 1–10. Springer, Heidelberg (2008)
[12] Xu, Z., Fu, Y., Mao, J., Su, D.: Towards the Semantic Web: Collaborative Tag Suggestions. In: Collaborative Web Tagging Workshop at 15th International World Wide Web Conference, Edinburgh, Scotland (2006)
[13] Basile, P., Gendarmi, D., Lanubile, F., Semeraro, G.: Recommending Smart Tags in a Social Bookmarking System. Bridging the Gap between Semantic Web and Web 2.0. In: SemNet 2007, pp. 22–29 (2007)
[14] Grineva, M., Grinev, M., Turdakov, D., Velikhov, P.: Harnessing Wikipedia for Smart Tags Clustering. In: Knowledge Acquisition from the Social Web (KASW 2008), Graz, Austria (2008)
[15] Begelman, G., Keller, P., Smadja, F.: Automated Tag Clustering: Improving search and exploration in the tag space. In: Collaborative Web Tagging Workshop at WWW 2006, Edinburgh, Scotland (2006)
[16] Girvan, M., Newman, M.E.: Finding and evaluating community structure in networks. Physical Review E 69(2), 026113 (2004)
[17] Hotho, A., Jaschke, R., Schmitz, C., Stumme, G.: Information Retrieval in Folksonomies: Search and Ranking. In: Sure, Y., Domingue, J. (eds.) ESWC 2006. LNCS, vol. 4011, pp. 411–426. Springer, Heidelberg (2006)
[18] Hotho, A., Jaschke, R., Schmitz, C., Stumme, G.: FolkRank: A Ranking Algorithm for Folksonomies. In: Workshop Information Retrieval (FGIR 2006), Germany (2006)
[19] Mika, P.: Ontologies Are Us: A Unified Model of Social Networks. Web Conference and Semantics, Web Semantics: Science, Services and Agents on the World Wide Web 5(1), 5–15 (2005)
[20] Specia, L., Motta, E.: Integrating Folksonomies with the Semantic Web. In: Franconi, E., Kifer, M., May, W. (eds.) ESWC 2007. LNCS, vol. 4519, pp. 624–639. Springer, Heidelberg (2007)
[21] Cattuto, C., Benz, D., Hotho, A., Stumme, G.: Semantic Analysis of Tag Similarity Measures in Collaborative Tagging Systems. In: 3rd Workshop on Ontology Learning and Population (OLP3), Patra, Greece (2008)

On the Development of Web-Based Argumentative Collaboration Support Systems

Manolis Tzagarakis[1], Nikos Karousos[1], and Nikos Karacapilidis[1,2]

[1] Research Academic Computer Technology Institute
26504 Patras, Greece
{tzagara,karousos,karacap}@cti.gr
[2] IMIS Lab, MEAD, University of Patras
26504 Patras, Greece
nikos@mech.upatras.gr

Abstract. Advanced argumentative collaboration support systems can rarely be found in today's World Wide Web. This can be partially justified by the fact that the current Web environment - its users and available data - differs significantly from the environments in which these systems were traditionally developed and used. Efforts to bring such systems to the Web must carefully consider how these need to change in order to be effective in the new environment. In this paper, we present how such concerns have been addressed in CoPe_it!, a tool that supports argumentative collaboration on the Web. Preliminary evaluation results show that the tool succeeds in meeting the challenges of today's Web environment without compromising its effectiveness.

Keywords: argumentative collaboration, incremental formalization, Web-based systems, CoPe_it!

1 Introduction

Argumentative collaboration support systems have a long history. Generally speaking, they offer sophisticated support for sense- and/or decision-making, and have been proven effective in addressing a wide range of concerns in various domains, such as engineering, law and medicine. The ability of argumentative collaboration support systems to explicate, share and evaluate knowledge makes them an important infrastructure component in the knowledge society.

In most cases, argumentative collaboration support systems have largely remained within the communities in which they originated, thus failing to reach a wider audience. When investigating how the advent of the World Wide Web affected them, the results are rather disappointing: only Web-based discussion forums, offering rather primitive support when compared to argumentative collaboration support systems, have successfully migrated to the Web. One key factor contributing to the wide adoption of these forums is their emphasis on simplicity. On the other hand, the formal nature of sophisticated argumentative collaboration support systems has been pointed out as an important barrier to their wide adoption, and as one factor that hinders them to make the step towards the World Wide Web. The new landscape of the Web, as

M.D. Lytras et al. (Eds.): WSKS 2009, LNAI 5736, pp. 306–315, 2009.

shaped by the so called Web 2.0 move, is considered as a rather hostile environment for traditional argumentative collaboration support systems. In this new environment, users are moving away from formal systems that prescribe their interactions, and favor applications that place the control of the formalization process in their hands. Nevertheless, attempts to bring advanced argumentative collaboration support systems into the new Web environment have already begun to appear [1].

In this paper, we use the Walker's concepts of *domesticated* and *feral* technology [2] to describe the current contradicting environments in which applications must operate, and present how CoPe_it! (http://copeit.cti.gr/), an innovative tool that supports argumentative collaboration on the Web, attempts to bridge the abovementioned gap. Our long term aim is to equip the Web with more powerful discussion tools, with which the complex problems of our society can be better addressed.

2 The Domestic Nature of Traditional Argumentative Collaboration Support Systems

Technologies have a long history in being repurposed, i.e. to be useful in ways they were not developed or evolved for. Aerosol was not designed to be used in graffiti or street art, the internet was designed for military and academic use and not to create today's social communication networks, the purpose of computers never included individual, everyday use [2]. In order to analyze, discuss and understand such evolutionary paths of technology in the context of hypertext, Walker coined the terms of domesticated and feral technologies [2].

The term domesticated technology refers to technology that is tamed, farmed and cultivated in carefully controlled environments and intended to be used in specific ways. Early hypertext systems, for instance, were developed to run on mainframe computers in research institutions, and provided the necessary means to address the problem of information organization in specific application domains. Such domestication shaped the form of these hypertext tools, which made distinctions between author and reader, provided fixed types of links and, generally speaking, enforced rules for guiding the process of creating hypertexts. They were intended to be used in carefully controlled environments.

The rapid spread of personal computers and the advent of the Web created a new environment in which hypertext existed in. The result was hypertext to become feral: to overcome the boundaries and constraints that guided its creation and use, giving birth to new forms of information organization paradigms. Wikipedia (http://www.wikipedia.org/) and the folksonomy of Flickr (http://www.flickr.com/) are characteristic examples of feral hypertext.

The above discussion may provide useful insights with respect to how technologies and applications change when their ecosystem changes and relocate to the Web. In general, while changes to their functionality are a rather expected implication, one alteration is common to all applications that are subject to such relocation: they give greater control to the user, attempting to minimize the constraints and rules that control the user's actions. The applications are characterized by greater flexibility compared to their domesticated counterparts.

Argumentative collaboration support systems are one particular class of applications that have recently begun to appear on the Web. Nevertheless, these systems have a long standing tradition in being developed and used in tightly controlled communities and environments, hence constituting examples of domesticated technologies. When argumentative collaboration support systems go feral on the Web, the question of how they need to change in order to successfully address the conditions of their new habitat is critical.

Argumentative collaboration support systems have offered advanced computational services to diverse types of groups and application areas. In general, these services are enabled through the prescribed semantics and methods of interaction that the respective systems impose to their users. One reason why these systems deploy such prescribed methods is because of their efforts to closely match their argumentation model to the vocabulary and needs of the groups and application areas in which they are used. These prescribed methods place control over the argumentative collaboration process on the system's side and make it impossible for users to escape it. Therefore, these systems exhibit a high degree of formality.

Considering the provided prescribed methods as mechanisms to constrain the argumentation model, existing argumentative collaboration support systems can be considered as domesticated technology. They attempt to 'tame' the argumentative collaboration process by being designed, developed and used in carefully controlled environments. In our context, this implies control over a number of factors, such as the type of user that is foreseen to use these tools and participate in argumentative collaboration, the structure of the community or group entering into argumentative collaboration, the type of the problem at hand, as well as the type of the media formats that the information must have in order to be admissible for a particular discussion.

3 The Untamed Nature of Today's World Wide Web

For argumentative collaboration support systems, which require carefully controlled environments for their proper use, the current form of the World Wide Web proves to be a rather hostile habitat. In general, the prescribed methods of argumentative collaboration support systems have received much criticism. The formal structures were the reason these systems were difficult to use, requiring great efforts from individuals [3, 4], and proved to be barriers rather than catalysts for collaboration, as they slow down the users' activities [5]. The formal structure imposed has been the leading cause for the failure of these systems with respect to their widespread adoption [6]. On the other hand, simpler online discussion tools, such as Web-based forums, gained exceptional adoption precisely because of their lack of sophisticated formal structures, and their emphasis on "naturalness of interactions" [6]. Towards their move to the Web, such observations are of great concern as today's so-called Web 2.0 environment – its users and available data – differs substantially from the controlled environments that traditional argumentative collaboration support systems were initially developed and used. Web users are moving away from applications that impose semantics and tightly control the formalization process, favoring applications that place such control into their hands.

In today's successful Web applications, semantics is an emergent and not a predefined property of the system. Folksonomies and Wikis are prominent examples in this regard. Users expect simple but powerful Web applications that provide easy-to-use and engaging collaboration environments. Social and awareness services are nowadays considered required functionalities. Moreover, users are familiar in re-using information and interlinking a wide range of media formats, from text to images and videos, which are available on the Web and not always under their control. Finally, they favor applications that are easy to use but at the same time provide powerful and advanced collaboration services.

Moving argumentative collaboration support systems to today's Web environments requires careful consideration with respect to how they need to change in order to keep their effectiveness but at the same time address the concerns of their new environment. CoPe_it! is a tool supporting argumentative collaboration aiming to be used in today's Web environment. In the next sections, we present some key aspects of this tool.

4 Argumentative Collaboration in CoPe_it!

CoPe_it! is a tool to support synchronous and asynchronous argumentative collaboration in today's Web environment. The tool aims at supporting the level of control that current Web users expect, while also providing advanced services that are traditionally associated with strongly formalized argumentation and decision making systems. CoPe_it! achieves this by opening up the semantics layer and introducing the notion of *incremental formalization* of argumentative collaboration. The tool permits a stepwise evolution of the collaboration space, through which formalization is not imposed by the system; instead, it is at the user's control. By permitting the users to formalize the discussion as the collaboration proceeds, more advanced services can be made available. Once the collaboration has been formalized to a certain point, CoPe_it! can exhibit an active behavior facilitating the decision making process. Our overall approach is the result of action research studies [7] concerning the improvement of practices, strategies and knowledge in diverse cognitively-complex collaborative environments. The research method adopted for the development of CoPe_it! has followed the design science paradigm [8].

4.1 Incremental Formalization of Discussions

In CoPe_it!, formality and the level of knowledge structuring is not considered as a predefined and rigid property, but rather as an adaptable aspect that can be modified to meet the needs of the tasks at hand. By the term formality, we refer to the rules enforced by the system, with which all user actions must comply. Allowing formality to vary within the collaboration space, incremental formalization, i.e. a stepwise and controlled evolution from a mere collection of individual ideas and resources to the production of highly contextualized and interrelated knowledge artifacts, can be achieved [9].

In our approach, *projections* constitute the 'vehicle' that permits incremental formalization of argumentative collaboration. A projection can be defined as a particular representation of the collaboration space, in which a consistent set of abstractions able to solve a particular organizational problem during argumentative collaboration is available. With the term abstraction, we refer to the particular data and knowledge items, relationships and actions that are supported through a particular projection, and with which a particular problem can be represented, elaborated and be solved. CoPe_it!

enables switching from a projection to another, during which abstractions of a certain formality level are transformed to the appropriate abstractions of another formality level. This transformation is rule-based; such rules can be defined by users and/or the facilitator of the collaboration and reflect the evolution of a community's collaboration needs. It is up to the community to exploit one or more projections of a collaboration space (upon users' needs and expertise, as well as the overall collaboration context).

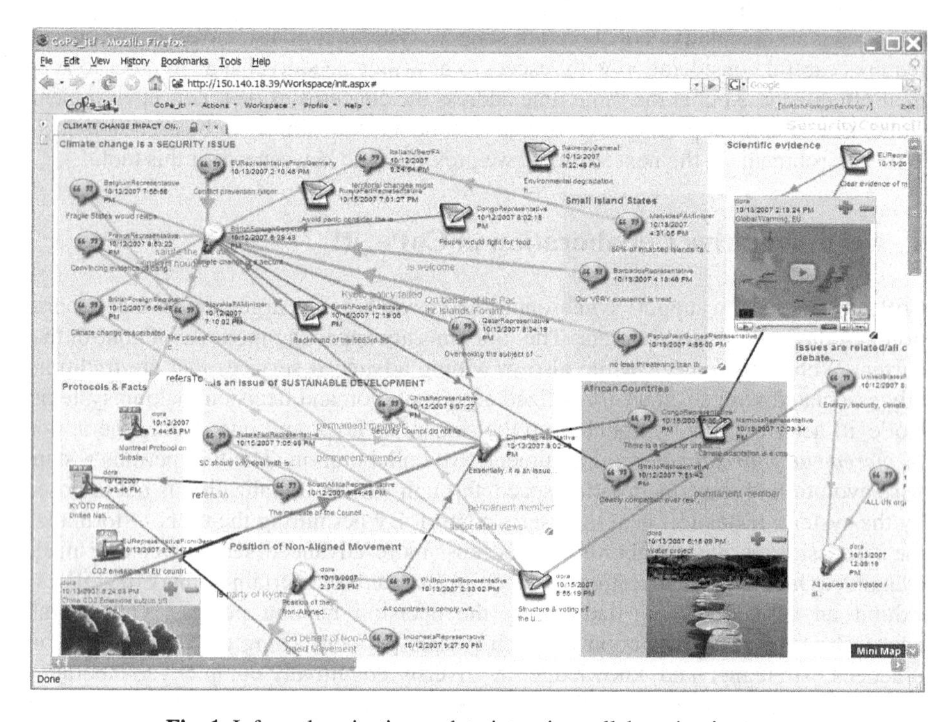

Fig. 1. Informal projection: a data-intensive collaboration instance

Each projection of the collaboration space provides the necessary mechanisms to support a particular level of formality. The more informal a projection is, the more easiness-of-use is implied; at the same time, the actions that users may perform are intuitive and not time consuming (e.g. drag-and-drop a document to a shared collaboration space). Informality is associated with generic types of actions and resources, as well as implicit relationships between them. However, the overall context is more human (and less system) interpretable. As derives from the above, the aim of an informal projection of the collaboration space is to provide users the means to structure and organize data and knowledge items easily, and in a way that conveys semantics to them. Generally speaking, informal projections may support an unbound number of data and knowledge item types. Moreover, users may create any relationship among these items; hence, relationship types may express agreement, disagreement, support, request for refinement, contradiction etc. (Figure 1)[1].

[1] The collaboration instances illustrated in Figures 1, 2 and 3 correspond to a detailed example of use of CoPe_it!, which can be found in [10].

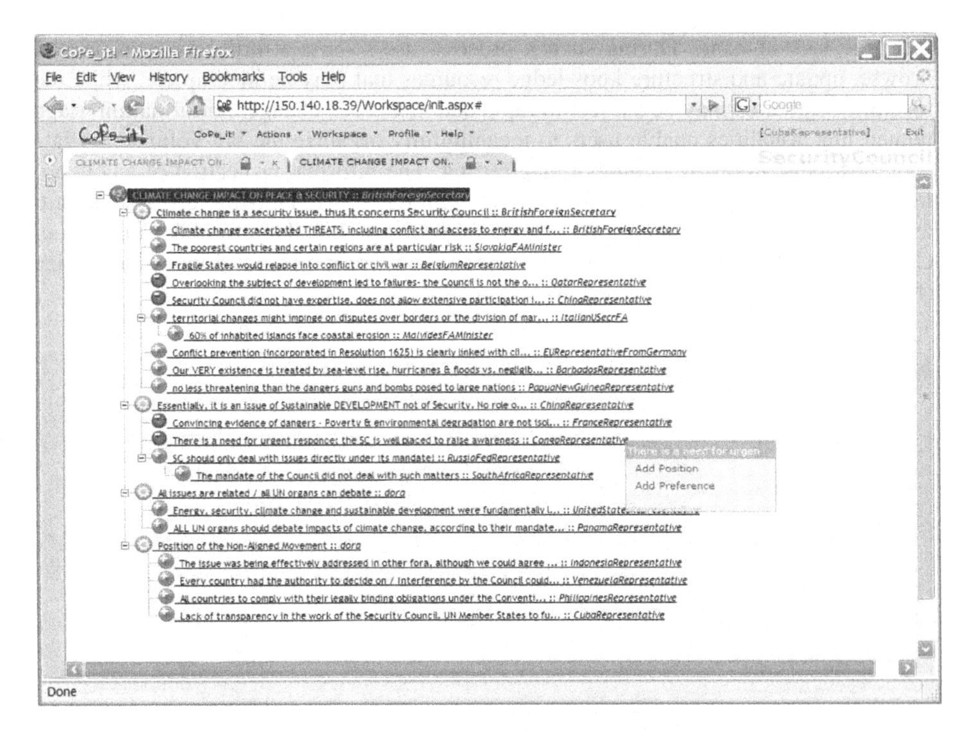

Fig. 2. Formal projection: a collaboration instance

While such a way of dealing with data and knowledge resources is conceptually close to practices that humans use in their everyday environment, it is inconvenient in situations where support for advanced decision making processes must be provided. Such capabilities require resources and structuring facilities with fixed semantics, which should be understandable and interpretable not only by the users but also by the tool. Hence, decision making processes can be better supported in environments that exhibit a high level of formality. The more formal projections of a collaboration space come to serve such needs. The more formal a projection is, easiness-of-use is reduced; actions permitted are less intuitive and more time consuming. Formality is associated with fixed types of actions, as well as explicit relationships between them. However, a switch to a more formal projection is highly desirable when (some members of) a community need to further elaborate the data and knowledge items considered so far. Such functionalities are provided by projections that may enable the formal exploitation of collaboration items patterns and the deployment of appropriate formal argumentation and reasoning mechanisms. A switch to a projection of a higher level of formality disregards less meaningful data and knowledge items, resulting to a more compact and tangible representation of the collaboration space (Figure 2). This effect is highly desirable in data-intensive situations.

4.2 Information Triage

Our solution builds extensively on the *information triage* process [11], i.e. the process of sorting and organizing through numerous relevant materials and organizing them to meet the task at hand. During such a process, users must effortlessly scan, locate, browse, update and structure knowledge resources that may be incomplete, while the resulting structures may be subject to rapid and numerous changes. Information triage related functionalities enable users to meaningfully organize the big volumes of data and knowledge items in a collaborative setting. The informal projection of a collaborative workspace in CoPe_it! is fully in line with the above. Drawing upon successful technologies coming from the area of spatial hypertext [11], the informal projection of CoPe_it! adopts a spatial metaphor to depict collaboration in a 2-dimensional space (Figure 3).

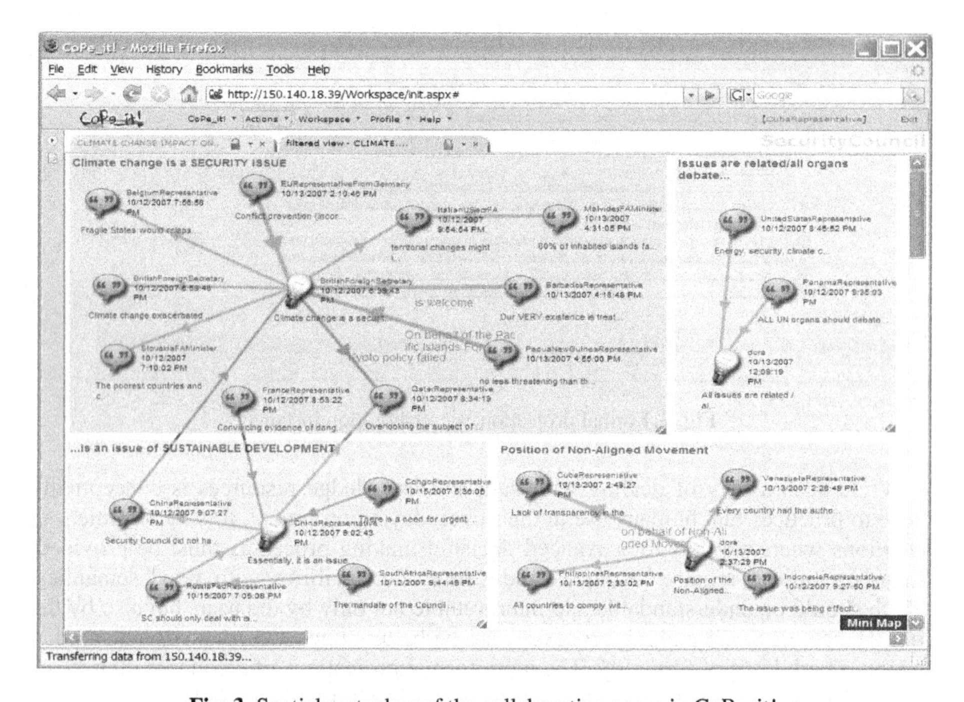

Fig. 3. Spatial metaphor of the collaboration space in CoPe_it!

In an informal projection, users are incrementally processing information and are not forced to predefined structural commitments. The related features and functionalities of CoPe_it! enable users to create and organize information by making use of spatial relationships and structures, giving them the freedom to express relationships among information items through spatial proximity and visual cues. Such cues are related to the linking of collaboration items (e.g. coloring and thickness of the respective links) and the drawing of colored rectangles to cluster related items.

As highlighted above, the informal projection of a collaborative workspace in CoPe_it! permits an ordinary and unconditioned evolution of data and knowledge structures. This projection also provides abstraction mechanisms that allow the creation of new abstractions out of existing ones. Abstraction mechanisms include: (i) annotation and metadata (i.e. the ability to annotate instances of various knowledge items and add or modify metadata); (ii) aggregation (i.e. the ability to group a set of data and knowledge items so as to be handled as a single conceptual entity); (iii) generalization/specialization (i.e. the ability to create semantically coarse or more detailed knowledge items in order to help users manage information pollution of the collaboration space); (iv) patterns (i.e. the ability to specify instances of interconnections between knowledge items of the same or a different type, and accordingly define collaboration templates). Information triage related activities can be conducted in CoPe_it! either collaboratively (a moderator may be required in some cases) or individually.

5 Evaluation Issues

CoPe_it! has been already introduced in diverse collaborative settings (from the management, engineering and learning domains) for a series of pilot applications. The results of the first evaluation phase were very encouraging. So far, the tool has been evaluated by 67 users. The above evaluation was conducted through questionnaires that contained: (i) two sets of closed-ended questions, aiming at evaluating the tool's 'perceived usefulness' and 'perceived ease-of-use', and (ii) a number of open-ended questions, through which users were asked to comment on the tool's advantages, disadvantages and/or limitations, as well as to suggest areas of improvement. A typical five-level Likert item (*strongly disagree, disagree, neither agree nor disagree, agree, strongly agree*) was used for the closed-ended questions.

With respect to the questions related to the tool's perceived usefulness, the percentage of the positive answers (sum of the 'strongly agree' and 'agree' percentages) in the majority of the questions was very promising. More specifically, 66.1% of the users confirmed that the tool helped them organize the collaboration efficiently, 73.6% that the tool was easy to learn, 71.1% that it was easy to use, 72.5% enjoyed its use, while 66.1% admitted that it was worth the effort. Furthermore, users have admitted that it stimulates interaction (63%), makes them more accountable for their contributions (64.3%), while it aids them to conceive, document and analyze the overall collaboration context in a holistic manner, by facilitating a shift from divergence to convergence (59.4%). However, users were skeptical about whether they will definitely consider the tool as their first choice for supporting their future collaboration (37.3%). Having further elaborated their answers to this last issue, we concluded that this was due to the change of the way they were accustomed to work.

Similar results were obtained for the questions concerning the tool's perceived ease-of-use. 82.6% of the users answered positively that were able to easily understand the tool's features and functionalities, 79.3% found it easy to use all available options, while 75% agreed that the achieved results (after a user's action) were clear. Nevertheless, only 52.3% could easily understand the contents of a workspace (we identified that this happens in data-intensive situations; efforts to provide more intuitive workspace icons are underway).

The open-ended questions revealed that users considered the ability of the tool to represent and manipulate the structure of an argumentative collaboration, along with its various visualization options, as its strongest features, setting it apart from Web-based forums. Respondents also commented positively on the tool's ability to provide multiple views of a particular collaborative session. When asked for the tool's disadvantages, respondents mentioned the cluttering of the workspace (basically due to the numerous arrows that appear in some workspaces), and the inability to make references from a workspace to another. With respect to improvements, most comments were around the need of providing awareness mechanisms that can inform on changes that happen within a workspace, the ability to reuse items between workspaces (by copy-pasting), and the integration of video/audio conference tools in order to enhance real time collaboration.

6 Conclusions

In this paper, we have presented how CoPe_it! provides advanced argumentative collaboration support on the Web. Drawing upon the understandings of how domesticated technologies mutate when they go feral on the Web, CoPe_it! opens up the semantics layer giving control of the formalization process to users participating in the collaboration. In CoPe_it! semantics are emergent and not predefined. By incrementally formalizing the discussion via projections, CoPe_it! is able to provide not only emergent semantics but also advanced decision making services. These aspects make the tool suitable for being deployed in today's challenging Web environment. Future work directions include the extensive evaluation of CoPe_it! in diverse contexts in order to shape our minds towards the development of additional means that are useful when such systems are deployed on the Web.

Acknowledgments. Research carried out in the context of this paper has been partially funded by the EU PALETTE (Pedagogically Sustained Adaptive Learning through the Exploitation of Tacit and Explicit Knowledge) Integrated Project (IST FP6-2004, Contract No. 028038).

References

1. Buckingham Shum, S.: Cohere: Towards Web 2.0 Argumentation. In: Proceedings of COMMA 2008, Toulouse, France, May 28-30, pp. 97–108 (2008)
2. Walker, J.: Feral Hypertext: when Hypertext Literature Escapes Control. In: Proceedings of the 16th ACM Conference on Hypertext and Hypermedia, Salzburg, Austria, September 06 – 09 (2005)
3. Grudin, J.: Evaluating Opportunities for Design Capture. In: Moran, T.P., Carroll, J.M. (eds.) Design Rationale: Concepts, Techniques and Use. Lawrence Erlbaum Associates, Mahwah (1996)
4. Hurwitz, R., Mallery, J.C.: The Open Meeting: A Web-Based System for Conferencing and Collaboration. In: Proceedings of the 4th International World Wide Web Conference, Boston, MA, December 11-14 (1995)
5. Buckingham Shum, S.: Design Argumentation as Design Rationale, The Encyclopedia of Computer Science and Technology, pp. 95–128. Marcel Dekker, Inc., New York (1996)
6. Nam, K., Ackerman, M.S.: Arkose: Reusing Informal Information from Online Discussions. In: Proceedings of the 2007 international ACM conference on Supporting Group Work, Sanibel Island, Florida, USA, November 04-07 (2007)

7. Checkland, P., Holwell, S.: Action Research: Its Nature and Validity. Systemic Practice and Action Research 11(1), 9–21 (1998)
8. Hevner, A.R., March, S.T., Park, J., Ram, S.: Design Science in Information Systems Research. MIS Quarterly 28(1), 75–105 (2004)
9. Karacapilidis, N., Tzagarakis, M.: Supporting Incremental Formalization in Collaborative Learning Environments. In: Duval, E., Klamma, R., Wolpers, M. (eds.) EC-TEL 2007. LNCS, vol. 4753, pp. 127–142. Springer, Heidelberg (2007)
10. : Karacapilidis, N., Tzagarakis, M., Karousos, N., Gkotsis, G., Kallistros, V., Christodoulou, S., Nousia, D., Mettouris, C., Kyriakou, P.: CoPe_it! - Supporting collaboration, enhancing learning. In: Proceedings of the 2009 International Conference on Information Resources Management (Conf-IRM 2009), Dubai, UAE, May 21-23 (2009)
11. Marshall, C., Shipman, F.M.: Spatial Hypertext and the Practice of Information Triage. In: Proc. 8th ACM Conference on Hypertext, pp. 124–133 (1997)

Semantic Service Search Engine (S3E): An Approach for Finding Services on the Web

Lemonia Giantsiou[1,2], Nikolaos Loutas[1,2], Vassilios Peristeras[1], and Konstantinos Tarabanis[2]

[1] National University of Ireland, Galway, Digital Enterprise Research Institute
firstname.lastname@deri.org
[2] Information Systems Lab, University of Macedonia, Thessaloniki, Greece
{lgiantsiou,nlout,kat}@uom.gr

Abstract. Currently, the Web is an important part of people's personal, professional and social life and millions of services are becoming available online. At the same time many efforts are made to semantically describe Web Services and several frameworks have been proposed, i.e. WSMO, SAWSDL etc. The Web follows a decentralized architecture, thus all the services are available at some location; but finding this location remains an open issue. Many efforts have been proposed to solve the service discovery problem but none of them took up. In this work, a lightweight approach for service discovery is proposed. Our approach comprises of three main phases. Firstly, during the crawling phase the semantic service descriptions are retrieved and stored locally. Afterwards, in the homogenization phase the semantics of every description are mapped to a service meta-model and the resulting triples are stored in a RDF repository. Finally, at the search phase, users are enabled to query the underlying repository and find online services.

Keywords: semantic, service, search, Web, meta-model.

1 Introduction

Currently a significant number of good practices and success stories exist, which prove that Service Oriented Architectures (SOAs) can be efficiently applied at the business level [1]. Nowadays, efforts to implement SOA at a Web scale are increasing. Towards this purpose, the term Web-Oriented Architecture (WOA) was coined to describe a lightweight version of SOA for the Web [2].

Web service standards, such as WSDL and UDDI, support SOA and enable many use cases. UDDI registries play the role of the service brokers in SOA environments. However, the service descriptions that are stored in UDDI registries are simple textual descriptions that lack rich expressivity. There are a lot of doubts of whether and how UDDI registries can be applied to the Web. According to [3], "it is unreasonable to assume that there would be a single global registry containing all of the information required to find and interact with businesses throughout the world". Additionally, the shutdown of the UDDI public registry by Microsoft [4], along with IBM and SAP,

M.D. Lytras et al. (Eds.): WSKS 2009, LNAI 5736, pp. 316–325, 2009.

reinforces this argument. Both the fact that the service providers didn't publish their services in centralized registries and the poor descriptions of the published services forced the research community to come up with other, less centralized, ways for publishing and storing services on the Web.

Currently, according to the Web service search engine Seekda [5] there are more than 27,000 Web services from more than 7,000 service providers available on the Web. The main problem that arises and needs to be resolved is how a user can find a service without having to know exactly where this service is available from. Hence, a solution for services' discovery that takes into consideration the Web's decentralized architecture is required. Such a solution should address the following requirements:

- *No need for publishing the services.* The approach should not require from the service providers to publish their services neither in a specific way nor in a centralized registry. As mentioned earlier, such practices did not take up.
- *Rich service descriptions.* Specific service processes, such as service discovery and composition, should be facilitated by the provision of service descriptions enriched with semantics.

The need for a rich and formal definition of a service including semantics led to the emerging and growth of Semantic Web Services (SWS) models. The system that is proposed in this paper enables the collection, processing and analyzing of Semantic Web Services. The prototype that is developed as a proof-of-concept of our approach processes two types of services, namely SAREST [6] and SAWSDL [7] services. We chose these types of services mainly for their strong connection with already accepted standards, such as WSDL, RDFa [8] and GRDDL [9]. More specifically, the system facilitates the collection of SAWSDL and SAREST services, the extraction of the semantic information, the translation of this information to a service meta-model and, finally, the analysis of this information.

The remainder of this paper is organized as follows. In Section 2 we present briefly the related efforts and provide the main points that differentiate our work. Section 3 discusses the proposed service meta-model. In Section 4 we present the overall system architecture and we focus on the back-end subsystem presentation. Section 5 highlights the decisions that were made for the testing and evaluation processes. Finally, Section 6 concludes the paper and discusses the future work.

2 Related Work

Our work has been mainly influenced by related research efforts in the fields of Web service search engines and search engines that process semantic annotations and microformats. Thus, this section is organized accordingly. At the end of this section we discuss how our work differentiates from and builds upon these related efforts.

2.1 Web Service Search Engines

The search engines that aim at facilitating Web services' discovery can be classified into three main categories:

- Search engines that are based on the crawling of the Web services' descriptions;
- Search engines that base their discovery mechanism on the use of ontologies;
- Search engines that use general-purpose search engines for finding services.

With respect to the first category, in [17], the authors propose a WSDL-based Web service discovery mechanism that retrieves relevant Web services from the Web. Their solution exploits well-established technologies from research fields like Information Retrieval and Nature Language Processing. There are concrete issues that led previous attempts to set up a Web service brokerage system to fail, such as the failure of current technologies to successfully support the publishing and finding procedure of the Web services at the Web level. Towards this direction, in [18] the authors argue that registry-based approaches, like UDDI, create a number of problems, which may complicate the sharing of Web services. Thus, the adoption of a single format that is used by a specific registry is considered to be of high risk.

Seekda [5] facilitates the on-demand use of services over the Web providing the means for crawling Web services. Thus, there is no need for publishing Web services at a central place. Actually, the service providers are only required to have their services available somewhere on the Web and afterwards these are collected by Seekda's crawler. However, Seekda does not take into consideration possible semantic information embedded in the service description. Woogle [19] is an online directory of Web services and Web service resources, which crawls the Web in order to find WSDL service descriptions. Afterwards, by processing these service descriptions it clusters the services according to their domain (i.e. weather, sports) and the place from where they are available.

As mentioned before, there are approaches where an ontology is used to support the services' discovery. In [20], Bin et al. propose a method for searching Web services based on a domain travel ontology aiming at finding domain related Web services. Following a similar approach, Condack et al. [21] present a software engineering tool (Swell) for searching Semantic Web Services.

In their work Daewook et al. [22] present a semantic search engine for Web services that can improve the recall and precision using both a clustering and an ontology approach. Additionally, Syeda-Mahmood et al. [23] use domain-independent and domain-specific ontologies in order to find matching service descriptions.

A potential for discovering Web services using search engines is explored in [24]. They identify key differences between Web services and Web pages. In [25] Song et al. investigate a similar approach where they use general-purpose search engines, like Google and Yahoo/Altavista, to discover Web services. The results showed that the general-purpose search engines can be used to find Web services on the Internet.

2.2 Semantic Web Search Engines

In [26] a categorization scheme for Semantic Web search engines is introduced and elaborated. There are two kinds of Semantic Web search engines; engines specific to the Semantic Web documents (Ontology Search Engines) and engines that try to improve the search results using Semantic Web standards and languages (Semantic Search Engines).

Swoogle [27] belongs to the first category. It is a crawler-based system for indexing and retrieving Semantic Web documents (SWDs) in RDF or OWL. The system is comprised of a set of crawlers for searching URLs that lead to SWDs. Sindice [28] also belongs to the first category as it crawls the Semantic Web and indexes the retrieved resources. It maintains a list of resources and the set of source files where these resources were found.

SWSE [29] belongs to the second category. It is based on an entity-centric model that allows for describing entities, such as people, locations and generally speaking resources, rather than just documents. Consequently, the search results are structured descriptions of entities. It scales upon billions of RDF statements and aims at providing search and query functionalities over data integrated from a large number of sources. We use in our work the SWSE's Fetch Module.

2.3 Our Approach

Summarizing, there are significant efforts for discovering Web services and Semantic Web content. However, the search mechanisms presented are based on the descriptions of the services without exploiting the availability of semantic information in them. On the contrary, our approach takes into consideration this information and aims at improving search and identifying potential relations between services.

More specifically, our approach differentiates in four main ways. Firstly, we aim at providing a system that detects Semantic Web Services and facilitates the extraction and the utilization of the semantic information. Furthermore, our architecture is independent of the semantic service models. Hence, the services that are crawled may follow any semantic service model.

Additionally, we propose a service meta-model, which emerges from the analysis of the common elements of the existing service models. The mapping of the semantic information to a service meta-model introduces a set of benefits, which will be further discussed in Section 3. Finally, we enrich this common model by adding elements that represent the end user's perspective of the service, namely social elements.

3 Service Meta-Model

The need for a rich and formal definition of a service including semantics led to the emerging and growth of Semantic Web Services (SWS) models, such as OWL-S (originally DAML-S) [10], the Semantic Web Services Framework (SWSF) [11], WSDL-S [12], the Web Service Modeling Ontology (WSMO) [13], the Internet Reasoning Service (IRS) [14], and WSMO-Lite [15]. These models target at enhancing Web services with rich formal descriptions of their behavioral and functional characteristics, such as service capabilities and interfaces, and service inputs and outputs. Nowadays, lightweight models, such as SAREST [6] and SAWSDL [7], are becoming very popular. These efforts allow the inclusion of semantics into service descriptions created using standards, such as WSDL or XHTML. The main reason that caused the emerging of these mechanisms is the need for lightweight approaches for describing a service on the Web. We aim at homogenizing these models by mapping them to a common service meta-model.

A service can be seen from different perspectives starting from a complex business process and going down to software component. Therefore, in order to develop a service meta-model, we studied the service models/ontologies being based on the five types of service contracts as these are presented in [15]. These are the Information Model, the Functional Descriptions, the Non-Functional Descriptions, the Behavioral Descriptions and the Technical Descriptions, and can be considered as complementary parts of a service description.

The service meta-model is heavily influenced by lightweight approaches for semantically annotating service, such as SAREST. In addition to the five service contracts, the service meta-model defines a sixth contract, which is that of the Social Contract, as well as a blank placeholder that can be used for possible extensions of the meta-model. An overview of the service meta-model is presented in Table 1.

Table 1. The Service Meta-Model

Information Model	Ontologies encoded in any language
Functional Descriptions	Input, Output, Operation
Non-Functional Descriptions	ServiceProvider, ServiceDomain, ReferenceClass
Behavioral Descriptions	*No explicit definition*
Technical Descriptions	LiftingSchema, LoweringSchema, Protocol
Social Contract	Folksonomy which represents the user-defined annotations
Blank Placeholder	*Placeholder for future use*

The Information Model can be encoded in any ontological language aiming at decoupling the service description from a specific ontological language. The Functional Descriptions of a service are mapped to three elements, namely the *Input*, the *Output* and the *Operation* which represent the inputs, the outputs and the operations of a service respectively. The value of these elements may be either a literal or a concept defined in an external ontology.

At this point, it is important to highlight the fact that the Behavioral Descriptions are deliberately omitted, as they add complexity and they do not really contribute to a lightweight discovery approach. The service meta-model provides three elements for the Non-Functional Descriptions. These are the *ServiceProvider*, the *ServiceDomain* and the *ReferenceClass*. The *ReferenceClass* element is very important as it facilitates the association of a URI with a concept of an ontology. The Technical Descriptions help to add features related to the execution of a service, such as the lifting and lowering schemata and the communication protocol.

We extend the main concepts defined in the current research efforts by adding the Social Contract [16]. We make the assumption, that in a time when everyone advocates active user participation on the Web, users will be enabled to add meaning and describe the services that they use, similarly to what they do for photos or videos. In that way the semantic descriptions of services would not come solely from the service providers, but they will emerge from the usage of the service, thus giving a social aspect to service annotation.

The service meta-model can be easily combined with technologies that facilitate the embedding of semantic information in XHTML pages, i.e. RDFa and GRDDL.

There are a lot of benefits from homogenizing the existing service models and mapping them to a common model. Some of these benefits that are applicable to our system are listed below.

- *Services' interoperability*. The existence of a service meta-model allows the interconnection and interaction of services that were developed using different technologies.
- *Potential execution sequences of services*. The mapping of the service descriptions to a common model enables the detection of output-input relations between services that were modeled following different frameworks. In that way, a potential execution sequence of services can be identified.
- *Services browsing according to the model's main elements*. The services can be browsed based on the elements that comprise their service description, i.e. browse service by input and/or output.
- *Services' clustering*. The application of similarity measures and consequently the identification of relatedness among different services are enabled. As a consequence, the formulation of clusters of services that are expressed in different languages is facilitated.

4 System Architecture

We propose an architecture for services' discovery that is modular, extendable and independent of the underlying semantic service models.

The system can be divided into two subsystems; the back-end subsystem and the front-end subsystem. The back-end subsystem is responsible for collecting, processing and analyzing the services, while the front-end subsystem enables the application of queries over this information. The two subsystems are connected via the repository component. In this paper, we will focus on the functionalities provided by the back-end system, and we will discuss particularly the Crawler. In Fig. 1 the overall system architecture is presented.

The back-end subsystem is responsible for the collection of the services' descriptions, the extraction of the semantic information, the translation of this information to the service meta-model and the analysis of this information. More specifically, the subsystem is comprised by the Crawler and the Analysis Component. In our prototype, we crawl service descriptions following the SAWSDL and SAREST models. In the section that follows we present the Crawler component.

4.1 Crawler

The crawler is the core component of the back-end subsystem as it encapsulates the fetching, detecting, transforming, storing and extracting processes. The crawler is conceptually organized in four layers, namely the WWW Layer, the Content Analysis Layer, the Service Meta-Model Layer and the Repository Layer.

WWW Layer. The WWW Layer consists of the Fetching and Extract Modules. These two modules handle a queue with the URIs of the Web pages that are going to be processed. Initially, this queue is populated with a set of URIs. We use a set of services in order to trigger the crawling process.

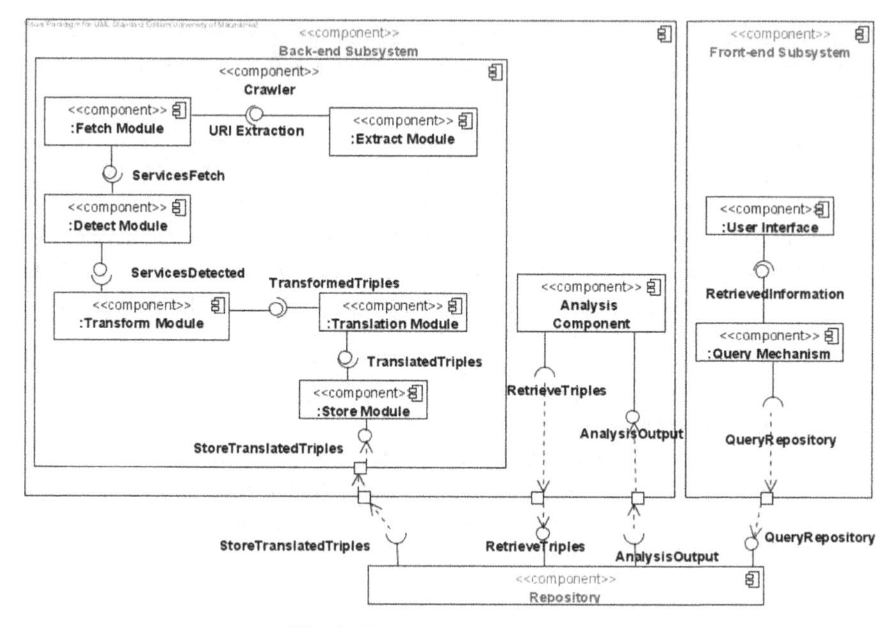

Fig. 1. The overall system architecture

The Fetching Module extracts a URI from the queue and checks the robots.txt file for this URI. The robots.txt file on a Website specifies whether all or part of the Website is accessible by Web crawlers. In case of absence of this file, then the Website is publicly available. This first check aims to specify whether the Web page can be processed. Afterwards, the Fetching Module fetches and stores locally the content and header information for this Web page. We use SWSE's Fetch Module [29] for this process.

The Extract module is responsible for extracting URIs from the Web page that is being processed and feed these URIs to the queue of the URIs. It has been developed in order to be able to take as parameters some restrictions regarding the URIs that will be added in the queue. For instance, one of these restrictions may be related to the extension of the extracted URIs (e.g. it may allow only URIs with .gov extension).

Content Analysis Layer. The Content Analysis Layer consists of the *Detect* and *Transform* modules. The Detect module is responsible for specifying the type of the file that is going to be processed and the existence of semantic information in it. At the moment, the system handles SAREST and SAWSDL services. Consequently, the file types that our system processes are (X)HTML and WSDL files.

The Detect module firstly tries to identify the type of the file from the extension of the URI. If this is not feasible, it uses the content information that was previously stored by the Fetching module and tries to detect the file type from the content type information. If the content type corresponds to HTML files, then the recognition is accomplished. If the file is an XML file, then we parse the file using a DOM parser to investigate whether this file is a WSDL file. When the file type recognition is completed, the Detect module parses the file, using an HTML parser for the HTML files or a DOM parser for WSDL files, and explores whether there is semantic information in it. If all checks are successful, then the file is passed to the Transform module.

The Transform module parses the file and extracts the semantic information in the format of RDF triples. These triples are formed based on the semantic service model that was used in the service description. For instance, if this model is SAREST then the predicates of these triples are the elements of the SAREST model. Afterwards, we translate these predicates, to the service meta-model by forwarding the triples to the Translation Module.

Service Meta-Model Layer. The Service Meta-Model Layer includes the *Translation* Module, which is responsible for the conversion of the predicates of the triples coming from the Transform Module to the elements that are specified in the service meta-model. The Translation Module depends on the service model specification. Actually, there is one Translation Module for each service, i.e. one for SAREST and one for SAWSDL, and it implements the mapping between the model's elements to the service meta-model's elements.

Repository Layer. The Repository Layer consists of the *Store* Module, which communicates with the RDF repository where the translated RDF triples are stored. All the triples that exist at any time in the RDF repository follow the service meta-model, namely their predicates refer to elements of the meta-model.

5 Evaluation and Validation

The existence of SAWSDL and SAREST services in the Web is limited. Both standards are relatively new, thus the services that are available are mainly sample services. However, both service models are supported by the research community, i.e. SAWSDL has already become a W3C Recommendation. W3C offers a SAWSDL test suite [30] and in [6] SAREST sample services are provided. Thus, in order to test and consequently evaluate our system we had to create our own dataset. Towards that purpose, apart from the sample services that are available online, we used a set of e-government services that were annotated using elements specified in an e-government domain ontology. Summarizing, in order to test and evaluate the back-end subsystem we used all the SAWSDL and SAREST services that are currently available on the Web along with the sample services that we have created. The crawler managed to process properly all of these services. Table 2 summarizes the tests that were performed and their results.

With respect to the validity of the proposed architecture, it is important to mention that it complies with the reference architecture for Semantic Web applications of [31]. More specifically, all of the system components are mapped to the components of the reference architecture, while the crawler is considered to be the core component enabling the discovery and the retrieval of remote data. Furthermore, the data and the user interface components of the reference architecture map to the Query Mechanism and the User Interface components of our architecture.

Table 2. Crawler's testing results

Module	Measurement Process	Results
Fetch	Check whether the Fetch Module manages to fetch and store locally the Web documents taking into consideration possible limitations coming from the robots.txt file.	All the URIs that are given as input to the module were fetched properly.
Detect	Check whether the detected file type is the same with the actual type.	Successful detection of file types.
Extract	Check if all the URIs are extracted by the Web document that is being processed.	The expected set of extracted URIs is the same with the actual set.
Transform/ Translate	Check if the transformation and the translation of the service descriptions into RDF triples is performed without any information loss.	Proper semantic information retrieval and conversion.
Store	Check whether the RDF triples are stored into RDF repository and whether their update after the crawling process is performed properly.	Both the storing and the updating processes are performed as expected.

6 Conclusions and Future Work

An approach for finding services on the Web, which does not require services to be published centrally and exploits the potential existence of semantic information in the service descriptions, has been presented in this work. Our Semantic Service Search Engine facilitates the collection, processing and analysis of Semantic Web Services. At the same time it remains independent of the various semantic service models.

In the context of our future work, we plan to extend the prototype by supporting more SWS models. Furthermore, we developed a front-end subsystem that serves as a user interface that supports searching and browsing of the collected services. We intend to extend and evaluate this front-end subsystem. Finally, scalability and performance issues remain open and will have to be researched.

References

1. Minglun, R., Xiaoying, A., Hongxiang, W.: Service oriented architecture for interorganizational IT resources sharing system. In: IEEE ICAL 2008, Qingdao, China (2008)
2. Hinchcliffe, D.: The SOA with reach: Web-Oriented Architecture (2006), http://blogs.zdnet.com/Hinchcliffe/?p=27
3. Papazoglou, M.P., Dubray, J.: A Survey of Web service technologies. Technical Report DIT-04-058, Informatica e Telecomunicazioni, University of Trento (2004)
4. Krill, P.: Microsoft, IBM, SAP discontinue UDDI registry effort (2005), http://www.infoworld.com/article/05/12/16/HNuddishut_1.html
5. Seekda (2009), http://seekda.com/
6. Sheth, A., Gomadam, K., Lathem, J.: SA-REST: Semantically Interoperable and Easier-to-Use Services and Mashups. IEEE Internet Computing 11(6), 91–94 (2007)
7. Kopecký, J., Vitvar, T., Bournez, C., Farrell, J.: SAWSDL: Semantic Annotations for WSDL and XML Schema. IEEE Internet Computing 11(6), 60–67 (2007)
8. RDFa Primer (2008), http://www.w3.org/TR/xhtml-rdfa-primer/
9. W3C GRDDL Working Group (2008), http://www.w3.org/2001/sw/grddl-wg/
10. Martin, D., et al.: OWL-S: Semantic Markup for Web Services. W3C Submission (2004)
11. Battle, S., et al.: Semantic Web Services Framework (SWSF). W3C Submission (2005)
12. Akkiraju, R., Farrell, J., Miller, J., Nagarajan, M., Schmidt, M., Sheth, A., Verma, K.: Web Service Semantics - WSDL-S. In: A joint UGA-IBM Technical Note, version 1.0 (2005)
13. Roman, D., et al.: Web Service Modeling Ontology. Applied Ontology 1(1) (2005)

14. Cabral, L., Domingue, J., Galizia, S., Gugliotta, A., Norton, B., Tanasescu, V., Pedrinaci, C.: IRS-III: A Broker for Semantic Web Services based Applications. In: Cruz, I., Decker, S., Allemang, D., Preist, C., Schwabe, D., Mika, P., Uschold, M., Aroyo, L.M. (eds.) ISWC 2006. LNCS, vol. 4273, pp. 201–214. Springer, Heidelberg (2006)
15. Vitvar, T., Kopecky, J., Fensel, D.: WSMO-Lite: Lightweight Semantic Descriptions for Services on the Web. In: CMS WG Working Draft (2008)
16. Loutas, N., Peristeras, V., Tarabanis, K.: Extending Service Models to Include Social Metadata. In: Proceedings of the WebSci 2009: Society On-Line, Athens, Greece (2009)
17. Chen, W., Chang, E.: Searching Services on the Web: A Public Web Services Discovery Approach. In: Third International IEEE SITIS 2007, Shanghai, China (2007)
18. Willmott, S., Ronsdorf, H., Krempels, K.H.: Publish and search versus registries for semantic Web service discovery. In: Proceedings of the IEEE/WIC/ACM International Conference on Web Intelligence, Compiegne University of Technology, France (2005)
19. Dong, X., Halevy, A., Madhavan, J., Nemes, E., Zhang, Z.: Similarity search for Web services. In: Proceedings of the Thirtieth international conference on Very large data bases 30, pp. 372–383 (2004)
20. Bin, X., Yan, W., Po, Z., Juanzi, L.: Web Services Searching based on Domain Ontology. In: IEEE Workshop on Service-Oriented System Engineering, Beijing, China (2005)
21. Condack, J., Schwabe, D.: Swell - Annotating and Searching Semantic Web Services. In: Third Latin American Web Congress (LA-WEB 2005), Buenos Aires, Argentina (2005)
22. Daewook, L., Joonho, K., SeungHoon, Y., Sukho, L.: Improvement of the Recall and the Precision for Semantic Web Services Search. In: 6th IEEE/ACIS International Conference on Computer and Information Science (ICIS 2007), Montréal, Québec, Canada (2007)
23. Syeda-Mahmood, T., Shah, G., Akkiraju, R., Ivan, A.-A., Goodwin, R.: Searching service repositories by combining semantic and ontological matching. In: IEEE International Conference on Web Services (ICWS 2005), Orlando, Florida, USA (2005)
24. Al-Masri, E., Mahmoud, O.H.: Discovering Web Services in Search Engines. IEEE Internet Computing 12(3), 74–77 (2008)
25. Song, H., Cheng, D., Messer, A., Kalasapur, S.: Web Service Discovery Using General-Purpose Search Engines. In: IEEE ICWS 2007, Salt Lake City, Utah, USA (2007)
26. Esmaili, K.S., Abolhassani, H.: A Categorization Scheme for Semantic Web Search Engines. In: IEEE International Conference on Computer Systems and Applications, Dubai/Sharjah, UAE (2006)
27. Swoogle (2007), http://swoogle.umbc.edu/
28. Oren, E., Delbru, R., Catasta, M., Cyganiak, R., Stenzhorn, H., Tummarello, G.: Sindice.com: A document-oriented lookup index for open linked data. International Journal of Metadata, Semantics and Ontologies 3(1) (2008)
29. Harth, A., Umbrich, J., Decker, S.: MultiCrawler: A Pipelined Architecture for Crawling and Indexing Semantic Web Data. In: Cruz, I., Decker, S., Allemang, D., Preist, C., Schwabe, D., Mika, P., Uschold, M., Aroyo, L.M. (eds.) ISWC 2006. LNCS, vol. 4273, pp. 258–271. Springer, Heidelberg (2006)
30. SAWSDL Test Suite (2007), http://www.w3.org/2002/ws/sawsdl/CR/testsuite.html
31. Heitmann, B., Hayes, C., Oren, E.: Towards a reference architecture for Semantic Web applications. In: Proceedings of the WebSci 2009: Society On-Line, Athens, Greece (2009)

Towards a Mediator Based on OWL and SPARQL

Konstantinos Makris, Nikos Bikakis, Nektarios Gioldasis, Chrisa Tsinaraki, and Stavros Christodoulakis

Technical University of Crete, Department of Electronic and Computer Engineering
Laboratory of Distributed Multimedia Information Systems & Applications (MUSIC/TUC)
University Campus, 73100, Kounoupidiana Chania, Greece
{makris,nbikakis,nektarios,chrisa,stavros}@ced.tuc.gr

Abstract. We propose a framework that supports a federated environment based on a Mediator Architecture in the Semantic Web. The Mediator supports mappings between the OWL Ontology of the Mediator and the other ontologies in the federated sites. SPARQL queries submitted to the Mediator are decomposed and reformulated to SPARQL queries to the federated sites. The evaluated results return to the Mediator. In this paper we describe the mappings definition and encoding. We also discuss briefly the reformulation approach that is used by the Mediator system that we are currently implementing.

Keywords: Information Integration, Semantic Web, Interoperability, Ontology Mapping, Query Reformulation, SPARQL, OWL.

1 Introduction

The Semantic Web community has developed over the last few years standardized languages for describing ontologies and for querying OWL [1] based information systems. Database implementations are striving to achieve high performance for the stored semantic information. In the future large databases, RDF [2] data will be managed by independent organizations. Federated architectures will need to access and integrate information from those resources.

We consider in this paper a Mediator based architecture for integrating information from federated OWL knowledge bases. The Mediator uses mappings between the OWL Ontology of the Mediator and the Federated site ontologies. SPARQL [3] queries over the Mediator are decomposed and reformulated to be submitted over the federated sites. The SPARQL queries are locally evaluated and the results return to the Mediator site.

We describe in this paper the mappings supported by our architecture as well as the SPARQL reformulation algorithms supported by our system.

Ontology Mapping in general, is a topic that has been studied extensively in the literature [13, 14, 15, 16, 17, 18, 19, 20, 21, 22, 23, and 24]. However, very few publications, in our knowledge, examine the problem of describing the mapping types that can be useful for the SPARQL query reformulation process and how they should be exploited. Only [7] deals with this subject but not directly, since it describes which mapping types cannot be used in the reformulation process.

M.D. Lytras et al. (Eds.): WSKS 2009, LNAI 5736, pp. 326–335, 2009.

Query Reformulation is a frequently used approach in *Query Processing* and especially in *Information Integration* environments. Much research has been done in the area of query reformulation; Up to now, though, limited studies have been made in the field of SPARQL query reformulation related to posing a query over different datasets. Some relevant work has been published for approximate query reformulation, based on an ontology mapping specification [12], using a part of OWL-DL, without specifying a particular query language. In addition, many approaches that deal with *optimization* [8, 9], *decomposition* [4, 5], *translation* [10, 11] and *rewriting* (in order to benefit from inference) [6] of a SPARQL query, have been published.

The rest of the paper is structured as follows: In section 2, we present a motivating example. Then, the patterns of the proposed correspondences and the mapping types are outlined in section 3, while the language that is used for mapping representation is discussed in section 4. The SPARQL query reformulation process is described in section 5 and the paper concludes in section 6.

2 Motivating Example

We present in this section a motivating example. In Fig. 1, we show the structure of two different ontologies. The source ontology describes a store that sells various products including books and cd's and the target ontology describes a bookstore.

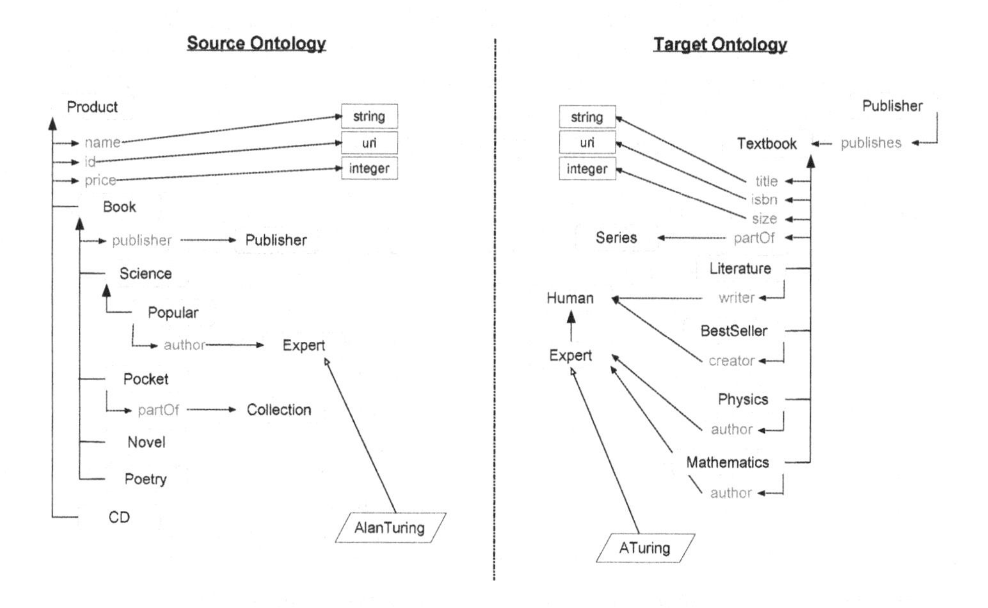

Fig. 1. Semantically Overlapping Ontologies. The notation is based on [24]. The rounded corner boxes represent the classes followed by their properties, the parallelogram boxes at the bottom express individuals, the rectangle boxes express the datatypes and finally, the arrows express the relationships between those basic constructs of OWL.

In Fig. 1, we observe some correspondences between the two ontologies. For example, the class "Book" in the source ontology seems to describe the same individuals with the class "Textbook" in the target ontology (equivalence relationship). In addition, correspondences like the one between the class "Collection" and the class "Series", or the one between the datatype property "name" and the datatype property "title" can be thought as more general (subsumption relationships).

Apart from the obvious correspondences, we observe more complex ones such as those between the class "Science" and the union of the classes "Physics" and "Mathematics", and the one between the class "Pocket" and the class "Textbook" restricted in its "size" property values.

3 Mapping Types

In this section we define the set of mapping types between the Mediator and the target ontologies. These mapping types are used for the SPARQL query reformulation process.

The basic concepts of OWL, whose mappings are useful for the reformulation process, are the classes (denoted as "*c*"), the object properties (denoted as "*op*"), the datatype properties (denoted as "*dp*") and the individuals (denoted as "*i*"). Since we deal with SPARQL queries, some mapping types may not be useful for the query reformulation process. For example, a mapping between an individual of the source ontology and a concatenation of two different individuals of the target ontology would be meaningless, since they cannot be represented in SPARQL. Such types of mappings are described in [7] and many of them could be useful for processing the query results but not during the query reformulation and query answering process.

In order to define the mapping types that can be useful for the reformulation process, we use the set of symbols presented in Table 1.

Table 1. The notation used to define the different mapping types and concept/role expressions

Symbol	Notation
$\sqsubseteq, \sqsupseteq$	inclusion operators
$\equiv$	equality operator
$\sqcap$	intersection operator
$\sqcup$	union operator
$\neg$	negation operator
$\emptyset$	empty set
$\mid$	logical or
.	used to specify a sequence of binded concepts/roles
domain(c)	property domain restriction to the values of a class c
range(c)	property range restriction to the values of a class c
inverse(p)	inverse of a property p

We consider that a class expression (denoted as "*CE*") from the source ontology can be mapped to a class expression from the target ontology.

$$\textit{Class Expression Mapping: } \text{CE rel CE, rel:=} \equiv \mid \sqsubseteq \mid \sqsupseteq \qquad (1)$$

As a *class expression*, we denote any complex expression between two or more classes, using disjunctions, conjunctions or both. Any class that participates in the class expression can be restricted by the value of one or more properties, attached to it directly or indirectly (directly in some of its own properties, or indirectly in some property of an associated class) using a path (provided that a path connecting the class and the desired property already exists).

$$\text{CE::= c} \mid \text{c.R} \mid \text{CE} \sqcup \text{CE} \mid \text{CE} \sqcap \text{CE},$$

$$\text{R::= R'} \mid \neg\text{R'} \mid \text{R} \sqcup \text{R} \mid \text{R} \sqcap \text{R},$$

$$\text{R'::= P opr V, opr:= !=} \mid = \mid \leq \mid \geq \mid < \mid >$$

R denotes the *restrictions* applied in a class, while *R'* stands for the restriction of a property path *P* in a possible value *V*. *V* can be either a data value or an individual. The operators that can be used in a property path restriction differ according to the type of *V*. In order to define a restriction on a data value all the above operators can be used (!=, =, $\leq$, $\geq$, $<$, $>$). But, in order to define a restriction on an individual only the !=, = operators can be used.

A *property path P* is a sequence of object properties (possibly empty) ending with a datatype/object property. It relates the class that we want to restrict with the property (object or datatype) in which the restriction should be applied.

$$\text{P::= P'} \mid \text{dp} \mid \text{P'.dp}$$

$$\text{P'::= } \emptyset \mid \text{op} \mid \text{P'.op}$$

Accordingly, an object property expression (denoted as "*OPE*") from the source ontology can be mapped to an object property expression from the target ontology.

$$\textit{Object Property Expression Mapping: } \text{OPE rel OPE} \qquad (2)$$

As an *object property expression*, we denote any complex expression between two or more object properties, using disjunctions, conjunctions or both. It is also possible for the inverse of an object property to participate in the object property expression of the target class. Any object property that participates in the object property expression can be restricted on its domain or range values, using the same type of restrictions with those described for the class expressions.

$$\text{OPE::= op} \mid \text{OPE} \sqcup \text{OPE} \mid \text{OPE} \sqcap \text{OPE} \mid \text{OPE} \sqcap \text{domain(CE)} \mid \text{OPE} \sqcap \text{range(CE)} \mid \text{inverse(OPE)}$$

Similarly, a datatype property expression (denoted as "*DPE*") from the source ontology can be mapped to a datatype property expression from the target ontology.

$$\textit{Datatype Property Expression Mapping: } \text{DPE rel DPE} \qquad (3)$$

As a *datatype property expression*, we denote any complex expression between two or more datatype properties, using disjunctions, conjunctions or both. Any datatype property that participates in the datatype property expression can be restricted on its domain values.

$$DPE ::= dp \mid DPE \sqcup DPE \mid DPE \sqcap DPE \mid DPE \sqcap domain(CE)$$

Finally, an individual from the source ontology (denoted as "i_s") can be mapped with an individual from the target ontology (denoted as "i_t").

$$\textit{Individual Mapping: } i_s \equiv i_t \quad (4)$$

We mention here that the equivalence between two different properties or property expressions denotes equivalence between the domains and ranges of those properties or property expressions. Similarly, the subsumption relations between two different properties or property expressions denote analogous relations between the domains and ranges of those two properties or property expressions.

4 Mapping Representation

The language that we use in order to represent the mappings between two overlapping ontologies has been defined in [20, 21]. It combines the alignment format [18], a format used to represent the output of ontology matching algorithms, and the OMWG mapping language [22] an expressive ontology alignment language. The expressiveness, the simplicity, the Semantic Web compliance (given its RDF syntax) and the capability of using any kind of ontology language are the key features of this language.

Below, we list a possible set of correspondences, using first-order logic expressions for the ontologies presented in Fig. 1 and afterwards we provide an example showing the mapping representation of a specific correspondence using the language that we mentioned above.

Type (1) mappings:

a. $\forall x, [Book(x) \Leftrightarrow Textbook(x)]$

b. $\forall x, [Publisher(x) \Leftrightarrow Publisher(x)]$

c. $\forall x, [Expert(x) \Leftrightarrow Expert(x)]$

d. $\forall x, [Product(x) \Rightarrow Textbook(x)]$

e. $\forall x, [Collection(x) \Leftarrow Series(x)]$

f. $\forall x, [Science(x) \Leftrightarrow (Physics(x) \lor Mathematics(x))]$

g. $\forall x, [Popular(x) \Leftrightarrow ((Physics(x) \lor Mathematics(x)) \land BestSeller(x))]$

h. $\forall x, [Pocket(x) \Leftrightarrow (Textbook(x) \land \exists y; [size(x, y) \land y \leq 14])]$

i. $\forall x, [(Novel(x) \lor Poetry(x) \Leftrightarrow Literature(x))]$

Type (2) mappings:

j. $\forall x, \forall y\ [publisher(x, y) \Leftrightarrow publishes(y, x)]$

k. $\forall x, \forall y\ [author(x, y) \Leftrightarrow (author(x, y) \wedge creator(x, y))]$ [1]

l. $\forall x, \forall y\ [partOf(x, y) \Leftrightarrow (partOf(x, y) \wedge \exists z; [size(x, z) \wedge z \leq 14])]$

Type (3) mappings:

m. $\forall x, \forall y\ [name(x, y) \Rightarrow title(x, y)]$

n. $\forall x, \forall y\ [id(x, y) \Rightarrow isbn(x, y)]$

Type (4) mappings:

o. AlanTuring = ATuring

The representation of mapping **f** (presented above) using the language that was discussed in this section is shown in Fig. 2.

```
<?xml version="1.0" encoding="utf-8" standalone="no"?>
...
<rdf:RDF xmlns="http://www.omwg.org/TR/d7/ontology/alignment/" ...>
  <Alignment rdf:about="http://example.com/ ">
    <xml>yes</xml>
    ...
    <onto1>
      <Ontology rdf:about="&Source;"/>
      ...
    </onto1>
    <onto2>
      <Ontology rdf:about="&Target;"/>
      ...
    </onto2>
    <map>
      ...
      <Cell>
        <entity1>
          <Class rdf:about="&Source;Science"/>
        </entity1>
        <entity2>
          <Class>
            <or rdf:parseType="Collection">
              <Class rdf:about="&Target;Physics"/>
              <Class rdf:about="&Target;Mathematics"/>
            </or>
          </Class>
        </entity2>
        <relation rdf:resource="&omwg;equivalence"/>
      </Cell>
      ...
    </map>
  </Alignment>
</rdf:RDF>
```

Fig. 2. The representation of mapping **f**

[1] The object property "author" in the target ontology consists of two different domains. According to the semantics of OWL, this means that the domain of the object property "author" is actually the union of the classes "Physics" and "Mathematics".

5 SPARQL Query Reformulation

In this section we provide an overview of the SPARQL query reformulation process, using a predefined set of mappings that follows the different mapping types described in section 3.

The SPARQL query reformulation process is based on the query's *graph pattern* reformulation and is consequently independent of the query type (*Ask, Select, Construct, Describe*). The SPARQL solution modifiers (*Limit, Offset, Order By, Distinct, Reduce*) are not taken into consideration since they do not affect the reformulation process.

In order to reformulate the *graph pattern* of a SPARQL query using *1:N* mappings (mappings between a basic OWL concept of the source ontology and a complex expression, among basic OWL concepts, of the target ontology), the reformulation algorithm traverses the execution tree in a bottom up approach, taking each triple pattern of the *graph pattern* and checking for existing mappings of the subject, predicate and object part in the predefined mappings set. Finally, it reformulates the triple pattern according to those mappings.

In case of *N:M* mapping utilization (mappings between a complex expression of the source ontology and a complex expression of the target ontology), the total graph pattern must be parsed in order to discover the predefined complex mapping and then, the algorithm must produce the required combination of triple patterns, based on this mapping.

The SPARQL graph pattern operators (*AND, UNION* and *OPTIONAL*), do not result in modifications during the reformulation process.

The reformulation of the *FILTER* expressions is performed by reformulating the existing *IRIs* that refer to a class, property, or individual, according to the specified mappings. The SPARQL variables, literal constants, operators (*&&*, ||, *!*, =, *!=*, >, <, >=, <=, +, -, *, */*) and built-in functions (e.g. *bound, isIRI, isLiteral, datatype, lang, str, regex*) that may occur in a *FILTER* expression remain the same during the reformulation process.

Finally, in case that more than one mappings are specified for a given class or property expression of the source ontology, the reformulation algorithm chooses the one that produces the most efficient reformulated query. Moreover, for efficiency reasons, a graph pattern normalization step is applied in parallel to the reformulation process, similarly with the one that is described in our SPARQL2XQuery [11] framework.

5.1 Reformulation Examples[2, 3]

We briefly present in this section the SPARQL reformulation process, using a set of examples due to the space limitation. We assume that an initial SPARQL query is

---

[2] The graph pattern normalization step is not included in the examples presented here, in order to make the reformulation process more easily understandable.

[3] For the SPARQL query examples presented in this subsection we use the following prefixes:

```
PREFIX s: <http://example.com/Source.owl#>
PREFIX t: <http://example.com/Target.owl#>
PREFIX rdf: <http://www.w3.org/1999/02/22-rdf-syntax-ns#>
```

posed over the source ontology presented in Fig. 1 and is reformulated to a semantically equivalent query in order to be posed over the target ontology of Fig. 1, using the mappings specified in section 4.

Example 1: Consider the query posed over the source ontology: *"Return the titles of the pocket-sized scientific books"*. The SPARQL syntax of the source query and the reformulated query is shown in Fig.3. During the reformulation process the mappings *f), h)* and *m)* from section 4 are used.

Source Query:

```
SELECT ?name
WHERE{
  ?x s:name ?name.
  ?x rdf:type s:Science.
  ?x rdf:type s:Pocket.
}
```

Reformulated - Target Query:

```
SELECT ?name
WHERE{
  {?x t:title ?name.}
  {{?x rdf:type t:Physics.}
   UNION
   {?x rdf:type t:Mathematics.}}
  {?x rdf:type t:Textbook.
   ?x t:size ?size.
   FILTER (?size ≤ 14)}
}
```

Fig. 3. The source and reformulated queries of Example 1

Example 2: Consider the query posed over the source ontology: *"Return the titles of books that belong to the poetry or novel category"*. The SPARQL syntax of the source query and the reformulated query is shown in Fig.4. During the reformulation process the mappings *i)* and *m)* from section 4 are used.

Source Query:

```
SELECT ?name
WHERE{
  {?x s:name ?name.}
  {{?x rdf:type s:Poetry.}
   UNION
   {?x rdf:type s:Novel.}}
}
```

Reformulated - Target Query:

```
SELECT ?name
WHERE{
  ?x t:title ?name.
  ?x rdf:type t:Literature.
}
```

Fig. 4. The source and reformulated queries of Example 2

6 Conclusions

In this paper we presented the formal definition and the encoding of the mappings of a semantic based mediation framework (based on OWL/RDF knowledge representations and SPARQL queries) that we are currently developing. The framework is based on a set of mapping types that can be useful in the context of SPARQL query reformulation. Thus, a SPARQL query that can be posed over a source ontology, is reformulated, according to the mappings, in order to be capable of being posed over a target ontology. We have also outlined the SPARQL query reformulation process and presented examples of query reformulation in a motivating example.

This work is part of a framework that we are pursuing, which aims to provide algorithms, proofs and middleware for the support of transparent access to federated heterogeneous databases across the web in the Semantic Web environment.

References

1. McGuinness, D.L., van Harmelen, F. (eds.): OWL Web Ontology Language: Overview. W3C Recommendation, February 10 (2004), http://www.w3.org/TR/owl-features
2. Manola, F., Milles, E. (eds.): RDF Primer. W3C Recommendation, February 10 (2004), http://www.w3.org/TR/rdf-primer
3. Prud'hommeaux, E., Seaborne, A. (eds.): SPARQL Query Language for RDF. W3C Recommendation, January 15 (2008), http://www.w3.org/TR/rdf-sparql-query/
4. Benslimane, S.M., Merazi, A., Malki, M., Amar Bensaber, D.: Ontology mapping for querying heterogeneous information sources. INFOCOMP (Journal of Computer Science) 7(2), 44–51 (2008)
5. Quilitz, B., Leser, U.: Querying Distributed RDF Data Sources with SPARQL. In: Bechhofer, S., Hauswirth, M., Hoffmann, J., Koubarakis, M. (eds.) ESWC 2008. LNCS, vol. 5021, pp. 524–538. Springer, Heidelberg (2008)
6. Jing, Y., Jeong, D., Baik, D.-K.: SPARQL Graph Pattern Rewriting for OWL-DL Inference Query. In: Proceedings of the 2008 Fourth International Conference on Networked Computing and Advanced Information Management (2008)
7. Euzenat, J., Polleres, A., Scharffe, F.: Processing ontology alignments with SPARQL (Position paper). In: International Workshop on Ontology Alignment and Visualization, CISIS 2008, Barcelona, Spain (March 2008)
8. Stocker, M., Seaborne, A., Bernstein, A., Kiefer, C., Reynolds, D.: SPARQL basic graph pattern optimization using selectivity estimation. In: Proceedings of WWW 2008 (2008)
9. Hartig, O., Heese, R.: The SPARQL Query Graph Model for Query Optimization. In: Franconi, E., Kifer, M., May, W. (eds.) ESWC 2007. LNCS, vol. 4519, pp. 564–578. Springer, Heidelberg (2007)
10. Bizer, C., Cyganiak, R.: D2R Server, http://www4.wiwiss.fu-berlin.de/bizer/d2r-server/index.html
11. Bikakis, N., Gioldasis, N., Tsinaraki, C., Christodoulakis, S.: Querying XML Data with SPARQL. In: Proceedings of the 20th International Conference on Database and Expert Systems Applications (DEXA 2009) (2009)
12. Akahani, J., Hiramatsu, K., Satoh, T.: Approximate Query Reformulation for Ontology Integration. In: Proc. of the Semantic Integration Workshop Collocated with the Second International Semantic Web Conference, ISWC 2003 (2003)
13. Choi, N., Song, I.-Y., Han, H.: A survey on ontology mapping. SIGMOD Record 35(3), 34–41 (2006)
14. Kalfoglou, Y., Marco Schorlemmer, W.: Ontology Mapping: The State of the Art. Semantic Interoperability and Integration (2005)
15. Ghidini, C., Serafini, L.: Mapping Properties of Heterogeneous Ontologies. In: Dochev, D., Pistore, M., Traverso, P. (eds.) AIMSA 2008. LNCS (LNAI), vol. 5253, pp. 181–193. Springer, Heidelberg (2008)
16. Ghidini, C., Serafini, L.: Reconciling Concepts and Relations in Heterogeneous Ontologies. In: Sure, Y., Domingue, J. (eds.) ESWC 2006. LNCS, vol. 4011, pp. 50–64. Springer, Heidelberg (2006)

17. Ghidini, C., Serafini, L., Tessaris, S.: On Relating Heterogeneous Elements from Different Ontologies. In: Kokinov, B., Richardson, D.C., Roth-Berghofer, T.R., Vieu, L. (eds.) CONTEXT 2007. LNCS (LNAI), vol. 4635, pp. 234–247. Springer, Heidelberg (2007)
18. Euzenat, J.: An API for ontology alignment. In: McIlraith, S.A., Plexousakis, D., van Harmelen, F. (eds.) ISWC 2004. LNCS, vol. 3298, pp. 698–712. Springer, Heidelberg (2004)
19. Scharffe, F., de Bruijn, J.: A language to specify mappings between ontologies. In: SITIS 2005, pp. 267–271 (2005)
20. Euzenat, J., Scharffe, F., Zimmermann, A.: D2.2.10: Expressive alignment language and implementation. Knowledge Web EU-IST Project deliverable 2.2.10 (2007)
21. Scharffe, F.: PhD thesis: Correspondence Patterns Representation, `http://www.scharffe.fr/pub/phd-thesis/manuscript.pdf`
22. Scharffe, F., de Bruijn, J.: A language to specify mappings between ontologies. In: Proc. of the Internet Based Systems IEEE Conference, SITIS 2005 (2005)
23. Scharffe, F.: Omwg d7: Ontology mapping language (2007), `http://www.omwg.org/TR/d7/`
24. Euzenat, J., Shvaiko, P.: Ontology matching. Springer, Heidelberg (2007)

STOWL: An OWL Extension for Facilitating the Definition of Taxonomies in Spatio-temporal Ontologies

Alberto Salguero, Cecilia Delgado, and Francisco Araque

Dpt. of Computer Languages and Systems, CITIC-UGR, University of Granada, 18071 Granada (Andalucía), Spain
{agsh,cdelgado,faraque}@ugr.es

Abstract. The spatio-temporal information is becoming more popular day by day. Well known web services (google maps, virtual earth...) make this type of information available to the public. On the other hand the Web is tending to the Semantic Web. The OWL language, the ontology description language the Semantic Web is based on, lacks some of the desired features for representing spatio-temporal information. We have developed an OWL extension which tries to encompass this issue. In this work we focus on how the OWL language has been extended in order to consider properties for defining taxonomies.

Keywords: Ontology, Geography, Information System, OWL, Taxonomy, Tourism.

1 Introduction

Ontologies are being developed as specific concept models by the Knowledge Management community. They can represent complex relationships between objects, and include the rules and axioms missing from semantic networks [5]. OWL is the language adopted by the W3C for defining ontologies and supporting the Semantic Web. A growing number of Semantic Web applications, which can interact between them and cooperate to find better solutions, is being developed [7]. It is a common belief that Semantic Web technology would significantly impact the use of the Web, essentially in terms of increased task delegation to intelligent software agents, and a subsequent amelioration of the information overload effect [8].

On the other hand, the generalization of geographical information has produced a proliferation of applications such as Geographic Information Systems, GPS navigators and other mapping tools. The problem is that the OWL language does not support some of the desired characteristics of a spatio-temporal data model.

Some of these missing features have to do with the possibility of defining complex taxonomies. Taxonomies are increasingly being used in object oriented design and knowledge management systems [17], [6] to indicate any grouping of objects based on a particular characteristic. It is common when dealing with geographic information to find this kind of structures. Although the W3C on its own has described how to represent these types of relations using OWL, the resulting code is long and difficult to be followed by a person. The exhaustive decompositions and partitions, this work

M.D. Lytras et al. (Eds.): WSKS 2009, LNAI 5736, pp. 336–345, 2009.

is focused on, cannot be expressed directly in OWL. They are usually represented combining other constructions. When there is a need, like when dealing with geographical information systems, of representing huge quantity of this type of relations the ontology description becomes practically not understandable by humans. It also creates a maintenance problem. We have developed an extension of the OWL language in order to directly support the common primitives found in spatio-temporal data models. We call this extension STOWL [14].

In this paper we focus on the extension of the OWL language in order to support the definition of exhaustive decompositions and partitions. Actually, these kinds of constructions are widely used when designing information repositories, even those which do not have to do with geographic information. Furthermore, there are specific situations in which the usage of detailed taxonomies can ease the knowledge management. The case where there is information coming from different sources and should be integrated in a common repository is one of those situations. It is sometimes difficult to propagate the sources' scheme underlying knowledge to the integrated repository because usually the sources are independent (different enterprises, departments…). The original taxonomies in the sources maybe do not represent the same knowledge in the integrated repository.

STOWL is build on top of OWL language. Most of its features rely on OWL features, like the possibility of defining partitions or exhaustive decompositions. On the other hand, there are spatio-temporal features which cannot be expressed in the OWL language because to the OWL ontology does not support those desired features. In this case the modification of the OWL ontology should be made in order to incorporate those features and the OWL language, consequently, has also to be modified in order to reflect those changes (giving STOWL as result).

In the former situation STOWL can be seen as a software layer that transform the new defined spatio-temporal features, which can be expressed straightforwardly in STOWL, in an more complex and equivalent definition in OWL. Due to this transformation the results continue being an ontology description expressed in OWL, so all the OWL tools can be used as usual (reasoners, editors...). This is the case of the proposed extension in this paper. Exhaustive decompositions and partition in STOWL are transformed to their equivalent in OWL.

The resultant ontology-based data model can be used as a common data model to deal with the data sources schemes integration [15]. Due to the fact that the data sources we work with contain many geographic information we early noticed that the OWL ontology lacks some of the desired spatio-temporal features and an extension focused on this type of information should be developed [13]. Although it is not the first time the ontology model has been proposed for this purpose [16], [12], [1], [2], to our knowledge, this is the first time the metadata storage capabilities of some ontology definition languages has been used in order to improve the Data Warehouse data refreshment process design.

The remainder of the paper is organized as follows. In Section 2 the spatio-temporal features the OWL language lacks and which have been incorporated to STOWL are depicted. Section 3 describes the modification made to OWL in order to consider the exhaustive decompositions and partitions. Finally, in section 4 the conclusions are presented.

2 Differences between OWL and STOWL Ontologies

It has been commented previously that STOWL, the common data model which is proposed in this work to describe the schemes of the different data sources, is an extension of the OWL language. In this point, the differences between both languages are explained.

Both are languages which allow the description of ontologies based on the OWL ontology. The OWL language can be used for describing spatio-temporal data repositories. The main problem is that this language lacks some desired features which makes difficult to express certain type of knowledge which is common when dealing with spatial and temporal information. Those missing features are what we have incorporated to the STOWL language. Following are described those features. The reasons why it can be useful to incorporate them to a spatio-temporal data model are also presented.

- *Description of exhaustive decompositions.* As well as in the DAML+OIL ontology (the OWL ontology derive from), in the OWL ontology is possible to express exhaustive decompositions. For this is used the primitive *owl:one of.* In spite of that, it is important to note that the classes described using this primitive are exhaustive decompositions of instances of other classes. The problem is that, in certain cases, it is necessary to describe exhaustive decompositions of classes, not specific instances. In the tourism or transport fields, for instance, is often to have certain transport stations where different types of means of transport are met (they are usually known as "transport interchangers"). The existence of this type of stations implies that the type of transport stations (train, bus, metro…) cannot be disjoints but, actually, there cannot be a transport station of another type of mean of transport. This situation cannot be described easily in OWL. We would need to define each possible combination of means of transport (bus-metro, bus-metro-train, metro-train…) and define them as a partition. This solution is not applicable sometimes due to effort needed in order to contemplate all the possible combinations. Because this situation is encountered often when dealing with geographic information in STOWL the method for defining exhaustive decompositions has been made easy.
- *Description of partitions.* Unlike the DAM+OIL model, in OWL is not possible to express partitions of concepts. This is because, in the case it is needed, a partition can be expressed combining the primitives *owl:disjointWith* and *owl:unionOf.* The former primitive express disjoint knowledge whereas the latter describes decompositions. Although valid, the code needed for describing partitions with this approach is long and complex to follow. Due to the fact that it is common to have partitions when dealing with spatial (administrative regions…) or temporal (months…) information the OWL code representing the ontology becomes ugly. As well as with the case of exhaustive decomposition, STOWL is defined in such a way that this type of knowledge is easier to describe.
- *N-ary relations.* OWL only allows the definition of binary relations. In case of higher arity relations need to be defined, the W3C suggests the creation of classes (or concepts) for representing those relations. These types of relations

are often used when dealing with spatial information. When defining topological relations is usual to define a property for connecting two elements by mean of a specific link. This is a basic ternary relation which cannot be expressed directly using OWL. An artificial concept should be created in order to binary relate those three elements. As well as with the previously described properties, although it is possible to express n-ary relations using the OWL language the resultant code is difficult to follow. The STOWL language has been defined to solve this issue.

- *N-ary functions*. OWL has not been designed for supporting functions. In case of necessity, the primitive *owl:FunctionalProperty* can be used as a mechanism for defining binary relations. It is not possible to express functions in OWL with higher arity. It is an important problem when dealing with spatial information. In fact, it is therefore not possible to define layered information, a common type of data managed by the geographical information systems (terrain height, terrain usage…). The STOWL language can manage this type of knowledge.
- *Formal axioms*. None of the markup-based languages for describing ontologies, including OWL, support the definition of formal axioms. The inclusion of this kind of knowledge in a spatio-temporal model is very interesting. Following with the example provided when talking about the n-ary relations, it would be desirable in a spatio-temporal model to be able to specify that the linked nodes can be the same or not. This kind of knowledge is incorporated to STOWL.
- *Rules*. They are a special type of formal axioms, so they cannot be expressed in OWL. They allow to infer new knowledge. Geographic information system can use this rules for performing complex queries.
- *Integrity constraints*. The integrity constraints make the maintenance of semantic consistency of data easier. OWL only considers the one type of integrity constraint: the functional dependency, which can be expressed by mean of the primitive *owl:FunctionalProperty*. There are other types of integrity constraint which would be useful to consider in a spatio-temporal model: unique integrity constraint (avoiding duplicities), and assertions (for expressing, for instance, that the total sum of all the areas in a region should be equal to the area of that region). Actually, the integrity constraints can be expressed by mean of formal axioms, so if the latter are present in STOWL the integrity constraints can also be expressed in STOWL. The integrity constraints are related to the quality of data. The stronger the integrity constraints the lower the possibility the data contains error.

The rest of this work is focused in how the OWL language has been extended in order to consider the two former features in the previous list.

3 Taxonomies in STOWL

There are many methods and automatic tools for structuring the knowledge. One of them is the definition of taxonomies [3][11][4], which allows the filtering of the underlying information in order to produce useful knowledge [9].

But let suppose the following situation. We have two independent data sources which contain some taxonomy decompositions. On the integrated repository the original decompositions may lose some of their knowledge. Consider, for instance, the integration of a exhaustive decomposition set of concepts in one source with the base concept, not decomposed, in the other source. The resultant decomposition may not be considered a exhaustive decomposition depending on the specific problem.

In this point the solutions adopted for the extension of OWL with the features related to the definition of taxonomies are described: exhaustive decompositions and partitions.

3.1 Exhaustive Decompositions

As it has been explained before, it is not possible to express disjoint and exhaustive decompositions directly in OWL. The W3C consortium, aware of this problem, defines a set of case of use patterns, for representing this type of knowledge [10]. These solutions include the definition of new symbols which are incorporated to the OWL diagrams. An exhaustive decomposition can be expressed visually using the new symbol illustrated in figure 1.

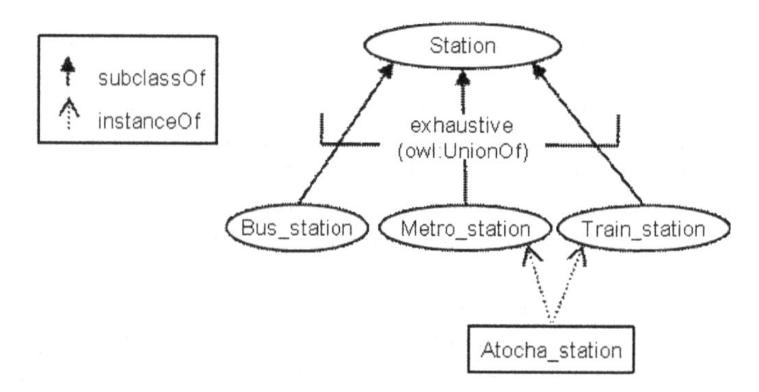

Fig. 1. The new symbol introduced by the W3C for representing exhaustive decompositions

This new symbol can be used only in the diagrams. It makes the diagrams human-friendly. When implementing the ontology by mean of the OWL language there is no direct translation for this new primitive. Instead, the exhaustive decomposition is represented using the primitives *owl:equivalentClass* and *owl:unionOf* in conjunction. The former primitive indicates that the instances belonging to the class which is being described are the same as those represented by the classes inside the *owl:equivalentClass* primitive. The primitive *owl:unionOf* represents the logic conjunction. Figure 2 illustrates how can be combined to represents the exhaustive decomposition depicted in figure 1.

It would be possible to have instances of the class *Station* which do not belong to any of the three subclasses defined in figure 1 if only using the *owl:unionOf* primitive. All of the instances of the class *Station*, when using a combination of the *owl:equivalentClass* and *owl:unionOf*, have to be instances of any of the classes defined inside the *owl:unionOf* primitive.

```
<owl:Class rdf:about="Station">
      <owl:equivalentClass>
            <owl:Class>
                  <owl:unionOf
                rdf:parseType="Collection">
                        <owl:Class
                      rdf:about="Bus_station"/>
                        <owl:Class
                      rdf:about="Train_station"/>
                        <owl:Class
                      rdf:about="Metro_station"/>
                  </owl:unionOf>
            </owl:Class>
      </owl:equivalentClass>
</owl:Class>
```

Fig. 2. Example of an exhaustive decomposition combining the *owl:equivalentClass* and the *owl:unionOf* primitives

Because this situations appears often when dealing with spatio-temporal information, in the STOWL data model the definition of the exhaustive decompositions has been simplified. A new primitive has been added to de model: *stowl:exhaustiveUnionOf*. This primitive represents the exhaustive decompositions and it has been defined as a sub-property of the property *owl:unionOf*, as it can be seen in figure 3. From the syntactic point of view both primitives are identical. From the semantic point of view, the difference between them is that the classes in the list of classes representing the range of the property have to include all the possible subclasses of the class representing the domain of the property.

```
<rdf:Property rdf:ID="exhaustiveUnionOf">
      <rdfs:label>exhaustiveUnionOf</rdfs:label>
      <rdfs:subPropertyOf rdf:resource="&rdf;UnionOf"/>
      <rdfs:domain rdf:resource="#Class"/>
      <rdfs:range rdf:resource="&rdf;List"/>
</rdf:Property>
```

Fig. 3. Definition of the exhaustiveUnionOf primitive in STOWL

Obviously, none of the tools capable of managing OWL ontologies will understand this new property. At this point we consider two possibilities:

- Develop a software layer which makes the translation between the ontologies described in STOWL, where the new primitive can be used, into the same ontologies but using only OWL valid primitives.
- Modify the applications (editors, inference engines…) in order to consider the new property.

Although the second alternative seems to be the best option, sometimes are not possible to make those required changes. The source code is not always available or its modification requires a huge effort. Also, the modified applications have to be maintained over time. For this reasons we have opted for the former approach (see figure 4). We also opted for that alternative because the inclusion of other of the features described in the previous section (n-ary relations, forma axioms…), out of the scope of this work, requires the implementation of that software layer. The translation, in the case of the exhaustive decompositions, consists on replacing the *stowl:exhaustiveUnionOf* occurrences by a combination of the *owl:equivalentClass* and *owl:unionOf* primitives as described in figure 2.

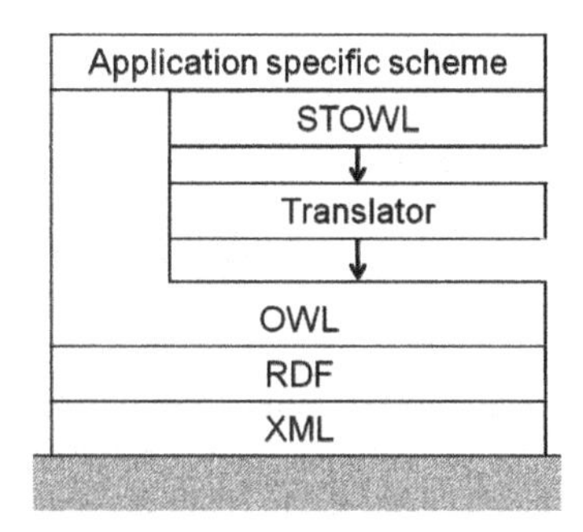

Fig. 4. Ejemplo de descomposición exhaustiva

3.2 Partitions

As well as with the exhaustive decompositions, it is not possible to express partitions directly in OWL. The W3C consortium suggest to combination of the *owl:unionOf* and the *owl:disjointWith* primitives for expressing this type of knowledge. Figure 5 illustrates how to define partitions by mean of the combination of such properties. Three subclasses are defined in that example, each of them representing the possible locations according to the country they belong to (considering only North American countries for the sake of simplicity).

As it can be seen in figure 5, the definition of partitions in OWL is a tedious task. For each involved class in the partition it should be specified that is disjoint with respect to the rest class forming the partition. Due to the fact that it is very common to find partitions when dealing with spatio-temporal information a simplified process is needed. The same idea proposed for the exhaustive decomposition is applied here.

A new property is defined in STOWL named *stowl:partitionOf* for representing this type of knowledge. It is defined as a subproperty of *stowl:extaustiveUnionOf* (see figure 6). The classes forming the range of this property should be mutually disjoint, as well as form an exhaustive decomposition.

Partitions expressed by means of the new property look like the one shown in figure 7. It can easily be seen that the effort needed for expressing partitions in this form suppose much less effort. The problem is that, as well as with the *stowl:exhaustiveUnionOf* property, none of the OWL tools can directly work with this new primitive. The ontologies expressed in STOWL should be translated to plain OWL. In this case, all the partitions expressed by means of the property *stowl:partitionOf* are transformed into a combination of the *owl:unionOf* and the *owl:disjointWith* primitives, as illustrated in figure 5.

```
<owl:Class rdf:ID="Location">
      <owl:unionOf rdf:parseType="Collection">
            <owl:Class
          rdf:resource="#CanadianLocation"/>
            <owl:Class
          rdf:resource="#MexicanLocation"/>
            <owl:Class rdf:resource="#USALocation"/>
      < /owl:unionOf >
</rdf:Class>
...
<owl:Class rdf:ID="CanadianLocation">
      <rdfs:subClassOf rdf:resource="#Location" />
      <owl:disJointWith
    rdf:resource="#MexicanLocation" />
      <owl:disJointWith rdf:resource="#USALocation" />
</owl:Class>
<owl:Class rdf:ID="MexicianLocation">
      <rdfs:subClassOf rdf:resource="#Location" />
      <owl:disJointWith
    rdf:resource="#CanadianLocation" />
      <owl:disJointWith rdf:resource="#USALocation" />
</owl:Class>
<owl:Class rdf:ID="USALocation">
      <rdfs:subClassOf rdf:resource="#Location" />
      <owl:disJointWith
    rdf:resource="#MexicanLocation" />
      <owl:disJointWith
    rdf:resource="#CanadianLocation" />
</owl:Class>
```

Fig. 5. Example of partition expressed in OWL by means of the combination of the *owl:unionOf* and the *owl:disjointWith* primitives

```
<rdf:Property rdf:ID="partitionOf">
      <rdfs:label>partitionOf</rdfs:label>
      <rdfs:subPropertyOf
    rdf:resource="&stowl;exhaustiveUnionOf"/>
      <rdfs:domain rdf:resource="#Class"/>
      <rdfs:range rdf:resource="&rdf;List"/>
</rdf:Property>
```

Fig. 6. Definition of the *stowl:partitionOf* property

```
<owl:Class rdf:ID="Location">
      <stowl:partitionOf rdf:parseType="Collection">
            <owl:Class
         rdf:resource="#CanadianLocation"/>
            <owl:Class
         rdf:resource="#MexicanLocation"/>
            <owl:Class rdf:resource="#USALocation"/>
      < /owl:partitionOf >
</rdf:Class>
```

Fig. 7. Partition example in STOWL

4 Conclusions

In this paper we have presented an extension of the OWL language focused in making easier the description of spatio-temporal information. This interest emerges because OWL, the language chosen to be the ontology description language for supporting the Semantic Web, lacks some of the desired features for a spatio-temporal data model. Because of the fact that the geographic information is becoming more and more popular day by day, this is an issue which should be addressed.

Actually, this work is focused in the part of the extension related with the description of exhaustive decompositions and partitions. This kind of knowledge is difficult to be expressed in OWL. The resultant code is long and difficult to be followed by humans. By mean of a software layer we have made easier the specification of exhaustive decompositions and partitions. The equivalent code is shorter and easier to be followed by human.

In general, the proposed extension allows the definition of more detailed taxonomies in OWL. This fact eases some common tasks which take part in the organization's knowledge processes. The most significant of those tasks is possibly the creation of integrated repositories from independent information sources.

Acknowledgments. This work has been supported by the Spanish Research Program under project PSI2008/00850 and by the Erasmus European Research Program under project GERM 2008 - 3214 / 001 - 001.

References

1. Ale, M.A., Gerarduzzi, C., Chiotti, O., Galli, M.R.: Organizational Knowledge Sources Integration through an Ontology-Based Approach: The Onto-DOM Architecture. In: Lytras, M.D., Carroll, J.M., Damiani, E., Tennyson, R.D. (eds.) WSKS 2008. LNCS (LNAI), vol. 5288, pp. 441–450. Springer, Heidelberg (2008)
2. Arch-Int, N., Sophatsathit, P.: A Semantic Information Gathering Approach for Heterogeneous Information Sources on WWW. Journal of Information Science 29, 357–374 (2003)
3. Cimiano, P., Staab, S., Tane, J.: Deriving concept hierarchies from text by smooth formal concept analysis. In: GI Workshop "Lehren-Lernen-Wissen-Adaptivität" (LLWA), Fachgruppe Maschinelles Lernen, Wissenentdeckung, Data Mining, Karlsruhe, Germany, October 6-8, pp. 72–79 (2003)

4. Heylighen, F.: Structuring Knowledge in a Network of Concepts. In: Workbook of the 1st Principia Cybernetica Workshop, pp. 52–58 (1991)
5. Hodge, G.: Systems of Knowledge Organization for Digital Libraries: Beyond Traditional Authority Files. Council on Library and Information Resources (2000)
6. Hsu, C., Kao, S.: An OWL-based extensible transcoding system for mobile multi-devices. Journal of Information Science 31, 178–195 (2005)
7. Kolas, D., Dean, M., Hebeler, J.: Geospatial semantic Web: architecture of ontologies. In: Rodríguez, M.A., Cruz, I., Levashkin, S., Egenhofer, M.J. (eds.) GeoS 2005. LNCS, vol. 3799, pp. 183–194. Springer, Heidelberg (2005)
8. Lytras, M.D., García, R.: Semantic Web applications: a framework for industry and business exploitation – What is needed for the adoption of the Semantic Web from the market and industry. International Journal of Knowledge and Learning 4(1), 93–108 (2008)
9. Mertins, K., Heisig, P., Alwert, K.: Process-oriented Knowledge Structuring. Journal of Universal Computer Science 9(6), 542–550 (2003)
10. Rector, A.: Representing Specified Values in OWL: "value partitions" and "value sets". W3C Working Group (2005), `http://www.w3.org/TR/swbp-specified-values/`
11. Rhodes, B.J.: Taxonomic knowledge structure discovery from imagery-based data using the neural associative incremental learning (NAIL) algorithm. Information Fusion 8(3), 295–315 (2007)
12. Yang, S.Y.: How Does Ontology help Web Information Management Processing. WSEAS Transactions on Computers. 5(9), 1843–1850 (2006)
13. Salguero, A., Araque, F.: Tourist trip planning applying Business Intelligence. In: 1st World Summit on the Knowledge Society, WSKS (2008)
14. Salguero, A., Araque, F., Delgado, C.: Using ontology meta data for data warehousing. In: 10th Int. Conf. on Enterprise Information Systems (ICEIS), Barcelona, Spain (2008)
15. Samos, J., Araque, F., Carrasco, C.A.R., Delgado, C., Garví, E., Ruíz, E., Salguero, A., Torres, M.: Ontology-based spatio-temporal DW models research project. TIN2005-09098-C05-03, Spain (2006)
16. Skotas, D., Simitsis, A.: Ontology-Based Conceptual Design of ETL Processes for Both Structured and Semi-Structured Data. International Journal on Semantic Web and Information Systems 3(4), 1–24 (2006)
17. Whittaker, M., Breininger, K.: Taxonomy Development for Knowledge Management. In: World library and information congress: 74th ifla general conference and council. Québec, Canada, August 10-14 (2008)

Learning by Sharing Instructional Segments

Edmar Welington Oliveira[1,3], Sean Wolfgand M. Siqueira[1], Maria Helena L.B. Braz[2], and Rubens Nascimento Melo[3]

[1] Department of Applied Informatics (DIA/CCET), Federal University of the State of Rio de Janeiro (UNIRIO): Av. Pasteur, 458, Urca, Rio de Janeiro, Brazil, 22290-240
{edmar.oliveira,sean}@uniriotec.br
[2] DECivil/ICIST, IST, Technical University of Lisbon: Av. Rovisco Pais, 1049-001 Lisbon, Portugal
mhb@civil.ist.utl.pt
[3] Computer Science Department, Pontifical Catholic University of Rio de Janeiro (PUC-Rio): R. Marquês de São Vicente, 225, Gávea, Rio de Janeiro, Brazil 22453-900
{eoliveira,rubens}@inf.puc-rio.br

Abstract. Workers in the Knowledge Society need to think creatively, solve problems, and make decisions as a team. Therefore, the development and enhancement of skills for teamwork through collaborative learning is one of the main goals of nowadays education. The term "collaborative learning" refers to an instructional approach in which students work together in groups toward a common goal. The present research was designed to evaluate the effectiveness of a collaborative learning approach where groups of students were involved in the creation of instructional content through the reuse of existing videos that could be segmented according to their needs. The experience was based on the use of a prototype supporting the collaborative activities and facilitating the creation, sequencing and presentation of video segments. Student's perceptions of the learning experience were evaluated and findings are presented as well as suggestions for further research.

Keywords: Learning content, Content segmentation, Content sharing, Collaborative Learning.

1 Introduction

The explosion of available information not only implies using systems that require new skills for accessing, organizing and retrieving it but also implies structuring the information in order to facilitate knowledge construction and acquisition. Nowadays' society depends on information, in particular structured information which is easier to process, but non-structured information is also very important and it is still a difficult problem to represent and extract the embedded knowledge.

The work presented in this paper focuses on a collaborative approach to handle non-structured information. In fact, the collaboration could be useful not only in order to acquire and organize knowledge, but also in order to disseminate it. In other words, it would be possible, through a collaborative environment, to structure knowledge,

M.D. Lytras et al. (Eds.): WSKS 2009, LNAI 5736, pp. 346–355, 2009.

making it easier to be understood. The proposed approach was applied to educational scenarios, but the idea can be used in other contexts of the knowledge society.

The collaborative learning approach has been referenced by many authors as a useful teaching method. In instructional environments based on this approach, there is a focus on providing rich interactions among learners in order to increase the knowledge transfer between the participants [1]. Some researches emphasize the collaborative effectiveness when applied to learning processes. According to [2], the development of the students' critical sense is highly improved by learning processes that are supported by collaborative strategies. Collaborative learning creates the conditions, in which learners can build knowledge, for instance from artefacts created/used by other learners [3]. In fact, in collaborative learning, the learners not only absorb new information but also create knowledge as they consider and analyse other learners' assumptions and points of view.

Another aspect to notice is related to the knowledge representation, as in the knowledge society, the real value of the products (e.g. instructional material) is the knowledge embedded in it. Nowadays there is a movement in technology enhanced learning towards representing learning content through Learning Objects (LOs), which aims at providing reusable portions of instructional content. The most cited definition of LO is provided by IEEE LTSC [4]: "a LO is any entity, digital or non-digital, that can be used for learning, education and training."

When considering digital LOs, they can be understood as reusable multimedia files with instructional content. Associated with LOs are their respective metadata, which are descriptions for enabling better retrieval, documentation and access of these objects. As in the work presented in this paper the focus is on digital LOs and their possible segments, the architecture proposed at [5] as well as the systematic for structuring the segments were adapted to consider collaborative scenarios. Such segmentation approach was evaluated and the results were presented at [6], but this evaluation did not consider collaborative scenarios.

According to the processes and collaborative strategies in an educational scenario, the participants must be encouraged to work together for developing knowledge. Although the individual work is important to build knowledge, the learning process (as well as the knowledge work) gains more amplitude and dimension when collaborative work also happens.

According to [7], knowledge society is generally defined as an association of people with similar interests who try to make use of their combined knowledge. Furthermore, when an educational group invests considerable effort toward sharing and producing new knowledge, then it contributes to the knowledge society. So, the work presented in this paper follows the idea that a learning process executed by a learner can also be useful to others. Therefore, participants can search for LOs, segments and composed objects, make new segments, compose segments in a new object, and share their content (LOs, segments and composed objects) in learning scenarios. It follows the fundamental idea that the knowledge is increasingly a collective artifact which requires the development of collaboration skills by learners.

Considering the taxonomy presented at [8], the work presented in this paper lets the pedagogy issues to the instructor while using metadata to describe the learning context and focuses on tasks such as assimilative, information handling and communicative.

Next section presents the architecture adaptation for sharing learning content. Section 3 describes a case study and its evaluation. Finally, section 4 presents some final remarks.

2 Architecture for Sharing Learning Content

In order to allow participants to share content in different aggregation levels (LOs, segments and composed objects) it is necessary not only to provide processes for manipulating these levels, but also managing user profile and group formation as well as learning scenarios.

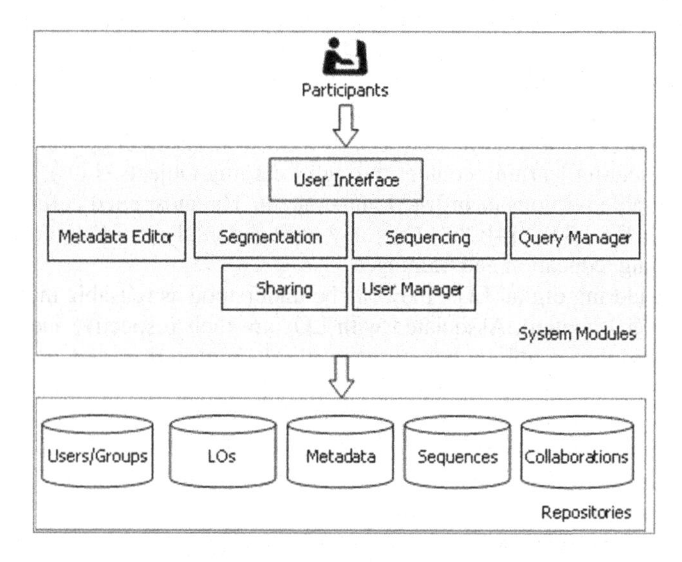

Fig. 1. The Architecture for Sharing Learning Content

The participant accesses the system through a **user interface**, which is responsible for managing user interactions with the computer. The following functions were provided: search, segment, sequence, and share.

The participant can search for learning content through the **query manager** module. The **query manager** gets a query from the user or another module and according to the intended use (provided by the learning scenario or input data) accesses the metadata repository looking for LOs, segments or composed objects. In order to provide the answer to a query, the **query manager** also considers the user privileges and learning background.

The **segmentation** module activates the query manager to search for LOs and once one LO is selected for segmentation, it is necessary to specify the segment as well as its context and sharing restrictions. According to the proposal presented in [5] for LO segmentation, it was also used a MPEG-7 description for specifying the segment. The developed prototype works with video files and therefore it is necessary to define starting time of the segment in the video as well as duration time. The **metadata editor** module is used in order to provide the learning context.

To provide the construction of semantically richer objects, it was also developed a **sequencing** module, in which it is possible to define sequences of content objects (LOs, segments or other composed objects). It is necessary to make searches for learning content, selecting the ones of interest and specify the sequences. The sequences are defined according to the metadata ids of the content objects. It uses SCORM specification [9] as described in [5]. SCORM is a group of specifications and patterns for content, technologies and services, associated to e-learning. Among other things, this model defines rules for the description of elements as well as in which way such elements can be grouped to form "content packages" (LOs). For the resulting composed object it is necessary to make a description of its usage context, through the use of the **metadata editor** module.

The **metadata editor** module defines the context of learning contents through the use of IEEE LOM descriptions. IEEE Learning Object Metadata (LOM) [10] items were created to facilitate search, assessment, acquisition and use of learning content and it is undoubtedly the most used and referenced standard for describing learning resources. In addition, it is also necessary to specify access restrictions.

The **user manager** module is important to control user accesses to the system as well as learning profiles of participants and groups formation. PAPI Learner was used for describing learners' profile. IEEE Public and Private Information (PAPI) for Learners (PAPI Learner) [11] defines and/or references elements for recording descriptive information about: knowledge acquisition, skills, abilities, personal contact information, learner relationships, security parameters, learner preferences and styles, learner performance, learner-created portfolios, and similar types of information. This Standard allows different views of the learner information (perspectives: learner, teacher, parent, school, employer, etc.) and substantially addresses issues of privacy and security.

The **sharing** module manages the content sharing as well as communication tools focused on content. For each learning scenario, if the user selects or creates content, it is possible to provide annotations about the content, to recommend it to other participants, to suggest new learning contexts (complementing the metadata description) and to inquire more information about the content from the authors or other participants.

3 Case Study and Its Evaluation

A prototype was developed using Java programming language, version 1.6. The Integrated Development Environment was Eclipse, version 3.4 and the application server was Apache Tomcat 5.5. All the database repositories (multimedia content - LOs, composition structures - sequences, metadata descriptions, learner profile, group formation and learning scenarios - collaborations) were developed on PostgreSQL version 8.3.3.

The prototype focused on dealing with video LOs. The user can search for stored videos, exploring LOM metadata elements and segment them by specifying initial and final time. Different types of audiovisual resources could have been used in the prototype and case study. However, videos were used in order to focus on one media type, simplifying the implementation effort and because there is now an increasingly number of digital videos that can be reused in learning scenarios.

To collect data for the case study, the implemented prototype was used in a class of the "Introduction to Information Systems" Course – in the Federal University of the State of Rio de Janeiro (UNIRIO). Twenty-four learners participated of the study. They were distributed in eight groups (A1, A2, B1, B2, C1, C2, D1 and D2) - each one with three members. This configuration was chosen in order to make possible the distribution of tasks, allowing their simultaneous execution, mainly considering a collaborative environment. The number of groups and the number of members were defined based on the infrastructure available to carry out the case study and the desired investigation scenarios.

Twenty-three videos were selected from the Web: 12 related to "Strategic Planning" topic, 11 related to "Object-Oriented Programming". These topics were chosen because both of them are subjects of the course and were interesting to the students. The number of videos was bounded by time restrictions on the considered tasks.

The case study was configured to be performed in four stages. In each one, the groups should execute a specific task - within a 20-minutes timeframe. Moreover, after the completion of each task, evaluation questionnaires were applied in order to collect data.

In the first stage, the groups created segments – through the segmentation process - using the prototype. Moreover, they should be related to the "Strategic Planning" topic. These groups should access the available videos (previously stored in the multimedia database repository) and choose - according to their interest – the ones that would be segmented. It was defined that all groups should create at least three segments and analyze all the involved processes.

In the second stage the groups A1, A2, B1 and B2 created a presentation related to "Strategic Planning" in video format, but, this time, using the segments created by groups C1, C2, D1 and D2 during the first stage. Meanwhile, the groups C1, C2, D1 and D2 shared the created segments and analyzed them. The experience on sharing learning content using segments created by other learners was evaluated. It was also considered and evaluated if the segments' sharing could favor the exchange of ideas, allows to develop the critical thinking skills and to consider different points of view, and if it allows to explore more learning materials.

In the third stage, all groups should create semantic segments. Just as in the other stages, it was established that each segment should take, at most, two minutes. However, now, they should be related to the "Object-Oriented Programming" topic. These groups should access the videos available for this case study (previously stored in the multimedia database repository) and choose according to their interest – the ones that would be segmented. It was defined that all groups should create at least three segments and analyze all the involved processes.

Finally, in the fourth stage, the groups C1, C2, D1 and D2 should create a presentation related to "Object-Oriented Programming" in video format using the segments created by anyone in the third stage.

Some questionnaires were applied in the case study. The collaborative learning was evaluated according to segments sharing. It was analyzed the learners' interest on using segments from other learners for learning a specific subject and some groups analyzed segments created by other groups.

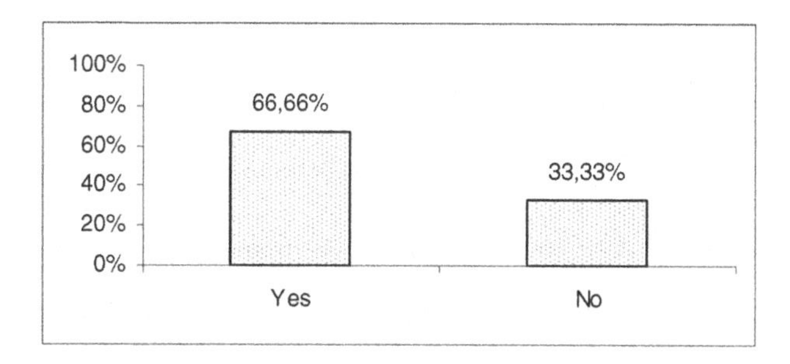

Fig. 2. Segments are useful in learning the subject

According to Fig. 2, the evaluated segments helped in learning the subject for 66,66% of learners because the segments provide an overview of the subject although presenting specific aspects. In addition, they explained and exemplified well concepts and definitions, facilitating understanding. However, the evaluated segments did not help learning for 33,33% because their duration was too short to enable a good understanding of the subject. It was also described that the segments' content were too generalist.

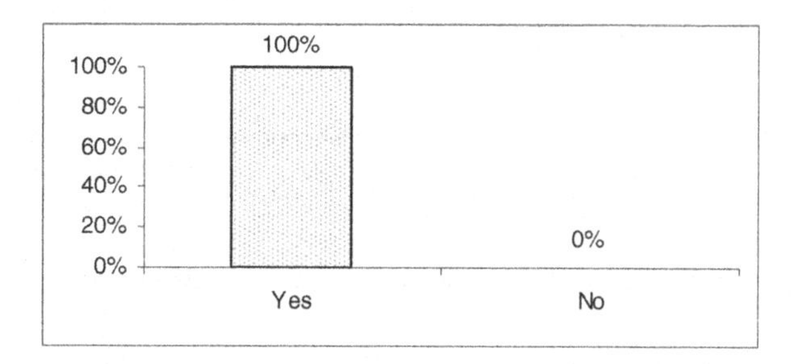

Fig. 3. Sharing segments promotes the exchange of ideas

Although some learners considered the evaluated segments not adequate to promote learning, everybody agrees – as presented in Fig. 3 – that sharing segments favors the exchange of ideas and it is an interesting approach.

Fig. 4 presents the evaluation results for the segments' sharing approach as an alternative for promoting the critical thinking skills of students and for considering different points of view (opinions presented by other learners).

As presented in Fig. 4, most learners (66,66%) totally agrees that sharing segments can be seen as an approach for promoting critical thinking and for considering different points of view. According to the learners, such strategy allows improving the understanding of the subject. A great advantage that was mentioned is the possibility of analyzing information that would not have been taken into account at first but that is brought to consideration by different groups or learners that are studying the same subject. In addition, new points of view can be considered so it is possible to review the personal opinion. However, for 33,33% of learners, this capability is possible but only depending on how the segments from other learners were structured.

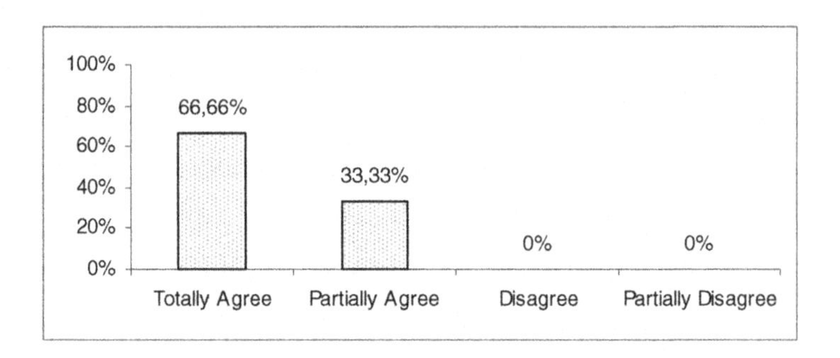

Fig. 4. Sharing segments improves critical thinking and noticing different points of view

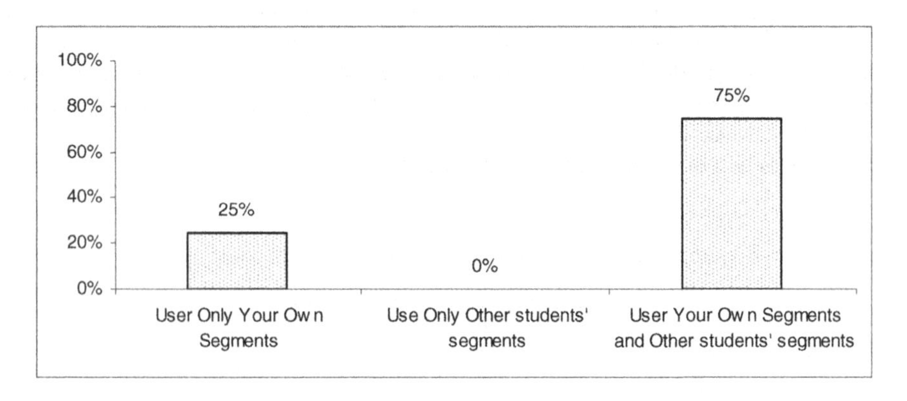

Fig. 5. Choice of authorship of segments in learning activities

Other important result is presented in Fig. 5, in which 75% of learners would choose to use segments created by themselves as well as segments created by other learners and groups. For 25% of evaluated students, the choice would be for using only the segments created by themselves.

When comparing segments created by themselves and those created by others, 75% of the learners find those segments similar as shown in Fig. 6. By the other side, 25%

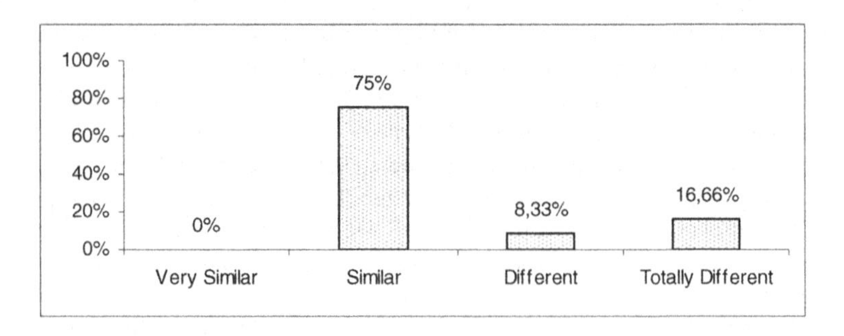

Fig. 6. Comparing their own segments with those created by others

think the segments are different or very different. Although most of learners think the evaluated segments were considered similar, it is important to notice the reduced time of the experiment. However, even though most of the segments were considered similar by learners, they were not considered useless. As presented in Figure 2, most of the learners (66,66%) believe that the evaluated segments (created by other groups) are useful in learning processes.

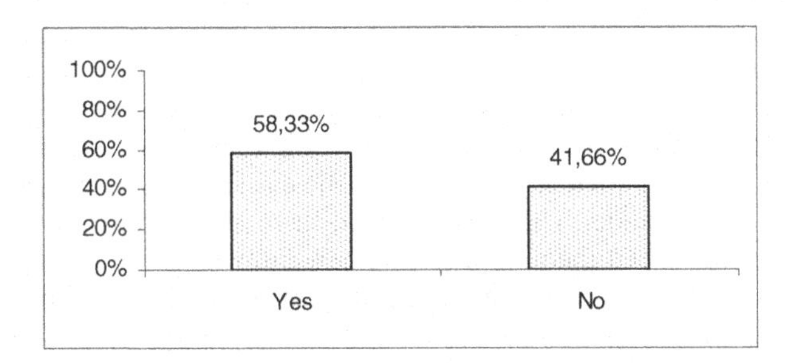

Fig. 7. Evaluated group explored different parts not explored by evaluator group

Fig. 7 presents that more learners (58,33%) believe that the group under evaluation explored parts of videos that they had not explored in their segmentation processes. However, as presented in Fig. 8, 91,66% of learners think their groups explored parts of videos that the evaluated group did not explore.

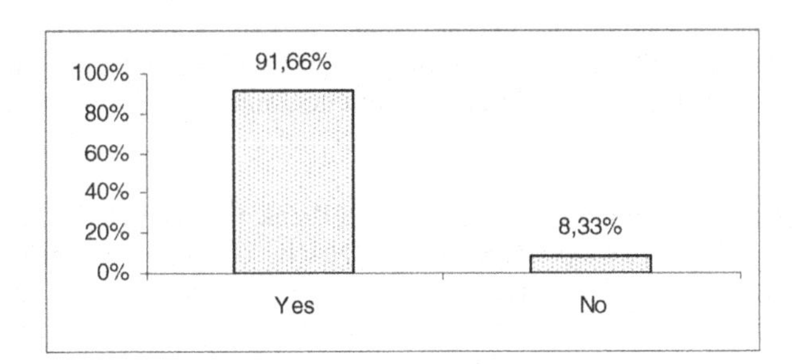

Fig. 8. Evaluator group explored different parts not explored by evaluated group

Considering the case study presented in this paper, using the collaborative approach allows exploring more learning content. Using the segmentation processes it is possible to select the most important parts of each LO. Then, through the sharing of segments, there is a considerable gain on exploring LOs in less time and with added benefits (such as those presented in these results), promoting knowledge construction.

In addition, knowledge acquired by a learner in his learning process can be reused to guide the learning processes of other learners (i.e., segments created by one participant can be useful for providing learning to another participant). In fact, sharing the segments can promote the discussion of ideas, the analysis of opinions, argumentation and counter-argumentation of points of views etc., improving knowledge construction.

4 Conclusion

This paper describes the architecture considered for exploring the segmentation approach presented in [5] and [6] in a collaborative educational environment. In fact, it presents a case study where groups of students used a prototype implementing the main parts of the architecture to build new content by performing several collaborative tasks centered on the segmentation approach. Finally some results obtained from this case study were presented and discussed. In fact, these results seem very encouraging as they have reinforced the idea that the segmentation of learning materials is an interesting and simple technique that is easily mastered by students and that can be used to propose collaborative tasks centered on the creation of instructional content by students. The created content (segments) in the case study was considered interesting and useful by the students involved, which have also expressed their agreement with the idea that it helped them in acquiring new knowledge, in identifying and analyzing different points of view and in increasing their motivation to work in groups.

As future work it is expected to extend the developed prototype in order to be able to deal with different media and apply it in different learning contexts supporting additional tasks. It would also be desirable to increase the number and diversity of participants in the evaluation process possibly building groups considering the learner's profile and studying how the profiles influence the results. In addition, it would be very interesting to study how to automatically propose different segments to students according to their specific needs, interests or characteristics recorded in their user profile when they look for multimedia materials. The architecture and the prototype are also going to be improved. Works such as [12] and [13] can motivate the development of richer compositions and tools for dealing with composing segments and multimedia materials. Monitoring knowledge building and interactions, such as described in [14] can enhance learning experience and should also be considered in future works.

Acknowledgments. The authors would like to thank the Knowledge and Reasoning Representation research group (RCR) from UNIRIO and the E-learning Technology research group from ICIST/DECivil who actively participated in the research discussions. Acknowledgments to the FCT Portugal, through the ICIST funding, for the financial support, making possible the accomplishment of this work.

References

1. Kumar, V.S.: Computer-Supported Collaborative Learning: Issues for research. In: 8th Annual Graduate Symposium on Computer Science, University of Saskatchewan, Canada (1996)
2. Miyake, N.: Constructive Interaction and the Iterative Process of Understanding. Cognitive Science 10, 151–177 (1986)
3. Smith, B.L., MacGregor, J.T.: What is collaborative learning? In: Collaborative Learning: A Sourcebook for Higher Education, National Center on Postsecondary Teaching, Learning, and Assessment, Pennsylvania State University (1992)
4. IEEE Learning Technology Standards Committee (LTSC), Draft standard of learning object metadata, IEEE 1484.12.1 (2002)

5. Oliveira, E.W., Siqueira, S.W.M., Braz, M.H.L.B.: Structuring Segments of E-Learning Multimedia Objects with LOM and MPEG-7. In: IASTED International. Conference on Computers and Advanced Technology in Education, Greece, pp. 353–358 (2008)
6. Oliveira, E.W., Siqueira, S.W.M., Braz, M.H.L.B.: Evaluating the Reuse of Learning Content through a Segmentation Approach. In: The 9th IEEE International Conference on Advanced Learning Technologies, Latvia (to appear, 2009)
7. Anderson, R.E.: Implications of the Information and Knowledge Society for Education. In: Voogt, J., Knezek, G. (eds.) International Handbook of Information Technology in Primary and Secondary Education Series. Springer International Handbooks of Education, vol. 20 (2009)
8. Conole, G.: Capturing practice, the role of mediating artefacts in learning design. In: Lockyer, L., Bennett, S., Agostinho, S., Harper, B. (eds.) Handbook of Research on Learning Design and Learning Objects: Issues, Applications and Technologies. IGI Global, Hersey (2008)
9. Advanced Distributed Learning (ADL), Sharable Content Object Reference Model SCORM 3rd Edition Documentation Suite (2004)
10. IEEE LTSC.: IEEE P1484.12.3, Draft 8: Draft Standard for Learning Technology - Extensible Markup Language (XML) Schema Definition Language Binding for Learning Object Metadata Draft P1484.12.3/D8, http://ltsc.ieee.org/wg12/files/IEEE_1484_12_03_d8_submitted.pdf
11. ISO/IEC JTC1 SC36: IEEE P1484.2.1/D8, PAPI Learner - Core Features (2002), http://metadata-stds.org/metadata-stds/Document-library/Meeting-reports/SC32WG2/2002-05-Seoul/WG2-SEL-042_SC36N0175_papi_learner_core_features.pdf
12. Watanabe, T., Kato, K.: Computer-supported interaction for composing documents logically. International Journal of Knowledge and Learning (IJKL) 4(6), 509–526 (2008)
13. Rodrigues, D.S.S., Siqueira, S.W.M., Braz, M.H.L.B., Melo, R.N.: A Strategy for Achieving Learning Content Repurposing. In: Lytras, M.D., Carroll, J.M., Damiani, E., Tennyson, R.D. (eds.) WSKS 2008. LNCS (LNAI), vol. 5288, pp. 197–204. Springer, Heidelberg (2008)
14. Caballé, S., Juan, A.A., Xhafa, F.: Supporting Effective Monitoring and Knowledge Building in Online Collaborative Learning Systems. In: Lytras, M.D., Carroll, J.M., Damiani, E., Tennyson, R.D. (eds.) WSKS 2008. LNCS (LNAI), vol. 5288, pp. 205–214. Springer, Heidelberg (2008)

Infrastructural Analysis for Enterprise Information Systems Implementation

Guoxi Cui[1,2] and Kecheng Liu[1,2]

[1] School of Management and Economics, Beijing Institute of Technology,
Beijing 100081, China
guoxicui@bit.edu.cn
[2] Informatics Research Centre, University of Reading,
Reading RG6 6WB, UK
k.liu@reading.ac.uk

Abstract. The success of enterprise information systems implementation depends more on how well an organization is prepared and organizational aspects are integrated than the pure technical systems. This reveals the soft issues of software engineering. Organizational situations for enterprise information systems implementation are often complex and soft, and it is difficult to get a satisfactory answer from hard system engineering methods. This paper addresses organizational situations by improving the Problem Articulation Method. This method inspects an enterprise information systems implementation from different perspectives of people, activities and systems, and delivers a structure for the analyst to systematically understand and arrange the implementation. We have applied the improved method in a physician workstation system and showed a preliminary result.

Keywords: System Implementation, Enterprise Information Systems, Problem Articulation Method.

1 Introduction

Enterprise Information Systems (EIS) are complex socio-technical systems. As developing rapidly in modern organizations, EIS have been more and more taking over in operational activities and taking part in organizational issues nowadays. In the modern knowledge society, it is a common scenario that people and EIS co-ordinate to jointly perform an activity. This requires EIS to integrate not only technical and operational procedures, but also organizational and cultural knowledge. It is the organizational and social issues that make EIS "soft". Dealing with "soft" problems raises a challenge in EIS. It also moves EIS towards a new research agenda featured as an alternate paradigm in the knowledge society.

Implementing an EIS is the first puzzle. One may begin a problem analysis with some need to be met, implying a new or altered system to provide the solution. If we call the total problem, undifferentiated and unstructured, "the total system", it includes the normal operating system that is to solve the problem plus all the additions

M.D. Lytras et al. (Eds.): WSKS 2009, LNAI 5736, pp. 356–365, 2009.

needed to create it, service it, and maintain it. The technical levels are well catered for, so we began to look for precise solution to organizational problems on the human level in the shape of a formalism for analyzing and designing information systems. Such solutions should ideally support the automatic generation of a default version of an application system if deemed useful but without pre-empting technically more efficient solutions. It therefore calls for an approach to analyzing such organizational infrastructures.

Existing methods for Implementing EIS are with less capability for analyzing total systems. Information systems planning deals with part of infrastructural analysis, but planning approaches are initially introduced as a solution to help the industry organize its resources and secure its operations. Many unsuccessful systems, including those lack of satisfaction and acceptance, fail in ignoring the total requirements.

Problem articulation concerns with the macro-structure of an EIS lifecycle [1], it starts from a vague or soft problem and finds out appropriate words and other signs, as well as exploring the cultural and physical constraints within which to frame the norms. The Problem Articulation Method (PAM) intends to give the implementation a structure, and explore the cultural and physical constraints within which to frame the norms [1, 2]. In this paper, we aim to identify a structure for vague problem from the perspectives of collateral tasks, stakeholders, and sub-problems.

This paper is structured as follows. Section 2 briefly reviews related work in this area. Section 3 improves the PAM method on the linkage between its techniques. We illustrate this improved method in a hospital information system in Section 4, and a discussion on the application is followed in Section 5. Section 6 gives conclusion and future work.

2 Related Work

Traditional hard systems methodology are most suited to address structured problems but the organizational and social problems generally are half-structured or unstructured. Soft Systems Methodology has emerged for solving complex and messy problem in social situations [3]. The SSM points out a sequence of activities for dealing with problems of ambiguity and change. The ideas of "soft problems" and "systems thinking" have shown considerable practice value. But it is not catered for information systems development.

Many authors have outlined success and failure factors based on cases. Al-Mudimigh et al. propose an integrative framework for ERP implementation based on the factors of top management commitment, business case, change management, project management, training, and communication [13]. These are essential elements that contribute to success in the context of ERP implementation. However, it does not provide a method on how to implement an EIS. Kansal proposes a framework for the selection process of EIS, identifying critical factors for EIS implementation [14]. He identifies five categories of constraints that influence the EIS implementation, which are technical, organizational, human, financial and time constraints. These aspects need further formalization as guidance for the analyst.

Decomposition is relatively straightforward for solving a complex problem in requirements engineering. This is also common in project planning. Such methods usually use divide-and-conquer strategy. They break up the whole problem into smaller sub-problems and thereof solve each sub-problem separately, either parallel or sequentially. The Work Breakdown Structure (WBS) is typical method of this type [4]. WBS is a tool that defines a project and groups the project's discrete work elements in a way that helps organize and define the total work scope of the project. It adopts a tree structure, which shows a subdivision of effort required to achieve an objective. The WBS starts with the end objective and then successively subdivides it into manageable components. Similar methods include product breakdown structure [5], and project breakdown structure. All approaches aim to define and organize the total scope of a project. The WBS is organized around the primary products of the planned outcomes instead of the work needed to produce the planned actions, which is different with PAM.

Stamper's seminal work on PAM aims to offer a structure to a vague problem situation [1, 9]. To start a problem, they developed the five techniques for PAM, each clarifying the problem from a different perspective. Following Stamper's framework, Liu et al. make further improvements on the individual techniques [10, 11]. They separate stakeholder identification as a single technique and sharpen the collateral analysis technique. Simon et al discusses a practical application of PAM to the development of enterprise information systems planning. However, PAM is still not mature for practice. The linkage between techniques is open.

3 Structuring Enterprise Information Systems Implementation

Inspired by the "soft" characteristics of SSM, the Problem Articulation Method (PAM) is employed to cope with the complexity and ambiguity in social systems. This method treats EIS implementation as a total problem, and attempts to produce a structure for solving the problem. The output of this method is an enterprise infrastructure model, from which the analyst can derive business requirements [11]. It is beneficial in the initial stage of a complex EIS implementation project. In this section we improve PAM with the interconnection of its three techniques, namely, unit system definition, stakeholder identification, and collateral analysis.

3.1 Unit System Definition

To understand the total problem of EIS implementation, PAM first defines the scope of the problem. PAM attempts to subdivide a problem into small units and organize them in a structured way. Each unit is called a unit system. A unit system is a collection of organized activities performed by people or automata to achieve certain objectives [11]. Kolkman adopts the notion of "course of action" to define a unit system [1]. A unit system has an action course as its kernel and, in turn, this course may have one or more sub-activities. An action course can be a simple activity by one person or a complex set of tasks by a group. Using the notion of a unit system, a problem situation is analyzed as a constellation of action courses about each of which one needs to say only a few things. This gives us a start for structuring the problem situation.

3.2 Collateral Analysis

The Collateral Analysis technique attempts to recognize all auxiliary tasks that support the unit system and provides a collateral structure. In this technique, the analyst selects a unit system as the focal system, and then begins to analyze all other systems that serve this system; we call them collateral systems. All collateral systems with the focal system constitute the collateral structure of the unit system. The focal system is the kernel of a collateral structure, determining the particular focal interest of the problem situation. In this technique we strip away all details of how to bring the focal system into operation and keep it going, but consider only its interrelated situation.

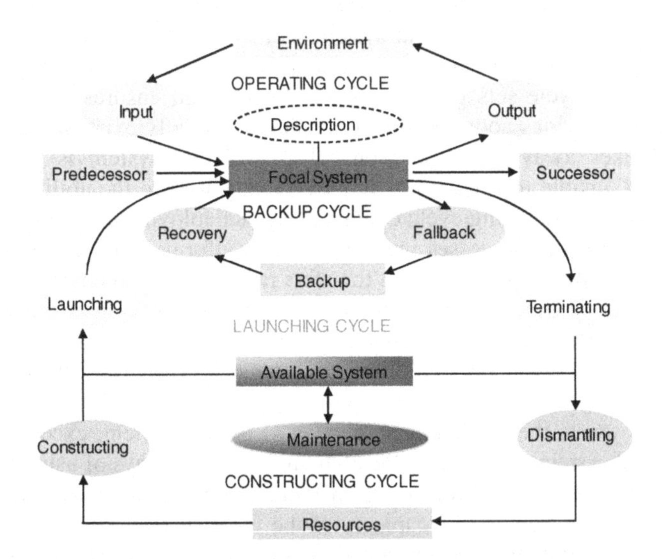

Fig. 1. Collateral Structure (adapted from [11])

Once having positioned a focal system, next step is to identify the related collateral systems. Two closely interrelated classes have been classified: object and service systems [11]. Object systems are the unit systems that are required, and instances include available system, resources, predecessor, and successor, as shown in rectangles in Fig.1. Service systems, as shown as ellipses in Fig.1, provide services and operations to object systems. A service system takes an object system as input and produces another object system as output. In other words, it transforms an object system from a previous state into its next state. Both object and service systems make up the focal system's context and dependencies, enabling it to function as required. Structured by the life-cycle of the system, collateral analysis considers the focal system's history line and then starts with the constructing, launching, operating, and backup cycles.

The history line is a linear progression from an earlier solution. It looks at the relationships between the predecessor, successor and focal system. The history line contains three object systems and no service systems. The predecessor, the focal system, and the successor are linked one by one because there may be a causal or relation between them. As there is no substantial link, the service systems are absent. We have to take this history and future into account because the materials, knowledge, people and other resources from predecessor may well be needed in focal system [1]. It may give some tips or provide motivating forces to think of a successor system. Certainly the people outside the system, who supply it or use its outputs, may be disconcerted if there is no continuity.

The constructing cycle builds up an available system from resources. The resources are regarded as object systems, including raw materials, energy, human resources, and information. The constructing system provides the focal system with the ability to have the required functions. By inputting energy to this cycle, resources go through a transition from a state of disorder to order. It is an entropy decreasing process. The dismantling system brings the focal system's function ability to an end. The maintenance system makes sure the focal system retains its functions while it needs services.

The launching cycle sets up an available system and ensures it operational. The launching system brings about the functions that potentially exist, while the terminating system takes away these functions. The available system is a system to be commissioned, unable to function as required but will have the ability to have those functions. Tasks in launching cycle contain systems deployment, staff training, etc.

The operating cycle processes input and output information, and makes an impact on the environment. The input system transfers resources in the environment to enable the focal system to function as required, while the output system transfers material from the focal system to the environment. The input and output are, in fact, patterns not single instances and these have to be considered [2]. Environment system consists of material, energy and information. There are sources and sinks in the environment [1]. In most respects the environment is given it is just a client, except that any new version of focal system may require the client to learn new ways of using this function.

The backup cycle deals with emergent situations and launches a backup system. The recovery system repairs the capacity of the focal system, while the fallback system terminates the operations of the focal system. Backup system is a normal operational function in some required measure. Activities in backup cycle include planning, informing, backup, etc.

3.3 Stakeholder Identification

Stakeholder identification technique attempts to build up a structure from an organizational perspective. It helps to find out stakeholders and take into account of their interests regarding the problem. Stakeholder identification takes a selected unit system and produces a categorized list of stakeholders. Each unit system is associated with a set of stakeholders. Liu et al. has identified six types of stakeholders, labeled by roles [11]:

$$\text{Role} = \{\text{actor} \mid \text{client} \mid \text{provider} \mid \text{facilitator} \mid \text{governing body} \mid \text{bystander}\} \qquad (1)$$

An actor is an agent (individual or group) who takes actions that directly influence the unit system. Actors perform action of course and provide service. They have a direct link with the system, e.g., the physicians. A client is an agent who receives the consequences or outcomes of the unit system, e.g., the patients. A provider is an agent who creates the conditions (supplies, permission, design, and funds) to facilitate the deliverables of the unit system, e.g. the equipment providers. A facilitator is the agent who solves conflicts and ensures continuality, steers the team towards the goals, e.g. the financial institutions. The governing body is the agent who enforces of the legal rules, policies, and regulations, e.g. city hall, ministry of employment. Bystanders are those who have other interests in the unit systems. Bystanders usually are not part of the project itself, but may have interests in the outcome and have influence, e.g. press agencies.

Responsibilities are captured and described to structure problem and its related issues. Stakeholders and their responsibilities make up an organizational structure of a unit system. The unit system describes an agent who is responsible for an action course of which the exact spatial and temporal boundaries are determined by a group of people who hold a stake in it. For instance, some agent will determine and govern the existence of the unit system and some agent will be responsible for the tasks being carried out.

3.4 Applying Sequences and Infrastructural Analysis Architecture

To implement an innovative EIS, the analyst starts from the objectives of a unit system. Though with a vague problem situation, he can identify at least one objective from his customer, for example, "EIS in operation".

With the objective, the analyst can use collateral structure as a guidance to conduct unit system decomposition, as it is a general structure for all EIS. As a result of collateral analysis, a set of collateral (unit) systems can be recognized, each representing corresponding work required in the implementation. Recursive decompositions apply to these new unit systems whenever necessary. The criterion is if the analyst feels confident about the unit system and can find a skilled person or company to accomplish this work without risks.

For a unit system that there is no need to conduct a further collateral analysis, stakeholder identification will be applied. Activities for unit systems can be exposed at the same time as identifying the stakeholders' responsibilities. The stakeholders and activities, organized in a certain way, are the main tasks in a unit system to achieve the objectives. These activities, covering from EIS selection and purchase, to staff trainings and system configuration, to business process reengineering and data preparation, and to EIS backup and maintenance, will be assigned with explicit responsible individuals or groups to ensure they are successfully implemented.

Having identified collateral systems, stakeholders and their activities that support the focal system in operation, the analyst has revealed the infrastructure of the EIS to be implemented. Therefore, for a unified representation, a unit system can be specified in the format of:

$$\text{Unit System} = \{\text{stakeholder, role, responsibility, representatives}\} \quad (2)$$

Fig 2 illustrates the applying sequence of PAM techniques to get infrastructure for an EIS.

For a single unit system, only a few things need to be considered about a unit system. Besides the objectives mentioned above, a unit system has a required start and finish. Authorities who start or finish this unit system also need to be made clear. This information tells the lifespan of a unit system, e.g., when to take into and remove

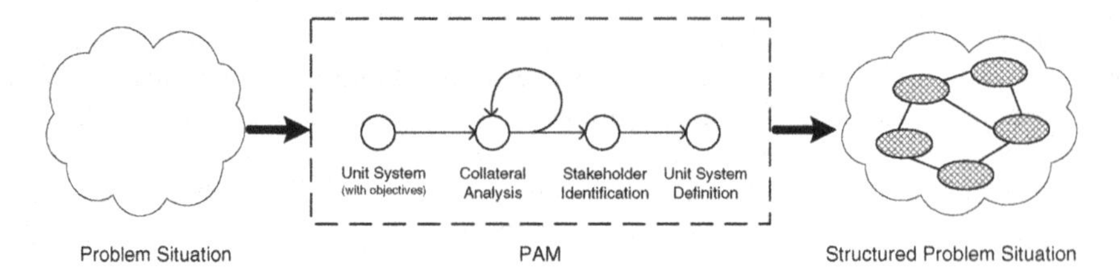

Fig. 2. PAM Techniques Applying Sequence

from consideration a unit system, and who holds the overall responsibility over the unit system. It is useful to manage all unit systems from a global view. For the ease of navigation, unit systems are organized in a hierarchical structure. A unit system map can be drawn to show all unit systems in one site. In this map, all collateral systems are treated as its sub-unit systems.

4 Example: Implementation for Physician Workstation System

We take a physician workstation system in hospital as an illustration. The focus is on the infrastructural analysis before implementing a physician workstation system and to offer a holistic structure of the problem situation.

The problem situation is fairly vague. The hospital keeps saying that it wants to catch up with Electrical patient record trend, and build up a digital hospital in its area. Therefore, we start with an overall objective statement that the hospital wants to have a physician workstation system in operation.

Collateral analysis helps to generate a collateral structure. Before this workstation, the medical orders are manually prescribed by physicians, and then taken over by the nurses. We therefore need to know about the procedures carried out behind. The environment is the physicians. It produces "raw" medical orders as input to the workstation system, and the latter processes medical orders as required as output. We also have to take into account its backup system in case the workstation system falls down. For constructing service, the hospital has decided to purchase a set of software from a healthcare service provider, and needs to consider the issues like market investigation, functionality selection, and making a purchase. The resources are the procedures, rules and money for buying an available system. Launching service needs system configuration, interface customization, and staff training. The latter needs special attention because the training of physicians would be a difficult task, and therefore a second round of collateral analysis against staff training is recommended.

Stakeholder analysis produces an organizational structure of the target system. By applying the stakeholder identification technique at the top level unit system, we have identified physicians and nurses as actors, patients as clients, the healthcare service supplier as provider, project management board as facilitator, hospital management board as governing body, and other departments in hospital as bystander. Among them, project management board takes the responsibility of allocating resources, coordinating conflicts. The patients are the beneficiary because the system improves their work efficiency directly. The hospital management board steer the entire project, and determine the existence of the system. Other departments in hospital are the bystanders of this system but any mistakes of the system may an impact to them. The healthcare service supplier is responsible to deliver a workstation system that meets all the requirements agreed. Fig 3 shows the result of collateral analysis and stakeholder identification.

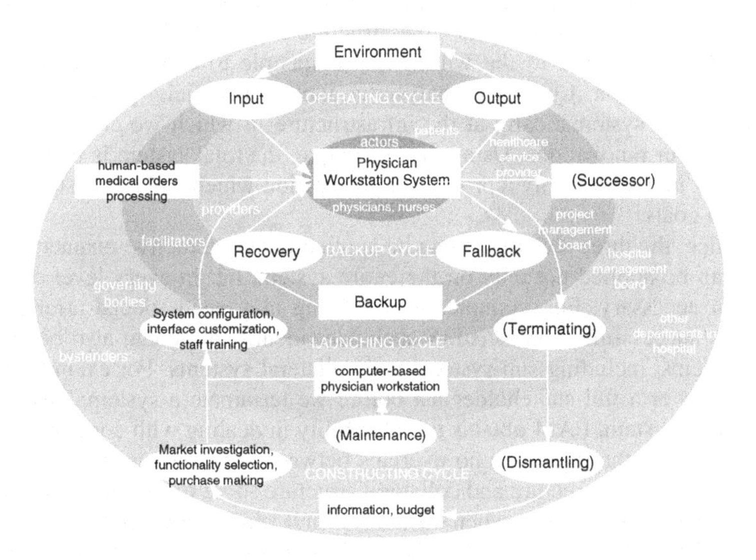

Fig. 3. Physician Workstation System: Stakeholders and Collateral Systems

For a unit system at lower levels that there is no need to conduct further collateral analysis, we can identify its stakeholders and activities involved to achieve the objective of this unit system. For instance, the constructing service has a specific objective of purchasing a physician workstation system. It also has a specific start and finish time authorized by the hospital management board. The actors involve IT department, who is responsible for market investigation and functionality selection, and financial department, who is responsible for making the purchase once the IT department has successfully finished its mission. All these activities will be facilitated by the project management board. Therefore, the constructing service can be specified as:

$U_{\text{Market Investigation}}$= {IT Department, Actor, Select a Physician Workstation System }

$U_{\text{Functionality Selection}}$= {IT Department, Actor, Choose Functionality}

$U_{\text{Purchase Making}}$= {Financial Department, Actor, Purchase the System}

At present, the output of the PAM analysis is an infrastructure model for the problem situation, making up by sub-system structure, organizational structure and collateral structure. Due to the space limit, we cannot give a representation of all the details.

5 Discussion

PAM is a method for senior staff or project managers at the initial stage to inspect clearly of the total problem in EIS implementation. Following the three techniques, it produces systemic, organizational and collateral structures. All make up the infrastructure of the target systems. Problem decomposition is a regular practice in requirement analysis, and it gives us a preliminary structure of the problem. Understanding the responsibilities associated with the various stakeholders involved in the problem is also one of the tasks in dealing with the soft part of the requirements analysis. This defines the relevant goal structure and relates it to the mores of the community, in particular to the willingness of people to conform to norms and to exchange information. Collateral analysis is another beneficial practice in PAM. It leads us to look systematically at the infrastructure in which we create our systems and perform our innovations. Using this technique, the total system is partitioned into a network of unit systems by a process or articulation which isolates systems with the same broad goals.

In practice, the three techniques can be applied in an interactive manner. Collateral analysis can be applied not only on the entire system, but in every level of unit systems when necessary. For example, staff training may need special arrangement to ensure it is successfully conducted. Stakeholder identification can also be applied in all unit systems, including sub-systems and collateral systems. For example, we may need to discover a full stakeholder list before we terminate a system. Thanks to the notion of unit system, PAM obtains this flexibility in dealing with complex problems. It is worth noting that there are no overlaps between sub-systems and collateral systems, as sub-systems structure and collateral structure serve for different purposes.

It is also worth mentioning when the analyst first applies collateral analysis. Once a definition of the focal system has been established the analyst can relate the focal system to the existing infrastructure. In defining these systems the analyst should try to capture all aspects of the problem and to concentrate on the interactions between systems that describe different parts of the problem solution. This makes for a very complex analysis. In an environment with a rich infrastructure many of the necessary collateral systems will already exist and be run by others. But in a less developed environment, the need to organize and run collateral systems will ensure that even a simple problem will be a considerable managerial task [1]. Collateral analysis helps considering the building and running of an infrastructure in which many diverse innovations can proceed smoothly. The difference between innovating in a developing country and in sophisticated socio-economic environment is made dramatically clear by collateral analysis.

As discussed, PAM positions itself in the initial stage of the project, and it helps in clarifying a vague and complex problem. The activities in PAM can also contribute to the information systems planning. Similar to information systems planning, PAM is dedicated to providing uncertain and unstructured problems with a structure, meanwhile considering global, long-term, and key issues. Therefore, we must be aware that PAM lies in the context of the whole organization, and align with organization management issues. PAM is not catered for concrete business problems, but for the system goal, strategy, architecture, and resource planning. From this point of view, the process of applying PAM is a decision-making process, which involves both technical and management issues. As a result, the targeted system's profile is roughly depicted.

6 Conclusion and Future Work

This paper discussed a preliminary infrastructural analysis for EIS implementation and the approach of Problem Articulation Method to achieve this. The main contribution is on the improvement and linkage of PAM techniques. We proposed a unified specification for unit systems, which connects three techniques in PAM. We also carried out an infrastructural analysis for physician workstation system, and obtained a holistic structural view of the total requirements. This helps in understanding the organizational aspect of IS development.

Currently, a software tool to support PAM is under development. In future, further research on the unit system modeling is in our schedule. Its identification, organization, and specification require more work. Also, improvement on collateral analysis is still needed to give a clear guideline to the analyst.

References

1. Kolkman, M.: Problem Articulation Methodology: PhD thesis, University of Twente, Febo, Enschede (1993) ISBN 90-9005972-5
2. Liu, X.: Employing MEASUR Methods for Business Process Reengineering in China: PhD thesis, University of Twente, Febo, Enschede, The Netherlands (2000) ISBN 90-3651618-8
3. Checkland, P., Scholes, J.: Soft systems methodology in action. John Wiley & Sons, Inc., New York (1990)
4. Haugan, G.T.: Effective Work Breakdown Structures: Management Concepts (2002)
5. OGC, Managing Successful Projects with PRINCE2, 5th edn.: The Stationary Office (2005)
6. Ross, J.W., Weill, P., Robertson, D.C.: Enterprise Architecture as Strategy, vol. 1. Havard Business School Press, Boston (2006)
7. Zachman, J.A.: A Framework for Information Systems Architecture. IBM Publication G321-5298. IBM Systems Journal 26 (1987)
8. TOGAF, The Open Group Architecture Framework, TOGAF 8.1.1 (2006)
9. Stamper, R., Kolkman, M.: Problem articulation: a sharp-edged soft systems approach. Journal of applied systems analysis 18, 69–76 (1991)
10. Liu, K., Sun, L., Bennett, K.: Co-Design of Business and IT Systems. Information Systems Frontiers 4, 251–256 (2002)
11. Liu, K., Sun, L., Tan, S.: Modelling complex systems for project planning: a semiotics motivated method. International Journal of General Systems 35, 313–327 (2006)
12. Tan, S., Liu, K.: Requirements Engineering for Organisational Modelling. In: 7th International Conference on Enterprise Information Systems, Porto, Portugal (2004) (presented)
13. Al-Mudimigh, A., Zairi, M., Al-Mashari, M.: ERP software implementation: an integrative framework. European Journal of Information Systems 10, 216–226 (2002)
14. Kansal, V.: The Enterprise Systems Implementation: An Integrative Framework. The Review of Business Information Systems 10(2), 2nd Quarter (2006)

Enterprise Resource Planning: An Applications' Training Aid Based on Semantic Web Principles

Aristomenis M. Macris

Dept. of Business Administration, University of Piraeus,
80 Karaoli & Dimitriou str., 185 34 Piraeus, Greece

Abstract. An integrated ERP system is an asset for any organization using it, but since its full deployment requires increased cooperation between business units, there is a need to provide the users involved with appropriate training material, so that they can effectively and efficiently exploit the business processes, which very often change in a dynamic business environment. Existing training materials fail to represent effectively the implicit business knowledge in order to help the users understand the underlying structures and relationships. This paper proposes a prototype model for the design and development of ERP training material, where both the multimedia objects used in training scenarios and the knowledge built into them are captured and fully reusable. The proposed approach helps trainees understand: (i) which are the building blocks of an ERP application, (ii) how they relate with each other and (iii) how they can be used in order to solve business specific problems.

Keywords: ERP ontology, ERP training, Ontology-based training, Semantic web ERP Training.

1 Introduction

World economy has been transformed into a knowledge economy. In that economy the application of knowledge is the main means of production and has become more important than traditional resources, such as labour, capita or base materials. Traditional economy that was primarily driven by transformational activities (turning raw product into finished product, or turning data into information) has been transformed into knowledge economy where the highest-value activities are complex interactions between people and systems. This shift from transformation activities to interactions represents a broad shift in the nature of economic activity. Economic success and most productivity gains in the future are going to be in interactions. Hence, enterprises are beginning to realize that strategic advantage becomes less focused on ownership of distinctive stocks of knowledge. Instead, strategic advantage resides in the institutional capacity to get better and faster the most promising flows of knowledge and in the rapid integration of the knowledge acquired from these flows into the enterprise activities [1].

M.D. Lytras et al. (Eds.): WSKS 2009, LNAI 5736, pp. 366–375, 2009.

1.1 Training Knowledge Workers for the Knowledge Economy

Knowledge workers can be described as highly professional, highly competent individuals with an excellent education, globally mobile and even independent from specific national restrictions. They are global "knowledge players" with a main function to act as "knowledge brokers".

On the other hand there are candidate knowledge workers. Those individuals own qualifications on a high level too and are capable to welcome constantly new working tasks. However, they are less virtuous and qualified than knowledge workers.

One of the key issues for the successful integration of knowledge flows into the enterprise activities is active user participation. This is achieved through continuous interactions between experts (knowledge workers) and trainees (candidate knowledge workers). In essence, active user participation, which is enabled through user training, is considered a knowledge-creation spiral that emerges when the interaction between tacit and explicit knowledge is elevated dynamically from lower to higher ontological levels, i.e. from the individual, to the group, to the organization, to the inter-organization level [2], [3].

The approach is outlined through a case study involving the development of a training scenario for the applications' training of Enterprise Resource Planning (ERP) users.

1.2 Enterprise Resource Planning Systems Training

An Enterprise Resource Planning (ERP) system is complex business software, as it optimizes business processes using enterprise resources, depending on the availability of internal or external enterprise objects or conditions. It consists of many integrated subsystems (modules) that are linked together in order to satisfy unique enterprise needs [4]. End-user training is a key success factor in ERP implementations, since the subsystems' customization and integration is not achieved through programming (as it used to be in the past) but through end-user parameterization (setting correctly the appropriate set of parameters) [5], [6].

Most of the existing automated training aids for ERP applications essentially manipulate collections of multimedia objects (text, images, videos, etc.) [7]. These multimedia objects are usually grouped hierarchically (e.g. in units and sub-units), indexed and combined, through hyperlinks, in order to formulate training materials that will support specific training needs. However, the training material developed using these aids, although it allows the user to examine specific constructs making up an ERP system or subsystem, is not designed in a way that will allow him/her to understand the underlying structures and relationships between these constructs. The reason being that most training aids only provide for manipulating and restructuring multimedia objects and not for externalizing the underlying logic for the knowledge domain under consideration. Hence, in order to help the user get an in-depth understanding of each construct making up an ERP system and subsystem, how it is used and how it relates to other constructs, this knowledge must be externalized, made explicit and therefore become diffused and reusable. To accomplish this, a method that considers the fundamental building blocks of the perception process is needed.

Perception is the process of building a working model that represents and interprets sensory input (mosaic of percepts) into a more abstract part (conceptual graph) [8], [9]. A conceptual graph is made of concepts (the simplest possible self-contained entities) and the relations between them. Therefore, when a trainee is asked to understand the training material accompanying a training process, the act of consuming this material can be modeled as a two stage process: (i) the analysis process, where the material is broken down into concepts and (ii) the synthesis process where concepts are linked to other concepts (found in the training material on hand and other related material that the trainee has already analyzed before) in order to form more complex structures (conceptual graphs). Therefore, meaning is not discovered but constructed, and training material has meaning only in relation to other material, being interconnected to each other as codes and systems in the culture and in the minds of trainees.

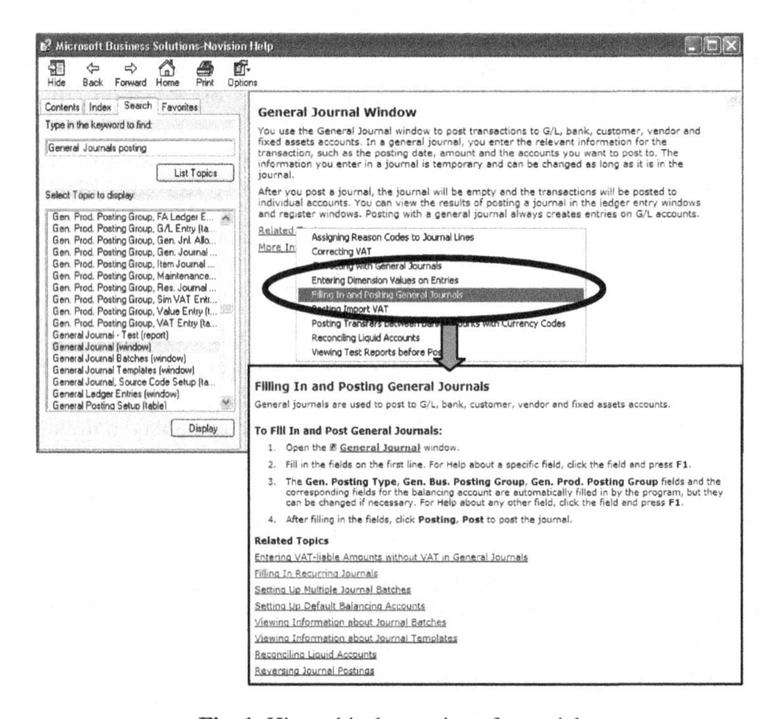

Fig. 1. Hierarchical grouping of material

One of the major issues in ERP systems' training is that the trainees are introduced with hundreds of new constructs, and they have problems understanding how each construct relates to the theoretical and practical knowledge that they have already developed and how constructs relate to each other. For example a "General Journal Posting" (construct) is a group of balanced General Ledger transactions that (a) is characterized (relation) by a "General Journal Template" (such as payments, receipts etc.) and (b) updates (relation) General Ledger accounts' statements, balance of accounts and journals. This knowledge is difficult to be extracted when the material is organized textually (Figure 1), but it can be easily extracted when the material is organized semantically (Figures 2 and 3). The main objective of this paper is to describe a method for developing training materials that capture that knowledge in a way that will help the trainees understand the underlying structures and relationships between these constructs and hence communicate it and make it reusable, while still allowing hands-on context specific learning that will help trainees bridge the gap to practice. This method is also highly transferable to the development of training materials for any kind of systems and users and can produce similar gains in learning and understanding in any domain of knowledge.

2 Design of an Ontology-Based Training Tool

The proposed model is developed on the bases of the principles set by various initiatives like the semantic web, all having in common the focus on extending current web technology with machine-understandable metadata [10]. Those metadata are stored in ontologies [11] and play an essential role in semantic web, since they provide the shared conceptualizations expressed in a logical form. The semantic web vision has been combined with the principles of knowledge transformation in order to provide a theoretical model of e-learning processes [2], [12], [13], [14] thus enhancing the Knowledge-creating company towards the vision of the Semantic Learning Organization (SLO) [15].

In designing an ontology-based training aid, the main objective is to capture and represent the knowledge, which is implicit in the application domain so that it can be made reusable. Thus, domain experts record their knowledge on the particular field under consideration in terms of an ontology, which is recorded in the *ontology repository* and, hence, better communicate it and make it reusable. Therefore each ontology construct is recorded only once and can be made available to every training scenario using it. In addition, relevant supportive material (either existing or created), in the form of multimedia objects (e.g. text, image, video and animation), is used in order to develop a collection of reusable multimedia objects that are related to the knowledge domain under consideration [16], [17]. This collection of multimedia objects comprises the *content repository*. The ontology and content repositories are then used to create knowledge networks, each corresponding to a training scenario, which are recorded in the *knowledge repository*.

Contrary to traditionally designed training scenarios which are based on mere user navigation to multimedia objects, training scenarios that are based on the proposed approach are enhanced and empowered in that they allow users to navigate into the domain knowledge which has been represented in the form of a knowledge network. Thus, the user of the training scenarios is guided either through a semantic search followed by a navigation to the knowledge network, or directly through navigation to the knowledge network. To enhance his/her understanding of each ontology construct included in a knowledge network, the user can access relevant supportive material in the form of multimedia objects and identify the relation of the particular construct with other relevant constructs.

3 The ERP Applications' Training Tool

For illustrative purposes a sample ontology and one training scenario were built based on the proposed approach. The scope of the training scenario is to help trainees understand what is the meaning and use of the basic general ledger constructs and how constructs relate to each other. The ontology-based knowledge network that follows was constructed using the SemTalk2[1] ontology editor for MS-Office 2003, a Visio 2003 add-in that provides all the modeling functionality needed to create ontologies complying with the standards set by W3C's (the World Wide Web Consortium) recommendation OWL[2].

The ontology built for the needs of this research was based on: (a) the ontological analysis of Sowa [18], (b) the resource-event-agent (REA) model for enterprise economic phenomena [19], (c) the work of Geerts and McCarthy [20] on a domain ontology for business enterprises, based on the REA model and (d) the Enterprise Ontology [21].

In the text that follows, the General Ledger (GL) transactions scenario is provided using ontology concepts (shown in italics) and ontology relations (shown in single quote enclosures). Higher level concepts appear in parenthesis, right after each lower level concept. In the diagram, concepts are represented as rounded rectangle nodes and relations as lines connecting them.

3.1 The GL (General Ledger) Transactions Scenario

The GL Transactions scenario shows what is the meaning and use of each general ledger construct and how constructs relate to each other.

Figure 2 shows the GL Transactions scenario. New transactions are entered through *General Journals (Data Entry Screen)*. When the user first enters the screen a 'choice' appears in *General Journal Template List (List of Values)* that 'refers to' *General Journal Template (Document Type)* and the user must choose the type of document he/she wishes to enter (payments, receipts, clearing transactions, etc.). Each *General Journal Template (Document Type)* 'uses for autonumbering' *Number of Series (Record ID)* as each new document takes automatically by the system a new ID. *General Journals (Data Entry Screen)* 'updates journal entry' *General Ledger Posting (Journal Entry)* that 'uses for autonumbering' *Number of Series (Record ID)* as each new journal entry must have a unique ID. When *General Ledger Posting (Journal Entry)* is posted it 'updates transactions' *GL Transaction (Transaction)*. Each *GL Transaction (Transaction)* 'refers to' an existing *GL Account (Account)* and 'updates balance' of *GL Account (Account)*. On the other hand *GL Account (Account)* 'balance is analyzed to' *GL Transactions (Transaction)*. For VAT calculations, each *GL Account (Account)* 'use for VAT' *General Business Posting Group (Data Entry Screen)* and *General Product Posting Group (Data Entry Screen)*. Each *GL Transaction (Transaction)* 'refers to' a *GL Account (Account)* and receives in order to 'use for

---

[1] http://www.semtalk.com

[2] Web Ontology Language. OWL facilitates better machine interpretability of Web content than that supported by other languages like XML, RDF, and RDF Schema (RDF-S) by providing additional vocabulary along with formal semantics.

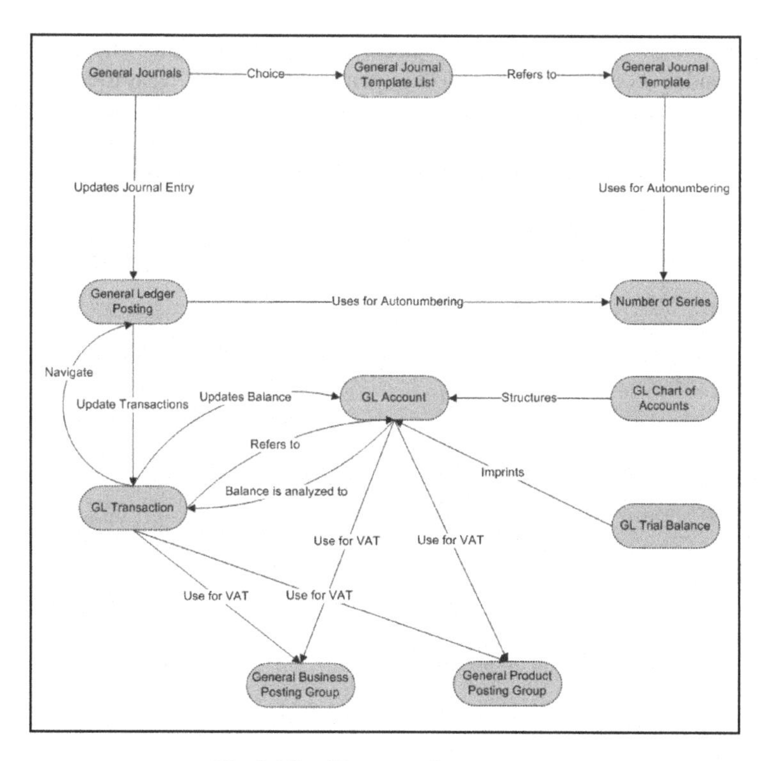

Fig. 2. The GL transactions scenario

VAT' *General Business Posting Group (Data Entry Screen)* and *General Product Posting Group (Data Entry Screen)* whose values can be modified per *GL Transaction (Transaction)*, if needed. The *GL Chart of Accounts (Chart of Accounts)* 'structures' *GL Account (Account)* and *GL Trial Balance (Balance of Accounts Report)* 'imprints' *GL Account (Account)* balances.

This scenario aims to show to the trainee (1) the steps that a user must go through in order to post a transaction into the GL subsystem (select template, enter journal entry and post transactions), (2) how setup is involved with the process (number of series, templates, GL accounts, VAT parameterization) and (3) how posted transactions relate to accounts and how accounts' balance is calculated from the transactions.

3.2 The User Interface

Figure 3 shows the users' interface of the scenarios. The user has two options: (1) chose one of the scenarios and navigate through concept instances and relations or (2) search for a concept instance, in which case the system displays all occurrences of the concept instance in all scenarios and then select a specific scenario to navigate. When selecting an instance (for example *GL Chart of Accounts*) the system displays its properties and all supportive multimedia associated with the specific instance. So the user can discover in the properties window (on the left): (1) all the scenarios (pages) where the specific instance appears, (2) comments explaining its use in the

ERP application, (3) the ontology-defined concept that relates the term used by the ERP vendor (concept instance) to the generally accepted term used in the literature (concept) and (4) all the relations defined for the specific instance in relation to other instances. The user can also navigate into the supportive multimedia (in the specific example an adobe acrobat document) and find additional information about the instance under consideration.

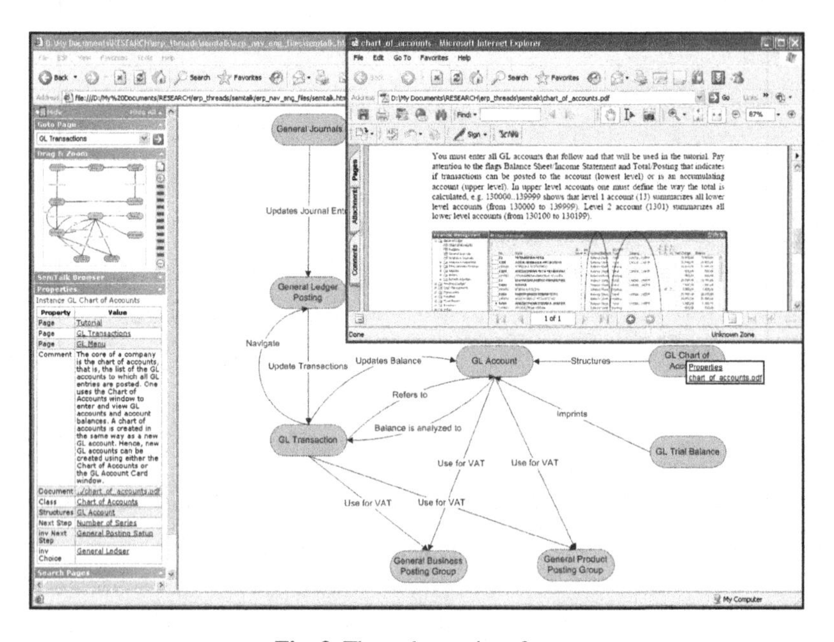

Fig. 3. The web user interface

4 Discussion and Concluding Remarks

The approach proposed in this paper is mainly concerned with capturing and representing the knowledge found in the logic, the structure and the ways of use of ERP systems as ontology-based knowledge networks, i.e. training scenarios serving a specific training needs. The ontology contains all the relative concepts and the relations between them. The knowledge network relates the basic entities defined in the ontology with the various multimedia objects, which are supportive for better understanding the ontology constructs. Thus, the user of the resulting training material is enabled to search for an ontology construct (for example an ERP task) and understand its meaning and usage with the help of the supportive multimedia. Furthermore, the user can navigate to associated ontology constructs in order to acquire an in depth knowledge about ERP processes, the data and control flows between them and how they can be combined in order to solve specific real-life problems.

With regard to the trainee, the main advantages of the proposed model are the following: a) Semantic search - This allows to search ontology or knowledge constructs semantically instead of textually putting emphasis on matching the content and the real meaning of each relevant concept searched [22]; b) Knowledge or conceptual navigation - This allows the use of browsing and navigation capabilities in order to identify the ontology or knowledge constructs as they are recorded into the knowledge repository and used in the training scenario and involves: (i) navigation into the knowledge domain under consideration, (ii) navigation into the business context and (iii) navigation into the structure of the educational scenario [22]; and c) Knowledge dissemination – This is an important function of any kind of training activity that can only be achieved if the trainee is provided with the ability to extract the knowledge implicit in the problem domain, as opposed to the mere presentation of facts and disconnected information which, in most cases, is not adequate. With the proposed model, knowledge is made explicit in order to assist the trainees' combination (from explicit to explicit) and internalization (from explicit to tacit) knowledge transformation processes [2].

In addition, significant advantages of the proposed model could be identified for the creator of the training material as well. The most important of them can be grouped around its three main components (repositories): a) Ontology repository - There are many benefits when using ontologies that have already been recognized in the learning technology community [23], [24]; b) Content repository - Content reusability is a key issue in the literature [16], [17]; and c) Knowledge repository - A knowledge network is a self contained entity that serves a specific training need in a specific knowledge domain, in a specific business context and has a specific structure [22]. Reusability of knowledge recorded into training scenarios is also achieved as knowledge constructs instilled into older scenarios can be used into new scenarios in order to meet new training needs.

Additionally, the expected deployment of the semantic web, which provides semantic meaning for concepts, allowing both users and machines to better interact with the information, will allow the combination and multiple exploitation of dispersed training ontologies through the internet: (i) from writers of training material, since training ontologies and scenarios will be available at various sites around the world and (ii) from web services that will be able to process the knowledge built into ontologies and knowledge networks for various purposes.

It must be noted, however, that the proposed model enhances and empowers existing methodologies by allowing the semantic representation of knowledge so that to enable trainees navigate into the underlying knowledge of the application domain under consideration. Thus, the model can combine the existing multimedia material with ontology constructs, using knowledge-based multimedia authoring tools, in order to build user training scenarios and satisfy specific training needs. Hence, in addition to the existing multimedia objects, the knowledge built into both the ontology and the training scenarios is fully reusable. Finally, due to the encouraging features of the approach described, it is intended to evaluate it extensively using more elaborate implementation tools and more complex ERP business processes.

References

1. Tome, E.: IC and KM in a macroeconomic perspective: the Portuguese case. International Journal of Learning and Intellectual Capital 5(1), 7–19 (2008)
2. Nonaka, I., Takeuchi, H.: The Knowledge-Creating Company. Oxford University Press, New York (1995)

3. Macris, A., Georgakellos, D.: A Business Plans Training Tool Based on The Semantic Web Principles. In: Lytras, M.D., Carroll, J.M., Damiani, E., Tennyson, R.D. (eds.) WSKS 2008. LNCS (LNAI), vol. 5288, pp. 225–233. Springer, Heidelberg (2008)
4. Bingi, P., Sharma, M., Godla, J.: Critical issues affecting and ERP implementation. Information Systems Management 16(3), 7–14 (1999)
5. Macris, A.: Enterprise Resource Planning (ERP): A Virtual lab implementation for managers' and users' training. 'SPOUDAI' Journal, 13–38 (October-December 2004)
6. Soffer, P., Golany, B., Dori, D.: ERP modelling: a comprehensive approach. Information Systems 28, 673–690 (2003)
7. Mahapatra, R.K., Lai, V.S.: Intranet-based training facilitates on ERP system implementation: a case study. In: Hoadley, E.D., Benbasat, I. (eds.) Proceedings of the Fourth Americas Conference on Information Systems, Baltimore, MD, pp. 1070–1072 (1998)
8. Novak, J.D., Gowin, D.B.: Learning How to Learn. Cambridge University Press, New York (1984)
9. Sowa, J.: Conceptual Structures, information processing in mind and machine, ch. 3. Addison-Wesley Publishing Company Inc., Reading (1984)
10. Berners-Lee, T., Hendler, J., Lassila, O.: The Semantic Web. Scientific American, 28–37 (May 2001)
11. Gruber, T.R.: A translation approach to portable ontology specifications. Knowledge Acquisition 5(2), 199–220 (1993)
12. Naeve, A., Yli-Luoma, P., Kravcik, M., Lytras, M., Lindegren, M., Nilsson, M., Korfiatis, N., Wild, F., Wessblad, R., Kamtsiou, V., Pappa, D., Kieslinger, B.: A conceptual modelling approach to studying the learning process with a special focus on knowledge creation. Deliverable 5.3 of the Prolearn EU/FP6 Network of Excellence, IST 507310 (June 2005)
13. Yli-Luoma, P.V.J., Naeve, A.: Towards a semantic e-learning theory by using a modeling approach. British Journal of Educational Technology 37(3), 445–459 (2006)
14. Collazos, C.A., Garcia, R.: Semantics-supported cooperative learning for enhanced awareness. International Journal of Knowledge and Learning 3(4/5), 421–436 (2007)
15. Sicilia, M.-A., Lytras, M.D.: The semantic learning organization. The Learning Organization 12(5), 402–410 (2005)
16. Chebotko, A., Deng, Y., Lu, S., Fotouhi, F., Aristar, A.: An Ontology-Based Multimedia Annotator for the Semantic Web of Language Engineering. International Journal of Semantic Web and Information Systems 1(1), 50–67 (2005)
17. Steinmetz, R., Seeberg, C.: Meta-information for Multimedia e-Learning. In: Klein, R., Six, H.-W., Wegner, L. (eds.) Computer Science in Perspective. LNCS, vol. 2598, pp. 293–303. Springer, Heidelberg (2003)
18. Sowa, J.: Knowledge Representation - Logical, Philosophical and Computational Foundations, ch. 2, Brooks/Cole, USA (2000)
19. McCarthy, W.E.: The REA accounting model: a generalized framework for accounting systems in a shared data environment. Accounting Review 57, 554–578 (1982)
20. Geerts, G.L., McCarthy, W.E.: An ontological analysis of the economic primitives of the extended-REA enterprise information architecture. International Journal of Accounting Information Systems 3(1), 1–16 (2002)
21. Uschold, M., King, M., Moralee, S., Zorgios, Y.: The Enterprise Ontology. The Knowledge Engineering Review 13(1) (1998), developed in the Enterprise project (IED4/1/8032), supported by the UK's Department of Trade and Industry under the Intelligent Systems Integration Program, pp. 31–89

22. Stojanovic, L., Saab, S., Studer, R.: eLearning based on the Semantic Web. In: Proceedings of the World Conference on the WWW and the Internet (WebNet 2001), Orlando, FL, USA (2001)
23. Sampson, D.G., Lytras, M.D., Wagner, G., Diaz, P.: Guest editorial: ontologies and the Semantic Web for e-learning. Educational Technology and Society 7(4), 26–28 (2004)
24. Lytras, M., Tsilira, A., Themistocleous, M.: Towards the semantic e-learning: an ontological orientated discussion of the new research agenda in e-learning. In: Proceedings of the 9th Americas Conference on Information Systems, Tampa, FL, USA, pp. 2985–2997 (2003)

Exploring the Potential of Virtual Worlds for Enquiry-Based Learning

K. Nadia Papamichail, Amal Alrayes, and Linda A. Macaulay

Manchester Business School, University of Manchester, Booth Street East, Manchester, M15 6PB, U.K.
nadia.papamichail@mbs.ac.uk, amal.alrayes@postgrad.mbs.ac.uk, linda.macaulay@mbs.ac.uk

Abstract. Nowadays academic institutions seek to equip students with interpersonal and project management skills by enhancing their learning experience through the use of new collaborative technologies. This paper presents an enquiry based learning initiative at Manchester Business School. A virtual world environment was introduced to facilitate group project work. The setting was the B.Sc. Information Technology Management for Business (ITMB) programme, a new degree designed to meet the needs of major employers in the business-led IT sector. This paper discusses how the project acted as a vehicle for increasing the involvement of employers in the programme and achieving the objectives of greater student creativity, productivity, engagement, participation and productivity in team work.

Keywords: Virtual worlds, Second Life, EBL, team work.

1 Introduction

Virtual worlds are 3D environments that combine game-based graphical capabilities with social interaction systems [5]. They allow educators to organize pedagogical events and explore new innovative methods for teaching and learning [12]. They provide a sophisticated environment where users can interact, socialize and collaborate [17].

It has been shown (see for example [10] and [16]) that enquiry based learning (EBL) approaches increase creativity and participation in student teams by encouraging students to 'learn by doing'. In this paper, we describe an EBL based project where students are encouraged to be creative through the use of 'Second Life', one of the most well known virtual worlds. In such a virtual world, learners' collaborative interactions can occur across distance using avatars (i.e. users' computer representations). Interpersonal dynamics amongst avatars facilitate learning activities in a manner rather different than typical face-to-face collaborative meetings [8]. It should be noted however, that virtual worlds are not limited in supporting distance education but they can also be used to enhance classroom activities [9].

Apart from academic institutes, businesses are increasingly deploying virtual worlds for business collaboration and seek to build a virtual presence for themselves.

M.D. Lytras et al. (Eds.): WSKS 2009, LNAI 5736, pp. 376–385, 2009.

Recent research suggests that "slowly, companies are leaving the physical world behind to cut costs, improve communication, and find new ways to collaborate" [20].

ITMB (IT Management for Business) is a UK undergraduate programme that was designed by leading IT employers such as Accenture, BBC, BT, Capgemini, Cisco, Deloitte, Logica, HP, IBM, Morgan Stanley, Procter & Gamble and Unilever. It seeks to equip students not only with IT management but also with interpersonal and project management skills (more details about ITMB can be found at www.e-skills.com/itmb). Manchester Business School runs an ITMB programme that involves employers through events, guest lectures, guru lectures delivered via video conferencing, and a mentoring scheme. Employers contribute to the curriculum by offering business problems and case studies and by organizing career and onsite events for the students.

This work discusses an EBL based project involving employers, tutors and ITMB students at Manchester Business School. The paper starts with an overview of the key concepts associated with virtual worlds and EBL. This is followed by a description of the EBL based project. The outcomes of the project are discussed and assessed through the perspectives of the main three stakeholders: employers, students and tutors. Group performance results are presented. The paper concludes with a summary of the work.

2 Virtual Worlds

Virtual worlds are computer programs which allow users to connect and interact with each other in real time in a shared visual space. They possess key features of interactivity, physicality, and persistence [2] that enable the creation of a feeling of being with others. They are essentially 3D graphical environments, which were originally developed from the field of computer games, and can be accessed over the Internet [29]. They allow a large number of users to interact synchronously. Such environments have evolved into stimulating, dynamic and collaborative settings [29]. Apart from education communities considering the use of virtual worlds, commercial enterprises such as IBM are also exploring the use of these spaces for improving leadership and strategic thinking skills [11].

As Salt et al [29] point out, Second Life is a popular virtual world that allows users to create complex environments and objects. It is a 3D immersive world with chat facilities and social spaces where users can reconstruct the virtual space. It provides sophisticated graphics and has a relatively low cost of entry. It has not been specifically designed as a pedagogical instrument for supporting teaching and learning processes. Its creator Philip Rosedale, launched Second Life as an objective-orientated game back in 2003. Later on, the concept evolved into a user-created environment, in which the in world residents are creators of their own environments [2].

Kay and Fitzgerald [19] suggest that 3D virtual worlds allow for rich experiences, experiential learning activities, role-play and simulation. They are platforms for data visualisation and offer collaborative opportunities. Kay and Fitzgerald also explain that Second Life is used to support educational needs and suggest a list of educational activities that can take place including: self-paced tutorials, displays and exhibits, data visualisations and simulations, historical re-creations and re-enactments, living and immersive archaeology, machine construction, treasure hunts and quests, language and cultural immersion and creative writing. Most of these activities can be used to encourage experiential learning. Discipline areas represented in Second Life include computer programming and artificial intelligence, literature studies, theatre and performance art, language teaching and practice, politics, commerce, architectural design and modelling and urban planning.

A large number of educational institutions around the world, including Manchester Business School, are now investigating and using Second Life as an educational platform. As the following statistics by the Gronstedt Group [13] suggest, an increasing number of individuals, communities and businesses are now using Second Life:

- 80 percent of Internet users will be active in non-gaming virtual worlds such as Second Life by the end of 2011.
- IBM is investing in a number of Second Life islands, and other major companies are following suite e.g. Sun, Dell, Intel, Adidas, and Toyota.
- Hundreds of universities, including Harvard and INSEAD, have set up classes in Second Life.

3 Enquiry-Based Learning (EBL)

EBL is described as an environment in which learning processes are approached by learners' enquiries. In such environments generated knowledge is easily retained as it was previously experienced by the learner. EBL paves the way for learners to be problem solvers and knowledge creators in their future working environments [10]. The theory of experiential learning specifies that "knowledge is created through the transformation of experience" [21] – see the four modes of learning in Fig. 1. Experiential learning, can be conversational, whereby, "learners construct meaning and transform experiences into knowledge through conversations" [3]. These conversations could be held electronically through the use of virtual worlds.

"Without the opportunity for experiential learning of some sort, management education programmes will continue to produce students unprepared for the realities of organizational life" [15]. The ITMB programme at Manchester Business School aims to equip students with academic as well as interpersonal and project management skills, by engaging students in project group activities and involving employers. In this paper, one example of engaging students in experiential work is presented.

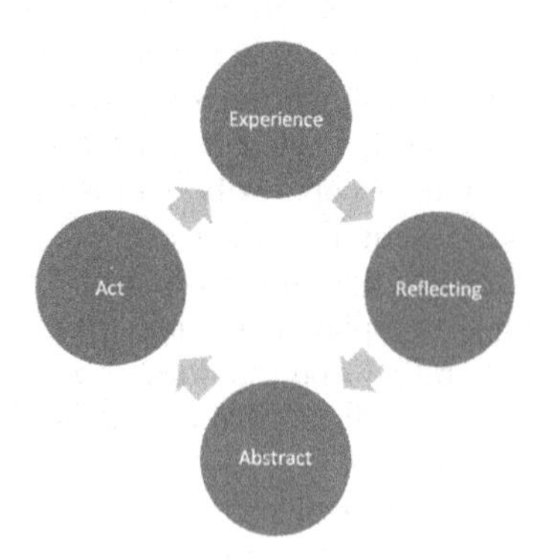

Fig. 1. Experiential Learning Model (source: Kolb [21])

4 An EBL Project

This section discusses the task, team arrangement, and structure of the EBL project that was undertaken at Manchester Business School. The setting was the Business Team Project course unit, whose main learning objective was to practice an EBL project with a business problem set by an industry client. E-skills UK, a not-for profit organization whose aim is to advance technology skills in the UK, and IBM provided the problem. The task was to create a virtual world environment where ITMB students, prospective employers, school teachers and university academics could meet for purposeful activity. The deliverable was to identify stakeholders and requirements, design a proof of concept ITMB environment and organize a virtual world event. As Manchester Business School had a presence in Second Life, it was decided that Second Life would be the preferred virtual world environment.

The students had access to a wide range of rooms and facilities available on the Manchester Business School island (i.e. virtual space). Thirty eight second year undergraduate students were divided into seven teams of five or six. Each team had to identify the needs of ITMB stakeholders e.g. students, employers, and tutors and organize an event in Second Life to address some of their requirements. The taught part of the Business Team Project involved a one-hour lecture and a two-hour tutorial session each week throughout the academic year (20 weeks of teaching). Discussion boards on Blackboard were set up in the second semester to provide technical support.

The lectures covered a range of topics including project management, research methodologies, stakeholder analysis, facilitation techniques, business report writing, organization of events and presentation skills. Guest lectures with speakers from companies on topics such as teamworking skills were arranged. The aim of the tutorials was to enable students to acquire technical skills (administration of virtual spaces) and encourage them to adopt an EBL approach. During the first semester students met with tutors during the tutorial hours to discuss ideas, define the scope of the project, identify requirements, and set up a plan. In the second semester, students undertook Second Life tutorials to familiarize themselves with the environment and design

Table 1. Second Life events organised on the MBS island

Team no.	Event description	Employers involved
1	ITMB careers event	HP
2	Collaborative meetings in Second Life	Deloitte, e-skills UK
3	ITMB troubleshooting session offering advice	--
4	Presentations by employers and Second Life tutorials	P&G, e-skills UK
5	Facilitation of group activities	--
6	Mock interviews with employers	Careers Service
7	Informal networking bar & lecture facilities	Accenture, e-skills UK

suitable areas for their event. Most of the teams invited employers and organized mock events. All of them used specialized software to video their actual events. Table 1 provides descriptions of the events organized. All groups produced video clips to describe their efforts and some of them were posted on YouTube:

www.youtube.com/watch?v=bzvRTByVeAw
www.youtube.com/watch?v=kYXiMZUxFGc

5 Results

In order to assess the EBL project, we applied a collaborative learning framework that consists of the following five concepts [24]:

- *Participation*: level of teamwork, participation in discussions
- *Productivity*: level of achievement, quality of outcome
- *Creativity*: level of contribution of ideas, novelty
- *Engagement*: level of motivation, passion for their work, enthusiasm
- *Understanding*: level of understanding of the problem, and application of theory to practice

The above key concepts have been found to be important by a number of studies on teamwork: participation ([1], [14], [16], [18], [25], [27]), productivity ([4], [18], [25], [27]), creativity ([10], [16]), engagement ([7], [10], [18], [22], [26]), and understanding ([6], [16], [23]).

The students' team work was assessed from three different perspectives (i.e. students, employers and module tutors) taking into account the above criteria. In the case of the students, we adapted the framework to establish whether Second Life improved their performance (students' perspective).

5.1 The Students' Perspective

In order to establish the students' attitudes towards the project, a questionnaire was administered. The aim of the questionnaire was to collect feedback on the use of Second Life and to establish the attitudes of the students towards the virtual world environment and rate their overall experience. A total of 17 responses were received. The analysis of the findings in terms of level of participation, creativity, productivity, engagement and level of understanding are shown in Fig. 2.

The results show that in terms of participation in Second Life events (i.e. equal contributions from teams' members, honest opinions on ideas and questioning of ideas) 50% of the students rated their participation as "weak", 6% as "average" and 44% as "very good". In terms of productivity i.e. whether the use of Second Life reduced social loafing and production blocking, 56% of the students reported productivity as "satisfactory", 13% as "good", 25% as "very good" and only 6% as "excellent". In terms of generating new ideas and knowledge (creativity), 38% of the students rated Second Life as a "very good" environment, 31% as a "good" environment, 25% as "satisfactory" and only 6% rated the environment as "poor". In terms of engagement i.e. whether team members applied more effort to the task when engaged in Second Life activities, 44% stated that engagement was "satisfactory", 19% stated it was "good" and 38% stated it was "very good". In terms of understanding (i.e. whether Second Life guided the students into a deeper level of understanding of theory such as IT development, project management and group theory through practice) 6% reported that understanding was "poor", 25% reported a "superficial understanding", 25% reported "average", and 44% reported the environment to lead into a "deep understanding".

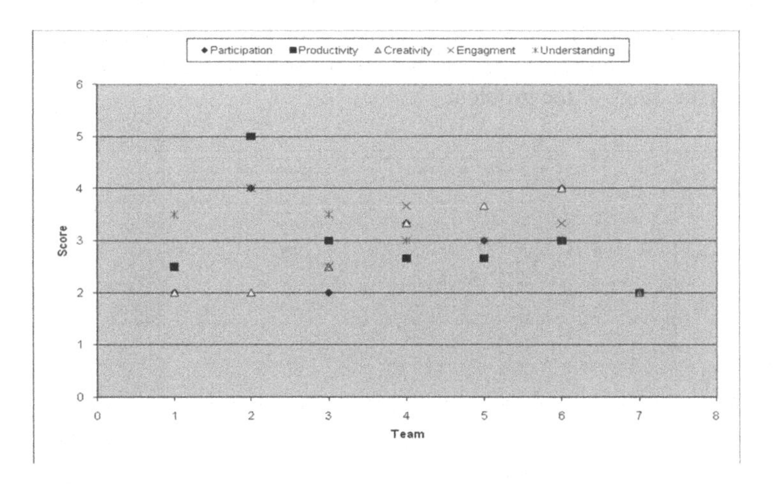

Fig. 2. Students' attitudes towards Second Life

The students also provided qualitative feedback. Some of their concerns regarding SL are highlighted in the following statements:

"I think we still prefer more face-to-face meeting than doing it in SL. Because in SL, people who are SL active users might tend to be distracted from participating and less interested people are not keen to do anything in SL."

"People do not take meetings seriously in SL it is treated like a game."
"People are unsure of its uses, so don't know how creative they can be."

5.2 The Employers' Perspective

The employers' assessment of the students' work was carried out at an ITMB employers' event. Such events are regularly organized at Manchester Business School to help ITMB students develop their presentation skills. They have been shown to increase the students' confidence and their ability to converse with business representatives, which will hopefully increase their performance at job interviews. During the ITMB event, all the groups were given the opportunity to demonstrate their work using posters and video clips. Several employers from a range of companies such as IBM, Accenture, Deloitte, e-Skills UK, P&G, Unilever, BT and Informed Solutions, visited the stands of the groups and assessed their project work by completing a feedback form.

Fig. 3 shows the employers' assessment results based on the level of participation, productivity, creativity, engagement and understanding of the groups. In terms of level of teamwork and participation in discussion, two out of the seven student teams were considered as "average", and five of them were considered "very good". In terms of level of achievement and quality of solutions, two out of the seven teams were regarded as "average" while the rest were rated as "very productive". In terms of level of contribution of own ideas and novelty, five of the teams were rated as "very creative" while the rest as "average". In terms of level of motivation, passion for their work and enthusiasm, five of the teams were thought of having a "very good" level of engagement. In terms of level of understanding of the problem, and application of theory to practice, the results showed that all of the teams were regarded as having a "deep understanding" of the problem.

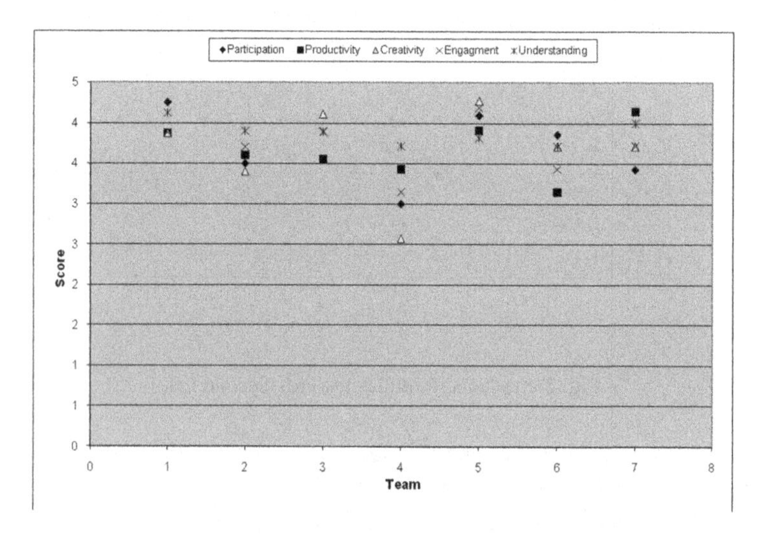

Fig. 3. Employers' assessment results

The employers also provided qualitative feedback to the groups. The following comments highlight the employers' priorities regarding the group project work:

"clearly defined roles, team worked together plus supported one another"
"highly innovative idea which given more time is likely to become something impressive"
"good understanding of content plus needs of employers"

Employers gave higher marks to those groups who were able to articulate the business problem well and provide a solution that addressed the needs of several stakeholders. They were also looking for high quality solutions. For example, group 1 had thoroughly explored the business problem and had a contingency plan in place (in case their Second Life event was not successful), group 2 conducted several trial events and group 5 developed a technically superior solution.

Those employers who participated in Second Life events and engaged in activities found these quite stimulating. They felt that their Second Life experience was more satisfactory to that of a teleconference meeting. They also stated that the current financial crisis, which has imposed a strain over commercial travel budgets, has made Second Life events an attractive alternative to more traditional settings.

5.3 The Tutors' Perspective

The Second Life group project was rather intensive. Students had to work closely together in their groups. Their first task was to make sense of the business problem presented to them (i.e. organize an event in Second Life) and develop necessary skills so as to formulate solutions. As in many other group projects (see Rickards et al. [28]) they went through emotions of anxiety before optimizing their performance. It was therefore important that the level of participation in team work activities and discussions was equal from all members. Peer assessment results obtained twice during the academic year showed that the contributions of group members to the project were equal in most groups.

All the groups were productive. They performed very well and achieved very high marks. The opportunity to showcase their work in front of employers and academics boosted their confidence and made them have a sense of achievement at the end of the project. The project outcomes were all of high quality. Half of the projects were rated at distinction level with the remaining projects achieving a 'very good' mark.

The tutors were impressed by the level of creativity shown by the students. Each group came up with a different concept and solution to the problem. The EBL setting and the 'learning by doing' approach allowed the students to develop their own ideas and devise their own solutions based on their own creativity.

The interactions with employers and the incentive of three financial prizes by IBM, increased the level of engagement of the students and energized their work. A learner's motivation has been shown to have an impact on the quality of the learning experience [6]. All groups showed considerable enthusiasm and passion for their work during their presentations to employers and tutors.

The EBL setting encouraged the students to integrate different pieces of information from the course unit such as research methodologies and scripting in virtual worlds as well as other course units such as business application design and development, human computer interaction, database design and development and project management. This allowed students to arrive at a deeper understanding of theory through practice.

6 Conclusions

This paper discusses an EBL initiative that has been undertaken at Manchester Business School. Students of an IT Management for Business undergraduate programme were presented with a business problem and were encouraged to 'learn by doing in groups'. Each group generated different ideas and produced different solutions to the same problem. The solutions were developed in Second Life, a virtual world environment. The project outputs were presented to both employers and tutors. A framework of five concepts (participation, productivity, creativity, engagement and understanding) was applied to assess the overall experience of the students in the virtual world (students' perspective), as well as the outcomes of the project (employers and tutors' perspectives). The students had to integrate theory from different modules and develop new skills such as navigating, interacting and designing new spaces in a virtual world as well as interpersonal skills such as presentation, project management and teamwork skills. The involvement of employers in the EBL project evidently stimulated active engagement and boosted students' self confidence. The group projects will continue next year with a different business problem presented to the students.

Acknowledgments

We would like to thank all those employers who participated in the Second Life events organized by the ITMB students and assessed the students' efforts. We are grateful to Bob Clift and Liam Hoult from e-skills UK, for providing the business problem of the project. We are also thankful to IBM for offering prizes to the best ITMB teams.

References

[1] Aiken, M., Govindarajulu, C.D., Horgan, D.: Using a Group Decision Support System for School-based Decision Making. Education 115(3), 420–425 (1995)
[2] Au, W.J.: The Making of Second Life: Notes from the New World. Collins Business, New York (2008)
[3] Baker, A.C., Jenson, P.J., Kolb, D.: Conversational Learning an Experiential approach to Knowledge creation. Quorum Books, Connecticut (2002)
[4] Carroll, J.: HCI Models, Theories, and Frameworks. Morgan Kaufmann, Elsevier Science (2003)
[5] Castranova, E.: Virtual Worlds: A First-Hand Account of Market and Society on The Cyberian Fronteir. CESifo Working Paper Series No. 618 (2001), http://ssrn.com/abstract=294828 (Retrieved April 27, 2009)
[6] Cocea, M., Weibelzahl, S.: Can Log Files Analysis Estimate Learners' Level of Motivation? In: 14th Workshop on Adaptivity and User Modeling in Interactive Systems, Hildesheim, October 9-11 (2006)
[7] Coch, L., French, J.: Overcoming Resistance to Change. Human Relations 1(4), 512–532 (1948)
[8] Dede, C.: Emerging technologies and distributed learning. American Journal of Distance Education 10(2), 4–36 (1996)
[9] Dillenbourg, P., Schneider, D., Synteta, P.: Virtual Learning Environments. In: Proceedings of the 3rd Hellenic Conference on Information and Communication Technologies in Education, pp. 3–18 (2002)
[10] Edelson, D.C., Gordin, D.N., Pea, R.D.: Addressing the Challenges of Inquiry-Based Learning Through Technology and Curriculum Design. The Journal of Learning Science 8(3&4), 391–450 (1999)
[11] GIO, Virtual Worlds, Real Leaders (2007), http://domino.watson.ibm.com/comm/www_innovate.nsf/pages/world.gio.gaming.html (Retrieved April 6, 2009)
[12] Good, J., Howland, K., Thackray, L.: Problem-based learning spanning real and virtual worlds: a case study in Second Life. ALT-J, Research in Learning Technology 16(3), 163–172 (2008)
[13] Gronstedt, A.: Welcome to Web 3-D! (2007), http://commons.iabc.com/branding/category/second-life/ (Retrieved April 5, 2009)

[14] Guzzo, R., Dickson, M.W.: Teams in Organizations: Recent Research on Performance and Effectiveness. Annual Review of Psychology 47, 307–338 (1996)
[15] Hyde, P.: Integrating experiential learning through 'live' projects: A psychodynamic account. In: Reynolds, M., Vince, R. (eds.) Experiential Learning and Management Education, pp. 291–308. Oxford University Press, Oxford (2007)
[16] Kahn, P., O'Rourke, K.: Understanding Enquiry-based Learning (EBL), Handbook of Enquiry & Problem Based Learning. In: Barrett, T., Mac Labhrainn, I., Fallon, H. (eds.) pp. 1–12. CELT, Galway (2005), http://www.nuigalway.ie/celt/pblbook/ (last accessed on February 17, 2009)
[17] Kahai, S., Carroll, E., Jestice, R.: Team Collaboration in Virtual Worlds. The DATA BASE in Advances in Information Systems 38(4), 61–68 (2007)
[18] Karau, S.J., Williams, K.D.: Social loafing: A meta-analytic review and theoretical integration. Journal of Personality and Social Psychology 65, 681–706 (1993)
[19] Kay, J., FitzGerald, S.: Educational Uses of Second Life (2007), http://sleducation.wikispaces.com/educationaluses (Retrieved April 5, 2009)
[20] King, R.J.: It's a Virtual World (2009), http://www.strategy-business.com/li/leadingideas/li00121 (Retrieved April 27, 2009)
[21] Kolb, D.: Experiential learning: Experience as the source of learning and development. Prentice-Hall, Englewood Cliffs (1984)
[22] Kravitz, D., Martin, B.: Ringelmann Rediscovered: The Original Article. Journal of Personality and Social Psychology 50, 936–941 (1986)
[23] Lee, V.S., Greene, D.B., Odom, F., Schechter, E., Slatta, R.W.: What is Inquiry-Guided Learning? In: Lee, V.S. (ed.) Teaching and Learning Through Inquiry: A Guidebook for Institutions and Instructors. Stylus Publishing, Sterling (2004)
[24] Macaulay, L.M., Tan, Y.L.: Enhancing Creativity through Group Intelligence Software (2009) (Working paper)
[25] Manning, L.M., Riordan, C.A.: Using Groupware Software to Support Collaborative Learning in Economics. Journal of Economic Education, 244–252 (2000) (Summer)
[26] Markus, M.L.: Power, Politics, and MIS Implementation. Communication of the ACM 26(6), 430–444 (1983)
[27] Nunamaker, J., Briggs, R., Mittleman, D., Vogel, D., Balthazard, P.: Lessons from a Dozen Years of Group Support Systems Research: a Discussion of Lab and Field Findings. Journal of Management Information Systems 13(3), 163–207 (1996)
[28] Rickards, T., Hyde, P.J., Papamichail, K.N.: The Manchester Method: A critical review of a learning experiment. In: Wankel, C., DeFillippi, R. (eds.) Educating Managers through Real World Projects, pp. 239–254. Information Age Publishing, Greenwich (2005)
[29] Salt, B., Atkins, C., Blackall, L.: Engaging with Second Life: Real Education in a Virtual World (2008), http://slenz.files.wordpress.com/2008/12/slliteraturereviewa1.pdf (Retrieved June 20, 2009)

A Model for Mobile Multimedia Capturing

Stavros Christodoulakis[1] and Lemonia Ragia[2]

[1] Laboratory of Distributed Multimedia Information Systems and Applications
Technical University of Crete (MUSIC/TUC)
Chania, Crete, Greece, 73100
stavros@ced.tuc.gr
[2] Advanced Systems Group, Centre Universitaire d'Informatique
University of Geneva
24 rue General-Dufour, 1211 Geneva 4, Switzerland
lemonia.ragia@cui.unige.ch

Abstract. The recent years mobile multimedia information gain considerable demand. We propose a model which can be easily addressed in different, interesting applications. In this paper we describe two applications more detailed in order to demonstrate functionality and usefulness of our model. The basic component is the description of real world events and the management of different data types like images, audio, video and texts.

Keywords: model, multimedia information, capturing, multimedia integration, engineering.

1 Introduction

Mobile multimedia capturing is an important issue for several applications. We propose to help the one man crew with software systems that will help him to do quality capturing of multimedia information, as well as systematic capturing and structuring of the multimedia information at the time of capturing for achieving among others fast post-processing and systematic integration of the captured multimedia information with the workflows and the data processing applications of the organizations.

We present a model that is based on the classification of events to be captured into **EventTypes**. The Event Types describe important types of complex events, some aspects of which have to be captured with multimedia information (video, sound, etc) and they have particular types of metadata that have to be captured as well. The elementary events are captured by shots. The system suggests shot types appropriate for capturing particular event types. The Event Type designer provides systematic methods for capturing the metadata for the event types, as well as for doing quality multimedia capturing (video capturing, etc.) based on the site geometry, the movements of actors and the camera, etc., based on cinematography principles.

The model also describes event instances which are instances of events of particular types as described above. Event instances are captured by event shot instances. Information captured at the instance level also includes camera parameters and spatial

M.D. Lytras et al. (Eds.): WSKS 2009, LNAI 5736, pp. 386–394, 2009.

information which can be used for better visualization of the captured multimedia, as well as for better contextual understanding.

The scenarios for event capturing of a particular event type are described by event capturing workflows. Workflows are modeled after the capabilities of the UML Activity Model. An Event Capturing Workflow instances describe what the state is at a certain point in time. The Event Capturing Workflows are used for describing in a systematic way a process of capturing certain events that may occur in a manner that depends on the event type.

Multimedia capturing applications have increased rapidly in importance the recent years. The most common task is the audio/video capturing. One example is described by Abowd et. all. [1]. It is a system which is specialized in one application, the capturing of lecture information and accessing on-line lectures via the Web. Another approach describes a system that captures and presents multimedia context to students [7]. A system for capturing and presentation of various types of shared communication experiences, which is information exchange provided by non-electronic (e.g. viewgraphs) and electronic devices is described in [3]. The management and integration of multimedia information has been also discussed in different scientific documents. A distributed system to interconnect multimedia data bases and to provide multimedia query processing is proposed by Berra et. all [2]. An interface for a multimedia data base linking information of different media types is presented in [4].There is considerable amount of work done in the research area of multimedia information retrieval. An approach proposes a model which integrates text and images using similarity based methods with semantic based retrieval [5]. Other paper presents the MPEG-7 Query Language for querying MPEG-7 multimedia content description [6]. Our approach is different in that it emphasizes generic models for multimedia information capturing in different application environments, and a framework for their system implementation.

2 Mobile Multimedia Applications

There are a lot of applications that have some emphasis on the multimedia content capturing. We describe here two of them.

2.1 Construction Engineering

Consider any construction site. For the site there is always an approved plan. The work in progress is monitored informally with inspections to direct the work and relate the construction to the building plan and the time plan. These inspections produce materials in reports and photos which are mainly useful to accomplish the construction in time and according to plan.

A mobile multimedia application kit can serve in a number of purposes:

- As documentation for the work done or the work in progress.
- As documentation for the history of a construction and all details that have been done.
- As a “real model” of the construction. This means that we can have a 3d model with all the parts and their connection. In addition, every part can be

accompanied by textual description, video description and a list of pictures. Using also GPS we can have absolute coordinates and import all this material in the office environment used by the construction company.

- It gives a realistic timetable with all the processes and events happening in the construction.
- It helps the communication between different people working for specific aspects of the construction. These can be the construction company engineers and subcontractors or owners.
- All this information can be put in a centralized data base for data storage and data management.
- It is a guide for the future for streamlining the construction or for a new construction.

All this can be accomplished provided that the kit is robust, the Mobile Multimedia report production is non intrusive and easy and the post production and use is efficient and effective. It points to a careful analysis of the critical points and time for the construction including what is critical, necessary or auxiliary to capture. The persons follow a script and shoot the material accordingly.

We note here that the engineering construction has a mobile part, it supports workflows at the distributed sites (order in which work has to proceed in the site) as well as workflows at the central office site (assessing progress, scheduling, reallocating personnel, ordering materials, etc.).

It can first be observed that it is clearly a multimedia application, not just a video application. There can be 2D and 3D representations of the building organized in floors, rooms, etc. in order to allow for the user to find where he is and to associate his multimedia information with the 2D or 3D representations. The interfaces allow simultaneous viewing of the 2D or 3D structures and the video, taking into account the location from where the video was taken, the direction of the video etc. The video is not the only multimedia information needed. There is a need for textual and/or voice information associated with the video to explain the observations such as omissions, damages, etc. There can also be graphics capabilities to identify the specific area where a damage may be. The graphics have to be associated with the video itself (to identify as a separate layer on top of the video the possible damage of the structures, or it may be associated with other multimedia representations (such as the floor plan).

Metadata can be stored with the video. They clearly identify the context, as well as other parameters. The context is possibly identified by location, direction as well as task performed or relevant task. The metadata are useful both for searching as well as for browsing. Browsing of the information can be done in a hierarchical manner from floor to floor and from room to room. However the inspectors and supervisor may want to search and browse in many different ways the information. For example searching where construction work of a specific task is taking place should produce an indication of locations in the floor plans, which would allow subsequently the supervisor to look at the related videos.

Annotation of the information is typically done asynchronously by different workers or inspectors on their mobile devices. It is however often seen synchronously and in a collaborative manner, for example in the communication of the construction personnel with their managers or with the headquarters. There is a requirement for synchronous viewing and browsing of multimedia information. In addition, since the headquarters and the mobile workers may have devices with different capabilities there may be a requirement for multimedia adaptation. Note also that there may be some need for synchronous annotations. This is the case for example where the headquarters cooperatively define the plan of action for an emergency or for next day on top of the maps.

Finally all the multimedia information is visible by all the construction personnel. Mechanisms to reduce the amount of information visible based on the needs of the specific task that a worker is involved are desirable. In addition, the workflows in the headquarters will have to select which multimedia information is needed to be maintained for the long run, and which tasks will handle it. This is done based on the type, the context and the other metadata associated with the multimedia information.

2.2 Real Estate Applications

There are important business applications of multimedia that require geospatial references and mapping information. One of them is real estate agents selling houses or property to remote or local users. The real estate market is important in the Mediterranean countries, not only for the locals, but also for North Europeans that want to buy houses in Mediterranean for spending their vacation and/or retirement. Real estate property and houses are expensive in Crete. In some cases the houses are preconstructed and sold. In other cases the land exists and a selection of house models which can be built exist. Several Real Estate Agents sell through the internet "houses in Crete", "Villas in Crete", etc.

A proper support of this application should allow the footprints of the property or the houses to be reflected on a map, together with videos and pictures of the property. The video presentation and the related metadata have to depend on the kind of the property. In addition, the information system may have to maintain information regarding the near by properties, as well as the laws governing the area (capability to build high rise, commercial, etc.).

For remote buyers the application should support navigation with at least 2D floor plans on the premises of a house, and simultaneous video playing showing the corresponding room. Several alternative navigation plans for the video playing within the building should be supported. Interactive control of the viewer should be also supported. Simultaneous viewing and interactive navigation and discussion between the buyer and the seller should be supported.

Editing notes in terms of text and graphics related to proposed and accepted changes for design personalization should be supported. Access to picture and video repositories that will facilitate the selection of choices for the personalization (such as the selection of the look and the decorative aspects of the fire places of the house or other internal space and external space decoration aspects) should be supported. The video navigation through the house should be able to accommodate the transparent synthetic video integration of the navigation video and the decoration video clips in order to give a good feeling of the final look to the buyer. To do that appropriate selections of scenarios of the presentations of the decorative elements should be defined to match the navigational scenarios of the house. The final decisions should be incorporated in the design model that will be given to the construction teams in terms of both, engineering diagrams as well as synthetic videos.

It is often the case that real estate agents own land in various locations with various views of the surrounding environment. They also have a selection of house models that they can build in agreement with the customer. The selection of the model and the placement on a specific location on the ground is a very important aspect for the buyer. One aspect is the view of the house from external viewers from different locations, and its match with the environment. Another aspect is its protection from the sun and the winds. A third aspect is the view that the owner will have from the various rooms and windows of the house. Some of those aspects can be supported with mixing of pictures and video only.

The video and the pictures can not be completely pre-constructed since they involve a variety of house models and a variety of house locations on the land. They should be assembled by mobile workers according to workflows and video scripts that are guided by the house model (at least floor plan, or 3D designs if available, location of windows and inside and outside sitting spaces in the design, orientation of the covered outside spaces, etc.), as well as the GPS footprint of the house on the ground. An interesting aspect, at least in the Mediterranean countries where the summers are continuously sunny and hot is to simulate the sun and shadow provided by the house and its orientation at various parts of the year. Back in the office, the video information captured can be assembled in a systematic manner with the videos navigating through the house for a total video presentation of the house and the views to the surrounding environment. These videos can be shown to the potential buyer locally or remotely. Based on the views shown small or larger changes in the design may be interactively agreed.

This kind of functionality in an extensive form that allows completely flexible building design (not just a set of preexisting models), also supported by 3D models, is also useful for Architecture houses and Internal and External Decorators. Architecture companies, Decoration companies, Furniture construction companies may cooperate to support high- end remote customers that are interested in shopping for personalized architecture, decoration and furniture.

3 The Model

In this section we will describe the models that we have developed for the support of the mobile multimedia applications.

3.1 The Complex Event Type

The model describes real world events and their capturing by shots. In cinematography a **shot is an uninterrupted video capture.** The figure 1 shows the requirements model for the complex event type design.

The types of events of the real world are described by a **ComplexEventType** which has a name, a complex event type Id and a description. For example a Complex Event Type may be used to describe events of the type “marriage”. In this case the name of the Complex Event Type will be marriage. The Complex Event Types have

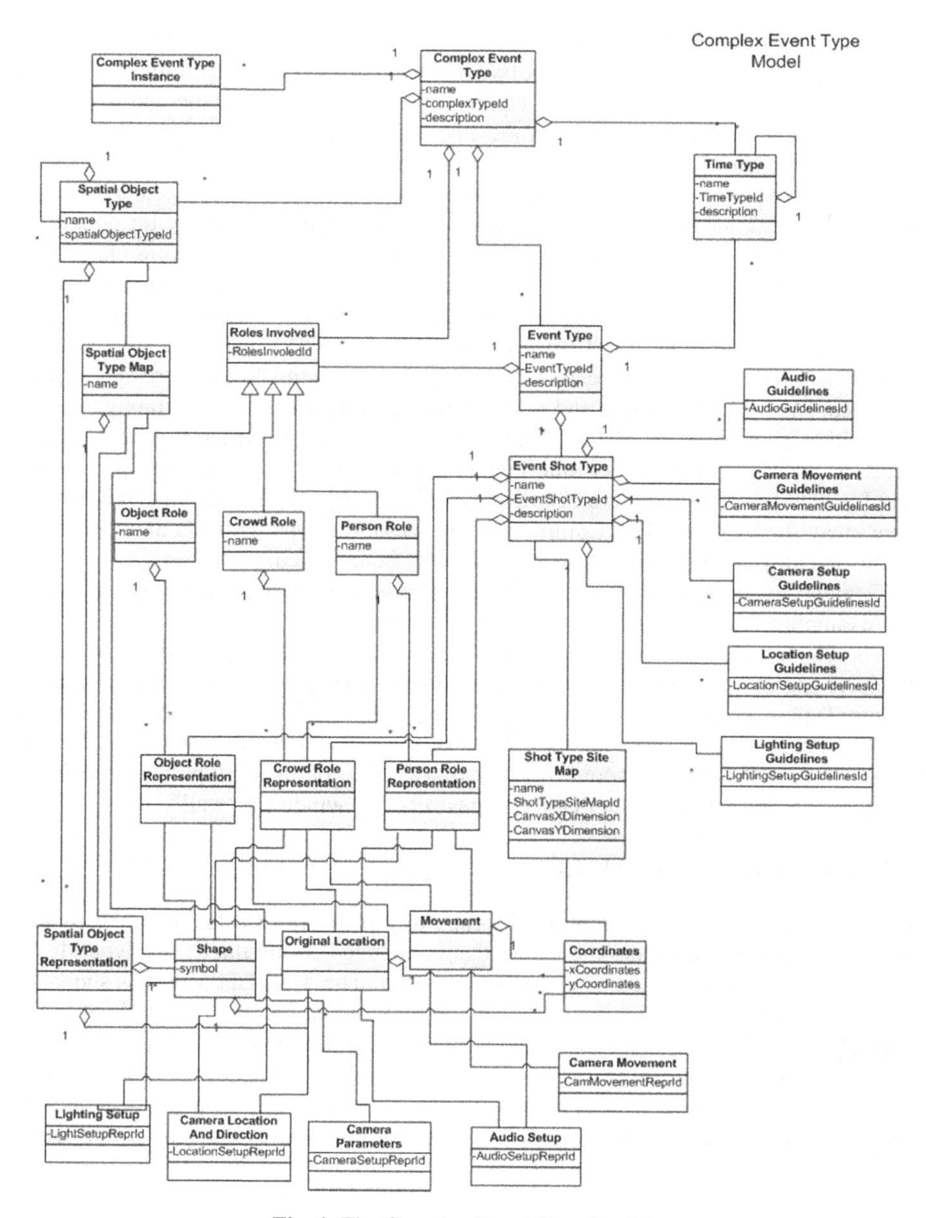

Fig. 1. The Complex Event Type Model

ComplexEventInstances. For example an instance of the complex event type marriage will be a specific marriage of two persons. Complex Event Type Instances will be described in a later section.

Complex Event Types are associated with certain roles of persons or objects that participate in the Complex Event Type. The **RolesInvolved** class is used to model such roles. A specialization of the RolesInvolved class is the class **PersonRole** which is used to model the roles of persons that participate in a ComplexEventType. For example person roles in the Complex Event Type marriage may be "bride", "groom", "priest", "father of bride", "friend", etc. The PersonRoles have name and a description in text. A Person Role name in the marriage type could be groom. Note that a specific PersonRole may be played by several different people in a specific ComplexEventType. For example the PersonRole "friend" could be played by several persons in a ComplexEventInstance. All those persons will be friends of the couple.

Another specialization of the RolesInvolved class is the **CrowdRole** class. It is used to model larger concentrations of people that play a role in a ComplexEventType without referring to each one in particular. For example in a ComplexEventType of the type "demonstration" the CrowdRoles "demonstrator crowd" and "police force crowd" may be modeled. In a class teaching environment the "student crowd" may be modeled. Crowd location, role and movement is important in cinematography.

Another specialization of the RolesInvolved class is the **ObjectRole** class. This class is used to model other non-human object types. For example in a ComplexEventType of the type "car accident event" the ObjectRole "car involved in accident" may be modeled. In a ComplexEventType of the type "product announcement event" the ObjectRole "product announced" may be involved.

ComplexEventTypes have a **TimeType** which is used to model time interval types. For example a TimeInterval type could be "evening".

The ComplexEventTypes are also associated with **SpatialObjectTypes**. A SpatialObjectType may be composed of other SpatialObjectTypes. For example a SpatialObjectType may be a "one floor bulding". The one floor building SpatialObjectType may be composed of a SpatialObjectType of the type "corridor", and some SpatialObjectTypes of the type "room". A SpatialObjectType of the type room could be further decomposed into SpatialObjectTypes "door", "window", "wall", etc.

3.2 The Event Metadata Model

One objective of the mobile multimedia capturing software is to facilitate the capturing of multimedia metadata. The model that is included within the Event Type Model described above is based on an event based model. The Metadata Model is shown in Figure 2. The dominant semantic model for multimedia content descriptions is the Mpeg-7.. It is also based on events, and it has been shown by us how to accommodate ontologies with its semantic primitives (although in a complex way). The semantic metadata model that we present here is simpler but compatible with the structures of Mpeg-7. We anticipate that most applications will be satisfied with the event model presented for the metadata capturing.

3.3 Event Capturing Workflows

The **Main Actor** for the **Creation of the Event Capturing Workflow Use Case** is the **Workflow Type Designer**. This Actor may be the same Actor as the Event Type Designer described earlier.

The purpose of the Event Capturing Workflow is to model the possible sequences of Event occurrences of a given Complex Event Type. In some cases the users will have control over what events are to be captured and in what sequence. In this case

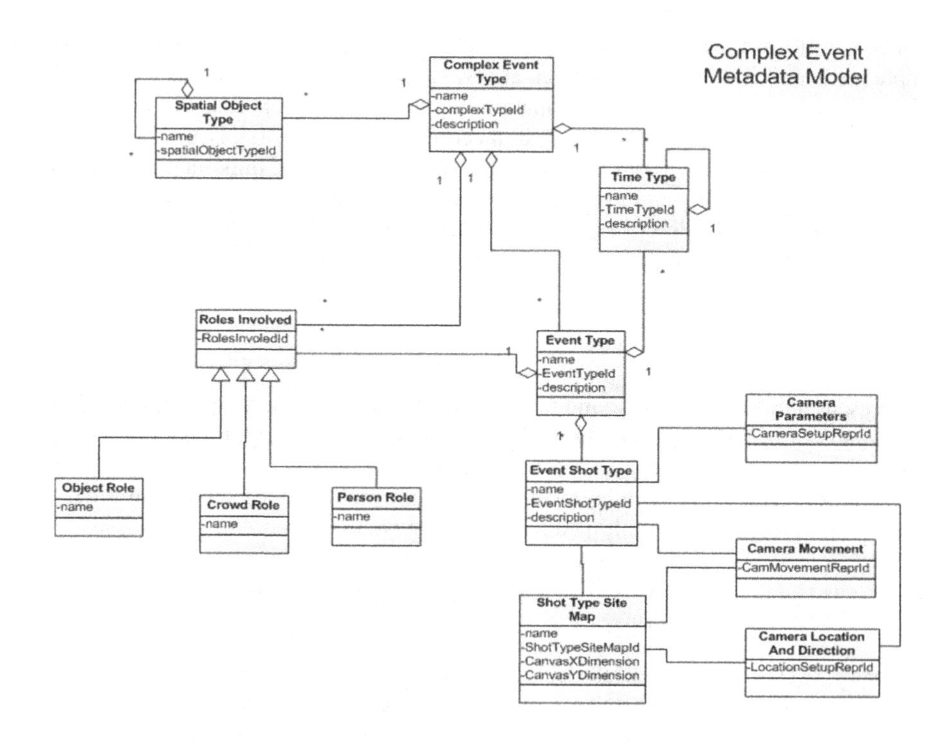

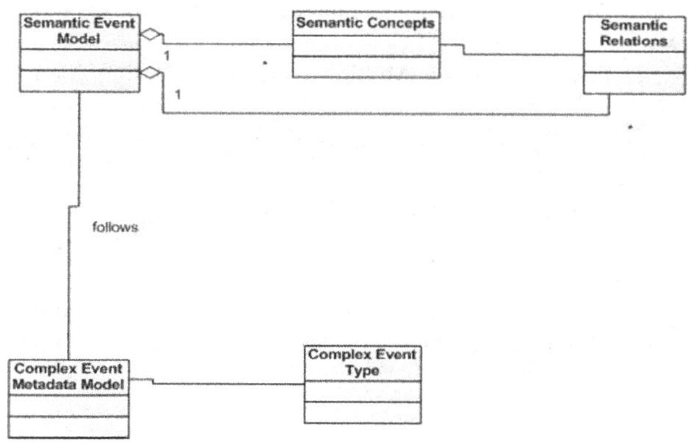

Fig. 2. The Complex Event Metadata Model

the workflow guides them on what to capture, and when. In other cases the events occur in a predetermine sequence that depends on the Complex Event Type. The workflow will help the user to be prepared for upcoming important events, giving also an indication on when they will occur.

The workflow aims also to remind the user on what information is missing (for example metadata) and has to be captured at various points in time. It has also to describe other activities, not related to information capturing, like preparing the equipment, packing and unpacking the equipment in a certain order, and possibly transmitting the captured information at various points in time (depending on the importance of the shot captured or the event captured, the urgency, the need for further editing, etc). The Workflow Model describes therefore Elementary Actions that can be reused to synthesize alternative workflows for the capturing of mobile multimedia events.

4 Conclusions

We present a model for mobile multimedia context for different kind of applications. The goal is to provide easy and powerful access to all kind of multimedia types. The model describes real world events which has a name, a complex event type ID and a description. The events can be taken by different multimedia methods. In order to incorporate more complex multimedia information we involve

References

1. Abowd, G.D., Brotherton, J., Bhalodia, J.: Classroom 2000: A System for Capturing and Accessing Multimedia Classroom Experiences. In: CHI 1998 Conference Summary on Human Factors in Computing Systems. CHI 1998, pp. 20–21. ACM, New York (1998)
2. Berra, P.B., Chen, C.Y., Ghafoor, A., Lin, C.C., Little, T.D., Shin, D.: Architecture for Distributed Multimedia Database Systems. Computer Communications 13(4), 217–231 (1990)
3. Cruz, G., Hill, R.: Capturing and Playing Multimedia Events with STREAMS. In: Proceedings of the Second ACM international Conference on Multimedia, pp. 193–200. ACM, New York (1994)
4. Laurel, B., Oren, T., Don, A.: Issues in Multimedia Interface Design: Media Integration and Interface Agents. In: CHI 1990 Proceedings, pp. 133–139 (1990)
5. Meghini, C., Sebastiani, F., Straccia, U., Nazionale, C.: A Model for Multimedia Information Retrieval. Journal of ACM 48(5), 909–970 (2001)
6. Tsinaraki, C., Christodoulakis, S.: An MPEG-7 query language and a user preference model that allow semantic retrieval and filtering of multimedia content. Multimedia Systems 13(2), 131–153 (2007)
7. Volker, M.M., Schillings, V., Chen, T., Meinel, C.: T-Cube: A multimedia Authoring System for eLearning. In: Proceedings of E-Learn, pp. 2289–2296 (2003)

Innovation Management in the Knowledge Economy: The Intersection of Macro and Micro Levels

Aljona Zorina and David Avison

ESSEC Business School, BP 50105, Avenue Bernard Hirsch, 95021,
Cergy Pontoise Cedex, France
AljonaZorina@yandex.ru, avison@essec.fr

Abstract. Although the ideas of innovation management have been developed and indeed implemented widely at the macro and micro levels, the intersection between the two has not been studied to the same extent. We argue that innovation management can only be fully effective through paying attention to this intersection, which is free of biases inherent in each individually. Our research looks at the macro and micro levels of innovation management in Denmark, Sweden, USA, India, Russia and Moldova. It suggests that differences in the success of innovation management between countries lie at the intersection level and in particular at a two-way mediation process between the micro and macro. Paying attention to this aspect of innovation management can develop further the knowledge society.

Keywords: innovation management, knowledge society, government, strategic programs.

1 Innovation Management: The Macro and Micro Levels

What makes innovation management so important for sustainable development of the knowledge economy? Does the answer lie at the macro or at the micro level? We suggest that the answer comes most of all from the intersection of these levels.

Theoretical aspects of innovation and knowledge management at the macro and micro levels have been studied by many authors including Porter, Nonaka and Takeuchi, Dovila, Senge, Bell, Toffler and Frydman. We will refer to some of their contributions in this paper. However, the intersection of these two levels, the mechanism of their connection and its influence, has been scarcely studied in the context of knowledge society development. We argue that this is often the reason why even the best innovation strategies and the most ambitious government projects underachieve. This paper studies the link between the macro and micro levels of innovation management. The conceptual scheme of the research is presented in figure 1.

The definition of innovation management depends on the level to which it is applied. In this paper we refer to innovation management at the macro level as a macro economic framework realized in various national and international innovation programs and projects and to the development of a knowledge society. Innovation management at the micro level deals with the firm's capacity for innovation seen in the

M.D. Lytras et al. (Eds.): WSKS 2009, LNAI 5736, pp. 395–404, 2009.

firm's patents, R&D investment and pioneering of its own new products and processes. Generally, the framework of both macro and micro levels of innovation management can be divided into two domains: theoretical and practical. We will refer to the main ideas of these domains as well as their common basis to see where the framework of the intersection level lies (see table 1).

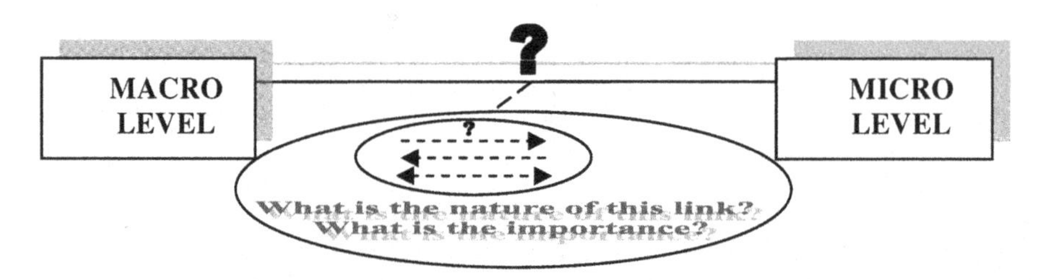

Fig. 1. Innovation management in the knowledge society

(1) Macro level. The idea that new knowledge and innovation creation is necessary for successful economic development has been proposed by many authors. Thus, Schumpeter [17] put the first stepping stone towards the development of Innovation Theory. He argued that the fundamental impulse for economic growth comes from entrepreneurs, who, looking for irrational ways of making profit, create innovations and technological change. Kuznets [8] argued that increasing knowledge and expanding its application make the essence of economic growth. Drucker [4] and Machlup [9] also suggest that expanding knowledge use is one of the most effective resources of the firm in the area of production and management. Such investment in knowledge leads to major change in the economy and society as a whole, turning the first into a "knowledge economy" and the second into a "knowledge society" [4], [9]. The analysis of world economic development made by Porter [15] suggests that knowledge is the main success factor in the competitive situation of different countries. All these theories indicate a dependence of economic growth at the macro level to innovation development and effective knowledge use.

The *practical domain* of innovation management at the macro level becomes apparent in many international and national strategic programs. Thus, the main goal of the United Nations *Information and Communications Technologies Task Force* lies "in addressing core issues related to the role of information and communication technology in economic development and eradication of poverty and the realization of the Millennium Development Goals" [6]. Since 2000 the creation of the economy based on knowledge has been announced by the European Union. This development in strategic programs is referred to as the *Lisbon Strategy* [5], aimed at world innovation leadership of Europe, and *Electronic Europe (2000-2010),* with a budget of about 100 million Euros. It consists of a number of separate programs aimed at R&D, computerization in the spheres of economy and management, an increase in the level of education and the adoption of lifelong learning [5]. Similar programs also exist at national levels in other countries, being a part of their strategic programs of sustainable socio-economic development. These include the *National Innovation Initiative* [11] in the USA; *Strategy of the development of information society in the Russian Federation* and *Electronic Russia*

[19] in Russia; and the *National Strategy on Information Society Development – e-Moldova* – and *E-Governance* in Moldova [12].

International organizations and projects that have this macro knowledge element include the World Summit on the Information Society (WSIS) [26], World Summit on the Knowledge Society (WSKS) [27], World Economic Forum (WEF) [22], United Nations Commission for Science and Technology Development (UNCSTD) [21], Association for Information Systems (AIS) [1], and many others. These organizations help to develop international standards and provide support to international and national initiatives.

(2) Micro level. The *theoretical domain* of innovation and knowledge management at the micro level has been studied by a number of authors. Thus Nonaka and Takeuchi [13] studied human capital and the transformation of information into knowledge, which may lead to a firm's competitive advantage. Porter [14] argued that a firm has two sources of competitive advantage: cost reduction and product differentiation. In both, innovation management can be of great use. Dixon [3] studied how knowledge management turns to prosperity and Rodriguez-Ortiz [16] described the Quality model of knowledge management for R&D organizations.

Theoretical studies suggest that the *practical domain* of innovation and knowledge management at the micro level lies in the firm's innovation strategy and its ability to transform R&D investment, human capital and information into competitive advantage. This ability can be seen in part in the number of patents for new products and processes which a firm owns.

All in all, we can see that the results of the analysis, presented in table 1, suggest that at the macro level innovation and knowledge creation are regarded as necessary for sustainable economic development of the knowledge society and the creation process demands special strategic programs. At the micro level innovation management participates in a firm's competitive advantage creation. Both these are complementary and mutually dependent: the macro level of innovation management creates the right business environment, whilst the micro level of innovation management creates the innovation and new knowledge necessary for the development of the knowledge society as a whole. This mutual dependence forms the intersection level of innovation management.

Table 1. Macro and micro levels of innovation management

Levels and domains of innovation management		Common basis of the ideas
Macro	**Theoretical domain:** Schumpeter, Kuznets, Bell, Keynes, Porter, Drucker	• Innovation and knowledge at the macro level are important and even necessary for sustainable economic development. • The process of innovation management and knowledge creation needs special strategic action, initiative and standards.
	Practical domain: Lisbon strategy, Agenda for action, National programs of information society creation; projects and actions of the WSIS, WSIK, AIS, UNCSTD.	
Micro	**Theoretical domain:** Nonaka, Takeuchi , Sullivan, Dosi, Drucker, Senge, Milner, Dovila	• There is a strong coherence between innovation management and a firm's competitive advantage. • Innovation strategy, human resource management, information strategy and R&D creates a firm's competitive advantage and defines its success
	Practical domain: Firm's innovation strategy, human resource management, information strategy and R&D	

2 The Intersection between the Macro and Micro levels

2.1 The Nature of the Intersection Level

The question about the nature of the economic society and the relationship between its macro and micro levels is, perhaps, one of the most ancient in the history of economic thought. Its philosophical roots go back to the contradictory ideas of Smith [2] and Spencer [18] on the one side, arguing that there should be no intervention into the economy at the macro level, and Comte [10] and Keynes [7] on the other side, insisting on the particular importance of strategic planning and government regulation of a national economy. This suggests different views and paradigms explaining the nature of the intersection level of macro and micro levels of innovation management. We propose a classification explaining its nature by referring to the possible types of relationship which can exist between the two levels. Theoretically we can define four different types of intersection link between the macro and micro levels of innovation management.

Type 1. There is no link between macro and micro levels of innovation management, or the link is very weak (see figure 2). In fact we would not expect to have this situation in a modern economy, especially at the level of innovation management (IM) development, as the intersection of macro and micro environment is both strong and inter-dependent.

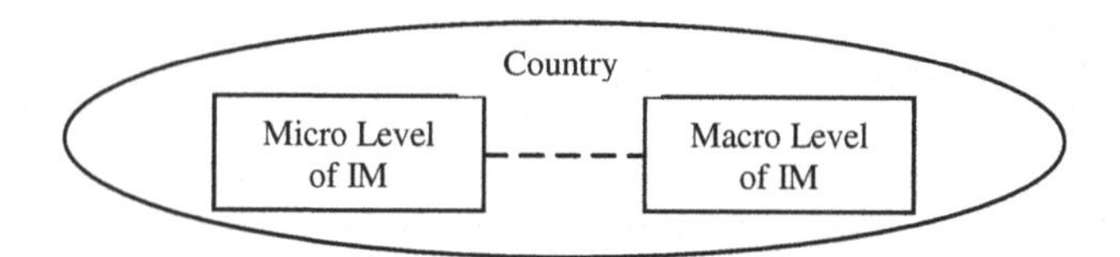

Fig. 2. Weak link between macro and micro levels of innovation management

Type 2. The interaction of macro and micro levels of innovation management comes mostly from the macro level (see figure 3).

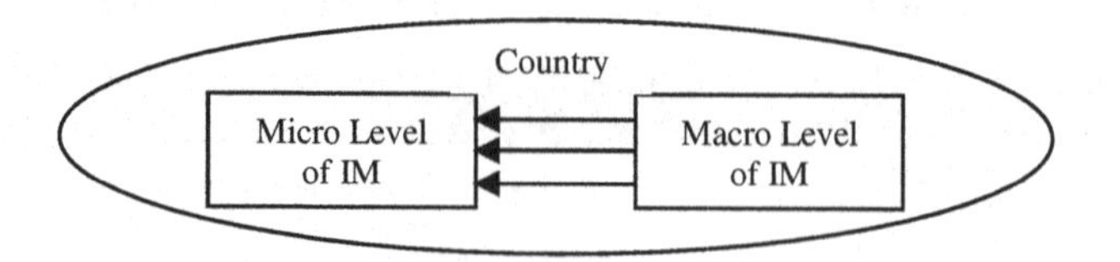

Fig. 3. The link comes from macro level of innovation management

In this situation there exists a strong motion coming from government and/or international organizations to create national programs of information and knowledge society creation, initiatives to develop knowledge communities, etc. On the other hand, individual economic and social players do not have a developed level of innovation management. This may occur because of some ideological restrictions which can exist in the society, because of the inappropriate economic, political, tax or legal conditions, or when participants of the micro level have low level of education, are not interested in the development of innovation management, etc. An example of this type of interaction could be seen in the Soviet Union when government had well developed programs of innovation creation in the military and technological spheres; but at the same time, market conditions and separate programs of innovation management for firms were not developed. Another example of this type of link may be seen in the developing countries (Belarus, Moldova, and Lithuania, along with Russia) which have only recently accepted national programs of information society creation but where the conditions for full and comprehensive implementation of innovation management at the micro level are not developed enough and need further elaboration.

Type 3. The interaction of macro and micro levels of innovation management comes mostly from the micro level (see figure 4). This type of interaction means that the economic model of the country is close to the pure market economy, the so-called "laissez faire" market. In this model, government's initiatives are weak and innovation management at the macro level does not exist or exists only in paper form. Innovation management at the micro level comes most from firms and organizations which are monopolies or oligopolies.

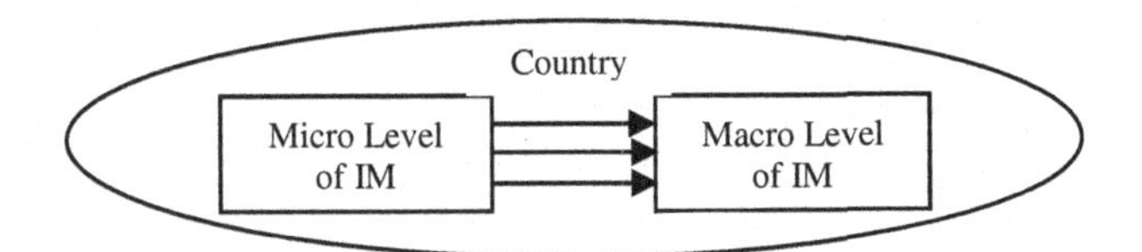

Fig. 4. The link comes from micro level of innovation management

In this model innovation management will not develop fully. In those spheres where there is no visible profit for firms and organizations (healthcare, public education, environmental security, safe product utilization) and which are usually controlled by government functions, innovation management either will not exist at all or will exist at a very elementary level.

Type 4. The interaction of macro and micro levels of innovation management comes from both macro and micro levels (see figure 5). In this situation innovation management is well developed both at the macro level (in the form of active national and international programs, creating special conditions for their implementation) and at the micro level (in the form of successful and comprehensive innovation strategies of firms).

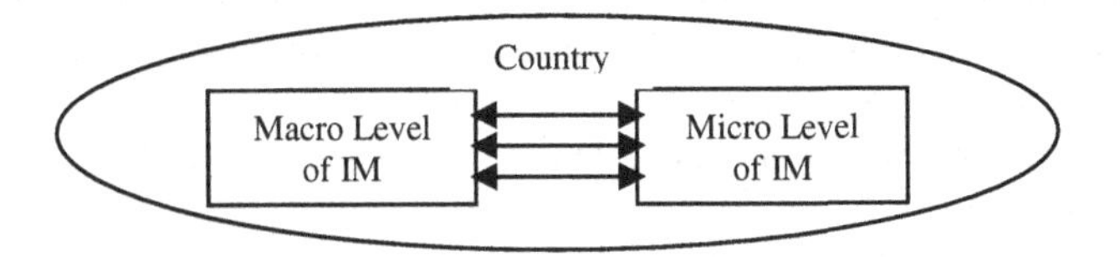

Fig. 5. The intersection comes from both micro and macro levels

In the next section we will apply this classification to the comparison of the degree of knowledge society creation in different countries. Our objective is to see whether the nature of the intersection level of the innovation management changes at different degrees of knowledge society development.

2.2 Research Model

To compare the degree of knowledge society creation in different countries we will refer to some particular indicators and indexes:

- ***The Networked Readiness Index (NRI)***: measures the level of the country's ICT development in terms of 67 parameters and is an important indicator of a country's potential and ability for development [20]. The index has been published since 2002 by the World Economic Forum and INSEAD. The 2007-2008 values of the Networked Readiness Index for Denmark, Sweden, United States, India, Russia and Moldova are presented in table 2.

Table 2. The Networked Readiness Index 2007-2008 for some countries

Rank	Country	Index
1	Denmark	5.78
2	Sweden	5.72
4	United States	5.49
50	India	4.06
72	Russian Federation	3.68
96	Moldova	3.21

In this paper we take the NRI as a suitable measure for characterizing the degree of knowledge society development at the macro level for the majority of countries.

- ***National and international programs of information society creation***: characterizes macro level innovation management and shows whether the knowledge society and innovation development is taken into account in national strategies and programs.
- ***Characteristics of macro and business environment:*** includes GDP, number of Internet users per 100 inhabitants, laws relating to ICT, efficiency of legal framework, e-government readiness index, venture capital availability, financial market sophistication and availability of latest technologies. The characteristics show how effective the above macro programs work in reality and how they can directly influence firms' abilities to create innovation programs at the micro level.
- ***Companies spending on R&D:*** this index on the micro side comes from the World Economic Forum Executive Opinion Survey 2006-2007 and shows the capacity of national firms to invest in research and development (R&D) which create innovations [24].
- ***Capacity for innovation:*** the index also comes from the World Economic Forum Executive Opinion Survey 2006-2007 [25].

The summarized comparison of the countries is presented in the table 3 ([5], [11], [12], [19], [20], [23], [24], [25]).

Table 3. Summary table for defining the intersection level type of innovation management in different countries

Country	Macro Level		Intersection Level	Micro Level		Type of the Inter-section
	N RI	Programs of knowledge society creation	Characteristics of macro and business environment *	Companies spending on R&D	Capaci-ty for innov.	
1.Denmark	1	"Electronic Europe" "Lisbon Strategy", The framework programs for research and technological development	GDP – 36920,4 (PPP US$) Intnt^{+} users (per 100 inhabitants) 58.23 Laws relating to ICT 1 Efficiency of legal frame-work 1 E-government readiness index 2 Venture capital availability 8 Financial market sophistication 13 Availability of latest technologies 5	5.47	5.54	4
2.Sweden	2	"Electronic Europe" "Lisbon Strategy", The framework programs for research and technological development	GDP – 34734,9 (PPP US$) Intnt users (per 100 inhabitants) 76.97 Laws relating to ICT 5 Efficiency of legal framework 5 E-government readiness index 1 Venture capital availability 7 Financial market sophistication 7 Availability of latest technologies 1	5.71	5.88	4
3.United States	4	"Agenda for action", "E-Government Strategy"	GDP – 43223,5 (PPP US$) Intnt users (per 100 inhabitants) 69.1 Laws relating to ICT 12 Efficiency of legal framework 30 E-government readiness index 4 Venture capital availability 1 Financial market sophistication 5 Availability of latest technologies 6	5.81	5.44	4
4.India	50	"National policy for ICT development", Government programs of support of ICT development	GDP – 3802 (PPP US$) Intnt users (per 100 inhabitants) 5.44 Laws relating to ICT 36 Efficiency of legal framework 34 E-government readiness index 91 Venture capital availability 29 Financial market sophistication 33 Availability of latest technologies 31	4.15	4.01	2 coming to 4
5.Russia	72	"Strategy of the development of information society in the Russian Federation", "Electronic Russia"	GDP – 12177,7 (PPP US$) Intnt users (per 100 inhabitants) 18.02 Laws relating to ICT 82 Efficiency of legal framework 103 E-government readiness index 57 Venture capital availability 60 Financial market sophistication 86 Availability of latest technologies 96	3.42	3.4	2-3
6.Moldova	96	"National Strategy on Information Society Development – "e-Moldova", E-Governance, National Program for Schools Informatization	GDP – 2869,1 (PPP US$) Intnt users (per 100 inhabitants) 17.36 Laws relating to ICT 90 Efficiency of legal framework 110 E-government readiness index 82 Venture capital availability 108 Financial market sophistication 105 Availability of latest technologies 125	2.48	3.01	2

*In cases where it is not mentioned specifically, the indexes show the position of the country among 127 countries which were included in the Networked index annually survey.

Intnt^{+} - Internet

Table 3 presents the research model of the paper. The *macro level* of innovation management is described by the country's national and international strategies of information and knowledge society creation and by its position according to the NRI ranking. The *micro level* of innovation management is described by ranking companies of the country according to spending on R&D and their capacity for innovation according to the 7 step scale. For the R&D index, "1" means that companies do not spend money on research and development, "7" means that companies "spend heavily on research and development relative to international peers" [24]. For capacity for innovation, the indicator ranks companies obtaining technologies from "1"- licensing or imitating foreign companies to "7"- conducting formal research and pioneering their own new products and processes [25]. The *intersection level* of innovation management is presented by the characteristics of the countries' business environment, which shows the real effectiveness of these programs. The analysis of the research model presented in table 3 is presented in the next session.

3 Findings and Discussion: Analysis of the Tendencies

Table 3 shows that Denmark, Sweden and the USA have the 4th type of intersection proposed in the previous section. These countries have a highly developed intersection level of innovation management, through having different national programs of innovation creation and different company structure of spending on R&D. The macro level of innovation management in these countries shows active national and international programs of innovation and knowledge society development. The micro level of innovation management in these countries is characterized by high level of firms' capacity for innovation and R&D.

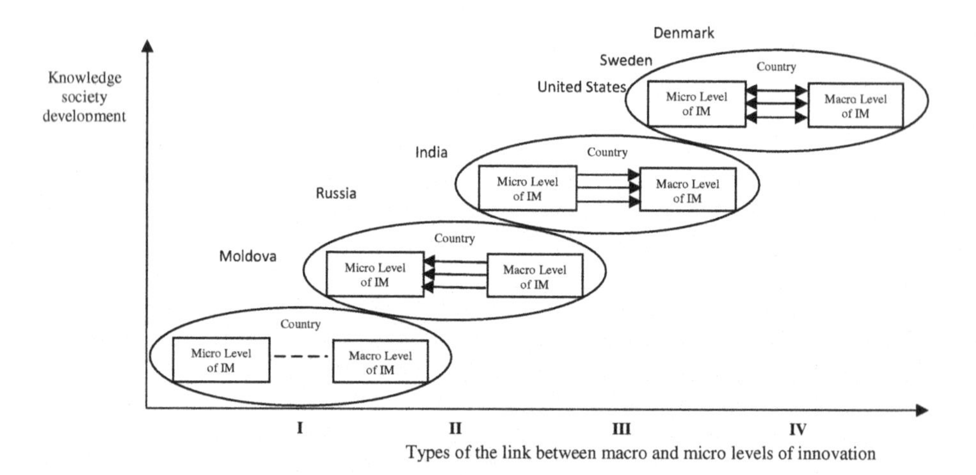

Fig. 6. Intersection level at the different degrees of the knowledge society development

The intersection level of innovation management in the countries which have low and middle levels of knowledge society development according to the NRI (India, Russia and Moldova) tend towards a "one-direction" link between macro and micro levels of innovation management, which corresponds to the types 1-3 of the proposed models.

The indicators of the macro and the micro levels of Denmark and Sweden are very close to each other. At the macro level these countries even share many international programs. Nevertheless the *intersection level* of the innovation management in Sweden and Denmark shows different indicators for the number of Internet users, laws related to ICT, efficiency of legal framework, etc. This suggests that it is the intersection level that constitutes *real* effectiveness of the innovation management at the country.

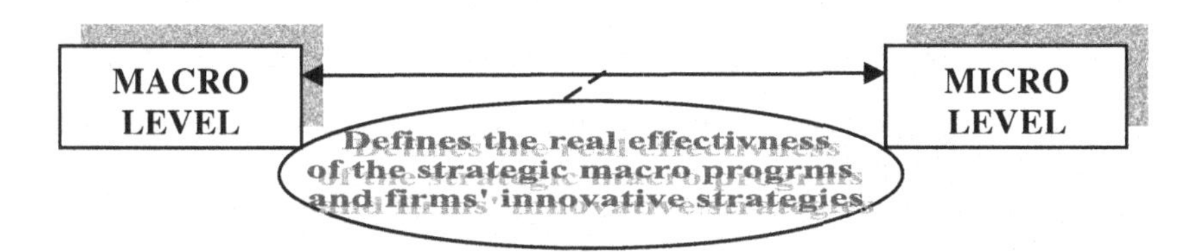

Fig. 7. Innovation management in the knowledge society

4 Conclusions: The Importance of the Intersection Level

The analysis discussed in the previous sections showed that the successful development of any society, especially one based on knowledge, is possible only if the intersection level of innovation management is well developed. The intersection level of innovation management is important because it combines strategic macro programs of innovation creation with firms' individual creation of competitive advantage.

The intersection level of innovation management shows the *real effectiveness* of a country's macro and micro levels. It can serve as a good indicator of the degree of innovation management development in the knowledge economy as it is free of the possible biases which can occur while estimating innovation management only at the macro or micro levels. Further, it does not depend on the traits of different strategies and models of knowledge society creation applied in different countries.

The analysis of the macro and the micro levels of innovation management realized in Denmark, Sweden, USA, India, Russia and Moldova show the trends in the development of the intersection level of innovation management in these countries, some of which are leaders of knowledge society development and those countries where only "one-way" intersection of innovation management is found.

The findings described in the article will assist in estimating the real effectiveness of strategic programs aimed at the innovative environment and knowledge society creation.

References

1. Association for Information Systems (AIS), `http://home.aisnet.org/associations/7499/files/Index_Markup.cfm`
2. Bryce, J.: The Glasgow Edition of the Works and Correspondence of Adam Smith. Liberty Fund, Indianapolis (1982)
3. Dixon, N.: Common Knowledge: How Companies Thrive by Sharing What They Know. Harvard Business School Press, Boston (2000)
4. Drucker, P.: Management Challenges for the 21st Century. Harper Business, New York (1999)

5. Europe's Information Society. I2010 in context: ICT and Lisbon Strategy, http://ec.europa.eu/information_society/eeurope/i2010/ict_and_lisbon.htm
6. Information and Communication and Technologies Task Force. UN Substantive session (2006), http://www.unicttaskforce.org/perl/documents.pl
7. Keynes, J.: The General Theory of Employment, Interest and Money (1936), http://www.marxists.org/reference/subject/economics/keynes/general-theory/
8. Kuznets, S.: Modern Economic Growth: Rate, Structure, and Spread. Yale University Press, New Haven (1966)
9. Machlup, F.: The Production and Distribution of Knowledge in the United States, Princeton (1962)
10. Martineau, H.: The Positive Philosophy of Auguste Comte. Thoemmes Press (2001)
11. National Innovation Initiative Interim Report, Innovative America (2004), http://www.goalqpc.com/docs/reports/NIIInterimReport.pdf
12. National Strategy on Information Society Development "e-Moldova", http://www.mdi.gov.md/info21_en/
13. Nonaka, I., Takeuchi, H.: The Knowledge Creating Company. Oxford University Press, New York (1995)
14. Porter, M.: Competitive Advantage: Creating and Sustaining Superior Performance. Free Press, New York (1998)
15. Porter, M.: The Competitive Advantage of Nations. Free Press, New York (1998)
16. Rodriguez-Ortiz, G.: Knowledge management and quality certification in a research and development environment. Computer Science, 89–94 (2003)
17. Schumpeter, J.: The Theory of Economic Development. Oxford University Press, New York (1961)
18. Spencer, H.: The Man versus the State, with Six Essays on Government, Society and Freedom. In: Mack, E. (ed.) Liberty Classics, Indianapolis (1981)
19. Strategy of the development of information society in the Russian Federation (2007), http://www.kremlin.ru/text/docs/2007/07/138695.shtml
20. The Networked Readiness Index (2007-2008), http://www.weforum.org/pdf/gitr/2008/Rankings.pdf
21. United Nations Commission for Science and Technology Development (UNCSTD), http://www.unesco.org/webworld/telematics/uncstd.htm
22. World Economic Forum (WEF), http://www.weforum.org/en/index.htm
23. World Economic Forum, Global Information Technology Report (2007-2008), http://www.insead.edu/v1/gitr/wef/main/analysis/choosedatavariable.cfm
24. World Ranking for Companies Spending on R&D, Global Information Technology Report (2007-2008), http://www.insead.edu/v1/gitr/wef/main/analysis/showdatatable.cfm?vno=5.21
25. World Ranking for Capacity for Innovation, Global Information Technology Report (2007-2008), http://www.insead.edu/v1/gitr/wef/main/analysis/showdatatable.cfm?vno=8.15
26. World Summit on the Information Society (WSIS), http://www.itu.int/wsis/index.html
27. World Summit on the Knowledge Society (WSKS), http://www.open-knowledge-society.org/summit.htm

An Eclipse GMF Tool for Modelling User Interaction

Jesús M. Almendros-Jiménez[1], Luis Iribarne[1], José Andrés Asensio[1], Nicolás Padilla[1], and Cristina Vicente-Chicote[2]

[1] Dpto. de Lenguajes y Computación, Universidad de Almería, Spain
{jalmen,jacortes,liribarn,padilla}@ual.es
[2] Dpto. de Tecnologias de la Información, Universidad Politécnica de Cartagena, Spain
Cristina.Vicente@upct.es

Abstract. Model-Driven Development (MDD) has encouraged the use of automated software tools that facilitate the development process from modelling to coding. User Interfaces (UI), as a significant part of most applications, should also be modelled using a MDD perspective. This paper presents an Eclipse GMF tool for modelling user-interaction diagrams –an specialization of the UML state-machines for UI design– which can be used for describing the behaviour of user interfaces.

1 Introduction

The adoption of *Model-Driven Development* (MDD) [16] in the design of *User Interfaces* (UI) allows software architects to use declarative and visual models for describing the multiple perspectives and artifacts involved in UI development. In the literature there are several works [3,22,21,7,20,19,18,14,6,8] whose aim is to apply the MDD approach to the development of user interfaces. In most cases they consider the following models: *task*, *domain*, *user*, *dialogue* and *presentation* models. The task model specifies the tasks that the user will carry out on the user interface. The domain model describes the domain objects involved in each task specified in the task model. The user model captures the user requirements. The dialogue model allows to model the communication between the user and the user interface. Finally, the presentation model describes the layout of the user interface.

In a previous work [2], we have proposed a MDD-based technique for user interface development. It involves the adoption of several UML models, which have been adapted to UI development. In particular, our UI-MDD technique can be applied to the modelling of *WIMP* (*Windows, Icons, Menus*, and *Pointers*) user interfaces. Our proposal mainly uses three models: (1) a dialogue model, which makes use of the so-called *user-interaction diagrams* (a special kind of UML state-machines for UI design); (2) a task model, which makes use of the so-called *user-interface diagrams* (an specialization of the UML use case diagram for UI specification); and, (3) a presentation model, which uses the so-called *UI-class diagrams* (UML class diagrams describing UI objects).

M.D. Lytras et al. (Eds.): WSKS 2009, LNAI 5736, pp. 405–416, 2009.

The main goal of our proposed UI-MDD technique is to allow designers to model the user interfaces with UML. The proposed technique is intended to be useful for rapid prototyping. Our technique has been extensively used by our students in the classroom. However, students need to manually carry out most of the tasks due to the lack of a tool for supporting our technique. In particular, one of the most tedious tasks that they have to manually carry out is the modelling of user interfaces by means of user-interaction diagrams.

In this paper we present $\mathcal{ALIR}$, a graphical modelling tool that implements the *core* of our UI-MDD modelling technique, described in [2]. $\mathcal{ALIR}$ allows the development of user-interaction diagrams. This tool has been implemented using some of the MDD-related plug-ins provided by the *Eclipse platform*, namely: the *Eclipse Modelling Framework* (EMF)[5] and the *Eclipse Graphical Modelling Framework* (GMF)[4]. Some further details about these tools will be provided in Section 3.

The structure of the paper is as follows. Section 2 presents the user-interaction diagrams. Section 3 describes the implementation of the Eclipse GMF Tool called $\mathcal{ALIR}$. Section 4 compares our work with existent proposal. Finally, Section 5 presents some conclusions and future work.

2 User-Interaction Diagrams

User-interaction diagrams are an specialization of the UML state-machine for describing the user interaction with the user interface. User-interaction diagrams include *states* and *transitions*. The states represent *data output/request actions*, that is, how the system responds to user interactions showing/requesting data. The transitions are used to specify how the user introduces data or interacts with the system, and how an *event* handles the interaction. Transitions can be *conditioned*, that is, the event is controlled by means of a *boolean condition*, which can either specify *data/business logic* or be associated to a *previous user interaction*. User interaction diagrams can include states with more than one outgoing transition, and which of them is executed depends on either data/business logic or the previous interactions of the user. The initial (resp. final) state is the starting (resp. end) points of the diagrams.

From a practical point of view, it is convenient to use more than one user-interaction diagram for describing the user interface of a software system, since the underlying logic of the user windows is usually too complex to be described in a single model. For this reason, a user-interaction diagram can be deployed in several user-interaction diagrams, in which a piece of the main logic is described in a separate diagram. Therefore, user interaction diagrams can include states which do not correspond to data output/request, rather than they are used for representing a sub-diagram. In this case, the states are called *non-terminal states*; otherwise, they are called *terminal states*.

Our proposed modelling technique provides a mapping between UI models and the *Java Swing package*. The wide acceptance of the *Java* technology: *applets*, *frames*, and *event-handlers*, etc for UI design guarantees the practical application of our proposal. The mapping to Java can be viewed as a *concrete modelling* of a more *general modelling technique*. In such mapping, UI components can be classified as *input* (e.g. a *button*) or *output/request* components (e.g. either a *text field* or a *list*). *Input/output* components are labelled by means of *UML stereotypes*. For instance, the `<<JTextField>>` and`<<JList>>` stereotypes are used in states, while the stereotype `<<JButton>>` is used in transitions.

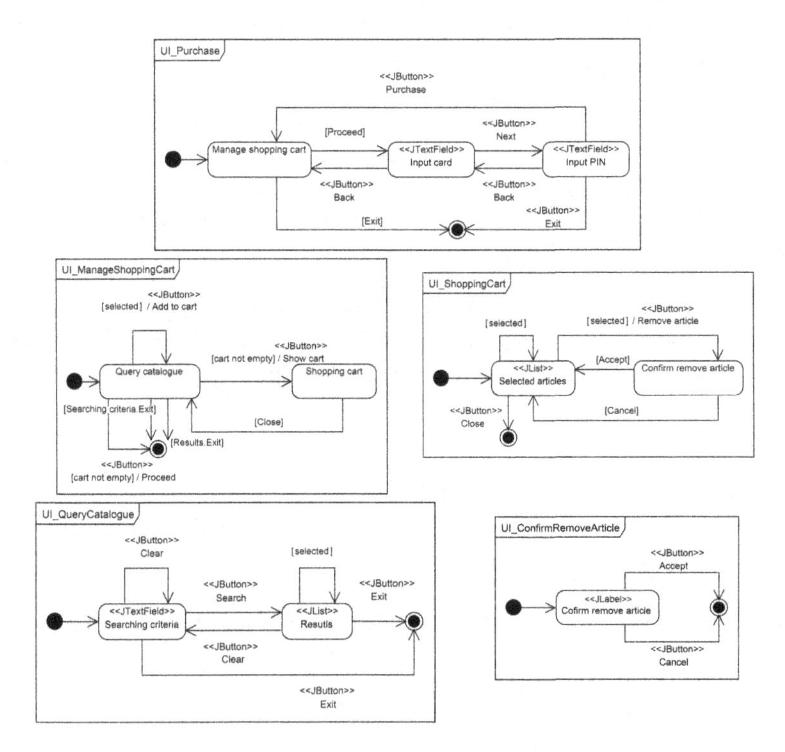

Fig. 1. The dialogue model of the purchase task

As running example, we will consider an *Internet Shopping System*. Figure 1 shows examples of user-interaction diagrams. They describe the *purchase* task. The *main* user-interaction diagram is `UI_Purchase`. This diagram includes non-terminal states (i.e. non-stereotyped states), which are described by means of *secundary* user interaction diagrams. They are sub-diagrams of the main diagram. Both the name of the non-terminal states and the name of its associated diagrams must be the same. The running example shows how the customer carries out the purchase task by querying from a catalogue and by adding or removing articles to/from a shopping cart. After, the shopping system requests the customer a card number and a PIN to carry out the order. The boolean conditions occurring in transitions specify the requirements to be fulfilled for state change. For instance, the *"cart not empty"* condition, in the `UI_ManageShoppingCart` diagram, means that the shopping cart can be only reviewed whenever it is not empty.

2.1 User-Interaction Diagrams Metamodel

User interaction diagrams can be considered as an specialization of UML state-machines in which states and transitions can be stereotyped by means of UI component names, but also states can represent subdiagrams. In order to implement in our tool user-interaction diagrams, we have to move to the UML metamodel. Figure 2 shows the user-interaction diagram metamodel (a *Platform-Independent Model* (*PIM*) perspective). The metamodel can be mapped to the elements of user-interaction diagrams (a *Platform-Specific Model* (*PSM*) view). Figure 3 summarizes the elements of user-interaction diagrams, and Table 1 describes the mapping of the elements defined in Figure 3 and the metamodel described in Figure 2. Due to lack of space, we have described the mapping of a subset of the elements of the meta-model in a subset of the elements of user-interaction diagrams. They are enough for the running example. The metamodel of Figure 2 includes the following elements:

States. They necessarily fall into one of the two following categories: *terminal states* (TS) and *non-terminal states* (NTS). A terminal state is labelled with a *UML stereotype*, representing a *Java* data output/request UI component. A non-terminal state is not labelled, and it is described by means of another *user-interaction diagram.*

Pseudo-states. They necessarily fall into one of the following categories: *initial* pseudo-state, *final* pseudo-state, *choice* states: which define alternative paths of type "OR" between two or more interactions and, finally, *fork/join* pseudo-states, which defines paths of type "AND". For instance, using a fork and a join, one can specify that the user can introduce the card number and the PIN in any order.

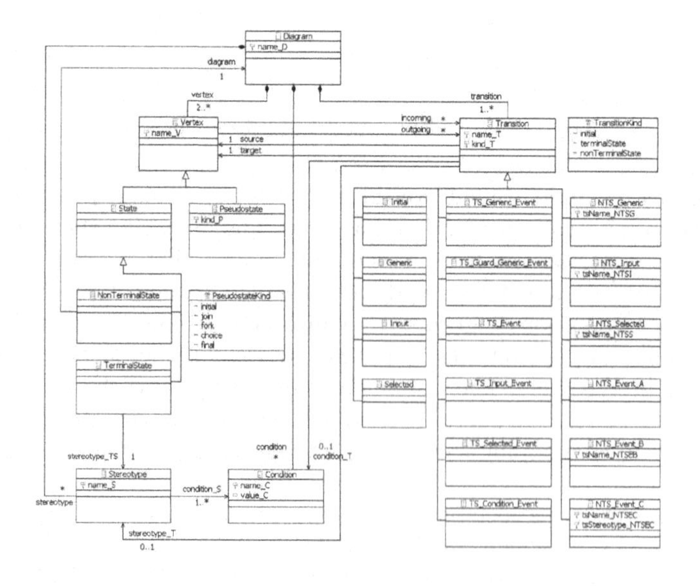

Fig. 2. The metamodel of user-interaction diagrams

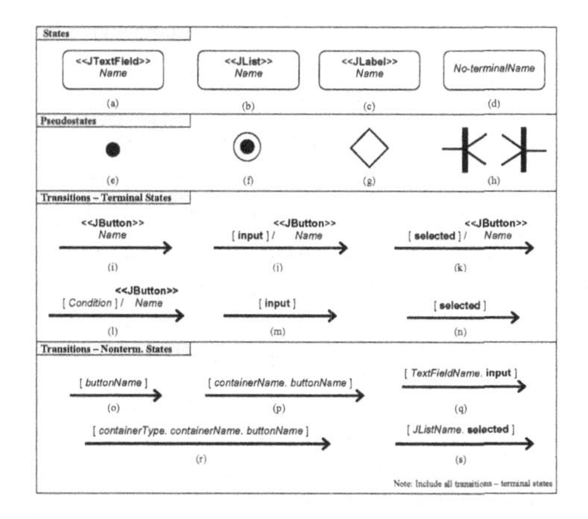

Fig. 3. Elements of the user-interaction diagrams

Table 1. Mapping between UI elements and Metamodel elements

#	Class (MM)	Parameters
(a)	TerminalState	(Name = Vertex.name_V) and (Stereotype.name_S = "JTextField")
(b)	TerminalState	(Name = Vertex.name_V) and (Stereotype.name_S = "JList")
(c)	TerminalState	(Name = Vertex.name_V) and (Stereotype.name_S = "JLabel")
(d)	NonTerminalState	Name = Vertex.name_V
(e)	Pseudostate	kind_P = "PseudostateKind::initial"
(f)	Pseudostate	kind_P = "PseudostateKind::final"
(g)	Pseudostate	kind_P = "PseudostateKind::choice"
(h)	Pseudostate	(kind_P = "PseudostateKind::join") or (kind_P = "PseudostateKind::fork")
(i)	TS_Event	(Name = Transition.name_T) and (Steretype.name_S = "JButton")
(j)	TS_Input_Event	(Name = Transition.name_T) and (Steretype.name_S = "JButton") and (Condition.name_C = "input")
(k)	TS_Selected_Event	(Name = Transition.name_T) and (Steretype.name_S = "JButton") and (Condition.name_C = "selected")
(l)	TS_Condition_Event	(Name = Transition.name_T) and (Steretype.name_S = "JButton") and (Condition = Condition.value_C)
(m)	Input	Condition.name_C = "input"
(n)	Selected	Condition.name_C = "selected"
(o)	NTS_Event_A	buttonName = Transition.name_T
(p)	NTS_Event_B	(containerName = NTS_Event_B.tsName_NTSEB) and (buttonName = Transition.name_T)
(q)	NTS_Evemt_C	TextFieldName = NTS_Input.tsName_NTSI
(r)	NTS_Input	(containterType = Stereotype.name_S) and (containerName = NTS_Event_B.tsName_NTSEB) and (buttonName = Transition.name_T)
(s)	NTS_Selected	JListName = NTS_Selected.tsName_NTSS

Vertex. States and Pseudo-states are special cases of the class *Vertex* included in the metamodel.

Transitions Coming from Terminal States. They may connect either TS to TS or TS to NTS. Transitions in the metamodel are classified according to the kind of *event* they represent. Events can include *input events* and *conditions.* Conditions can represent either *previous user choices* (i.e. previous events) or *business/data logic* (in particular, UI component status).

In the running example, transitions can be labelled with a button name (stereotyped with `<<JButton>>`) (case (i) of Figure 3). The button represents an input event. But they can also be conditioned (cases (j), (k) and (l) of Figure 3). Transitions can be conditioned to a *previous user interaction* (cases (j) and (k) of Figure 3), or can be conditioned to *business/data logic* (case (l) of Figure 3). Finally, they can be conditioned to a previous user interaction but they are not associated to a input event (cases (m) and (n) of Figure 3). Boolean conditions about *business/data logic* are used, for instance, for checking the status of UI components. This is case of the user introduces a text into a *text field* (case (m) of Figure 3) or selects an element from a *list* (case (n) of Figure 3).

Transitions Coming From Non-Terminal States. They may connect either NTS to NTS or NTS to TS. They are transitions with boolean conditions about the user interaction in the non-terminal state (i.e. sub-diagram). There are five kinds of transitions. In the case of transitions of type (o), (p) and (r) of Figure 3 the boolean condition is related to the *"exit condition"* in a subdiagram. In other words, they check which button has been pressed as exit transition of the sub-diagram.

For instance, the terminal state *Query catalogue* of Figure 1 has two *"exit"* buttons, qualified as follows: [Results.Exit] and [Searching criteria.Exit]. They describe the action to be achieved when the user clicks the *"exit"* button and closes the Query catalogue window. However, the "exit" button can be pressed from two states: Results and Searching criteria. This is the reason boolean conditions are qualified in the main diagram. There are more cases in Figure 1 of *"exit"* conditions: [Close], [Cancel], [Accept] or [Proceed]. In these cases the *"containerName"* is not specified because there is only one button with this name in the sub-diagram.

The cases (q) and (s) of Figure 3 check in the main diagram the user interaction in the sub-diagram. They consider two cases: the introduction of a text into a *text field* (case (q)) and the selection from a list (case (s)).

For instance, the [selected] condition, in the transition triggered from the Query catalogue state of Figure 1, is not an *"exit condition"*, rather than it is an *"internal transition"* of the UI_QueryCatalogue diagram, which is checked from outside of the sub-diagram. This makes possible that UI_ManageShoppingCart controls the user interaction in the UI_QueryCatalogue window. The keyword ''selected'' is associated to the JList container. The keyword ''input'' is associated to the JTextField container. The transition [selected] / <<JButton>> Remove article of Figure 1, means that the button *"Remove article"* will be enabled/visible whenever the user selects one element of the container *"Selected articles"*. An input event can also represent that the mouse is placed over a *label* or that the mouse is focused on a *text field.* For this reason we need the following generalization.

Generalization. The Generic transition in the metamodel are transitions of type [String] —an abstraction of the *"Input"* (case (m) of Figure 3) and *"Selected"* (case (n) of Figure 3) transitions—. TS_Generic_Event transition in the metamodel are transitions of type <<Name>> Name (where *Name* is an string). This kind of transition is similar to TS_Event (case (i) of Figure 3, i.e.,

<<JButton>> Name). Finally, TS_Guard_Generic_Event transitions are transitions of type [Guard] / <<Name>> Name (where *Guard* and *Name* are strings). This kind of transition is similar to TS_Input_Event (case (j) of Figure 3), TS_Selected_Event (case (k) of Figure 3), and TS_Condition_Event (case (l) of Figure 3).

However, specific transitions, for instance, <<JButton>> Name, are used by the EMF/GMF implementation of the $\mathcal{ALIR}$ tool to draw an specific element of the toolbar, for instance, *"JButton"*.

3 Implementing an Eclipse GMF Tool for User-Interaction Diagrams

Now, we would like to present the implementation of an Eclipse GMF tool, called $\mathcal{ALIR}$, for the modeling of user-interaction diagrams. Such tool allows the designer to model user-interaction diagrams with the help of a toolbar in which the elements of user-interaction diagrams are available. In addition, the tool is able to check some of the constraints imposed to user-interaction diagrams.

Our Eclipse GMF tool can be integrated with our UI-MDD technique described in [2] as follows.

- The first step of our UI-MDD technique consists in the design of the layout of each window of the system. The windows have to include its UI components (i.e buttons, text fields, labels, etc). We have used the *Eclipse Window Builder-Pro Plug-in* for modelling the layout of the windows and for adding UI components. The developer can obtain prototypes of the windows by means of the code generation (step 0 of Figure 4).
- Secondly, the developer can store the layout in XMI format (step 1). The Eclipse GMF tool accepts XMI format as input and extracts (step 2 of Figure 4) the UI components (i.e., buttons, text fiels, labels, etc.), which will be used as elements in the tool (step 3 of Figure 4).
- Thirdly, the developer uses the Eclipse GMF tool to draw user-interaction diagrams.
- Next, the new models (i.e. user-interaction diagrams) are stored in XMI format (step 4 of Figure 4).
- Finally, the XMI file can be imported from the *Eclipse Window Builder-Pro Plug-in* (step 5 of Figure 4) to test the behaviour of the windows.

Both the metamodel discussed in Section 2 and the $\mathcal{ALIR}$ modeling tool presented next, have been developed using some of the MDD facilities provided by the *Eclipse* platform. This free and open-source environment provides the most widely used implementation of the *OMG standard Meta-Object Facility* (MOF) [10], called *Eclipse Modelling Framework* (EMF) [5]. Although EMF currently supports only a subset of MOF, called *Essential MOF* (EMOF), it allows designers to create, manipulate and store both models and metamodels using the OMG *XML Metadata Interchange* (XMI) [12] standard format. Many MDD-related initiatives are currently being developed around Eclipse and EMF. Among them, and directly related to our tool, it is worth mentioning the following ones:

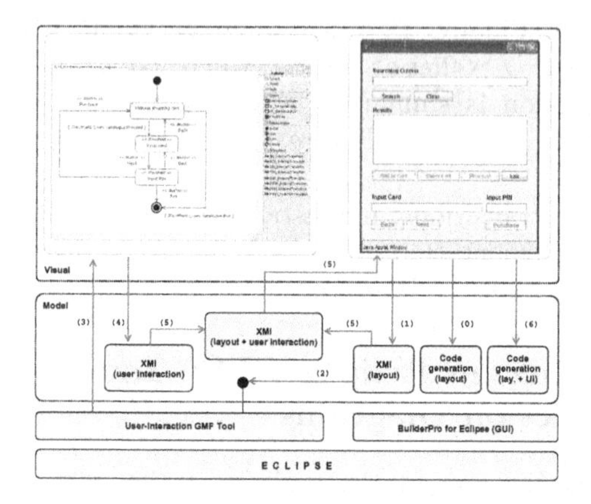

Fig. 4. UI-MDD Technique Steps

- The *Eclipse Graphical Modelling Framework* (GMF) [4], which allows designers: (1) to create a graphical representation for each domain concept included in the metamodel, (2) to define a tool palette for creating and adding these graphical elements to their models (see Figure 5), and (3) to define a mapping between all the previous artifacts, i.e. the metamodel elements, their graphical representations, and the corresponding creation tools.
- The *Object Constraint Language (OCL)* [9] facilities provided by the *Eclipse Modeling Framework (EMF)*. This plug-in can be used, together with GMF, to enable the definition and evaluation of OCL queries and constraints in EMF-based models.

It is worth mentioning that both the metamodel and the Eclipse GMF tool implemented as part of this work, have been developed as an extension and modification of StateML+ [1]: a set of MDD tools aimed at designing hierarchical state-machine models (and automatically generating thread-safe Ada code from these models).

3.1 The $\mathcal{ALIR}$ Tool

Figure 5 shows an snapshot of the $\mathcal{ALIR}$ tool. It has been used for drawing the `UI_ShoppingCart` user-interaction diagram of Figure 1. Figure 6 shows the instance of the metamodel of Figure 2 in the case of the `UI_ShoppingCart` user-interaction diagram.

The $\mathcal{ALIR}$ tool is able to validate user-interaction diagrams according to some semantic and syntactic rules. More than 150 *OCL* constraints have been implemented to support the model validation. The most relevant ones are those whose aim is to check whether the boolean conditions about user interactions, that is, about "exit conditions" and "internal transitions", in a main diagram have an associated event in sub-diagrams. In Table 2 we show some of the constraints the $\mathcal{ALIR}$ tool is able to check. The reader can find more detailed information about the $\mathcal{ALIR}$ tool in our Web page: `http://indalog.ual.es/mdd/alir`.

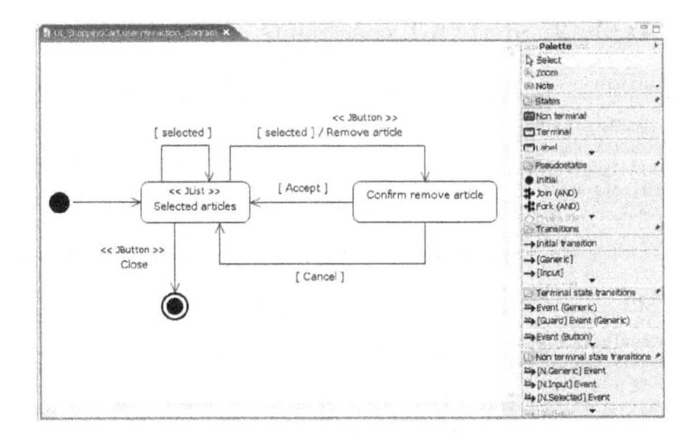

Fig. 5. Snapshot of the Eclipse GMF tool

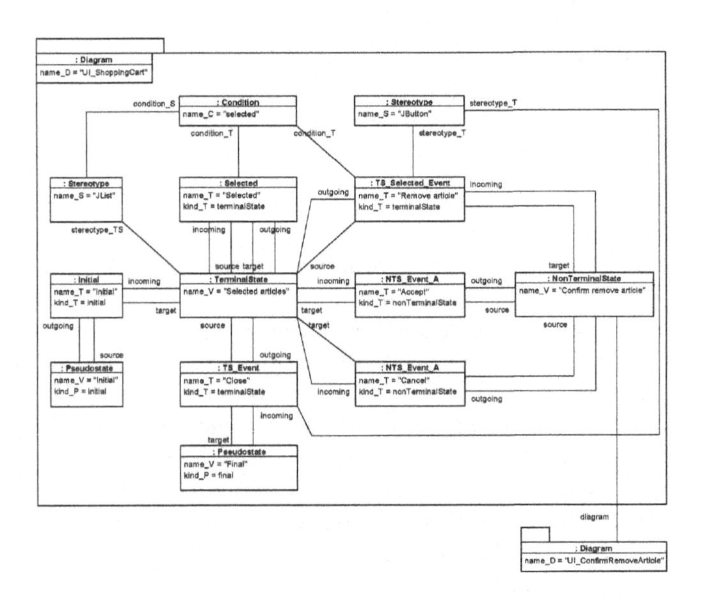

Fig. 6. Instance of the Metamodel for the `UI_ShoppingCart` diagram

4 Related Work

In the literature, there are several works about MDD and UI design, and most of them are supported by a tool. The most relevant ones are *TRIDENT*, *TERESA*, *WISDOM*, *UMLi*, *ONME* and *IDEAS*. TRIDENT (*Tools foR an Interactive Development ENvironmenT*) [3] proposes an environment for the development of highly interactive applications. User interfaces are generated in a (quasi-) automatic way. There is a CASE tool called SEGUIA (*System Export for Generating a User Interface Automatically*) [22] for generating the user interface from TRIDENT models. This approach can be considered a non-UML compliant proposal. However, most of the concepts developed in TRIDENT were later included

Table 2. Some OCL constraints of the $\mathcal{ALIR}$ tool

NTS_Event_C transition

```
self.source.oclAsType(NonTerminalState).diagram.vertex->
    one(v | (v.name_V=self.tsName_NTSEC) and
             (v.oclAsType(TerminalState).stereotype_TS.name_S=self.tsStereotype_NTSEC) and
             (v.outgoing->
                one(t | (t.name_T=self.name_T) and
                          ((t.target.oclIsTypeOf(Pseudostate)) and
                          (t.target.oclAsType(Pseudostate).kind_P=PseudostateKind::final)))))
```

NTS_Input transition

```
self.source.oclAsType(NonTerminalState).diagram.vertex->
    one(v | (v.name_V=self.tsName_NTSI) and
             (v.outgoing->
                one(t | (t.condition_T.name_C=self.condition_T.name_C) and
                           ((t.target.oclIsTypeOf(Pseudostate)) and
                           (t.target.oclAsType(Pseudostate).kind_P=PseudostateKind::final)))))
```

Input transition

```
if (self.source.oclIsTypeOf(NonTerminalState)) then
    self.source.oclAsType(NonTerminalState).diagram.vertex->
        one(v | (v.outgoing->
            one(t | (t.condition_T.name_C=self.condition_T.name_C) and
                    ((t.target.oclIsTypeOf(Pseudostate)) and
                    (t.target.oclAsType(Pseudostate).kind_P=PseudostateKind::final)))))
else
    true
endif
```

in UML proposals. TERESA (*Transformation Environment for InteRactivE System RepresentAtions*) [21] is a tool aimed to generate user interfaces for multiple platforms from task models designed with the so-called *ConcurTaskTrees*. Partially based on UML, WISDOM (*Whitewater Interactive System Development with Object Models*) [7] is a proposal for user interface modeling. WISDOM uses UML but the authors have also adopted the ConcurTaskTrees. Fully based on UML, UMLi [20,19,18] proposes an extension of UML for user interface modeling. UMLi aims to preserve the semantics of existing UML constructors since its notation is built by using UML extension mechanisms. Like our technique, the scope of UMLi is restricted to WIMP interfaces. UMLi is supported by ARGOi [17]. UMLi is very similar to our proposal in the adoption of UML diagrams for user interface modeling, although it supports different extensions. Another case of MDD-based UI development is the OO-Method [14]. The authors have developed the ONME tool (*Oliva Nova Model Execution*, http://www.care-t.com), an implementation of the *OO-Method*, which complies with the MDA paradigm by defining models at different abstraction levels. Finally, IDEAS (*Interface Development Environment within OASIS*) [6,8] is also a UML-based technique for the specification of user interfaces based on UML and on the OASIS (*Open and Active Specification of Information Systems*) specification language [13].

5 Conclusions and Future Work

This paper presents the first step towards the development of a tool for user interface design based on MDD. User interaction diagrams are one the main elements of our technique, since they are used for dialogue modelling. However, our modelling technique requires also task and presentation modeling. The presentation model is partially handled by means of the UI design tool (like *Windows Builder-Pro of Eclipse*). However we have to integrate it with our Eclipse GMF tool by means of XMI. It is considered as future work. In addition, the presentation modelling in our proposal is also achieved by means of the so-called *UI-class diagrams*, which are an specialization of UML class diagram for UI components. An extension of the tool for supporting UI-class diagram modelling is planned to be implemented in the near future. Finally, task modelling is achieved in our UI-MDD technique by means of *user-interface* diagrams which are an specialization of use case diagrams for UI specification. We would also like to extend our tool for modeling such diagrams in order to be able to fully support in $\mathcal{ALIR}$ our UI-MDD technique.

Acknowledgements

This work has been partially supported by the EU (FEDER) and the Spanish MEC under grants TIN2008-06622-C03-03, TIN2006-15175-C05-02 (MEDWSA), and TIN2007-61497 (SOLERES).

References

1. Alonso, D., Vicente-Chicote, C., Pastor, J.A., Álvarez, B.: StateML+: From Graphical State Machine Models to Thread-Safe Ada Code. In: Kordon, F., Vardanega, T. (eds.) Ada-Europe 2008. LNCS, vol. 5026, pp. 158–170. Springer, Heidelberg (2008)
2. Almendros-Jiménez, J.M., Iribarne, L.: An Extension of UML for the Modeling of WIMP User Interfaces. Journal of Visual Languages and Computing 19, 695–720 (2008)
3. Bodart, F., Hennebert, A.-M., Leheureux, J.-M., Sacré, I., Vanderdonckt, J.: Architecture Elements for Highly-Interactive Business-Oriented Applications. In: Bass, L.J., Unger, C., Gornostaev, J. (eds.) EWHCI 1993. LNCS, vol. 753, pp. 83–104. Springer, Heidelberg (1993)
4. Eclipse Graphical Modeling Framework (GMF), `http://www.eclipse.org/modeling/gmf/`
5. Eclipse Modeling Framework (EMF), `http://www.eclipse.org/modeling/emf/`
6. Lozano, M., Ramos, I., González, P.: User Interface Specification and Development. In: Proceedings of the IEEE 34th International Conference on Technology of Object-Oriented Languages and Systems, pp. 373–381. IEEE Computer Society Press, Washington (2000)
7. Nunes, N.J., Falcao e Cunha, J.: WISDOM - A UML Based Architecture for Interactive Systems. In: Palanque, P., Paternó, F. (eds.) DSV-IS 2000. LNCS, vol. 1946, pp. 191–205. Springer, Heidelberg (2001)
8. Molina, J., González, P., Lozano, M.: Developing 3D UIs using the IDEAS Tool: A case study. In: Human-Computer Interaction. Theory and Practice, pp. 1193–1197. Lawrence Erlbaum Associates, Mahwah (2003)

9. OMG Object Constraint Language (OCL) Specification, version 2.0, http://www.omg.org/technology/documents/formal/ocl.htm
10. OMG Meta-Object Facility, http://www.omg.org/mof/
11. OMG OMG Unified Modeling Language (OMG UML), Superstructure, V.2.1.2, http://www.omg.org/spec/UML/2.1.2/Superstructure/PDF
12. OMG XML Metadata Interchange (XMI), http://www.omg.org/spec/XMI/
13. Pastor, O., Hayes, F., Bear, S.: OASIS: An Object-Oriented Specification Language. In: Loucopoulos, P. (ed.) CAiSE 1992. LNCS, vol. 593, pp. 348–363. Springer, Heidelberg (1992)
14. Pastor, O.: Generatig User Interfaces from Conceptual Models: A Model-Transformation based Approach. In: Chapter in Computer-Aided Design of User Interfaces, CADUI, pp. 1–14. Springer, Heidelberg (2007)
15. Paternò, F.: Model-Based Design and Evaluation of Interactive Applications. Springer, Berlin (1999)
16. Selic, B.: UML 2: A model-driven development tool. IBM Systems Journal 45(3) (2006)
17. Paton, N.W., Pinheiro da Silva, P.: ARGOi, An Object-Oriented Design Tool based on UML. In: Tech. rep. (2007), http://trust.utep.edu/umli/software.html
18. Pinheiro da Silva, P., Paton, N.W.: User Interface Modelling with UML. In: Information Modelling and Knowledge Bases XII, pp. 203–217. IOS Press, Amsterdam (2000)
19. Pinheiro da Silva, P., Paton, N.W.: User Interface Modeling in UMLi. IEEE Software 20(4), 62–69 (2003)
20. Pinheiro da Silva, P.: Object Modelling of Interactive Systems: The UMLi Approach, Ph.D. thesis, University of Manchester (2002)
21. Berti, S., Correani, F., Mori, G., Paternó, F., Santoro, C.: TERESA: A Transformation- based Environment for Designing and Developing Multi-device Interfaces. In: Proceedings of ACM CHI 2004 Conference on Human Factors in Computing Systems, vol. II, pp. 793–794. ACM Press, NY (2004)
22. SEGUIA, System Expert Generating User Interfaces Automatically, http://www.isys.ucl.ac.be/bchi/research/seguia.htm

The Effectiveness of e-Cognocracy

Xhevrie Mamaqi and José María Moreno-Jiménez

Zaragoza Multicriteria Decision Making Group (http://gdmz.unizar.es)
Faculty of Economics, University of Zaragoza, Spain
mamaqi@unizar.es, moreno@unizar.es

Abstract. This paper presents an empirical analysis of the effectiveness of the democracy model known as e-cognocracy [1-7]. The paper considers the extent to which the goals of e-cognocracy (transparency, control, participation, knowledge democratisation and learning) are achieved in an e-discussion process in a cognitive democracy using a collaborative tool (forum). Students on the Multicriteria Decision Making course at the Zaragoza University Faculty of Economics were surveyed with the aim of identifying their opinions with regards to the *Gran Scala* leisure complex project which involves locating the biggest leisure complex in Europe in the *Los Monegros* area of Aragon, north-east Spain. The results of the survey, which focuses on different dimensions of the problem (quality of the web site, quality of information available and attributes for effectiveness), have been analysed using structural equations. This approach has been formulated as a general framework that allows an empirical evaluation of citizens' e-participation in electronic governance.

Keywords: e-cognocracy, e-democracy, e-government, e-participation, knowledge society, effectiveness, structural equation model (SEM).

1 Introduction

Electronic Government can be understood as the application of information and communications technology (ICT) in the field of public administration. Electronic Government could include the provision of services via Internet (e-administration), e-voting, e-discussion, e-democracy and even the most complicated issues concerning the governance of society.

In a world of growing complexity, the objective of scientific decision making [8] not only involves the search for truth, as suggested by traditional science, but the formation of the individual (intelligence and learning), the promotion of relationships with others (communication and coexistence), the improvement of society (quality of life and cohesion) and building the future (development). All this can be seen as a part of the 'Knowledge Society' – for the advancement of knowledge and human talent.

In the context of democracy (considered as the most widespread and accepted representative model among western societies), Moreno [1] proposed a new democratic model known as *e-cognocracy* [1-7]. This new model combines the preferences

M.D. Lytras et al. (Eds.): WSKS 2009, LNAI 5736, pp. 417–426, 2009.

of political parties with the preferences of the people, selecting the best alternative from a discrete set and providing the arguments that support the decisions taken. E-cognocracy also seeks to improve the transparency of the democratic system by increasing the citizen's control over government through greater participation in the running of society.

In general, taking the correct decision can be understood as effectiveness. E-cognocracy is an effective democratic model if it allows the creation and diffusion of knowledge related to the scientific resolution of the public decisional problem that is being considered.

An evaluation of the effectiveness of e-cognocracy was undertaken by means of an experimental study consisting of a questionnaire completed by students of Multicriteria Decision Making at the Zaragoza University Faculty of Economics. The questionnaire was based on the proposed "*Gran Scala*" leisure complex development - the construction of one of the world biggest entertainment centres with an estimated investment of 17,000 million euros, 25 million visitors per year and the creation of 65,000 jobs in the 70 hotels, 32 casinos and five theme parks.

The electronic survey provided the information necessary for the Structural Equation Model (SEM) empirical analysis. The following dimensions in the study of e-cognocracy behaviour were considered: "web service quality", "quality of available information" and "knowledge transfer".

Section 2 offers a brief introduction to e-cognocracy and its objectives. Section 3 deals with the experimental research. Section 4 summarizes the empirical results and Section 5 presents the main conclusions of the study.

2 Conceptual Framework and Hypothetical Model

Whilst traditional democracy refers to government of the people, cognocracy refers to governance of knowledge (wisdom). Cognocracy does not refer to knowledge provided exclusively by 'wise men', as suggested by Plato [6]; it is a new model of democracy which combines representative democracy (political parties) and participatory democracy (the citizen) in accordance with weightings that depend on the type of problem that is dealt with. The general objective of e-cognocracy is the creation and dissemination of knowledge associated with the scientific resolution of public decision making problems related to the governance of society. The specific objectives of e-cognocracy are: (i) to improve the transparency of the democratic system, (ii) to increase the control of citizens and (iii) to encourage the participation of citizens in the governance of society [3,5,9].

The creation and dissemination of knowledge is achieved through internet discussion in which political parties and citizens put forward arguments that support decisions and justify preferences. Transparency is increased because the political parties are forced to express their viewpoint before the resolution of the problem; the citizen's control is increased because the political parties must win the vote. Citizens and opposition political parties may request that a specific problem be resolved by means of e-cognocracy and this could involve a vote of censure on a specific political measure. Public participation in the governance of society enables citizens to be directly implicated in the final decision on an issue, not just in the election of representatives (representative democracy).

E-Cognoracy is closely linked to the Internet [1-7]. The Internet is used to deliver judgments and to facilitate the exchange of information and opinion of the participants. The application of e-cognocracy includes two fundamental dimensions, the Internet service and its available information. The Internet service is related to multiple factors that characterise the citizen's participation in the decision making process (adequate website design, accessibility, security and ease of use). Clear instructions to enable access to information related to the debate are also of vital importance to the development of e-cognocracy.

First hypothesis: H_1: *The quality of the Internet service has a direct and positive effect on the effectiveness of e-cognocracy.*

Moreno et al (2003) have suggested the use of e-information and e-voting through the Internet as a means to enhance the transparency of the democratic process and to increase the participation of citizens in their own government. In e-cognocracy, the dissemination of information and knowledge is aimed at improving policy formulation within the democratic process, giving the citizen the right to know about key aspects of public administration which ensures integrity, objectivity and justice. People are able to contribute their knowledge and opinions to the voting process of decision making through discussions in the forums.

The availability of information on the Internet rapidly increases communication between participants and their knowledge about a specific problem. It is therefore important that the information provided through the Internet in the development phase of discussions and decisions is accessible and transparent. The contents must be clear, comprehensive, objective and appropriate for the level of the user [21-22].

Second hypothesis: H_2: *The quality of information available has a direct and positive effect on the effectiveness of e-cognocracy*

The third hypothesis concerns the dissemination of information and the exchange of opinions between participants in the Internet forum [5]. The exchange of information and opinions between participants in the forum increases the knowledge of the actors involved in the resolution of the problem.

Third hypothesis: H_3: *The "transfer of knowledge and opinion" has a direct and positive effect on the effectiveness of e-cognocracy.*

To evaluate the effectiveness of e-cognocracy as a latent construct, multiple objectives (such as increasing the degree of trust, responsibility and commitment between government and citizens, legitimacy through participation and representation of the citizens in decision making), must be considered [24]. All these aspects are reflected in the items included in the electronic survey. The three hypothesis are reflected in the model structure, as shown in Figure 1.

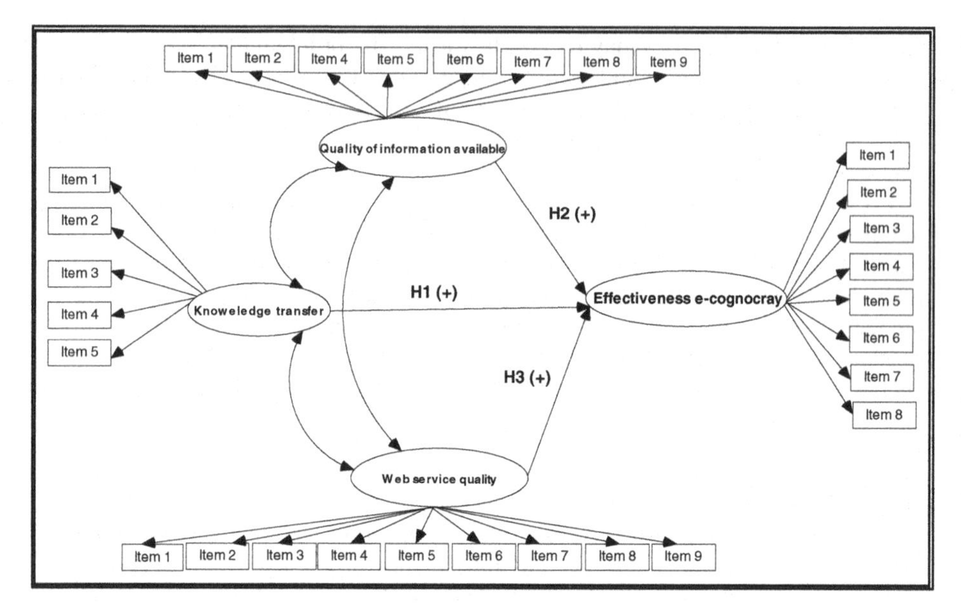

Fig. 1. Hypothetical Structural Model

3 Experimental Research

The survey was conducted in April 2008 among 4th year students of Multicriteria Decision Making at Zaragoza University. From February 27 to March 4, 2008, the students participated in an e-discussion (in a forum) on the proposed location of a leisure and entertainment complex in the area of *Los Monegros* in Aragon [23]. In the forum discussion, participants submitted 77 messages (19 on advantages, 13 on costs, 22 on risks and 23 on opportunities) and 257 comments. To enhance the participation of students in solving the problem, the analytic hierarchy process, AHP [11-12] was used as methodological support, the Internet as the communication tool and the forum for the public debate.

At the end of the forum, the participants responded to an online survey, made up of 6 parts and 36 questions (34 closed and 2 open): I. Identification (3 issues); II. Web Service Quality (9 items); III. Quality of information available (9 items); IV. Knowledge Transfer (5 items); V. Impact of Knowledge (1 issue) and VI. Effectiveness of e-cognocracy (9 items). The scale used in measuring the variables of the study was 1 to 5 (1-not important, 5-very important).

Structural Equation Modelling (SEM) was used as a method for measuring relationships among latent variables [12-14]. The graphical representation for SEM and estimations were obtained using AMOS 7 [15].

The model structure is made up of two parts: The Measurement Model (or Confirmatory Factor Analysis - CFA) that tests the relationship between items and latent variables, and The Full Structural Model that consists of joint estimation equations of the Measurement and Structural Model.

4 Case Study

Table 1 shows the statistical average of punctuation items for the latent constructs. For "Web Service Quality", the highest marked items are "Easy web debate" and "Web presence information" with 3.87. In the case of the construct "Quality of information available", the indicator with the highest average (3.45) corresponds to "Update information" and the lowest average score is "Specific information". For the "Knowledge Transfer" construct, the highest rated indicators are: "Interaction among participants" (4.16) and "Known other alternatives" (4.13). Finally, for the "Effectiveness" construct, the highest ranked are: "legitimacy" (3.58) and "amplification" (3.66).

Table 1. Statistics of the Items

ITEMS*	Statistical					
	Average score	Standard deviation	Standard error	Skewnees	Curtosis	Response rate
Latent Variable "Web service quality"						
Visual design						
I 1. Easy web access	3,79	0,811	0,132	-0,230	-0,327	38
I 2. Web usability	3,63	0,970	0,157	-0,299	-0,793	38
I 3. Web availability	3,47	1,059	0,172	-0,648	1,785	38
Degree of personalization						
I 4. Easy web debate	3,87	1,143	0,185	-1,449	2,850	38
I 5. Web security	3,50	1,225	0,199	-1351	2,129	38
I 6. Web e-voting	3,45	1,245	0,202	-0,937	0,388	38
Availability of information						
I 7. Easy access to information	3,55	1,155	0,187	-1,356	3,108	38
I 8. Web management information	3,47	1,084	0,176	-1,743	4,066	38
I 9. Web presence information	3,87	1,189	0,193	-1,259	1,640	38
Latent variable "Quality of information available"						
Accessibility to information						
I 1. Transparent information	3,13	1,256	0,204	-0,261	-0,313	38
I 2. Specific information	2,79	1,166	0,189	-1,077	0,662	38
I 3. Attractive information	3,37	0,942	0,153	-1,235	3,192	38
I 4. Update information	3,45	1,005	0,163	-0,859	2,508	38
Comprehensiveness of the information						
I 5 Accuracy of information	3,24	1,051	0,170	-0,800	1,429	38
I 6 Comprehensiveness	3,50	1,180	0,191	-0.886	0,865	38
I 7 Objectivity of information	3,11	1,181	0,192	-0,526	1,081	38
I 8 Clarity of information	3,53	1,109	0,180	-0,196	-0,794	38
I 9 Apropiate level of information	3,34	0,966	0,157	0,001	-0,118	38
Latent variable "Knowledge transfer"						
I 1. Enrichment of knowledge	3,71	,867	0,141	-0,697	1,370	38
I 2. Hear other views	3,97	1,102	0,179	-0,968	0,184	38
I 3. Interaction	4,16	,945	0,153	-0,534	-1,227	38
I 4. Known other alternatives	4,13	,906	0,147	-1,193	2,233	38
I 5. Transfer alternatives	3,76	1,076	0,175	-0,596	-0,281	38
Latent variable "Effectiveness e-cognocracy"						
I 1. Effectiveness legitimacy	3,58	1,056	,171	-1,091	2,237	38
I 2. Effect. amplification	3,66	,815	,132	-,226	-,272	38
I 3. Effect representativeness	3,42	,889	,144	,008	-,650	38
I 4. Effect trust	3,18	1,159	,188	-,379	,264	38
I 5. Effect commitment	3,24	1,195	,194	-,686	,539	38
I 6. Effect government efficiency	3,08	1,302	,211	-,696	,117	38
I 7. Effect common goals	3,03	1,423	,231	-,761	-,203	38
I 8. Effect common commitment	3,11	1,410	,229	-,562	-,395	38

***measurement scale for items: 1-nothing important; 2-unimportant, 3-relatively important, 4-important, 5-very important**

This analysis confirms that many indicators present acceptable skewness and kurtosis levels. For indicators that presented a slight deviation from normality, a square root transformation to reduce its deviation was applied.

To check the dimensionality of the latent constructs, a principal component factor analysis (not reported here) was implemented. This analysis revealed that the latent construct items in "Web service quality" saturate in three factor loadings (or dimensions) called "Visual design", "Degree of personalization" and "Availability of information". The analysis also shows that there are two factors for the latent construct

"Quality of information available", which are called "Information accessibility" and "Comprehensiveness of the information." Consequently, for both constructs, the measurement model has been designed as a second order confirmatory factor analysis.

The results of the measurement model and the reliability indices for each individual item are presented in Table 2. For the majority, the rate of reliability is $\alpha \succ 0{,}5$ (widely accepted by the relevant bibliography). Only four indicators, three in the "Web service quality" construct and one in the "Transfer knowledge" construct, present an alpha value below this limit.

Table 2. Loadings and cross loadings of the measurement model

Items	λ	t	p	standarized
Web service quality (exogenous latent variable)				
Visual design				
Easy web access<--- Visual design	1			,431
Web usability <--- Visual design	0,637	2,528	***	,660
Web avialibility<--- Visual design	0,717	6,111	***	,627
Degree for personalization				
Easy web debate<--- Degree for personalization	1			,359
Web security<--- Degree for personalization	1,946	3,169	**	,540
Web e-voting<--- Degree for personalization	1,966	3,171	**	,579
Availability of information				
Easy access to information <--- Availability of information	1			390
Web management information<--- Availability of information	1,201	0,248	**	,249
Web presence information <--- Availability of information	1,111	0,641	**	,239
Scale reliability overall items $\alpha_{cronbach}$ = 0,56				
Quality of information available (exogenous latent variable)				
Accessibility to information				
Transparent information<---Accessibility to information	1			,724
Specific information<--- Accessibility to information	0,331	1,692	**	,540
Attractive information<--- Accessibility to information	1,132	4,486	***	,531
Update information<--- Accessibility to information	1,047	3,195	***	650
Comprehensiveness of the information				
Transparent information<--- Comprehensiveness information	1	1		,656
Comprehensiveness <--- Comprehensiveness information	0,856	3,339	***	,738
Objectivity of information<--- Comprehensiveness information	1,058	3,786	***	,633
Clarity of information<--- Comprehensiveness information	1,249	4,243	***	,790
Appropriate level<--- Comprehensiveness information	0,869	3,623	***	,702
Scale reliability overall items $\alpha_{cronbach}$ = 0,75				
Knowledge transfer (exogenous latent variable)				
Enrichment of knowledge<--- Knowledge transfer	1			,575
Hear other views<--- Knowledge transfer	0,876	0,153	*	,334
Exchange viewpoint<--- Knowledge transfer	1,876	2,673	***	,670
Known other solution<--- Knowledge transfer	1,785	0,987	*	,339
Known new alternative<--- Knowledge transfer	1,984	3,111	***	,726
Scale reliability overall items $\alpha_{cronbach}$ = 0,73				
Effectiveness e-cognocracy (endogenous latent variable)				
Legitimacy<--- Effectiveness e-cognocracy	1			602
Amplification<--- Effectiveness e-cognocracy	0,981	3,995	***	798
Representativeness <--- Effectiveness e-cognocracy	1,045	3,756	***	,632
Trust<--- Effectiveness e-cognocracy	0,984	3,965	***	,789
Commitment<--- Effectiveness e-cognocracy	0,893	4,653	***	,817
Government efficiency<--- Effectiveness e-cognocracy	1,117	3,875	***	,827
Common goals<--- Effectiveness e-cognocracy	1,327	4,675	***	,840
Common commitment<--- Effectiveness e-cognocracy	1,011	4,328	***	,866
Scale reliability overall items $\alpha_{cronbach}$ = 0,78				
Findings for goodness-of-fit indices: CFA model RMR (Root Mean Squared Residual) = 0,187 GFI (Goodness of Fit Index) = 0,687 AGFI (Ajusted Goodness of Fit Index) = 0,889 = 789,3, df=582, p=0,000				
*** p value <0,001 (t=1,96); **p<0,05 (t=2,576); * p<0,10 (t=3,291)				

The joint reliability (Alpha Cronbach) of each latent construct offers positive results. For three latent constructs, this coefficient is above 0.70. For the latent "Quality web service" construct, the value is 0.563. The estimated path for the relationships between observable items and their respective latent construct is above the $\lambda \succ 0{,}42$, recommended by the literature as a good value. The fit of the model were examined through the RMSE and GFI tests. The results of this analysis showed an excellent performance of all latent constructs.

Figure 3 shows the hypothesized Full Structural Model on "E-cognocracy effectiveness". It contains the direct relationships between the latent variables and their respective items and also those among the three latent exogenous variables and the endogenous variable.

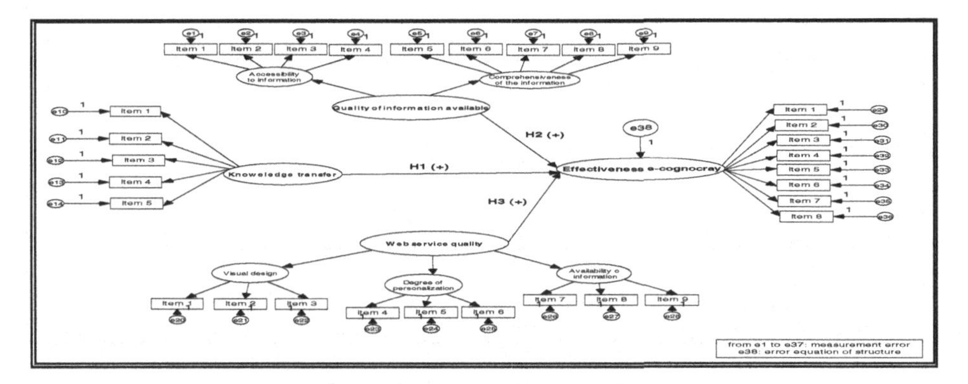

Fig. 3. Hypothesized Full Structural Model

4.1 Analysis and Results of the Full Structural Model

The results of the analysis are shown in Figure 4. The maximum likelihood (ML) estimation technique was used but other estimation methods to verify the goodness of fit were also applied [17]. The indicators χ^2 = 545,759 (p = 0.000 df = 365.1), RMSR= 0,157, AGFI = 0,918 and RMSE = 0,072 show a good fit. The estimated coefficients are significant at 1% and 5% for the relationships between the observable items and their latent variables.

The three hypotheses on the relationships between "Quality web service", "Transfer knowledge", "Quality of available information and "E- cognocracy effectiveness" were confirmed by the values of the coefficient path estimated as statistically significant and positive. The relationships between "Knowledge transfer", "Quality web service" and "E- cognocracy effectiveness" have been confirmed with a confidence level of 5%.The influence of the latent construct. "Quality of available information" on "E-cognocracy effectiveness" was confirmed by a result of minor significance at 1%. The high significance between "Quality of information available" and "E- cognocracy effectiveness" indicates that this effectiveness depends to on the quality of information.

The results of the simultaneous regressions analysis suggest that the indicators 'legitimacy', "trust", "representativeness", "efficiency", "commitment", "web service", "quality of available information" and "transfer of opinion/knowledge", are the best for the presentation of the constructs 'e-cognocracy effectiveness' and "service web quality", respectively.

The importance of "E- cognocracy effectiveness" depends on the degree of the importance of indicators such as easy web access, web comfort, web security, and e-voting. For web discussion, exchange opinion, expand and transfer web knowledge available, comprehensive, clear and adequate information is vital and therefore it is clear that knowledge playa a positive role in decisions involving citizens.

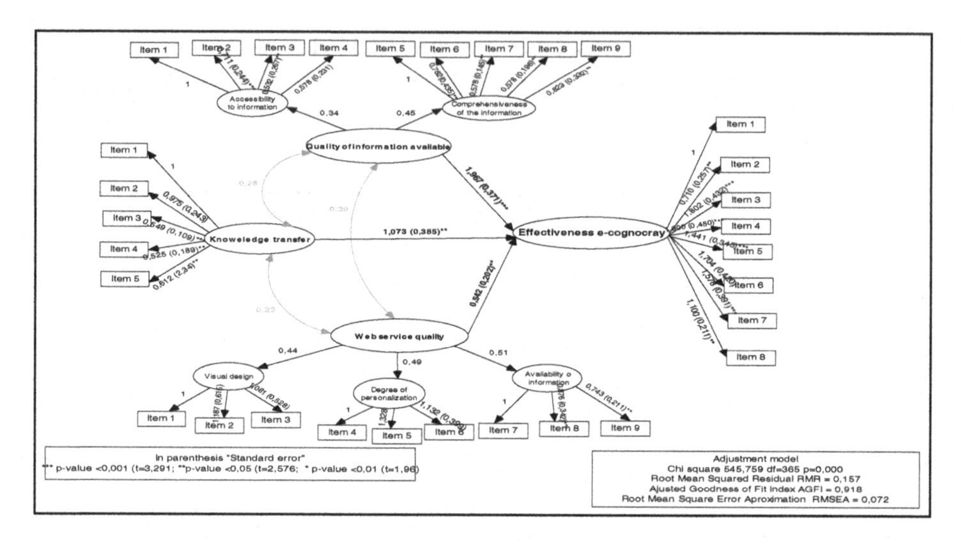

Fig. 4. Standardised estimated coefficient

5 Discussion and Conclusions

Conclusions based on the empirical results:

- "E-cognocracy effectiveness" is a concept that involves multiple issues such as legitimacy, trust, commitment and shared responsibility between citizens and government in making decisions. It is a theoretical concept that can not be observed directly and must be treated as a latent construct.
- The quality of information is an important factor in the implementation of e-cognocracy. Estimations showed that it has a direct impact and plays a significant role on "e- cognocracy effectiveness".
- As characteristics of web-information, "transparency", "accuracy", "comprehensiveness", "objectivity" and "a appropriate level" received high ratings.
- Internet "transfer of knowledge" is an issue related to the possibility of interaction and exchange of viewpoints between participants. It is positively related with the latent variable "Effectiveness of e-cognocracy". There is a similar relationship between "Quality of internet service" and "Effectiveness e-cognocracy". "Comfort", "security" and "choice of different alternatives" were also highly rated in relation to the latent variable "Quality of internet service".

However, the results of the study are empirically limited, therefore:

- The model should be applied to other cases in order to achieve further improvements and generalise the methodological approaches.
- A broader database to overcome limitations related to the response rate is recommended.

Acknowledgments

This work has been partially funded by: "*E-participation, Security and Knowledge Democratization*" (Ref. PM2007-034), the Government of Aragon and "*Collaborative Decision Making in e-Cognocracy*" (Ref. TNI2008-06796-C04-04), the Spanish Ministry of Science and Innovation. We would also like to thank David Jones for his help with the English edition of the paper and our colleagues Juan Aguarón and Alberto Turón for their collaboration in the development of the electronic forum and survey.

References

1. Moreno-Jiménez, J.M.: Las Nuevas Tecnologías y la Representación Democrática del Inmigrante. IV Jornadas Jurídicas de Albarracín. Consejo General del Poder Judicial (2003)
2. Moreno-Jiménez, J.M., Polasek, W.: E-democracy and Knowledge. A Multicriteria Framework for the New Democratic Era. Journal Multicriteria Decision Analysis 12, 163–176 (2003)
3. Moreno-Jiménez, J.M.: E-cognocracy y Representación Democrática del Inmigrante. Anales de Economía Aplicada (2004)
4. Moreno-Jiménez, J.M., Polasek, W.: E-cognocracy and the participation of immigrants in e-governance. In: Böhlen, et al. (eds.) TED Conference on e-government 2005. Electronic democracy: The challenge ahead, vol. 13, pp. 18–26. University Rudolf Trauner-Verlag, Schriftenreihe Informatik (2005)
5. Moreno-Jiménez, J.M.: E-cognocracy: Nueva Sociedad, Nueva Democracia. Estudios de Economía Aplicada 24(1-2), 559–581 (2006)
6. García Lizana, A., Moreno Jiménez, J.M.: Economía y Democracia en la Sociedad del Conocimiento. Estudios de Economía Aplicada 26(2), 181–212 (2008)
7. Moreno-Jiménez, J.M., Escobar, M.T., Toncovich, A., Turón, A.: Arguments that support decisions in e-cognocracy: A quantitative approach based on priorities intensities. In: Lytras, M.D., et al. (eds.) The Open Knowledge Society. Communications in Computer and Information Sciences, vol. 19, pp. 649–658 (2008)
8. Moreno-Jiménez, J.M.: Los Métodos Estadísticos en el Nuevo Método Científico. In: En Casas, J.M., Pulido, A. (eds.) Información económica y técnicas de análisis en el siglo XXI. INE, pp. 331–348 (2003)
9. Moreno-Jiménez, J.M., Aguarón, J., Escobar, M.T., Turón, A.: Philosophical, Methodological and Technological Foundations of E-cognocracy. In: Proceeding on the TED 2006 Conference on Towards e-Democracy: Participation, Deliberation, Communities, Mantova, Italia (2006)
10. Saaty, T.L.: Multicriteria Decision Making: The Analytic Hierarchy Process. Mc Graw-Hill, New York (1980); 2ª Impression RSW Pub., Pittsburgh (1990)
11. Saaty, T.L.: Fundamentals of Decision Making. RSW Publicat. (1994)
12. Jöreskog, K.G.: A general method for estimating a linear structural equation system. In: Goldberger, A.S., Duncan, O.D. (eds.) Structural Equation Models in the Social Sciences, pp. 85–112. Seminar Press, New York (1973)
13. Bollen, K.A.: Structural equation with latent variable. John& Wile, New York (1989)
14. Fornell, C., Larcker, D.F.: Evaluating structural equation models with unobservable variables and measurement errors. Journal of Marketing Research 18, 39–50 (1981)
15. Arbouckle, L.J.: AMOS 7 User's guide. Amos Development Corporation. Copyright, Chicago-USA (2007), http://amosdevelopment.com

16. Heck, R.H.: Factor Analysis: Exploratory and Confirmatory Approaches. In: Marcoulides, G.A. (ed.) Modern Methods for Business Research, pp. 177–216. Lawrence Erlbaum Associates, Mahwah (1998)
17. Ullman, J.B.: Structural equation Modeling. In: Tabachnick, B.G., Fidell, L.S. (eds.) Using Multivariate Statistics, pp. 653–771. Allyn and Bacon, Boston (2001)
18. Brown, M.W., Cudeck, R.: Alternative ways of assessing model fit. In: Bollen, K.A., Long, J.S. (eds.) Testing structural equation models, pp. 136–161. Sage Publication, Thousand Oaks (1993)
19. Held, D.: Modelos de Democracia. Alianza Editorial (2003)
20. Moreno-Jiménez, J.M., Piles, J., Ruiz, J., Salazar, J.L.: E-cogning: the e-voting process of e-cognocracy. Rios's International Journal on Sciences of Industrial and Systems Engineering and Management 2(2), 25–40 (2008)
21. Henderson, M., Henderson, P.: E-democracy evaluation framework (2005), `http://www.getinvolved.qld.gov.au/share_your_knowledge/documents/pdf/eval_framework_summaryfinal_200506.pdf`
22. Coleman, S., Macintosh, A., Schneeberger, A.: e-Participation Research Direction based on barriers, challenge and need. Demonet deliverable D12.3 (2008), `http://www.demo-net.org`
23. Mamaqi, X., Moreno-Jiménez, J.M., Muñoz, L.G.: Efectividad de la e-cognocracia en el gobierno electrónico de la sociedad. In: Comunicación presentada en la XXII-Congreso de la Economía Aplicada (ASEPELT), Junio, Barcelona, España (2008)
24. Tolbert, C.J., Karen, M.: The effects of e-government on trust and confidence in government. Public Administration review 66(3), 354–368 (2006)

Arguments That Support Decisions in e-Cognocracy: A Qualitative Approach Based on Text Mining Techniques

José María Moreno-Jiménez[1], Jesús Cardeñosa[2], and Carolina Gallardo[2]

[1] Zaragoza Multicriteria Decision Making Group (http://gdmz.unizar.es)
Faculty of Economics, University of Zaragoza, Spain
[2] Validation and Business Applications Research Group (http://www.vai.dia.fi.upm.es/)
Faculty of Informatics, Universidad Politecnica de Madrid, Spain
moreno@unizar.es, carde@opera.dia.fi.upm.es,
carolina@opera.dia.fi.upm.es

Abstract. E-cognocracy [1-5] is a new democratic system that tries to adequate democracy to needs and challenges of Knowledge Society. This is a cognitive democracy oriented to the extraction and democratization of the knowledge related with the scientific resolution of public decision making problems associated with the governance of society. It is based [3,6] on the evolutionism of live systems and it can be understood as the government of the knowledge and wisdom by means of the information and communication technology (ICT). E-cognocracy combines the representative and the participative democracies by aggregating the preferences of the political parties with those of citizens and by generating knowledge from the conjoint discussion of the arguments that support their own positions. This paper presents a qualitative approach based in text mining tools to identify these arguments from the analysis of the messages and comments elicited by political parties and citizens through a collaborative tool (forum). The proposed methodology has been applied to a case study developed with students of the Faculty of Economics at the University of Zaragoza and related with the potential location at the region of Aragón (Spain), of the greatest leisure project in Europe (Gran Scala).

Keywords: e-cognocracy, e-democracy, knowledge society, knowledge extraction, text mining.

1 Introduction

This paper continues with the research line about e-cognocracy [1-5] followed by the Zaragoza Multicriteria Decision Making Group (GDMZ) during the last five years. It extends and complements a previous paper of the GDMZ [7] presented in the First World Summit on Knowledge Society (WSKS08). In that case, following a quantitative approach based on the preference intensities of the political parties and citizens, we identify the messages and comments that support the different positions or patterns of behaviour of the actors involved in the resolution of the problem.

M.D. Lytras et al. (Eds.): WSKS 2009, LNAI 5736, pp. 427–436, 2009.

In this paper, a qualitative approach based on text mining tools have been used to identify the arguments embedded in these messages that support the different positions of the actors. This approach has been developed in collaboration with the Validation and Business Applications research group of the Universidad Politecnica de Madrid who has been working on knowledge extraction for more than ten years [11].

The methodology followed in the qualitative approach is based on the analysis of the different patterns that an expert defines in order to classify a message. These patterns are implemented in a text mining system that tries to emulate the lines of reasoning of the expert. The implemented text mining system is backed in the linguistic knowledge codified in a tailor-made grammar and a specific lexical resource. Its main purpose is to ascribe each comment to the different positions identified in the resolution of the problem.

The final and main aim of e-cognocracy is the democratisation of the knowledge associated with the scientific resolution of the problem. This way, we collaborate in the construction of a social knowledge and wisdom in accordance with the cognitive process that characterise the evolution of live systems [3,8]. The individual and social knowledge added value related with the scientific resolution of problems, the associated citizens' education and the improvement of transparency, participation and control will be some aspects that increase the individual and social welfare.

This work is structured as follows: Section 2 briefly presents e-cognocracy and the stages of its participation process; Section 3 presents the qualitative approach proposed for identifying arguments; Section 4 illustrates this methodology in a case study related to the construction of a leisure complex near the city of Zaragoza (Spain) and, finally, Section 5 highlights the most important conclusions.

2 E-Cognocracy: Definition and Participation Process

E-cognocracy [1-8] is a new democratic model that, based on the evolution of living systems, focuses on the extraction and social diffusion of the knowledge derived from the scientific resolution of highly complex problems associated with public decision making related with the governance of society.

This cognitive democracy seeks to convince citizens by means of arguments and not to defeat them (e-democracy) by means of votes [6,7]. To this end, e-cognocracy identifies the arguments that support the decisions made by the political parties and the citizens from the messages both group includes in the collaborative tool used to implement the discussion phase of its methodology. Obviously, some of these messages can incorporate strategic opinions that do not correspond with the final preferences.

The key idea of this democracy of the knowledge society (e-cognocracy) is to educate people (intelligence and learning), promote relations with others (communication and coexistence), improve society (quality of life and cohesion) and construct the future (evolution) in a world of increasing complexity [5].

The stages followed in the e-cognocracy process [7,9] are:

Stage 1: *Problem Establishment.* Using the web, this stage identifies the relevant aspects of the problem: context, actors and factors (mission, criteria, subcriteria, attributes and alternatives), as well as their interdependencies and relationships.

Stage 2: *Problem Resolution*. This stage provides the priorities of the alternatives being compared by using one of the most extended multicriteria decision making approach, AHP [10].

Stage 3: *Model Exploitation*. This stage of the resolution process derives the patterns of behaviour of the actors involved in the resolution process.

Stage 4: *Discussion*. Using any media or collaborative tool (forum in our case), the citizens' representatives (through their respective political parties) and the citizens themselves (through the network) give their motives and justify the decisions. From these comments and messages, the arguments that support the alternatives, as well the attributes that are more relevant in the resolution process are identified.

Stage 5: *Second round in problem resolution*. After updating the individual preferences with the explicit knowledge derived in the previous resolution stage, the priorities of the alternatives in a second round are obtained.

Stage 6: *Knowledge Extraction and Democratisation*. Using the information on the preferences of the two rounds and the quantitative information included in the comments elicited in the discussion stage, the comments that support each pattern of behaviour and each preference structure are identified.

Next, by means of text mining techniques, we consider in more detail the fourth stage: searching for the arguments that support decisions. Finally, integrating this qualitative approach with the quantitative one proposed in [7] and the individual changes in preferences, we will be able to identify the social leaders (the actors whose opinions provoke these changes) and the arguments taken into account by the citizens.

The procedure concludes with the social democratisation of the knowledge through the web. This final step is the main objective of the cognitive democracy we propose to deal with public decision making.

3 A Qualitative Approach for Searching for Arguments

This section identifies the comments and messages included in the discussion stage of e-cognocracy that support the different patterns of behaviour and changes in preferences that appear in the problem resolution stage. The qualitative approach follows a knowledge-based methodology. First, an expert performs a manual classification of the messages taking into account the presence of specific assertions that allows the expert to reasonably infer the participant's position underlying in the message. Then, the set of identified assertions are analyzed and codified into some linguistic patterns that are implemented in a text mining system. The design and implementation of the whole process is described in the following subsections.

3.1 The Expert Task

The expert has at his disposal a set of 332 messages extracted from a forum about the construction of a leisure complex in the area of Los Monegros (near Zaragoza). He then classifies each message according to the following categories:

- **A1:** In favour of the implementation of the project.
- **A2:** Support for the majority's position.
- **A3:** Against the implementation of the project.
- **A1_A2:** Position is not clear but surely the subject won't vote against (not A3)
- **A2_A3:** Vote is not clear but surely the subject won't vote in favour (not A1)

Messages are decontextualized so that it is avoided any other source of information (even the discourse thread is missed). The rationale to avoid contextual information is to define patterns merely based on the content of the message.

The expert then reads the messages, classifies each according to the five postulated categories, and marks up the expressions that support the proposed classification. Table 1 shows some examples of tagged messages; supporting arguments are underlined.

Table 1. Example of messages and supporting arguments

MESSAGE	A1	A2	A3
I think you are right, it is not logical that this year in expo water is treated as a very important problems and that in the next X years we will like to build an aquatic park in Monegros, where <u>water will be seriously wasted</u>.		✓	✓
I think that the installation of Gran Scala <u>will bring lot of opportunities</u> that will favour people's continuity in the area, specially young people. Besides, it is another way to <u>promote tourism</u>.	✓		

3.2 Patterns Extraction

Once the expert has classified the messages and identified the supporting expressions, common features and regularities have to be looked for. In general, simple categories (A1, A2 y A3) show clearer patterns than intermediate ones (A1&A2 and A2&A3). So it can be stated that:

- **Category A1** is characterized by the presence of verb phrases, noun phrases and words with positive connotation (that is, they refer to realities that are commonly considered as good or positive).
- **Category A3** is characterized by the presence of noun phrases and words with negative connotation and verb phrases that describe a negative reality (for example, *doesn't respect the bases of sustainable development*).
- **Category A2** is identified by the absence of significant expressions.
- In **categories A1_A2 and A2_A3**, negative and positive expressions are intermingled along with the so-called *hedges*.

As can be seen, the key aspect to identify patterns is the presence of expressions with either positive or negative connotation. Thus, this work requires lexical units (in particular, adjectives, nouns and verbs) to be classified as positive, negative, or neutral. *Hedges* are expressions that lessen the degree of assertiveness of the speaker's intention. Modal verbs like "should", "could" or adverbs like "possibly" or "perhaps" will be used for the identification of intermediate categories.

Thus, the system of automatic classification mainly seeks noun and verb phrases and then assigns a global connotation to the message according to the nature of the identified expressions. The identification of phrases will be done using shallow parsing [12,13].

3.3 Creation of the Lexical Resource

The set of messages to be mined presents two crucial features: a) they are framed in a closed domain, namely, the settling Gran Scala; and b) they show a high degree of informality and all sorts of errors. This implies that a robust parser is required, able to extract maximum information even with insufficient lexical and grammatical information.

The unique information registered in the dictionary is the grammatical category of the word. The postulated categories for this work are for open categories: verb, noun, adjective, and adverb; for functional categories: determinant, conjunction, preposition, and pronoun; and finally two ad hoc categories for the domain: hedge, and quantifier.

This classification is a simplification of the categories present in natural languages, but it is sufficient for this work where linguistic accuracy is not a priority. Words belonging to open categories are marked with either positive or negative connotation. This is the only semantic feature allowed in the dictionary. Dictionary entries have the following appearance:

```
mafia: SUS,negative.                         [mafia]
puestos de trabajo:   SUS,positive.          [job]
riesgos: SUS,negative.                       [risks]
```

In order to avoid the excessive dependence of the dictionary on the corpus of available messages, we started from an existent resource that was adapted to the requirements of the system. In particular, we consulted the Spanish Wordnet (Spanish lexical resource developed at the EuroWordNet project) [14] in order to obtain a general purpose dictionary as well as it was the source to infer the connotation of open categories words from the information codified in the synset (set of synomyns that define a concept) and in the definitions. Undoubtedly, this resource constitutes a raw version of the final lexical resource, refined and tuned during the testing phase of the system.

3.4 Pattern Codification and Generation

Once obtained the first version of the lexical resource, it is required to define a grammar that can identify simple phrases and that assigns a positive/negative connotation to the phrase as a whole. Thus, the grammar is restricted to the identification of noun, verb, and adjective phrases. As an example, let's see the treatment of noun phrases.

The following rule implements a pattern for a generic positive noun phrase:

```
DET? AP? N AP* PP*
```

Where

- DET is the class of determiners, AP stands for Adjective Phrase, N stands for Noun, and PP stands for Prepositional Phrase.

- The symbol ? denotes an optional element and * indicates that the element can be repeated zero o more times.
- At least one of these elements must have positive connotation except for the PP that can present a negative connotation, like in the following examples:
  `creación de [puestos_de_trabajo]`$_{pos}$ *[creation of jobs]*
  `utilización`$_{pos}$ `[de tierras infertiles]`$_{neg}$ *[use of waste lands]*

3.5 Assigment of Category Labels

Category labels are ascribed to a message according to the following rules which constitute the core of the automatic classification:

- If the message does not contain any expression, or the number of positive expressions equals the number of negative one, **then assign A2.**
- If only one positive expression has been found, or there are hedges and there are more positive than negative expressions, then **assign A1_A2.**
- If only one negative expression has been found, or there are hedges and there are more negative than positive expressions, **then assign A2_A3.**
- If there are more positive expressions than negative, **then assign A1**.
- If there are more negative expressions than positive, **then assign A3**.

4 Case Study

Gran Scala will be the largest leisure complex ever built in Europe. The project statistics are staggering: 17,000 million Euros in investment; 1,000 million in State and 677 million Euros in Aragon Regional Government revenues through taxation. Gran Scala is expected to receive 25 million visitors per year, create 65,000 direct and indirect jobs at its 70 hotels, 32 casinos, five theme parks, museums, golf courses and racetracks. In other words, the project would transform the area into a town with a population of 100,000.

Since its inception, the project has caused much debate and controversy in Aragonese society. For these reasons, the project was selected for an electronic voting experiment with students from the Multicriteria Decision Making course of the 4th year of the Bachelor of Science in Business Administration at the University of Zaragoza. The experiment was intended to elicit the opinion and preferences of the students regarding the implementation of the Gran Scala Project in the "Los Monegros" district and to assist in the construction of the ontology used in the final stage of the methodology, the identification of arguments that support the decisions by means of text mining techniques.

The stages of the experiment are the following:

1. Modelling, assessment and prioritization. This stage is based on the Analytic Hierarchy Process (AHP) methodology and the outcomes of it are the initial priorities for all the decision-makers.
2. Discussion, argumentation and feedback. In this stage a debate is implemented to improve the quantity and quality of the information available.
3. Final prioritization and conclusions. In this stage, a new assessment is made and the final priorities for the decision makers are obtained.

4.1 First Stage: Modelling

In this case, the hierarchy was directly introduced to the students and its design was not part of the decision making process. The hierarchy is composed by the goal (G), four criteria and three alternatives, as can be seen in Figure 1. The goal consists of determining the best course of action concerning the implementation of the Gran Scala project. The alternatives are evaluated with respect to four criteria in terms of their contribution to the goal: Benefits (B), Costs (C), Opportunities (O) and Risks (R). The alternatives considered in the process are: Implementation of the project (A1), Support for the majority's position (A2), and Rejection of the project (A3). The second alternative was introduced to provide an intermediate option and to serve as the recipient of potentially indifferent or undecided decision-makers. In theory, if the debate and discussion stage is carried out in a proper manner, part of those uncommitted or indifferent decision-makers will adopt a clear position as a result of a better understanding of the problem.

The factors that affect the decision are grouped in four main criteria (B, C, O, R). This general scheme was further detailed in order to facilitate a clear comprehension of the problem. We provided some economic, social and environmental ideas for benefits and costs (positive and negative short term effects), and opportunities and risks (positive and negative long term effects). The previous information was only used only to initiate the decision making process and accelerate the global comprehension of the situation under consideration. Students were completely free to make their own decisions and form their own opinions.

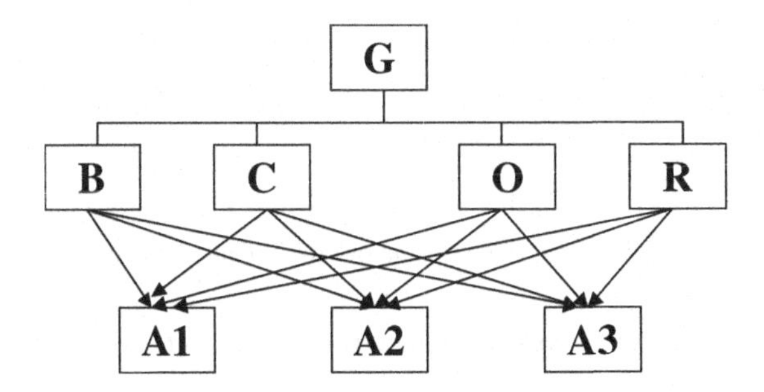

Fig. 1. Hierarchy considered in the *Gran Scala* problem

Once the hierarchy has been presented, it is possible to start the assessment process that will provide the priorities for the alternatives. This process was implemented using a web platform that allowed the full participation of the students. The data are then synthesized to obtain the priorities for the alternatives. This process is carried out by means of specific software that performs the calculations required by the AHP methodology.

4.2 Second Stage: Discussion

In this stage, a debate session was initiated after the end of the first round of voting. The implementation of the process was performed through a web interface in the form of a discussion forum (Figures 2a,b,c,d). The students posted messages grouped in four categories associated to the criteria of the problem. Different threads of discussion were created and, in general, several comments related to the initial messages of the threads were also posted. When a message is posted on the forum the author of a message posts it on the forum he/she is required to express his/her appreciation concerning the importance of the message using a scale of 1 to 10 (1 is minimum, 10 is maximum importance).

The discussion process was prolonged during ten days (24th February until 4th March 2008) in order to facilitate the sharing of viewpoints, information and feedback necessary to achieve an appropriate level of maturity in the participants' judgments.

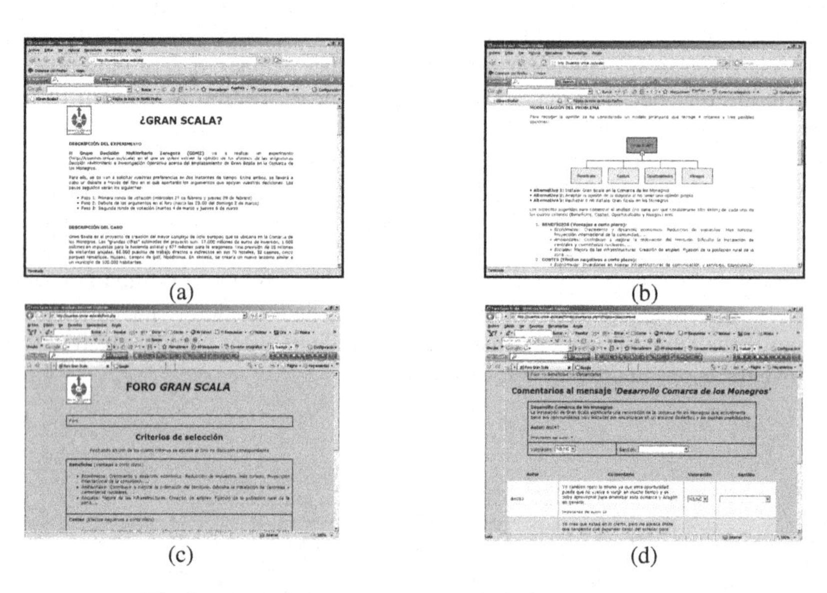

Fig. 2. Web interface for the discussion and prioritization

4.3 Comparative Analysis of the Expert and the Automatic Classification

Table 2 shows the results of the manual (based on the expert) and the automatic (based on text mining techniques) approaches. The first column (#Expert) shows the number of messages that the expert has classified for the different Categories (A1 to A3) whereas the second column (#Automatic) indicates the number of messages classified by the text-meaning system for each category. Although the figures may look similar, it has to be remarked that the expert has identified 177 messages out of 332 as A1; whereas the automatic procedure has assigned category A1 to 172 messages. However, only 128 comments are the same in both sets (column #Coincide). Thus the degree of coincidence of both approaches is 128/177 (taking the expert as the reference), that is, 72.31% (column % Coincidence). For each category, it is shown the total number and rate of coincidence between the expert and automatic classification.

Table 2. Comparative analysis of the expert's and automatic classification

Category	# Expert	# Automatic	# Coincide	% Coincidence
A1	177	172	128	72.31 %
A1_A2	31	48	9	29.03 %
A2	93	73	33	35.48 %
A2_A3	21	22	3	14.28 %
A3	10	17	4	40.00 %
Total	332	**332**		

Coincidence rate is significant for category A1 but rather low for the rest. A possible reason for this is that the positive tendency is much more marked and salient than the others within the group of participants. It is also possible that both the expert and the automatic procedure detect the positive expressions more accurately. What is clear is that the mixed categories (A1_A2 and A2_A3) are an artificial construct of the expert used to represent *uncertainty areas*, so that it may be reasonable to omit them from computation. This would force both the expert and the automatic procedure to choose between simple categories (A1, A2 or A3) or even pose a binomial classification just considering tags A1 and A3. This can be done since A2 is a category that implies the acceptance of the majority and not vagueness between A1 and A3.

Summing up, these results suggest that supporting expressions in favour of the implantation of Gran Scala are more precise than the arguments against it. Besides, they also point at the necessity of revision of the rules of the automatic procedure as well as the intensity of the arguments against the implantation of the project, which the expert finds weaker.

Taking into account this interpretation, it could be known which arguments provoke more accurate and intense assertions, uncovering the concrete aspects that a participant considers as more relevant, aiming at the basis of an e-cognocracy system.

5 Conclusions

The paper presents a method for identifying the messages that supports decisions and that lead to the change of preferences of the decision makers which participate in a decision making process in the context of e-cognocracy, the democracy of the knowledge society. This method does not depend of the application field and can be applied in any other multi-actor decision making context (e-democracy, e-participation, marketing…).

By using these techniques, the outstanding ideas and opinions contained into the messages can be detected and related with the changes observed in the preferences, so the actors whose arguments provoke those changes (social leaders) can be identified.

Finally, the arguments that support the different patterns of behaviour of the actors involved in the decision making process and the knowledge associated with the scientific resolution of the problem would be shared through the network (democratisation of knowledge). This would allow the citizens to experience a real learning procedure in accordance with the evolution of living systems.

E-cognocracy allows the direct implication of citizens, and takes advantage of their insights, in the construction of a new and better world in a context of increasing complexity. The social learning provided by e-cognocracy will help citizens to improve their quality of live and cohesion in a Global Knowledge Society.

References

1. Moreno-Jiménez, J.M.: Las Nuevas Tecnologías y la Representación Democrática del Inmigrante. IV Jornadas Jurídicas de Albarracín. Consejo General del Poder Judicial (2003)
2. Moreno-Jiménez, J.M.: E-cognocracia y Representación Democrática del Inmigrante. Anales de Economía Aplicada (2004)
3. Moreno-Jiménez, J.M.: E-cognocracia: Nueva Sociedad, Nueva Democracia. Estudios de Economía Aplicada 24(1-2), 559–581 (2006)
4. Moreno-Jiménez, J.M., Polasek, W.: E-democracy and Knowledge. A Multicriteria Framework for the New Democratic Era. Journal Multicriteria Decision Analysis 12, 163–176 (2003)
5. Moreno-Jiménez, J.M., Polasek, W.: E-cognocracy and the participation of immigrants in e-governance. In: Böhlen, et al. (eds.) TED Conference on e-government 2005. Electronic democracy: The challenge ahead, vol. 13, pp. 18–26. University Rudolf Trauner-Verlag, Schriftenreihe Informatik (2005)
6. García Lizana, A., Moreno Jiménez, J.M.: Economía y Democracia en la Sociedad del Conocimiento. Estudios de Economía Aplicada 26(2), 181–212 (2008)
7. Moreno-Jiménez, J.M., Escobar, M.T., Toncovich, A., Turón, A.: Arguments that support decisions in e-cognocracy: A quantitative approach based on priorities intensities. In: Lytras, M.D., et al. (eds.) The Open Knowledge Society. Communications in Computer and Information Sciences, vol. 19, pp. 649–658 (2008)
8. Moreno-Jiménez, J.M., Piles, J., Ruiz, J., Salazar, J.L., Sanz, A.: Some Notes on e-voting and e-cognocracy. In: Proceedings E-Government Interoperability Conference 2007, Paris, France (2007)
9. Moreno-Jiménez, J.M., Piles, J., Ruiz, J., Salazar, J.L.: E-cognising: the e-voting tool for e-cognocracy. Rio's Int. Jour. on Sciences of Industrial and Systems Engineering and Management 2(2), 25–40 (2008)
10. Saaty, T.: The Analytic Hierarchy Process. McGraw-Hill, New York (1980)
11. VAI Group homepage, `http://www.vai.dia.fi.upm.es/ing/areas.htm`
12. Abney, S.: Parsing By Chunks. In: Berwick, R., Abney, S., Tenny, C. (eds.) Principle-Based Parsing. Kluwer Academic Publishers, Dordrecht (1991)
13. Special Issue on Shallow Parsing. J. Machine Learning Research 2(4) (2002)
14. `http://www.illc.uva.nl/EuroWordNet/`

An Integrated eGovernment Framework for Evaluation of Taiwan Motor Vehicles Driver System

Chi-Chang Chang[1], Pei-Ran Sun[1], Ya-Hsin Li[1], and Kuo-Hsiung Liao[2]

[1] Chung Shan Medical University, No.110, Sec. 1, Chien-Kuo N. Rd., Taichung, Taiwan
threec@csmu.edu.tw
[2] Yuanpei University, No.306, Yuanpei St., HsinChu, Taiwan
liao@mail.ypu.edu.tw

Abstract. The "Electronic Government" movement is sweeping across almost all the world in the last decades. This movement represents a new paradigm for public services. As we know, depending on the advantages provided by internet, the traditional public services can be improved in many aspects. According to the literatures, we found many studies were only focused on how to technically establish the web sites which allow citizens access government information more appropriately. However, few studies paid attention to explore the relationship management among the different stakeholders of e-government. Therefore, the objective of this paper intends to integrate the relationship management among the three groups of stakeholders, which include government itself, its citizens and employees. In this paper, we examine the literature regarding to the underlying rationale of a successful e-government. Also, a framework which supports the relationship management among citizens and government's employees, and public services are developed and empirically tested.

Keywords: Electronic government, customer relationship management, public delivery service.

1 Introduction

The pervasive spreading of the ICTs has created a tremendous opportunity for providing services over internet. In the last decades, researchers and practitioners from different fields investigate various issues of public administration for virtual processes. Much of current research on e-government focuses on improving efficiency and increasing performance within public administration. But e-government is definitely more than just redesigning citizen services, and using state-of-the-art-IT. According to the recent survey of University of Manchester's Institute for Development Policy and Management (2009) found around 35% are total failures, 50% are partial failures, only 15% are successes. [15] Most of the failure results in the unmet of the design and reality gap. For example, citizens showed little demand for such information and did not have the presumed level of skills, which further inhibited their involvement, and the heterogeneity of systems, processes and cultural background etc. Since,

M.D. Lytras et al. (Eds.): WSKS 2009, LNAI 5736, pp. 437–451, 2009.

e-government face serious challenges and most important is to offer two-way-communication services for transactions between administrations and their partners (citizens, companies and other administrations).

In a word, a successful implementation of e-government services requires developing both as regards their internal reorganization and their external relations with customers in a coordinated way. To provide the benefits of such transaction services, a customer-centric solution is necessary. Yet, a comprehensive approach is still not found. [6,13] The purpose of this study was proposed a framework for customer-centered online citizens' public deliver service support. As a result of various web technologies, the functionality and utility of web technologies in public management can be broadly divided into two categories: "internal" and "external": Internally, the website and other technologies hold promise potential as effective and efficient managerial tools that collect, store, organize, and manage an enormous volume of data and information. [1] Government also can transfer funds electronically to other governmental agencies or provide information to public employees through an intranet or internet system. Thus, government so can do many routine tasks more easily and quickly. On the other hand, externally, website technologies also facilitate governmental linkages with citizens and businesses. Information and data can easily be shared with and transferred to external stakeholders. In addition, some website technologies enable the government to promote public participation in policy-making processes by posting public notices and exchanging messages and ideas with the public. According to reviewing the e-government research [8,11,17], even with the use of advanced IT, e-government plan still left some key problems which can be discussed as follow.

1.1 The Problem of Heterogeneous Systems

The delivered services electronically assume that the public sector functions as an integrated environment. One of the major problems that governments will encounter is that their data are usually hard to reach, distributed in disparate and inaccessible systems across various departments. The problem becomes harder when it is related to cultural resistance for information dissemination. Adequate working environments must be established, where employees can access information easily, evaluate it and share their emerging knowledge with fellow colleagues. Therefore, government employees should be able to make the most of document management, workflow and intranet tools to assist them in this difficult task.

1.2 The Lack of Citizens' Viewpoint

Web-based consumer services are generally perceived as being successful, but there has been little evaluation of how well the web meets its users' primary information requirements [10]. In other word, the important point is what the citizens' expect, want and need, and the way they perceive, accept and judge the services of the administration. For example, "citizens are not directly connected into back offices", " back office partners are not directly connected into the front office", "citizens are directly connected into the front office", "the front office is directly integrated into the back offices", "the back office is directly connected into back office partners". [14]

Thus, the challenge of today's e-government is to carefully integrate the technological advancements for citizens' reactions.

1.3 The Lack of Employees' Viewpoint

Before the internet emerged in the late 1980s, the government was already actively pursuing information technologies to improve operating efficiency and to enhance internal communication. [18,12,16] However, the focus of e-government in that era was primarily internal and managerial. The internet gradually has matured into user-friendly platform for employees to communicate directly with citizens and to deliver massive quantities of information to the public. While the lack of internal customer's perspective, it make hard to design, evolving employees expectations and integrate interdependent networks coordinated by governance.

In regard to these problems, Coulthard (2004), and Dabholkar et al. (2005) have proposed e-government can be seen from four perspectives: citizens and customers, process (reorganization), (tele)cooperation, and knowledge. Besides, from the practical considerations, Hiller and Bélanger(2001) have proposed a five-stage of e-government, which reflect the degree of technical and interaction with users. In stage 1 is the most basic form of e-government and uses IT for disseminating information, simply by posting information or data on the website for constituents to view. Stage 2 is two ways communication characterized as an interactive mode between government and constituents. In stage 3, the government allows online service and financial transactions by completely replacing public servants with "web based self-services". In stage 4, the government attempts to integrate various government services vertically (inter-governmental integration) and horizontally (intra-governmental integration) for the enhancement of efficiency, user friendliness, and effectiveness. This stage is a highly challenging task for governments because it requires a tremendous amount of time and resources to integrate online and back-office systems. Both vertical and horizontal integrations push information and data sharing among different functional units and levels of governments for better online public services. [4] Finally, in stage 5 involves the promotion of web-based political participation, in which government website include online voting, online public forums, and online opinion surveys for more direct and wider interaction with the public. While the previous four stages are related to web-based public services in the administrative arena, the fifth stage highlights web-based political activities by citizens.

However, the above-mentioned, the framework just simply provides an exploratory conceptual tool that helps one understand the evolutionary nature of e-government. [9] On the contrary, a crucial problem, not yet completely addressed, is the challenge of developing an evolutionary architecture to integrate large heterogeneous systems and to meet the requirements of customer. Frequently, crafting such architecture involves not only reengineering technological systems according to government's needs, but also reengineering the administrative and business processes that provide services to customers. Recently, CRM has become a strategy asset for government, a critical resource of government's adaptation and survival. In order to avoid the failures of CRM, Zachman (2005) proposed a framework of six perspectives (what-how-where-who-when-why) to define what aspects of CRM need to be studied. In order to communicate with the potential customers through WWW electively, a well-designed webpage is needed. [2] Yet, the factors that affect customer's perception about the acceptance of a web site are unclear.

2 Research Design

Based on the above discussion, the challenge for today's e-government is the need to integrate the technological advancements for the citizens and employee's benefit. In order to develop a measure framework for e-government, a systematic approach is used. Fig 1 show the triple-diamond website measure framework of an e-government, which consists of three functional orientations: IT strategy orientation, citizen orientation and administration orientation which offer transport basic and cooperative services with each other.

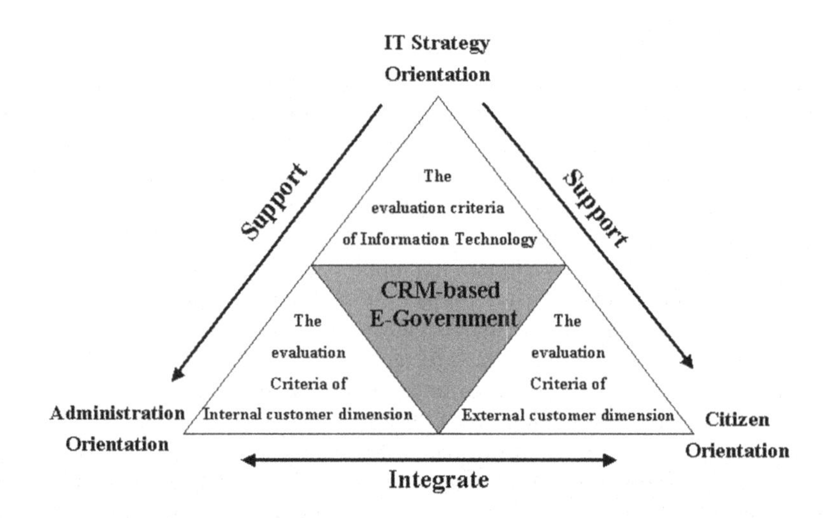

Fig. 1. The Triple-Diamond website Measure Framework of e-government

2.1 IT Strategy Orientation

Basically, the central issue of an e-government is applying IT in information activities, in order to redefine and improve existing administration services. IT is used in a broad sense of information resource configurations referring to IT handling techniques (storage, processing, transport, capturing and presentation) of data including text, sound and visual images and knowledge. It is an intermediate technology or communication gateway, enabling electronic interaction between actors. Therefore, the sector needs to understand the role of the new technology in information service delivery, aiming at developing inter-organization cooperative information systems supporting e- government actions.

2.2 Citizen Orientation

E-government support and assistance should be available anytime to aid the potential visitor as a valuable citizen, reflecting the fact that a government respects citizens. However, referring to potential customer as citizen reflects the attitude that a government must have when it interacts with citizens. It is meant to acknowledge that issues such as citizen s satisfaction, fulfillments of needs, quality of service etc. should be part of an e-government's mentality and practice.

2.3 Administration Orientation

This study suggests that government administrations face serious challenges and requirements both as regards their internal reorganization and their external relations with customers. A successful IT implementation of e-government services requires the internal and external components to be developed in a coordinated way. Therefore, their tasks can be better structured, and employees can concentrate on improving the quality and performance of their process workflows. For this purpose, the employee-oriented design goes one step further by categorizing information and services on the web according to the needs of different departments or work groups.

Therefore, e-government has to rethink their existing services and possibly create new ones. These services must be developed in a way accessible to all potentials customers, easy to use and based on delivery channels assisted by advanced technological means, requiring an evaluation of the service processes. When citizens perceive that the information meets their needs and requirements, they are willing to criticize the value of each product or service based on their purchase decision criteria. Thus, determining customers' perception of information delivery quality on the website is a primary stage in assessing their potential behavior. Citizens' attitudes toward accepting the government's website was constructed with the combined effect of three measurement scales: citizen, administration and IT. For this reason, an established scale was used, and a citizens-focus performance evaluation scale system for the government's website was measured by using 22 positively worded items from the e-government performance evaluation scale (see Table 1). In order to develop an employees-focus performance evaluation scale for a government's website. We firstly define 12 dimensions for a government information delivery service from a literature review and experts' opinions. Further, the 51 items of the 12 defined dimensions are named "ease of use," "ability," "reliability," "communication," "safety," "understanding," "form," "classification & frame," "approachability," "information quality," "useable" and "maintainable." With growing interest in the evaluation of internet information resources, many sets of criteria have been proposed for the evaluation of websites (Marche & McNiven, 2003; Seilheimer, 2004). In order to develop the criteria used for the evaluation of an e-government's website. This study has adapted the related literatures and concluded the IT measurement, which identified five dimensions (e.g. security, the constructional capacity of the network, data processing, performance, and database system) and 32 attributes (Table 2).

Table 1. The Criteria for Citizens Delivery Service Quality and Measure

No	Content
1.	I'm satisfied with the query system.
2.	The e-government website informs me of the status of business processing.
3.	I can find the information to meet my needs via the e-government website anytime.
4.	It can actively accomplish all the related applications to various offices under my authorization.
5.	All of the functions & service items can be operated normally.
6.	The web site is equipped with a confidential service.
7.	The web site is equipped with a plan for identification.
8.	There is no need to worry about personal information being used illegally by others.
9.	There is no need to worry about e-government not having received my register.
10.	The Q&A column of e-government web site answers common questions.
11.	Information offered by e-government website meets my needs.
12.	From the image introduction of e-government website it is easy to find information.
13.	Information provided by e-government is organized and classified.
14.	The e-government website provides me with the latest news of e-government.
15.	The image introduction of e-government website helps me to classify related businesses.
16.	Information on e-government web site is easily understood.
17.	The e-government website plans teaching activities to help me access to the service of e-government website.
18.	The e-government website provides me with a channel to express myself.
19.	Through letters, ads, and activities etc., e-government informs me of the network service.
20.	I know that I have an equal opportunity to access the service provided by e-government website.
21.	There are staffs to respond to my opinions and questions on the website.
22.	There is no difficulty for me to find the connection to e-government website.

Citizen-focus performance evaluation scale system for e-government's website (Cronbach α=0.9064).

Next, for the external customer, this study organized a panel of coders composed of 112 citizens to conduct structured content evaluation of the website of Hsin-Chu Motor Vehicles Driver Office (HMVD) in Taiwan, to define the concepts perceived from citizens' perspectives. Subjects were given a 60 min training session to familiar themselves with the websites of HMVD. Scales were developed for measuring each of the attributes in the Tables 1 and 3. For each scale, citizens were asked to express the degree to which they agree with the state on a 1-4 scale with "1" representing disagree completely and "4" agree completely. Ninety-eight subjects handed in their responses and all of them were valid. Second, for the internal employees, we conducted 75 structured interviews with administrative employees of HMVD. A employees' perception of information quality is presumed to affect their intention and acceptance to use web information positively. In this part, attitude measures were developed by asking respondents to rate each attribute on a 4-point semantic differential scale by ascending from "not at all agree" to "very agree." Finally, IT quality was represented by the five constructs: Security, Network capacity, Data Processing, Operating performance and Database system. In order to discriminate the different levels of each construct, they were marked with "yes," "no" or "N/A (Not Available)."

Table 2. The Measurement of Website Delivery Service Quality Dimension Attribute Index Result

Dimension	Attribute	Index	Result
	1 Fire wall	Y/N	N/A
	2 Backup and Load equation	Y/N	N/A
	3 Routine test and Monitor	Y/N	N/A
	4 Standard Operation Procedure	Operating manual	N/A
		Obey communication protocol	N/A
	5 System monitor software	User access log analysis	YES
		User route log analysis	N/A
		User flow analysis	N/A
	6 Privacy	Selectivity	N/A
		Fire wall function	N/A
Security		Privacy announce	NO
		Accessible	NO
		Encryption	N/A
	7 Authority	Public key recognize system	NO
		Independent supervise faculty	NO
		Copyright announce	YES
	8 Completeness	Relief regulation	NO
		Data–guard supervisor system	NO
		Version declare	YES
	9 Undeniable	Digital Signature	NO
		Id/No. Check	YES
Network capacity	1 Performance analysis	Low bandwidth Support	YES
	2 Quality service	Proxy server provide	NO
	1 Reliable	Renew update	YES
		Full linkage execute	NO
	2 Rapid	Provide pure-text page	YES
		Download file within 15 seconds	YES
	3 Managerial efficiency	Show the message of wrong linkage	NO
		Backup function	N/A
	4 Free from suffering	Make a recovery plan	N/A
		24x7service	NO
	5 Search	Keyword search	NO
Data Processing		Advanced search	NO
	6 Publish	Classify function	NO
		Download function	YES
		Print function	YES
	7 Related data linkage	Basic data support	YES
		Site map	NO
		News	YES
	8 Index	Subject matter	NO
		Outline all function	NO
		Provide the interpretation	NO

Table 2. (*continued*)

Dimension	Attribute	Index	Result
Operating performance	1 Speed	Reasonable average response time	22 Sec.
		Fit in with 3 clicks	NO
		E-mail respond within 2 days	N/A
	2 Consistence	Webpage name accord with content	YES
		Same format	YES
		Right information	YES
	3 Elasticity	Multi-formats download files	NO
		Linkage to homepage	YES
		Support NETSCAPE browser	YES
	4 Malfunction service	Phone Service	YES
		Linkage to webmaster	NO
	5 Communication	Audio clip	NO
		Video clip	NO
		E-mail Service	YES
		Chat	NO
		FAQ	NO
		E-newspaper support	YES
		Community or not	NO
	6 Effortless	Site map	NO
		Website guild	NO
		URL accord with content	YES
		No complex URL	YES
		Use the ranking record	NO
		Content search function	NO
	7 Linkage portal	Y/N	YES
Database System	1 Usability	Click No. vs. Access No.	NO
		Several file format for download	NO
	2 Expandable	FTP function support	NO
		Whether customize or not	NO
	3 Multilanguage support	Y/N	NO
	4 English support	Y/N	YES
	5 Easy backup	Y/N	N/A
	6 Update function	Y/N	YES

3 Data Analysis

Of the 112 external citizens, as shown in Table 5, citizens' intention to use an e-government website was directly and positively affected by their perceptions of "Communication", "Understanding" received a lower average score, followed by the attributes, "Response", "Friendliness", "Safety". Second, for the internal customers, of the 75 respondents, 32% had technical jobs, and 68% were in administrative positions. The results are shown in Table 6. In the light of the IT quality, we study the features that are available at HMVD website. Regarding this criterion, the results are shown in Table 4.

The coefficient alpha estimation of customer-focus performance evaluation scale system was 0.9014. These 22 items were divided into 3 dimensions, which were respectively named "systemic ability," "website design" and "promotion." In order to verify the employee-focus performance evaluation scale for the website, we invited experts to examine the content validity, and conducted a field study in HMVD website. Based on the results of factor analysis, we found that three dimensions, which are respectively named "usability," "responsiveness," and "reliability," can represent the proposed scale (see Table 7). A Cronbach α analysis revealed that each dimension and the scale have high reliability, ranging from 0.7567 to 0.8832 (see Table 8). The Cumulative Variance explained was 0.68717. Finally, there are 14 employee-focus performance evaluation scales were developed to assess the website. The individual questionnaire items used to construct the scales in the analysis are shown in Table 3.

Table 3. The Criteria for Employees Delivery Service Quality and Measure at HMVD Website

No	Content
1.	This website improves my working quality.
2.	This website saves me working time.
3.	I feel that this website can improve my working efficiency.
4.	The information provided by the web site meets my working needs.
5.	Through the website, I can respond to the questions and opinions from citizens.
6.	Through the web site, I can transfer and organize information from various organizations and citizens.
7.	The e-government handles my problems correctly.
8.	I can contact the Webmaster easily to inform duly the website problems.
9.	Fast response from the customer service to my query.
10.	During the use of website, I have clear ideas of the steps and expectations of coming results.
11.	The titles and contents are consistent and they are connected efficiently.
12.	I won't see the wrong content and information.
13.	All of the characters and illustrations can be presented normally during my access to the web site.
14.	I trust the content provided by the website.

Employee focus performance evaluation scale system for e-government's website (Cronbach α=0.9251).

As shown in Table 9, the results of analyzing the existing website of HMVD website are divided into three parts. The most common disagreements observed on the website are lack of navigation support, design inconsistency, overly long reaction times, lacking a foreign language version, orphan pages, security issues, and lack of biographies. Many of the above mistakes are interrelated and symbolize the failures of coordination between the different phases of the website development. Most importantly, all of them could cause serious integrated problems on the usability and endanger the effort of this. Hence, to ameliorate these problems, a formal strategic planning needs to be implemented and suitable IT measures need to be adopted.

Table 7. The results of the exploratory factor analysis (Internal employees)

Dimension	Items	Communality	Factor Loadings	Cumulative Variance explained
Usability	23	0.672	0.743	26.694%
	34	0.644	0.705	
	11	0.640	0.617	
	14	0.768	0.878	
	28	0.703	0.804	
	37	0.613	0.767	
Responsiveness	26	0.603	0.742	52.220%
	24	0.688	0.735	
	47	0.683	0.705	
	25	0.793	0.856	
	27	0.724	0.782	
Reliability	48	0.684	0.788	68.717%
	15	0.724	0.804	
	43	0.663	0.707	

Table 8. The results of reliability test (Internal employees)

Dimension	Reliability Coefficients (Alphas)	Total-Scale Reliability
Usability	0.8832	0.8959
Responsiveness	0.8686	
Reliability	0.7567	

Table 9. The results of analysis of the website of HMVD website

Citizen orientation	Administration orientation	IT orientation (selected No)
- Channel (75%)	- Perceived ease of use (75%)	- Security (privacy,
- Reaction (73.5%)	- Trustworthy (65.6%)	authorization)
- Connective (68.7%)	- Ability (59.4%)	- The constructional capacity of
- Trustworthy (68.4%)	- Classification (59.4%)	network (proxy server offer)
- Perceived ease of use(61.7%)	- Approachability (59.4%)	- Data processing (linkage,
- Ability (59.4%)	- Information quality (59.4%)	24x7 service, search engine,
- Safety (59.4%)	- Format (50%)	Web map, index)
- Search (53.1%)	- Maintainable (50%)	- Performance (format,
- Equitable (53.1%)	- Communicable (43.8%)	communicable, perceived ease
- Friendly (44.1%)	- Usefulness (40.6%)	of use)
- Information quality (43.8%)	- Perceived ease of use (31.3%)	- Database system (usefulness,
	- Safety (28.1%)	expandable function,
		customized, multi-lingual)

4 Discussion

In order to identify the success of the e-government action, a triple-diamond measurement framework was proposed as in Fig 2 and described as below. *External Customer*: IT holds great potential to improve the interface between e-government and customers. The information may comprise simple "where-to-go," detailed information regarding delivery service, support for customers searching processes or even general everyday information. For example, customers who are not familiar with the logic of administrative thinking will need active help in finding the information. Therefore, by way of external customer criteria for the evaluation of website, the goal is to move constituents to this new channel while continuing to provide excellent service through the internet. *Internal Customer*: In e-government domain, for information sharing over the network, the internal customer criteria for the evaluation of an e-government's website will play a crucial role for realizing the expected revolution. Similar to the customers, the employees do not only search and access information, but also communicate and operate to fulfill the administrative processes with others through the website. The websites also support employees' demands when accomplishing their tasks. *IT Strategy*: For the purpose of co-operation, autonomous machine agents should be available. This will support employees' demands when accomplishing their tasks. In addition, these IT elements need to be tightly integrated and need effective analysis by the information technology criteria for the evaluation of websites. The successful elements are highly dependent on the effectiveness of the other elements.

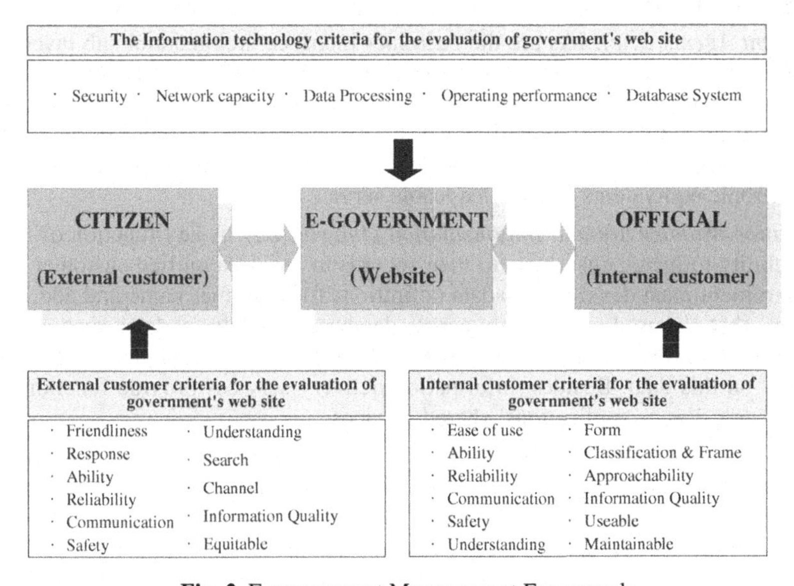

Fig. 2. E-government Measurement Framework

These elements must be integrated in a seamless fashion. While each of these IT elements has obvious advantages, it is the relationship among these elements that provides e-government with the potential to effectively interact with its customers. When an e-government's information strategically links together with these key elements, it produces an atmosphere of customer interaction where the product is greater. As mentioned above, e- government is a new management concept that relies heavily on technology and process automation to create its environment. However, to create such an environment will entail change. Berry (1995) referred to technology's role in customer delivery service as 'high touch through high tech'. Information technology can be used in both manual and automated customer interactions. Figure 4 shows the e- government website function model, which consists of two components offering transport, basic and co-operative services.

4.1 Automated Interaction

The key to the automated service encounter is to pass the control of interaction process to the customer (see part I of Fig 3). Technical infrastructure is a key consideration when an e-government website designs its automated interaction strategy. This will consist of a telecommunication network and terminal equipment and can be internal or external to the government.

Customer Use Interface (CUI): IT holds great potential to improve the interface between government and customers. The information may comprise simple 'where-to-go', detailed information regarding delivery service, support for customers searching processes or even general everyday information. Therefore, the Web-Interface presence is well suited to new efficiencies afforded by e- government, and the goal is to move constituents to this new channel while continuing to provide excellent service through the internet.

Intelligent Agents (IAs): IAs are the customer interface architectural sub layer, as are sets of customer-facing function logic that assist customers in conducting complex, multistage transactions or sets of transactions. Generally, these transactions represent events that have known, typical lifecycles; IAs assists the customer through the life cycle, over application boundaries, and over time (stop and start) and channels. IAs is the electronic equivalents of knowledgeable service clerks.

Integration Middleware and Intermediation Hub (IM/IH): A key function of this part is the ability to match customers to their records to ensure a unified customer history. E-government must develop metadata definitions for customer name and address data elements that are used consistently across databases. Ideally, updates such as change of address should be dealt with one time, with revisions made available to all systems that rely on this data. Besides, information delivery services provide functionality to multiple agencies or applications' shared services.

Customer Information System (CIS): IT's impact on customer services has been studied based on the MIS and marketing disciplines. With respect to IT, there have been those who have focused on how data integration and customer support activities can be a foundation for improving an e-government's ability to serve customers effectively. Within the content of this part, CIS is defined as the acquisition, storage, and distribution of customer information. In addition, the desired end result is to increase profitability and customer satisfaction by getting the right campaign information, order information, and interaction information to targeted customers. In e-government, the goal is about getting the right information to the customer, while the analytical focus is on improved delivery service and more efficient internal operations.

4.2 IT-Assisted Interaction

IT-assisted interaction is predominantly a manual process that uses IT to enhance the relationship between the service provider and the customer (see part II of Fig 3). This will consist of a customer interaction center and employee use interfaces, which can be mentioned as follows:

Customer Interaction Center (CIC): Providing customer service over the web is a given, but as expectations expand, customers' demand for self-service options will cross over to non-Internet channels. Therefore, the complexities of anticipating and responding to customers increase and robust customer interaction centers become even more essential. The CIC should concentrate on providing a framework for supporting applications that solve administration issues, rather than satisfying the urge to deploy point solutions.

Employee Use Interface (EUI): In an e-government domain, for information sharing over the network, the EUI plays a crucial role in realizing the expected revolution. Similar with the CUI (Customer Use Interface), through the website, employees not only search and access information, but they also communicate and operate to fulfill the administrative process with others. For the purpose of co-operation, autonomous machine agents should be available that will support officials on demand when accomplishing their tasks.

The previous discussion highlights the fact that these IT elements need to be tightly integrated. The success of either element is highly dependent on the effectiveness of the other element. It is our contention that for e-government to maximize its ability to interact with its customers, these elements must be integrated in a seamless fashion. While each of these IT elements has obvious advantages, it is the relationship among these elements that provides e-government with the potential to effectively interact with its customers. When e-government strategically links together each of these key elements, it produces an atmosphere of customer interaction where the product is greater that the sum of its parts.

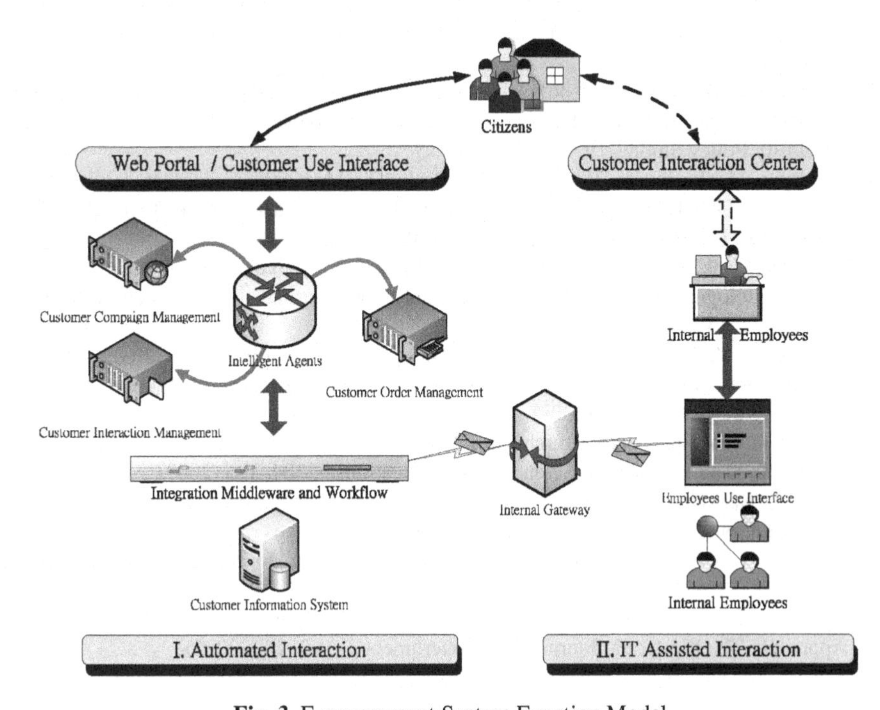

Fig. 3. E-government System Function Model

5 Conclusion

This paper examines the emerging issue of e-government in a web-based system function framework. By investigating local HMVD in Taiwan, the survey results show that customers' requirements are significant factors in the implementation and development of e-government. According to the results of this study, the data also raises a number of issues relevant to e-government policymakers and practitioners. This study also proposes an evaluation framework for an online customer information delivery service support. In this evaluation framework, there are obvious interrelationships that are necessary for government to realize the benefits of becoming an e-government. We suggest that future research should assess the extent to which this model is validated by e-government reality, in particular the way in which the model develops from initial rhetorical intentions through strategic planning, systems development, integration to final transformation. Such innovations may well change e-government as we know them today. In addition, the result can be used by decision-makers as a guidance and direction for architecture development, to reduce the complexity of the progression of e-government initiatives, to communicate changes to the rest of the organization and to provide milestones to evaluate and control cost of architecture development.

References

1. Abanumy, A., Al-Badi, A., Mayhew, P.: e-Government Website Accessibility: In-Depth Evaluation of Saudi Arabia and Oman. The Electronic Journal of e-Government 3, 99–106 (2005)
2. Jaeger, P.T.: User-centered policy evaluations of section 508 of the rehabilitation act: Evaluating e-government web sites for accessibility for persons with disabilities. Journal of Disability Policy Studies 19, 24–33 (2008)
3. Landrum, H., Prybutok, V., Zhang, X., Peak, D.: Measuring is system service quality with SERVQUAL: Users' perceptions of relative importance of the Five SERVPERF dimensions. Informing Science 12, 17–35 (2009)
4. Van Den Haak, M.J., De Jong, M.D.T., Schellens, P.J.: Evaluating municipal websites: A methodological comparison of three think-aloud variants. Government Information Quarterly 26, 193–202 (2009)
5. Omachonu, V., Johnson, W.C., Onyeaso, G.N.: An empirical test of the drivers of overall customer satisfaction: Evidence from multivariate Granger causality. Journal of Services Marketing 22(6), 434–444 (2008)
6. Lin, J.: A consumer support architecture for enhancing customer relationships. WSEAS Transactions on Information Science and Applications 6, 384–396 (2009)
7. Hiller, J., Bélanger, F.: Privacy Strategies for Electronic Government. E-government Series. Pricewaterhouse Coopers Endowment for the Business of Government, Arlington (2001)
8. Hutto, D.H.: Recent literature on government information. Journal of Government Information 28, 185–240 (2001)
9. Janssen, M., Veenstra, A.F.: Stages of Growth in e-Government: An Architectural Approach. The Electronic Journal of e-Government 3, 193–200 (2005)
10. Sandoz, A.: Design principles for e-government architectures. Lecture Notes in Business Information Processing 26, 240–245 (2009)

11. Badri, M.A., Alshare, K.: A path analytic model and measurement of the business value of e-government: An international perspective. International Journal of Information Management 28, 524–535 (2008)
12. Palkovits, S., Wimmer, M.A.: Processes in E-government - A holistic framework for modeling electronic public services. In: Traunmüller, R. (ed.) EGOV 2003. LNCS, vol. 2739, pp. 213–219. Springer, Heidelberg (2003)
13. Robert, M., Davison, C.W., Ma, L.C.K.: From government to e-government: a transition model. Information Technology & People 18, 280–299 (2005)
14. Lee, K.C., Kang, I., Kim, J.S.: Exploring the user interface of negotiation support systems from the user acceptance perspective. Computers in Human Behavior 23, 220–239 (2007)
15. The University of Manchester's Institute for Development Policy and Management: Success and Failure Rates of eGovernment in Developing/Transitional Countries: Overview, http://www.egov4dev.org/success/sfrates.shtml
16. Traunmuller, R., Wimmer, M.A.: e-Government at a decisive moment: Sketching a roadmap to excellence. In: Traunmüller, R. (ed.) EGOV 2003. LNCS, vol. 2739, pp. 1–14. Springer, Heidelberg (2003)
17. Chircu, A.M.: E-government evaluation: Towards a multidimensional framework. Electronic Government 5, 345–363 (2008)
18. Brusa, G., Caliusco, M.L., Chiotti, O.: Enabling knowledge sharing within e-government back-office through ontological engineering. Journal of Theoretical and Applied Electronic Commerce Research 2, 33–48 (2007)
19. Zachman, A.: The framework for enterprise architecture, http://www.intervistainstitute.com/resources/zachman-poster.html

Adapting the SPMSA (Software Project Management Supported by Software Agents) Model to PMBOK2004 Guidelines

Rita Nienaber and Elmé Smith

School of Computing, University of South Africa, Pretoria, South Africa
nienarc@unisa.ac.za, smithe@unisa.ac.za

Abstract. Numerous software development projects either do not live up to expectations or they fail outright. The scope, environment and implementation of software projects are changing due to globalisation, advances in computing technologies as well as the deployment of software projects in distributed, collaborative and virtual environments. As a result, traditional project management methods fail to address the added complexities found in this ever-changing environment. The authors proposed the software project management model, entitled SPMSA (Software Project Management Supported by Software Agents) that aims to enhance software project management by taking the unique nature and changing environment of software projects into account. The SPMSA model supports the entire spectrum of software project management functionality, supporting and enhancing each key function with a team of software agents. In this paper the authors adapt the SPMSA model to incorporate PMBOK2004 guidelines. The SPMSA model makes a fresh contribution to enhance software project management by utilising software agent technology.

Keywords: Software projects; software project management; software agent technology.

1 Introduction

With the advent of global enterprises and virtual organisations, the environment impacting on traditional software project management (SPM) has changed. The traditional single project, which was commonly executed at a single location, has evolved into distributed, collaborative projects deployed in distributed and collaborative environments.

Research furthermore implies that traditional project management methods are unable to address the added complexities found in a distributed environment. Consequently, tools are required for the effective sharing of information among project contributors, as well as for efficient task scheduling, tracking and monitoring. High levels of collaboration, task interdependence and distribution have become essential across time, space and technology [1]. There is a clamant need for managing software projects in such a way that this complex distributed environment is addressed and supported optimally.

M.D. Lytras et al. (Eds.): WSKS 2009, LNAI 5736, pp. 452–461, 2009.

Literature reveals that ongoing research aims to address the shortcomings and impact of managing Information technology projects [2], [3]. Practitioners have attempted to apply several software engineering principles to SPM processes [4]. They have explored standard structured analysis and design methods and incorporated object-oriented approaches to overcome the aforementioned shortcomings [5]. Different standard project management approaches exist, which are applicable to different areas of software project management, such as PRINCE 2 [5].

The authors proposed the SPMSA model – entitled Software Project Management Supported by Software Agents – to enhance SPM processes by incorporating a software agent technology framework – See [6]. The purpose of this paper is to adapt the SPM model to align with the PMBOK Guide Third edition. The first part of this paper highlights the unique features and changing nature of the SPM environment. In the second part an overview of software agents is given. The third part is devoted to a discussion of the adaptation of the SPMSA model and the agent framework supporting it. The paper culminates in a discussion of the verification of the adapted SPMSA model against the ISO standard 10006:2003.

2 Software Project Management Environment

SPM is defined as the process of planning, organising, staffing, monitoring, controlling, and leading a software project [7]. Figure 1 illustrates the key elements in SPM identified by *The Project Management Body of Knowledge (PMBOK)* [8].

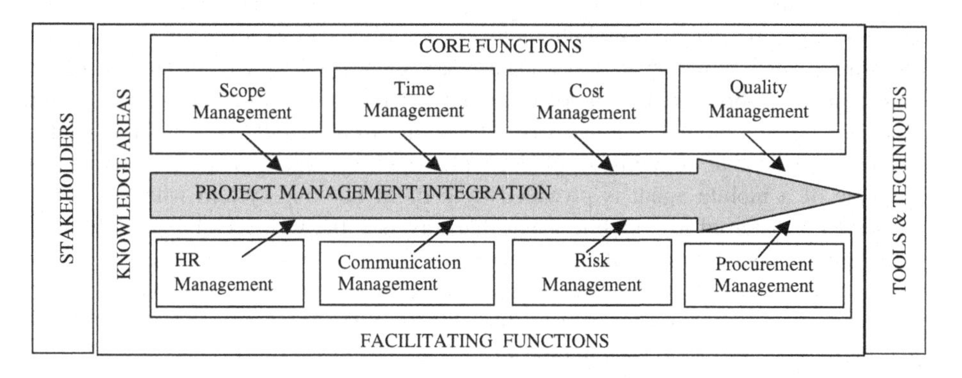

Fig. 1. Software Project Management framework (adapted from Schwalbe [9])

The Software project management knowledge areas include the key competencies, categorised as core and facilitating functions. The core functions, namely scope-, time-, cost- and quality management lead to specific project objectives and are supported by the facilitating functions. The facilitating functions represent the means through which different objectives are to be met and include human resource management, communication-, risk-, and procurement management.

Software project management (SPM) differs from general project management as certain inherent characteristics are unique to software development. These characteristics are invisibility, complexity, conformity and flexibility [10]. SPM practices has changed due to factors such as globalisation and advances in computing technology [11]. Nowadays projects are deployed in distributed and collaborative environments. SPM is also characterised by its unique and dynamically changing nature.

Since the operational environment of SPM has changed, new methods are needed to enhance and support standard SPM practices.

3 An Overview of Software Agent Technology

A software agent is a computer program that is capable of autonomous (or at least semi-autonomous) actions in pursuit of a specific goal [12]. The autonomy characteristic of a software agent distinguishes it from general software programs. Autonomy in agents implies that the software agent has the ability to perform its tasks without direct control, or at least with minimum supervision, in which case it will be a semi-autonomous software agent. Software agents can be grouped, according to specific characteristics, into different software agent classes [13]. Literature does not agree on the different types or classes of software agents. As software agents are commonly classified according to a set of characteristics, different classes of software agents often overlap. For the purpose of this research, we distinguish between two simple classes, namely stationary agents and mobile agents. **A stationary agent** can be seen as a piece of autonomous (or semi-autonomous) software that permanently resides on a particular host. An example of such an agent is one that performs tasks on its host machine such as accepting mobile agents, performing specific computing tasks, and so forth. A well known example of a stationary agent is Clippie, the Microsoft Office Assistant. **A mobile agent** is a software agent that has the ability to transport itself from one host to another in a network. The ability to travel allows a mobile agent to move to a host that contains an object with which the agent wants to interact. An example of a mobile agent is provided by a flight booking system where a logged request is transferred to a mobile agent that traverses the web seeking suitable flight information quotations as well as itineraries.

The computational mechanisms of agent systems are extremely suitable to address the ever-changing organizational structures of the SPM environment, as will be further highlighted in Table 2 (paragraph 4).

4 The SPMSA Model

The authors compiled a comprehensive model of SPM functionality to be supported by software agent technology. The SPMSA model enhances and supports *all* core and facilitating functions of SPM by utilizing an agent framework to enhance the SPM processes. This model thus addresses the *entire* spectrum of SPM [14]. This model (version 1) is based on PMBOK2000 guidelines [8] and is adapted to comply with PMBOK2004 guidelines in this article (version 2). The reader is referred to [6] for a detailed discussion on the first version of the SPMSA model.

4.1 Conceptual View of the SPMSA Model

Having delineated all SPM processes, the authors defined the SPM core and facilitating functions, and adapted these according to PMBOK2004 guidelines as depicted in table 1.

Table 1. SPM Core and Facilitating functions

Intergration management	Develop charter	Develop scope statement & management plan	Manage project execution	Monitor and control	Integrated change control	Close project
Scope management	Scope planning	Scope definition	Creating WBS	Scope verification	Scope control	
Time management	Activity definition	Activity sequencing	Activity resource estimation	Activity duration estimation	Schedule development	Sche-dule control
Cost management			Cost estimating	Cost budgeting	Cost control	
Quality management		Planning	Assurance	Control		
HR management	HR Planning	Acquiring the team	Developing project team	Managing the team		
Communica-tion management	Planning		Information Distribution	Performance Reporting	Managing stakeholders	
Risk Manage-ment	Risk manage-ment planning	Risk Identification	Risk Analysis	Risk Response planning	Risk monitor and control	
Procure-ment management	Procure-ment purchases	Planning contracting	Request seller response	Select sellers	Administer contract	Closing contract

When examining Table 1 closely, *overlapping phases* can be identified, as executed in each of these functions. An abstraction of these functions may be mapped to a generic model of software development, containing the overlapping phases depicted in figure 1, for each function (or process) of SPM. These constitute the second version of the SPMSA model which reflects one added core key function, namely Integration management. Furthermore various phases have changed in this second version. The discussions in the remainder of this paper focus on the adapted second version of the SPMSA model.

Figure 2 aims to explain the concept of this process of support as it consists of the phases of software development for each SPM key function.

The primary goal of the SPMSA model is to support the teams and individual team members in the SPM environment while they are executing their tasks and, in so doing, to enhance the complete SPM environment. This support is enabled by an agent framework as depicted in figure 5. Teams and individual team members will be supported during each process of SPM, utilizing an agent framework to simplify the environment, enhance communication and implement dynamic changes in the system.

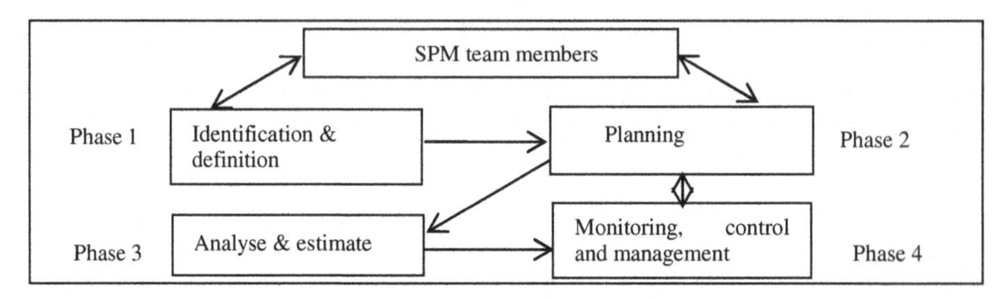

Fig. 2. Phases of Software Development

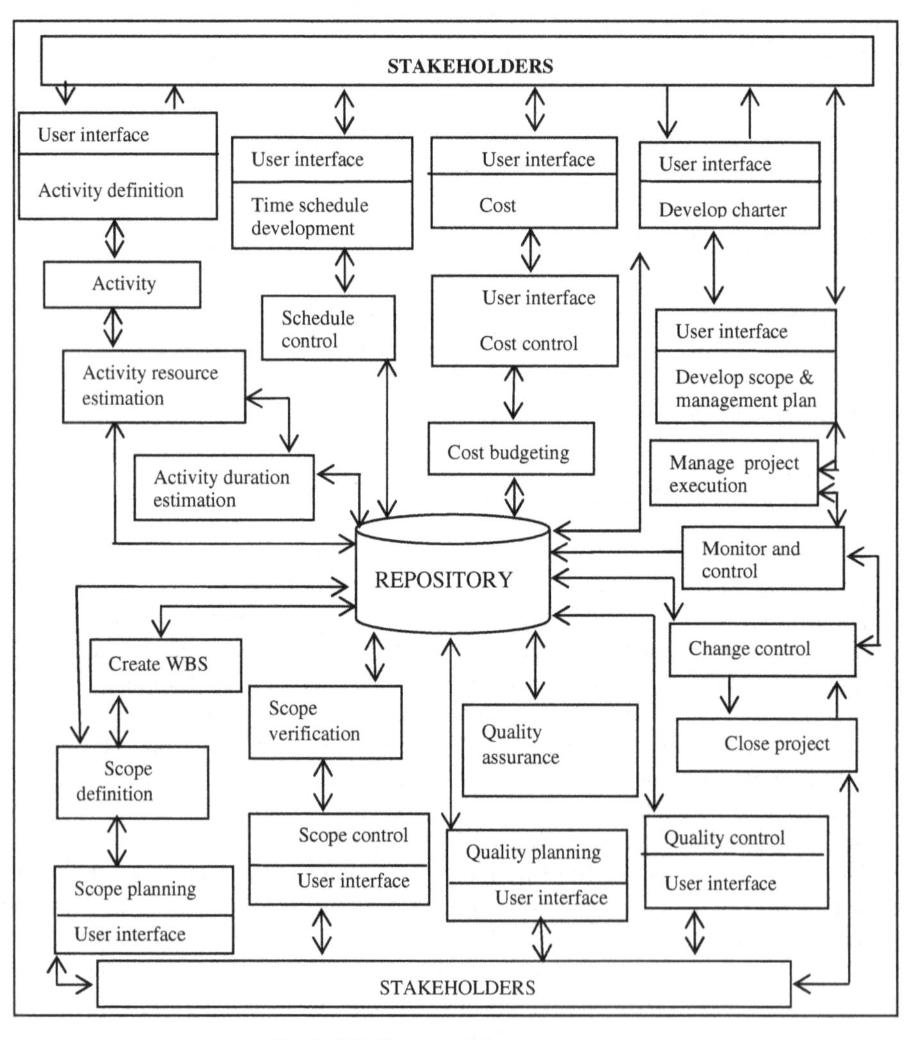

Fig. 3. SPMSA model for core functions

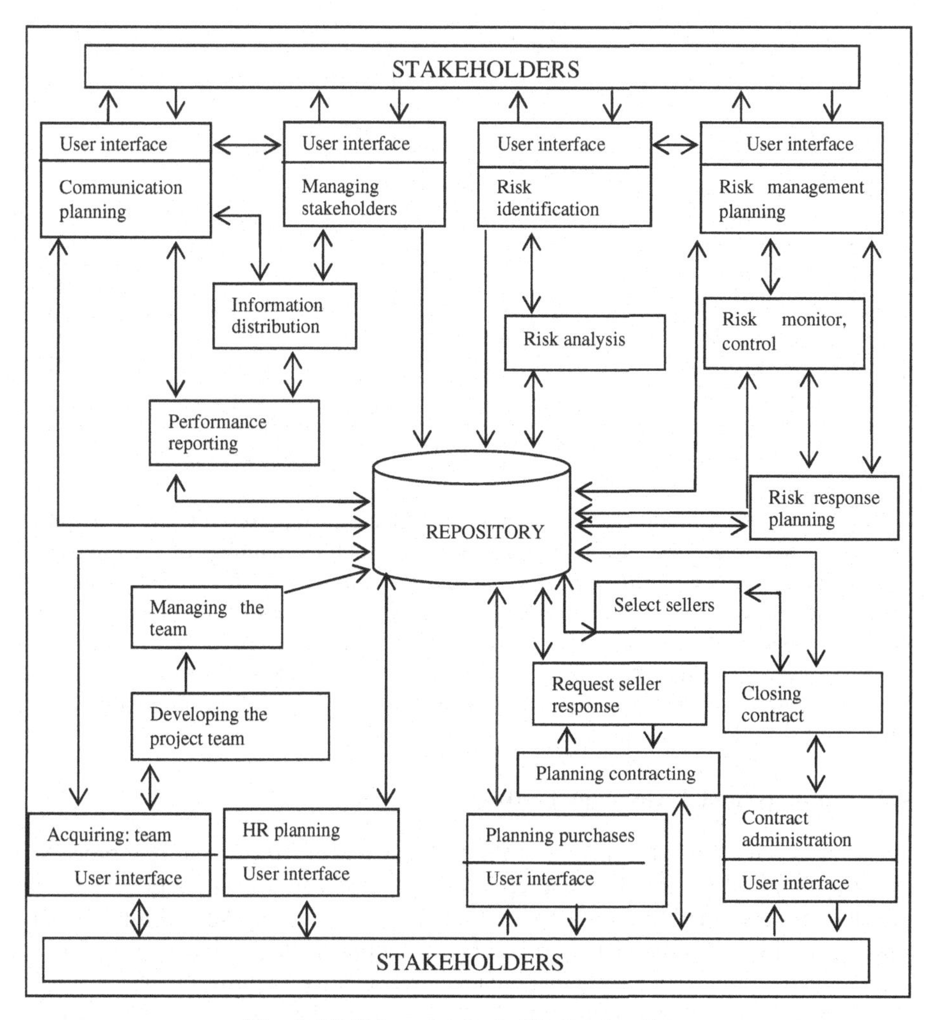

Fig. 4. SPMSA model for facilitating functions

The phases for each of the SPM key functions, as illustrated in table 1, were investigated in detail to compile the comprehensive SPMSA model. The SPMSA model comprises all processes of SPM as supported by an agent framework, which are illustrated in figures 3 and 4. Due to space limitation the core and facilitating functions are presented separately.

4.2 Agent Support Framework for the SPMSA Model

Each of the key processes discussed in Table 1 could successfully be addressed by following a black box approach that is based on software agent technology. According to this approach, we use multiple (simple) agents, each with a particular objective, rather than fewer (complex) agents which each has a long list of tasks to accomplish - see Figure 5.

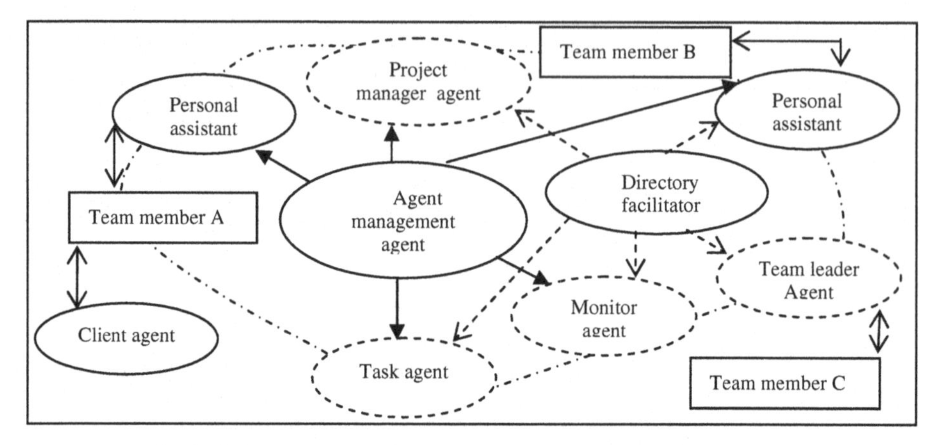

Fig. 5. Software Agent framework to support each SPM key function

Agents are not commonly used in SPM applications and are typically constrained to one or two of the core or facilitating functions such as planning or scheduling. It has not been applied to *all core and facilitating functions of SPM*. Supporting and enhancing the whole spectrum of SPM processes by a software multi-agent system could provide software project managers with significant advantages over using contemporary methods [15]. The potential advantages that will result from this approach become clear through the increasing number of deployed agent applications in other application areas [16].

4.3 Prototype and Technology Platform

A prototype (JPMPS) of a section of the SPMSA model (risk-monitoring processes) was implemented as 'proof of concept' and was tested in a real-life project i.e. the Corridor Sensor Web Application (CSWA) [10]. As Java contains most of the required technologies to implement software and mobile agents such as multithreading, remote method invocation, portable architecture, security features, broadcast support and database connectivity, it was viable to implement the risk management function area of the proposed model in Java [14].

JADE can be considered as agent middleware that implements an agent platform and sustains a development framework. JADE facilitates mobile agent application development, providing key features for distributed network programming. Using the JPMPS prototype enabled the project leader to monitor risk probability and consequence, as well as the status of tasks and deliverables of the project. It was therefore concluded that the JPMPS succeeded in supporting and enhancing the SPM processes during risk management.

A summary of the advantages of how the SPMSA model addresses the limitations of current SPM approaches through software agents is given in table 2 on page 459.

Table 2. Limitations of SPM addressed by agent technology

Limitations of SPM approaches and tools	**Agent enhancement**
Environmental factors	
Stakeholders in virtual teams may have different goals [1]	Virtual teams, supported by automated agents work toward a *similar goal.*
Support homogeneous environments	Agents support *heterogeneous* environments [13]
Executes synchronously and must be connected to be execute [13]	Agent system executes asynchronously and autonomously.
Human interaction/automated control	
Documents distributed: humans [17]	*Automated workflow management*
Team member interaction dependent on human action [18]	*Automates team member interaction* by regular prompting improving productivity
Execution without specific process coordination measures [17]	*Automates process coordination:* will improve programmer productivity and minimise errors
Tasks	
Complexity of tasks and environment -indusive to failure	Complexity of tasks is minimised by automated support, i.e. *automated calculations*
Difficult to maintain consistently [17]	*Maintenance and change control is automated*
No progress management support	*Management of progress status automated* [3]
Risks commonly identified at the start of project, only 50% followed-up [17]	Agents monitoring risk will automate the *continuous monitoring of risks.*
Passive reporting of SPM [1]	Continuous ***dynamic reporting***
Data and tasks will be sent over a network to execute at the user.	Tasks embedded into agent behaviour - lessens *communication overheads* and network load
Quality measures selected by team [18]	Continuous automated input *quality control.*
Intelligent support	
SPM tools - no intelligent support [18]	Agents with intelligence may provide *knowledge base support*
Current SPM tools are static and do not support dynamic simulation [14]	Agent systems support *dynamic simulation* concerning planning, i.e. resource allocation.
All interaction by stakeholders [11]	*Personal assistant agent* support [16]

5 Verification of the SPMSA Model

The SPM phases of the SPMSA model were compared with the processes in the ISO 10006:2003 standard (targeting projects specifically) to determine the relevance of the SPMSA model regarding SPM processes. This comparison was done in detail for the first version of the model and may be accessed in Nienaber & Smith [6].

The ISO 10006:2003 standard consists of a full and extensive list of clauses. The ISO 10006:2003 standard consists of 8 main clauses, 27 sub-clauses and 61 sub-sub clauses. The key areas in the second version of the SPMSA model are all reflected in the standard. The additional process namely integration management, contains 6 processes, which are all reflected in the ISO 10006:2003 clauses, namely clauses 7.2.2, 7.2.3, 7.2.4, 7.2.5, 7.3.3 and 7.6.4 . The first version of the SPMSA model contained ten processes not reflected in the ISO model, whereas the second version of the SPMSA model contains only 6 processes not reflected in the ISO model. Furthermore the majority of processes in the second version of the model correlate with processes in ISO 10006:2003.

The above comparisons clearly indicate that the SPMSA model conforms to the ISO 10006:2003 standard and, as such, can justifiably be applied to the SPM area.

6 Conclusion

It is clear from this research study that traditional methods and techniques of SPM do not meet the requirements posed by this dynamically changing platform. Software agent technology is ideally suited to meeting these new challenges.

In this paper we proposed an approach to using software agent technology to address these challenges. The SPMSA model was compiled to enhance standard SPM practices and also to address challenges encountered due to the unique and changing environment of SPM. The SPMSA model is specifically tailored to address each of the unique features, namely *complexity*, *flexibility*, *conformity* and *invisibility,* through the agent framework.

We believe that our solution is innovative and contribute to the cross-disciplinary field of IT research, disseminating new scientific ideas relevant to international research agendas. By research collaboration in today's hyper-complex world agent technology can be utilized in various scientific, industrial and business applications. This research is aimed at Information technology practitioners and software developers, but will also be beneficial to researchers engaging in dialogue towards a better world. Information system projects supporting crucial business activities may worldwide be supported by agent technology to attain a competitive advantage for that organization.

References

1. Chen, F., Nunamaker, J.F., Romano, N.C.J., Briggs, R.O.: A Collaborative Project Management Architecture. In: 36th Hawaii International Conference on System Sciences. IEEE, Hawaii (2003)
2. Arigliano, F., Ceravolo, P., Fugazza, S., Storelli, D.: Business Metrics Discovery by Business Rules. In: Lytras, M.D., Carroll, J.M., Damiani, E., Tennyson, R.D. (eds.) WSKS 2008. LNCS (LNAI), vol. 5288, pp. 395–402. Springer, Heidelberg (2008)
3. The Standish Group International. Latest Standish Group Chaos Report. Chaos Chronicles. Massachusets, `http://www.standishgroup.com`
4. Lethbridge, T.C., Laganiere, R.: Object-oriented Software Engineering: Practical Software development using UML and Java. McGraw-Hill, London (2001)
5. Hughes, B., Cotterell, M.: Software Project Management, 4th edn. McGraw-Hill, London (2006)
6. Nienaber, R.C., Smith, E.: Enhancing and supporting SPM: the SPMSA model. In: International Conference on Business Information Management. Paris (2009)
7. IEEE Standards Board. IEEE Standard for SPM Plans. IEEE, Los Alamitos (1987)

8. Project Management Institute (PMI). The Guide to the Project Management Body of Knowledge (PMBOK), 3rd edn., http://www.pmi.org
9. Schwalbe, K.: Information Technology Project Management. Thomson, Canada (2006)
10. Nienaber, R.C., Smith, E., Barnard, A., Van Zyl, T.: Software Agent Technology supporting Risk Management in SPM. In: IADIS International Conference on Applied Computing (IADIS 2008), Algarve, Portugal (2008)
11. Romano, N.C., Chen, F., Nunamaker, J.F.: Collaborative Project Management Software. In: 35th Hawaii International Conference on System Sciences. IEEE, Hawaii (2002)
12. Krupansky, J.W.: What is a software agent? http://agtivity.com/agdef.htm
13. Wooldridge, M.: Multi Agent Systems. John Wiley, England (2002)
14. Nienaber, R.C., Barnard, A.: A Generic Agent Technology Framework to Support the Various Software Project Management Processes. In: International Conference on Issues in Informing Science and Information Technology (INSITE), Slovenia (2007)
15. Jennings, N.R.: An Agent-Based approach for building Complex Software Systems. Communications of the ACM 44, 35–39 (2001)
16. Gawinecki, M., Kruszyk, M., Paprzycki, M., Ganzha, M.: Pitfalls of agent system development on the basis of a Travel Support System. In: Abramowicz, W. (ed.) BIS 2007. LNCS, vol. 4439, pp. 488–499. Springer, Heidelberg (2007)
17. O'Connor, R., Jenkins, J.: Using Agents for Distributed Software Project Management. In: 8th International Workshop on Enabling Technologies: Infrastructures for Collaborative Enterprises, pp. 54–60. IEEE, Stanford (1999)
18. Verner, J.M., Cerpa, N.: Australian Software Development: What Software Project Management Practices Lead to Success. In: 2005 Australian Software Engineering Conference (ASWEG). IEEE, Los Alamitos (2005)

Transformational Government in Europe: A Survey of National Policies

Konstantinos Parisopoulos, Efthimios Tambouris, and Konstantinos Tarabanis

University of Macedonia, 156 Egnatia street,
54006 Thessaloniki, Greece
{konparis,tambouris,kat}@uom.gr

Abstract. Transformational government (t-Gov) is often defined as the stage of eGovernment evolution characterized by the radical restructuring of the public sector towards efficiency. In this paper, we investigate the level of t-Gov sophistication in Europe by assessing the eGovernment policies of 18 EU Member States. First, we attempt to understand t-Gov. By reviewing the relevant literature, we suggest that t-Gov embraces nine defining elements, namely user-centric services, joined-up government, one-stop government, multi-channel service delivery, flexibility, efficiency, increased human skills, organizational change and change of attitude of public servants, and finally, value innovation. Based on this understanding, we examine 18 national eGovernment policy documents to determine the degree of t-Gov sophistication. The results suggest that most strategies have set objectives referring to the development of user-centric services, increased efficiency, breaking out of silos and creating joined-up government structures. However, to exploit the full potential of t-Gov, a multi-perspective approach is needed.

Keywords: Transformational government; National policies; Public sector; Efficiency; Value innovation.

1 Introduction

In March 2000, the Lisbon European Council set the goal for the Union to become the most competitive and dynamic knowledge-based economy in the world, capable of sustainable economic growth with more and better jobs and greater social cohesion. Scientific and technological research is essential in providing new knowledge and consequently forms one of the key elements in meeting the Lisbon Agenda.

In addition, the rapid evolution of Information and Communication Technologies (ICT) and particularly the Internet has tremendously affected the way public and private organizations respond to their customers' needs. In the new millennium an increasing number of public sector bodies and agencies are embracing the concept of electronic government (eGovernment) through the introduction of national policies and initiatives [35]. Such initiatives recognize the incremental role of ICT towards the transformation of rigid, bureaucratic governance models to eGovernment models

M.D. Lytras et al. (Eds.): WSKS 2009, LNAI 5736, pp. 462–471, 2009.

where services are offered according to the customer's needs in a more democratic and efficient way that will eventually create a real knowledge society [16, 31]. To achieve this transformation, all developed countries, including the European Union Member States, have now implemented some form or level of eGovernment policies and initiatives [1, 3] and the developing countries are now closely following this example [35].

In this context, the concept of transformational government (t-Gov) has been introduced. T-Gov refers to the radical restructuring and re-engineering of the internal and external business processes of the public sector in order to achieve severe cost cuttings and increased efficiency [26]. In fact, literature shows that the very essence of eGovernment is about transforming the internal and external processes of government using ICT to provide efficient and user focused services to citizens, businesses and other stakeholders [5, 9].

The main objective of this paper is to assess how governments in Europe understand and strategically pursue t-Gov by examining the national eGovernment policies of the European Union Member States. In order to make such an assessment, we first attempt to better understand and scope t-Gov based on the relevant literature. We thereafter conduct a survey by reviewing the national eGovernment policy documents of 18 EU Member States. This enables us to identify how Europe perceives t-Gov, which t-Gov definition elements are more popular and to demonstrate that there is a common set of beliefs that inspire eGovernment initiatives in these countries which influences policy-makers towards the possibilities of institutional innovation and governmental transformation. This cross-comparison will shed light on some of the factors that affect t-Gov as well as on what more needs to be done in order to realize the Lisbon objectives and transform Europe into a real ICT innovation inventory.

The rest of this paper is structured as follows: Section two describes our methodology while section three presents our understanding of t-Gov. Section four outlines the results of our survey. Finally, section five draws the conclusions and offers recommendations about t-Gov in Europe.

2 Methodology

In order to assess how Europe perceives and is strategically progressing towards t-Gov, we conducted a desktop literature survey so as to better understand how t-Gov is defined and what the main elements involved in the relevant definitions are. For this purpose, the literature review methodology proposed by Webster and Watson [34] was followed.

We should note however that this review was not restricted to gathering definitions explicitly referring to t-Gov. Instead, we reviewed the broader literature for important eGovernment elements that in our view are relevant to the ideas behind t-Gov. As a result, we concluded that t-Gov is a wider and more complex concept compared to existing definitions in the literature. We further concluded that t-Gov is constituted by more elements than what the current relevant definitions imply. In this respect, we provide our own understanding of t-Gov, which consists of nine defining elements and which is used as an assessment basis.

Desktop research was employed to collect the national eGovernment policy documents of the EU Member States, as these are normally published on the website of each national authority responsible for eGovernment. The survey was carried out in autumn 2007 and was updated in summer 2008. Three languages were used for the survey (English, French, and Greek). The survey revealed that eighteen out of the 27 EU Member States have published a national eGovernment policy document online in one of the three afore-mentioned languages. In some cases, there seemed to be no eGovernment policy online. We should note however that, in some other cases, an English summary of the original document was provided online. However, we decided not to include such summaries in the study since the complete document was deemed essential for the purposes of our study.

Under these limitations, the documents examined include the national policies of Austria, Bulgaria, the Czech Republic, Denmark, Estonia, Finland, France, Greece, Ireland, Luxembourg, Malta, The Netherlands, Poland, Romania, Slovenia, Sweden and the UK. These documents are the complete original policy documents or the complete English translations of the original policy documents.

After collecting the documents, we reviewed them by reading them thoroughly and cross-examining them with our t-Gov understanding in order to identify which of the nine defining elements are included in each national policy. In addition, qualitative analysis [23] has been conducted in order to identify and understand each country's approach towards t-Gov, taking into account the different contextual factors that influence the country-specific approaches as well as the implementation. Furthermore, we employed StratML forms [2] for each strategy in order to identify all the core elements of each document and to conduct homogeneous comparison that eliminates structural or semantic differences of each text. This method enables us to examine how Europe perceives t-Gov and whether the approach followed by each Member State is fragmented or not.

In addition, by reviewing the documents, we were able to examine which t-Gov elements are most common across the EU and which are considered more important. Finally, we compared the objectives laid down in each document to the actual implementation of eGovernment, as the latter is measured by Cap Gemini [11] on behalf of the European Commission. This comparison allows us to see how eGovernment implementation has evolved since the publication of each strategy.

3 Understanding Transformational Government

In the past decade, a variety of definitions have been offered to describe eGovernment, including "the use by government agencies of information technologies (such as Wide Area Networks, the Internet, and mobile computing) that have the ability to transform relations with citizens, businesses, and other arms of government" [33], "the use of information and communication technology in public administrations combined with organizational change and new skills in order to improve public services and democratic processes and strengthen support to public policies, "government's use of technology, particularly web-based Internet applications to enhance the access to and delivery of government information and service to citizens, business partners, employees, other agencies, and government entities" [24] and others.

In literature, it is now accepted that eGovernment is an evolutionary phenomenon and therefore eGovernment initiatives should be accordingly derived and implemented. In this respect, various frameworks have been developed to explain the evolutionary stages of eGovernment. For example, Layne and Lee [24] propose four stages of eGovernment growth: (1) cataloguing, (2) transaction, (3) vertical integration, and (4) horizontal integration. The United Nations eGovernment Readiness Knowledge Base [33] proposes a five-stage model of eGovernment evolution: (1) emerging presence, (2) enhanced presence, (3) interactive presence, (4) transactional presence and (5) networked presence. Finally, Cap Gemini [11], in collaboration with the European Commission, proposes the following five stages: (1) information, (2) one-way interaction, (3) two-way interaction, (4) transaction and (5) personalization.

An underlying principle of eGovernment definitions and frameworks is the *transformation* the public sector has to undergo in order to delivery efficient eGovernment services [18]. 'Transformation' itself is not new conceptually, since it was at the heart of the introduction of information systems into government in the 1980s, and it was central to organizational reform in the 1990s [30] which looked at the use of entrepreneurial approaches to service delivery. It does however, reinvigorate challenges. As Bonham et al. [8] state "at its most advanced level, eGovernment could potentially reorganize, combine, and/or eliminate existing agencies". This view is further supported in literature; researchers such as Mansar [25] and Layne & Lee [24] highlight that business process reengineering (BPR) is particularly important when eGovernment projects reach the later stages of eGovernment development such as the transformational stage where all services are centralized in a one-stop-shop environment.

In this respect, Murphy [27] argues that t-Gov is about fundamentally changing the way government does what it does. In fact, Murphy defines transformational government as the stage of eGovernment evolution which implies radically changing the way government conducts its business internally and externally. The transformational phase of eGovernment should primarily focus upon cost savings and service improvement through back-office process and information technology (IT) change. Ultimately, the objective of the transformational stage of eGovernment implies that process reengineering is needed to rethink the value propositions of the government and how they function in serving citizens more efficiently and effectively. A major goal therefore is to also change the behavior and culture of government.

In literature many academics and practitioners have tried to analyze the different stages of eGovernment evolution and have referred to the final stage by different names such as transformation [6], horizontal integration [24], transforming government [5, 27] and fully integrated or single point of access [24]. Despite the variance of names, one can argue that many scholars have agreed on the purpose of this final stage of eGovernment implementation [36] i.e. that it leads to integrated government services which are accessible from a single point.

Although we agree with Murphy's [27] approach of fundamental BPR for the public sector, we feel that changes implemented on the path towards t-Gov should not be designed as a one-time thing. Given the highly volatile socio-technological environment of eGovernment, durability and sustainability of t-Gov benefits are also necessary. In this context, achieving durable efficiency and keeping up with the pace of changes, requires a deeper look in the organizational and process restructuring of the public sector. We are inclined to believe that apart from offering joined up, seamless, flexible and personalized one-stop-shop services, it is also necessary for governments to ensure that service delivery value is constantly created for users.

In order to effectively execute value creation strategies for achieving t-Gov, governments will need to transform their organizational architectures appropriately. Core business processes may need to be rethought and redesigned, new governance and organizational forms that foster collaboration and partnering may need to be developed, and human resource and reward systems may need to be redesigned [13]. Furthermore, given the volatility of the modern eGovernment environment, government architectures will have to be designed for dynamic stability [17]. Dynamic stability implies a continuous transformation as conditions change and new opportunities arise.

Dynamic stability will transform the organization towards a competitive-advantage-oriented approach. The only way to sustain competitive advantage, however, is through launching new value concepts and continuously re-inventing the way customer value is created and delivered [26]. The recommendations made by numerous authors [19, 22] can guide managers in these endeavors. Value innovation implies breaking free from taken-for-granted assumptions about the environment, and the intra- and inter-organizational ways of working; as in the private sector, governments must deviate from the dominant bureaucratic, rigid public sector recipe.

Value innovation acts as a catalyst in transforming organizations for pursuing durable efficiency and sustainable competitive advantage and therefore, we consider it as a defining concept and important element of t-Gov.

All things considered, t-Gov implies much more than the efficient-driven restructuring of the public sector for the provision on one-stop government services. In our view, transformational government is that stage of eGovernment development in which joined-up public administration is breaking out of silos and offers in a one-stop government environment seamless, personalized, user-centric services with increased flexibility and improved efficiency. In t-Gov, public services are available through a wide range of channels (Internet, mobile phones, telephony, contact points etc.) and the public administration human resources have an incremental role to play in transformational service delivery. However, the most important element of t-Gov is the concept of value innovation enabled by constant organizational and behavioral change; restructuring and reshaping of the business processes (front-office along with back-office) should be seen as the means towards modernizing administration and renewing institutions rather than as threats.

Constant value creation, value innovation and business process transformation will ensure that governments are alert towards an ever-changing technological environment. For the public sector and the successful implementation of t-Gov, it is important to be aware of the new technologies, to update infrastructure on regular basis and to integrate new or emerging technologies so as to keep government service delivery efficient and competitive. A holistic t-Gov approach can ensure that new technologies are always looked into and integrated in the governmental processes without the need to undergo any complex or time-consuming (re-)formulation of policies, action plans and implementation road maps.

In summary, based on published literature, we identify nine defining elements for t-Gov:

- User-centric services [27, 37]
- Joined-up, seamless government which offers public services [15]
- Breaking out of silos and offering one-stop access [32, 35]
- Multi-channel delivery [27]
- Flexibility of service delivery [27]
- Efficiency [27]
- Important role of public sector HR in facilitating transformation [27, 28]
- Organizational change and change of attitude of public servants [27]
- Value innovation and constant value creation in service delivery [26]

We believe that addressing together all these elements within an eGovernment strategy provides a holistic approach to t-Gov. This is achieved by focusing on all the factors that can facilitate a deep, real, thorough, innovative and value-oriented transformation of government.

4 Survey Results

The examination of the National Policies for eGovernment of eighteen EU Member States demonstrate that at European level, focus is shifting from initial service automation approaches to issues such as inclusion; effectiveness, for example through quality of service; organizational strategies, such as the 'transforming' of government organizations; and in working together with the private and the third sectors so that the strengths of each are used strategically to deliver 'public value' to citizens. The overall eGovernment brand has therefore matured from an early focus on the 'production' of eGovernment, to one of relevance and value to citizens - the 'consumption' of eGovernment.

As far as the nine t-Gov defining elements are concerned, our survey has shown that the two most important elements for all 18 strategies are the user-centric services and the improved efficiency. More specifically, the vast majority of documents set priorities for single-window access to services (16 out of 18 strategies) as well as to creating seamless, joined-up government structures (14 out of 18). Twelve out of 18 documents aim to provide services via various channels while only ten documents place some focus on the importance of providing new skills to governmental human resources. The concepts of flexibility and organizational change, despite their importance, have gained very limited attention by policy makers, as they appear in only 7 and 4 out of the 18 documents respectively.

What is striking is that none of the eighteen policies put any emphasis on constantly creating value for the users of the public services by pursuing value innovation. Despite the importance of value innovation as means for sustainability and durability of the t-Gov benefits, this definition element lacks focus. Figure 1 below illustrates our findings.

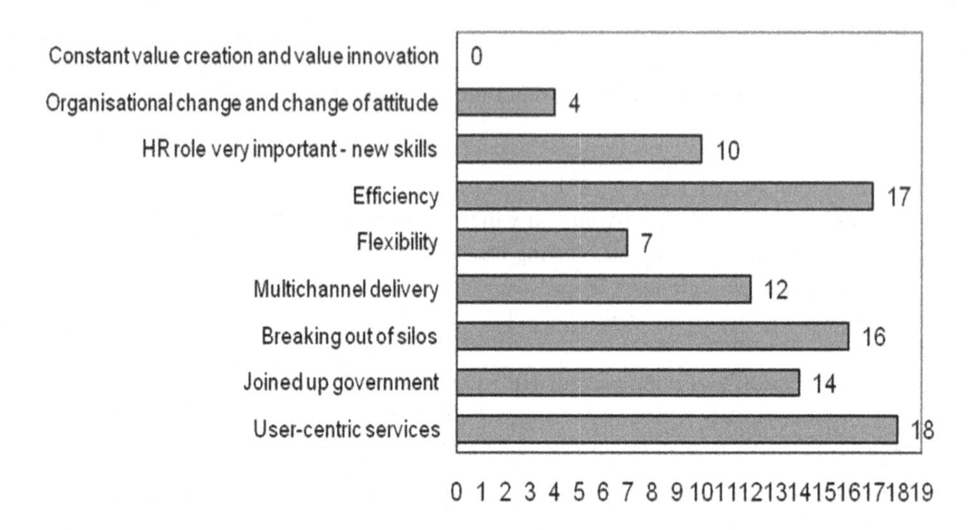

Fig. 1. Most Common t-Gov Definition Elements

5 Conclusions

In this paper, we have reviewed the literature in order to understand and better scope the concept of transformational government (t-Gov) which is becoming increasingly popular among researchers. In addition, we have analyzed the national eGovernment policy documents of eighteen EU Member States. This analysis has shown that although an agreement among researchers as well as among policy-makers as to how t-Gov is defined exists, this definition is not comprehensive enough as it does not incorporate some indispensable elements that will facilitate real, deep, innovative and sustainable government transformation.

More specifically, apart from restructuring the internal and external business processes of the public sector to deliver user-centric services with increased efficiency, emphasis should also be given on the constant value creation and value innovation of public agencies, the organizational change and the change of attitude among public employees, the retraining of public servants with new IT-oriented skills, the increased flexibility of public bodies and the multichannel provision of online public services. To reach these ends, it is essential for governments to escape from their siloed, bureaucratic and fragmented approach and to create a modern, joined up, seamless and efficient structure.

In this endeavors more long-term focus is required in order to ensure that public investments on ICT do deliver the highest value for money, that eGovernment is comprehensively implemented and that thorough transformation is achieved. Constant value creation and value innovation should be embedded and actively incorporated in the managerial approaches and the transformational efforts of the public sector.

In addition, measurement and evaluation frameworks that cover all elements of t-Gov need to be established. So far, the European Commission has been evaluating the development of eGovernment by measuring the online availability and sophistication of public services. However, more indexes need to be developed along with more generic evaluation models that incorporate both quantitative and qualitative data on elements such as efficiency, multichannel delivery, value creation, HR training etc.

Transforming organizations to deliver citizen-centricity will continue to benefit from flexible strategies at the European level, particularly ones that help to understand the complexity and diversity of the t-Gov landscape, which promote constructive sharing of good and bad experiences, the building of measures of deeper transformation and restructuring as well as of constantly creating public value, and overall, which focus on the strategically important processes of the consumption of governance, rather than its technological production. As Blakemore [7] puts it, in the European context of citizen, organizational, and business heterogeneity should the EU structure stakeholder dialogue differently to achieve greater impact? What can Member States realistically learn from each other's strategies and experiences? And, are citizens and businesses really concerned whether it is eGovernment, t-Gov or something else that delivers them seamless and one-stop-shop services? All these questions remain highly relevant and open for further work.

References

1. Accenture: Leadership in Customer Service: New Expectations, New Experiences (2005), http://www.accenture.com
2. Association for Information and Image Management (AIIM): Strategy Markup Language (StratML) project, http://www.xml.gov/stratml/index.htm#Forms
3. Baacke, L., Fitterer, R., Mettler, T., Rohner, P.: Transformational Government – A Conceptual Foundation for Innovation in Public Administrations. In: Proceedings of the Eighth European Conference on e-Government, Lausanne, Switzerland, p. 43 (2008)
4. Baden-Fuller, C., Pitt, M.: Strategic innovation: an international casebook on strategic management, London. Routledge, New York (1996)
5. Balutis, A.: E-government strategy 2001, Part 1: understanding the challenge and evolving strategies. The Public Manager 30(1), 33–37 (2001)
6. Baum, C., Di Maio, A.: Gartner's four phases of e-government model (2001), http://gartner3.gartnerweb.com/public/static/hotc/00094235.html
7. Blakemore, M.: Think Paper 4: eGovernment strategy across Europe - a bricolage responding to societal challenges (2006), http://www.ccegov.eu/
8. Bonham, M.G., Seifert, J.W., Thorson, S.J.: The Transformational Potential of e-Government. Maxwell School of Syracuse University (2003), http://www.maxwell.syr.edu/maxpages/faculty/gmbonham/ecpr.htm
9. Burn, J., Robins, G.: Moving Towards e-Government: A Case Study of Organizational Change Processes. Logistics Information Management 16(1), 25–35 (2003)
10. Cameron, K., Quinn, R.: Diagnosing and changing organizational culture: Based on the competing values framework. Organizational Development Series. Prentice Hall, Englewood Cliffs (1999)
11. Cap Gemini: Benchmarking the Supply of Online Public Services (2007), http://www.capgemini.com/resources/thought_leadership/benchmarking_the_supply_of_online_public_services
12. Di Maio, A.: Moving from e-government to government transformation, Business Issues (2006), http://www.gartner.com

13. ElSawy, O.A., Malhotra, A., Gosain, S., Young, K.M.: Intensive Value Innovation in the Electronic Economy: Insights from Marshall Industries. Management Information Systems Quarterly 23(3), 305–335 (1998)
14. European Commission, Communication from the Commission to the Council, The European Parliament, The European Economic and Social Committee and the Committee of the Regions, i2010 – A European Information Society for growth and employment, COM(2005) 229 final, Brussels (2005)
15. Fagan, M.H.: Exploring city, county and state e-government initiatives: an East Texas perspective. Business Process Management Journal 12(1), 101–112 (2006)
16. Fountain, J.: The paradoxes of public sector customer service. Governance 14(1), 55–73 (2001)
17. Ghemawat, P., Ricart, I.C., Joan, E.: The Organizational Tension between Static and Dynamic Efficiency. Strategic Management Journal 14, 59–73 (1993)
18. Gupta, M.P., Jana, D.: E-government Evaluation: A Framework and case study. Government Information Quarterly 20, 365–387 (2003)
19. Hamel, G.: Strategy innovation and the quest for value. Sloan Management Review 39(2), 7–14 (1998)
20. Kim, J.K., Pan, G., Pan, S.L.: Managing IT-enabled transformation in the public sector: case study on e-government in South Korea. Government Information Quarterly 24(1), 338–352 (2007)
21. Kim, C.S.: Diagnosing And Changing Organizational Culture based on The Competing Values Framework. Series in Organizational Development. Prentice Hall, New Jersey (1999)
22. Kim, C.W., Mauborgne, R.: Value innovation: The strategic logic of high growth. Harvard Business Review 75(1), 102–115 (1997)
23. Layce, A., Luff, D.: Qualitative Data Analysis, Trent RDSU (2007)
24. Layne, K., Lee, J.: Developing fully functional e-government: A four stage model. Government Information Quarterly 18(2), 122–136 (2001)
25. Mansar, S.L.: E-Government Implementation: Impact on Business Processes, pp. 1–5. IEEE, Los Alamitos (2006)
26. Matthyssens, P., Vandenbempt, K., Berghman, L.: Value innovation in business markets: Breaking the industry recipe. Industrial Marketing Management 35, 751–761 (2005)
27. Murphy, J.: Beyond e-government - the world's most successful technology-enabled transformations, Executive Summary, MP Parliamentary Secretary Cabinet Office Report (2005), http://www.localtgov.org.uk
28. O'Donnell, O., Boyle, R., Timonen, V.: Transformational aspects of e-Government in Ireland: Issues to be addressed. In: The Proceedings of the Third Conference on E-Government. Trinity College, Dublin (2003)
29. Organization for Economic Cooperation and Development: The e-government imperative (2003)
30. Osborne, D., Gaebler, T.: Reinventing Government: How the Entrepreneurial Spirit is Transforming the Public Sector. Addison Wesley, New York (1992)
31. Peristeras, V., Tarabanis, K.: Towards an enterprise architecture for public administration using a top-down approach. European Journal of Information Systems 9, 252–260 (2000)
32. Tambouris, E.: An Integrated Platform for Realizing One-Stop Government: The eGOV project. In: E-Government Workshop within DEXA 2001, pp. 359–363. IEEE Press, Los Alamitos (2001)
33. United Nations E-government Readiness Knowledge Base, Global E-Government Readiness Reports and Survey, http://www2.unpan.org/egovkb/global_reports/08report.htm

34. Webster, J., Watson, R.T.: Analyzing the past to prepare for the future: Writing a Literature Review. MIS Quarterly 26(2), 13–23 (2002)
35. Weerakkody, V., Baire, S., Choudrie, J.: E-Government: A Case for Process Improvement in the Public Sector. In: Proceedings of the Hawaii International Conference on Systems Sciences (Hicss-39). Computer Society Press, Washington (2006)
36. Weerakkody, V., Dwivedi, Y.K., Dhillon, G., Williams, M.D.: Realizing T-Government: A UK Local Authority Perspective. In: Proceedings of the Fifth International Conference on Electronic Government. Inderscience Publishers, Geneva (2007)
37. West, D.: E-government and the transformation of service delivery and citizen attitudes. Public Administration Review 64(1), 15–27 (2004)
38. World Bank: Definition of eGovernment, `http://go.worldbank.org/M1JHE0Z280`

A Framework to Evaluate the Impact of e-Government Services in the Improvement of the Citizens' Quality of Life

Ourania I. Markaki, Dimitris E. Charilas, Yannis Charalabidis, and Dimitris Askounis

National Technical University of Athens, Department of Electrical & Computer Engineering, Heroon Polytechneiou 9, Zographou, 15773, Athens, Greece
omarkaki@epu.ntua.gr, dcharilas@mobile.ntua.gr, yannisx@epu.ntua.gr, askous@epu.ntua.gr

Abstract. Governments around the world are embracing the digital revolution to enhance services for their citizens. However, the development of quality electronic services and delivery systems that are efficient and effective is only one way to improve the citizens' quality of life. The essence of e-government lies as well in engaging citizenry into the use of e-government services. As a result, this paper builds on the elementary concept of the time that citizens spend on their transactions with the Public Administration to provide a) an evaluation of e-government services in terms of the benefits that their use involves for citizens compared to the use of conventional ones b) an evaluation of the actual utilization of e-government services by the citizens. To overcome the impediment of subjectivity and uncertainty that is innate in the citizens' estimations regarding the time spent, fuzzy triangular numbers are adopted. The proposed framework is then applied for the case of Greece using actual data.

Keywords: E-government Evaluation, Sophistication Stage, Quality of Life, Fuzzy Sets Theory, Fuzzy Triangular Numbers.

1 Introduction

As e-government transformation is no longer just an option but a necessity for countries aiming at improving services for their citizens, e-government performance evaluation becomes more and more important in guiding and inspecting e-government's development. A variety of models have been proposed in the literature to assess e-government performance [1][2][3]. Such models have an unquestionable usability; nevertheless their use in isolation leaves the impression that e-government is solely about delivering government services over the internet and that a nicely designed, user-oriented web site is all that matters. Such an assumption however ignores the substantial investments that are needed in people, tools and policies.

Even Quality of Service (QoS), a topic that has already gained attention in multiple application domains, is presented in [4] as an emerging promising approach to promote the development of e-government services. Quality measurements may provide quantitative measures of specific quality aspects of delivered services and allow the

M.D. Lytras et al. (Eds.): WSKS 2009, LNAI 5736, pp. 472–482, 2009.

Public Administration to determine whether the activities and related investments are valuable. Attention has to be drawn therefore to the fact that Quality of Service does not necessarily prove quality of life.

An innovative indicators' system is proposed in [5] as a measurement and evaluation tool of the national strategic objectives regarding the improvement of the administrative capacity of the Greek Public Administration: the proposed system includes mainly indicators of impacts, rather than indicators of flows, meaning that the actual focus is on the final recipients of the actions undertaken to support these strategic objectives i.e., the citizens, the enterprises and the state itself; however a complete framework in terms of measurement methodology and base values to be used is not suggested.

Still, the potential impact of e-government transformation in the improvement of the citizens' quality of life is only one side of the problem. The statistical information provided by Eurostat, the Statistical Office of the European Communities [6], is rather disappointing, since only a small percent of the population of some member states exploit currently the e-government services. At the normative level, concerns have already been expressed about the "digital divide" and whether e-government will exacerbate inequities among citizens [7]. Addressing this concern and finding an amicable way out should also constitute a part of the effectiveness of governments.

Bringing public services online is thus only one way to enhance the quality of public service delivery. The next challenge that e-government has to tackle is, as implied in [3], to shift the focus from "availability" and "quality" of e-government services, through a transition stage of "use", that refers to e-government services' take-up and user satisfaction, to a desired end-state of "impact", which concerns long-term results of e-government initiatives on society.

As a consequence, the evaluation framework proposed in this paper is oriented towards two main axes: a) the evaluation of e-government services in terms of the benefits that their use involves for citizens compared to the use of conventional ones b) the evaluation of the actual utilization of e-government services by the citizens. More specifically, this paper builds on the elementary concept of the time that citizens spend on transactions with the public authorities to provide a quantitative indication of the impact of the modernization of Public Administration in the improvement of the citizens' quality of life, taking as well into account the current sophistication stage of online public services. The scheme proposes the use of questionnaires to collect the raw data with regard to the time spent, which are then converted to fuzzy numbers to be further processed, in order to incorporate the uncertainty factor which is innate in the imprecise answers provided by the citizens. On an upper level, statistical information is exploited in order to perform e-government evaluation in terms of e-government services take-up.

The paper is structured as follows: Section 1 is an introduction to the problem of e-government evaluation. Section 2 presents the e-service sophistication model used to benchmark the sophistication of online public services. Section 3 provides an insight to Fuzzy Sets Theory and linguistics, which constitute important tools in the evaluation attempted, while Section 4 introduces the evaluation framework along with the two axes of analysis. In Section 5 the case study of Greece is presented, to illustrate the proposed scheme. Finally, Section 6 summarizes conclusions.

2 Sophistication of e-Government Services

In order to measure the sophistication of online public services, an e-service sophistication model is used in [8]. This model illustrates the different degrees of sophistication of online public services going from "basic" information provision over one-way and two way interaction to "full" electronic case handling. As seen in Table 1, Stage 1 corresponds solely to the online availability of the information required to start the procedure to obtain a specific public service. Stage 2 indicates that the publicly available website offers the possibility to obtain in a non-electronic way, thus by downloading forms, the paper form necessary to commence the procedures related to the specific public service. Stage 3 represents as well the possibility of an electronic intake of an official electronic form in order to obtain the service, implying the existence of an authentication mechanism for verifying the identity of the person requesting the service, while Stage 4 is indicative of full electronic delivery of services, where no other formal procedure is necessary for the applicant via "paperwork". Finally, the 5th level of sophistication, built around the concepts of pro-activity and personalization [8], gives an indication of fully integrated electronic procedures that help reduce "red tape" and improve data consistency. Note that both stages 4 and 5 represent full electronic case handling.

Table 1. The five levels of sophistication in e-government

Stage 1:	Information
Stage 2:	One way interaction (downloadable forms)
Stage 3:	Two-way interaction (electronic forms)
Stage 4:	Transaction (full electronic case handling)
Stage 5:	Personalization (pro-active, automated service delivery)

3 Introduction to Fuzzy Numbers

3.1 Linguistics and Fuzzy Sets Theory

There are some situations in which information may be hard or even impossible to quantify due to its nature, and thus, it can only be expressed in linguistic terms (e.g., when evaluating the comfort or design of a car, terms like "good", "fair", "poor" can be used). In other cases, precise quantitative information cannot be provided because it is either unavailable or the cost for its computation is too high and an approximate value can be tolerated. Such qualitative information can be mathematically modeled through the use of Fuzzy Sets Theory. The latter handles fuzziness and represents qualitative aspects as linguistic variables, i.e. variables whose values are not numbers but words or sentences according to a natural or artificial language. More specifically, qualitative information expressed through linguistic terms (e.g. "low", "medium", "high") can be converted to fuzzy numbers using a suitable conversion scale. Processing of the relevant information takes place using these fuzzy numbers, which are finally converted to crisp numbers through the process of defuzzification. Obviously, the same linguistic terms in different conversion scales can have different crisp values.

3.2 Fuzzy Numbers

The fuzzy sets theory, introduced by Zadeh (1965) [9] as a means of dealing with vagueness, imprecision and uncertainty in problems, has been used as a modeling tool for complex systems that can be controlled by humans but are hard to define precisely. A fuzzy set is one that assigns grades of membership between 0 and 1 to objects using a particular membership function $\mu_A(x)$. The membership function of a triangular fuzzy number is defined by three real numbers, expressed as (l, m, u), where l is the lower limit value, m is the most promising value and u is the upper limit value.

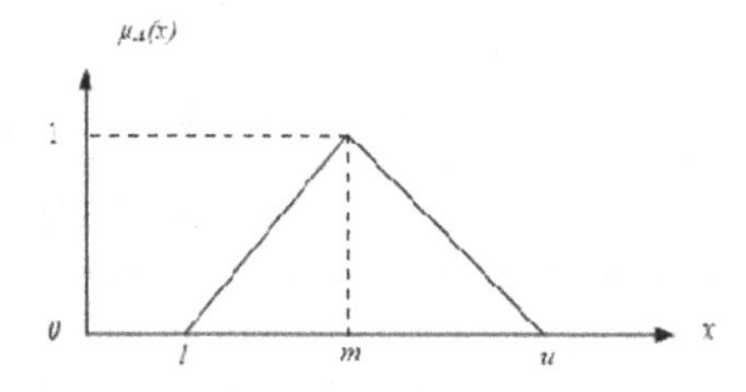

$$\mu_A(x) = \begin{cases} (x-l)/(m-l), & l \le x \le m \\ (u-x)/(u-m), & m \le x \le u \\ 0 & otherwise \end{cases}$$

Fig. 1. Fuzzy Triangular Number

3.3 Defuzzification Methods

Defuzzification is the process of producing a quantifiable result in fuzzy logic; in other words the extraction of a crisp value that represents effectively a given fuzzy number. The following common methods of defuzzification may be used:

Centre of Gravity (COG):
It is one of the most commonly used defuzzification techniques and is based on the determination of the centre of the area under the combined membership functions, i.e. the area under the curve. The centroid or centre of gravity of the area is calculated as:

$$F_{COG}^{-1} = \frac{\int_x \mu_A(x) \cdot x\,dx}{\int_x \mu_A(x) \cdot dx}$$

Bisector of Area (BOA):
The bisector is the vertical line that divides a region into two sub-regions of equal area. It sometimes coincides with the centroid line.

$$\int_a^{x_{BOA}} \mu_A(x)dx = \int_{x_{BOA}}^{\beta} \mu_A(x)dx$$

4 Evaluation Framework

4.1 Fuzzification of Time

Common approaches for e-government service evaluation rely on questionnaires. The citizens are requested to express their overall experience of a service or their opinion on a specific feature on a predefined satisfaction scale [10], that in several cases consists of a series of linguistic values, as described in Section 3.1. Final conclusions are then usually reached through statistic processing of these values. Such an approach is adopted here as well in order to obtain information on the time allocated by the citizens on interactions with the Public Administration; however in our approach the raw data obtained from the questionnaires are converted to fuzzy numbers to be further processed.

As a consequence, this section describes how Fuzzy Sets Theory can be exploited and thus how fuzzy numbers can be extracted so as to model the time required for the completion of a public service. At this point attention must be drawn to the fact that even if time is a countable quantity, it is considered extremely difficult or even impossible to record accurately the time spent by the citizens for interactions with the Public Administration: citizens cannot be constantly monitored and neither can they provide with absolute precision the necessary information when asked to do so. The use of fuzzy instead of crisp numbers is therefore adopted as a means of ensuring that the uncertainty factor that is innate in subjective evaluations/estimations is incorporated into the analysis and more reliable results are produced.

The citizens are able to evaluate each one of a pre-selected package of services ("basket of services"), the use of which is suggested for practical reasons, in terms of the time they spend on transactions with the Public Administration using either conventional or electronic services, based on a qualitative seven-level scale of linguistic variables that range from "Very Low" to "Very High". To exclude the factor of subjectivity involved, due to the fact that different people may have a different understanding of expressions such as "Very Low", "Low" etc., an indication of the respective amount of time, defined according to the authors' personal experience is as well provided, as in Table 2. The latter shows the equivalence between the linguistic variables and the respective fuzzy sets.

Table 2. Relations between Linguistics and Fuzzy Numbers

Linguistic	Abbreviation	Fuzzy Set (minutes)	Fuzzy Number
Very Low	VL	1-20	(1,10,20)
Low	L	10-30	(10,20,30)
Medium Low	ML	20-40	(20,30,40)
Medium	M	30-50	(30,40,50)
Medium High	MH	40-60	(40,50,60)
High	H	50-70	(50,60,70)
Very High	VH	60-80	(60,70,80)

In this way a citizen may describe for example the amount of time required to obtain a birth certificate, using the respective electronic service, as "Very Low", which means that the time needed does not exceed 20 minutes and it is approximately around 10 minutes. The concept of fuzziness is expressed through the term "approximately". Note that in the framework of this study symmetric triangular fuzzy numbers are considered. Optimality of results may be investigated through a variety of conversion scales; this issue however lies outside the interests of the current work. Using the Matlab Fuzzy Logic Toolbox, the adopted conversion scale is illustrated as shown in Figure 2. Using the selected conversion scale, the observations from the questionnaires for both conventional and electronic services are transformed into fuzzy numbers and average values are calculated for each one of the pre-selected public services. A frequency is also assigned to each service as well, providing an indication of the number of times that the service is used by a citizen during a specific period of time.

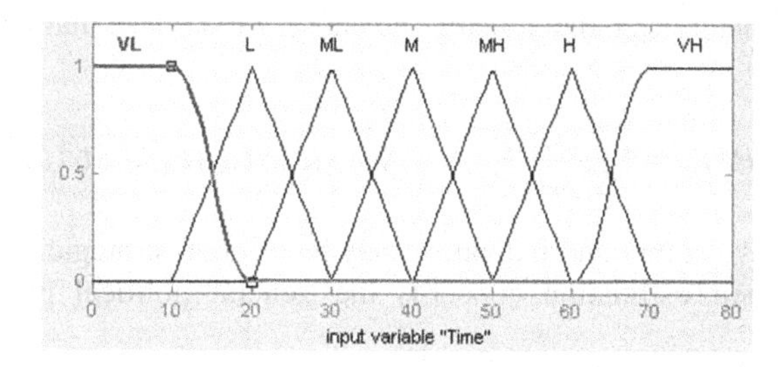

Fig. 2. Linguistics to minutes conversion scale

4.2 Axis I

The evaluation of the impact of e-government transformation in the improvement of the citizens' everyday life is performed in terms of the time that citizens save because of the implementation of the relevant transactions with electronic services. A set of composite indicators are suggested in this frame:

- The average time (%) that citizens may gain exploiting the current sophistication stage of electronic services as opposed to conventional methods (Average Gained Time I) is defined as:

$$\mathrm{AGT_I} = 100\% \cdot \frac{\mathrm{AT_{cs}} - \mathrm{APT_s}}{\mathrm{AT_{cs}}} \tag{1}$$

where AT_{cs}, is the average time required per service using conventional services and is calculated as:

$$\mathrm{AT_{cs}} = \frac{\sum_{i=0}^{n} f_i t_{i,c}}{\sum_{i=0}^{n} f_i} \tag{2}$$

and APT_s, is the average time potentially required per service exploiting the current sophistication stage, being calculated as

$$APT_s = \frac{\sum_{i=0}^{n} f_i t_{i,s}}{\sum_{i=0}^{n} f_i} \tag{3}$$

Note that $t_{i,c}$ and $t_{i,s}$ are the fuzzy numbers (mean values) that derive from questionnaires and cover the n pre-selected services and f_i are the corresponding frequencies.

- Similarly, the average time (%) that citizens could potentially gain, if there were fully transactional services may be defined as:

$$AGT_{II} = 100\% \cdot \frac{APT_s - APT_{fs}}{APT_s} \tag{4}$$

where APT_s is calculated as above and APT_{fs} is the average time potentially required if there was full electronic case handling. As it is assumed that not all public services of the pre-selected package have reached a stage of transactional maturity, APT_{fs} can only be approximately calculated using the subset of services that belong to stages 4 and 5 as:

$$APT_{fs} = \frac{\sum_{i=0}^{n} f_i t_{i,s}}{\sum_{i=0}^{n} f_i} \qquad , \forall i: s_i \epsilon (Stage_4 \cup Stage_5) \tag{5}$$

Note that the above compound indicators may be of great strategic and political significance, since their estimation concerns the general political planning of Public Administration.

4.3 Axis II

The second axis of the analysis refers to the evaluation of the actual utilization of e-government services by the citizens. The goal this time is to estimate the average time (%) that citizens could potentially save exploiting the current sophistication stage, which however is left unexploited since not all citizens make use of e-services and even if they do, they do not necessarily exploit the maximum sophistication stage that is currently available.

- The metric in question may be defined as:

$$AGT_{III} = 100\% \cdot \frac{ART_s - APT_s}{ART_s} \tag{6}$$

where APT_s is calculated as above and ART_s, which denotes the actual mean time that is currently allocated by the citizens on their interactions with the Public Administration, may be determined, using relevant statistical information, i.e. the percentage of citizens that actually benefit from the available sophistication level of a service.

Assuming that S_i ($Stage_i$) denotes the subset of services that have currently reached the i_{th} stage of sophistication, we define the Average Real Time per service as:

$$ART_s = \frac{1}{\sum_{i=1}^{n=4} F_i} \cdot \sum_{i=1}^{n=4} \left[F_i \cdot \left[p_i T_i + \sum_{j=1}^{m=i-1} (p_j - p_{j+1}) T_j + (1 - p_1) T_{i,c} \right] \right] \tag{7}$$

where

T_i is the average time required for the completion of an i-stage service
$T_{i,c}$ is the average time required for the completion of an i-stage service with traditional methods
F_i is the frequency assigned to the i_{th} sophistication stage, as $F_i = \sum_{i=0}^{n} f_i, \forall i: s_i \epsilon Stage_i$

We also assume that stage i is the maximum stage that is being used by percentage p_i of citizens. In equation (7) $p_i T_i$ indicates that a percentage p_i of citizens exploit the maximum sophistication stage available and spend therefore T_i time for the completion of an i-stage service. The sum $\sum_{j=1}^{m=i-1}(p_j - p_{j+1})T_j$ denotes the time allocated on transactions with the public authorities by the rest of the citizens who make use of e-government services but do not exploit the maximum sophistication stage offered. Finally, a percentage $(1 - p_1)$ of citizens do not rely on electronic services at all and therefore the time they spend is respectively$(1 - p_1)T_{i,c}$.

5 Case Study: Greece

5.1 Sophistication Stage of e-Services in Greece

In this section the evaluation framework presented in this paper is applied for Greece using the basket of twelve basic public services for citizens that has been composed by Capgemini for the European Commission. The services under evaluation along with their sophistication stage, as this is defined in the latest version of the country's e-Government factsheet [11] are summarized in Table 3. Table 4 presents the actual utilization of e-government services per sophistication stage, according to Eurostat [6]. Since no information is provided on the use of fully transactional services (stages 4 and 5), we assume that the corresponding percentage is equal to zero.

Table 3. Sophistication stage of selected services

	Services	Sophistication Stage
s1.	Income taxes: declaration, notification of assessment	5/5
s2.	Job search services by labour offices	4/4
s3.	Social security benefits (unemployment benefits, child allowances, medical costs, student grants)	1/5
s4.	Personal documents (passport and driver's license)	1/4
s5.	Car registration (new, used, imported cars)	4/4
s6.	Application for building permission	1/4
s7.	Declaration to the police (e.g. in case of theft)	1/3
s8.	Public libraries (availability of catalogues, search tools)	3/5
s9.	Certificates (birth, marriage): request and delivery	4/4
s10.	Enrolment in higher education /university	2/4
s11.	Announcement of moving (change of address)	2/4
s12.	Health related services (availability of services in different hospitals, appointments for hospitals)	2/4

Table 4. Population percentages

Sophistication Stage	Percentage
Stage 1	8.8%
Stage 2	4.0%
Stage 3	3.6%
Stage 4,5	0.0%

5.2 Results of the Proposed Framework

Table 5 summarizes the results of questionnaires for both electronic and conventional services. The frequencies assigned correspond to the number of times that a specific service is used by a citizen during a period of thirty years. Note that any moderate or excessive judgments are counterbalanced by the fact that the same persons evaluate both conventional and electronic services. Note also that all metrics in question (see equations 1, 4, and 6) are fuzzy numbers. Given two fuzzy numbers $\tilde{A}$= (a_1, a_2, a_3) and $\tilde{B}$ = (b_1, b_2, b_3), the fuzzy number $\frac{\tilde{B}-\tilde{A}}{\tilde{B}}$ is calculated based on operations on triangular fuzzy numbers as $\frac{\tilde{B}-\tilde{A}}{\tilde{B}} = (\frac{b1-a3}{b3}, \frac{b2-a2}{b2}, \frac{b3-a1}{b1})$. Although this equation possesses mathematical correctness, it has no physical sense. In order therefore to obtain the average gained time, a defuzzification method has to be applied. In this study, both the COG and BOA defuzzification values are extracted (see section 3.3), using the fuzzy operations of Matlab. The fuzzy and crisp values for all metrics described are finally presented in Table 6. As a general conclusion it can be stated that BOA provides more moderate crisp values compared to COG.

Table 5. Linguistic values and frequencies for all services

	Current Soph. Stage	Conventional Services	Freq.
s1	Very Low	High	30
s2	Very Low	Medium Low	18
s3	Medium	Medium High	115
s4	Very High	Very High	7
s5	Low	High	4
s6	High	High	1
s7	Medium	Medium High	8
s8	Very Low	Medium	90
s9	Very Low	Medium	10
s10	High	High	1
s11	Medium High	High	2
s12	Very Low	Low	90

The result for AGT_I indicates that thanks to the current sophistication stage, there is a gain of 55% for citizens as far as transaction time is concerned. However, the value of AGT_{III}, which incorporates the actual utilization of e-government services seems to counterbalance this gain and provides strong evidence that the authorities responsible should take into account the information and technology literacy of citizens as well as the challenges they encounter with regard to the use of e-government services due to physical impairments. It is not the case that all citizens are familiar with computing and Internet-based technologies nor does the design of web sites currently encompass all different disability types (associated with aging, visual, auditory, speech, motor and cognitive deficiencies). Furthermore, issues related to security and privacy continue to be a challenge: users must be confident that the web sites they visit and transactions they complete are safeguarded against theft, fraud and unauthorized access. On the other hand the higher percentage that derives for AGT_{II} illustrates in an explicit way that the actual benefits of e-government transformation are located in the upper stages of sophistication.

Table 6. Fuzzy and crisp values of proposed metrics

	L	M	U	COG	BOA
AT_{cs}	30.58511	40.5851	50.58511		
APT_s	12.14894	21.516	31.51596		
APT_{fs}	1.580645	10.6452	20.64516		
ART_s	29.90319	39.88611	49.89568		
AGT_I	-1.84017	46.98558	125.6696	56.9386	55
AGT_{II}	-26.9585	50.52434	246.4027	89.9906	83
AGT_{III}	-3.23229	46.05652	126.2299	56.3510	54

6 Conclusions

This paper proposed an evaluation framework to project an aspect the impact of the modernization of the Public Administration in the citizens' quality of life through the adoption of a series of metrics with regard to the time required for the completion of transactions with the public authorities. The framework was applied for the case of Greece, pointing out that any benefits associated with e-government transformation are counterbalanced by the fact that only a small fraction of the population make use of electronic services and suggesting that effort should be put in raising public awareness around e-government services, so that the citizens realize the envisaged e-government benefits.

References

[1] Wang, L., Bretschneider, S., Gant, J.: Evaluating web-based e-government services with a citizen-centric approach. In: Proceedings of the 38th Hawaii International Conference on System Sciences, January 2005, vol. 5, p. 1292, Island of Hawaii (2005)

[2] Panopoulou, E., Tambouris, E., Tarabanis, K.: A framework for evaluating web sites of public authorities. In: Aslib Proceedings: New Information Perspectives, vol. 60(5), pp. 517–546. Emerald Group Publishing Limited (2008)

[3] Wauters, P.: Benchmarking e-government policy within the e-Europe programme. In: Aslib Proceedings: New Information Perspectives, pp. 389–403. Emerald Group Publishing Limited (2006)

[4] Corradini, F., Marcantoni, F., Polzonetti, A., Re, B.: A Formal Model for Quality of Service Measurement in e-Government. In: 29th Int. Conf. on Information Technology Interfaces, Croatia (June 2007)
[5] Deliverable, Indicators for the quantification of G.S.P.A & E.G.'s objectives with regard to eGovernment. The use of Information and Communication Technologies in the Public Sector (eGovernment) (February 2008), http://www.observatory.gr
[6] EUROPA – Eurostat – Home page, http://epp.eurostat.ec.europa.eu
[7] Hubregtse, S.: The digital divide within the European Union. New Library World 106(1210/1211), 164–172 (2005)
[8] Capgemini, The User Challenge: Benchmarking the Supply of Online Public Services (September 2007), http://www.uk.capgemini.com
[9] Zadeh, L.A.: Fuzzy sets. Inform. and Control 8, 338–353 (1965)
[10] Abhichandani, T., Horan, T.A.: Toward A New Evaluation Model of E-Government Satisfaction: Results of Structural Equation Modeling. In: 12th America's Conference on Information Systems, Mexico, August 4-6 (2006)
[11] eGovernment in Greece, Country Factsheets (December 2007), http://epractice.eu

e-Government: Challenges and Lost Opportunities*

Jaroslav Král[1,2] and Michal Žemlička[1]

[1] Charles University, Faculty of Mathematics and Physics
Malostranské nám. 25, 118 00 Praha 1, Czech Republic
kral@ksi.mff.cuni.cz, zemlicka@ksi.mff.cuni.cz

[2] Masaryk University, Faculty of Informatics,
Botanická 68a, 602 00 Brno, Czech Republic
kral@fi.muni.cz

Abstract. The e-government systems are the largest software systems ever used. They influence almost all aspects of the life of a society. The analysis of the capabilities they provide indicates that e-government does not provide enough tools for the strategic aspects of the development of states and societies. Examples include control and optimization of education system, crisis prediction and control, and analysis of administrative processes. It is amplified by the delayed application of modern software architectures and e-government. We show that the precondition of any solution of these issues is a proper treatment of data collection and a reasonable data security policy in e-government systems. It should lead to an architecture of e-government having three well-developed virtual tiers: user, application, and data tier. Current systems have almost no data tier. The condition for it is an explicit formulation of measurable aims regarding strategic aspects of the use of e-government systems.

1 Introduction

Software systems for e-government are very large and very complicated and the most expensive software systems in every country. The development of software for e-government (SWeG) is therefore quite complex and expensive. One would expect that such a process is based on proven principles of software engineering. It seems that it is not so.

There is an evidence that standard software engineering methods are not properly applied in SWeG. It can be argued that SWeG is so specific problem that standard software engineering methods and processes are not applicable in the development of SWeG.

Such an opinion is based on the fact that states (government systems) are organizations with "professional bureaucracy" [1] being different from the "machine bureaucracy" of enterprises and other organizations. It may be the reason why e.g. Enterprise Service Bus is in SWeG rarely used, if ever.

* This work has been partially supported by the Program "Information Society" under project 1ET100300517 and by the Czech Science Foundation by the grant number 201/09/0983.

M.D. Lytras et al. (Eds.): WSKS 2009, LNAI 5736, pp. 483–490, 2009.

Although there are differences, main software engineering principles are common for all software systems, SWeG inclusive:

1. It is always necessary to formulate clear visions – why a system is to be developed. A special issue is the absence of vision of the strategic effects like the prediction of economical turbulences, control/supervision/assessment of the quality of education, etc.
2. The structure of a state is a network of rather autonomous entities – offices and other majorities having many features common with the division [1] organization of large decentralized enterprises. Common techniques can and should be therefore applied.
3. Many technical tools (pragmatic service-oriented techniques, databases, middleware, prototyping) are applicable in all large organizations, entire states inclusive.

We will show that these points are not properly taken into account. It unnecessarily increases the development expenses of SWeG and reduces the possible benefits and useful capabilities for citizens, businessmen, politicians, for state administration itself, and for researchers.

This paper is based on the personal experience of the authors with e-government in Czech Republic. They collected the experience as IT experts, scholars, and citizens using the existing e-government systems.

There are, however, strong indications that our conclusions are common at least for almost all SWeG in European countries and North America. The results of the researches of world-wide personal agency Manpower [2] are an indirect evidence that the situation is unlikely to be substantially different in Asia or South America. The researches indicate that demands for experts are similar all over the world. It in turn indicates the common problems with education and, maybe, issues of the reward system. The problems are caused by many reasons enabled by the deficiencies of SWeG. The deficiencies of SWeG support public prejudices and hidden wrong effects of education systems reforms. Similar effects exist in other domains (e.g. health care) in many societies.

2 Operative Level and Strategic Level of e-Government

Contemporary SWeG support almost exclusively the administrative operations like various registers, operations related to ownership and its changes, etc. It corresponds to the operation tier in enterprise resource planning (ERP) systems. The individual operations of this type require as a rule small collections of data of the guaranteed quality, the semantic of the operations is quite fixed and intuitively clear. The operations can be nodes of a network defining user-oriented processes. It is known for ERP systems that the operation tier does not usually bring substantial long term advantages. Similar properties are probably present in the operation tier in e-government. The tier speeds up and simplifies the communication between citizens and SWeG. It is nice but it has e.g. not detected the coming contemporary crisis of economies.

Strategic level in enterprise systems requires varying actions over large data bodies. The data come from different sources are of a varying quality and are used in dynamic "processes" producing information, compare OLAP in an ERP as a simple example. The aims are the detection of trends, analysis, etc. as a support of long-term (strategic) decisions. Different users can use/design different information producing applications according to their instantaneous needs. The needed data can be in local databases or can be looked for elsewhere, e.g. over the web.

Similar opportunities should available be in e-government systems too. The users are in this case citizens, enterprises, and institutions. The data analysis could be used in research of social or economic phenomena. The problem is that the needed data are often sensitive (personal, business, secret, etc.) and there are bans limiting their use.

The fact that there can be links between data regarding a given individual from distant time moments implies that the data cannot be fully anonymized. There are satisfactory technical solutions (see below) of the problem but the solution requires a proper legislative and above all the vision and understanding that strategic tier and strategic aims are highly desired and that the bans can be in fact very expensive.

Current e-government tend to be rather a virtual two-tier system (user/application where both tiers are implemented by networks of software services) than as a virtual three-tier system (users/applications/data stores). The limited capabilities of the data tier reduce the extent of possible capabilities of the system.

3 e-Government and Data

The above conclusions imply that, strictly speaking, the SWeG lack some important capabilities of information systems in common sense. SWeG aims (visions) are not formulated so that they can enable the "measurement" whether the developed system fulfills the requirements and whether no components were developed unless capabilities were required.

The main issue therefore is that SWeG are not developed and used with any proper data tier and with almost no useful capabilities based on data provided by the data tier. Such capabilities are often not required and even not mentioned in visions/aims of the developed systems. It is the consequence of two main factors:

- The data in e-government are of varying quality and are scattered over various components of e-government and often outside it as well. It implies some conceptual and technological problems if we want to enable access all the data and to use the data properly.
- The data are often sensitive (personal, secret, business, etc.). Much interesting information generable from them is not sensitive and they should be published. The process of the information generation must be properly defined and supported by legislative.

The first issue can be solved if we properly use modern software design principles, especially the principles borrowed from service-oriented (SO) technologies. The principles should be, at least for some time, used and applied in a pragmatic way [3]. Current practice in e-government does not seem to apply SO properly – not to speak about the application of technologies allowing integration of distributed heterogeneous databases. Some standards and laws disallow in fact the use of SO in e-government properly.

The problem of data security is in fact solved in a very straightforward way. Although stated otherwise the main principle can be with some simplification stated as "disallow the access and the use of any data if it is imaginable that the data can be misused." It can be characterized as a "huge virtual data shredding".

4 Benefits of Data Tier

It is clear that issues similar to the ones discussed above must be solved if we want to increase the quality of health systems, municipal systems, environment control systems, etc. In all these cases the benefits of properly used data tier are of the following types:

1. Generally accessible snapshots of current situation in the society.
2. Prediction of further development using standard statistical methods like time-series analysis.
3. Research based on the analysis of the history based on the data analysis.
4. Quantitative research of social and economic processes. The research methods can use powerful simulation experiments based on large data bodies provided by that e-government data tier.

It is not difficult to see that all these aspects make sense in all the above mentioned domains.

We can hope that it could reduce the probabilities of such failures like wrong predictions of rising economic or social crisis. It could enable to analyze and to review and evaluate various market regulation attempts, and so on. It is not sure whether we will ever be able to do it precisely enough. It is sure, however, that without it we will be blind in the primeval forests of economic phenomena.

There is a small chance that it could help in building macro-economy as a real experimental science.

5 From Sensitive Data to Public Information

Information needed for e-government partners (citizens, institutions, enterprises, etc.) to make decisions like "what schools are the best for my children?" is typically open; it is not sensitive. An example is the average salary of the graduates/alumni of a school. The needed data (e.g. the salary of a person) are sensitive and should not be directly accessible for the querying people.

The solution can be based on the fact that the data can be used by tools (applications) producing the information provided that the application is maintained and its outputs (i.e. the information) controlled by a trusted (accredited) body.

The body can use the data store of sensitive data and information producing tools. The datastore can be distributed especially if a pragmatic variant of service-oriented architecture (SOA) is used. A solution can be based on the following principles (see Fig. 1).

The data store accepts the data to store them or to establish links to other datastores.

- It cleans and filters the data it uses.
- It uses only accredited information producing tools after it evaluate their applicability. Users can write and accredit the tools they need.
- The outputs of data queries and of information producing applications are logged and tested whether they do not break security rules.

We believe that it, if designed and used properly, provides satisfactory data security.

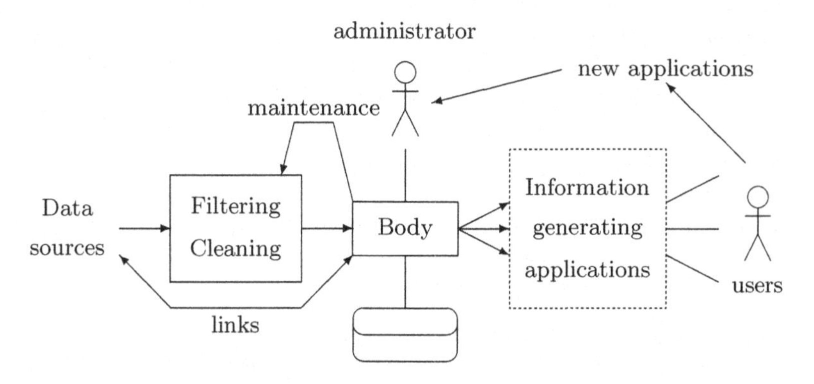

Fig. 1. Data handling scheme

It can be argued that our proposal decreases the data security in comparison to current data security practices as it can provide a new data leak tunnel. It is partially true but not too significant. Why? There are many ways of (e.g. personal) data leaking now: various registers, sample researches, land registries, social software, etc. Full closing of one tunnel does not change the situation substantially as the probability of the data leak through the tunnels being not closed (social software, passport procedures, e-commerce, various registers, etc.) is not small.

6 e-Government and Education System

The missing proper data tier and missing capabilities of long-term data analysis and predictions in *e*-government have crucial consequences. Let us discuss the most obvious case, the case of the assessment of education systems.

The quality of education is considered to be of a crucial importance. If we want to test whether an education system is "good" and if we want to improve it, then the evaluation of the education systems should start from the evaluation of schools. We must, however, take into account that quality is a subjective concept depending on the requirements of users [4].

It follows that the evaluation process must be dynamically adaptable to the needs of evaluators, e.g. parents looking for the best schools for their children.

The majorities should be able to test the quality not only of schools but also of study programs. The criterion of quality should be the success of graduates/alumni in real life. It all can be achieved if we build a database of data on professional positions of graduates. Such data does exist in many countries but they cannot be used due the unbalanced data security policy mentioned above.

It has the following consequences:

1. Parents looking for a good school for their children must use unreliable information like reputation of the school among their friends, i.e. rumors must be used. It is especially wrong in post-communistic countries where school systems and pedagogic processes are changing very (maybe too) quickly.
2. The rumors prefer non-STEM (STEM stands for science, technology, engineering, and mathematics) education. It is one of the causes of the chronic lack of people having professions requiring STEM knowledge and skills (see the research of Manpower [2] and the gossips on engineering education). It is sometimes felt as a threat for national security and prosperity. Compare [5,6,7,8,9,10].
3. The education institutions are not induced to increase the quality of their study programs – especially the STEM-oriented ones. The education institutions are misusing this opportunity.

A proper information system on education and school systems accessible for the broad collection of users can solve many problems, provided that the delay caused by the fact that we query current situation and judge future needs does not matter substantially.

Note that the existence of a good information system enabling evaluation/assessment of education can have substantial research possibilities in the domain of pedagogical sciences and other humanities.

There are technical issues. It can be quite difficult to find an effective implementation allowing scalable and customizable solution with agile features.

7 e-Government and SOA

We will understand as a service-oriented architecture (SOA) any collection of collaborating autonomous software entities communicating in asynchronous way and forming a virtual peer-to-peer network. Such a system can be constructed in a pragmatic way. For example it is good to use as few standards as possible.

Pragmatic SOA simplifies the communication of SWeG with software systems of enterprises, municipalities, and with electronic devices of citizens. SOA simplifies insourcing and outsourcing, autonomy of institutions and incremental development and modernization/maintenance. These challenges in SWeG have not been fully taken into account yet. The reasons for such an attitude are:

- Existing standards do not meet some critical requirements.
- The existing standards of many XML-based formats and languages recommended to use for e-government communication and data exchange in general are typically cumbersome and quickly changing.
- The SWeG developers have insufficient experience with pragmatic application of service orientation.

SOA can simplify the integration of data stores e.g. to implement the education assessment discussed above.

The main problem is that SOA is a system of specific paradigms. As such it is difficult to accept it, to use it efficiently, and to be properly taken into account in standards. Let us give an example. ITIL [11] standard is based on the concept of (infrastructure) services. ITIL posses only very weak tools enabling to build a pragmatic software architecture and composite services. It makes the reuse of existing systems (like the systems of autonomous institutions) more difficult than necessary. A proper design of SOA moreover enables an easy construction of software prototypes, easy logging, and so on [12].

8 Crucial Problems of e-Government

The application of above proposals and attitudes is not easy from the technical point of view. The data to be used are of different quality and the data sets are very large and scattered over various technical frameworks. Technical obstacles are, however, not the main barrier of the development of e-government towards full three-tier information system.

Main obstacles in the modernization of e-government and of application of our proposals discussed above are:

- e-government is felt as a potential threat for the positions of the state administration officers;
- (slow and repeated) implementation of e-government systems is a wonderful opportunity to make a lot of money;
- if the state administrative uses "proper" implementation it can be locked to a particular software vendor; the laws of economy force the vendors to behave so;
- it is better for some subjects if the evidence and evaluation in their domain remain so weak as it is now (for example an improved evaluation of schools can imply a higher effort of the schools in teaching their students what can be in many cases undesirable for the school owners – they therefore lobby against such changes);

Hence any technical solution leading to real effective implementation is in fact often blocked by some involved parties.

9 Conclusion

The current e-government systems have brought many excellent advantages. Current e-government systems are, however, in fact operation supporting systems.

The capabilities known in enterprise information systems as management information systems are almost missing in e-government systems.

It is the consequence of the fact that the data storing and providing capabilities are rather limited, sometimes almost forbidden. It has substantial consequences. The consequences are, however, rarely properly taken into account.

This issue is the consequence of a broader and deeper snag. Current data management and the information providing practices in e-governance imply that the modern opportunities enabled by modern software are often wasted. Examples are knowledge society, supervision and prediction of social processes and state institutions like education systems. It is a great obstacle of the proposals from [13,14].

To avoid these limitations the overall attitude to data security and use in e-government must be changed. Otherwise the e-government will be unable to support or collaborate properly with many important systems, and to provide capabilities enabling the analysis and the control of long-term social and economic processes. It can lead to fatal consequences, sometimes even to civilization collapses (compare [15]).

References

1. Mintzberg, H.: Mintzberg on Management. Free Press, New York (1989)
2. Manpower: Talent shortage survey: 2007 global results (2008), `http://www.manpower.com/research/research.cfm`
3. Král, J., Žemlička, M.: Pragmatic web-based service-oriented systems. In: ICWS 2009 (submitted, 2009)
4. International Organization for Standardization: ISO 9000:2005 Quality management systems – fundamentals and vocabulary (2005), `http://www.iso.org/iso/catalogue_detail?csnumber=42180`
5. Spellings, M.: Answering the Challenge of a Changing World. U.S. Department of Education (2006)
6. Fortenberry, N.L., Sullivan, J.F., Jordan, P.N., Knight, D.W.: Engineering education research aids instruction. Science 317, 1175–1176 (2007)
7. Mervis, J.: Congress pases massive measure to support reearch and education. Science 317, 736–737 (2007)
8. Mervis, J.: A new bottom line for school science. Science 319, 1030–1033 (2008)
9. Mervis, J.: U.S. says no to next global test of advanced math, scicence students. Science 317, 1851 (2007)
10. Jamieson, L.: Engineering education in a changing world. The Bridge (Spring 2007)
11. OGC – the Office of Government Commerce (UK): IT infrastructure library (2007)
12. Král, J., Žemlička, M.: Software architecture for evolving environment. In: Kontogiannis, K., Zou, Y., Penta, M.D. (eds.) Software Technology and Engineering Practice, pp. 49–58. IEEE Computer Society, Los Alamitos (2006)
13. Moreno-Moreno, P., Yáñez-Márquez, C.: The new informatics technologies in education debate. In: Lytras, M.D., et al. (eds.) WSKS 2008. CCIS, vol. 19, pp. 291–296. Springer, Heidelberg (2008)
14. Korres, G.M., Tsamadias, C.: Looking at the knowledge economy: Some issues on theory and evidence. In: Lytras, M.D., et al. (eds.) WSKS 2008. CCIS, vol. 19, pp. 712–719. Springer, Heidelberg (2008)
15. Diamond, J.M.: Collapse: How Societies Chooses to Fail or Succeed. Viking, New York (2005)

The Influence of Customer Churn and Acquisition on Value Dynamics of Social Neighbourhoods

Przemysław Kazienko[1,2], Piotr Bródka[1], and Dymitr Ruta[2]

[1] Institute of Computer Science, Wrocław University of Technology
Wyb.Wyspiańskiego 27, 50-370 Wrocław, Poland
[2] BT Innovate, British Telecom Group, Intelligent Systems Research Centre (ISRC),
Orion 1/12G, Adastral Park, IP5 3RE Ipswich, UK
kazienko@pwr.wroc.pl, piotr.brodka@pwr.wroc.pl,
dymitr.ruta@bt.com

Abstract. The customers of modern telecommunication service providers implicitly create an interactive social network of individuals, which both depend on and influence each other through various complex social relationships grown on friendship, shared interests, locality, etc. While delivering services on the individual basis, the social network effects exerted from customer-to-customer interactions remain virtually unexplored and unexploited. The focus of the paper is on customer churn and acquisition, where social neighbourhood effects are widely ignored yet may play a vital role in revenue protection. The key assumption made is that a value loss or gain of a churning or new customer extends beyond the revenue stream and directly affect interaction within local neighbourhoods. This influence is evaluated experimentally by direct measurements of the total neighbourhood value of the churning customer taken before and after the churn event.

Keywords: telecommunication social network, customer churn and acquisition, social value, social neighbourhood, social network analysis, network dynamics.

1 Introduction

A social network is one of the many possible representations of a human community, in which people interact and get into relationships with one another. These relationships can be very complex and usually involve our emotions and feelings. Besides, associations within the social network may result from family dependencies or work cooperation. Moreover, a social network continuously evolves and changes its structure. Every second some new communities arise while the others disappear, some relationships reinforce while the other vanish [15]. In the everyday world, people relay on each other. Thus, their choices and behaviour also influence choices and behaviour of the others [5]. This is the fundamental concept of recommender networks [13, 16] or recommender systems [17] and enacts a significant role in marketing [14], in which people spread information and opinion about products through their mutual, personal contacts. Capability to predict changes and their consequences is crucial in every business. Apparently, dynamic analysis within the customer network

M.D. Lytras et al. (Eds.): WSKS 2009, LNAI 5736, pp. 491–500, 2009.

especially in the telecommunication social network is very important. General concept of analysis of dynamic social networks was presented in [1]. In order to forecast such changes and investigate the evolution of social networks even physics and molecular modelling can be utilised [8]. In some other approaches, clustering [4], statistical analyses and visualizations [1] or multi agent systems [2, 19] are used to get an insight into network dynamics. Daspupta *et al.* tried to predict churn based on the analysis of relationship strength in the mobile telecommunication social network [3], whereas Gopal and Meher used typical prediction method – regression to estimate churn time and tenure for the same domain [7].

This paper addresses the question: how much our behaviour, as the customers, influences the others and are we able to evaluate this influence based on the available data about mutual contacts or not? In particular, we analyse the influence of churning or recruited customers on their neighbourhoods after the churn or acquisition, respectively. These questions appear to be very important in the general analysis so-called knowledge society, in which people influence one another through their contact and interactions facilitated by telecommunication channels.

2 Telecommunication Social Network

Telecommunication data like voice calls (including residential, mobile and VOIP) contains enormous amount of information about customer activities. Moreover, each phone call can be treated as the evidence of mutual relationship between two subscribers [12]. A telecommunication social network *TSN* is the tuple *TSN*=(*M*,*R*) that consists of the finite set of members (customers, nodes) *M* and the set of relationships *R* that join pairs of distinct members: $R=\{r_{ij}=(x_i,x_j): x_i\in M, x_j\in M, i\neq j\}$. Relationships in *TSN* are directed, i.e. $r_{ij}\neq r_{ji}$. In other words, one member corresponds to one phone number, which, in turn, is assigned to one social entity – a human, group of people or an organisation.

3 Node and Neighbourhood Social Values

Measures are one of the social network analysis tools to describe human characteristic, specific for the given social network and to indicate personal importance of individuals in the community.

Some simple measures and one a bit more complex were used during experiments. All express the social value of a network member. In particular, they are: a total number or duration of phone calls initialized (*Out calls* and *Out duration*), a total number or duration of both received or initialized phone calls (*In+Out calls* and *In+Out duration*), Fig. 6 and 7. Another complex measure is Social Position measure, which has been proposed and developed in [9, 11]. It can be evaluated both for duration and number of calls in an iterative way:

$$SP^{(n+1)}(x) = (1-\varepsilon) + \varepsilon \cdot \sum_{y\in M} SP^{(n)}(y) \cdot C(y \to x)\,, \tag{1}$$

where:

$SP^{(n+1)}(x)$, $SP^{(n)}(x)$ – social position of node x after n+1th or nth iteration;

ε – the fixed coefficient from the range (0;1);

$C(y \rightarrow x)$ – the commitment function which expresses the strength of the relation from member y to x.

The constant ε represents the openness of human social position on external influences, in other words high ε means that the social position is highly influenced by others and low ε means that the social position is more static while others' influence is week [10, 11].

The set $N(x)$ of members y_i which are directly connected to member x is called x's neighbourhood. In real world, it is a set of members, with which member x maintains the closest relationships – nearest neighbours or first-level neighbours. We assume that these closest members from $N(x)$ have the biggest influence on member x and in opposite, member x has big influence on them.

Social value of the neighbourhood of member x – $SVN(x)$ is the sum of social values $SV(y)$ of all x's neighbours y:

$$SVN(x) = \sum_{y \in N(x)} SV(y) \cdot \tag{2}$$

Note that the neighbourhood does not include member x. Furthermore, also other churning or just acquired nodes y are excluded from set $N(x)$, see Fig. 1.

In the telecommunication business, members and separately their neighbours can belong to various classes. Two of them are usually distinguished: residential (individuals and their families, acquaintances, friends, etc.) and business (a company, department in the organisation, a position or single employee in the organization).

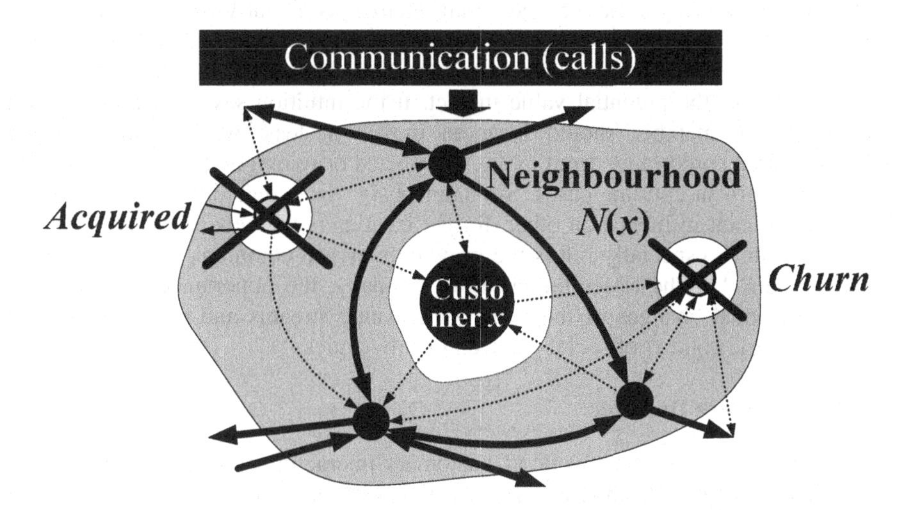

Fig. 1. Churning and acquired nodes from the neighbourhood as well as the central node itself are excluded from social value calculation

4 Social Effect of Customer Churn and Acquisition

Today's global telecommunication market environment can be characterized by the strong competition among different telecoms and a decline in growth rate due to maturity of the market. Furthermore, there is a huge pressure on those companies to make healthy profits and increase their market shares. Most telecom companies are in fact customer-centric service providers and offer to its customers a variety of subscription services. One of the major issues in such environment is customer churn known as a process by which a company loses a customer to a competitor. Recent estimates suggest that churn rates in the telecom industry could be anything between 25% and 50% [6]. Moreover on average it costs around $400 to acquire a new customer which takes years to recoup [6]. These huge acquisition costs are estimated to be between 5 to 8 times higher than it is to retain the existing customer by offering him some incentives [20]. In this competitive and volatile environment, it makes therefore every economic sense to have a strategy to retain customers which is only possible if the customer intention to churn is detected early enough [18].

4.1 Customer Churn and Acquisition

Nowadays client churn is one of the most important problems in many companies like telecommunication and internet providers [7]. Some analysis indicates that possibility of customer churn strongly depends on the number of neighbours which have already churned from the network. It is extremely challenging task to predict customer churn and prevent it or at lest be able to predict how much this churn affect others network members [3] and how much the company may lose because of particular member's churn.

This paper is trying to deliver the initial intelligence about the impact of customer churn on the dynamics of a service value within a social neighbourhood of the churning customer, such that a decision to retain or rescue a churning customer can be better aligned to the potential value impact. It the intuition says that a churn of an active network member should have an impact at least on his direct network neighbours that could range from fading, redirected or reinvigorated activity up to the follow-up churn in extreme cases. Simultaneously, the acquisition of customers can have a significant influence on other, former customers the new client gets into relationships with. It especially refers the growth or loss in communication between the old customers. For the telecommunication company, the importance of this change lies in individual changes of the neighbours' value streams and can be considered within a generic context of social value and its dynamics.

4.2 Social Values Dynamic

Let us consider a social network of customers interacting through telephone calls. Each such customer established his local social network consisted of customers whom he called or who called him at least once during his lifetime. Members of such local social network are customers' first-level neighbours as depicted in Fig 1. Each customer generates a dynamic value consisting of a value of his outbound calls as well as value added network component stemming from the fact that his presence drives inbound calls from his neighbours. We refer to such network value added component as social value of a customer.

While a dynamics of customer value stream is explicitly evident in his outbound calls that translate into telephone bills, the social value remains implicit and is hidden from direct observations. One naïve way of estimating the social value of a customer is to periodically measure the value of his inbound calls. The problem with this method is that it is unclear to which degree the calls made by customer neighbours are driven by the customer presence, or in other words it is not clear if under customer absence his neighbours would call less, redirect calls from the absent customer to other customers or perhaps even get stimulated to grow their neighbourhood and call more.

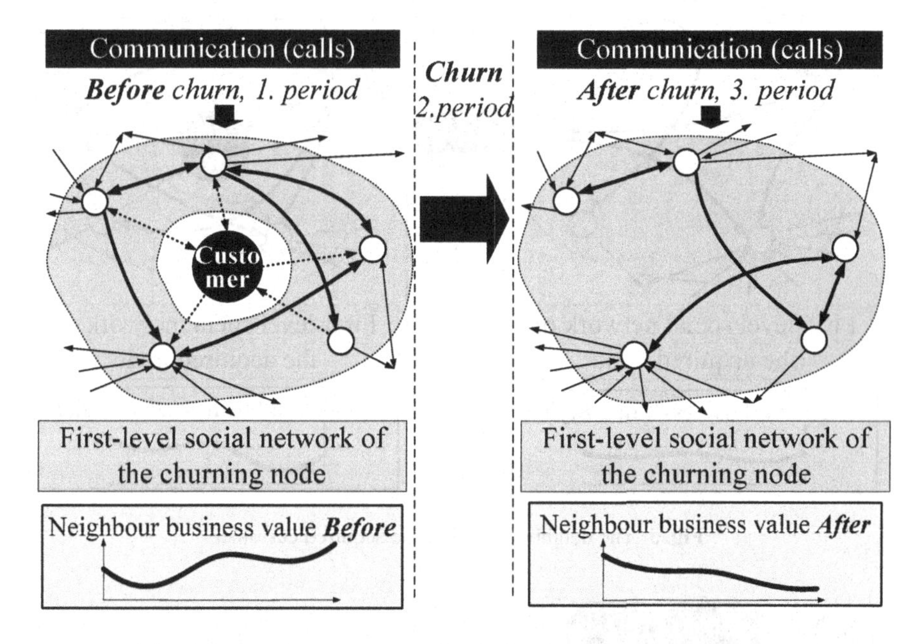

Fig. 2. A churning customer and their first-level (nearest) neighbours

Customer churn gives a realistic opportunity to evaluate the social value of a customer and explore its dynamics over time. By comparing the value of customer neighbourhood before and after the churn one can truly estimate the impact of churning customer on the change in neighbourhood value which is equivalent o the social value of churning customer, as shown in Fig. 2. A similar case occurs for acquired customers. A new element in the community can influence not only on the communication with this node but also on the information exchange between the old customers, Fig. 3.

It is important to remember that the neighbourhood value may continue to change at different rates well after the churn or acquisition event until the new equilibrium is achieved. Moreover, it might be very difficult to extract a direct impact of particular customer's churn/acquisition on his neighbourhood value as there may be many other concurrent drivers of value dynamics like other customers' churn, acquisition, customer moves and other significant network events. From the global perspective, all these additional processes impacting social value dynamics happen continuously anyway and are part of the ongoing network value fluctuations; hence their impact should be statistically similar before and after the churn/acquisition event.

Note that the process of analyzing the impact of churn on social value dynamics is directly reverse to the process of analyzing the impact of customer acquisition hence the two can be analyzed together possibly even sharing similar observations and conclusions. A diagram illustrating such process is shown in Fig. 4.

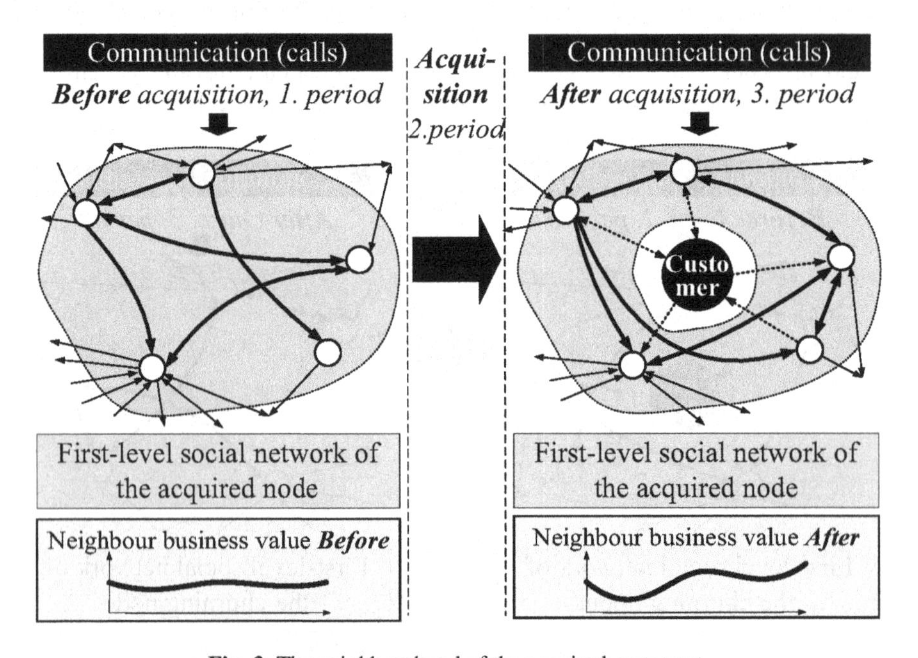

Fig. 3. The neighbourhood of the acquired customer

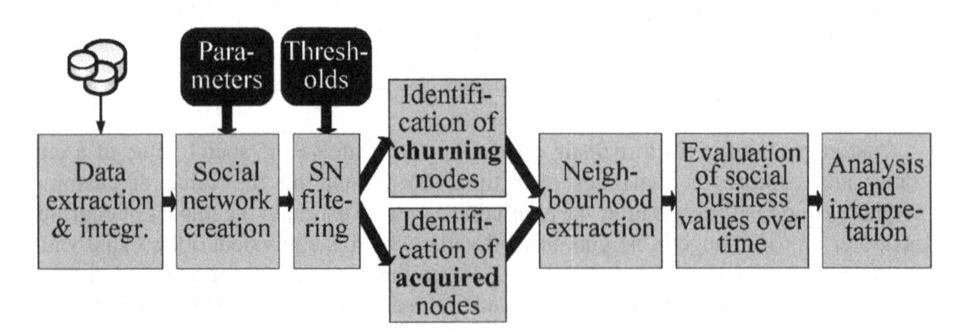

Fig. 4. Process of analysis of social neighbourhood for churning and acquired customers

The first part of this process is the identification of relationships in social network which allows establishing the neighbourhood of any particular customer. The next step is finding customers for which we want to analyse the social value dynamics, and those would be the customers who churn or are acquired preferable during the middle part of the period the analysis is conducted for. Then the key part involves establishing the neighbourhoods of such customers (prior- for churn and post- for acquisition) and measure the time series of their total business values from before the event until the point after the event for which the neighbourhood value time series attains again stationary. Note that as illustrated in Fig. 1, the value of churned or acquired customer is excluded from the neighbourhood value both before and after the churn/acquisition event and thus social value of a customer is a measure of customers ability to drive business value from the rest of the social network.

5 Experiments

The experiments were performed for two real telecommunication social networks only for churning customers, Fig. 2. The first data set with several dozens of thousand of residential customers (*Residential*) and a few hundred of churning customers. The second one with several hundreds of thousands of both business and residential clients (*Busines & Residential*) and a few thousand of churning clients and in line with the process shown in Fig. 4. The data came from 1-month period that has been split into three 10-day slots. The second slot was used to identify churning members, the first one to extract their neighbourhoods and calculate neighbourhood values "*Before*", whereas the last slot was exploited to evaluate values "*After*" and the relative change: "*After*" compared to "*Before*". During the experiments, measures described in section 3 were utilized.

The average social position of the churning nodes turned out to be about 40% for the *Residential* network and 57% for the *Business&Residential* network lower than the average social position of all other network members, Fig. 5. It suggests that churning customers lower their activities within the network before they leave it. Moreover, weaker social position of the churning customer also affects his

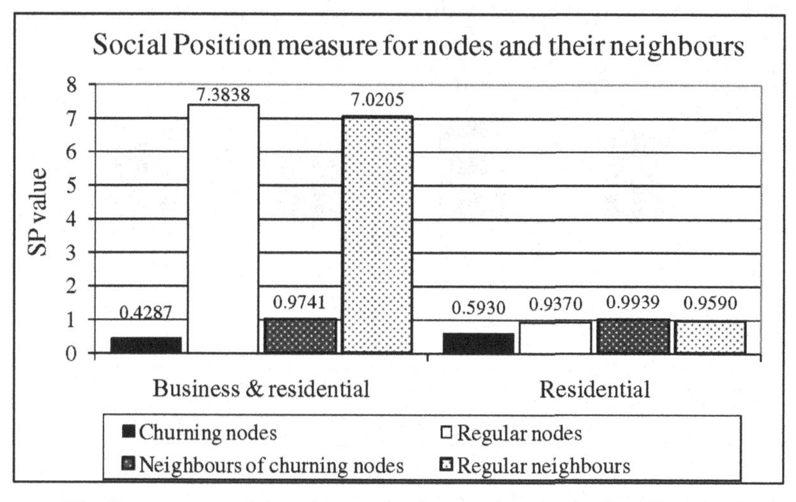

Fig. 5. Average social position value for members in the first time slot

neighbourhood although the average neighbours' social position is only 2.3% smaller than the average neighbour of the non-churning customer. However this is happening only in the *Residential* network, in case of the *Business & Residential* network average social position is about 5% higher than the average neighbour. Anyway, if we look into the dynamics of social position over the three time slots, the impact of churn on the entire neighbourhood is becoming more apparent.

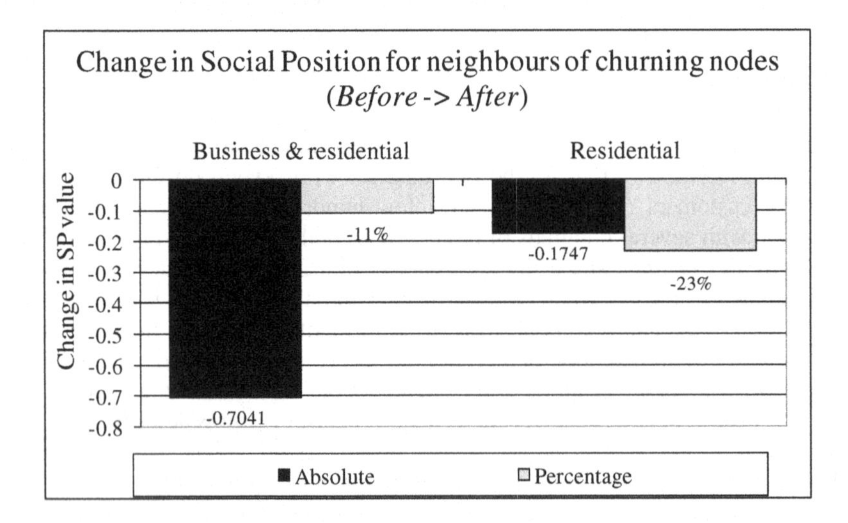

Fig. 6. Average change of neighbourhood social position value for churning members compared to regular members; the third time slot (*After*) compared to the first one (*Before*)

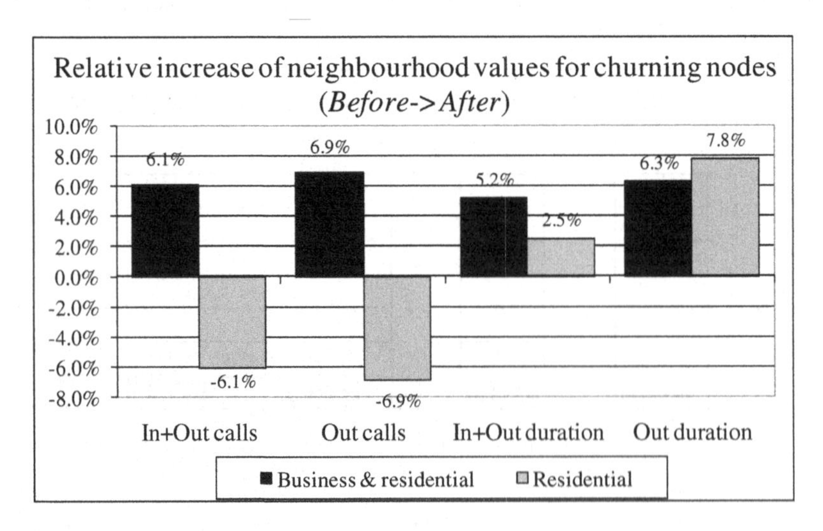

Fig. 7. Average change of neighbourhood value for churning members compared to regular members; the third time slot (*After*) compared to the first one (*Before*)

The average social position of churner's neighbourhood decreased by 23% for the *Residential* network and 11% for the *Business & Residential* network after the churn event compared to regular members as shown in Fig. 6. It shows how big influence the customer churn has on his neighbourhood. Because of customer churn, the network loses both the member and big part of the member's neighbour's activities. In order to present churn influence the trend, describing general change of customer activities between the third and the first period, was removed.

The similar studies were carried out using four other measures described in section 3 and the conclusions are the same.

6 Conclusions and Future Work

People influence one another and this principle can be used to analyse and understand customer behaviour especially in retail companies. This impact can be observed by means of social network analysis and changes in social value of nearest neighbours.

The preliminary experiments presented in the paper revealed that the churning customers influence their neighbourhoods. In particular, social position of the neighbours drops significantly after the churn. Smaller values of social position also point to the churning customers before they churn.

However, to indicate why it happens and to verify the extent of the presented phenomena and to build a dynamic profile of the social value change additional studies on larger data set stretching along longer periods is necessary. Once these observations are validated and formally described the next step could be to try to predict the magnitude of the change in network's activity based on the local properties of the node that triggers the change.

References

[1] Berger-Wolf, T.Y., Saia, J.: A Framework for Analysis of Dynamic Social Networks. In: Proceedings of the 12th ACM SIGKDD international conference on Knowledge discovery and data mining, Philadelphia, PA, USA, August 20-23 (2006)

[2] Bocalettia, S., et al.: Complex networks: Structure and dynamics. Physics Reports 424, 175–308 (2006)

[3] Dasgupta, K., Singh, R., Viswanathan, B., Chakraborty, D., Mukherjea, S., Nanavati, A.A.: Social ties and their relevance to churn in mobile telecom networks. In: Proc. of the 11th International Conference on Extending Database Technology: Advances in Database Technology, EDBI 2008, Nantes, France, March 25-30, pp. 668–677. ACM Press, New York (2008)

[4] Ebel, H., Davidsen, J., Bornholdt, S.: Dynamics of social networks. Complexity 8(2), 24–27 (2002)

[5] Fowler, J.H., Christakis, N.A.: Dynamic spread of happiness in a large social network: longitudinal analysis over 20 years in the Framingham Heart Study. BMJ 337, a2338 (2008)

[6] Furnas, G.: Framing the wireless market. The Future of Wireless, WSA News:Bytes 17(11), 4–6 (2003)

[7] Gopal, R.K., Meher, S.K.: Customer Churn Time Prediction in Mobile Telecommunication Industry Using Ordinal Regression. In: Washio, T., Suzuki, E., Ting, K.M., Inokuchi, A. (eds.) PAKDD 2008. LNCS (LNAI), vol. 5012, pp. 884–889. Springer, Heidelberg (2008)
[8] Juszczyszyn, K., Musiał, A., Musiał, K., Bródka, P.: Molecular Dynamics Modelling of the Temporal Changes in Complex Networks. In: IEEE Congress on Evolutionary Computation, CEC 2009, Trondheim, Norway. IEEE Computer Society Press, Los Alamitos (2009)
[9] Kazienko, P., Musiał, K., Zgrzywa, A.: Evaluation of Node Position Based on Email Communication. Control and Cybernetics 38(1) (in press, 2009)
[10] Kazienko, P., Musiał, K.: Assessment of Personal Importance Based on Social Networks. In: Gelbukh, A., Kuri Morales, Á.F. (eds.) MICAI 2007. LNCS (LNAI), vol. 4827, pp. 529–539. Springer, Heidelberg (2007)
[11] Kazienko, P., Musiał, K.: On Utilising Social Networks to Discover Representatives of Human Communities. International Journal of Intelligent Information and Database Systems, Special Issue on Knowledge Dynamics in Semantic Web and Social Networks 1(3/4), 293–310 (2007)
[12] Kazienko, P.: Expansion of Telecommunication Social Networks. In: Luo, Y. (ed.) CDVE 2007. LNCS, vol. 4674, pp. 404–412. Springer, Heidelberg (2007)
[13] Kempe, D., Kleinberg, J.M., Tardos, E.: Maximizing the spread of influence through a social network. In: The Ninth ACM SIGKDD International Conference on Knowledge Discovery and Data Mining, KDD 2003, Washington, DC, USA, August 24 - 27, pp. 137–146. ACM Press, New York (2003)
[14] Leskovec, J., Adamic, L.A., Huberman, B.A.: The dynamics of viral marketing. ACM Transactions on the Web 1(1) (2007)
[15] Leskovec, J., Backstrom, L., Kumar, R., Tomkins, A.: Microscopic evolution of social networks. In: Proc. of the 14th ACM SIGKDD International Conference on Knowledge Discovery and Data Mining, Las Vegas, Nevada, USA, August 24-27, pp. 462–470. ACM Press, New York (2008)
[16] Leskovec, J., Singh, A., Kleinberg, J.M.: Patterns of Influence in a Recommendation Network. In: Ng, W.-K., Kitsuregawa, M., Li, J., Chang, K. (eds.) PAKDD 2006. LNCS (LNAI), vol. 3918, pp. 380–389. Springer, Heidelberg (2006)
[17] Musiał, K., Kazienko, P., Kajdanowicz, T.: Social Recommendations within the Multimedia Sharing Systems. In: Lytras, M.D., Carroll, J.M., Damiani, E., Tennyson, R.D. (eds.) WSKS 2008. LNCS (LNAI), vol. 5288, pp. 364–372. Springer, Heidelberg (2008)
[18] Ruta, D., Adl, C., Nauck, D.: Data Mining Strategies for Churn Prediction in Telecom Industry. In: Wang, H.-F. (ed.) Intelligent Data analysis: Developing New Methodologies Through Pattern Discovery and Recovery, pp. 218–235. IGI Global, New York (2008)
[19] Schweitzer, F.: Brownian Agents and Active Particles – Collective Dynamics in the Natural and Social Sciences. Springer Series in Synergetics, New York (2007)
[20] Yan, L., Miller, D.J., Mozer, M.C., Wolniewicz, R.: Improving prediction of customer behaviour in non-stationary environments. In: Proc. of International Joint Conference on Neural Networks, IJCNN 2001, vol. 3, pp. 2258–2263 (2001)

The Impact of Positive Electronic Word-of-Mouth on Consumer Online Purchasing Decision

Christy M.K. Cheung[1], Matthew K.O. Lee[2], and Dimple R. Thadani[3]

[1] Department of Finance and Decision Sciences, Hong Kong Baptist University
ccheung@hkbu.edu.hk
[2] Department of Information Systems, City University of Kong Kong
ismatlee@cityu.edu.hk
[3] Department of Information Systems, City University of Kong Kong
dimplet@student.cityu.edu.hk

Abstract. Despite the extensive use of online reputational mechanism such as products reviews forum to promote trust and purchase decisions, there has been little empirical evidence to support the notion that positive word-of-mouth (eWOM) plays a role in impacting trust and purchase intentions. Using the belief-attitude-intention framework as a foundation, we suggest that positive eWOM reinforces consumers' original belief and attitude towards vendors in the aspect of trust. Through a laboratory experiment, we investigate the moderating effect of positive eWOM on the relationships among consumers' belief (i.e. cognitive trust - competence & integrity), attitude (i.e. emotional trust), and behavioral intention to shop online. Results show that positive eWOM strengthens the relationship between consumers' emotional trust and their intention to shop online, as well as the relationship between consumers' perceived integrity and attitude. Implications for the current investigation and future research directions are provided.

Keywords: Electronic word-of-mouth, online consumer behavior, trust, online shopping, electronic commerce, e-marketing, virtual community.

1 Introduction

With the emergence of Web 2.0 technologies which place an emphasis on online collaboration and sharing among users, consumer-generated product reviews have proliferated online and had a profound impact on electronic commerce [7, 15]. It is reported that 85 percent of the world's online population has used internet to make a purchase [1] and 77 percent of online shoppers in US reported using consumers-generated reviews and rating to aid their purchase decisions [14a], [14b]. Electronic word-of-mouth (eWOM) communication has become a dominating channel that influences consumers buying decisions online [13].

With the explosion of consumer generated media over the past few years, information on products including consumption-related advices, regardless of positive or negative, are made highly accessible. Online discussion forum, electronic bulletin board systems, and newsgroup are considered to be important sources of information

M.D. Lytras et al. (Eds.): WSKS 2009, LNAI 5736, pp. 501–510, 2009.

influence that facilitate information exchange among consumers [2]. These media have generated spheres of influence that encompass millions of consumers [17].

Although consumer purchase decisions are influenced by both positive and negative information about products they obtain from fellow consumers, positive messages are more likely to encourage a purchase decision. Positive eWOM communication has been recognized as a valuable vehicle of promoting products and services. A recent survey conducted by AC Neilson [1] found that user-generated reviews, especially recommendations, play an enormous role in selecting websites on which to shop. Past research on e-commerce has mainly focused on building trust in the online environment by alleviate risks associated with shopping online. There exists little research on the role of positive eWOM in impacting trust as well as consumer purchase decisions. Driven by this notion, the purpose of this paper is to examine how positive online consumer reviews affect consumer emotional trust towards the retailer, as well as online purchasing decision.

2 Theoretical Background and Hypotheses Development

2.1 Online Consumer Behavior and Belief-Attitude-Intention Framework

Consumers' lack of trust constitutes a major psychological barrier to the adoption of electronic commerce. Prior studies [5], [9], [20], [21] have demonstrated, with empirical evidence, the importance of trust in online purchasing. Komiak and Benbasat [16] further built on belief-attitude-intention framework and proposed a trust model of electronic commerce adoption.

The belief-attitude-intention framework [6] that relates belief, attitude, and behavioral intention has been widely used in the study of online shopping adoption [4], [7]. This framework suggests that the attitude toward a particular object depends on the direct effects of beliefs about the object, while attitude has a direct positive impact on behavioral intention toward the object. Komiak and Benbasat [16] distinguished two types of trust, namely cognitive trust and emotional trust. Cognitive trust basically comprises of the beliefs of online shopping, while emotional trust reflects the trusting attitude. This framework is adopted in the current study of consumer online purchasing behavior. As shown in Figure 1, consumer trusting beliefs (perceived competence and perceived integrity) determine their attitudes (emotional trust) toward online shopping, and the emotional trust formed, in turn affects consumer intention to shop online.

In addition to these basic variables, Monsuwe et al. [19] suggested that there exist exogenous factors moderating the relationships between the core constructs in the belief-attitude-intention framework of online shopping, such as consumer traits, situational factors, product characteristics, and previous online shopping experiences. As discussed before, in most circumstances, prospective online consumers usually get information regarding online shopping and the product they are interested to buy online before they take the action. Electronic word-of-mouth (eWOM) in the form of online consumer review is believed to play an important role in determining consumer purchasing decision.

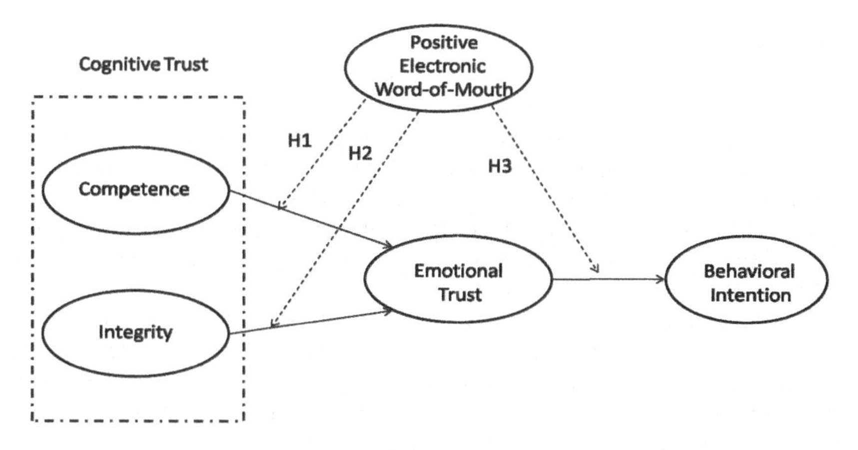

Fig. 1. Research Model and Hypotheses

2.2 Electronic Word-of-Mouth

Given the connectivity nature of the Internet, consumers can easily interact and exchange shopping experiences with other consumers using online discussion forums or any other social network technologies. Online consumer review represents a new form of electronic word-of-mouth. Similar to traditional word-of-mouth communication, eWOM refers to any positive or negative statement made by potential, actual, and former customers about a product or a company [12].

eWOM is especially important to online purchasing decision because of the amount of perceived risk involved. Many potential Internet shoppers tend to wait and observe the experiences of others. Previous research suggested that information from external sources (such as online consumer reviews) can enhance consumer's confidence in their beliefs or attitudes toward some object through internalization process. For instance, Spreng and Page [22] suggested that the more confidence an individual is in his or her belief, the more likely it is the belief will influence attitude formation as well as later behavior toward the object.

In the current study, it is believed that if a potential online shopper finds the online consumer review supports what he/she has already believes about online shopping in a particular online vendor, his/her confidence about the beliefs would be enhanced and exhibited a stronger impact on his/her attitude, as well as later behavior. Therefore, the following hypotheses are postulated:

Hypothesis 1: *Positive eWOM would strengthen an existing positive relationship between perceived competence and emotional trust toward an online vendor.*

Hypothesis 2: *Positive eWOM would strengthen an existing positive relationship between perceived integrity and emotional trust toward an online vendor.*

Hypothesis 3: *Positive eWOM would strengthen an existing positive relationship between emotional trust toward an online vendor and intention to shop online.*

3 Methodology

A controlled laboratory experiment was conducted to test the above hypotheses. The laboratory situation was devised in an effort to emulate an actual purchase situation. To control for the confounding effects of brand features and other marketing mix features that are difficult to capture in an experimental study, we confine ourselves to exploring the effects of positive online consumer reviews on consumer online purchasing decision given that only one particular product (watch) was to be bought.

One hundred university students, evenly divided between male and female, participated in the study on a voluntary basis. An invitation was sent to students via email broadcasting, posters, and flyers inside the campus in a local university in Hong Kong. To avoid potential biases in evaluations, only those students who have not visited the experimental website were invited to participate in the study. Upon successful completion of the experiment, each participant would receive a monetary compensation (US$7) for their time spent.

3.1 Design and Procedure

There were one experimental condition and one control condition in the experiment. Two sessions (one control group and one treatment group) were held at one time and each session held in a computer laboratory had around 20 participants. The session lasted for around 45 minutes. Each participant was randomly assigned to participate in an experimental session and each of them was randomly assigned to a computer set in the computer laboratory in that session. After a ten-minute introduction of the task by the experiment administrator, the participants were requested to decide whether to make online purchases in a real UK watch selling website (www.easywatch.com) under a hypothetical scenario: *"You friend is studying overseas and her birthday is coming. You are planning to use US$40 to buy her a watch as a birthday present."*

For the control group, participants were asked to view the watch website for about 15 minutes and decide whether they would make online purchases via that website. The participants then completed an online questionnaire containing measures of research variables and demographic information. For the treatment group, apart from browsing the watch website, participants had to login and browse through the online consumer discussion forum for 10 minutes before deciding whether they would make online purchases. To ensure that participants browsed the watch website before the online consumer discussion forum, participants were required to click a "confirmation" button after they finished browsing the watch website, and we would then directed them to the online discussion forum. Similarly, they were required to click a "confirmation" button before we directed them to the online questionnaire.

The online discussion forum and the messages were created by three research assistants. Each message have been checked and amended by the group to ensure that the messages disseminate positive online purchasing experience in the tone of perspective users of the website and are realistic and trustworthy. The same administrator conducted all sessions in the study to ensure consistency in the instructions given to the participants. No communication was allowed between subjects during the experiment. Subjects were arranged to sit in every other seat so as to increase the difficulty for them to view the computer screens of the subjects sitting next to them. Log files were checked after the experiment to ensure that participants did not browse through other websites. We also audited the responses with respect to the time they spent on the experimental websites, the online discussion forum (for the treatment group), and the completion and subsequent online questionnaire.

3.2 Measures

The constructs in the research model were measured by using multiple-item scales adopted from previous studies with minor modifications to ensure contextual consistency. The scale items used seven-point Likert scale. Table 1 lists the measures of all the constructs and their sources.

Table 1. Measures

Cognitive Trust in Competence (COM) [18]	
COM1	Easywatch.com is competent and effective in offering high-quality watch.
COM2	Easywatch.com performs its role of offering high-quality watch very well.
COM3	Overall, Easywatch.com is a capable and proficient online watch store.
COM4	In general, Easywatch.com is very knowledgeable about the watch.
Cognitive Trust in Integrity (INTEG) [18]	
INTEG1	Easywatch.com is truthful in its dealings with me.
INTEG2	I would characterize Easywatch.com as honest.
INTEG3	Easywatch.com would keep its commitments.
INTEG4	Easywatch.com is sincere and genuine.
Emotional Trust (ET) [16]	
ET1	I feel comfortable about relying on Easywatch.com for my shopping decision.
ET2	I feel content about relying on Easywatch.com for my shopping decision.
Behavioral Intention (BI) [10], [23]	
BI1	I am very likely to buy watch from the Easywatch.com.
BI2	I intend to use the Easywatch.com to buy watch.
BI3	I intend to use the Easywatch.com frequently to buy watch.
BI4	It is likely that I am going to buy from the Easywatch.com.

4 Analysis and Results

PLS-Graph (Partial Least Squares) version 3.0 [3] was chosen to perform the analysis in this study. This technique allows the estimation of multiple and interrelated dependence relationships, has the ability to represent unobserved concepts in these relationships, and accounts for measurement errors in the estimation process [11]. Before conducting the hypotheses testing, the manipulation checks were first performed.

4.1 Manipulation Checks

The participants were asked to indicate the extent to which they agreed with the several statements about online consumer reviews. For instance, we would expect the treatment groups who were exposed to the eWOM were more likely to agree that the online consumer discussion forum displays online consumer reviews about the watch websites. Our manipulation checks suggested that the experimental manipulation between the treatment group and control group was successful.

4.2 PLS Analysis – Measurement Models

Both the convergent validity and discriminant validity of our measures were examined. Convergent validity indicates the extent to which the measuring items of a scale that are theoretically related should be related in reality. A composite reliability (CR) of 0.70 or above and an average variance extracted (AVE) of 0.50 or above are the recommended level of convergent validity [8]. Table 2 summarizes the item loadings, composite reliability, and average variance extracted of the measuring items for the control group and treatment group. All items have significant path loadings at the 0.01 level and they all fulfill the recommended levels of the composite reliability and average variance extracted.

Table 2. Convergent Validity of the Measures

Competence (COM)		**Control Group**	**Positive Group**
Factor loading	COM1	0.80	0.87
	COM2	0.87	0.92
	COM3	0.87	0.81
	COM4	0.59	0.79
Composite Reliability		0.87	0.91
Average Variance Extracted		0.63	0.72
Integrity (INTEG)			
Factor loading	INTEG1	0.87	0.77
	INTEG2	0.91	0.86
	INTEG3	0.86	0.74
	INTEG4	0.77	0.86
Composite Reliability		0.91	0.88
Average Variance Extracted		0.73	0.65
Emotional Trust (ET)			
Factor loading	ET1	0.89	0.88
	ET2	0.90	0.89
Composite Reliability		0.89	0.88
Average Variance Extracted		0.80	0.79
Behavioral Intention (BI)			
Factor loading	BI1	0.91	0.94
	BI2	0.93	0.95
	BI3	0.91	0.86
	BI4	0.80	0.91
Composite Reliability		0.94	0.95
Average Variance Extracted		0.79	0.84

Discriminant validity involves checking whether the measuring items measure the construct in question or other related constructs. Discriminant validity is examined with the squared root of the average variance extracted for each construct higher than the correlations between it and all other constructs [8]. As shown in Table 3, each construct shares greater variance with its own block of measures than with the other constructs representing a different block of measures. Overall, the results provide strong empirical support for the convergent validity and discriminant validity of the measures of our research model.

Table 3. Discriminant Validity of the Measures

Control Group	**COM**	**INTEG**	**ET**	**BI**
Competence (COM)	0.79			
Integrity (INTEG)	0.47	0.85		
Emotional Trust (ET)	0.62	0.47	0.89	
Behavioral Intention (BI)	0.61	0.38	0.60	0.89
Positive Treatment Group	**COM**	**INTEG**	**ET**	**BI**
Competence (COM)	0.85			
Integrity (INTEG)	0.68	0.81		
Emotional Trust (ET)	0.77	0.70	0.89	
Behavioral Intention (BI)	0.76	0.65	0.71	0.92

4.3 PLS Analysis – Structural Models

Figures 2 and 3 present the results of our study with the overall explanatory power, the estimated path coefficients, and the associated t-value of the paths for the control group and treatment group respectively. Tests of significance of all paths were performed using the bootstrap resampling procedure.

Figure 2 shows the structural model of the control group. The structural model explains 37% of the variance. Emotional trust (β= 0.60, t=6.46) has a significant effect

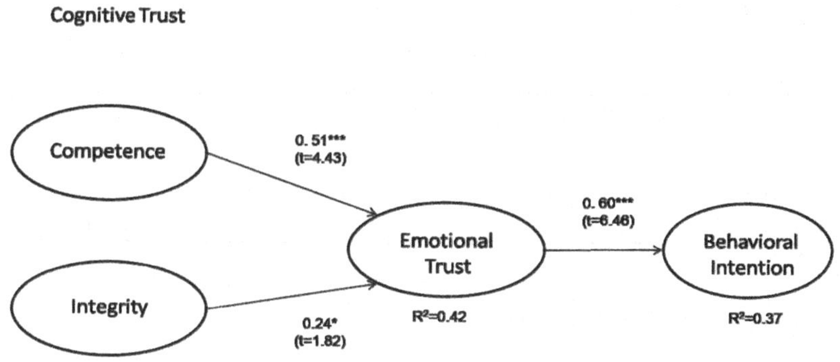

Fig. 2. PLS result of the Control Group

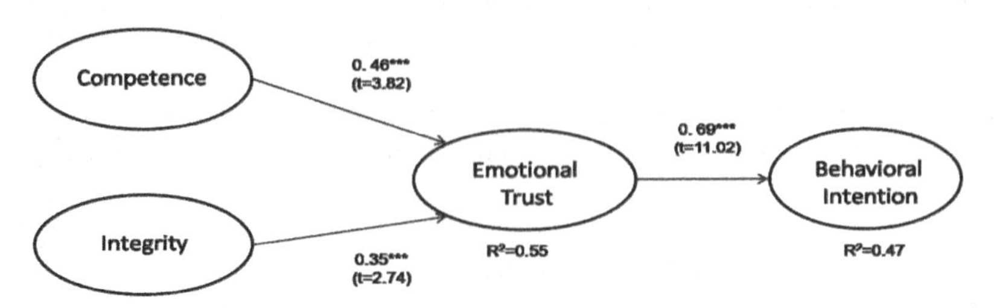

Fig. 3. PLS result of the Positive Treatment Group

on behavioral intention to purchase online. Both trusting beliefs, perceived competence (β= 0.51, t=4.43) and perceived integrity (β= 0.24, t=1.82), exhibit significant impact on emotional trust to shop online.

Figure 3 shows the structural model of the positive treatment group. The structural model explains 47% of the variance. Emotional trust (β= 0.69, t=11.02) has a significant effect on behavioral intention to purchase online. Both trusting beliefs, perceived competence (β= 0.46, t=3.82) and perceived integrity (β= 0.35, t=2.74), exhibit significant impact on emotional trust to shop online.

Hypotheses on the impact of positive treatment group can be tested by statistically comparing corresponding path coefficients between the two structural models. The statistical comparison was carried out using the procedure as stated in Appendix A. Table 4 summarizes the comparisons. Results show that positive electronic word-of-mouth (eWOM) significantly enhances the relationship between respondents' emotional trust toward online shopping and their intention to shop online. The path coefficient of emotional trust to behavioral intention of the positive treatment group is significantly stronger than the corresponding path of the control group. The path coefficient of perceived integrity to emotional trust of the positive treatment group is also significantly stronger than the corresponding path of the control group. However, it is interesting to find that the path coefficient of perceived competence to emotional trust of the positive treatment group becomes slightly weaker than the corresponding path of the control group.

Table 4. Path Comparisons between the Control Group and the Treatment Group

Path	Control Group	Positive Treatment Group	Conclusion
ET-> BI	0.60	0.69	t-statistics = 76.92
COM-> ET	0.51	0.46	t-statistics = -18.87
INTEG-> ET	0.24	0.35	t-statistics = 32.54

5 Discussion and Conclusion

This study aims at exploring the moderating effect of positive electronic word-of-mouth on the relationships among consumers' beliefs, attitudes, and behavioral intention to shop online. Our findings show that positive eWOM strengthens the relationship between consumers' emotional trust and their intention to shop online, as well as the relationship between consumers' perceived integrity and attitude. Between the two trusting beliefs, perceived integrity of an online vendor is more difficult to judge than its perceived competence, especially for those who do not have any prior experience with the online vendor. For instance, online shoppers can judge the competence of an online vendor based on its professional website design, however, they cannot easily judge the credibility of the online vendor. Therefore, online consumer reviews (positive eWOM) should exhibit a more significant impact on the relationship between perceived credibility and emotional trust than that between perceived competence and emotional trust.

The main contribution of this study is that while past research on online shopping have focused largely on the relationships between beliefs, attitudes, and behavioral intentions, this study goes further and investigates how an exogenous variable, positive electronic word-of-mouth, could affect consumer adoption of online shopping. eWOM is postulated as a moderator in the belief-attitude-intention framework. By taking a contingency approach, there is a significant increase in the amount of variance explained for the positive treatment group.

While this study raises interesting implications for researchers, it is also relevant for practitioners, especially for marketers. Web 2.0 applications encourage users and consumers to create and share opinions with others in online consumer opinion platforms. As this activity continues to expand, it is becoming increasingly necessary for marketers to understand and harness this phenomenon in order to remain in touch with their consumers. This study showed that positive eWOM in the form of online consumer reviews significantly affects their trusting beliefs, emotional trust, and online purchasing decision. Marketers should adopt procedures to habitually monitor and encourage consumers' opinions.

To conclude, this study has raised many interesting implications for eWOM and many additional avenues for research. This study is expected to trigger additional theorizing and empirical investigation aimed at a better understanding of eWOM and online purchasing decision.

Acknowledgment

The work described in this article was partially supported by a grant from the Research Grant Council of the Hong Kong Special Administrative Region, China (Project No. CityU 145907).

References

1. Nielsen, A.C.: Trends in Online Shopping a Global Nielsen Consumer Report (February 2008)
2. Bickart, B., Schindler, R.M.: Internet Forums as Influential Sources if Consumer Information. Journal of Interactive Marketing 15(3), 31–40 (2001)
3. Chin, W.W.: PLS Graph Manual (1994)
4. Cheung, C.M.K., Chan, G.W.W., Limayem, M.: A Critical Review of Online Consumer Behavior: Empirical Research. Journal of Electronic Commerce in Organizations 3(4), 1–19 (2005)

5. Cheung, C.M.K., Lee, M.K.O.: Understanding consumer trust in Internet shopping: A multidisciplinary approach. Journal of the American Society for Information Science and Technology 57(4), 479–492 (2006)
6. Fishbein, M., Ajzen, I.: Belief, Attitude, Intention, and Behavior: An Introduction to Theory and Research. Addison-Wesley, Reading (1975)
7. Forman, C., Ghost, A., Wiesenfeld, B.: Examining the Relationship Between Reviews and Sales: The Role of Reviewer Identity Disclosure in Electronic Markets. Information Systems Research 19, 291–313 (2008)
8. Fornell, C., Larcker, D.F.: Evaluating structural equation models with unobservable variables and measurement error. Journal of Marketing Research 18(1), 39–50 (1981)
9. Gefen, D., Karahanna, E., Straub, D.W.: Trust and TAM in online shopping: An integrated model. MIS Quarterly 27(1), 51–90 (2003)
10. Gefen, D., Straub, D.W.: The relative importance of perceived ease-of-use in IS adoption: A study of e-commerce adoption. Journal of the Association for Information Systems 1(8), 1–30 (2000)
11. Hair, J.F., Black, W.C., Babin, B.J., Anderson, R.E., Tatham, R.L.: Multivariate Data Analysis. Prentice-Hall, Englewood Cliffs (2006)
12. Hennig-Thurau, T., Gwinner, K.P., Walsh, G., Gremler, D.D.: Electronic word-of-mouth via consumer opinion platforms: What motivates consumers to articulate themselves on the Internet. Journal of Interactive Marketing 18(1), 38–52 (2004)
13. Hu, N., Liu, L., Zhang, J.: Do online reviews affect product sales? The role of reviewer characteristics and temporal effects. Info. Tech. Management. 9, 201–214 (2008)
14. Jupiter Research (2006a), http://www.ratepoint.com/resources/industrystats.html; Jupiter Research, Reluctant Shoppers Increase Due Diligence (2008b)
15. Korfiatis, N., Rodriguez, D., Sicilia, M.-A.: The Impact of Readability on the Usefulness of Online Product Reviews: A Case Study on an Online Bookstore. In: Lytras, M.D., Carroll, J.M., Damiani, E., Tennyson, R.D. (eds.) WSKS 2008. LNCS (LNAI), vol. 5288, pp. 423–432. Springer, Heidelberg (2008)
16. Komiak, S.Y.X., Benbasat, I.: The Effects of Personalization and Familiarity on Trust and Adoption of Recommendation Agents. MIS Quarterly 30(4), 941–960 (2006)
17. Musial, K., Kazenko, P., Kajdanowicz, T.: Social recommendations within the multimedia sharing systems. In: Lytras, M.D., Carroll, J.M., Damiani, E., Tennyson, R.D. (eds.) WSKS 2008. LNCS (LNAI), vol. 5288, pp. 364–372. Springer, Heidelberg (2008)
18. McKnight, H.D., Choudhury, V., Kacmar, C.: Developing and validating trust measures for e-commerce: An integrative typology. Information Systems Research 13(3), 334–359 (2002)
19. Monsuwe, T.P., Dellaert, B.G.C., Ruyter, K.d.: What drives consumers to shop online? A literature review. International Journal of Service Industry Management 15(1), 102–121 (2004)
20. Pavlou, P.A.: Consumer Acceptance of Electronic Commerce - Integrating Trust and Risk, with the Technology Acceptance Model. International Journal of Electronic Commerce 7(3), 101–134 (2003)
21. Pires, G., Stanton, J., Eckford, A.: Influences on the Perceived Risk of Purchasing Online. Journal of Consumer Behaviour 4(2), 118–131 (2004)
22. Spreng, R.A., Page, T.J.: The Impact of Confidence in Expectations on Consumer Satisfaction. Psychology & Marketing 18(11), 1187–1204 (2001)
23. Taylor, S., Todd, P.A.: Understanding information technology usage: A test of competing models. Information Systems Research 6(2), 144–176 (1995)

Information Technology Leadership in Swedish Leading Multinational Corporations

Lazar Rusu, Mohamed El Mekawy, and Georg Hodosi

Stockholm University/Royal Institute of Technology, Stockholm, Sweden
lrusu@dsv.su.se, moel@dsv.su.se, hodosi@dsv.su.se

Abstract. This paper presents a comparative study of the Chief Information Officer (CIO) role, responsibilities, profile and the IT impact in five leading multinational corporations (MNCs) in Sweden. The first part of the paper comprehends a research review regarding the CIO role, responsibilities and barriers that the CIO is facing today in MNCs with references to the European business environment. After that in the second part of the paper a comparative analysis is provided regarding the CIO role, responsibilities and CIO profile using the Sojer et al. model [25] in the Swedish leading MNCs. Moreover the paper is providing in this part an analysis upon if the CIO is adding value by using Earl and Fenny profile [7]. In the last part of the paper an analysis is done concerning the IT impact in the Swedish MNCs by using McFarlan strategic grid framework [20] together with IT strategies used by the CIOs to support the business operations and strategies. The results of this analysis provide a detail overview that can be usefully for Swedish IT executives or top managers about the development of the CIO role, responsibilities and profile.

Keywords: Chief Information Officer (CIO), Swedish Multinational Corporations, CIO role and responsibilities, CIO profile, IT impact.

1 Introduction

Today's highly competitive global market requires sophisticated IT strategies and resources management for multinational corporations (MNCs) to be able to sustain their competitive advantage in their businesses. Moreover IT has proofed to be a driver or an enabler for changing businesses of almost all types of organisations. Therefore, it is very important for an organization to manage the IT systems that span its entire business to increase productivity and to establish and maintain a global business. This reveals the importance of a Chief Information Officer (CIO) in an organization since he/she is the one who should plan, develop, implement and control the local versus global IT strategy and policies. This allows an organization to leverage economic power, realize huge economies of scale, and gain a global view of customers and operations which in turn makes the organization capable of operating effectively and efficiently. Moreover, a CIO needs to think about a renewable organisation of IT structure, architecture and strategies for his organisation considering increasingly global factors such as cultural and economic spatial powers. A very important

M.D. Lytras et al. (Eds.): WSKS 2009, LNAI 5736, pp. 511–522, 2009.

aspect for a CIO or an IT manager effectiveness is according to Brown et al. [2] related on "their abilities to work closely with key business managers in their organizations". The importance of managing the IT-business relationship asset is as important as the management of the others assets: human and technology assets that are forming the organization's IT resources [2].

2 Research Methodology

Our research mainly concerns the analysis of CIO's role and responsibilities for evaluating the CIO profile and if the CIO's is adding value in leading Swedish MNCs. Apart from these IT leadership aspects we have analysed the impact of IT in these MNCs and the IT strategies used by the CIOs to support the MNC's business operations and business strategies. There are at least two reasons that can motivate our choice for adopting such an empirical study in our research. Firstly, most of the research and work done till now in this area has not investigated in a comparative approach the CIO role and responsibilities in these leading Swedish MNCs. Secondly, there is a need of identifying the Swedish CIOs profile that can be later on used in on-job training for future CIOs' development.

For collecting the data semi-structured interviews (that has predetermined questions, but the order can be modified based upon the interviewer's perception of what seems most appropriate) were performed with the CIOs from five Swedish leading multinational corporations. The interviews were transcribed and used later on for analysis. The data obtained from these interviews has been collected in the last three consecutives years 2006, 2007 and 2008. Apart from the interviews other reliable sources like internal reports and published case studies has been used in or research. The main research question we have addressed in our paper is: What is the CIO role, responsibilities and profile and his contribution on the IT impact on business operations and business strategy in a Swedish Leading MNC? The study is an empirical one and has as limitation the fact that is reflecting the situation from the Swedish business environment. The research methodology is a case study one that is the most appropriate for studying a contemporary set of events, over which the investigators has little or no control [27].

3 Research Background

3.1 Role and Responsibilities of CIO

The Chief Information Officer (CIO) as a position in organizations has emerged in the 80's of the past century. Since that date, it has gained acceptance in practice and attracted much academic interest. As the impact of information technology (IT) on business has increased, it has become important in today's organizations that the CIO should delivers effectively and rapidly on the premise IT makes to business. Hence as the business functions became heavily dependent on IT at strategic and planning levels, most of the issues that primary had impacts on CIOs' roles and responsibilities are factors within the organisation [6][19][22][23]. Only by the end of the 1990's, researchers started to examine the external factors that affect the strategic role of CIOs in managing the firm's information systems [3][8].

In the 2000's, especially the last seven years, the CIO has become a key controller of any organisation. Like others, European market has witnessed a number of business developments that have led to a tremendous change in CIO's role and responsibilities [3][4][26]:

- Management of IT resources has become a necessity for business executives.
- Emerging of a new executive role of "Chief Technology Officer" that has increased the importance of business-IT alignment from a view of architecture and infrastructure.
- New regulations, such as the Sarbanes-Oxley and Health Insurance Portability and Accountability Act of 1996 (HIPPA) have caused different changes in the governance and IT resources activities.
- An extreme increase in outsourcing activities.
- Organisations today prefer to buy IT module packages and applications rather than to build them in-house.

Theses changes have led to that the CIO has come to be responsible for adjusting the IT infrastructure and capabilities into the business strategies and operations. Additional to that, different organisations have contributed to the package of CIOs' roles and responsibilities based on their experiences.

Many researchers have focused to define the CIO role among them Gottschalk [9] that has proposes "a model of leadership roles" with 6 different CIO roles based on study of IT leadership in Norway. In his opinion for e.g. when the CIO works with data processing then it has two roles: for his own organization he is technology provocateur and for the rest of organization he is acting as a coach. When the CIO is working with information systems than he is product developer within its organisation and a change leader for the rest of organization and finally when the CIO works with networks then he is a chief architect for its own organization and chief operating strategist for the rest of organization. Other researchers have investigated some of these newly added responsibilities in organisations. Among the responsibilities added to the CIO role we have: budget control [16][24], IT architecture [13], security [24], supply chain management [15] and governance [10]. In the last years researchers like Hoving [11] have stated that the IT leaders need to have many talents to succeed like for example: natural intuition to know which projects are going to pay off, ability to manage a diverse set of internal and external resources and business knowledge to provide business value: with measurable benefits. On the other hand Johnsson & Lederer [12] stated that today "CEO perceives the role of CIO to be more strategically important and also views the impact of the IT projects more favourable then the CIOs". As the position of CIO has followed a natural evolution, his role and responsibilities have followed the same line of development passing through mainframe and distributed towards the current web-based era. From a responsible for '*processing information*' within the company in the 1980s, he has become the main responsible for '*managing information resources*'. One of the reasons for this change is the increase of IT usage within organisations. This has created the need for better holistic view of what IT can do for an organisation [17]. Furthermore, IT usage has been extended outside organisations forming an advanced networked environment of vendors, customers and suppliers. In such environment, CIO has become responsible for more inter-organisational and global view of an organisation. He has to present strategies for how to manage, implement and merge different inter-organisational system. If there are no local CIOs for subsidiaries, a global CIO has to manage local resources for every region including resources for e-business and e-commerce. Additional to that, the CIO has become the main responsible for business-IT alignment to support business executives to develop and determine appropriate strategies for their business [17]. In fact in according to Luftman and Kempaiah [18] business-IT alignment is still the top IT management concern among the IT executives.

3.2 Barriers That CIOs Are Facing Today in Managing IT

In our business today where vast investments with innovations are required, a number of barriers or challenges face the top IT executives of any organisation. These barriers can be divided in two main groups. The first group, which refers to IS/IT management barriers, consists of the general challenges of managing operations and processes of IS/IT. The second group, which refers to environmental barriers, consists of those challenges appear at the global level when making global business. This group concerns more with dynamic aspects in different market places and the inter-organisational environment. Many researchers argue that in the 21st century, as organisations started globalising more challenges have become similar all over the world [26][21][4]. Moreover Palvia et al. [21] have described and analyzed different IS/IT management barriers that have been identified during the last few years in different regions of the world. Based on wide surveys and studies, their research gives an extensive summary of IS/IT management barriers in US companies while other researches like Varajão et al. [26] explained the barriers that CIOs are facing in Portuguese companies. These results represents a small part about how the barriers can be and are varied from one region or country to another as well as ranking in the same region or even in the same country. For example for the last five years, the three top IS/IT management barriers in the US were: IT and business alignment, IT strategic planning, and security and privacy. However, the Canadian list three years earlier has included: building a responsive IT infrastructure, improving IS project management practices and planning, and managing communication networks [21]. Different researchers argue that IS/IT management barriers are tightly connected to a specific country or context where the marketplaces' conditions are similar. However, environmental barriers can be encountered anywhere around the globe. By simply considering the diversity in culture, economy, politics and technology these formulate significant challenges that CIOs should consider in managing IT. These challenges; firstly do not allow a uniform IS or an IT applications around the world; and secondly, create a number of environmental risks [26]. As the American companies, the European companies have witnessed a stretch in the list of barriers from internal obstacles, politics and difficulties in regional coordination to cultural differences [5][4].

4 Case Studies Analysis in Five Swedish Leading Multinational Corporations

4.1 Companies' Profile

For analyzing the CIO role, responsibilities, profile and the IT impact we have selected five Swedish leading multinational corporations. Due to the confidentiality we have agreed on with these companies we can not mentioned their names neither the type of their business. However a short description of the companies' profile is presented below.

Company A is over 100 years old and had at the glance over 100.000 employees and still has about 40% of the world market in one of its brand, and is a global leader in the other areas. The company's strengths are a high Research &Development (R&D) budget and many patens and intellectual properties' on their products.

Company B is newer then A. The main characteristics are: innovative design of products to the end-users, own style, competitive prices and global presence around the world with well over 100.000 employees.

Company C is over 100 years old and has about 40.000 employees. It belongs to the largest European companies in their area with a strong brand. R&D and production are their competitive edge and are today strongly affected by the financial crises.

Company D is more than 100 years old and has today more than 100.000 employees. High tech is their business area and is the global leaders within several areas. The company has high R&D and global presence.

Company E is near 100 years old and has about 50.000 employees and R&D and production are the competitive edge on a global market.

4.2 CIO Role, Responsibilities and Profile

The role and responsibilities of the five CIOs are diverse in the five leading MNCS and are summarized in the table 1. The characteristics of the CIO role are coming out from our interviews with these five IT executives and the information presented in table 1 will be later on used for the evaluation of the CIO profile in the case of our MNCs.

Based upon the information presented in table 1, we have evaluated the role of CIO using Sojer et al. model [25] that is based on McFarlan's strategic grid [20]. In their model Sojer et al. argue that the role of CIO can be defined by two variables: strategic importance of running IT, and the strategic importance of changing IT [25]. Their

Table 1. The Main Role and Responsibilities of the CIOs from five Swedish Leading MNCs

	Role and Responsibilities
Company A	- Making sure about having reliable enterprise systems infrastructure. - Providing efficient and competitive ICT tools. - IT services outsourcing and follow up service level quality. - Maintaining relationships with suppliers and customers. - Support IT for suppliers, internal/external manufacturing and customers via e-business and e-commerce. - Development of integrated IT architecture.
Company B	- Coordinating all IT-related issues. - Development of the long-term IT direction. - Keeping relationships with external strategic partners. - Organisational development. - Procurement and costs. - Build of relations with different stakeholders. - Planning and structuring of infrastructure policies. - Assessment of the global risks.
Company C	- Business-IT alignment is a priority of the business goals. - Building up the IT structure. - Supporting core business and rapid growth of company. - Establishing a leadership environment in the organisation. - Cutting down the cost of non business-driven projects. - Development of future vision for IT. - Supporting the organisation development at global level.
Company D	- IT Leader. - Relationship manager. - Marketer of business and products. - Open IT systems-oriented. - Business-IT alignment at strategic and operational level.
Company E	- To stay close to business policies rather than to the IT policies. - To gave reviews in strategic decisions. - In charge with IT strategies and the plan of implementation. - Designing of IT metrics that meet the business metrics. - Management of resources for acquired companies around the world. - Management of global resources and subsidiaries' CIOs. - Management of IT outsourcing projects.

model consists of 4 different areas based on the two variables mentioned before that are defining the IT usage and the role of CIO (see Figure 1) and are the followings:

1. Supporter: in these companies neither the current nor the future IT has any significant impact on business and the role of CIO is to support current business processes.
2. Enabler: In these companies IT has a high strategic relevance, however the future IT systems are not expected to have the same significant impact on business and therefore the CIO role is to enable current business.

3. Cost Cutter or a Project Manager: in these companies the impact IT is low now but in the future the impact of IT will be higher, and therefore the CIO role "is either a Cost Cutter or a Project Manger who drives the future oriented measures within IT department".
4. Driver: in these companies IT has now and in the future a high strategic importance and the CIO is a driver with a significant impact in the whole organization.

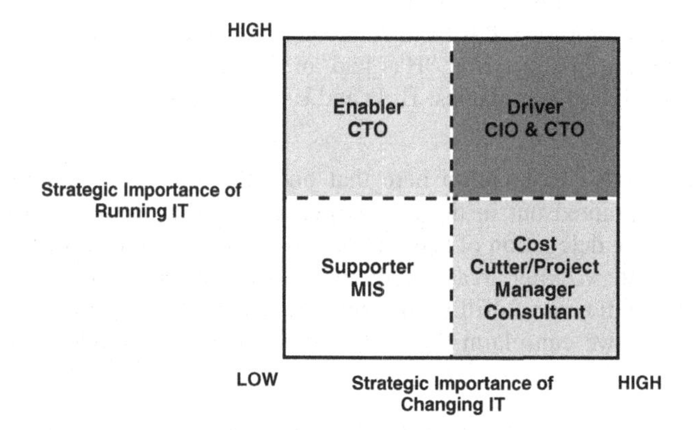

Fig. 1. Role and Title Assignment for IT Executives [25]

The result of the analysis of the CIO regarding the role and title assignment for IT executives [25] is the followings:

1. Supporter: none of the CIOs.
2. Enabler: none of the CIOs.
3. Cost Cutter/Project Manager Consultant: none of the CIOs.
4. Driver CIO & CTO: all 5 CIOs.

The figures above show that our selected CIO's are according to the model of Sojer et al. [25] very homogenous grouped and they "will have significant impact beyond the organizational boundaries of the IT department".

On the other hand as Earl and Feeny [7] stated, it is extremely important that a CIO is able to deliver value. In their research the authors have described several cases where CEO could not show value and therefore have been replaced and also cases where CEO was satisfied with the value that has been added by CIO. Furthermore Earl and Feeny [7] argue that the profile of the CIO who adds value has to fulfill the following criteria:

1. Behavior: "is loyal to the business and is open".
2. Motivation: "is oriented towards goals, ideas, and systems".
3. Competences: "is a consultant/facilitator, is a good communicator and has IT knowledge".
4. Experience: "has had an IS function analyst role".

In the case of our Swedish MNCs we have analyzed the profile of a CIO who adds and how our studied CIOs fulfill these criteria from Earl and Feeny model [7]. The results of this analysis are described below.

1. Regarding first criterion "Is loyal to the business and is open": we have found that this criterion it is relevant for all 5 CIOs.
2. For the second criterion "Is oriented towards goals, ideas, and systems" we have found that it is relevant for all 5 CIOs.
3. For the third criterion "Is a consultant/facilitator, is a good communicator and has IT knowledge" we have found that regarding consultant/facilitator and good communicator it is relevant for all 5 CIOs. But regarding IT knowledge: the CIO of company E definitively not, in case of the CIOs of company B and D this is partly true, and for the CIOs from A and C is relevant.
4. Concerning the last criterion "Has had an IS function analyst role" definitively not for the CIOs of companies: B, D and E but relevant for CIOs of companies A and C.

In summary we have to mention here that our interpretation of "competences and experiences" as pointed out in the last two criteria is that in the Swedish culture is used frequently the delegation of tasks and team works that enable the CIOs to use the best experts for all working areas. Furthermore, we can see from this case study that two out the five CIOs are fulfilling the entire criteria according to the model of Earl and Feeny [7] so we can claim that only these two CIOs are adding value to their companies.

4.3 The IT Impact in the Swedish Leading MNCs

For studying the impact of IT in or five leading MNCs we have used McFarlan's strategic grid [20] and analyze how the IT projects and IT initiatives impact the business operations and business strategies. But because this framework doesn't explain the use of IT in these MNCs we have firstly analyzed their IT strategies used by the CIOs and the way they are supporting the business operations and business strategies. The results of our analysis are described for each of the companies in table 2.

As it can be seen from the table 3, the IT strategies are supporting both the business operations and the business strategies in the case of the five MNCs analysed here. On the other hand to get information about the impact of IT in these MNCs we have analysed the IT project portfolio in these MNCs. As we knew the IT projects in a company are changing overtime and this will affect the IT impact too. It is also important to mention that in all five MNCs the IT projects are in different phases now and some of them are having a higher impact versus the other IT projects that are having a lower impact on business operations and business strategies. For assessing the impact of an organization's portfolio of IT initiatives and projects we have used McFarlan's strategic grid [20] which is a framework with four quadrants (support, turnaround, factory and strategic) looking on two dimensions: (1) the impact on business operations and (2) the impact of strategy and having two values: low and high. By using McFarlan's strategic grid [20] the executives from a company could assess how "the approaches for organizing and managing IT" are placed on the strategic grid and get an indication of the alignment of IT to the strategic business goals [1].

Table 2. IT Strategies used in the five Swedish Leading MNCs

	IT strategies used
Company A	- Minimize the number of the different databases and adapt to future business needs. - Adaptations of the increasing strategically information flow to the national needs and lower the lead-time for the distribution. - Develop the multicultural collaboration and use it for best performance 24 hours/day. - Efficiently respond to business changes regarding number of employee, new supplier, customers and services.
Company B	- Providing precise and up-to-date information about every manufacturing level. - Relevant customised information systems. - Own developed ERP-like system and other subsystems. - Centralised strategic decisions and development. - Centralised IT and customisations to local environments. - E-commerce applications for suppliers. - E-commerce for customers.
Company C	- In-house IT department. - Service-Oriented IT for cutting the cost and being specific. - Security of daily operations. - Modularisation of IT technologies. - Keeping records of all manufactured and exported units.
Company D	- "Clean-up" strategy for outsourcing IT solutions. - Reducing the number of ERP application for more standardisation. - Outsourcing the whole infrastructure. - Periodical for IT infrastructure. - Security plans for data.
Company E	- Outsourcing IT manufacturing processes to low-cost countries. - Reducing the number of ERP applications. - Own developed manufacturing system. - Centralised IT strategies and architecture but local implementation.

In case of our five Swedish MNCs the results of the IT impact among the two dimensions business operations and business strategy from the four quadrants of McFarlan's strategic grid [20] will be presented below. The results are the followings:

Support: None of the five Swedish MNCs have IT projects or IT initiatives that have little impact on the business strategy or business operations.

Factory: All five Swedish MNCs have their own production; however company A has partly outsourced his production. Moreover Company B has a big manufacturing, which previously has been outside but later on they have in-sourced a big part of this production. Regarding Company C this has a large production in several manufactories while Company D has its own production with relative small series, but complex,

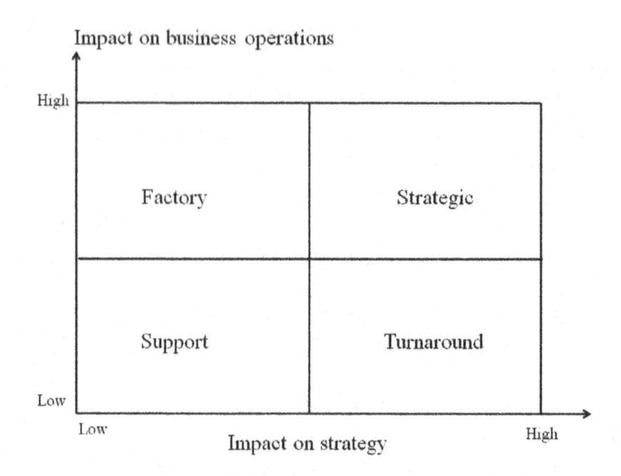

Fig. 2. McFarlan's Strategic Grid [1]

proprietary products that can only be produced in-house. Concerning Company E this has a large production with large series. As we have noticed for all companies a "zero defects" of IT is essential. All companies, except company B has a large Research and Development department with several thousands of employee that would be affected in case of failure of IT systems. Therefore for company B that has a huge logistics around the world an important requirement is for a high IT reliability. Furthermore all of the IT projects in these companies have to be further developed in order to cut the costs and improve their performance.

Turnaround: All five Swedish MNCs exploit the emerging strategic opportunities and innovations and therefore time to market and cost savings are fundamentals for these companies.

Strategic: All five Swedish MNCs are committed to use IT for enabling core business operations and business strategies. The companies A, C, D and E have long term strategic plans with their product development, which takes several years to design. Furthermore the product development is IT intensive in all phases and this requires tide coordination between IT organization and the developers' group. For these companies with such a long strategic planning the IT impact is very high on business strategy and operations. On the other hand for company B the IT impact is lower on business operations than the others but the IT impact on business strategy is higher due to the planning, logistics etc. which is extremely important for this company.

5 Conclusions

The results of the comparative analysis of IT leadership in five Swedish MNCs have brought detailed information about CIO role, responsibilities and profile. Furthermore this analysis has detailed the profile of the five IT executives that adds value to their organizations. The results in this direction has pointed out that in only two of the five Swedish MNCs the CIO is adding value more precisely in case of companies A and C. Concerning the role played by the CIOs we have found that in all five MNCs he/she is a Driver and the strategic importance of running IT is high. On the other hand as King [14] stated "to be effective in strategic business context, the CIOs must be strategic change agents". In our case studies as we have seen none of the CIOs is a Cost Cutter/Project Manager Consultant therefore the strategic important of changing IT is low. Last but not the least we have found that in companies A, C, D and E the impact of IT strategy on business strategy is high due to the fact that these MNCs have long term strategic plans. In conclusion the CIO role, responsibilities and profile analysed in the case of the five leading Swedish MNCs are important factors that have a contribution to the strategic impact of IT in these MNCs therefore the comparative analysis provided in this paper will help the IT executives and top mangers to the development of the CIO profile in this business environment.

References

1. Applegate, L.M., Austin, R.D., McFarlan, F.W.: Corporate Information Strategy and Management –Text and Cases, 7th edn. McGraw-Hill/Irwin (2007)
2. Brown, C.V., DeHayes, D.W., Hoffer, J.A., Martin, E.W., Perkins, W.C.: Managing Information Technology, 6th edn. Pearson Education Inc., London (2009)
3. Chun, M., Mooney, J.: CIO Roles and Responsibilities: Twenty-Five Years of Evolution and Change. In: Proceedings of AMCIS 2006, Paper 376 (2006), http://aisel.aisnet.org/amcis2006/376 (accessed on March 7, 2009)
4. Daum, M., Haeberle, O., Lischka, I., Krcmar, H.: The Chief Information Officer in Germany - Some Empirical Findings. In: Proceedings of the European Conference on Information Systems, Turku, Finland, June 14-16 (2004)
5. Deresky, H.: International Management. Managing Across Borders and Cultures – Text and Cases, 6th edn. Pearson Education Inc., London (2008)
6. Davenport, T.H.: Saving IT's Soul: Human-Centered Information Management. Harvard Business Review, 119–131 (March-April 1994)
7. Earl, M.J., Feeney, D.F.: Is your CIO adding value. Sloan Management Review 35(3), 11–20 (Spring 1994)
8. Feeny, D.F., Willcocks, L.P.: Core IS capabilities for exploiting information technology. Sloan Management Review 39(3), 9–21 (1998)
9. Gottschalk, P.: Information Systems Leadership Roles: An Empirical Study of Information Technology Managers in Norway. Journal of Global Information Management 8(4), 43–52 (2000)
10. Holmes, A.: The Changing CIO Role: The Dual Demands of Strategy and Execution, CIO Magazine (2006), http://www.cio.com/article/16024/State_of_the_CIO_The_Changing_CIO_Role (accessed on March 14, 2009)
11. Hoving, R.: Information Technology Leadership Challenges – Past, Present and Future. Information Systems Management 24, 147–153 (2007)
12. Johnson, A.M., Lederer, A.L.: The Impact of Communication between CEOs and CIOs on their Shared Views of the Current and Future Role of IT. Information Systems Management 24, 85–90 (2007)
13. Kaarst-Brown, M.L.: Understanding An Organization's View of the CIO: The Role of Assumptions About IT. MIS Quarterly Executive 4(2), 287–301 (2005)
14. King, W.R.: Including the CIO in Top Management. Information Systems Management 25(2), 188–189 (2008)
15. Kohli, R., Devaraj, S.: Realizing the Business Value of Information Technology Investments: An Organizational Process. MIS Quarterly Executive 3(1), 53–68 (2004)

16. Leidner, D.E., Beatty, R.C., Mackay, J.M.: How CIOs Manage IT During Economic Decline: Surviving and Thriving Amid Uncertainty. MIS Quarterly Executive 2(1), 1–14 (2003)
17. Luftman, J.N., Bullen, C.V., Liao, D., Nash, E., Neumann, C.: Managing the Information Technology Resource. Leadership in the Information Age. Pearson Prentice Hall, Upper Saddle River (2004)
18. Luftman, J., Kempaiah, R.M.: The IS Organization of the Future: The IT Talent Challenge. Information Systems Management 24, 129–138 (2007)
19. Mata, F.J., Fuerst, W.L., Barney, J.B.: Information Technology and Sustained Competitive Advantage: A Resource-Based Analysis. MIS Quarterly, 487–506 (December 1995)
20. McFarlan, F.W., McKenney, J.L., Pyburn, P.: The information archipelago – plotting a course. Harvard Business Review, 145–156 (January-February 1983)
21. Palvia, P., Palvia, S.C.J., Harris, A.L. (eds.): Managing Global Information Technology: Strategies and Challenges. Ivy League Pub. (2007)
22. Rockart, J.F., Earl, M.J., Ross, J.W.: Eight Imperatives for the New IT Organization. Sloan Management Review 38(3), 43–55 (1996)
23. Ross, J.W., Feeney, D.F.: The Evolving Role of the CIO, CISR WP. No. 308 (1999), `http://web.mit.edu/cisr/working%20papers/cisrwp308.pdf` (accessed on March 14, 2009)
24. Ross, J.W., Weill, P.: Six IT decisions your IT people shouldn't' make. Harvard Business Review 80(11), 84–91 (2002)
25. Sojer, M., Schläger, C., Locher, C.: The CIO – hype, science and reality. In: Proceedings of the European Conference on Information Systems, Goteborg, Sweden (2006)
26. Varajão, J., Trigo, A., Bulas-Cruz, J., Barroso, J.: Biggest Barriers to Effectiveness in CIO Role in Large Portuguese Companies. In: Lytras, M.D., Carroll, J.M., Damiani, E., Tennyson, R.D. (eds.) WSKS 2008. LNCS (LNAI), vol. 5288, pp. 479–488. Springer, Heidelberg (2008)
27. Yin, R.K.: Case Study Research: Design and Methods, 3rd edn. Sage Publications, Thousand Oaks (2003)

Government Online: An E-Government Platform to Improve Public Administration Operations and Services Delivery to the Citizen

Athanasios Drigas and Leyteris Koukianakis

NCSR 'Demokritos',
Institute of Informatics and Telecommunications,
Net Media Lab,
Agia Paraskevi, 153 10, Athens, Greece
{dr,kouk}@iit.demokritos.gr

Abstract. E-government includes fast and improved citizen service from a quantitative and qualitative point of view, as well as the restructuring and reengineering of organizations and their services, through the increased usage and exploitation of the capabilities and services of ICT's and the Internet. The escalation of the e-government services begins with easy access to governmental information and passes through the e-transactions between citizens and the public organization and reaches the electronic delivery of the requested document. A prerequisite in order to support the aforementioned e-government services "layers" is the development of an electronic system, which supports e-protocol, e-applications/e-petitions and internal organizational function of the public organization. In addressing the above context, this article presents an e-government structure which supports and provides the aforementioned e-government services "layers" in order to provide public information dissemination, accept electronic document submissions, manage them through e-protocol and support the operations through the appropriate electronic structure.

Keywords: e-government, e-protocol, e-transactions, governmental functions, e-tools, e-applications, e-petitions, e-delivery, ICT.

1 Introduction

E-government constitutes an extensive area of knowledge, principles and policies, and thus among others includes the following ideas. Firstly, user centric services are designed from the perspective of the user. This implies taking into account the requirements, priorities and preferences of each type of user. Efficient, high quality public services for all are fundamental for economic growth, more and better jobs and affordable solidarity in Europe.

E-government should now realize its promise of measurably more efficiency, more effectiveness for the users, higher quality services, full accountability, better democratic decisions, and inclusive services for all. Widespread modernization and

M.D. Lytras et al. (Eds.): WSKS 2009, LNAI 5736, pp. 523–532, 2009.

innovation of public administrations must become a reality. The focus is to move from readiness to impact and transformation. The emphasis should shift from online availability to achieving impact and wider user take-up, using more comprehensive benchmarking. Organisational innovation includes the improvement of human resources and skills. These "human" factors are essential to be able to make progress on organisational innovation but – most importantly – they should also be a driving force to turn organisational innovation into a continuous process. E-government refers to the federal government's use of information and communication technologies (such as Wide Area Networks, the Internet, and mobile computing) to exchange information and services with citizens, businesses, and other arms of government. It originates from penetration of ICTs within the governmental domain. E-government transforms the traditional and well known shape of governmental structure, services and operations to a new figure which affects strongly the e-citizen transactions with the governmental services. More and more governments are using information and communication technologies and especially the Internet or web-based applications, to provide services among governmental agencies and citizens, businesses, employees and other nongovernmental organizations [1],[2]. Just as e-learning [3],[16],[17], e-health and e-commerce [4], e-government represents the introduction of a great wave of technological innovation as well as government reinvention. E-government uses the most innovative information and communication technologies, particularly web-based applications, to provide citizens and businesses with access to governmental information and services, to improve the quality of the services and to develop and provide greater opportunities to citizens to participate in democratic institutions and processes [5],[6]. This includes transactions between government and business, government and citizen, government and employee, and among different units and levels of government like justice, taxation, welfare, social security, procurement, intra-governmental services etc [7],[8],[9],[10],[11],[12]. All these require technical policies and specifications for achieving interoperability, security and information systems coherence across the public sector [13],[14],[15].

The above context constitutes a basic body of knowledge for the design and development of e-government applications. On this basis, and towards a modular design of the electronic transactions, we analytically specified, designed, and developed a generic e-government environment that is based on a highly interactive, user-case model (citizen, employee, and administrator) and a flexible-interoperable scheme of assistive communication tools.

2 Structure of the E-Government Environment

The e-government environment consists of three systems: A web portal, the e-protocol system and finally the e-applications/e-petitions system. The last two will be described as one, since the latter may be considered as an extension of the e-protocol system. The governmental organization consists of six departments (planning, havoc compensation, housing, protocol, finances and research). Each department has one director and a large number of employees.

2.1 Web Portal's Environment Tools

The web portal environment includes tools that offer flexibility and adaptability depending on their use. The design of these tools is based on web services, such as discussion forums, chat, message box, e-libraries, which are widespread in the public web community. These tools are distinguished into two groups: "informative" and "communicative". On the one hand, the "informative" tools include services related to the information of governmental functions and their presentation. On the other hand, the "communicative" tools include services that allow the communication of users belonging to the same or different group (session level). The web portal environment enables the management of these tools according to the user groups' permission. More explicitly, the "informative" tools are the following: announcements, frequently asked questions (F.A.Q.) and e-libraries. Respectively, the "communicative" tools are: discussion forums, message boxes, chat and e-requests. Finally, it must be noted that the environment relates the tools according to the specific user level permissions. These levels are analyzed in the sections to follow.

2.2 User Levels

Seven user levels are distinguished (Fig.1) in the web portal environment. Different supporting tools exist in each one of them.

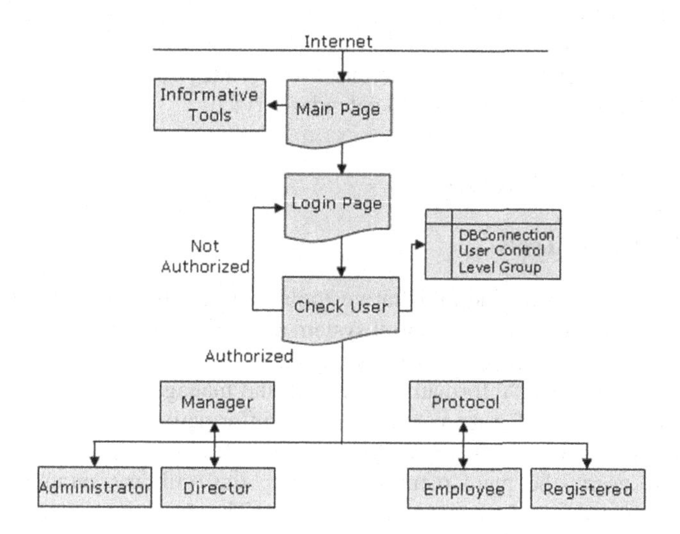

Fig. 1. User Levels

Depending on the corresponding use, these levels have also a different role: Administrator, Manager, Director, Employee, Protocol Administrator (Employee), Registered (Authorized) User and Unauthorized User (Guest). Each of them interacts with the other through the "informative" and "communicative" tools related to each level.

The administrator coordinates and manages the e-government application through the administrative tools. The administrator determines which user level-group has the permission to use the corresponding ‘‘informative’’ and ‘‘communicative’’ tools. Moreover, the administrator can communicate with the other user levels in order to solve issues and has the privilege of updating the system. Finally, the administrator decides about the preparation, design and diffusion of the electronic content to the citizens. Through user friendly and interactive ICT web tools, the administrator authors the governmental content.

The Manager, Director and Employees user levels are described together, as they incorporate many similarities. The manager decides about the preparation, design and diffusion of the electronic content. Moreover, through the communicative tools, the employees cooperate with the directors, the directors with the manager and the manager with the administrator with respect to discussing solutions to problems and to exchange ideas for the better functionality of the system. Finally, these three user levels play an important and diverse role in the e-protocol chain, which will be described later.

The Protocol Administrator (Employee) is responsible for the e-protocol system. Besides the “informative” and “communicative” tools, he/she has the ability to view, change (under conditions) and add applications/petitions to the e-protocol system. The applications/petitions are fully categorized and new categories can be created. The Registered (Authorized) Users have the ability to see and change specific information regarding their account, can view the progress of their applications/petitions and finally, they can make new applications/petitions that are supported by the e-application/e-petition system.

Finally, Unauthorized Users (Guests) can enter and search the data structure as a means of gathering important information. Finally, they may be informed about the news and events through the news and calendar service.

3 User Tools and Services

3.1 Administrative Tools

The environment provides administrative tools that are divided into two groups as follows: Management of the web portal system and management of the e-protocol and e-applications/e-petitions system. The management of the web portal system incorporates management of the “informative” services and management of the “communicative” services (Fig. 2). The management of the informative services is an important issue, as through it the administrator has the flexibility to manage the following ontologies: The users, the main menu description, the e-library, announcements and finally, the frequently asked questions (F.A.Q.). The environment tools enable the administrator to organize the informative content.

Correspondingly, the communicative services group consists of interactive forms through which the administrator manages chat session, the discussion forum and finally, the message box.

Management of the e-protocol and e-applications/e-petitions systems incorporates management of the petitions, their categories and their deadlines. The transactions executed in each group concern retrieval, insertion and update of the corresponding data. All web requests/responses are carried out through interactive and user-friendly forms.

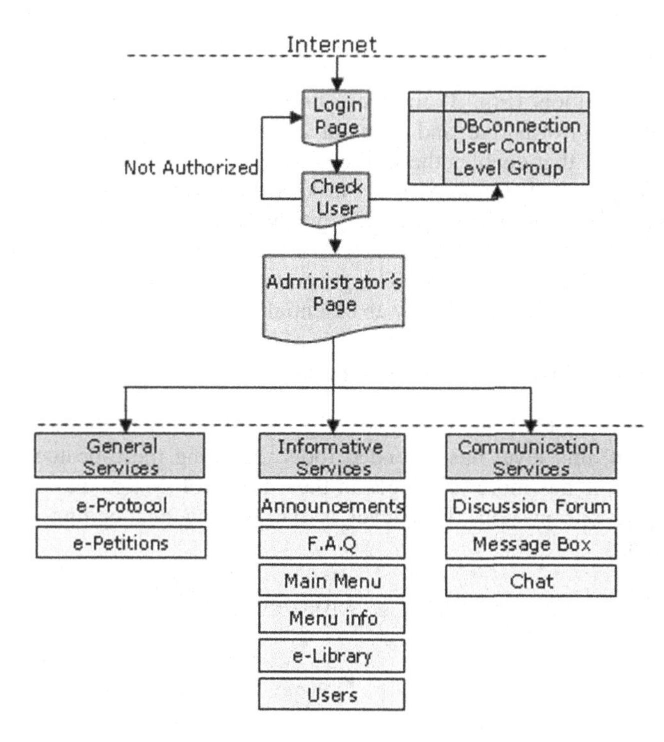

Fig. 2. Administrative Tools

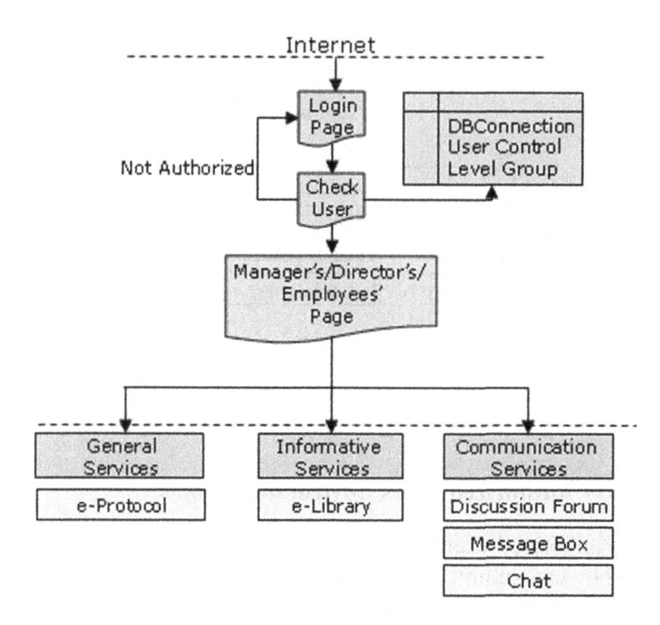

Fig. 3. Manager / Director / Employees Environment Tools

3.2 Manager-Director-Employees Environment Tools

The environment tools (Fig. 3) for these user levels are divided into three groups: Communicative, Informative and General Services. The group of communicative services is the one that enables these three user levels to communicate with the other user levels. The tools that employ these tasks are: the discussion forum, the message box, and chat. The second group of the informative services consists of tools that enable the fast access and management of the electronic content. This content cannot be accessed by unauthorized users. The general services group includes tools that are different for each user level and play an essential role in the e-protocol chain.

3.3 Protocol Employee Environment Tools

The environment tools of this user level (Fig. 4) are similar to the ones mentioned in 3.2. In addition, this level has extended tools regarding the e-protocol system. The protocol employee has more privileges in the e-protocol system and can also interact with the e-petitions system. This level is the starting and ending point in the e-protocol chain.

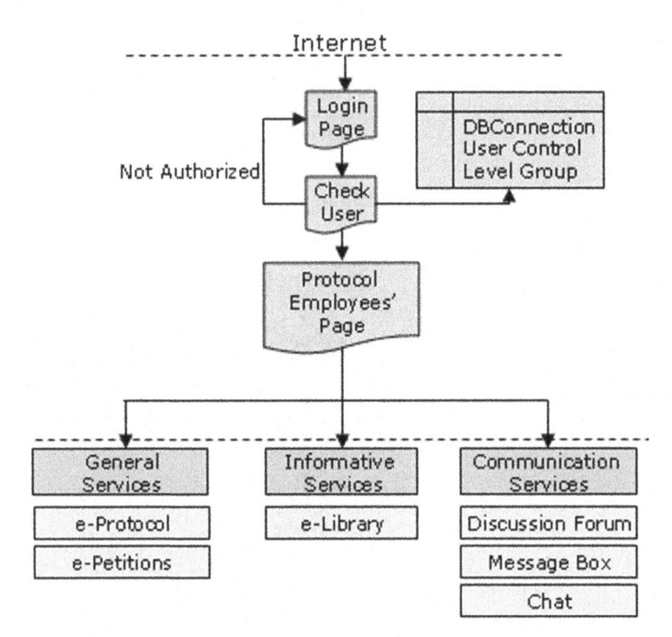

Fig. 4. Protocol Administrator Environment Tools

3.4 Registered - Authorized Users Environment Tools

The registered-authorized users have permission to interact with the e-Petitions system (Fig. 5). They can submit an application to the agency, as long as it is supported by the system. Moreover, the registered-authorized users have the ability to track the status of the applications they had submitted in the past. Finally, they can view and change some of their account information.

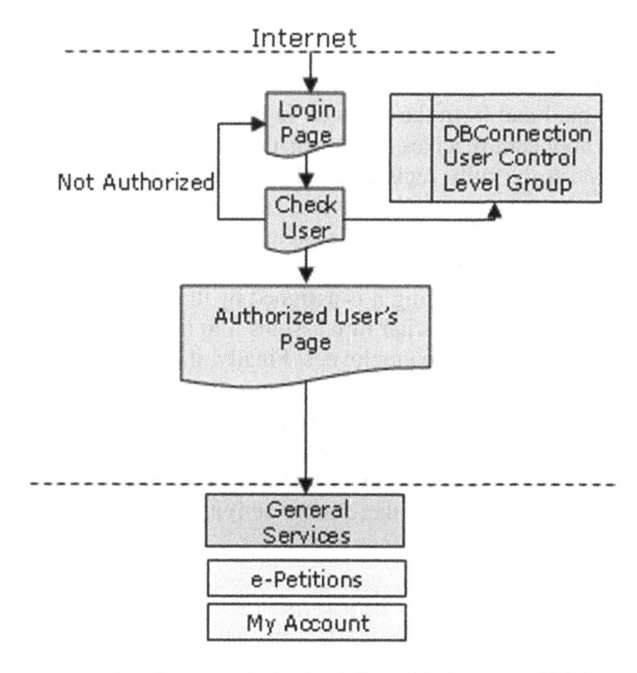

Fig. 5. Registered - Authorized Users Environment Tools

3.5 Guest - Unauthorized Users Environment Tools

The Guests – Unauthorized Users, on the other side, can browse the web portal in order to obtain valuable information regarding the agency and/or the issue(s) they wish to apply for. In order to apply, the guests-unauthorized users have to create an account (register) and interact with the e-Petitions system.

4 Structure Presentation

4.1 General Description

The presented environment is used as the web portal of the Earthquake victims' Compensation Agency. The application serves as a means for the electronic collaboration of the agency's employees as well as for the general informing of citizens regarding the e-services. The basic contribution is the application of the communicative services (discussion forum, chat, message box) as a means of central-based communication of the agency with its employees and with the citizens. The main objective of the developed infrastructure is the diffusion of information from the agency to everyone and the improvement of the e-services to the citizens. The portal's contribution with respect to information and valorization is the diffusion of the agency's information and services to the simple Internet user.

4.2 The Core of the e-Protocol System

The e-Protocol system accepts petitions from various sources such as deposits, faxes, standard mail, e-mail and from the Internet. In the case where the petition's source is the Internet, the applicant receives a confirmation number and directions in order for his/her application to be fully registered. This mechanism is intended to avoid fake applications entering the e-protocol system.

Once a new petition has entered the system, the procedure shown in Fig.6 is applied. The application is regarded as a new task that must be assigned to someone in order to process it. In the beginning it is assigned by the Protocol Administrator to the Department Manager, who in his/her turn assigns it to one or more Department Directors and the latter to one or more employees. Finally, it reaches the Protocol Administrator who completes it and sends it to the Correspondence Office. All steps are automated and the system has been designed so as to minimize the need of human intervention. For instance, the users are notified by the system when a new task is assigned to them. The transactions are made under secure communications (SSL) and there is an idle timeout of 20 minutes. If there is no activity during this period, the system automatically logs out the user.

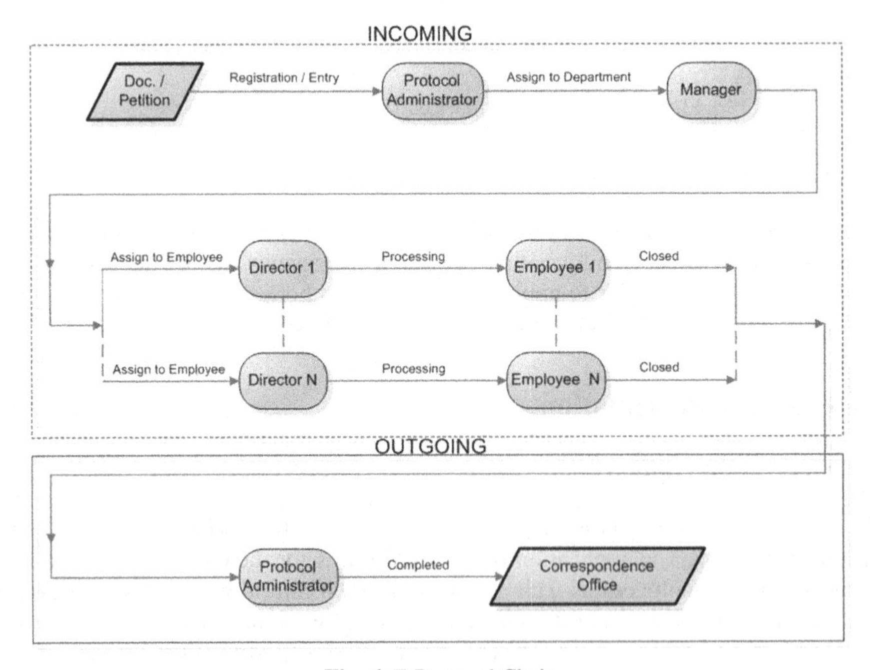

Fig. 6. E-Protocol Chain

5 Conclusions

E-government is the challenge for European Governments, to simplify, accelerate and improve, the delivery of the governmental circle services and facilities to the citizen.

According to Lisbon's Ministerial Meeting, EU groups of policies and initiatives, known as E-Europe, among other policies and priorities for the Information Society, give emphasis on Government online. This ensures that citizens have easy access to government information, services and decision-making procedures online. The big issue for e-government is to use Information and Communication Technologies (ICT's) to develop efficient services for European citizens. In particular, the basic aim is to develop Internet-based services to improve public access to information and public services, to improve public administration transparency through the Internet and to ensure that citizens have access to essential public data to allow them to take part in the decision-making process. Moreover, another goal is to ensure that electronic technology is fully used within public administration, with secure software and usage of secure layers in order to guarantee and reach some security standards.

E-government refers to electronic access to government information and services. As it has been mentioned before, the e-government idea includes fast and improved citizen service from a quantitative and qualitative point of view, as well as the restructuring and reengineering of the providing organization and its services, through the increased usage and exploitation of the capabilities of the information and communication technologies and especially through the facilities and services of the Internet. The escalation of the e-government services begins with the easy dissemination and easy access of the citizen to the governmental information and passes through the electronic transactions between the citizen and the public organization and reaches the electronic delivery of the requested document by the public organization to the citizen. An obvious prerequisite, in order to support the above "layers" of the e-government services is the development of an electronic infrastructure which is able to support e-protocol, e-applications/e-petitions and internal organizational function of the public organization. In addressing the above context, this article presented an e-government structure which supports and provides the aforementioned "layers" of the e-government services. This e-government structure, which introduces the notion and practicalities of electronic technology into the various dimensions and ramifications of government, has been developed by Net Media Lab of N.C.S.R. "Demokritos" for a Hellenic Public Organization, in order for the latter, to provide public information dissemination, accept electronic document submissions and manage them via e-protocol and support all the operations via the appropriate electronic structure which supports easy communication among the organization's departments as well as robust and user-friendly document management, storage, search, retrieval, handling and delivery.

Summing up, this paper presented an e-government environment based on information and communications tools. Our contribution is based on the proposal of a generic electronic scheme that enables distant collaboration of the agency's employees and the e-citizen. The included tools serve communicational and informative governmental functions through a user-friendly, interoperable and distributed web-based architecture. It must be noted that two basic axes are served. The first is the communication of the employees and the e-citizen, 24 hours a day, and the e-content development for different user levels. The second axe includes the delivery of e-services to the citizen.

References

1. Metaxiotis, K., Psarras, J.: E-Government: New Concept, Big Challenge, Success Stories. Electronic Government: An International Journal 1(2), 141–151 (2004)
2. Fang, Z.: E-Government in Digital Era: Concept, Practice, and Development. International Journal of the Computer, the Internet and Management 10(2), 1–22 (2002)
3. Drigas, A., Vrettaros, J., Kouremenos, D., Stavrou, P.: E-learning Environment for Deaf people in the E-Commerce and New Technologies Sector. WSEAS Transactions on Information Science and Applications 1(5), 1189 (2004)
4. Drigas, A., Koukianakis, L.: A Modular Environment for E-Learning and E-Psychology Applications. WSEAS Transactions on Information Science and Applications 6(3), 2062–2067 (2004)
5. Xenakis, A., Macintosh, A.: G2G Collaboration to Support the Deployment of E-Voting in the UK: A Discussion Paper. In: Traunmüller, R. (ed.) EGOV 2004. LNCS, vol. 3183, pp. 240–245. Springer, Heidelberg (2004)
6. Macintosh, A., Robson, E., Smith, E., Whyte, A.: Electronic Democracy and Young People. Social Science Computer Review 21(1), 43–54 (2003)
7. Kelly, E.P., Tastel, W.J.: E-Government and the Judicial System: Online Access to Case Information. Electronic Government: An International Journal 1(2), 166–178 (2004)
8. Barnes, S.J., Vidgen, R.: Interactive E-Government Services: Modelling User Perceptions with eQual. Electronic Government: An International Journal 1(2), 213–228 (2004)
9. Henman, P.: E-Government and the Electronic Transformation of Modes of Rule: The Case of Partnerships. Journal of Systemics, Cybernetics and Informatics 2(2), 19–24 (2004)
10. Verma, S.: Electronic Government Procurement: A Legal Perspective on the Indian Situation. Electronic Government: An International Journal 1(3), 328–334 (2004)
11. Wild, R.H., Griggs, K.A.: A Web Portal/Decision Support System Architecture for Collaborative Intra-Governmental Planning. Electronic Government: An International Journal 1(1), 61–76 (2004)
12. Kanthawongs, P.: An Analysis of the Information Needs For E-Parliament Systems. WSEAS Transactions on Information Science and Applications 1(5), 1237–1242 (2004)
13. Borras, J.: International Technical Standards for E-Government. Electronic Journal of E-Government 2(2), 75–80 (2004)
14. Chen, Y.S., Chong, P.P., Zhang, B.: Cyber Security Management and E-Government. Electronic Government: An International Journal 1(3), 316–327 (2004)
15. Abie, H., Foyn, B., Bing, J., Blobel, B., Pharow, P., Delgado, J., Karnouskos, S., Pitkanen, O., Tzovaras, D.: The Need for a Digital Rights Management Framework for the Next Generation of E-Government Services. Electronic Government: An International Journal 1(1), 8–28 (2004)
16. Drigas, A., Vrettaros, J., Kouremenos, D.: An E-Learning Management System for the Deaf People. WSEAS Transactions on Advances in Engineering Education 1(2), 20–24 (2005)
17. Drigas, A.: E-Course Support and Delivery for E-Psychology. WSEAS Transactions on Advances in Engineering Education 1(2), 25–28 (2005)

Main Elements of a Basic Ontology of Infrastructure Interdependency for the Assessment of Incidents

Miguel-Ángel Sicilia[1] and Leopoldo Santos[2]

[1] Information Engineering Research Unit
Computer Science Department, University of Alcalá
Ctra. Barcelona km. 33.6 – 28871 Alcalá de Henares (Madrid), Spain
msicilia@uah.es
[2] Emergency Military Unit, CG J6
28850 Torrejón de Ardoz Air Base, Spain
lsantos@et.mde.es

Abstract. Critical infrastructure systems currently conform highly complex and interdependent networks. While simulation models exist for different infrastructure domains, they are not always available when incidents are unfolding, and in many cases they cannot predict the cascading effect of failures that cross domains, or they are not able to support the rapid situation assessment required in response phase. To address a high-level view of the incidents in a given situation, both expert and domain knowledge and also computational models cross-cutting infrastructure systems are required. This paper describes the main elements of a basic formal infrastructure incident assessment ontology (BFiaO) that factors out the main elements required for applications dealing with incident assessment. Such ontology is intended to be extended with the specifics of each domain and infrastructure network. Situation assessment knowledge including reasoning can be captured with rules that suggest potential risks given the interdependency patterns in the network, having the outcomes of these rules different status according to their sources. Examples of this kind of risk-oriented assessment are provided.

Keywords: Emergency management, critical infrastructures, ontologies, situation assessment, rules OWL, SWRL.

1 Introduction

Emergency management (EM) is the continuous process by which individuals, groups, and communities manage hazards in an effort to avoid or ameliorate the impact of disasters resulting from them. The process of emergency management involves four broad phases: mitigation, preparedness, response, and recovery. Situation assessment becomes critical while an emergency unfolds and response is being deployed, as effective decision making relies on the availability of rules that guide the rapid understanding of the situation (McLennan et al., 2007).

As incidents are complex and evolving, systems able to assess the vulnerability of infrastructures and anticipate consequences and side effects of adverse events play a

M.D. Lytras et al. (Eds.): WSKS 2009, LNAI 5736, pp. 533–542, 2009.

key role in effective response management (Ezell, 2007). Infrastructure systems are a central element in the assessment of potential hazard risks in the course of incident tracking, as they have direct impact in life conditions of diverse kinds. An important aspect of critical infrastructures such as electric power, water distribution, transportation and telecommunications is that they form highly *interdependent* networks (Brown, Beyeler and Barton, 2004), and in consequence, vulnerabilities in many cases cross infrastructure boundaries, as is known to occur in the case of earthquakes (Dueñas-Osorio, Craig and Goodno, 2007). Further, adverse events causing emergencies have in many cases a cascading effect in interrelated networks (Pederson et al., 2006; Zimmerman and Restrepo, 2006). For example, a failure in gas distribution in many cases puts electrical supply at risk.

Causal relations or failure cascading inside and between infrastructure systems can in some cases be anticipated by experts that have experienced similar incidents in the past or have sufficient knowledge to predict them, complementing simulation models (Ezell, 2007). Simulation requires in general human expertise and complex and demanding computing requirements. Further, simulation is aimed at predicting effects of some events by using models with considerable fidelity with the real systems, so it is typically used in the *preparedness* phase of emergency management. However, in the *response* phase, a higher-level support for situation assessment is required, and simulation often does not fit the requirements of quick decision-making. Further, operational units do not always count with simulation systems, and simulation models that include several infrastructure domains are often difficult to find. This entails that experts become a key source of reliable assessment on the potential courses of evolution of the incident. The problem of relying on experts is twofold. On the one hand, there is a requirement of availability of concrete people and information flow that is difficult to manage when the incident has just started. And on the other hand, knowledge is dispersed among experts of different domains, and some relations between infrastructural domains might be overlooked. This justifies the engineering of systems that integrate high-level expert knowledge, domain knowledge and computational models for the assessment of risks before and during emergency.

Systems dealing with emergencies are data-intensive and require a number of integrated data models, including geospatial information, inventories of available resources and of vulnerable facilities or populations. Proposed standards for the interchange of data related to EM have emerged in the last years, e.g. the Tactical Situation Object (TSO) is aimed at exchanging information during EM and the OASIS Emergency Management Technical Committee[1] has released the Common Alerting Protocol (CAP) and the suite of Emergency Data Exchange Language (EDXL) specifications with related aims. Also, systems supporting the sharing of information during emergencies have appeared, as the open source system SAHANA[2]. Nonetheless, while information sharing is essential in the process of building collaboration during emergencies, *situation assessment* requires a priori models and expert knowledge that cross-cuts several aspects of the affected area. Ontologies have been proposed as knowledge-based models for threats, infrastructures and other EM-related elements (Little and Rogova, 2009; Araujo et al., 2008), but in infrastructure interdependency modeling such representations require some sort of upper-level ontology that

[1] http://www.oasis-open.org/committees/emergency/
[2] http://www.sahana.lk/

integrates the common aspects and relationships that concern the different domains. This paper describes the first version of a Basic Formal Infrastructure Incident Assessment Ontology (BFiaO[3]) addressing those high-level aspects. The reasoning processes related to situation assessment as the incident unfolds is captured with SWRL rules[4]. As knowledge related to these situations is imperfect, a diversity of kinds of reliability in the results of inferential processes is considered, along with the modeling of potential events as possible worlds that might be confirmed by events actually occurring.

The rest of this paper is structured as follows. Section 2 describes the main elements of a basic formal infrastructure incident ontology. Then, Section 3 describes how reasoning on cross-domain situations can be implemented via SWRL rules, using examples. Finally, conclusions and outlook are provided in Section 4.

2 Main Elements of a Basic Formal Infrastructure Incident Assessment Ontology (BFiaO)

The BFiaO provides a basic model covering the common high-level aspects of infrastuctures (electical, gas, etc.), causes of incidents (be them natural or caused by humans) and the incidents themselves as situations that in some cases evolve or are qualified as emergencies. Infrastructure can be defined as the basic physical and organizational structures needed for the operation of a society or enterprise. The *Basic Formal Infrastructure Incident Assessment Ontology* (BFiaO) aims at providing a base model for the assessment of situations that involve a risk that might derive in an emergency.

2.1 Basic Domains and Network Model

A *domain* is in general a "sphere of activity or concern" (often referred to as "sectors" as electricity, gas, etc.). In our context, a domain is a sphere of activity for which specific infrastructures exist and provide services, capabilities or goods that are important to human activity. Domains are not reified as objects in the ontology, but are expected to be reflected in the modular structure of the ontology. A *network* is a concrete infrastructure covering a domain that is deployed (usually at a national level). A country may have several networks covering a single domain. A network is described in an abstract way as a collection of geolocalized `Nodes`[5] and a collection of physical or logical `Connections` among them (both kinds of elements are labeled `NetworkElements`). This represents a basic graph model where `Nodes` and `Connections` are infrastructure elements of any kind with `connectsTo` (inverse `isConnectedBy`) as the property to specify the nodes connected. A typical example of a node is an electrical generator, while an example of a connection is an electrical transmission line. It should be noted that different kinds of nodes and connections determine different incident assessment models. A transitive `partOf` property is used to define a mereology equivalent to the one in the Basic Formal Ontology (BFO[6]) for networks

[3] The BFiaO version 0.5 is currently available in Ontology Web Language (OWL) format.
[4] http://www.w3.org/Submission/SWRL/
[5] Ontology elements are provided in Courier font in this document.
[6] http://www.ifomis.org/bfo

(including eventually ovelaps). Geographical description in BFiaO is currently modeled in two variants: one reusing the SWEET spatial ontology[7] and a lighter one using a simple point-based geolocation definition. The geospatial location of every `Node` is required in the ontology.

The semantics of connexions are highly service-dependent, which requires models of service flow. For example, a `RailRoad` segment `a` can be modeled as a kind of `Connection` between two nodes (representing for example populations). Then, if we have the event `road:CarAccidentEvent(cae)`[8] (events are introduced later) with `blocksConnection(cae, a)` and `hasStartTime(cae, t)`, then a directed-graph traffic model is able to provide alternative routes while end time is not known, and perhaps predict the level of congestion in them. However, the failure of a gas conduction follow different semantics, as in that case, the target is providing additional input to the network from other gas sources. In gas conduction networks, some `Node` instances are sources of service (modeled as a `isConnectedBy` sub-property `providesInput)` to connections. The pressure level of the gas in each connection part can be measured from the nodes providing input.

2.2 Interdependencies

Interdependencies are of a various nature. Table 1 provides a classification of its major types, covering part of those defined by Dudenhoeffer, Permann. and Manic (2006). These interdependencies are the source of cause-effect short-term predictions that can be combined with other cause-effect (Hernández and Serrano, 2001) models.

Note that dependencies between networks are perfectly possible, e.g. there might be several companies providing some infrastructure service and they have some points of connection between them. Robert, De Calan and Morabito (2008) describe an assessment framework for any kind of interdependency based on modeling the resources used (and which of them are alternatives in case of failure) for each commodity or service delivered. That model can be represented as a specialization of the BFiaO in which alternative resources are represented as a kind of connection. It should be noted that `Connection` is not tied a priori to any kind of directionality, effect or valence between the nodes connected, these domain-specific aspects are left to subsumed concepts.

Interdependencies may be declared as instances a priori (e.g. when the dependency is known because of physical connectivity), but in other cases, they are tacit, as results of inference. This is the case of physical connections which can be modeled by SWRL rules as the following:

```
Trans:TrainAccidentEvent(?ae) ∧ geolocation(?ae, ?g)
   ∧ hasStartTime(?ae, ?t) ∧ Connection(?c)
   ∧ physicalLocation(?c, ?g2) ∧ overlaps(?g, ?g2)
→ PossibleEvent(?pe)  ∧
   negativelyAffectsConnection(?pe,?c) ∧ hasStartTime(?pe, ?t)
```

[7] http://sweet.jpl.nasa.gov/ontology/

[8] Ontology elements not belonging to the abstract upper BFiaO model are prefixed with namespace labels.

Table 1. Kinds of infrastructure interdependency and their nature

Interdependency kind	Nature	Identification
Physical	Physical elements lying close to each other are likely to be affected by the same physical incidents.	They can be identified among physical network elements by using geospatial search functions. However, the notion of "closeness" presents element-specific aspects, e.g. not all the incidents affect the same kinds of nodes. This requires the use of a mereotopology (Smith, 1996) and domain-specific knowledge.
Connectivity based	As nodes require inputs from other nodes that go through connections, failure in network elements potentially affect those directly or indirectly connected.	They can be identified from network structure, taking into account the high-level dynamics of the network.
Policy or procedural	The failures of some nodes that have a special role in the control of the system inhibit the operation of other nodes, even though the connections are not in failure.	These require network-specific knowledge of the configuration of the system.

Thus some interdependencies may be tacit instead of explicitly declared. The rule above is still too generic, as it does not consider the level of protection of physical connections. Also, only narrow geospatial location specifications are useful in that kind of rules, e.g. a geospatial location defining a relatively large area for an accident is not giving the appropriate precision for the kind of inference intended.

2.3 Events and Incidents

`Events` are any kind of spatiotemporal wholes that has differentiated beginnings and endings corresponding to real discontinuities. `Incidents` are a kind of event that are considered "adverse". *Adversity* in our context cannot be defined except by enumeration of event types. Further, the potentiality of becoming a threat of some event types is highly dependent on the context and previous knowledge, e.g. intense rainfall can be considered adverse for some places and situations and not for others. Then, incidents are modeled through incident type taxonomies, and the decision on which events become an incident will be application-determined. A taxonomy derived from the CEN Workshop Agreement (CWA) code CWA 15931-2 "Disaster and emergency management - Shared situation awareness - Part 2: Codes for the message structure" has been used as a point of departure. Concretely, the categories specified in the CWA as `/EVENT/TYPE/CATEGORY` can be mapped as subsumed by `Incident`, e.g. class `FIRA_Event` in the ontology maps the `/FIR/CLA` category for "class A" fire. Some categories determine means to be applied, e.g. class A fire requires the use of water. Incidents are also temporally and geographically determined. Geospatial localization is common to the one used for infrastructure nodes.

2.4 Possible Events

Interdependencies between different kinds of events are the central element of the BFiaO, which focuses on the short-term evolution of incidents in terms of new events. A critical element in emergency response answering the "*what may happen?*" question. `Events` are facts about the past, and they are "fixed" or necessary. However, assessing what may happen requires statements about the future, which are by nature not necessary. The `PossibleEvent` concept introduces a category of future events that serve as the basis for those short-term future statements. They should be interpreted in terms of epistemic possibility, i.e. "given the available information" such element is possible. Predictions in this context are of a diverse kind. Some of them that come from some kind of evidential or probabilistic model (Klir, and Wierman, 1998) are represented as `ModelBasedPossibleEvent` and often numerical probability or evidence levels are associated to them. A category of these events are `WeatherForecasts`, which come from analysis of meteorological conditions, and are considered reliable to some extent. Other possible events may come from the codification of expert knowledge, which have a very different epistemological status. Eventually, these different kinds of predictions will be combined in the reasoning process when crossing domains. Then, provenance of predictions is a critical element in decision making, helping in the differentiation of two very different epistemological categories. This is for now accomplished by the trace of the reasoning process.

An event occurring in a concrete situation that affects nodes or connections change in general the dynamics of the networks and the overall properties of the services provided by the infrastructures. The changes in some elements affect others (i.e., the interact with others), leading to a new status, eventually with a reconfiguration of the collection and level of criticity of the elements. The computation of the new status depends on the characteristics of the models for each network, and they might entail reasoning or numerical algorithms that are outside the BFiaO scope. Figure 1 shows a general transition diagram representing the main situation tracking elements. Events are received at any state, and eventually one of them is considered an incident, thus changing to a situation in which possible events are started to be considered. As possible events are temporal, they will be confirmed or disconfirmed, causing a dynamic.

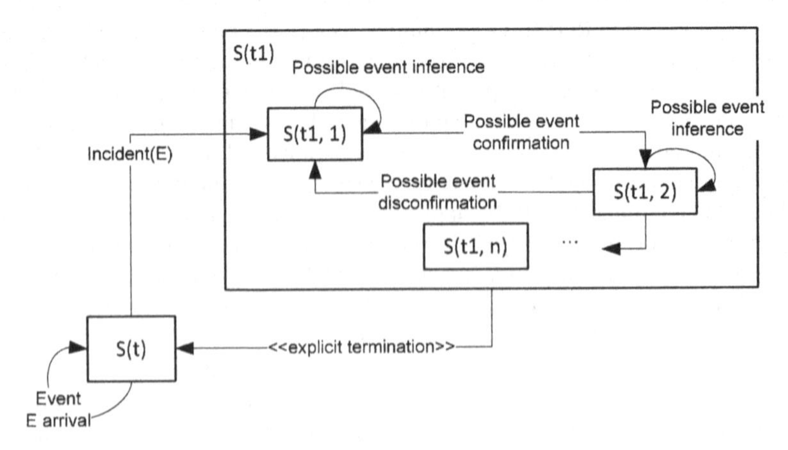

Fig. 1. Schematic dynamics of the assessment system system as a state-transition diagram

reconsideration of the present state. This requires a sort of non-monotonic reasoning in which the possible world considered are retracted and reconfigured as events occur (or as the predictions are not materialized at the time they were expected to occur).

Finally, incident situations have an indefinite duration. Since the modeling of situations is targeted to human assessment, it is expected that a decision maker determines when an incident situation has terminated.

2.5 Criticity as a Temporal Category

There is not an universal definition of a "critical node" or "critical connection", as criticity depends both on the objectives of the ongoing action and on the current situation, i.e. criticity is not an essential but an accidental feature of these elements. Then, criticity is assessed by means of `CriticalElementAssessments`, which are time-stamped records of criticity for a given element and situation. In some cases, criticity is interpreted as vulnerability, for which assessment methods have been proposed (Baker, 2005). Vulnerability can also be associated to an entire network (Holmgren, 2006).

3 Integrating Reasoning on Incident Assessment

In what follows, an example will be used to illustrate the use of the BFiaO to model emergency situation assessment cross-cutting domains. The starting point of the operation of a situation assessment episode is the arrival of an alert with effects that may qualify as incident[9]. Concretely, the starting point is that of the forecast of a severe storm localized in the area of the Barcelona port (see Figure 2). The forecast of the event specifies a time `t` and the geographical area affected, and uses the *Phenomena* SWEET ontology to specify the type `Storm`, and the associated quantitative strength measures. The forecast results in the creation of a new situation with the storm as the starting point. If the forecast is not confirmed by an event, the entire situation will be discarded.

If the event is confirmed, the elaboration of short term potential effects starts. The `StormEvent` is used to infer a potential closing of the port (label 1 in Figure 2), including the loading bays. If the port is modeled as a connection to several nodes in different infrastructure networks, then the following node generically (domain-independent) infers possible related effects in these nodes.

```
PossibleNodeClosingEvent(?nce) ∧ hasStartTime(?nce, ?t)
   ∧ blocksConnection(?nce, BCNport)
   ∧ connectsTo(BCNPort, ?n) ∧ ?Node(?n)
→    PossibleEvent(?pe) ∧
     blocksNode(?pe, ?n) ∧ hasStartTime(?pe, ?t)
```

Then, potential blockings (e.g. labels 2 and 3 in Figure 2) are interpreted in terms of domain-specific infrastructure networks. For example, the blocking of the port may

[9] The situation is completely fictitious, including infrastructure elements, conditions and the logic of interdependencies, it serves only as an illustration.

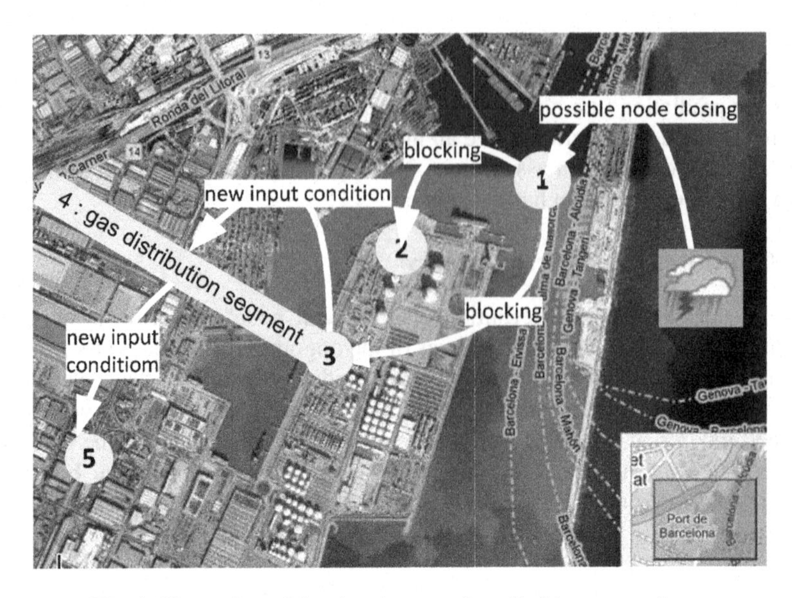

Fig. 2. Illustration of the situation area for a fictitious scenario

result in the blocking of the delivery of liquefied natural gas (usually transported in ships), which results in a decrease in the input to the gas distribution network affected. Having a specific kind of connection `NaturalGasConduction`, the following rule (elements specific to the gas domain are in namespace `ng`) triggers the computation of a possible gas conduction state (subsumed by `PossibleConnectionState`) related to the possible event. The possible state is computed by a `ng-alg:computeNew` algorithm. In this case, the resulting new pressure state (label 4 in Figure 2) could also be represented through rules, but in other domains numerical models are complex and bridging the rules with domain-specific algorithm implementations become a more efficient solution.

```
PossibleEvent(?pe) ∧ blocksNode(?pe, ?ge) ∧
   ng:GasProvidingStation(?ge) ∧ ng:NaturalGasConduction(?ngc)
   ∧ providesInput(?ge, ?ngc)
→   ng:PossibleGasConductionState(?x) ∧ causedBy(?x, ?pe)
    ng:possibleConductionPressure(?x, ng-alg:computeNew(?pe))
```

If the pressure in the new potential state computed goes below a threshold, a new possible event will be predicted, in this case affecting infrastructure elements fed by the gas conduction. These may include a combined cycle power plant, which will be affected by the lack of gas supply (label 5 in Figure 2). This in turn might lead to a loss of power supply in a part of the electrical network.

The resulting chained events will be time-located and a graph of possible events will serve as the trace of the short-term possible worlds predicted, they will then be timed-out, disconfirmed or in some cases confirmed by actual events. A problem with that chain of cascading events is placing a limit in the inference process. The inference engine can be provided with a maximum limit of chained inference steps, or on-demand chaining depth could be implemented to provide control to operators.

Associated to the above, measures of criticity might evolve, e.g. it might be that the vulnerability of the electricity network in the above example lead to a new `PossibleNetworkState` with an increased vulnerability, computed using a numerical metric as a clustering coefficient (Holmgren, 2006). That new state might be followed by a suggestion of special protection other electrical stations different from the one affected, that become critical as alternative sources in that situation.

4 Conclusions

Infrastructure systems play a central role in emergency management, and they are interdependent due to physical, operational and functional dependencies. High-level models of infrastructure components can be used to monitor the short-term evolution of incidents when crossing infrastructure domains. This paper has described the main elements of a Basic Formal Infrastructure Incident Assessment Ontology (BFiaO) serving as an upper model for representations combining infrastructure interconnections and situation assessment based on the notion of *incident*. As the model is aimed at being generic, it includes support for different kinds of epistemological accounts regarding the events that may happen considering an unfolding situation, including numerical models based on domain theories but also subjective expert knowledge representations. The modeling of short term predictions requires modeling possible worlds (in the epistemic sense), so that actual events confirm or invalidate previous hypothetical scenarios.

The BFiaO version described here is 0.5, covering the base model and examples coming from several domains, future versions are expected to be improved and modified as different domains are elaborated. Some of the aspects that are currently considered as extensions to the BFiaO include the merging of situations (i.e. combining two situations into a single one, provided that they are somewhat interrelated) and some conventions for interfacing the high-level causal models based on BFiaO with sources of data as CAP alerts or Geographical Information Systems (GIS).

References

Araujo, R.B., Rocha, R.V., Campos, M.R., Boukerche, A.: Creating Emergency Management Training Simulations through Ontologies Integration. In: Proceedings of the 11th IEEE International Conference on International Conference on Computational Science and Engineering (workshops), pp. 373–378 (2008)

Baker, G.H.: A Vulnerability Assessment Methodology for Critical Infrastructure Sites DHS Symposium: R&D Partnerships in Homeland Security, Boston, Massachusetts (April 2005), `http://works.bepress.com/george_h_baker/2`

Brown, T., Beyeler, W., Barton, D.: Assessing infrastructure interdependencies: the challenge of risk analysis for complex adaptive systems. International Journal of Critical Infrastructures 1(1), 108–117 (2004)

Dudenhoeffer, D.D., Permann, M.R., Manic, M.: CIMS: A Framework for Infrastructure Interdependency Modeling and Analysis. In: Proceedings of the Winter Simulation Conference. WSC 2006, pp. 478–485 (2006)

Dueñas-Osorio, L., Craig, J.I., Goodno, B.J.: Seismic response of critical interdependent networks. Earthquake Eng. Struct. Dyn. 36(2), 285–306 (2007)

Ezell, B.: Infrastructure Vulnerability Assessment Model (I-VAM). Risk Analysis 27(3), 571–583 (2007)

Hernandez, J., Serrano, J.M.: Knowledge-based models for emergency management systems. Expert Systems with Applications 20(2), 173–186 (2001)

Holmgren, A.J.: Using graph models to analyze the vulnerability of electric power networks. Risk Analysis 26(4), 955–969 (2006)

Klir, G., Wierman, M.: Uncertainty-Based Information: Elements of Generalized Information Theory. Springer, Heidelberg (1998)

Little, E., Rogova, G.: Designing ontologies for higher level fusion. Information Fusion 10(1), 70–82 (2009)

McLennan, J., Holgate, M., Omodei, M.M., Wearing, A.J.: Human information processing aspects of emergency incident management decision making. In: Cook, M., Noyes, J., Masakowski, Y. (eds.) Decision making in complex environments, pp. 143–151. Ashgate, Aldershot (2007)

Pederson, P., Dudenhoeffer, D., Hartley, S., Permann, M.: Critical infrastructure interdependency modeling: a survey of US and international research. Report INL/EXT-06-11464, Idaho Falls: Idaho National Laboratory (2006)

Robert, B., De Calan, R., Morabito, L.: Modelling interdependencies among critical infrastructures. International Journal of Critical Infrastructures 4(4), 392–408 (2008)

Smith, B.: Mereotopology: A Theory of Parts and Boundaries. Data and Knowledge Engineering 20, 287–303 (1996)

Zimmerman, R., Restrepo, C.: The next step: quantifying infrastructure interdependencies to improve security. International Journal of Critical Infrastructures 2(2/3), 215–230 (2006)

Cognitive Agent for Automated Software Installation – CAASI

Umar Manzoor and Samia Nefti

Department of Computer Science,
School of Computing, Science and Engineering,
The University of Salford, Salford,
Greater Manchester, United Kingdom
umarmanzoor@gmail.com, s.nefti-meziani@salford.ac.uk

Abstract. With the evolution in the computer networks, many companies are trying to come up with efficient products which can make life easier for network administrators. Silent automated installation is one category among such products. In this paper, we have proposed Cognitive Agent for Automated Software Installation (CAASI) which provides silent, automated and intelligent installation of software(s) over the network. CAASI will intelligently install any kind of software on request of the network administrator. CAASI uses an efficient algorithm to transfer the software installation setup on the network nodes. The system is fully autonomous and does not require any kind of user interaction or intervention and performs the task(s) over the network with the help of mobile agents.

Keywords: Automated Software Installer, Network Installation, Silent Installation, Cognitive Agents, Multi-Agent System.

1 Introduction

Computer technology plays an important role and become a necessity in current modern world. Computer networks have become core part of every organization in today's modern world. Hospital, software houses, offices, colleges, universities etc computer networks exists every where and have become the backbone of every organization. Computer networks have become very complex (network consisting of networks of networks) and managing, maintaining such networks has become a challenging task for network administrators [19]. Network administrator along with its team is responsible for maintaining and managing the network and they have to keep each and every node of the network up and running. The job includes monitoring of network nodes (i.e. monitoring of malicious applications on network), software deployment (i.e. installation /un-installation of software's) on network nodes and management of network (adding / removing resources on the network).

Software deployment on network sometimes becomes very tedious and time consuming task as the process needs human interaction. Traditionally, network administrator or their subordinates have to physically move to each computer in the network one by one and run the installation setup manually which initiate installation wizard. The installation wizard helps the user in the installation of the software and usually

M.D. Lytras et al. (Eds.): WSKS 2009, LNAI 5736, pp. 543–552, 2009.

consists of series of steps which requires user interaction [3]. However, some software setups provides silent installation feature (i.e. software can be deployed without user interaction) but all softwares do not provide this feature because it is dependent on the installer used for software deployment [2].

With the evolution in the computer networks, many companies are trying to come up with efficient products which can make life easier for network administrators. Silent automated installation is one category among such products. Remote installers [4, 5] are softwares which helps network administrator to install software remotely on the network nodes silently without user interaction. Remote installer has the following problem. 1) Remote client should be installed on network node which requires manual maintenance 2) It requires training 3) Installation on network node is homogenous 4) Not all software's support –s switch for silent installation.

Umar et al [3] generalize the process of silent unattended installation / un-installation and proposed Silent Unattended Installation Package Manager (SUIPM) which generates Silent Unattended Installation / Un-Installation Packages. SUIPM 1) supports heterogeneous setting on different nodes 2) does not require any client software on network nodes 3) support all kind of software's but SUIPM still requires 1) training, 2) generates its own silent unattended installation / un-installation setups and 3) has no intelligence. The size of the setups generated by SUIPM is almost double as compared to original setup size.

The need of the time is to develop a framework which supports silent unattended (autonomous) installation but should be intelligent (i.e. no training required) and use the original installation setup for silent installation. In this paper, we have proposed Cognitive Agent for Automated Software Installation (CAASI) which provides silent, automated and intelligent installation of software(s) over the network. CAASI framework is based on Agent paradigm and motivated by An agent based system for activity monitoring on network (ABSAMN) [1] architecture where Agents monitor network resources on the behalf of network administrator.

Agent is a software program which acts autonomously on some environment on the behalf of its user [6, 7]. If an agent has the capability of moving from one node to another network node autonomously and has intelligence, the agent is known as Intelligent Mobile Agent [8, 9]. In CAASI framework, Intelligent Mobile Agents are responsible for silent autonomous software installation / un-installation over the network. Because of flexibility, self recovering, fault tolerant and decentralized features of Agent paradigm, they have been used in many areas such as searching, network monitoring, file transfer, e-commerce, mine detection, financial sector, network installation and file synchronization [14, 1, 12, 8, 7, 10, 3, and 15].

CAASI does not require any specific kind of setup like MSI nor require any training for installation. CAASI will intelligently install any kind of software on request of the network administrator. CAASI not only provides installation services, it also enables network administrators to verify the installation of software over the network. CAASI uses an efficient algorithm to transfer the software installation setup on the network nodes. CAASI provides one click installation on demand of network administrator at anytime and there is no need for the network administrator or it's subordinate to use the traditional method of network installation.

2 How Installation Works on Microsoft Windows

Following are the basics that are required to understand how installation of software works on Microsoft Windows [17]. The following discussion is with reference to Microsoft Windows (any version released after year 2000).

- Every application has one or more windows with one as parent window and each window has unique identifier known as handle. Buttons (Next or Back) or controls (Text box, combo box etc) within the window are referred as child windows.
- Each application has a message queue and message loop. Occurred events (clicks, keystrokes, etc) are stored in message queue and message loop is responsible to retrieve the events from the message queue and dispatch (deliver) them to the appropriate window which has a procedure to process these messages.
- All the messages / events (keyboard input, mouse movement, signals etc) are stored in the System message queue and the System is responsible to forward these to the appropriate application message queue.

Let suppose user is running an application on some computer and a window appears on the screen which has "Next" button. If the user clicks the "Next" button following sequence of events will occur

- A click event will be stored in System message queue.
- System message queue forwards this event to Application message queue.
- Application picks this event using message loop and Translate the event then Dispatch the event and calls the specific window procedure.
- Window procedure performs the actual event i.e. in this case the code behind the next button will be executed.

3 System Architecture

Cognitive Agent for Automated Software Installation (CAASI) provides a complete solution for silent autonomous software installation over large networks. CAASI is a multi agent based application for silent autonomous software installation over the network. The system is fully autonomous and does not require any kind of user interaction or intervention. Network administrator is responsible for assigning the task(s) to CAASI and once assigned, it performs the task over the network autonomously with the help of mobile agents.

The system consists of the following five agents

- Server Agent
- Sub Server Agent
- File Transfer Agent
- Cognitive Installer Agent
- Verifier Agent

3.1 Server Agent (SA)

Server Agent (SA) is the main agent as it manages and initializes the whole system. In initialization SA performs the following tasks

- Load network configuration which contains hierarchy of the network, name and IP address of sub servers, range of the sub network IP Addresses etc from a pre-configured XML file. For the first time network administrator has to provide the network configuration, once the system is up and running, it will update and maintain the configuration autonomously.
- SA will create and initialize n Sub Server Agents (SSA) where n depends on the number of sub servers in the network. Each SSA is assigned one sub server and after intialization these agent move to the assigned sub server.

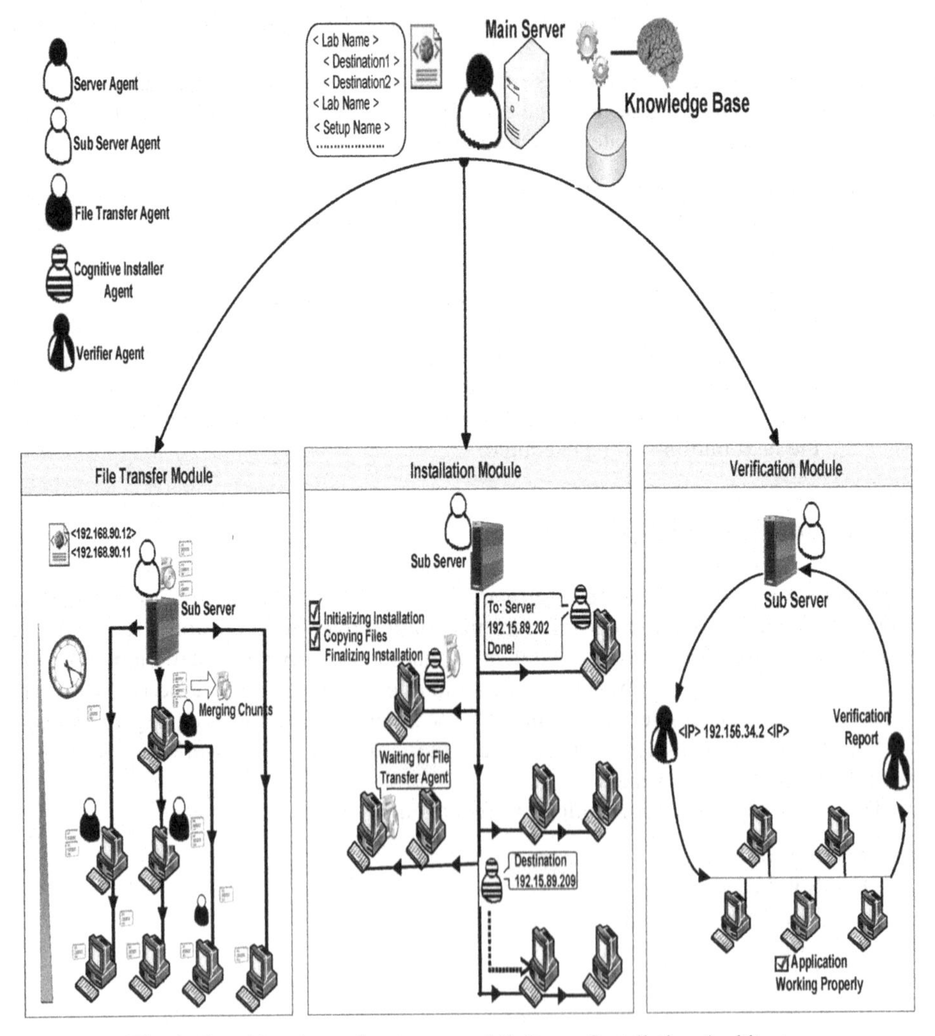

Fig. 1. Cognitive Agent for Automated Software Installation Architecture

In initialization SA transfers part of knowledge base, configuration of the sub network to SSA and it is responsible to execute any task(s) related to its sub network and report it back to SA.

- After initialization, SA waits for the task(s) from network administrator.

Network administrator can assign the task to SA through the online web based interface provided to administrators. The task generated by web based interface will be in the form of an XML file and contains all the necessary administrative information required to perform task. XML file will have the IP addresses of all the nodes on which operation has to be performed along with the product specific details. After receiving the task from the administrator, SA forwards the task to appropriate Sub Server Agents for execution of the task.

SA is also responsible for monitoring the progress, keeping logs and learning from tasks completed successfully or unsuccessfully. The learning process includes how to handle the errors generated during installation or verification, prediction of the events for the installer or generating the profiles of the newly installed software. The errors in installation can be low hard disk space, requirement of the dependent software and etc. SA is also responsible to manage the Knowledge Base which contains rules and profiles of the installed software. Initially Knowledge Base will have only rules, however, as new tasks are assigned by network administrator, CAASI will learn from the assigned tasks and Knowledge Base will be updated (i.e. new rules and profiles will be added or existing rules will be updated) which makes the execution of new tasks easy and efficient.

3.2 Sub Server Agent (SSA)

Sub Server Agent (SSA) is the created and initialized by Server Agent (SA) and after initialization SSA move to the destination (sub server) assigned and performs the following steps.

- SSA loads the network configuration which contains names and IP addresses of the machines in the sub-network.
- Loads the Knowledge Base
- SSA is responsible for the creation and initialization of File Transfer Agent, Cognitive Installer Agent, and Verifier Agent. Depending upon the request SSA dynamically create and initialize any (or all) of these agent. If the task assigned is installation of some software on sub-network, SSA will create a pair of FTA, one of which will move to File Server where the setup resides. Before transferring the software setup, FTA will divide the setup into n chunks where n depends on the size of the setup. After chunk creation, FTA start transferring chunks of software package from the File server to SSA machine. Once the setup transfer is complete, the setup is transferred to the network nodes using the File Transfer Module which uses exponential data transfer algorithm discussed later.
- Once FTA completes the transfer of chunks on some network node, it will merge the chunks into original setup and notify SSA about the completion of setup transfer. On receiving file transfer completion message, it will create Cognitive Installer Agent (CIA) and initialize it with destination(s), setup path on the destination machine, software profile if available (i.e. if the software

was installed before on the network, the complete information about previous installation {event, action, behavior} will be passed to CIA), software information (i.e. serial key, activation code etc) and the Knowledge Base.

- On receiving installation completion message from Cognitive Installer Agent, SSA will create Verifier Agent and initialize it with destination(s), path of the installed software, size of the software, file and registry XML files (Optional). File XML contains directories / file information and registry XML contains all the registry entries of the software. File and registry XML files will only be transferred if Verifier Agent requests these files.

SSA monitors the whole activity of File transfer, Installation, and verification and also keeps log of the activity running on its sub-network. SSA sends periodic logs which includes task status, activity details etc to SA.

3.3 File Transfer Agent (FTA)

File Transfer Agent (FTA) is responsible to transfer the software setup from source to destination(s) and uses exponential file transfer similar to one proposed by Koffron [11]. Exponential File Transfer is a method for optimized distribution of files from a file server to large number of client machines. Koffron has proposed a binary tree like logical structure of the network where each level receives the files from the higher level and transfers it to a lower level and file server will be root of the tree.

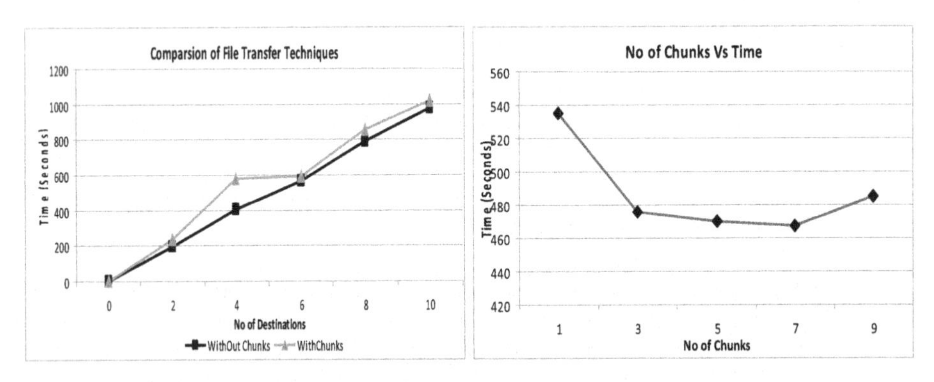

Fig. 5. a) Comparison of file transferring technique with and without chunks

Fig. 5. b) No of Chunks Vs Time

Instead of sending the complete setup file from one network node to another we divided the file into chunks and transfer these chunks from one node to another. Using our modification, as soon as one node receives the chunk, it will forward the chunk to the lower level. This change increases the level of parallelism in the file transfer and after the modification a visible improvement observed in file transfer time. Fig 5 (a) shows the comparison of file transfer time with and without chunk with reference to file size 8MB and 3 chunks (i.e. setup is divided 3 chunks). Fig 5 (b) shows the comparison of number of file chunks Vs time, the number of chunks plays an important role in decreasing the transfer time but up to a specific value after that the transfer time increases as shown in Fig 5 b).

3.4 Cognitive Installer Agent (CIA)

After initialization Cognitive Installer Agent (CIA) moves to the first node in destination list and loads the Software Profile, Knowledge Base and Software information passed by SSA. Before starting the installation process, CIA will run a Software Constraints Check to verify constraints like space available in a targeted drive, prerequisite requirement etc. Based on the result of Software Constraint Check, CIA will autonomously generate an appropriate action to find a solution by querying the Knowledge Base. The task of the CIA is to install the software autonomously and silently on the destination.

To automate the installation process CIA has to generate automatically the same events for the Software Setup which are generated when the user manually installs the application. Once the event is generated automatically, CIA has to trap the Software Setup window to send it the corresponding event. CIA will capture Software Setup window text and controls, store the information in vectors, filter the text vector using the keywords, pass the filtered text vector along with controls vector to Knowledge Base to generate the corresponding event. Once the event is generated, SendMessage API is used to send the event to the Installer Window.

CIA will generate two kinds of profile during the software installation. 1) Node based profile will be created on the node (where installation took place) which includes software installation directory, date and time of installation, size of the installed software etc, this information will help Verifier Agent to detect weather installed software is working properly or not. 2) A product based profile will be created which includes event sequence, actions generated against events etc. CIA will send this software profile to SSA and SA to update the Knowledge Base.

3.4.1 Test Case

CAASI has been evaluated on large number of software setups and results were very promising. The test case of AML Free Registry Cleaner [18] and the Cognitive Installer Agent working is given below.

Enumerated Text:
Welcome to the AML Free Registry Cleaner Setup Wizard
This will install AML Free Registry Cleaner 3.2 on your computer.
It is recommended that you close all other applications before continuing.
Click Next to continue, or Cancel to exit Setup.
Enumerated Control:
Next, Cancel

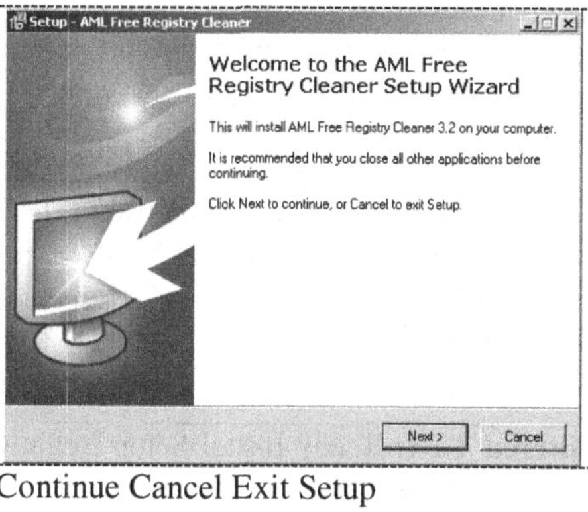

Filtered Text: Install Close Click Next Continue Cancel Exit Setup
Action Inferred Using KB: Click Next

Enumerated Text:
License Agreement
Please read the following important information before continuing.
Please read the following License Agreement. You must accept the terms of this agreement before continuing with the installation.
Enumerated Control:
I accept the agreement, I do not accept the agreement, Back, Next, Cancel

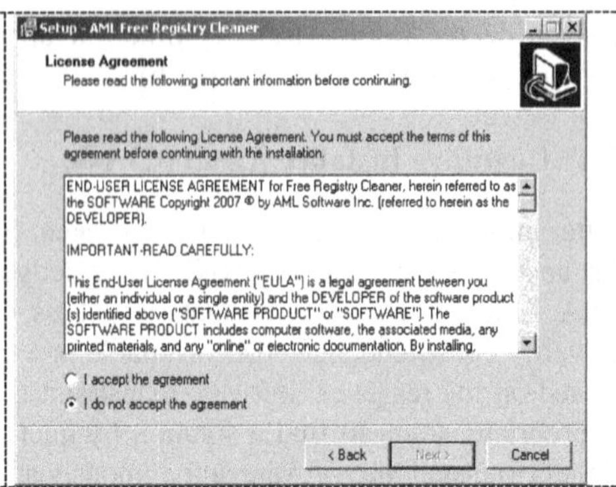

Filtered Text: License Agreement Read Accept Installation
Action Inferred Using KB: Select I Accept Control Then Click Next

Enumerated Text:
Select Destination Location
Where AML Free Registry Cleaner should be installed?
Setup will install AML Free Registry Cleaner into the following folder.
To continue, click Next. If you would like to select a different folder, click Browse.
C:\Program Files \ AML Products \ Registry Cleaner
At least 7.0 MB of free disk space is required.
Enumerated Control:
Back, Next, Cancel

Setup - AML Free Registry Cleaner
Select Destination Location
Where should AML Free Registry Cleaner be installed?
Setup will install AML Free Registry Cleaner into the following folder.
To continue, click Next. If you would like to select a different folder, click Browse.
C:\Program Files\AML Products\Registry Cleaner
Browse...
At least 7.0 MB of free disk space is required.
< Back
Next >
Cancel

Filtered Text: Select Destination Location Setup Install Folder Continue Next Select Click Browse 7.0MB Disk Space Required
Action Inferred Using KB: Change Drive (if space checks fail) then Click Next

Enumerated Text:
Ready to Install
Setup is now ready to begin installing AML Free Registry Cleaner on your computer.
Click Install to continue with the installation, or click Back if you want to review or change any settings.
Enumerated Control:
Back, Install, Cancel

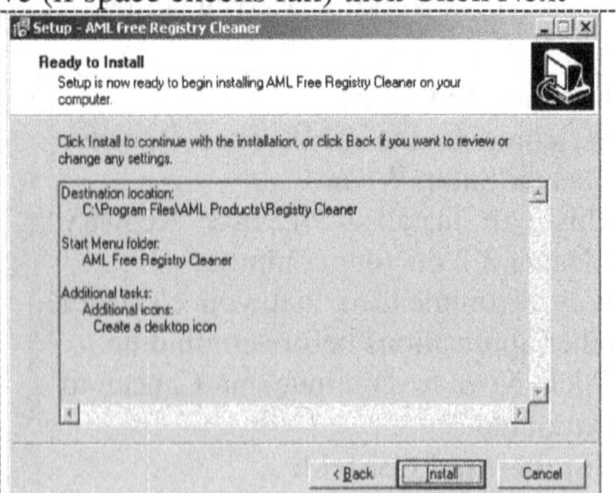

Filtered Text: Ready Install Setup Begin Click Continue Installation Back Review Change Setting
Action Inferred Using KB: Click Install

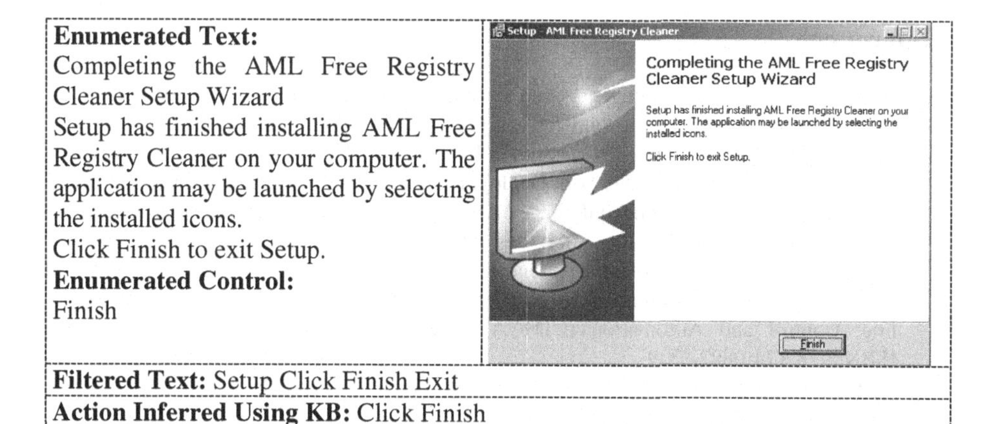

Enumerated Text: Completing the AML Free Registry Cleaner Setup Wizard Setup has finished installing AML Free Registry Cleaner on your computer. The application may be launched by selecting the installed icons. Click Finish to exit Setup. **Enumerated Control:** Finish	
Filtered Text: Setup Click Finish Exit	
Action Inferred Using KB: Click Finish	

3.5 Verifier Agent (VA)

After initialization Verifier Agent (VA) moves to the first node in destination list and loads the product information stored on the node during installation. VA is responsible to verify and check the product installed on the network node. VA will verify the size of the application, number of directories, number of files, and registry entries etc and create a verification log of the product. After verifying the product information, VA will run the application in the background. After delay of α, VA will close the application and checks the exit code of the application. Depending on the exit code Success / Failure will be updated in the verification log (i.e. if application return 0x0 it means success else failure). VA will notify SSA about the product verification result and application running status. After completing the task on the current node, VA will move to the next node in the destination list and repeat the same steps.

4 Conclusion

Software deployment on network sometimes becomes very tedious and time consuming task as the process needs human interaction. In this paper, we have proposed Cognitive Agent for Automated Software Installation (CAASI) for silent automated installation over the network. Proposed Framework support silent and automated installation over the network for any type and kind of software and is independent of the installer type used for deployment. CAASI does not require any specific kind of setup like MSI nor require any training for installation. CAASI will intelligently install any kind of software over the network with the help of mobile agents. The work can be extended in many directions. Repairing software feature and support for installing software patches over the network can be added in the existing framework.

References

1. Manzoor, U., Nefti, S.: An agent based system for activity monitoring on network – AB-SAMN. Expert Systems with Applications 36(8), 10987–10994 (2009)
2. Manzoor, U., Nefti, S.: Silent Unattended Installation / Un-Installation of Software's on Network Using NDMAS – Network Deployment Using Multi-Agent System. In: Proceeding of The Fourth European Conference on Intelligent Management Systems in Operations (IMSIO), Manchester, UK, July 7-8 (2009) (accepted for publication)
3. Manzoor, U., Nefti, S.: Silent Unattended Installation Package Manager – SUIPM. In: Proceeding of IEEE International Conference on Computational Intelligence for Modelling, Control and Automation (CIMCA), Vienna, Austria, December 10-12 (2008) ISBN: 978-174-0882-98-9
4. EMCO Remote Installer (2009), `http://www.emco.is/products/remote-installer/features.php`
5. Compulsion Software Remote Installer (2008), `http://www.compulsionsoftware.com/`
6. Ilarri, S., Mena, E., Illarramendi, A.: Using cooperative mobile agents to monitor distributed and dynamic environments. Information Sciences 178, 2105–2127 (2008)
7. Manzoor, U., Nefti, S., Hasan, H., Mehmood, M., Aslam, B., Shaukat, O.: A Multi-Agent Model for Mine Detection – MAMMD. In: Lytras, M.D., Carroll, J.M., Damiani, E., Tennyson, R.D. (eds.) WSKS 2008. LNCS (LNAI), vol. 5288, pp. 139–148. Springer, Heidelberg (2008)
8. Summiya, I.K., Manzoor, U., Shahid, A.A.: A Fault Tolerance Infrastructure for Mobile Agents. In: Proceeding of IEEE Intelligent Agents, Web Technologies and Internet Commerce (IAWTIC 2006), Sydney, Australia, November 29 – December 01 (2006)
9. Milojicic, D., Douglis, F., Wheeler, R.: Mobility: Processes, Computers, and Agents. ACM Press, New York (1999)
10. Manzoor, U., Ijaz, K., Shahid, A.A.: Distributed Dependable Enterprise Business System – DDEBS. In: Proceeding of Springer Communications in Computer and Information Science, Greece, Athens, September 24-28, vol. 19, pp. 537–542 (2008) ISBN 978-3-540-87782-0
11. Koffron Micah, A.: System and method for optimized distributed file transfer. In: Microsoft Corporation, WA, US (December 2006), `http://www.freepatentsonline.com/y2006/0282405.html`
12. Rajah, K., Ranka, S., Xia, Y.: Scheduling bulk file transfers with start and end times. Computer Networks 52, 1105–1122 (2008)
13. Manzoor, U., Nefti, S.: Agent Based Activity Monitoring System – ABAMS. In: Proceeding of IEEE International Conference on Tools with Artificial Intelligence (ICTAI 2008), Dayton, USA, November 3-5, pp. 200–203 (2008) ISSN: 1082-3409 ISBN: 978-0-7695-3440-4
14. Niazi, M., Manzoor, U., Ijaz, K.: Applying Color Code Coordinated LRTA* (C3LRTA*) Algorithm on Multiple Targets. In: Proceeding of 9th IEEE International Multitopic Conference (INMIC), pp. 27–32 (2005)
15. Niazi, M., Manzoor, U., Summiya, I.K., Saleem, H.: Eliminating duplication and ensuring file integrity in Multisync: A multiagent system for ubiquitous file synchronization. In: Proceeding of 9th IEEE International Multitopic Conference (INMIC), pp. 767–772 (2005)
16. Java Agent Development Framework (JADE), http://jade.tilab.com/
17. Microsoft Windows (2009), `http://www.microsoft.com/WINDOWS/`
18. AML Free Registry Cleaner (2009), `http://www.amltools.com/free-registry-cleaner-tutorial.html`
19. Matthieu, L., Walter, W.: Complex computer and communication networks. Computer Networks 52(15), 2817–2818 (2008)

A Formal Definition of Situation towards Situation-Aware Computing

Minsoo Kim[1] and Minkoo Kim[2]

[1] Graduate School of Information and Communication, Ajou University, Suwon, South Korea
visual@ajou.ac.kr
[2] College of Information Technology, Ajou University, Suwon, South Korea
minkoo@ajou.ac.kr

Abstract. Context-aware computing has emerged as a promising way to build intelligent and dynamic systems in overall computer science areas such as ubiquitous computing, multi-agent systems, and web services. For constructing the knowledgebase in such systems, various context modeling and reasoning techniques were introduced. Recently the concept of situation-awareness is focused beyond the context-awareness. In fact, the concept of situation is not a new one; McCarthy has introduced the theory of situation calculus in 1963 and Reiter et al. formalized it based on action theory. Recent works are trying to exploit the concept of situation for understanding and representing computing elements and environments in more comprehensive way beyond context. However, in such research, the situation is not defined and differentiated from the context clearly. Accordingly, the systems do not take advantage of the situation. In this paper, we provide a formal definition of the situation and differentiate it from the context clearly.

Keywords: Situation, Situation-Aware Computing, Context-aware Computing.

1 Introduction

As context-aware computing [1][2] becomes a promising way to build intelligent and dynamic systems in overall computer science and engineering areas [3][4][5], context research takes up much room in the knowledge society. Several beneficial features of context-awareness, such as dynamicity, adaptability and interoperability, work as basis of system operations and, moreover, context-awareness becomes one of essential requirements of recent systems.

In spite of many advantages from adopting context-awareness, some researchers tackled the problem of limitation of representation power [6]; the context is weak for giving comprehensive understanding of a phenomenon. In order to make up for the weakness, the concept of situation was introduced. In fact, the concept of situation is not a new one; McCarthy has introduced the theory of situation calculus in 1963 [7], Reiter et al. formalized it based on action theory [8], and Endsley proposed the theory of situation awareness. Based on these previous works, recently researchers are trying to exploit the concept of situation for representing computing environments beyond the context [6][9][10].

M.D. Lytras et al. (Eds.): WSKS 2009, LNAI 5736, pp. 553–563, 2009.

However, in such recent researches, the situation is not defined clearly and there are ambiguities of differentiating *situation* from *context*. To differentiate from the context, several information types, such as action history, triggering operation, and relations between, are inserted into the definition of situation. Accordingly, the situation seems to be composed of various information as well as context and it makes that situation has much expressive power than context. If we agree with that, then situation is a representation method like context and something can be represented in more comprehensive way using the situation rather than context.

The major motivation of this work is here. Even though the situation is a crucial issue nowadays computing, there are just weak notions and, accordingly, the systems do not take the advantages of the situation. In this paper, we argue that a situation is a concept not only to represent a phenomenon, but also to implicate further operations related to the phenomenon. A situation is a problematic and developing status of a computational element and it implicates understanding the reason of the problem and the methods for solving. Therefore, on a given situation, it is possible to understand and develop the current status by grasping the current context. This is a major advantage coming from the situation-awareness. A system can be represented in more comprehensive way with the concept of situation as well as the context.

Summarizing the analysis and motivations, we have following three contributions in this work:

1. We define the situation formally and differentiating it from context. From the upper analysis, we define the situation as a problematic state of computation element and it includes the plans for solving the problem. The context is used to define a situation, but a context cannot be a situation itself.
2. Base on this definition of situation, we introduce a situation flow that composes of a sequence of situations for achieving a certain goal. We define five types of situation and transition rules between situations. In addition, we introduce situation composition for flexible situation modeling.
3. As a final goal of this work, from the formalization of situation we will propose a situation-driven application development. We briefly describe situation-drive application development in last section.

2 Related Works

2.1 Brief History of Context Research

Attribute, context and situation are concepts representing knowledge around objects or systems and hence, at times it is difficult to clearly distinguish among them. Generally, an attribute is a specification that defines a property of an object or computing elements, for example size of file and age of person. The context is referred to as 'any information that can be used to characterize the situation of an entity (i.e. whether a person, place or object) that are considered relevant to the interaction between a user and an application, including the user and applications themselves' [2]. Therefore, any attribute of a system can be context directly and can be used for manipulating context. In other words, context is more conceptualized as a set of attributes. For example, an aggregation of two attributes temperature and humidity can manipulate context weather. We can say 'it is so hot' when temperature is 95F and humidity is 80%. Using the context, the status of system can be represented in a more comprehensive form and made more human understandable.

The issues of context have been broadly researched over the world and all areas of computer sciences. In special, as computing environment changes from distributed computing to mobile computing, and again from mobile computing to ubiquitous/pervasive environment [11], context has played a key role for designing and developing applications. In the literature the term context has been defined in several ways. Schilit who first introduces the term context-awareness defined the context as follows: *Three important aspects of context are: where you are, who you are with, and what resources are nearby* [1]. From the definition, only information about location identifies of people and objects, and changes to those objects can be regarded as context information. (This is why a location-aware system is treated as a context-aware system, even though those systems were using only the location information.) These kinds of context information are very useful for most distributed and mobile applications, but it is difficult to describe a situation of system sometimes. Schmidt et al. pointed out the problem and argued there is more to context than location [12]. Anind et al. pointed out the same problem in which these definitions are too specific and consequently proposed more general definition as follows: *Context is any information that can be used to characterize the situation of an entity. An entity is a person, place, or object that is considered relevant to the interaction between a user and an application, including the user and applications themselves* [2]. The definition makes it easier for an application developer to enumerate the context for a given application scenario. System developer or designer may use any kinds of information as context to express situations. Similar definition by Yau, who proposed a situation-aware application developing methodology, was introduced recently as follows: *Context is considered any detectable attribute of a device, its interaction with external devices, and/or its surrounding environment* [6].

These definitions may help to model context data relevant to systems in flexible way, but there is an ambiguity regarding application development. Any information, including attribute, action history, even context and situation itself, can be context of a system. From the definitions, it is obvious that context defines situation somehow, but it is not easy to find out the differences between them. For better understanding of situation-aware computing, we need to make sure what the situation is, what differences are between situation and context, and what situation-aware computing is.

2.2 Situation-Aware Computing

The research about situation has long history and now it has been re-focused to represent systems more comprehensively. The situation has defined in several different ways. First, McCarthy, the originator of the situation calculus, defines situation 'is the complete state of the universe at an instant of time' [7]. Reiter formalized the situation calculus and argues situation 'is a finite sequence of actions. It's not a state, it's not a snapshot, it's a history' [8]. In research and systems which adopted McCarthy's definition, a situation is described by using properties of systems such as variables, attributes and context. In contrary, if a situation is a history, for example in GOLOG [13] which is a programming language for describing situation calculus, the state of the world is described by functions and relations (fluent) relativized to a situation.

Recently, Yau et al. defined the situation in the perspective of pervasive environment as follows [6][14]: *The situation of an application software system is an expression on previous device-action record over a period of time and/or the variation of a set of contexts relevant to the application software on the device over a period of time. Situation is used to trigger further device actions.* This definition seems to just combine McCarthy' and Reiter' definitions, but it provides a more comprehensive understanding of situation. First, the concept of context is inserted into the definition. Second, they refer the purpose of situation which is to trigger the further actions of systems. It shows interoperation between context and situation, and direction of situation-aware computing. However, this definition does not show what differences are between context and situation and what meaning (semantic) of situation is (situation is not just expression).

One more issue related to the situation research is the theory of situation-awareness [15][16] which was significantly used in traffic and aircraft control. In general, situation awareness refers the perception of environmental elements, the comprehension of their meaning, and the projection of their status of the near future. Even though the definition of the situation did not discuss deeply, we can find it out from the definition of situation-awareness. At least, a situation implicates some meaning and catalyst which can make problems or trigger actions. This arbitrary definition seems like to make the situation more comprehensive and differentiate it from the context. Our definition of the situation starts from this idea will be discussed in next section.

3 Formal Definition of Situation

From the survey of existing context and situation researches, the situation is a concept not only representing a phenomenon, but also implicating an understanding and developing the phenomenon. Situation does not use to trigger actions, but on a given situation the actions are triggered by recognizing a set of context information. Therefore, situation-awareness means to recognize problems by context-awareness and to discover a sequence of actions for solving problem also by the context-awareness. In other words, at a certain point of time, a system which has abilities of situation-awareness as well as context-awareness understands its context, recognizes a problem, and makes plans to resolving the current situation.

In this sense, the situation and context are conceptually different terms. While the context is a concept more like to data, the situation is more like to a process including context-awareness. Therefore, situation-awareness is an integrated-awareness mechanism with recognizing problems and planning to further actions as well as context-awareness. Fig. 1 shows the notion of the situation-awareness in this perspective. One point we have to explain is to separate history information from the context. Although the history information can be represented as context information, the history of actions takes up much portion in the perspective of planning. For example, some actions should be performed as an exact order or certain two actions cannot accomplish at same point of time [13].

Fig. 1. The Notion of Situation-Awareness

Definition 3.1 (Situation). *Situation is a problematic and developing state of a computational element characterized by its context. 'Problematic' means that there is a problem of which a system should take care in a given situation and 'developing' means that a situation is changeable to another situation or state by system operations for solving the problem. Formally,*

$$\mathcal{S} = < \mathcal{C}, \mathcal{H}, \mathcal{P}, \mathcal{A} >, \textit{where}$$

$\mathcal{C}$ is context information, $\mathcal{H}$ is is history of actions, $\mathcal{P}$ is a problem, and $\mathcal{A}$ is a plan (set of actions).

We note that it may be not a same situation even though a same problem is recognized at different two points of time. It does not mean the situation is time-dependent. As shown in fig. 2, the system recognizes a same problem in both sub figures (a) and (b), but different plans are applied. For example, both actions A_1 and A_2 can solve problem P_1, but in the context A_1 or A_2 cannot be allowed to activate. In other words, a problem is specified by context and history information and a plan is also specified in same way.

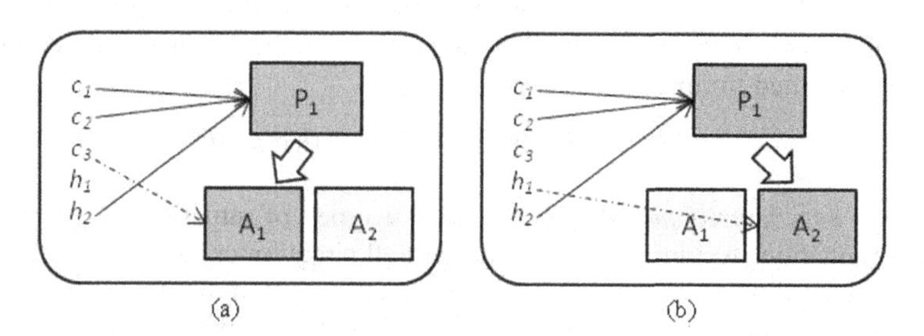

Fig. 2. Situation is a concept including problems and plans as well as context. Moreover, for a same problem, it is possible to plan in different ways depending on context. Both have the same problem in this figure, but each has different solution because A_2 is not available in the context (a) and A_1 is not available respectively (b). That is, (a) and (b) are not a same situation.

4 Situation Flow and Situation Composition

4.1 Situation Flow

The situation helps to design a process for solving a problem. In the case of a complex problem or a problem which requires several plans, a number of situations may be required and they make a flow of situations. In fact, a problem which is recognized in a certain situation should be solved out of the situation, because of the definition of situation does not represent the achieving condition of goal (state after problem is solved). Moreover, another situation is recognized immediately after problem is solved. For convenience for designing such flows as well as representing the 'out of the situation', we introduce following five types of situation.

Definition 4.1.1 (Null Situation φ). Nothing defines the null situation; no problems, no goals, and no plans. It is worth to be aware current situation and represent out of the situation, even though the null situation does not mean anything itself.

$$\varphi \overset{\text{def}}{=} \{\}$$

Definition 4.1.2 (Start Situation s_0). The start situation is a situation in which a situation flow starts. The start situation is defined by a problem, goal and plan.

$$s_0 \overset{\text{def}}{=} \{\exists \rho, \exists g, \exists \alpha \mid \rho \in \mathcal{P}, g \in \mathrm{G}, \alpha \in \mathcal{A}\}$$

Definition 4.1.3 (Goal Situation s_g). The goal situation is a situation in which a problem is solved. The goal situation is defined only by achieving condition.

$$s_g \overset{\text{def}}{=} \{\text{achieving condition}, \nexists \rho, \nexists g, \nexists \alpha \mid \rho \in \mathcal{P}, g \in \mathrm{G}, \alpha \in \mathcal{A}\}$$

Definition 4.1.4 (Ongoing Situation s_{on}). If a situation comes from the start situation or other ongoing situation, then the situation is ongoing situation. The ongoing situation is defined in same with the tart situation. In fact two types of situation equal logically, but two are distinguished in a situation flow. See the axiom 3.1.

$$s_{on} \overset{\text{def}}{=} \{\exists \rho, \exists \alpha, \exists g \mid \rho \in \mathcal{P}, g \in \mathrm{G}, \alpha \in \mathcal{A}\}$$

Definition 4.1.5 (Congest Situation s_c). If a situation has a problem and goal but no plan, then the situation is a congestion situation. The congestion situation is caused by either a plan is not defined or the plan cannot be executed because the condition of plan is not satisfied in run-time.

$$s_c \overset{\text{def}}{=} \{\exists \rho, \nexists \alpha, \exists g \mid \rho \in \mathcal{P}, g \in \mathrm{G}, \alpha \in \mathcal{A}\}$$

Axiom 4.1 lists key relationship between five types of situation in perspective of situation transition. Axiom (1) states that the null situation can move to only the start situation. According to axiom (2), the start situation can move to any other type of situation excepting the null situation. Axiom (3) indicates that the ongoing situation can move to any other type of situation excepting the null and start situation. Axiom (4) states that the goal situation can reach form the start situation or ongoing situation and, according to axiom (5), the goal situation automatically transits to the null situation. The congestion situation cannot move any type of situation by axiom (6).

The congestion situation can be moved from all types of situation excepting goal situation. The usefulness of the congestion situation comes from mutual complement between design-time and run-time. In design time, a number of congestion situations are designed by intent and the situations become unreachable. On the other hand, if a congestion situation is captured in run-time, then it goes to back to the system designer as a feedback and the situation is modified to be unreachable. This beneficial process makes situation flows to work correctly. The goal situation, in fact, is a null situation, but it makes a situation flow to be finished. Only one process which operates in the goal situation is to check that a goal of previous situation is achieved. The goal situation, and then, automatically is changed to the null situation. As we mentioned in definition 4.1.4, the start situation and ongoing situation cannot be differentiated by definition itself. The difference comes from the situation flow; the start situation can be reached only from the null situation. The reason why the start situation are defined separated from the ongoing situation is to explicitly know the point of time the problem is occurred and to trace the situation flow.

Axiom 4.1 (Situation Flow Rules). For the situation flow between five types of situation,

(1) $s_i \xrightarrow{trans} s_0 \vdash s_i \equiv \varphi$

(2) $s_0 \xrightarrow{trans} s_i \vdash s_i \not\equiv \varphi,$

$\varphi \xrightarrow{trans} s_i \wedge s_i \xrightarrow{trans} s_j \vdash s_i \equiv s_0$

(3) $s_{on} \xrightarrow{trans} s_i \vdash s_i \not\equiv \varphi \wedge s_i \equiv s_0,$

$s_0 \xrightarrow{trans} s_i \wedge s_i \xrightarrow{trans} s_k \vdash s_i \equiv s_{on} \wedge s_k \not\equiv \varphi$

(4) $s_i \xrightarrow{trans} s_g \vdash s_i \equiv s_0 \vee s_i \equiv s_{on}$

(5) $s_g \xrightarrow{auto-trans} s_i \vdash s_i \equiv \varphi$

(6) $s_c \xrightarrow{trans} s_i \vdash s_i \notin \mathcal{S}$

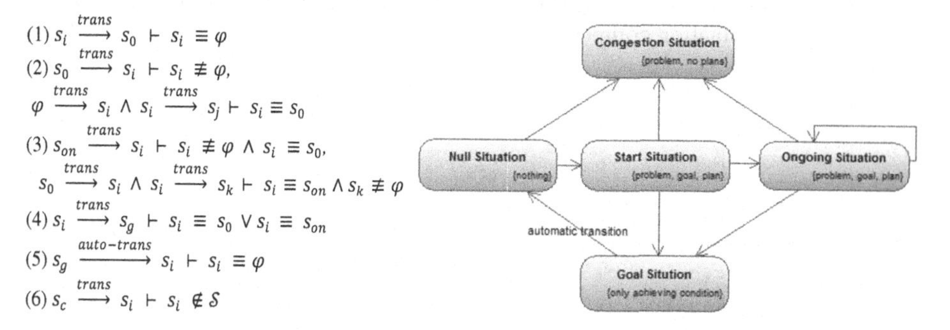

4.2 Situation Composition

According to the definition of situation, a situation is specified with exactly one problem (one goal) and one plan. When a problem has several solutions, therefore, a number of situations should be designed separately as the number of plans. In contrary, different problems can be solved by one plan depending on context. In such case, it also needs to design multiple situations as the number of problems. These problems make a situation design to be tedious and difficult. It is definitely true that all situations are needed if they have different problems or plans with each other. However, in order to provide convenience for the situation design, we introduce the situation composition. Several benefits can come from the situation composition such as reducing overhead associated with designing situation flow.

Definition 4.2.1. (Situation Composition). Given following components within a system,

$\mathcal{S}$: a set of situations,

$\mathcal{P_H}$: a problem hierarchy containing all problems, in special p_0 is null problem which is a root of the hierarchy (i.e. an ancestor of all other problems),

$\mathcal{A}_{\mathcal{H}}$: a plan (action) hierarchy containing all plans, in special a_0 is a null plan which is a root of the hierarchy (i.e. an ancestor of all other plans), and following three types of composition are defined.

$$S' \xrightarrow{P-composition} s_c^P \ iff,$$
$$\exists p_a(\forall s_i.p \preccurlyeq p_a) \mid S' \subseteq S, s_i \in S',\ p_a \in \mathcal{P}_{\mathcal{H}},\ p_a \not\equiv p_0,$$
$$S' \xrightarrow{A-composition} s_c^A \ iff,$$
$$\exists a_a(\forall s_i.a \preccurlyeq a_a) \mid S' \subseteq S, s_i \in S',\ a_a \in \mathcal{A}_{\mathcal{H}},\ a_a \not\equiv a_0,$$
$$S' \xrightarrow{PA-composition} s_c^{PA} \ iff,$$
$$\exists p_a \exists a_a(\forall s_i.p \preccurlyeq p_a \wedge \forall s_i.a \preccurlyeq a_a) \mid S' \subseteq S, s_i \in S, p_a \in \mathcal{P}_{\mathcal{H}}, a_a \in \mathcal{A}_{\mathcal{H}}, p_a \not\equiv p_0, a_a \not\equiv a_0.$$

According to the composition type, there are three types of composited situation. First, s_c^P composed by *P-composition* has one problem and more than one plans. The problem, obviously, is an ancestor problem of all problems of source situations. Second, s_c^A composed by *A-composition* has one plan and more than one plans. The s_c^{PA} composed by *PA-composition* has one problem and one plan like normal situation by definition 4.2.1. The problem is an ancestor problem of all problems of source situations and the plan is an ancestor plan, respectively. Fig. 3 illustrates an example of three types of situation composition.

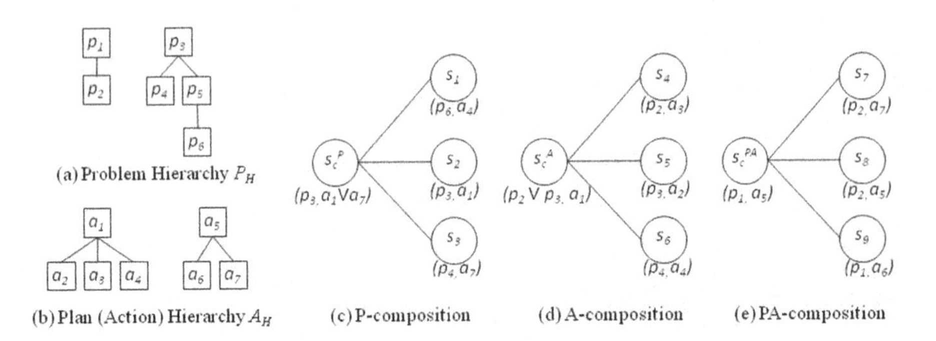

Fig. 3. Illustrative example of Situation Composition

The situation flow should be satisfied following conditions after composition. First, for the any type of composited situation, a set of previous situations (incoming edges) of the situation should include all previous situations of source situations for the composition. Respectively, a set of next situation (outgoing edges) should include all next situations of source situations.

The relation between ancestor and child in the problem hierarchy and plan hierarchy is the 'is-a' like the sub-class in ontology languages. It means that recognizing condition of child problem includes the condition of its ancestor. Even though ancestor problem is recognized, the child problem may not be recognized. Therefore, recognizing a composited situation does not mean that one of the source situations is recognized, but it is possible to develop to the source situations. In this sense, in order to simplify the flow, it is not a good way to composite situations having same previous situation or same next situations, since two situations have different types of plans as well as problems, the conditions of two situations may be different although they have same previous situation. In this case, even though two situations are composed to one situation, there are no benefits to be aware situations. Moreover, it is possible to make confusion about flow of situation.

5 An Example of Designing Situation-Flows

Scenario (Online Complaint). *John, a manager of customer service center of an online sale company, received a complaint from customer; the customer has ordered a product one month ago, but it was not delivered yet. The customer still wants to get that product. In order to solve the problem, first John plans to find out the reason. He thinks that the reason is whether the product was not delivered from the company or the delivery company did not deliver. He will let someone to re-deliver if the company or delivery company keeps the product. Otherwise, he will refund the customer.*

Figure 4 shows possible situation flows. There are one start situation, five ongoing situations, and one goal situation. We note again that any flow of situation cannot be determined but expected. Instead, depending on the context and history in run-time, one of following four flows will be occurred. In this scenario, there are four possible flows that can resolve online complaint; (a) $s_0 \rightarrow s_1 \rightarrow s_g$ *(The company re-deliveries), (b)* $s_0 \rightarrow s_1 \rightarrow s_4 \rightarrow s_g$ *(The company refunds), (c)* $s_0 \rightarrow s_2 \rightarrow s_3 \rightarrow s_g$ *(The delivery company re-deliveries), (d)* $s_0 \rightarrow s_2 \rightarrow s_3 \rightarrow s_5 \rightarrow s_g$ *(The delivery company refunds).*

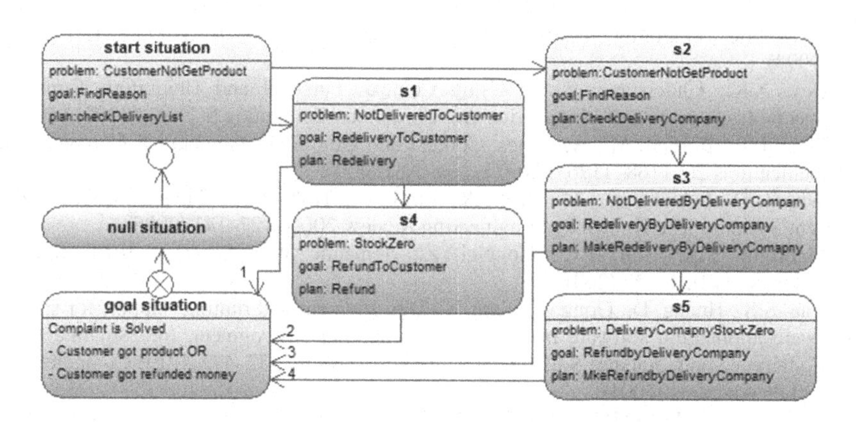

Fig. 4. Situation flow of example '*online complaint*'

6 Conclusion and Future Works

Context-aware computing becomes an essential requirement of nowadays computing and several notable works has proposed. Recently, situation-aware computing has emerged in order to make up for the weakness as well as to take advantage of context-aware computing. In the literature, we formalized the concept of situation and provided the notion of situation-awareness in this work. Situation is a concept of developing a state as well as understanding context. Situation is more like to a process, while the context is more like to a data. In this perspective, we introduced the concepts of situation flow and situation composition. It makes easy and efficient to design and implement dynamic systems based on the concept of situation.

There are several issues that need to be considered as future works.

- Knowledge representation scheme is required. Techniques that have been adopted in context modeling were in our consideration and, finally, we choose ontology-based approach to exploit the expressive power and reasoning ability. Now we are constructing situation ontology based on Web Ontology Language OWL.
- Scenario-based examples should be developed for showing which beneficial features advance nowadays computing through situation-aware computing. We did not show illustrative examples due to the limit of paper length in this work, but we continue to develop scenarios in diverse domain.
- Integrated development tool-kit is required in order to easy and fast development of situation-based applications. In particular, the tool-kit should support easy ways to model situation as well as context. Now we are on a design stage and we expect that the development tool-kit woks as an integrated development environment supporting all stage of development from modeling to deployment of applications.

References

1. Schilit, B., Adams, N., Want, R.: Context-aware Computing Applications. In: Workshop on Mobile Computing Systems and Applications, pp. 85–90. IEEE Press, Los Alamitos (1994)
2. Dey, A.K.: Understanding and Using Context. Personal and Ubiquitous Computing-Special Issue on Situated Interaction and Ubiquitous Computing 5(1), 4–7 (2001)
3. Prekop, P., Burnett, M.: Activities, Context and Ubiquitous Computing. Computer Communications 26, 1168–1176 (2003)
4. Weyns, D., Schumacher, M., Ricci, A., Viroli, M., Holvoet, T.: Environments in Multiagent Systems. The Knowledge Engineering Review 20(2), 127–141 (2005)
5. Mcllraith, S.A., Son, T.C., Honglei, Z.: Semantic Web Services. IEEE Intelligent Systems 16(2), 46–53 (2001)
6. Yau, S.S., Huang, D., Gong, H., Seth, S.: Development and runtime support for situation-aware application software in ubiquitous computing environment. In: 28th International Conference on Computer Software and Applications, pp. 452–472. IEEE Computer Society, Los Alamitos (2004)
7. McCarthy, J., Hayes, P.J.: Some Philosophical Problems from the Standpoint of Artificial Intelligence. Machine Intelligence 4, 463–502 (1969)
8. Reiter, R.: The situation Calculus Ontology. Electronic News Journal on Reasoning about Actions and Changes (1997), `http://www.ida.liu.se/ext/etai/rac/notes/1997/09/note.html`
9. Matheus, C.J., et al.: SAWA: an assistant for higher-level fusion and situation awareness. In: SPIE proceedings, vol. 5813, pp. 75–85 (2005)
10. Baumgartner, N., Retschitzegger, W.: A Survey of Upper Ontologies for Situation Awareness. In: Knowledge Sharing and Collaborative Engineering. ACTA press (2006)
11. Strang, T., Linnhoff-popien, C.: A Context Modeling Survey. In: 1st International workshop on Advanced Context modeling, Reasoning and Management at UbiComp 2004 (2004)
12. Schmidt, A., Beigl, M., Gellersen, H.W.: There is more to Context than Location. Computers and Graphics 23, 893–901 (1999)

13. Levesque, H.J., Reiter, R., Lespérance, Y., Lin, F., Scherl, R.B.: GOLOG: A logic programming language for dynamic domains. The Journal of Logic Programming 31(1), 59–83 (1997)
14. Yau, S.S., Wang, Y., Huang, D., In, H.P.: Situation-aware contract specification language for middleware for ubiquitous computing. In: Ninth IEEE Workshop on Future Trends of Distributed Computing Systems, pp. 93–99. IEEE Computer Society, Los Alamitos (2003)
15. Endsley, M.R.: Measurement of situation awareness in dynamic systems. Human Factors 37(1), 65–84 (1995)
16. Endsley, M.R.: Toward a theory of situation awareness in dynamic systems. Human Factors 37(1), 32–64 (1995)

Knowledge Artifacts Modeling to Support Work and Learning in the Knowledge Society

Fabio Sartori, Federica Petraglia, and Stefania Bandini

Dipartimento di Informatica, Sistemistica e Comunicazione
Università degli Studi di Milano–Bicocca
Via Bicocca degli Arcimboldi 8, 20126 Milano, Italy
{bandini,federica.petraglia,sartori}@disco.unimib.it

Abstract. This paper introduces a conceptual and computational framework for the design of Computer Supported Collaborative Learning systems devoted to support situated learning–by–doing within organizations that can be interpreted as a constellation Communities of Practice. The leading idea is that the way in which the learning requirements can be identified is to look at the artifacts that communities' members develop to support their work. Due to the extreme heterogeneity of knowledge, information and problem solving strategies that characterize them, an approach based on Knowledge Based System has been adopted. Moreover, the framework separates the representation of the core knowledge produced by Communities of Practice over the years from the problem solving strategies they adopt: in this way, more than one approach to problem solution can be incorporated in the target system.

1 Introduction

The understanding of the complex interconnections between work and learning makes it difficult to answer to the central question on "how can they be productively taken up together in future CSCW research" without taking into consideration the (classes of) specific contexts in which they are performed. Actually, these contexts characterize the nature of these interconnections. In this paper we will focus on work settings where people accomplish tasks that require technical core knowledge, typically as the design and the manufacturing of complex products. This framework, although restricted in relation to the variety of learning situations that could be considered, is anyhow broad and relevant enough since it encompasses a crucial need of a huge number of companies: to be competitive on the market with innovative products and to be adaptable to unforeseen contextual conditions. Innovation and flexibility are good stimuli for them to pay attention to continuous training and to invest in organizational solutions and technological tools that make them effective. From now on, we will refer, by default, to technical knowledge workers when speaking of people involved in work and training. In this perspective, the competencies of an individual are the result of the composition of basic skills possibly acquired in institutional educational programs, and the ability to apply them in the specific context of the organization the individual belongs to.

M.D. Lytras et al. (Eds.): WSKS 2009, LNAI 5736, pp. 564–573, 2009.

This combination is especially valuable for companies since it is a precondition necessary to let problem solving activities lead to solutions that are viable and sustainable for the company itself, both in case of design for innovation and in case of unforeseen events that require adapting the production process to the unpredictable conditions. From a more conceptual point of view, a growing number of studies, researches and information technology systems are focusing on the concept of *Community of Practice* (CoP), that combines organizational aspects with issues typically related to learning processes. A CoP may be characterized as *a group of professionals informally bound to one another through exposure to a common class of problems, common pursuit of solutions, and thereby themselves embodying a store of knowledge* (Hildreth et al., 2000). Originally born in the situated learning theory, the concept of CoP (Wenger, 1998) has been conceived to delineate the social and collective environment within companies where (core) knowledge is generated and shared. Therefore, CoPs may be considered as complementary organizational structures, made up of networks of people and built up to solve problems arising from common practice.

The concept of CoP is thus useful to name and describe a situation where work and learning are tightly related but, per se, is not useful to highlight the mechanisms by which this interconnection can be understood in operational terms so as to identify a technology able to support both of them. Again, it is difficult to give an answer to this problem in general terms: each working situation requires understanding its peculiarities and construct the solution around them. In this paper we describe how our experience in dealing with this problem in specific cases has allow developing a general framework for cultivating CoPs within organizations, especially from the educational point of view.

The next section describes how we used the notion of CoP to come up with the above requirements identification and how the latter allows the definition of a more general computational framework supporting learning within and across CoPs. In section 3, we motivate and discuss the overall architecture supporting our solution. Section 4 positions our approach in the framework of Computer Supported Collaborative Learning (CSCL), according to a classification proposed in (Kumar, 1996). Finally, conclusions about our approach and its developments are briefly pointed out together with lessons learned that can be used in the design of CSCW applications.

2 Learning Mechanisms in a Community of Practice

As alluded in the introduction, the notion of CoP is suggestive but too vaguely defined so as to be a conceptual tool practically useful to uncover the way in which knowledge is actually managed within organizations for sake of work and learning. This fact is testified by the several declination of the notion of community (sometimes called network or group) associated with qualifying items: e.g., epistemic community, community of interest or interest group, knowledge network, and so on. Hence, *the final effect appears to be that different names are applied to the same phenomenon and the same name appears to refer to different phenomena* (Andriessen, 2005). In such situation, we had to find our own way to uncover the existence of communities in the target reality, and more importantly to identify their boundaries. This was necessary not to build fictitious fences between them but to characterize them enough to be able to identify suitable supports to work and learning within the communities themselves.

Several factors make this characterization difficult. Communities of Practice (CoPs) are groups of people sharing expertise and flexibly interacting to deepen their knowledge in this domain (Wenger, 1998). As such they are of an informal nature: a CoP is typically transversal with respect to the formal structure of an organization. Since the membership is established on the basis of a spontaneous participation rather than on an official status, CoPs can overcome institutional structures and hierarchies, typically they span *across Business Units*. Since core knowledge is often distributed in different organization components, people working in cross–functional units can develop communication strategies that transcend product line fragmentation or other formal communication flows to share/maintain their expertise. On the one hand, the official organizational structure cannot help since CoPs are different from business or functional units and teams (Lesser and Storck, 2001); on the other hand, they are not simple networks: the existence of a CoP lies in the production of a shared practice and in the involvement of its members in a collective learning process. Last but not least, people are not always conscious to belong to a CoP and are very influenced by the formal structure of their company: hence, they often describe their working life in terms of the latter and of the involved processes, although they recognize the existence of a parallel cooperation structure.

The key to overcome these difficulties emerged from the investigation on how organization members use professional languages in their double role of locality and boundary (Smethurst, 1997). In so doing, first we recognized the existence of several professional jargons; second we were able to delineate the boundaries of the groups speaking them: both to cooperate by using deep and rich technical details (locality) and to interact with other groups in a more abstract way (boundary); third, we were able to check if these groups behaved like a CoP – that was the case - and to identify the main mechanisms they used to create, use and share knowledge in their daily work. More details on this process can be found in (Bandini et al., 2003a). Here we concentrate on the aspects of this discovery that have a more direct impact on the requirements of a learning functionality.

A second discovery, strictly related to the presence of professional jargons, was the identification of what we called *knowledge artifacts*: that is, conceptual, linguistic and/or modeling tools whose structure is shared by the members of the community to make their cooperation and problem solving activity more effective. This sharing can be explicit, e.g., reified in an information structure on a piece of paper, or implicit, e.g. in the way in which production phases are typically described and referred to. Anyhow, each knowledge artifact stratifies competencies and experience by means of a jargon that, irrespective of the level of detail of the represented knowledge, owns very precise syntactic and semantic rules and conventions. Now, exactly the mastering of these knowledge artifacts and the related jargons is what characterizes the members of the identified communities. According to CoP's feature named legitimate peripheral participation (Lave and Wenger 1991), these communities are naturally oriented to accept new or differently skilled people and share their practice and knowledge with them: here is where the learning issue comes upfront.

The primary quality of this training is to make people able to effectively use the jargon and the artifacts of the target community. In other words, the goal of the training is to shorten the time spent to cover the trajectory from the periphery to the center of the community. From the above considerations, it follows that training has to be concentrated on the practices typical of the community: in this view, learning–by-doing is a paradigmatic approach since it is based on well known pedagogical theories, like cognitivism and behaviorism, which emphasize the role of practice in humans' intellectual growth and knowledge improvement (Arrow, 1962). In particular, this kind of learning methodology states that the learning process is the result of a continuous interaction between theory and practice, between experimental periods and theoretical elaboration moments. According to this iterative view, the Learning–by–doing cycle (Figure 1) fits the needs of learning within a CoP, if doing is situated in its actual context. To emphasize this point we will speak of situated Learning–by-doing. In fact, the emergence of a new critical situation (e.g. a problem to solve, a new product to design and manufacture, and so on) causes a negotiation process (Wenger, 1998) among its members whose aim is to find a solution that, when agreed upon, will become a new portion of the community patrimony.

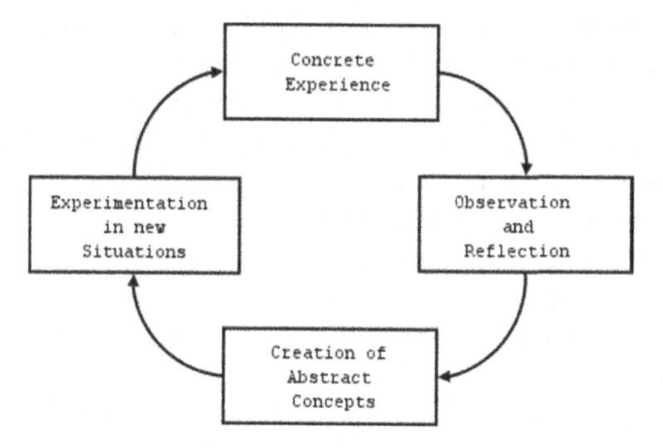

Fig. 1. An initial experience originates a reflection that ends with the discovery of a relation between the experience and its results (typically a cause–effect relation) and its generalization to a learned lesson that is applicable to new situations

The above considerations give an empirical evidence of the fact that in knowledge work there is an intimate connection between doing and learning, but more importantly that both use the same sources: the community experience about previous solutions and the community practice to create innovation. This means to develop the two basic functionalities, namely knowledge management and learning, as routed on the same source knowledge that was derived from the observation leading to the discovery of jargons and knowledge artifacts.

Unfortunately, the Learning–by–doing paradigm results very difficult to be implemented into a computational framework. Indeed the analysis of cause–effect relation between a problem and its solution and the generalization of it are very problematic, since it is sometimes impossible to explain how a given solution can be applied to a problem and solve it. What can be achieved is the memorization of previous experiences in terms of problem solution pairs and, when possible, of (recurrent) cause–effect relations to be applied in specific situations, with the aim to let professionals recover past solutions as the basis for new products or new production processes, and to check their adequacy against empirically based quality rules.

Thus, it is important to adopt methodologies and tools which allow to build an incremental model of the knowledge involved in the learning process or to bypass it if such model cannot be reasonably built up. These methodologies and tools are typical of the Knowledge Based Systems (KBS) paradigm that proved to be useful in the development of a computational framework to support work and learning within the identified communities of practice. What might look natural, if not obvious, a posteriori, was the outcome of a intensive activity characterized by subsequent refinements of the final knowledge based system. These refinements allowed us to identify an abstract architecture that reflects the needs of a generic CoP and that can be further specialized for specific instances of them.

3 A General Framework to Support Situated Learning–by–Doing

A few years ago the learning–by–doing approach was not so diffused: workers had typically to spend the first period of their professional life in reading general manuals about the new job and follow intensive courses promoted or organized by the management. Anyway, the limits of this kind of learning are evident.

For this reason, an increasing number of organizations are currently changing their view on the learning problem towards approaches more oriented to learning-by-doing. This change has an important impact on how the learning strategy can be effectively supported by computer–based tools and methodologies: while manuals and procedures are relatively simple to implement into a web-portal (although knowledge inside organizations is so variable from a day to the day after that manuals would have to be continuously updated), implicit knowledge must be externalized and modeled before it can effectively be part of learning system.

This is much more complex when the knowledge is developed within a CoP. In fact, a CoP is typically made of people having different roles in the decision making process about a problem. This is particularly true when the problem is so difficult that it has to be divided into more than one decisional steps, that is the typical situation when the considered CoP is involved in the design and manufacturing of complex products. When this happens, a newcomer enters a group characterized by heterogeneous knowledge, that is the result of a very complex process of experimentation and reification developed over many years. This knowledge is typically implicit or tacit and hence not reified: the only way for a beginner to quickly share this knowledge and contribute to increase it is to take part in actual problem solving as soon as possible. This is what situated learning-by-doing is about.

Here below we present our approach to the design and implementation of a framework for supporting situated learning–by–doing. The proposed framework is a three–tier architectural pattern based on an integration between a KBS approach and a component–based approach (Herzum and Sims, 2000): the architecture deals with the problems of knowledge Distributed access and maintenance through the modular architecture shown in Figure 2. The framework has been proposed for the development of Knowledge Management Systems supporting learning in wide organizations (Bandini et al., 2004).

The main feature of this three–tier architecture is the clear separation between application components implementing problem solving strategies that have to be experimented by newcomers and the ones providing other services. The problem solving strategy is distributed over a set of Knowledge Processing Modules (KPM), each one devoted to support a specific CoP role. KPMs are components of the framework and they are designed as KBSs. In this way different problem–solving strategies can be chosen according to their effectiveness in representing CoP decision making processes; moreover, heterogeneous knowledge–based modules based on different approaches to implement these strategies (e.g. rule–based, case–based, and so on) can coexist within the same system. The representation of knowledge and information needed by KPM is implemented into a Knowledge Repository (KR) component that can be viewed as a set of information sources (i.e. databases).

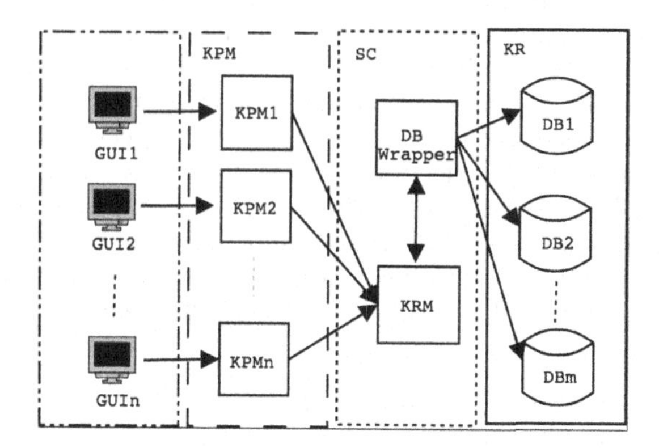

Fig. 2. A pattern architecture for the design of complex KBS

Since knowledge sources are usually heterogeneous and generally difficult to be accessed for several reasons (e.g., presence of confidential data, security policies, and so on), the framework incorporates a specific Middleware Service Component (MSC) to manage the communication between KPM and KR. This component is made of the Knowledge Repository Manager (KRM) and the DB Wrapper modules. KRM provides KPM with a set of functionalities to work on KR (e.g. access, visualization, updating), while the DB Wrapper interacts with the DBMS of the target database. When a KPM asks KRM for operating on KR, its request is interpreted by DB Wrapper that translates the query, sends it to the DBMS, receives the result, translates and returns it to KRM. In this way, potentially geographically distributed databases can be managed as a single and centralized knowledge source. From the knowledge maintenance point of view, this architecture allows to provide CoPs with a knowledge repository and a flexible tool to manage it. A centralized knowledge base facilitates all activities that require access to it and, at the same time, is an advantage also from the knowledge sharing among KPM components. In fact, they are provided with a specific view on the whole knowledge base they access to perform their tasks. Finally, advantages from the system extensibility point of view rely on the possibility to simply add new KBSs that exploit the same knowledge base content.

From the learning support point of view, the framework has many benefits:

- the centralized knowledge base acts as a comprehensive representation of the explicit knowledge produced by CoP over the years;
- the distributed intelligence allows to distinguish between different roles within a CoP. The KPM approach allows representing the knowledge involved in

specific problem solving strategies related to each role. This is particularly useful in case of CoPs involved in the design of complex products, where each step of the manufacturing process is managed by a well defined person or group. In this way, newcomers addressed to a specific phase of the decision–making process can be immediately introduced to it;
- moreover, the possibility to design and implement each KPM module independently from each other allows to choose the KBS approach that better describes the decision making process adopted by each CoP role. This is very important from the learning point of view, since a newcomer is immediately taught about the problem solving strategies he/she has been addressed;
- the division of system intelligence into more than one module allows to keep trace of the negotiation process among CoP roles. This allows to provide newcomers with a suitable support for dealing with process anomalies, occasional events that negatively affect the whole decisional process. In these situations, all CoP roles negotiate about solutions: the system memorize the negotiation process in terms of the different operation executed by its KPM part. In this way, it is possible for a CoP new member to retrieve what kind of operations are needed when problems are detected and who is necessary to contact inside the CoP in order to have suggestions.

The presented framework is general enough to be applied to domains characterized by heterogeneous knowledge.

4 Discussion: The Framework in the CSCL Context

The framework for supporting Learning-by-doing in organizations described above can be considered a sample of Computer-Supported-Collaborative-Learning.

Depending on the type of task to perform, a CSCL system could be employed to support the learning of concepts and notions, but also to support the learning of problem solving or designing activities (Blaye et al., 1991). The proposed framework takes care of all these aspects: in fact, concepts and notions are represented from the knowledge repository that is a comprehensive and easy to update archive of all the concepts involved in the target domain. Moreover, problem solving activities are represented by the presence of a specific KPM for each CoP role.

In (Kumar, 1996) a classification of CSCL systems according to seven dimensions is proposed: we use it to better set our approach in the CSCL scenario. The proposed dimensions are: control of collaborative interactions, tasks of collaborative learning, theories of learning in collaboration, design of collaborative learning context, roles of the peers, domains of collaboration and teaching/tutoring methodologies that inherently support collaboration.

From the control perspective, the proposed framework could be both active and passive. It depends on how it is implemented, although in most cases it will be meant passive, since it acts as a delivery vehicle for collaboration and it doesn't take any analysis or control action on it.

With reference to the tasks of the system, the framework supports collaborative concept–learning, problem–solving and designing tasks. In particular, when, the members of the overall community have a single goal represented by producing a given product that meets all the requirements. In this sense, according to (Smith, 1994), the result of the decisional process can be seen as the product of single effort made by the CoP. About the other two types of control (i.e. problem–solving and designing), they are implemented in by specific KPMs supporting each role in reaching their own objectives that are subgoals of the entire CoP process.

From the theory of collaboration standpoint, the proposed framework can be put in relation with shared cognition theory. This is implicit in the nature of learning–by–doing theorized by researchers in the situated learning field, like Lave and Wenger (1991). In a shared cognition theory collaboration is viewed as a process of building and maintaining a shared conception of a problem (Roschelle and Teasley, 1995), thus ensuring a natural learning environment. The division of the problem solving strategy into more than one KPMs is a direct realization of this: a newcomer for a specific CoP role can exploit its module to learn about the typical problem solving strategy adopted by its subgroup, as well as propose new solution to a problem that will become an asset for the entire CoP. In this way, a continuous training for all the CoP members is guaranteed, as well as a potentially never ending growth of core–knowledge patrimony.

The approach adopted in the design of collaborative learning context is knowledge-based, as KPMs have been implemented as Knowledge Based Systems. A KBS is more effective than other kinds of tools (e.g. knowledge portals, document repositories and so on) in supporting problem solving strategies and modeling core–knowledge. Moreover, the component–based paradigm allowed us to develop different KBS to implement different problem solving strategies into a uniform conceptual framework: a novice can learn as he/she is paired with an expert, according to Azmitia (1988) and Rogoff (1991), becoming an expert on his/her turn in a relatively short period of time.

By definition, a CoP is not characterized by the presence of different levels of skill and importance in its informal structure. Anyway, it is reasonable to think that, especially in enterprises, CoP will be made by *more expert people*, *less expert people* and *newcomers*. The way how our framework considers these roles depends on the specific domain. About the domains of collaborative, our approach has been developed thinking at designing and configuration problems (Puppe, 1996) in knowledge domains characterized by the presence of CoP.

Finally, the tutoring methodology behind a learning system developed according to our approach is situated learning, that is implied by the reference to the notion of CoP and of legitimate peripheral participation.

5 Conclusions and Future Work

This paper has presented a conceptual and computational framework for the development of a special kind of CSCL systems that is devoted to support learning in organizations where a constellation of interacting CoPs can be identified. From the technical point of view, the framework is based on the integration of approaches like knowledge–based systems and component–based design to represent heterogeneous core knowledge bases and guarantee a high degree of extensibility and maintainability.

An application of the framework, not described here due to the lack of space, is the P–Truck system, where three knowledge based systems have been created to support experts in truck tire manufacturing process. The P–Truck Training module of P–Truck exploits these KBSs to help novices to learn about their job according to the learning–by–doing paradigm. The implementation of P–Truck Training has allowed to demonstrate the suitability of the proposed approach in building this kind of CSCL systems, as presented e.g. in (Bandini et al., 2003). Part of the future work is devoted to verify if and with which effort the framework could be applied to other kinds of domains.

Another future line of development is the understanding of how knowledge management and learning systems can be integrated with CSCW systems, behind more or less sophisticated access to shared repositories of documents. The first unquestionable lesson learned (or better a confirmation of something that has been already pointed out by a number of researches) is that artifacts developed by people should be the core of the analysis and lead the development of collaborative applications since they incorporate a stratification of meanings that serves as the basis for both work and situated learning. Artifacts can be mainly oriented to coordination of practices around them (as discussed e.g., in (Schmidt and Simone, 1996)) or more oriented to represent core (technical) knowledge: in any case, they are the key to understand how coordination and knowledge are actually managed by people in their specific situations. The second lesson learned is that the development of applications supporting collaboration should consider in a joint way functionalities oriented to coordination, knowledge management and learning since the related practices are strongly interconnected, maybe at various degrees in different settings, in different steps of collaboration or for different purposes. This is a specific way to interpret flexibility of applications and raises challenging issues both at the conceptual and technological levels.

References

Andriessen, J.H.E.: Archetypes of Knowledge Communities. In: Proceedings of C&T 2005–2nd International Conference on Communities and Technologies, pp. 191–213. Kluwer Press, Milan (2005)

Arrow, K.: The Economic Implications of Learning by Doing. Review of Economic Studies 29(3), 155–173 (1962)

Azmitia, M.: Peer interaction and problem solving: When are two heads better than one? Child Development 59, 87–96 (1988)

Bandini, S., Simone, C., Colombo, E., Colombo, G., Sartori, F.: The Role of Knowledge Artifact in Innovation Management: the Case of a Chemical Compound Designers' CoP. In: Proceedings of C&T 2003–1st International Conference on Communities and Technologies, pp. 327–346. Kluwer Press, Amsterdam (2003)

Bandini, S., Colombo, E., Mereghetti, P., Sartori, F.: A General Framework for Knowledge Management System Design. In: Proceedings of ISDA 2004–4th IEEE International Conference on Intelligent Systems Design and Application, Budapest, pp. 61–66 (2004)

Blaye, A., Light, P.H., Joiner, R., Sheldon, S.: Joint planning and problem solving on a computer-based task. British Journal of Developmental Psychology 9, 471–483 (1991)

Herzum, P., Sims, O.: Business Component Factory. Wiley: OMG Press (2000)

Hildreth, P., Kimble, C., Wright, P.: Communities of practice in the distributed international environment. Journal of Knowledge Management 4(1), 27–38 (2000)

Kumar, V.: Computer-Supported Collaborative Learning: Issues for Research. In: Graduate Symposium. University of Saskatchewan, Sasktchewan (1996)

Lave, J., Wenger, E.: Situated Learning: Legitimate peripheral participation. Cambridge University Press, Cambridge (1991)

Lesser, E.L., Storck, J.: Communities of practice and organizational performance. IBM systems journal 40(1), 831–841 (2001)

Puppe, F.: A Systematic Introduction to Expert Systems. Springer, Berlin (1996)

Rogoff, B.: Social interaction as apprenticeship in thinking: Guided participation in spatial planning. In: Resnick, L., Levine, J., Teasley, S. (eds.) Perspective on Socially shared cognition, pp. 349–364 (1991)

Roschelle, J., Teasley, S.D.: Construction of shared knowledge in collaborative problem solving. In: O'Malley, C. (ed.) Computer-supported collaborative learning. Springer, New York (1995)

Schmidt, K., Simone, C.: Coordination Mechanisms: Towards a Conceptual Foundation of CSCW Systems Design. Computer Supported Cooperative Work: The Journal of Collaborative Computing 5, 155–200 (1996)

Smethurst, J.B.: Communities of Practice and Pattern Language. Journal of Transition Management (1997)

Smith, J.B.: Collective Intelligence in Computer-Based Collaboration. Lawrence Erlbaum Associate, Mahwah (1994)

Wenger, E.: Community of Practice: Learning, Meaning and Identity. Cambridge University Press, Cambridge (1998)

Image and Reputation Coping Differently with Massive Informational Cheating

Walter Quattrociocchi, Mario Paolucci, and Rosaria Conte

Laboratory of Agent Based Social Simulation - Institute of Cognitive Sciences and Technologies - CNR Rome, Italy

Abstract. Multi-agent based simulation is an arising scientific trend which is naturally provided of instruments able to cope with complex systems, in particular the socio-cognitive complex systems. In this paper, a simulation-based exploration of the effect of false information on social evaluation formation is presented. We perform simulative experiments on the RepAge platform, a computational system allowing agents to communicate and acquire both direct (image) and indirect and unchecked (reputation) information. Informational cheating, when the number of liars becomes substantial, is shown to seriously affect quality achievement obtained through reputation. In the paper, after a brief introduction of the theoretical background, the hypotheses and the market scenario are presented and the simulation results are discussed with respect to the agents' decision making process, focusing on uncertainty, false information spreading and quality of contracts.

1 Introduction

The functioning of communication based systems, as any socio-cognitive systems, requires individuals to interact in order to acquire information to cope with uncertainty. These systems relie on the accuracy and on the completeness of information. If agents need to be correctly informed, then the quality of information becomes a fundamental factor on the global system dynamics. The point is that in order to have a complete information, agents need a perspective, where all the relevant angles of looking are presented as squarely and objectively as possible. What is the relation between information credibility and opinion formation? What happens when the informational domains are not reliable and the trust level is poor? Can false information spread throughout a society to such an extent that it becomes a prophecy? In the real world, each decision needs strategies to reduce the level of uncertainty in the process of beliefs' formation and revision with respect to the decision's consequences. According to [14], there is uncertainty when an agent believes that there is a set of possible alternatives with equal probabilities of verification. The decision making process requires specific cognitive capacities enabling social agents to deal with uncertainty. Reputation has been shown to reduce uncertainty [7]: agents resort to external sources to improve their knowledge about the world, in particular to

M.D. Lytras et al. (Eds.): WSKS 2009, LNAI 5736, pp. 574–583, 2009.

improve their evaluations that will be applied, among others, to partner selection. In this paper, the dynamics of evaluations that are essential for intelligent agents to act adaptively in a virtual society was explored by means of simulation. A model for describing different types of evaluations spreading among agents endowed with different communicative attitudes (honest and liars) was applied to a computer simulation study of partner selection in a simulated market. Thanks to the RepAge Platform - a computational system allowing agents to communicate and acquire direct (image) and indirect and unchecked (reputation) information [11] - a number of experiments have been performed to investigate the role of this type of information on market efficiency. In our simulations, different cheating strategies are implemented. Agents can cheat (i) on the quality of products and sellers; (ii) on the quality of the other agents as informers. In this paper, we focus on the effects of the second, investigating how reputation and image can be used in coping with false information by means of simulations.

2 Social Evaluations

The dynamics of social systems are based on interactions among individuals. We will call informational domain the information available in a society spreading over the social network through gossip. People can populate their informational domain by communicating opinions and acquired information to other people and, in turn, getting information from the informational domain. This mechanism brings about a double process of information acquisition and production, which is reflected on beliefs' formation and revision. There are two sources of such beliefs, individual and collective. The individual source is based upon encounters, communications or observations, which lead the agent to form her opinions and evaluations about given matters of fact. This happens not only with direct experience, but also with indirect experience, whenever the recipient of information decides to trust the informer and share his views. Collective source, instead, generates information that is transmitted from one agent to another without the informers adopting it as truth. It is a kind of "meta-information", which agents form thanks to a special form of social intelligence, the capacity to form and manipulate in their minds the minds of other agents. Growing in quantity, this meta-information represents a sort of collective experience, the evaluations of the targets encountered from each individual, and passed on to other individuals within the same system. This meta-information may be inaccurate, both because a great deal of information around the evaluation (who is the evaluator, when and which factors contributed to build this evaluation) is lost, and because the information is passed on without being checked. Hence, there is a need for agents to reduce, metabolize and compensate uncertainty. The components of social intelligence dealing with this need are trust and reputation.

In [2] a fundamental distinction in social evaluations between *image* and *reputation* was introduced. Both are mental constructs concerning other agents' (targets) attitudes toward a given socially desirable behavior, and may be shared by a multitude of agents. Image is an evaluative belief and asserts that the target is "good" when it displays a certain behavior, and "bad" in the opposite case. Reputation, according to the theory in object, differs from image because it is true when the evaluation conveyed is actually widespread but not necessarily accurate. Reputation circulates within society through gossip [1], a communication stream which allows, among other effects, cheaters to be found out or neutralized. Gossip plays a basic role in social control and consequently in the formation and revision of social informational domains. To take place, reputation spreading needs four types of agents: evaluators, targets, gossipers, beneficiaries. As an example, the sentence "Walter is considered an honest person" includes a set of evaluators (who adopt the evaluation), a single targets (Walter), a set of beneficiaries (the people for which it is important to know that Walter is an honest person) and the set of gossipers (who refer the evaluation as such, independently from its adoption). Reputation is a belief about an evaluation circulating within the society. Hence, to acknowledge the existence of a reputation does not imply to accept the evaluation itself. For instance, an agent A might have a very good image of agent B as a seller, and at the same time recognise that there is a voice about agent B as a bad seller. Unlike ordinary communication and deception, reputation does not bind the speaker to commit herself to the truth-value of the evaluation conveyed but only to the existence of rumours about it. Hence, reputation implies neither personal commitment nor responsibility over the evaluation transmission and on its consequences. It represents the agents' opinions about a certain target's attitude with respect to a specific norm or standard, or her possession of a given skill.

3 Previous Works

Many other reputation models have been proposed in the field of agent based systems [13,9] (for a comparison of these models from the point of view of information representation, refer to [10]). However, in these models image and reputation collapse on the same type of social evaluation. Even when agents are allowed to acquire indirect information, this will not be represented as distinct from and interacting with own opinion. As a consequence, there is no way to distinguish known rumours from shared evaluations, a distinction which instead is central to our theoretical approach. Repage, which is the computational model implementing the cognitive theory of reputation from [2], is then the only system able to provide insight on how these processes may develop. Simulations based on Repage have already been presented in [12] and compared with the results obtained in a simpler system, implemented on NetLogo, and based on image only. Even using Repage, in [6] [7] and [8], the authors show that a market with exchange of reputation exhibits characteristics that are absent in a market with exchange of image only: (1) reputation produces a stabilizing effect in the discovery of critical resources; (2) more information circulates providing a better quality than with image only; (3) reputation reduces uncertainty and simplifies the agents' decision-making. Besides agents' societies, our reputation theory has been studied simulating the Internet of Services scenario [3] in which open global computing infrastructures providing services are made available to users through the Internet; the choice of a good service depends on the users' direct experience (image), and their ability to acquire information (reputation), which can be used to update their own evaluations. Even in that specialized application, reputation maintains its expected effect: (i) to reduce uncertainty and (ii) to facilitate the discovery of resources are still present.

4 Research Questions

This work addresses the relation between information trustworthiness and the cognitive processes of belief formation and revision with respect to the decision making process.

The experiments have been designed in order to stress the importance of information reliability. In particular, the more the market presents false information circulating, the more agents will need strategies to cope with uncertainty in order to select the best partner. On one hand, reputation has shown, in our previous works, to make the market more efficient through a faster information spreading; but on the other hand, reputation should amplify the effect of false information by influencing negatively the agents' decision making process. In particular, these negative effects of reputation transmitting false information should emerge when the market is provided with a poor offer and scarce quality levels. We will try to understand how does the faster false information spreading affects the agents performances, and which kind of information is needed to optimize the decision-making process under uncertain conditions.

5 Simulations

5.1 Computational Model

Repage. Repage [11] is a computational implementing a cognitive theory of reputation [2].

Unlike current systems, Repage allows agents to transmit their own image of a given target, which they hold to be true, or to report on what they have heard about the target, i.e. its reputation, whether they believe this to be true or not. Of course, in the latter case, they will neither commit to this information's truth value nor feel responsible for its consequences. Consequently, agents are expected to transmit uncertain information, and a given positive or negative reputation may circulate over a population of agents even if its content is not actually shared by the majority.

The main element of the Repage architecture is the memory, which is composed by a set of predicates. Predicates are objects containing a social evaluation, belonging to one of the main types accepted by Repage, the most important being image and reputation, or to one of the types used for their calculation (valued information, evaluation related from informers, and outcomes). The memory store allows for different evaluations - including image and reputation - for the same agent in the same role.

Predicates contain a tuple of five numbers representing the evaluation plus a strength value that indicates the confidence the agent has on this evaluation.

Each predicate in the Repage memory has a connection with a set of antecedents and a set of consequents; for example, an image is connected to the information used to calculate it - outcomes and valued information - and it will change with new outcomes, with new information, or with a change in the evaluation of an informer. If an antecedent is created, removed, or changes its value, the predicate is notied, recalculates its value and noties the change to its consequents.

This is not the place to proceed to a detailed explanation of Repage's mechanisms, for which we refer to [11] and [5].

The Market. The experiments were performed in a simulated marketplace in which buyer agents can purchase goods by selecting a seller. Agents can acquire information from other agents about the quality of sellers or about the quality of informers. Answers are provided as image or reputation evaluations. Simulations allow us to observe how the informational cheating affects the decision making process. The market has been designed with the purpose of reproducing the simplest possible setting where information is both valuable and scarce. In a simulation turn, a buyer performs one communication request and one purchase operation. In addition, the buyer answers all the information requests that it receives. Goods are characterized by a utility factor that we interpret as quality, with values between 1 and 100. Sellers have a fixed stock, that is decreased at every purchase, of goods with the same quality. Sellers exit the simulation when the stock is exhausted and are substituted by a new seller with similar characteristics but with a new identity (and as such, unknown to the buyers). The disappearance of sellers makes information necessary; reliable communication allows for faster discover of the better sellers. This motivates the agents to participate in the information exchange. With finite stock, even after having found a good seller, buyers should be prepared to start a new search when the good seller's stock ends.

5.2 Simulation Scenarios

There are six parameters that describe an experiment: the number of buyers (NB), the number of sellers (NS), the number of cheaters (C), the stock for each seller (S), the good sellers (GS), which are sellers providing the maximum quality, and the bad sellers (BS) providing minimum quality. Sellers that are neither good nor bad have a quality drawn from a uniform random distribution over the quality spectrum. We examine two experimental conditions: **L1** where there is only exchange of image and **L2** where both image and reputation circulate. The experiment explores 28 different market scenarios with a large amount of BS (50%), few (15%) GS and an increasing number of cheaters. For each scenario we have 25 sellers (NS) and 50 buyers (NB), the seller's stock is fixed to 30. All the scenarios are simulated in both experimental conditions; for each scenario ten simulation are performed.

5.3 Results

In this section the experimental results are reported. In both experimental settings, the agents' objective is to reach the maximum quality value in a contract under informational cheating. To observe the most relevant dynamics of factors,

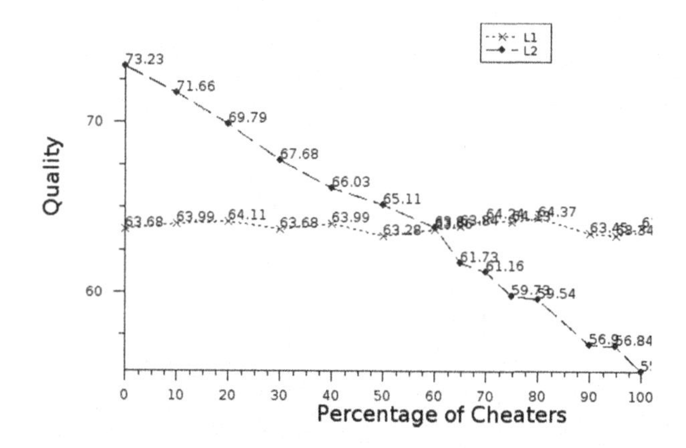

Fig. 1. Global quality with increasing number of cheaters. The curves represent quality in the stabilised regime for L1 and L2. Until cheaters remain below the threshold of 60% reputation allows for quality to reach higher values than happens in the complementary condition. The truth circulates faster providing social evaluations which are coherent with the reality. Coherence between information and reality, shared by a larger amount of agents in L2, increases the trust and the market is more efficient. Over the threshold of 60%, false information floods in, hence social evaluations, circulating faster with reputation, are often distorted with respect to reality.

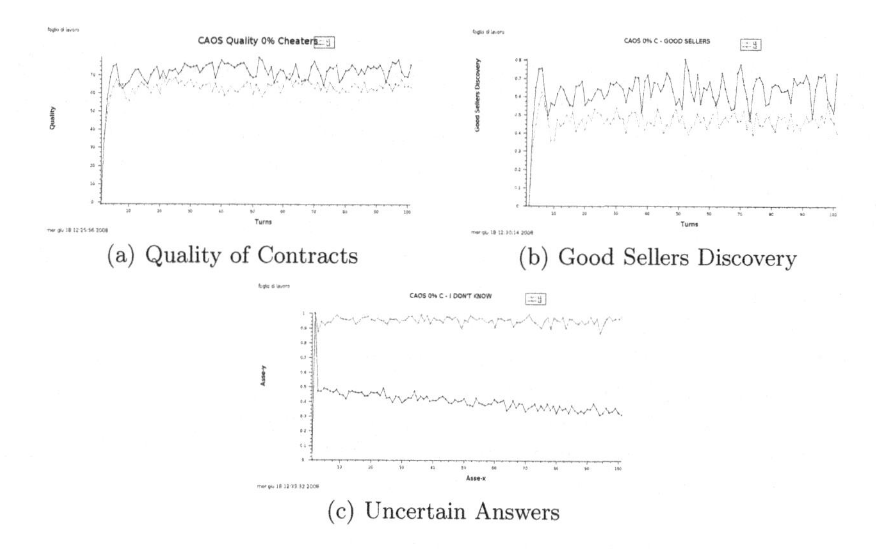

(a) Quality of Contracts

(b) Good Sellers Discovery

(c) Uncertain Answers

Fig. 2. Simulation Results in L1 and L2 with 0% of Cheaters for 100 turns

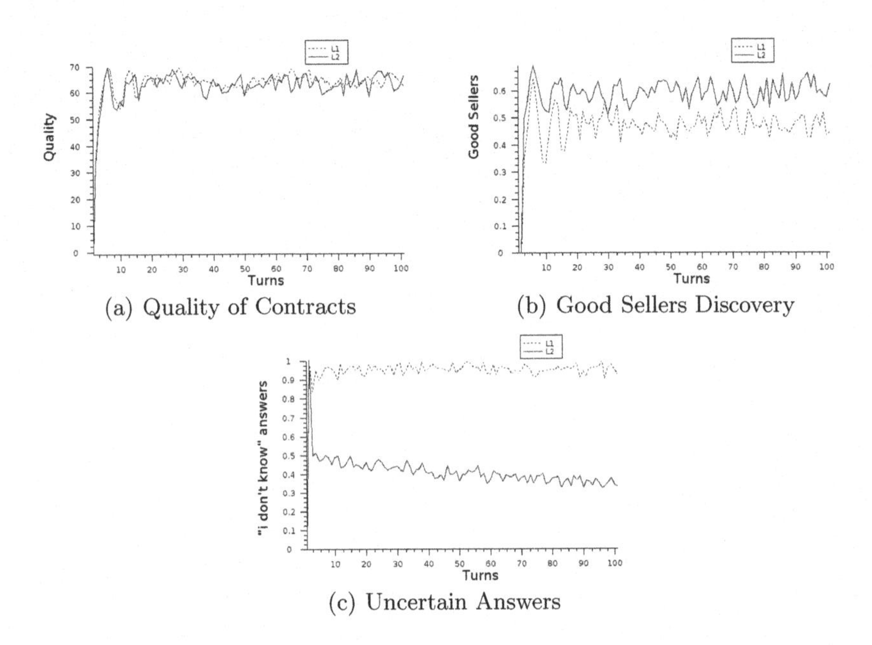

(a) Quality of Contracts

(b) Good Sellers Discovery

(c) Uncertain Answers

Fig. 3. Simulation Results in L1 and L2 with 60% of Cheaters for 100 turns. **(a)** Quality. The two curves represent Quality with Image (L1) and with both Image and Reputation (L2) when the number of cheaters inhibits the good effect of Reputation. Reputation and Image are on the same levels. **(b)** Good Sellers Discovered. The two curves are average values of Good Sellers found out for each simulation turn. **(c)** "i don't know" answers. The two curves are average values of "i don't know" answers for each simulation turn. The L2 curve shows the ability of reputation in reducing uncertainty.

we present a few selected charts from our analysis. In **Figure 1** we show the average market quality in the stabilised regime in (L1) and (L2) for a percentage of cheaters varying from 0 to 100% . We can observe the different performances of image and reputation at the increasing number of cheaters: reputation is more sensitive than image to the increase of cheating, but gives a better average quality performance than image does, until the rate of informing cheaters overcomes 60%. For a comparative analysis results are presented for the following critical values of the cheaters' number: at **0%** of cheaters, when only true information circulates, at **60%** when the positive effects of reputation disappears, and finally, at **100%** when all of the agents are informational cheaters. The following figures present the trends of quality, good sellers' discovery and uncertainty for each of these values. **Figure 2(a)** shows the trends of quality in absence of cheaters; here L2 curve is constantly higher than L1. In **Figure 2(b)** the trends of Good Sellers discovery in L1 and L2 are shown; here again L2 curve is higher with than L1, and the same pattern is shown by the uncertainty trend in **Figure 2(c)**.

Looking at **Figure 3(a)**, which shows the quality trend, we find that the L1 and L2 curves converge for all the simulation turns (100). **Figure 3(b)** shows

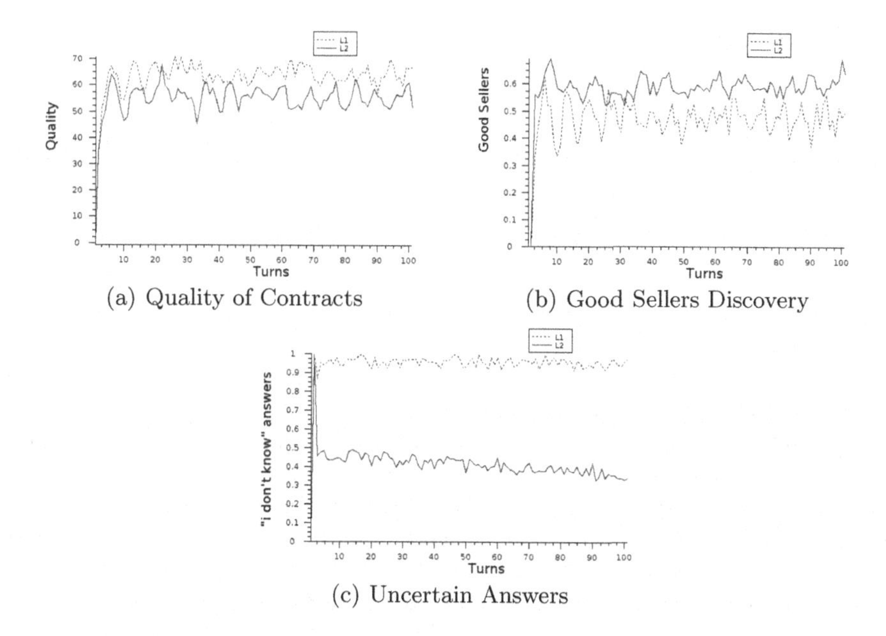

(a) Quality of Contracts

(b) Good Sellers Discovery

(c) Uncertain Answers

Fig. 4. Simulation Results in L1 and L2 with 100% of Cheaters for 100 turns. (a) Quality. The two curves represent Quality with Image (L1) and with both Image and Reputation (L2) when there are only cheaters. Image performs better.**(b)** Good Sellers discovered. The two curves shows the average values of Good Sellers found out for each simulation turn. In L2 more good sellers are discovered. **(c)** "i don't know" answers. The two curves are average values of "i don't know" answers for each simulation turn. The L2 curve shows the ability of reputation in reducing uncertainty also when the positive effects of reputation on the quality are neutralized and only false information circulates.

that with reputation more good sellers are found out, indicating that there is a higher consumption of resources. **Figure 3(c)** shows that reputation performs better at reducing uncertainty even with 60% of cheaters.

When information is unreliable, being the market populated only by cheaters, the different impact of image and reputation on quality is shown in **Figure 4(a)**, where L2 curve is lower than the L1 one. The GS discovery shown in **Figure 4(b)** presents higher values in L2 curve than in L1. The uncertainty trends in **Figure 4(c)** show that in L2 there still are less "i don't know" answers, although only false information circulates. It has been argued that reputation reduces uncertainty [7], but since bounded intelligent agents need communication to enlarge their knowledge about the world, the impact of reputation in reducing uncertainty depends on this information's reliability: if the information circulating is false the perception of reality will be fully distorted causing the market to collapse.

6 Conclusions

In this work, we have presented new simulation results obtained on the Repage platform. The simulations were aimed to support a social cognitive model of social evaluation, which introduces a distinction between image (the believed evaluation of a target) and reputation (circulating evaluation, without reference to the evaluation source). Reputation may be inaccurate, because there is a loss of information concerning the evaluation (who evaluates, when and which factors come to build this evaluation). What is the role of such inaccurate information? Why reputation is transmitted at all? As shown by our simulation data, in a market characterized by impervious conditions, i.e. increasing numbers of informational cheaters, the two types of evaluation produce interesting effects: the market benefits from reputation more than from image when information circulating is reliable, but when informational cheating is really really harsh, meaning that more than 60 percent of the total population of buyers is composed by liars, the quality of products decreases under the image setting. This phenomenon is due to the following factors: (1) when the number of cheaters grows, good evaluations are reported on more often than bad evaluations because cheaters transform the real conditions - which are harsh - into the opposite, making them better, (2) as soon as cheaters are found out, most information is not believed (average trust in informers collapses). As a consequence of both factors, although uncertainty decreases with reputation, the general level of trust does decrease as well. Hence, decisions (contracts) are taken randomly and the market collapses.

In future studies, we would like to investigate the performance of both types of evaluations under different market conditions, in at least two specific directions: (i) the interplay between informational and material cheating. What would happen with a variable number of good sellers, keeping constant informational cheaters? (ii) the interplay between informational cheating and the market settings. For example, how would image an reputation behave with a variable ratio between buyers and sellers (i.e. mitigating the resource scarcity)? Furthermore, we intend to use Repage as a test-bed for exploring some insightful theories of social phenomena, such as a subset of social contagion theories [4]. Some of them might reawaken, providing a scientific background for, the current debate on the role of media in opinion formation and dynamics: what is the respective role of media and other sources of information, such as reputation and gossip within a given population's informational domain? Are they independently formed and revised or do they reinforce each other? To these and to other more specific questions, we will turn in further applications of Repage and its underlying cognitive theory.

Acknowledgments

This work was supported by the European Community under the FP6 programme (eRep project CIT5-028575). A particular thanks to Jordi Sabater, Isaac Pinyol, Federica Mattei, Daniela Latorre, Elena Lodi and the Hypnotoad for support.

References

1. Coleman, J.: Foundations of Social Theory, August 1998. Belknap Press (1998)
2. Conte, R., Paolucci, M.: Reputation in Artificial Societies: Social Beliefs for Social Order. Springer, Heidelberg (2002)
3. Konig, S., Balke, T., Quattrociocchi, W., Paolucci, M., Eymann, T.: On the effects of reputation in the internet of services. In: ICORE 2009, Gargonza, Italy (2009)
4. Phillips, D.P.: The impact of mass media violence on u.s. homicides. American Sociological Review 48, 560–568 (1983)
5. Pinyol, I., Paolucci, M., Sabater-Mir, J., Conte, R.: Beyond accuracy. reputation for partner selection with lies and retaliation. In: Antunes, L., Paolucci, M., Norling, E. (eds.) MABS 2007. LNCS (LNAI), vol. 5003, pp. 128–140. Springer, Heidelberg (2008)
6. Quattrociocchi, W., Paolucci, M.: Cognition in information evaluation. The effect of reputation in decisions making and learning strategies for discovering good sellers in a base market. In: EUMAS HAMMAMET 2007, Tunisia (2007)
7. Quattrociocchi, W., Paolucci, M., Conte, R.: Dealing with uncertainty: simulating reputation in an ideal marketplace. In: AAMAS 2008 Trust Workshop Cascais Portugal (2008)
8. Quattrociocchi, W., Paolucci, M., Conte, R.: Reputation and Uncertainty Reduction: Simulating Partner Selection. In: Falcone, R., Barber, S.K., Sabater-Mir, J., Singh, M.P. (eds.) Trust 2008. LNCS (LNAI), vol. 5396, pp. 308–325. Springer, Heidelberg (2008)
9. Sabater, J., Sierra, C.: Who can you trust: Dealing with deception. In: Proceedings of the Second Workshop on Deception, Fraud and Trust in Agent Societies, Montreal, Canada, pp. 61–69 (2001)
10. Sabater-Mir, J., Paolucci, M.: On representation and aggregation of social evaluations in computational trust and reputation models. Int. J. Approx. Reasoning 46(3), 458–483 (2007)
11. Sabater-Mir, J., Paolucci, M., Conte, R.: Repage: REPutation and imAGE among limited autonomous partners. JASSS - Journal of Artificial Societies and Social Simulation 9(2) (2006)
12. Salvatore, A., Pinyol, I., Paolucci, M., Sabater-Mir, J.: Grounding reputation experiments. A replication of a simple market with image exchange. In: Third International Model-to-Model Workshop, pp. 32–45 (2007)
13. Schillo, M., Funk, P., Rovatsos, M.: Who can you trust: Dealing with deception. In: Proceedings of the Second Workshop on Deception, Fraud and Trust in Agent Societies, Seattle, USA, pp. 95–106 (1999)
14. Shannon, C.E., Weaver, W.: The Mathematical Theory of Communication. University of Illinois Press, Urbana (1949)

Distributed Cognitive Mobile Agent Framework for Social Cooperation: Application for Packet Moving

Umar Manzoor and Samia Nefti

Department of Computer Science,
School of Computing, Science and Engineering,
The University of Salford, Salford,
Greater Manchester, United Kingdom
umarmanzoor@gmail.com, s.nefti-meziani@salford.ac.uk

Abstract. In this paper we have proposed Distributed Cognitive Mobile Agent Framework for Social Cooperation to transfer packets from source to destination in randomly varying environment in the most efficient way with the help of mobile agents. We are considering Packet Moving problem in 2-D environment consisting of packets and destinations. The task of each mobile agent is to move the packet from the source to the destination. Simulated environment is dynamically generated with varying graph size (i.e. number of nodes in graph), package ratio and number of destinations. Packages and destinations are randomly placed in the graph and the graph is divided into societies to encourage an efficient and manageable environment for the Agents to work in parallel and share expertise within and outside the society. The proposed framework is based on layered agent architecture and uses decentralized approach and once initialized manages all the activities autonomously with the help of mobile agents.

Keywords: Social Cooperation, Automated Packet Moving, Cognitive Mobile Agents, Collaborated Agent System, Multi-Agent System.

1 Introduction

The unique characteristic of Agent paradigm makes it potential choice for implementing distributed systems. Agents are software programs which acts autonomous on the environment on the behalf of the user. Agents are assigned specific goals and before acting on the environment agent chooses the action which can maximize the chance to achieve their goal(s). Agents are of many types and in this paper we are interested in intelligent and mobile agents. Intelligent agents are software programs which have the capability of learning from the environment and respond to the dynamic events in timely and acceptable manner in order to achieve its objectives or goals [1, 6].

Mobile agents are software programs which have the capability of moving autonomously from one machine to another with its data and state intact [2, 8]. In Multi Agent System, two or more autonomous agents work together in order to achieve their common goal(s). These agents can collaborate and communicate together to solve complex problems which are outside single agent ability. In order to

M.D. Lytras et al. (Eds.): WSKS 2009, LNAI 5736, pp. 584–593, 2009.

improve its and overall efficiency, these agent(s) can learn from the environment and use the knowledge learned in making decisions.

The major characteristics of Multi Agent System are its flexibility and decentralized design which makes it ideal for implementing distributed application. Because of the decentralized design partial failures do not affect the overall performance of the system and these systems have the capability of recovery from partial failures automatically. Because of the tremendous capabilities of Multi Agent System, MAS have been used in the areas of organizational learning, network routing, network management, computer-supported knowledge work, mine detection and file transfer [5, 13, 4, 3, and 14].

1.1 Problem Definition

We are considering Packet Moving problem in 2-D environment consisting of packets and destinations. The task of each Mobile Agent is to move the packet from the source to the destination and the problem has the following constraints.

- Each Agent can carry one packet at a time.
- Each Agent can move in four (Left, Right, Top, and Bottom) directions.
- Agent can not pass through the packet or hurdle (obstacle) in the way.
- Agent can move the packet or hurdle in the way or can maneuver around it.
- Moving the hurdle or packet out of the way has two times the normal cost.
- Moving the packet from one state to another has some cost associated with it.
- Picking the packet or setting it down has no cost.

In [9] Guillaume et al proposed reinforcement learning algorithm (Q-Learning) for block pushing problem. In [10] Adam et al proposed STRIPS planning in multi robot to model block pushing problem. In this paper we have proposed Distributed Cognitive Mobile Agent Framework for Social Cooperation to transfer the packet from source to destination in randomly varying environment in the most efficient way with the help of mobile agents. The framework is based on layered agent architecture and uses decentralized approach. The framework once initialized with the domain knowledge manages all the system activities autonomously with the help of mobile agents. What makes proposed framework different from other approaches is its flexibility, decentralized design, and social cooperation.

1.2 Simulated System

In our simulated environment Goal Graph is dynamically generated with varying graph size (i.e. number of nodes in graph), package ratio and number of destinations. Packages and destinations are randomly placed in the graph and graph is divided into societies. Each society will have some Worker Agents and one Area Manager Agent, the society area will be managed its Area Manager Agent. The purpose of creating societies is to encourage an efficient and manageable environment for the Agents to work in parallel and share expertise within and outside the society.

We have generated graph of different sizes ranging from 10x10 to 100x100 with an increment of 10. For each size we have generated graphs with packet ratio ranging from 1 to 20 percent with an increment of 1 percent and destination ranging from 1 to 9 with an increment of 1. For each packet ratio, twenty different graphs were generated and for each run 10 trails are generated with varying range of destinations (i.e. from 1 to 9). For each twenty different mazes the number of agent's engaged in the search are changed from 2 to 5.

We have tested the framework on more than 4000 test runs with different graph sizes, packet ratios, number of destinations and number of agents. The real application of this multi-agent based software simulation would be for the Mobile Robots to drop the packets to the destination efficiently in an unknown and continuously varying environment.

3 System Architecture

Distributed Cognitive Mobile Agent Framework for Social Cooperation is a multi-agent based framework for moving the packets efficiently to destination with the help of mobile agents. Each Agent can move in four (Left, Right, Top, and Bottom) directions and each move has some cost. Moving the packet has cost associated with it but not with picking it or setting it down. Each agent can not pass through the packet or obstacle and have the capability of carrying only one packet at a time. While moving one packet to destination, if it encounters another package in the way, it can maneuver around it or move the package out of the way depending upon the situation. Moving the package out of the way might be beneficial but the cost of moving is high (two times the normal cost). The system consists of the following agents:

- Manager Agent (MA)
- Area Manager Agent (AMA)
- Worker Agent (WA)

3.1 Manager Agent (MA)

Manager Agent (MA) is responsible for the initialization of the system and in initialization it performs a set of tasks.

1) MA creates and initialize the grid using the Goal Graph Module (GGM) and place the packages at random positions in the grid.
2) MA divides the grid into n societies using Society Formulation Module (SFM) where n depends upon the resources.
3) Each society will have some Worker Agents and will be managed by Area Manager Agent.
4) MR is responsible for creation and initialization of Area Manager Agent (AMA). MA will create n AMA's where n is the number of societies and each AMA will be assigned one society. In initialization MA passes society goal graph (partial goal graph of the grid), number of Worker Agents, number of packets, and the number of destinations as arguments to AMA. The AMA of each society can communicate with other AMA's for information sharing, resource required, etc.

MA maintain the states of all the AMA's in state table using State Module (SM) and keep checking the state table after periodic interval. If AMA state is set to "IDLE", this means AMA has performed the task assigned and its resources are free.

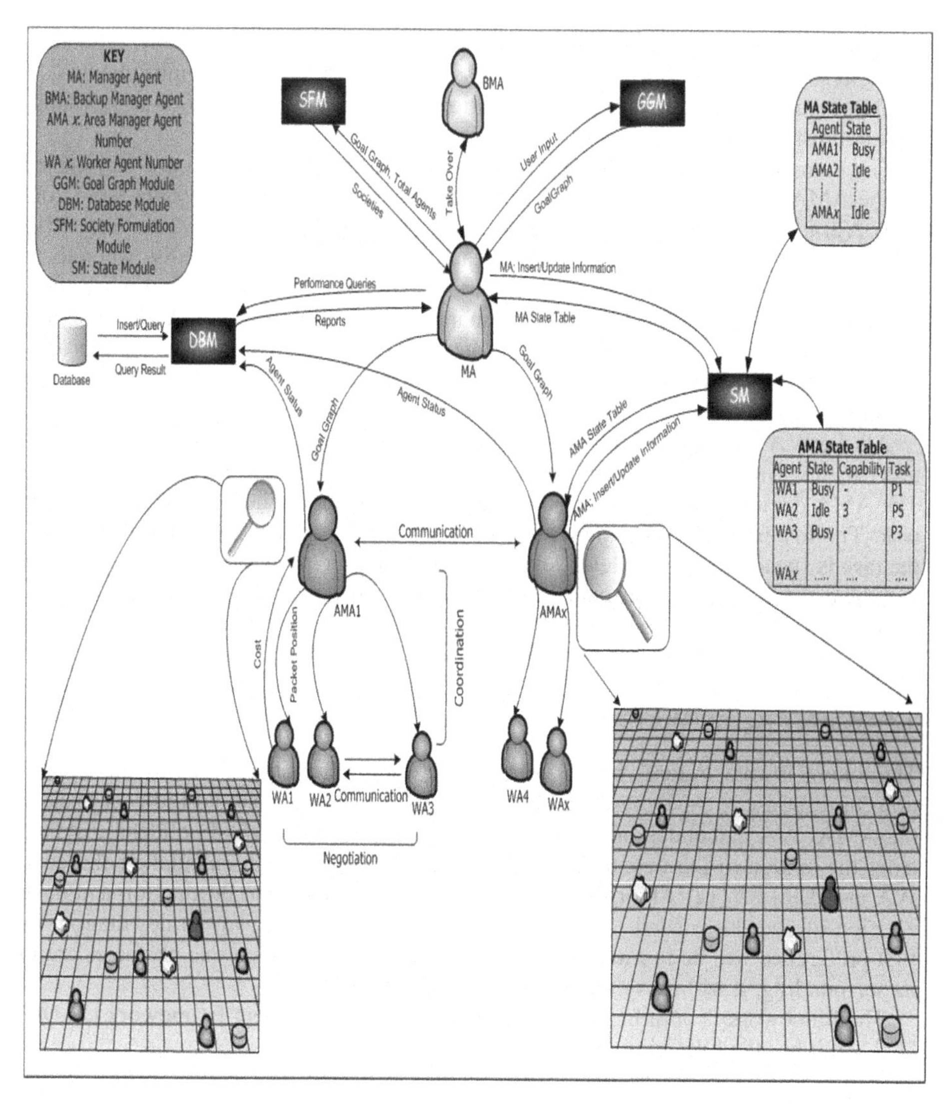

Fig. 1. Distributed Cognitive Mobile Agent Framework for Social Cooperation: Application for Packet Moving

MA can assign AMA and its resources to the neighboring society where packet to destination ratio is high. If all the AMA's are set to "IDLE", this means that all the tasks in the goal graph have been performed.

3.2 Area Manager Agent (AMA)

The role of Area Manager Agent (AMA) is to coordinate and manage a group of Worker Agent (WA) in a society. The coordination is done to ensure that there are no collisions and congestion among the WA. AMA also provides timely information to

WA so that the actions of the WA's are synchronized in order to avoid redundant problem solving. In order to manage its society each AMA set its commitments (i.e. handing over a specified task) and conventions (i.e. managing commitment in changing circumstances).

The commitments are implemented in the form of state tables that are maintained by the AMA of each particular society for managerial purposes. The state table contains Agent number, state (Busy or Idle), task (Packet Number) and their respective capabilities (in terms of cost). The AMA can change its role dynamically, when there are enough agents in an area to perform the task, it acts as a manager and when there are not enough agents AMA can act as manager as well as Worker Agent. It is the responsibility of the AMA to assign a particular task of the goal graph to an appropriate WA. AMA multicast's the information of different tasks to all the WA of the society.

All the WA are required to calculate the optimal cost (shortest path to perform the task) keeping in view their previous estimation of cost and report it back to the AMA. The AMA decides on the bases of bidding which WA should be assigned the specific task. The lowest bidder is assigned the task and AMA updates its state table. When the task is successfully completed, every WA is supposed to report the AMA, the start and the end time for that particular task. All the AMA's periodically submits the task table to the Data Base Module for insertion in database. When all the tasks on the AMA's society are completed, it reports to the MA about his idle state.

3.3 Worker Agent (WA)

The role of Worker Agent (WA) is to move the packet from the source to the destination. In response to AMA task multicast, WA estimates the cost and bid cost back to the AMA. WA calculates the cost for bidding purpose using the Euclidean distance from the source to the destination and ignores the obstacles in calculating the cost. WA uses LRTA* [12] Algorithm to move packet from source to destination and have a vision of one state ahead. If WA encounters obstacle in the way to destination, its first preference will be to move around the obstacle. If more then one obstacles are in its surrounding, it will move the obstacle out of the way.

The framework is built in Java using Java Agent Development Framework (JADE) [11] and it has the following modules.

- Goal Graph Module (GGM)
- Society Formulation Module (SFM)
- State Module (SM)
- Data Base Module (DBM)

3.4 Goal Graph Module (GGM)

The GGM is responsible for creating the goal graph with N Nodes where N is passed as arguments. Goal graph consist of nodes and edges and GGM places Packets, Destinations and Agents randomly on the nodes of the graph and also adds random cost to the edges of the graph. The goal graph represents the dependencies between the packet and destination and goal graph is used to calculate the shortest path to the packet or destination. Once the goal graph is divided by the RA, the particular part of the goal graph is passed to the AMA for task completion.

3.5 Society Formulation Module (SFM)

This module takes total number of agents, graph as input and calculates the best society formation keeping in view total number of agents and area to packet ratio. The purpose of creating societies is to encourage an efficient and manageable environment for the Agents to work in parallel and share expertise within and outside the society. Grid area will be divided into societies and each society will manage its own area. The AMA of each society can communicate with other AMAs for information sharing or resource sharing.

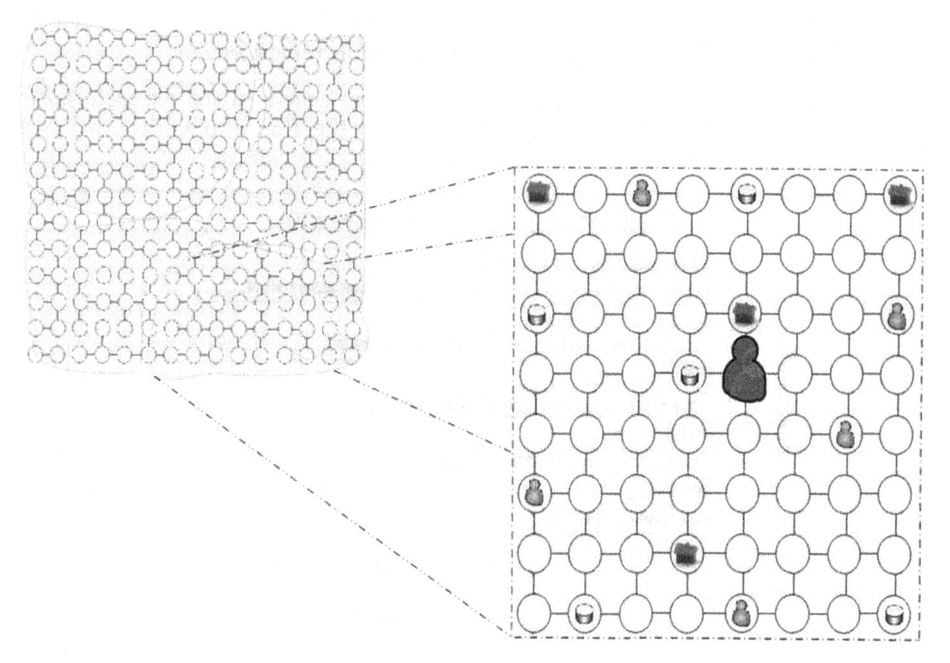

Fig. 2. Grid Division

3.6 State Module (SM)

The SM is responsible for maintaining two instances of hash table, one instance of Area Manager Agents (AMA) and other one for Worker Agents (WA). The AMA table is continuously monitored by the Manager Agent (MA) to get update on the status of tasks performed by AMAs. Similarly WA table is continuously monitored by AMA to get update on the state of tasks performed by WA. Agents state can be busy or idle and with each state extra information (i.e. task id, source, destination, cost, current position etc) is stored to help AMA or MA in planning.

3.7 Data Base Module (DBM)

Data Base Module contains all the necessary data base related functions. This module contains methods to Connect, Insert, Retrieve, Delete and Run Queries on the database. DBM will be used by different agents to insert experiment results in the database which is used to analyze the performance of the system.

4 Performance Analysis

We have evaluated this framework on randomly generated graph of different sizes ranging from 10x10 to 100x100 with an increment of 10. For each size different graphs were generated graphs with varying

- Packet ratio (ranging from 1 to 20 percent with an increment of 2 percent)
- Destinations (ranging from 1 to 9 with an increment of 2)

For each packet ratio, twenty different graphs were generated and for each run 10 trails are generated with varying range of destinations (i.e. from 1 to 9).

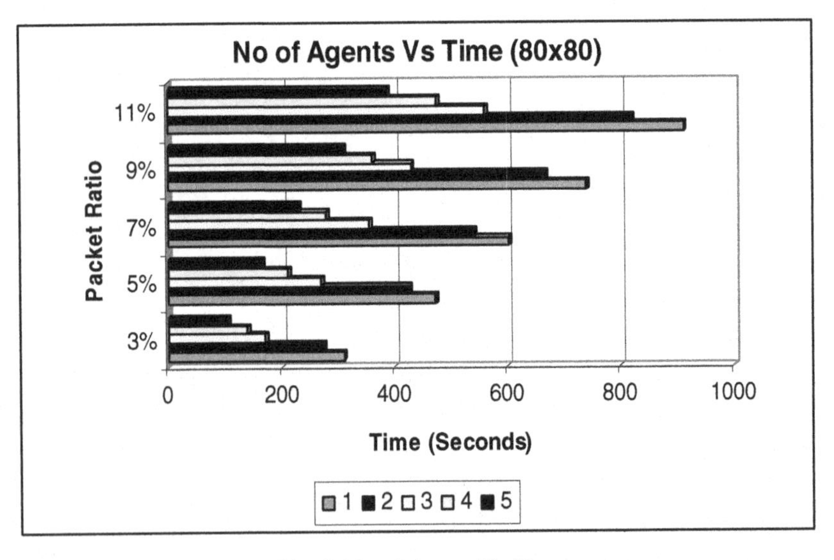

Fig. 3. No. of Agents Vs Time

For each twenty different mazes the number of agent's engaged in the search are changed from 2 to 5.

Figure 3 shows the time taken by different number of agents per society with varying number of packet ratio in 80x80 graph size. As we increase the number of agents in the society, time to complete the task reduces. However, as we keep on increases the number of agent per society the improvement in task time reduces because of dependency between the societies. As we increase the number of agents per society to 3, the time taken to complete the task reduces considerably but if we keep on increasing the numbers of agents the improvement is marginal (Figure 3).

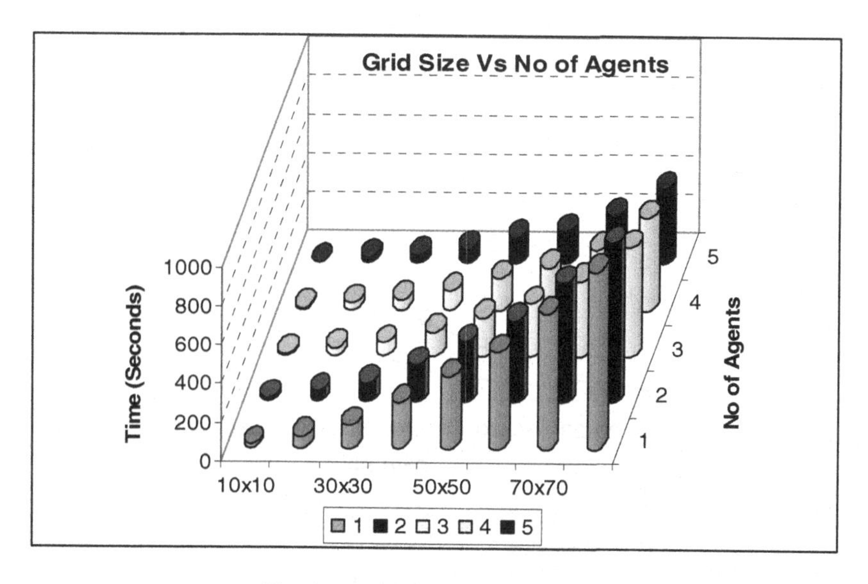

Fig. 4. Graph Size Vs No. of Agents

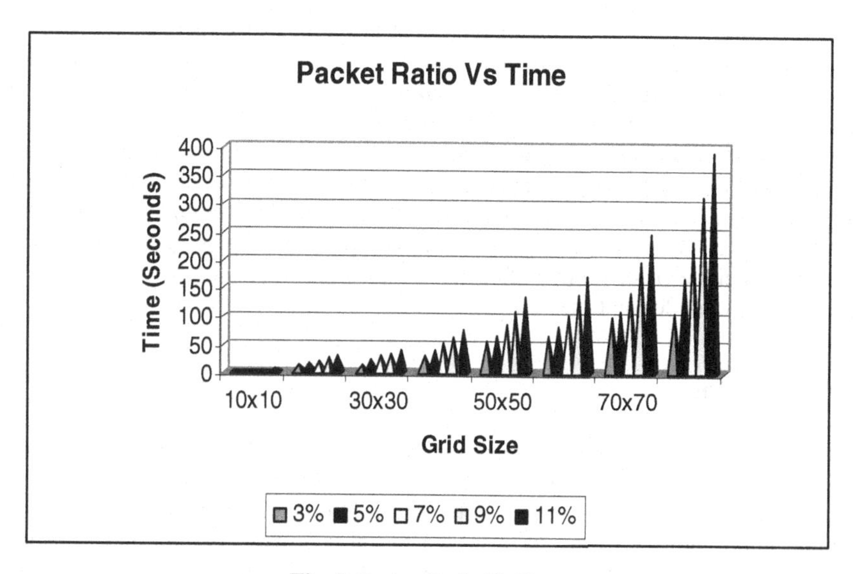

Fig. 5. Packet Ratio Vs Time

Figure 4 shows the comparison of Graph size with reference to number of agents with 11 percent packet ratio, increasing the number of agents in small graphs size does not improve the task completion time considerably however, as we increase the graph size number of agent plays important role in decreases the task time as shown in Figure 4.

Figure 5 shows the time taken by five agents to complete the task with varying graph size and packet ratio. For small graph size the completion time difference with

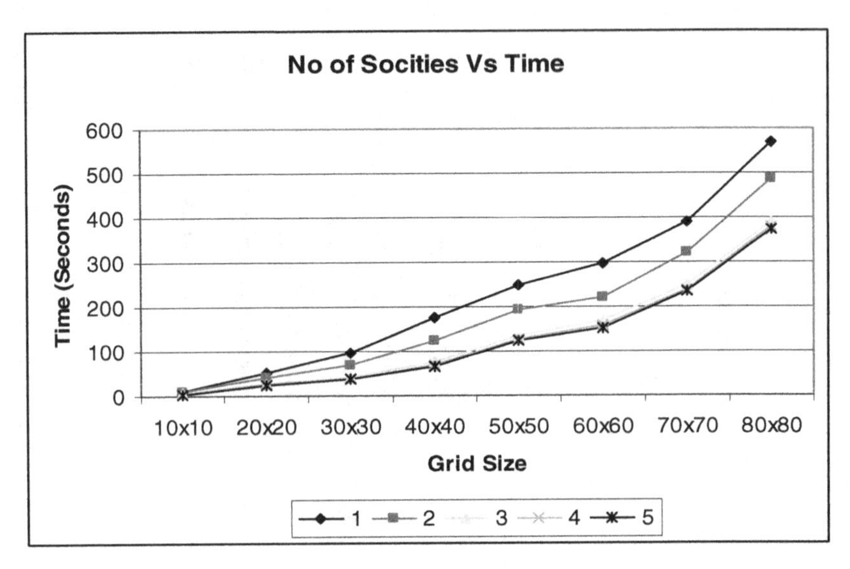

Fig. 6. No of Societies Vs Time

respect to varying packet ratio is insignificant however as we increase the graph size the completion time increase slightly. But if we consider the number of packets increased in the graph the difference between the times is ignorable and it demonstrates the efficiency of our system because of parallelism.

Figure 6 shows the effect of increasing the number of societies on the task completion time with reference to five agents and 11 percent packet ratio on varying graph size. Increasing the number of societies decreases the task completion but up to a specific value. As shown in Figure 6, the completion time is decreasing until the number of societies is equal to 3 after that the increase in the number of societies does not affect the completion time.

5 Conclusion

In this paper, we have proposed Distributed Cognitive Mobile Agent Framework for Social Cooperation for moving the packets efficiently to destination with the help of mobile agents. Mobile agents are software programs which have the capability of moving autonomously from one machine to another with its data and state intact. Packages and destinations are randomly placed in the graph and graph is divided into societies. Each Agent can move in four (Left, Right, Top, and Bottom) directions and can carry one packet at a time. While moving one packet to destination, if it encounters another package in the way, it can maneuver around it or move the package out of the way depending upon the situation. We have tested the framework on more than 4000 test runs with different graph sizes, packet ratios, number of destinations and number of agents. Results were promising and demonstrate the efficiency of the proposed framework. The real application of this multi-agent based software simulation would be for the Mobile Robots to drop the packets to the destination efficiently in an unknown and continuously varying environment.

References

1. Manzoor, U., Nefti, S.: An agent based system for activity monitoring on network – AB-SAMN. Expert Systems with Applications 36(8), 10987–10994 (2009)
2. Ilarri, S., Mena, E., Illarramendi, A.: Using cooperative mobile agents to monitor distributed and dynamic environments. Information Sciences 178, 2105–2127 (2008)
3. Watanabe, T., Kato, K.: Computer-supported interaction for composing documents logically. Int. J. Knowledge and Learning 4(6), 509–526 (2008)
4. Gavalas, D., Greenwood, D., Ghanbari, M., O'Mahony, M.: Hierarchical network management: a scalable and dynamic mobile agent-based approach. Computer Networks 38(6), 693–711 (2002)
5. Martínez León, I.M., Ruiz Mercader, J., Martínez León, J.A.: The effect of organisational learning tools on business results. Int. J. Knowledge and Learning 4(6), 539–552 (2008)
6. Greaves, M., Stavridou-Colemen, V., Laddaga, R.: Dependable Agent Systems. IEEE Intelligent Systems 19(5) (September-October 2004)
7. Manzoor, U., Ijaz, K., Shahid, A.A.: Distributed Dependable Enterprise Business System – DDEBS. In: Proceeding of Springer Communications in Computer and Information Science, Athens, Greece, September 24-28, vol. 19, pp. 537–542 (2008), ISBN 978-3-540-87782-0
8. Milojicic, D., Douglis, F., Wheeler, R.: Mobility: Processes, Computers, and Agents. ACM Press, New York (1999)
9. Laurent, G., Piat, E.: Parallel Q-Learning for a block-pushing problem. In: Proceeding of IEEE International Conference on Intelligent Robots and Systems, Hawaii, USA, October 29 – November 03 (2001)
10. Galuszka, A., Swierniak, A.: Translation STRIPS Planning in Multi-robot Environment to Linear Programming. In: Rutkowski, L., Siekmann, J.H., Tadeusiewicz, R., Zadeh, L.A. (eds.) ICAISC 2004. LNCS (LNAI), vol. 3070, pp. 768–773. Springer, Heidelberg (2004)
11. Java Agent Development Framework (JADE), http://jade.tilab.com/
12. Niazi, M., Manzoor, U., Ijaz, K.: Applying Color Code Coordinated LRTA* (C3LRTA*) Algorithm on Multiple Targets. In: Proceeding of 9th IEEE International Multitopic Conference (INMIC), pp. 27–32 (2005)
13. Qu, W., Shen, H., Sum, J.: New analysis on mobile agents based network routing. Applied Soft Computing 6(1), 108–118 (2005)
14. Manzoor, U., Nefti, S., Hasan, H., Mehmood, M., Aslam, B., Shaukat, O.: A Multi-Agent Model for Mine Detection – MAMMD. In: Lytras, M.D., Carroll, J.M., Damiani, E., Tennyson, R.D. (eds.) WSKS 2008. LNCS (LNAI), vol. 5288, pp. 139–148. Springer, Heidelberg (2008)

iPark: A Universal Solution for Vehicle Parking

Krishan Sabaragamu Koralalage and Noriaki Yoshiura

Department of Information and Computer Sciences,
Saitama University, Saitama, 338-8570, Japan
krishjp@gmail.com, yoshiura@fmx.ics.saitama-u.ac.jp

Abstract. Urbanization and drastic increase of vehicle usage has brought up a considerable problem in vehicle parking. Lack of efficient infrastructure facilities to manage increasing demand with available space and inability to provide intelligent, standard, and interoperable parking solutions are some of the main reasons to cause the above problem. Therefore, in this paper we introduced a novel parking solution called iPark which could solve almost all the existing issues using rich radio frequency chip design. iPark could realize the potential benefits in intelligent transportation systems (ITS) and allows real time, efficient, effective and hassle free parking. It also enables dynamic parking management, provides information on idling lots, allow prior reservations, provide navigation support towards idling or reserved lot, automate collecting of charges and enforce rules and regulations on illegal parking. Moreover, unlike other solutions, iPark works for both free entrance and controlled entrance parking lots by reducing the unnecessary burdens.

Keywords: Intelligent Transportation System, OTag, RFID, Parking.

1 Introduction

Parking is an essential part of transportation system. Vehicles must park at every destination. Therefore to provide the optimum solutions, parking system requires seamless integration of different areas: personnel, vehicle management, data processing, billing, and accounting in order to provide high quality, efficient services to the customers [1, 4, 5]. Similarly it is necessary to allow standardized usage for customers without adopting several proprietary systems.

There are two main types of parking lots: free entrance parking lots and controlled entrance parking lots. Free entrance parking lots do not have any entry control system or locking system to enforce the rules and regulations. Some of them even do not use any mechanisms to collect money. Parking lots in highways, along the main streets, hospitals and government service centers comes under this category. Mostly these types of parking lots are governed by the city offices or the government. Main concepts in this type of parking lots are to manage the traffic and allow smooth and efficient services to the consumers without expecting much economical benefits.

Controlled entrance parking lots are the parking lots that are mostly aiming to achieve the economic benefits. They are installed with one of the well known entry control mechanisms at the point of entrance and exit to monitor and control the

M.D. Lytras et al. (Eds.): WSKS 2009, LNAI 5736, pp. 594–604, 2009.

incoming and outgoing vehicles. Most of these parking lots are run by commercial companies. The main concepts of these parking lots are to achieve the economical benefits while providing customer convenience.

Like in automated toll collection, automated parking management systems offer great benefits to owners, operators, and patrons. Some of the main benefits include reduced cash handling and improved back-office operations, high scalability, automatic data capture and detailed reporting, improved traffic flow at peak hours, improved customer service, cash-free convenience, and also provinces to arrange special privileges for VIP customers such as volume discounts, coupons, and other discounts for daily parkers [1, 2, 5].

One of the best candidate technology used in parking systems is the Radio Frequency Identification (RFID) technology because it enables carrying data in suitable transponders. The non-contact short-range communication makes it suitable for implementing an automatic identification of vehicles. Hence the industry is working on RFID based solution to improve parking systems enabling automatic parking and toll applications minimizing delays and hassles while providing efficient service to customers [1, 2, 3, 4].

Once the unique ID in RFID tag is read the database system will check for its validity and allow vehicle to enter to the premises. When the vehicle leave parking premises vehicle is identified and billed to the pre-registered account by back office system. The advantages of this type of system include not only easy access for the customer, but the elimination of staffing at entry and exit points. Thus RFID enabled automated parking access control systems eliminates customers' need to fumble for change, swipe cards, or punch numbers into a keypad. Vehicles can move smoothly through controlled entrances, and more parkers can be accommodated, thereby increasing revenues. As there are no cards or tickets to read, the whole system enables a convenient, efficient, hassle free and easy vehicle parking [1, 2, 3].

Some of existing parking solutions are developed by Rasilant Technologies, EasyPark Solutions, PARKTRACK, Essen RFID, TagMaster North America, Inc., Macroware Information Technology, Infosys Solutions, TransCore, and ActiveWave. TransCore is one of the pioneers to provide parking access control systems with RFID technology. They uses proven eGo® and Amtech®-brand RFID technology to deliver reliable, automated parking access control solutions. TransCore's parking control systems facilitate airport security parking and parking at universities, corporations, hospitals, gated communities, and downtown parking facilities.

ActiveWave is another company who provide such systems using RFID technology. In their system, surveillance cameras or video recorders can be triggered whenever a vehicle enters or exits the controlled area.

After comprehensive literature review, we found that there is no standard solution developed to control both free entrance and controlled entrance parking systems instead the trend is to commercial parking systems to improve their management process and customer convenience to achieve the maximum economical benefits. Though commercial parking lot operators request customers to adopt some requirements to subscribe their services, similar requests cannot be made to enforce the rules and regulations of a country. Additionally the companies who provide RFID based parking management systems are proprietary, and provide no interoperability. Almost all the cases they only uses unique ID to identify and billing the vehicles. Furthermore, they are unable to provide interoperability. Thus customers have to have different devices installed for different subscriptions such as one for electronic toll collection and another for parking access. Additionally no existing system provides support to enforce the rule and regulations of the country. Hence, all the existing and previously proposed RFID based parking management systems has no support towards the standardized and interoperable parking management systems for both free entrance and controlled entrance parking lots while providing the facility of rules and regulation enforcement of the country.

Thus we proposed a novel architecture called OTag[6], its installation strategy and common communication platform to solve the issues arise in the intelligent transportation systems including the parking management system. In this paper we described how OTag and proposing composition helps to fulfill the future needs extensively without creating any inconvenience to the service provider or to the costumer when parking. We believe that this process should start at the time of vehicle manufacturing to enable standardized and interoperable systems. Therefore our architecture starts from the point of manufacturing.

2 OTag Architecture

Object tag (OTag[6]) is a radio frequency (RF) tag which can represent real world objects such as vehicles. Considering the concepts of Object Oriented Architecture, active RF tag is designed with its own attributes and methods to stand alone by itself. OTag may differ according to expected characteristics and behaviors of the real world objects. Each OTag can be considered as an instance of an object class. If a vehicle is the object class, tags attached to all vehicles are considered as instances of vehicle object class. Furthermore, each attribute of the OTag has got access modifiers to manage the access based on roles. They are categorized as public, friendly, protected and private.

Access modifiers can be considered as roles, and one of the four access modifiers is assigned to each attribute. With the help of those modifiers, OTag acts above four roles. Public means no security and any reader can read any attribute value of public area. Private, Protected, and Friendly attributes of the OTag need secure access. Furthermore the writing permission is also available for these areas, but that is controlled by both keys and memory types. For example write once memory is used for some attributes in friendly area to prevent updating fixed attribute values whereas rewritable memory is used for dynamically changing attribute values according to the behaviors of object class.

Role base accessing methods are implemented using access modifiers, memory types and encryption algorithms. Memory types used here are ROM, EPROM, and EEPROM. OTag controls writing permissions according to the process flow by using these three types of memories. ROM is used for read only data and they are created by time of fresh tag manufacturing. EPROM is used for one time writing data and they are all about product information. EEPROM is used for all the rewritable data and those data may be changed in any stage of the product lifecycle by relevant authorities. Figure 1 represents the logical structure of vehicle OTag. For other types of OTags such as road symbol OTag, attribute names, get method, and set method implementations are different from vehicle OTag. Additionally the oName attribute value should be class name, such as oName for road symbol will be "SYMBOL" instead of "VEHICLE". Therefore the other OTags can be developed using the same architecture only by changing the attribute name value pair and implementations of get and set methods according to the characteristics and behaviors of those classes.

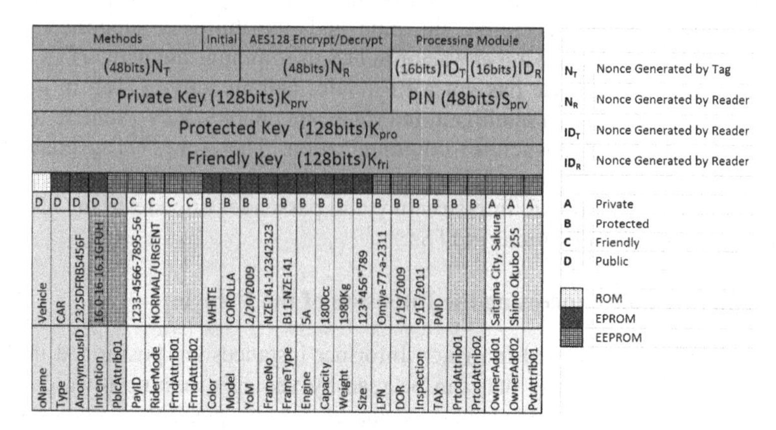

Fig. 1. Illustrate the Logical Structure of an OTag instance in Vehicle

Public area stores the object class name, type, anonymous ID, intention and customizable attribute. Information stored in this area can be used to understand object and its movements. Identifying those public information leads to create applications like collision avoidance, traffic signal management, vehicle to vehicle communication, road congestion management, road rule enforcements, driving assistant systems, etc.

Protected area stores the color, model, manufactured year, frame number and type, engine, license plate number, date of first registration, inspection validity, etc. Therefore, after first registration the data stored in this area can only be manipulated by government authority. Protected information can be read only by the owner or police. Inspection validity, insurance, tax, etc. help to identify the current status of the vehicle. Recognizing illegal, fake, clone, stolen or altered vehicles, issuing fines for inspection expired vehicles, verification of tax payments, management of carrying garage, temporary and brand new vehicles are some of the main applications using this area.

Friendly modifier allows several services to be catered to the user in effective manner. This area stores the pay ID or wallet ID, rider mode, two customizable attributes. Information stored in this area can be used to subscribe variety of services provided by companies. Electronic fee collection systems like toll collection and parking can use WalletID to collect the fee only after prior registration with relevant authorities. Furthermore, the emergency vehicles like ambulances can use this area to prioritize traffic signals by using the value of RiderMode attribute.

Private area stores owner name, address and one customizable attribute. Information stored in this area is to prove the ownership of the vehicle. No one can read or write data into this area without the permission of the owner. When the vehicle is sold the ownership information will be changed.

Unlike conventional RFID tag, OTag can manage its access depending on four roles: public, friendly, protected and private. Thus it provides role base access control mechanism. Additionally, OTag eliminates the necessity of accessing database by keeping its own data with it and thereby guarantees the stand alone capability. Similarly OTag assures ability of self-describing using the characteristics and behaviors of RF technology. Furthermore the interoperability is guaranteed in OTag by providing the common communication interface. Therefore the plug and playability is supported using the above three main points allowing any actor in ITS to communicate with OTag.

Suppose that each vehicle has got a RF reader and an active OTag in it. This arrangement makes vehicle an intelligent interface and enables V2V (vehicle to vehicle) and V2I (vehicle to infrastructure) communication. Therefore, OTag enables large array of applications to improve ITSs. External readers can read OTag in vehicles and the readers in vehicle can read infrastructure and passenger or pedestrian OTags opening up large array of applications in ITSs.

2.1 Identifying and Expressing the Intension of the Vehicle

There is an OTag class called Informer. Informer instances are positioned in the center line of the road in a way that can be read by vehicles running towards the both directions. Those tags contained the next target, current position, current route, and next crossing lane, etc. as shown in Figure 2. Once the immediate previous 16.1 Informer OTag position is passed by a vehicle, intended action "16.1GF" of that vehicle is sent to the 16.2 Informer OTag via the vehicle reader requesting the next target information. Then the Informer OTag sends the next target information to the vehicle. If there is no previous Informer, the reader in the vehicle requests yourPosition attribute of first Informer OTag to understand the current position.

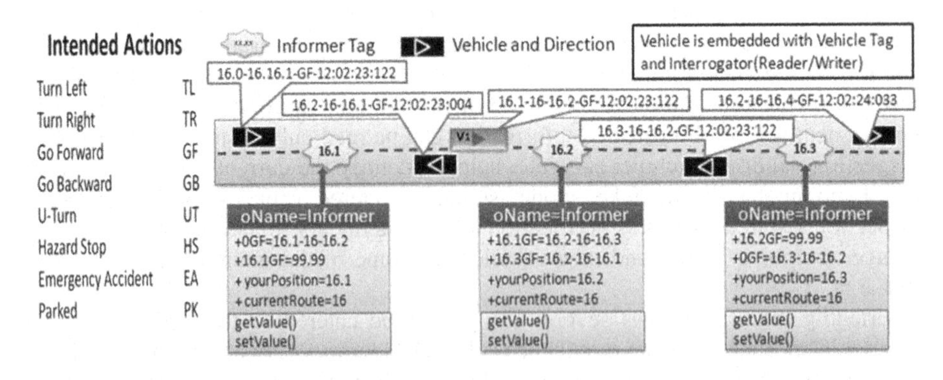

Fig. 2. Illustrate the identification layout of intended movements of vehicles

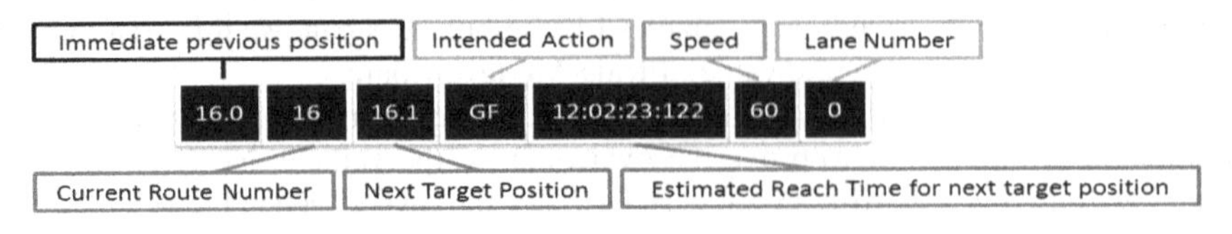

Fig. 3. Target message expressed by moving vehicle

When a vehicle passes the very first Informer, the interrogator in the vehicle gets to know the next immediate target. After that, the interrogator writes that information with intended action, running speed, and expected reach time of current immediate target and lane number if it exists, to the public area of the vehicle OTag. If there is no lane number, or previous position information before the Informer, the predefined not applicable dummy values will be used to compose the intention attribute. Once the vehicle OTag is filled with intention, it will start expressing the intended movements as shown in Figure 2. This process continues until the no Informer tag is found. Whenever changes happened to the values relevant to the parameters used to compose above message, recalculation is done and vehicle OTag starts retransmitting the new message. For instance, if the speed is changed the recalculation is done and retransmission starts. A target message interpretation is represented in Figure 3.

The message in Figure 3 is interpreted as a vehicle is moving over the route number 16 by passing the Informer position 16.0 and heading forward (GF) to 16.1. Current estimated reach time to the 16.1 is 12:02:23:122 and the vehicle is running at a speed of 60kmph in lane 0. Here the lane 0 means that the road 16 has only one lane for one side. Therefore as explained in the above message, one vehicle can understand the intended movements of surrounding vehicles and thereby collision avoidance can be achieved when crossing an intersection and merging lanes.

3 iPark: The Proposed Solution

Once the vehicle tag and interrogator is installed into vehicles, optimal automation could be achieved in parking management system while providing better services to the customers. Our aim was to develop a parking management system which works on both free entrance and controlled entrance parking lots. To enable such parking management system it is necessary to identify the arriving and departing vehicles, detect the idling parking lots, support navigation towards the idling or reserved parking lot, bill parked vehicles, allow inquiring availability and reserve parking lot and enable enforcing of parking rules and regulations.

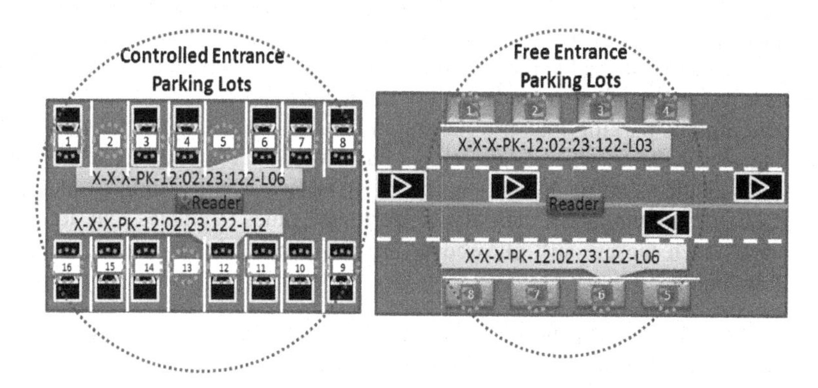

Fig. 4. Illustrate controlled and free entrance parking blocks with Lot tag installations

Figure 4 explained controlled entrance parking area and free entrance parking area respectively. The message in figure 4, "X-X-X-PK-12:02:23:122-L06" means that the car is parked (PK) in lot 6 and start time is 12:02:23:122. Both controlled entrance and free entrance parking should be equipped with Lot tag instances and Readers. Each Lot tag on each parking block is uniquely named. Since it is necessary to cope with the vehicles which do not have installed a vehicle tag and interrogator, sensing system is used to detect the occupancy of the parking lot. When a vehicle is parked to a parking lot, the sensor will check the occupancy and informed the control system that there is a car in its lot.

3.1 Identification of Arriving and Departing Vehicles

Automatic Vehicle Identification (AVI) is the solution to secure and convenient hands-free access control. In parking AVI is used to automatic identification of vehicles for safe access control and correct billing of parking fees. There can be two main types of vehicles entering to the parking premises: pre-registered vehicles (service subscribers) and other vehicles. To subscribe the services provided by the park operator, vehicle must be equipped with a vehicle tag and interrogator. Such vehicles become service subscribers when they are registered with the park operators. On the other hand, vehicles that are not subscribed but installed with above composition can use most of the services other than automated parking fee payment. Pre-registered vehicles can enter and leave the parking lot without any hassle whereas others need to pick a ticket at the entrance to enter the premises and made the payment when leaving the premises by inserting the parking ticket like in existing systems. A vehicle enter to the controlled entrance parking premises will first identify by its "AnonymousID" and check for the validation of the subscription and allowed to enter smoothly like in automated toll collection gates. If the vehicle is not a service subscriber, parking ticket will be issued before opening the entering gate. Similarly the vehicles that are not embedded with vehicle tag and readers are also issued such ticket to enter the premises.

When pre-registered vehicle leaves the parking premises fee will be collected automatically and allowed to pass the gate smoothly. When charging the fee, pre-registered vehicle tag's "WalletID" will be read securely by using the friendly key and request the confirmation from the driver. Once the confirmation is given the credit limit will be checked and charged accordingly. Note that the "WalletID" attribute value will be taken from a credit card or debit card.

In free entrance parking lots, pre-registered vehicles are charged then and there by requesting confirmation whereas the others have to pay the money to nearby control box. If they neglect the payment fine can be issued as it is possible to identify the vehicle using "anonymousID" and "type" attribute values. Vehicles that are not embedded with such tag are still allowed to use the parking lot but need to make the payment manually.

Lot Tag. Lot tag contains the following attributes to identify the lot number, occupancy and reservation information. Figure 5 illustrate some of the key attributes of Lot tag.

Oname	Lot
LotNumber	1
LotFloor	2
LotStatus	Idle/occupied/reserved
Occupant	AnonymousID
StartTime	2009-April-13 12:06:25 PM
EndTime	2009-April-13 12:42:20 PM
ParkerID	1230u22hsdfoIJHFI23

Fig. 5. Illustrate key attributes of Lot Tag

3.2 Detecting Idling Parking Lots

Detection should be possible in both free entrance and controlled entrance parking lots. As shown in figure 4 each parking lot should be tagged with a Lot tag with an infrared sensor. Both the free and controlled entrance parking lots have to have a sensor to detect the unregistered subscribers or untagged vehicles. Whenever a vehicle is parked to the parking lot LotStatus attribute value will be changed to occupied status by the sensor as shown in figure 5.

When a tagged vehicle is parked to a lot, the reader in the vehicle read the Lot tag and writes its position to the intended movement message for expressing its action as explained in the section 2.1 and the figure 2 and 3. Here the last pass Informer, route number, next Informer may be blank as shown in the figure 4 messages. Then the control system can identify the vehicle in a particular parking lot. Next, the interrogator in the center of the parking lot start reading the vehicle tag at the occupied parking lot and writes the start time to relevant Lot tag after verifying the vehicles identity by reading the "AnonymousID". Same way when the customer leaves the parking lot, end time of the Lane tag is updated by the interrogator in the parking block. Just after leaving the place, billing will be started. Therefore, using the sensor installed in the parking lot and Lot tag, it is possible to detect the occupancy of the lot and thereby the idling lots can be identified in both free and controlled entrance parking lots.

3.3 Navigation towards the Idling or Reserved Parking Lot

Navigation towards the idling parking lots or reserved parking lots can be provided by proposed solution. For that the One Informer tag at the point of entrance and several Position tags must be installed strategically according to the shape of the parking block in addition to the lot tags with sensors. Figure 6 illustrates a parking block with Informer, Position, and Lot tags and their placement strategy example for such shape.

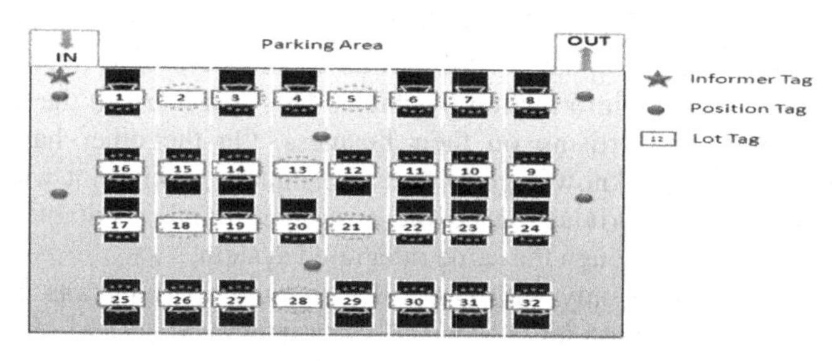

Fig. 6. Illustrate the setup for supporting navigation towards the reserved or idling parking lot

Just after passing the main entrance gate vehicle meets the Informer tag and read the destination details including the idling lot. After that the vehicle meets a position tag and asks for the directions. For instance when a vehicle get the idling lot number 21, direction provided by the position tag is go straight and turn after 2nd position tag. When the second position tag is passed the idling arrow to the left can be depicted with the log number 21. This way it is possible to arrange the navigation assistance system towards a given parking lot. When a vehicle needs to reach the reserved parking lot, reserved lot number must be given to the reserved vehicle by reading its "AnonymousID". Writing to Informer tag at the entrance gate must be completed just before the reserved vehicle come closer to the Informer tag.

Irrespective of the subscription, when a vehicle with vehicle tag and interrogator enters to the parking premises, guidance to the closest idling parking lot can be provided. Note that the reservations are only possible with controlled entrance parking blocks. Vehicles that are not having vehicle tag can not be guided instead the ticket can be printed with available parking lot numbers to assist finding a lot.

3.4 Billing of Parked Vehicles

Process of billing differs to two ways in controlled entrance parking blocks. If a customer is a service subscriber, parking management system can read the "WalletID" from their vehicle tag. Parking charge can be calculated according the time frame and the discount options available to that vehicle. Then the system read the "WalletID" and charged. Note that the customer should set the "WalletID" attribute value by inserting a credit card. Once the payment is successful gate will be released automatically enabling smooth and fast exit. Others must make the payment before leaving the exit gate by card or cash.

When it comes to the free entrance parking lots, everyone is treated same irrespective of the subscription. The billing process here works on the unique combination of the "AnonymousID" and "VehicleType". In these lots, there is no restriction applied before leaving since the billing will be done later in the month to the owner or to the driver if it is a rental vehicle. But if a rule is breached, a fine will also be issued with the bill. For patrons subscription of free entrance parking can also be done after prior registration.

3.5 Inquiring Availability and Reserving Parking Lots

Checking availability of a parking lot can be achieved by integrating the parking system to a web based system which can be access via system in vehicle, mobile phone, or over the web. Dynamic update of idling lots should be passed to such integrated system to make necessary information available. This is a merit to the park operator as it acts as sort of advertising on their business. On the other hand customers are provided with enough knowledge to find the suitable parking lot without any inconvenience even while driving. Once the customer found a suitable parking lot, reservation can be done through the same integrated system.

Reservation is allowed only with controlled entrance parking lots. Once the instruction to reserve a parking lot is received from integrated system, the "status" attribute of the Lot tag in a parking block will be filled with the value "reserved", with the expire time. If the time expires same vehicle cannot reserve another lot in the same company within the same day instead vehicle has to be present or there may be an option to extend the reservation by making an extra payment. On the other hand, Lot tag itself inform other vehicles about its reserved status by expressing value of "LotStatus" attribute value to "reserved" as shown in figure 5.

3.6 Enforcing of Parking Rules and Regulations

Enforcing rules and regulations had become a serious issue and big overhead to the government authorities. So far there were no automated system to enforce parking rules and regulations. Instead there are checkers or police officers to find the illegal parking. Difficulties arise in free entrance parking lots are greater than the controlled entrance.

Mainly it is necessary to prevent illegal parking. Illegal parking can be defined as parking in no parking areas and roadsides. By this system the vehicles can be informed if they stop in no parking area or no parking roadsides to prevent illegal parking. If they do not adhere to the rules, vehicles can be identified automatically and fined accordingly. In addition to that some vehicles may use some free parking lots beyond the permitted number of hours. Using proposed composition it is possible to detect such parking and fine accordingly. Other scenario is the misusing of reserved, elderly, and disabled parking lots. These scenarios can be detected through the proposed system and pre-inform or warn the drivers to avoid such misuse. This way the discrepancies can be eliminated and thereby unnecessary expenditures can be reduced.

3.7 Web Based Parking Management

Citywide parking blocks can be integrated into one central system to pass their idling lot information and price plans. At the same time there must be province to make the reservation. Once such integration is made a driver can made reservation desired or closest parking block to their destination. Each park operator must update their status whenever the changes happened attract more customers. Thus the drivers are provided with real time information to take better decisions by reducing the air pollution and unnecessary traffic congestions. In addition to this the web based system must provide support to take decision on multi mode transportation options to assist the drivers to take the best possible decision when reaching their destinations.

4 Concluding Remarks

A novel universal parking solution called iPark is proposed and explained. Using the OTag architecture and strategic composition almost all the issues arise in parking lot management could be solved. iPark can be implemented on both free entrance and controlled entrance parking lots and assist enforcing rules and regulations automatically while providing better services to the drivers, car park operators and also for the jurisdictions. Since this composition can solve the most unsolved issues in parking management and provide interoperable, intelligent and standardized usage providing ubiquitous usage, no doubt that iPark system become a better parking management system enhancing the future of ITS.

References

1. ITS: http://www.esafetysupport.org/, http://www.ewh.ieee.org/tc/its/, http://www.ertico.com/ (accessed on February 2009)
2. Active RFID Tag Architecture, http://www.rfidjournal.com/article/view/1243/1/1 (accessed on February 28, 2009)
3. RFID Parking Access, http://www.transcore.com/wdparkingaccess.html (accessed on March 02, 2009)
4. ITS Japan, http://www.mlit.go.jp/road/ITS/2006HBook/appendix.pdf (accessed on February 20, 2009); e-Plate and RFID enabled license plates, http://www.e-plate.com (accessed on July 2008)
5. Safety Applications of ITS in Europe and Japan, International Technology Scanning Program report, American Association of State Highway, and Transportation Officials NCH, Research Program
6. Koralalage, K.S., Yoshiura, N.: OTag: Architecture to Represent Real World Objects in RF Tags to improve future Intelligent Transportation Systems. Journal of Convergence Information Technology 4(2) (June 2009)

Author Index